Human Anatomy and Physiology Laboratory Manual

Cat Version

Updated Sixth Edition

Elaine N. Marieb, RN, PH.D.
Edison Community College
Holyoke Community College, Emeritus

PhysioEx Exercises authored by
Peter Z. Zao, North Idaho College

An Imprint of Addison Wesley Longman, Inc.

San Francisco • Reading, Massachusetts • New York • Harlow, England
Don Mills, Ontario • Sydney • Mexico City • Madrid • Amsterdam

Publisher/Producer: Daryl Fox
Senior Project Editor: Lauren Fogel
Managing Editor: Wendy Earl
Production Supervisor: Janet Vail
Art and Design Supervisor: Bradley Burch
Text Designer: Juan Vargas
Cover Designer: Yvo Riezebos
Channel Marketing Manager: Susan Winslow

Benjamin/Cummings gratefully acknowledges Carolina
Biological Supply for the use of numerous histology images
founds on the PhysioEx CD-ROM.

The Author and Publisher believe that the lab experiments
described in this publication, when conducted in conformity with
the safety precautions described herein and according to the
school's laboratory safety procedures, are reasonably safe for the
students to whom this manual is directed. Nonetheless, many of
the described experiments are accompanied by some degree of
risk, including human error, the failure or misuse of laboratory
or electrical equipment, mismeasurement, spills of chemicals,
and exposure to sharp objects, heat, bodily fluids, blood, or other
biologics. The Author and Publisher disclaim any liability arising
from such risks in connection with any of the experiments
contained in this manual. If students have any questions or
problems with materials, procedures, or instructions on any
experiment, they should *always* ask their instructor for help
before proceeding.

ISBN 0-8053-4919-7

2 3 4 5 6 7 8 9 10—VH—03 02 01 00 99

Credits appear following Appendix C.

The Benjamin/Cummings Series in Human Anatomy and Physiology

Textbooks

By R.A. Chase
The Bassett Atlas of Human Anatomy (1989)

By Kapit/Elson
The Anatomy Coloring Book, second edition (1993)

By Kapit/Macey/Meisami
The Physiology Coloring Book, second edition/2000

By E.N. Marieb
Human Anatomy and Physiology, fourth edition (1998)

Human Anatomy and Physiology, Study Guide,
fourth edition (1998)

*Human Anatomy and Physiology Laboratory Manual,
Main Version*, fifth edition (2000)

*Human Anatomy and Physiology Laboratory Manual,
Fetal Pig Version*, sixth edition (2000)

Human Anatomy Laboratory Manual, first edition
(1992)

Essentials of Human Anatomy and Physiology,
sixth edition (2000)

*The A&P Coloring Workbook: A Complete Study
Guide*, sixth edition (2000)

By E.N. Marieb and J. Mallatt
Human Anatomy, second edition (1997)

Library of Congress Cataloging-in-Publication Data

Marieb, Elaine Nicpon, 1936–
     Human anatomy and physiology laboratory manual —
Updated 6th ed., cat version.
          p.   cm.
     ISBN 0–8053–4919–7
     1. Human physiology Laboratory manuals.   2. Human
          anatomy Laboratory manuals.   3. Cats—Dissection
          Laboratory manuals.   I. Zao, Peter Z.
QP44.M34   2000
612′.0078—dc21   99-26244
                                                                  CIP

Benjamin/Cummings Science Publishing
1301 Sansome Street
San Francisco, CA 94111

Contents

Preface to the Instructor

The philosophy behind the sixth edition of this manual mirrors that of all earlier editions. It reflects a still-developing sensibility for the way teachers teach and students learn engendered by years of teaching the subject, and by listening to the suggestions of other instructors as well as those of students enrolled in multifaceted health-care programs. *Human Anatomy and Physiology Laboratory Manual: Cat Version* was originally developed to facilitate and enrich the laboratory experience for both teachers and students. This, its sixth edition, retains those same goals.

This manual, intended for students in introductory human anatomy and physiology courses, presents a wide range of laboratory experiences for students concentrating in nursing, physical therapy, dental hygiene, pharmacology, respiratory therapy, health and physical education, as well as biology and premedical programs. It differs from *Human Anatomy and Physiology Laboratory Manual, Main* version (fifth edition, 1999) in that it contains detailed guidelines for dissecting a laboratory animal. The manual's coverage is intentionally broad, allowing it to serve both one- and two-semester courses.

Basic Pedagogical Approach

The generous variety of experiments in this manual provides flexibility that enables instructors to gear their laboratory approach to specific academic programs, or to their own teaching preferences. The manual is still independent of any textbook, so it contains the background discussions and terminology necessary to perform all experiments. Such a self-contained learning aid eliminates the need for students to bring a textbook into the laboratory.

Each of the 47 exercises leads students toward a coherent understanding of the structure and function of the human body. The manual begins with anatomical terminology and an orientation to the body, which together provide the necessary tools for studying the various body systems. The exercises that follow reflect the dual focus of the manual—both anatomical and physiological aspects receive considerable attention. As the various organ systems of the body are introduced, the initial exercises focus on organization, from the cellular to the organ system level. As indicated by the Table of Contents, the anatomical exercises are usually followed by physiological experiments that familiarize students with various aspects of body functioning and promote the critical understanding that function follows structure. Homeostasis is continually emphasized as a requirement for optimal health. Pathological conditions are viewed as a loss of homeostasis;

these discussions can be recognized by the homeostasis imbalance logo within the descriptive material of each exercise. This holistic approach encourages an integrated understanding of the human body.

New Features and Changes

In this revision, I really tried to respond to reviewers' and users' feedback concerning trends that are having an impact on the anatomy and physiology laboratory experience, most importantly:

- the growing reluctance of students to perform experiments using living laboratory animals and the declining popularity of animal dissection exercises

- the increased use of multimedia (e.g., CD-ROMs, computers, etc.) in the laboratory, and hence the subsequent desire for more computer simulation exercises and advanced instrumentation.

- the ever-increasing computer literacy of our students

- the visual nature of today's students engendered by the advent of television and computer games

The specific changes implemented to address these trends fall neatly into two areas: multimedia and pedagogical. Changes made in each of these areas are described next.

Multimedia

1. PhysioEx
The PhysioEx CD-ROM, shrink-wrapped with every lab manual, is a fresh and exciting addition to the sixth edition of the manual. Unlike the typical tutorial-based computer supplements that usually target anatomy, the five easy-to-use physiology experiments on PhysioEx allow students to freely investigate various topics in physiology, and yet still have the structure and security of a written lab exercise guiding them through the process of discovery. Particularly advantageous is the fact that the student can conduct or review the experiments at home on their personal computer. PhysioEx also provides convenient "laboratory access" for students enrolled in Internet-based distance education courses.

PhysioEx topics include:

- Exercise 5B, *Cell Transport Mechanisms and Permeability.* Explores how substances cross the cell's membrane. Simple and facilitated diffusion, osmosis, filtration, and active transport are covered.

• Exercise 16B, *Skeletal Muscle Physiology.* Provides insights into the complex physiology of skeletal muscle. Electrical stimulation, isometric contractions, and isotonic contractions are investigated.

• Exercise 33, *Cardiovascular Dynamics.* Allows students to perform experiments that would be difficult if not impossible to do in a traditional laboratory. Topics of inquiry include vessel resistance and pump (heart) mechanics.

• Exercise 35B, *Frog Cardiovascular Physiology.* Variables influencing heart activity are examined. Topics include setting up and recording baseline heart activity, the refractory period of cardiac muscle, and an investigation of physical and chemical factors modifying heart rate.

• Exercise 40B, *Chemical and Physical Processes of Digestion.* Turns the student's computer into a virtual chemistry lab where enzymes, reagents, and incubation conditions can be manipulated (in compressed time) to examine factors that affect enzyme activity.

The PhysioEx exercises with the B designation cover the same basic material as the A versions, which use laboratory animals or the wet lab approach to experimentation.

2. Intelitool Instructions Update

The four physiological experiments (former Exercises 16i, 22i, 31i, and 37i) using Intelitool equipment and students as subjects have been removed from the text of this edition. Market research indicated that campuses are moving toward computerized laboratory experiences. However, these exercises have been updated and are now available in the *Instructor's Guide.* Instructors using Intelitool equipment in their laboratory may copy these exercises for student handouts.

3. New Videotapes

Two new videotapes are available to qualified adopters—*The Human Urinary System* and *The Human Reproductive Systems.* Produced by University Media Services and scripted by Rose Leigh Vines and Ann Motekautis of California State University, Sacramento, these excellent videotapes reinforce many of the understandings covered in Exercises 41–44 of this manual and will represent a valuable addition to any multimedia library.

Pedagogical

1. New Design

In response to complaints that the students often spend lab time reading descriptive material (much better read before the lab), thus delaying the actual hands-on lab experience, the manual now has cobalt blue *Activity* heads to alert the student that "this is the place" they get actively involved in their learning. Additionally, sections that entail dissection of isolated organs are heralded by a dissection logo and a kelly green *Dissection* head. The end of such sections is indicated by a colored block of the same hue. Besides the color-coded activity and dissection heads, essentially all diagrams and photos are now in full color. This edition is a colorful one to be sure—a pedagogical plus for enhancing the interest of the visually oriented students of today.

2. New Organization

A major change has been effected in the manual's organization as well. There are now two major sections in the manual. The first part of the lab manual includes the sequential 47 laboratory exercises, four of which have two (A and B) variations, followed by a unified lab review section pertaining to those exercises. Additionally, the lab review section, particularly those sections pertaining to bone and muscle anatomy, have been expanded.

All instructions for dissection of the major laboratory animal (the cat in this case) have been moved to a special section of the manual. In this section, the laboratory review sections are made part of the individual dissection exercises to encourage the student to look at the questions during the laboratory when the dissection specimen is still in front of them. Four-color photographs, formerly present in the Cat Anatomy Atlas, are now integrated in the dissection exercises and additional photographs have been taken to ensure that nearly all diagrams in this section are accompanied by a corresponding photo. This approach provides students with both a realistic (photographs in real color) and an idealized (diagrams in four-color) art program to guide their dissection experiences.

3. Expanded Art Program

Besides the new photographs and reworked art, several new cadaver photographs have been integrated into the anatomical exercises where appropriate, and a new 9-page Human Anatomy Atlas has been added. This atlas is indicated by its red-edged pages. Additionally, the Histology Atlas, which has always received good marks from our audience, has been updated and expanded. This atlas is recognized by its purple-edged pages.

4. Anatomical Terminology Updated

The anatomical terminology in this sixth edition has been updated to match that in *Human Anatomy and Physiology,* fourth edition (by this author), to avoid student confusion over conflicting terms.

Special Features Retained

Virtually all the special features appreciated by the adopters of the last edition are retained.

• The prologue, "Getting Started—What to Expect, The Scientific Method, Scientific Notation, and Metrics," explains the scientific method, the logical, practical, and reliable way of approaching and solving problems in the laboratory. It also reviews the use of exponents, metric units, and interconversions.

• Each exercise begins with learning objectives.

• Key terms appear in boldface print, and each term is defined when introduced.

• Illustrations are large and of exceptional quality. Full-color photographs and drawings highlight, differentiate, and focus student attention on important structures.

- Body structures are studied from simple-to-complex levels, and physiological experiments allow ample opportunity for student observation, manipulation, and experimentation.

- There are numerous physiological experiments for each organ system, ranging from simple experiments that can be performed without specialized tools to more complex experiments using laboratory equipment and instrumentation techniques.

- Tear-out laboratory review sheets, located toward the end of the manual, are designed to accompany each lab exercise. The review sheets provide space for recording and interpreting experimental results, and require students to label diagrams and answer multiple-choice, short-answer, and review questions.

- In addition to the figures, isolated animal organs such as the sheep heart and pig kidney are employed because of their exceptional similarity to human organs. If no major dissection animal is used, the laboratory manual version entitled *Human Anatomy and Physiology, Main Version,* fifth edition, is recommended.

- The 11-page Histology Atlas has 62 color photomicrographs, and the edges of this atlas are colored purple to allow it to be quickly located. The photomicrographs selected are those deemed most helpful to students because they correspond closely with slides typically viewed in the lab. Most such tissues are stained with hematoxylin and eosin (H & E), but a few depicted in the Histology Atlas are stained with differential stains to allow selected cell populations to be identified in a given tissue. Line drawings, corresponding exactly to selected plates in the Histology Atlas, appear in appropriate places in the text and add to the utility and effectiveness of the atlas. These diagrams can be colored by the student to replicate the stains of the slides they are viewing; thus they provide a valuable learning aid.

- All exercises involving body fluids (blood, urine, saliva) incorporate current Centers for Disease Control (CDC) guidelines for handling human body fluids. Because it is important that nursing students, in particular, learn how to safely handle bloodstained articles, the human focus has been retained. However, the decision to allow testing of human (student) blood or to use animal blood in the laboratory is left to the discretion of the instructor in accordance with institutional guidelines. The CDC guidelines for handling body fluids are reinforced by the laboratory safety procedures described on the inside front cover of this text and in the *Instructor's Guide.* The inside cover can be photocopied and posted in the lab to help students become well versed in laboratory safety.

- Appendix C correlates some of the required anatomical laboratory observations with the corresponding sections of A.D.A.M. Interactive Anatomy. Using A.D.A.M. to complement the printed manual descriptions of anatomical structures provides an extremely useful study method for visually oriented students.

- Four logos alert students to special features or instructions. These include:

 The dissection scissors icon appears at the beginning of activities that entail the dissection of isolated animal organs.

 The homeostasis imbalance icon directs the student's attention to conditions representing a loss of homeostasis.

 A safety icon notifies students that specific safety precautions are to be observed when using certain equipment or conducting particular lab procedures (for example, when working with ether a hood is to be used or when handling body fluids such as blood, urine, or saliva).

The A.D.A.M. "walking man" icon indicates where use of the A.D.A.M. software would enhance the study and comprehension of the laboratory topics.

Supplements

- The *Instructor's Guide* that accompanies all versions of the *Human Anatomy and Physiology Laboratory Manual* contains a wealth of information for those teaching this course. Instructors can find help in planning the experiments, ordering equipment and supplies, anticipating pitfalls and problem areas, and locating audiovisual material. The probable in-class time required for each lab is indicated by a clock icon. Other useful resources are the Trends in Instrumentation section that describes the latest laboratory equipment and technological teaching tools available and the directions for using Intelitool instrumentation. Additional supplements, all mentioned above, include these available free of charge to qualified adopters:

- *Selected Actions of Hormones and Other Chemical Messengers* videotape by Rose Leigh Vines and Juanita Barrena

- *Human Musculature* videotape by Rose Leigh Vines and Allan Hinderstein

- The *Human Cardiovascular System: The Heart* videotape by Rose Leigh Vines and University Media Services, California State University, Sacramento

- *The Human Cardiovascular System: The Blood Vessels* videotape by Rose Leigh Vines and University Media Services, California State University, Sacramento

- *Human Nervous System* videotape by Rose Leigh Vines and University Media Services, California State University, Sacramento

- *The Human Respiratory System* videotape by Rose Leigh Vines and Ann Motekautis

- *The Human Digestive System* videotape by Rose Leigh Vines and Ann Motekautis

- *The Human Urinary System* videotape by Rose Leigh Vines and Ann Motekautis

- *The Human Reproductive Systems* videotape by Rose Leigh Vines and Ann Motekautis

- Student Video Series Vol. I (0-8053-4110-2)

- Student Video Series Vol. II (0-8053-6115-4)

A.D.A.M.® Software

Available for purchase from Benjamin/Cummings Publishing to enhance student learning are the A.D.A.M.–Benjamin/Cummings Interactive Physiology CD-ROM Series and the *A.D.A.M. Interactive Anatomy Student Lab Guide* by Mark Lafferty and Samuel Panella (0-8053-4350-4). Modules in the CD-ROM series are:

- *Cardiovascular System* by Elaine Marieb and Marvin Branstrom
- *Muscular System* by Marvin Branstrom
- *Respiratory System* by Andrea Salmi
- *Nervous System I* by Susan Mitchell
- *Urinary System* by Marvin Branstrom
- *Fluid, Electrolyte, and Acid/Base Balance* by Martha Sanner and Christine Hooper
- *Nervous System II* by Susan Mitchell

Acknowledgments

I wish to thank the following reviewers for their contributions to this edition: Debra Barnes (Contra Costa College), Joseph Berger (Springfield College), Moges Bizuneh (Ivy Technical State College), William Cliff (Niagara University), James Constantine (Bristol Community College), Kim Cooper (Arizona State University), Diana Curley (Catonsville Community College), Gerald Dotson (Front Range Community College), Diane Dudzinski (Washington State Community College), Ron Dunn (Kapiolani Community College), Kathleen Flickinger (Iowa State University), Pamela Fouche (Walters State Community College), Daniel Gong (Seattle Central Community College), Susan Greimel (Lake City Community College), Clare Hays (Metropolitan State College of Denver), Mary Healey (Springfield College), Linda Heath (Xavier University), Jean Helgeson (Collin County Community College), S. Beth Howard (Rutgers University), James Huggins (Union University), Alison Jasson (Northwestern County Community College), Kevin Lien (Portland Community College), Jeri Lindsey (Tarrant County Junior College), Katherine Lockwood (University of New Hampshire), Anne Miller (Middlesex Community College), Patricia Munn (Longview Community College), Ruth Nash (College of Marin), Gisele Nasr (Brevard Community College), Leslie Nesbitt (Niagara University), Betsy Ott (Tyler Junior College), Jackie Reynolds (Richland College), Sharon Simpson (Broward Community College), Ann Smith (Joliet Junior College), David Smith (San Antonio College), Karin Van Meter (Des Moines Area Community College), Mark Wagner (Laramie County Community College), Linda Winkler (Oregon State University).

My continued thanks to my colleagues and friends at Benjamin/Cummings who worked with me in the production of this edition, especially Daryl Fox, the executive editor who got the project perking, and Lauren Fogel, the senior project editor who efficiently shepherded the manuscript every millimeter of the way, as she simultaneously managed the production schedule of PhysioEx. (Bravo, Lauren.) Speaking of PhysioEx, its excellence reflects the expertise of Peter Zao, my coauthor for this edition and friend, who generated the ideas behind the equipment graphics and envisioned the animations that would be needed; the programming genius of Mary Ann Tenorio; and the wonderful graphics of Steve McEntee. I complained about my schedule, but these people were really under the gun!

Kudos also to the folks at, or working with, Wendy Earl Productions, who did their usual great job. Janet Vail, my production editor for this project, got the job done seemingly effortlessly (but I know it wasn't) because of her mellow approach to everything. Bradley Burch changed hats this time from production editor to art and design coordinator. Brad was responsible for overseeing photography of the new cat dissection photos, and seeing that the corresponding art was aptly done, as well as obtaining any and all new photomicrographs and overseeing art modifications for this now totally digital version. Juan Vargas came up with our new crisp, clean, and "labby" (if you listen to Daryl) design. I think it's wonderful, and hope you do too. As usual, Yvo Riezebos produced a just-right cover image. Last but not least, I want to thank Anita Wagner, a very competent copyeditor who is now an old hand at my books, and Linda Kollett, who is preparing the corresponding *Instructor's Guide* for this edition, for her careful and thoughtful work.

As ever, preparations for the next edition begin well in advance of its publication. I invite users of this edition to send me their comments and suggestions for subsequent editions.

Elaine N. Marieb
Anatomy and Physiology
Benjamin/Cummings Science Publishing
2725 Sand Hill Road
Menlo Park, CA 94025

Preface to the Student

Hopefully, your laboratory experiences will be exciting times for you. Like any unfamiliar experience, it really helps if you know in advance what to expect and what will be expected of you.

Laboratory Activities

The A&P laboratory exercises in this manual are designed to help you gain a broad understanding of both anatomy and physiology. So, you can anticipate that you will be examining models, dissecting isolated animal organs, and using a microscope to look at tissue slides (anatomical approaches). You will also investigate chemical conditions or observe changes in both living and nonliving systems, manipulate variables in computer simulations, and conduct experiments that examine responses of living organisms, such as frogs or yourself, to various stimuli (physiological approaches).

Because some students question the use of animals in the laboratory setting, their concerns need to be addressed. Be assured that the commercially available major dissection animals and preserved organ specimens used in the A&P labs are *not* harvested from animals raised specifically for dissection purposes. Instead, cats brought to the SPCA or other animal shelters for humane extermination are sent to qualified biological supply houses, where they are prepared for laboratory use according to USDA guidelines. Organs that are of no use to the meat packing industry (such as the brain, heart, or lungs) are sent from slaughterhouses to biological supply houses for preparation.

Relative to using live animals for experimentation, every effort is being made to find alternative methods that do not use living animals to study physiological concepts. For example, new to this edition is the PhysioEx CD-ROM shrink-wrapped with your lab manual. The five simulation exercises on this CD employ the easy-to-use, colorful PhysioEx software that allows you to convert your computer (be it in the college laboratory or in your bedroom) into a virtual laboratory. You will be able to manipulate variables to investigate physiological phenomena. Such computer-based simulations provide you with alternatives to the more traditional exercises on muscle activity and electrocardiography that use frogs as experimental subjects.

There is little doubt that computer simulations offer certain advantages: (1) they allow you to experiment at length without time constraints of traditional animal experiments, in which fragile living tissues need to be kept alive and viable for the duration of the experiment; and (2) they make it possible to investigate certain concepts (such as isometric versus isotonic contraction in this case) that would be difficult or impossible to explore in traditional exercises. Additionally, they remove much of the time requirement of some traditional wet labs, typical of experiments that require samples to sit for a while or be incubated, as in investigations of membrane transport and enzyme activity. Yet, the main disadvantage of computer simulations is that the real-life aspects of experimentation are sacrificed. An animated frog muscle or heart on a computer screen is not really a substitute for observing the responses of actual muscle tissue. Consequently, living animal experiments remain an important part of the approach of this manual to the study of human A&P. However, wherever possible, the minimum number of animals needed to demonstrate a particular point is used. Furthermore, more instructor-delivered and videotaped demonstrations of live animal experiments are suggested.

If you use living animals for experiments, you will be expected to handle them humanely. Inconsiderate treatment of laboratory animals will not be tolerated in your A&P laboratory.

A.D.A.M.® Interactive Anatomy

If the A.D.A.M. CD-ROM software is available for your use, Appendix C will help you link the various laboratory topics with specific frames of the A.D.A.M. software to help you in your studies.

Icons/Visual Mnemonics

I have tried to make this manual very easy for you to use, and to this end two colored section heads and four different icons (visual mnemonics) are used throughout:

The new *Dissection* head is kelly green and is accompanied by the **dissection scissors icon** at the beginning of activities that require you to dissect isolated animal organs.

The new *Activity* head is cobalt blue. Because most exercises have some explanatory background provided before the experiment(s), this visual cue alerts you that your lab involvement is imminent.

 The **homeostasis imbalance icon** appears where a clinical disorder is described to indicate what happens when there is a structural abnormality or physiological malfunction, i.e., a loss of homeostasis.

 The **A.D.A.M. walking man icon** alerts you where the use of A.D.A.M. would enhance your laboratory experience.

⚠ The **safety icon** alerts you to special precautions that should be taken when handling lab equipment or conducting certain procedures. For example, it alerts you to use a ventilating hood when using volatile chemicals and signifies that you should take special measures to protect yourself when handling blood or other body fluids (saliva, urine).

Hints for Success in the Laboratory

With the possible exception of those who have photographic memories, most students can use helpful hints and guidelines to ensure that they have successful lab experiences.

1. Perhaps the best bit of advice is to attend all your scheduled labs and to participate in all the assigned exercises. Learning is an *active* process.

2. Scan the scheduled lab exercise and the questions in the review section in the back of the manual that pertain to it *before* going to lab.

3. Be on time. Most instructors explain what the lab is about, pitfalls to avoid, and the sequence or format to be followed at the beginning of the lab session. If you are late, not only will you miss this information, you will not endear yourself to the instructor.

4. Review your lab notes after completing the lab session to help you focus on and remember the important concepts.

5. Keep your work area clean and neat. This reduces confusion and accidents.

6. Assume that all lab chemicals and equipment are sources of potential danger to you. Follow directions for equipment use and observe the laboratory safety guidelines provided inside the front cover of this manual.

7. Keep in mind the real value of the laboratory experience—a place for you to observe, manipulate, and experience hands-on activities that will dramatically enhance your understanding of the lecture presentations.

I really hope that you enjoy your A&P laboratories and that this lab manual makes learning about intricate structures and functions of the human body a fun and rewarding process. I'm always open to constructive criticism and suggestions for improvement in future editions. If you have any, please write to me.

Elaine N. Marieb
Anatomy and Physiology
Benjamin/Cummings Science Publishing
2725 Sand Hill Road
Menlo Park, CA 94025

Getting Started—What to Expect, The Scientific Method, Scientific Notation, and Metrics

Two hundred years ago science was largely a plaything of wealthy patrons, but today's world is dominated by science and its technology. Whether or not we believe that such domination is desirable, we all have a responsibility to try to understand the goals and methods of science that have seeded this knowledge and technological explosion.

The biosciences are very special and exciting because they open the doors to an understanding of all the wondrous workings of living things. A course in human anatomy and physiology (a minute subdivision of bioscience) provides such insights in relation to your own body. Although some experience in scientific studies is helpful when beginning a study of anatomy and physiology, perhaps the single most important prerequisite is curiosity.

Gaining an understanding of science is a little like becoming acquainted with another person. Even though a written description can provide a good deal of information about the person, you can never really know another unless there is personal contact. And so it is with science—if you are to know it well, you must deal with it intimately.

The laboratory is the setting for "intimate contact" with science. It is where scientists test their ideas (do research), the essential purpose of which is to provide a basis from which predictions about scientific phenomena can be made. Likewise, it will be the site of your "intimate contact" with the subject of human anatomy and physiology as you are introduced to the methods and instruments used in biological research.

For many students, human anatomy and physiology is taken as an introductory-level course; and their scientific background exists, at best, as a dim memory. If this is your predicament, this prologue may be just what you need to fill in a few gaps and to get you started on the right track before your actual laboratory experiences begin. So—let's get to it!

The Scientific Method

Science would quickly stagnate if new knowledge were not continually derived from and added to it. The approach commonly used by scientists when they investigate various aspects of their respective disciplines is called the **scientific method.** This method is *not* a single rigorous technique that must be followed in a lockstep manner. It is nothing more or less than a logical, practical, and reliable way of approaching and solving problems of every kind—scientific or otherwise—to gain knowledge. It comprises five major steps.

Step 1: Observation of Phenomena

The crucial first step involves observation of some phenomenon of interest. In other words, before a scientist can investigate anything, he or she must decide on a *problem* or focus for the investigation. In most college laboratory experiments, the problem or focus has been decided for you. However, to illustrate this important step, we will assume that you want to investigate the true nature of apples, particularly green apples. In such a case you would begin your studies by making a number of different observations concerning apples.

Step 2: Statement of the Hypothesis

Once you have decided on a focus of concern, the next step is to design a significant question to be answered. Such a question is usually posed in the form of a **hypothesis,** an unproven conclusion that attempts to explain some phenomenon. (At its crudest level, a hypothesis can be considered to be a "guess" or an intuitive hunch that tentatively explains some observation.) Generally, scientists do not restrict themselves to a single hypothesis; instead, they usually pose several and then test each one systematically.

We will assume that, to accomplish step 1, you go to the supermarket and randomly select apples from several bins. When you later eat the apples, you find that the green apples are sour, but the red and yellow apples are sweet. From this observation, you might conclude (*hypothesize*) that "green apples are sour." This statement would represent your current understanding of green apples. You might also reasonably predict that if you were to buy more apples, any green ones you buy will be sour. Thus, you would have gone beyond your initial observation that "these" green apples are sour to the prediction that "all" green apples are sour.

Any good hypothesis must meet several criteria. First, *it must be testable.* This characteristic is far more important than its being correct. The tests may prove the hypothesis incorrect; or new information may require that the hypothesis be modified. Clearly the accuracy of a prediction in the green apple example or in any scientific study depends on the accuracy of the initial information on which it is based.

In our example, no great harm will come from an inaccurate prediction—that is, were we to find that some green apples are sweet. However, in some cases human life may depend on the accuracy of the prediction. Take the case of testing drugs for their effectiveness in treating disease. If one set of observations erroneously indicates that the drugs are risky but very effective, such a conclusion could lead to the death of the subsequent drug recipient(s). This illustrates two points: (1) Repeated testing of scientific ideas is important, particularly because scientists working on the same problem do not always agree in their conclusions. The studies on the use of saccharin and amino acid sweeteners are only two examples. (2) Conclusions drawn from scientific tests are only as accurate as the information on which they are based; therefore, careful observation is essential, even at the very outset of a study.

A second criterion is that, even though hypotheses are guesses of a sort, *they must be based on measurable, describable facts. No mysticism can be theorized.* We cannot conjure up, to support our hypothesis, forces that have not been shown to exist. For example, as scientists, we cannot say that the tooth fairy took Johnny's tooth unless we can prove that the tooth fairy exists!

Third, a hypothesis *must not be anthropomorphic.* Human beings tend to anthropomorphize—that is, to relate all experiences to human experience. Because man is a social animal influenced by culture, these two characteristics tend to promote biased thinking. Whereas we could state that bears instinctively protect their young, it would be anthropomorphic to say that bears love their young, because love is a human emotional response. Thus, the initial hypothesis must be stated without interpretation.

Step 3: Data Collection

Once the initial hypothesis has been stated, scientists plan experiments that will provide data (or evidence) to verify or disprove their hypotheses—that is, they *test* their hypotheses. Data are accumulated by making qualitative or quantitative observations of some sort. The observations are often aided by the use of various types of equipment such as cameras, microscopes, stimulators, or various electronic devices that allow chemical and physiologic measurements to be made.

Observations referred to as **qualitative** are those we can make with our senses—that is, by using our vision, hearing, or sense of taste, smell, or touch. The color of an object, its texture, the relationship of one part to another, and its relative size (large versus small) may all be part of a qualitative description. For some quick practice in qualitative observation, compare and contrast* an orange and an apple.

Whereas the differences between an apple and an orange are obvious, this is not always the case in biological observations. Quite often a scientist tries to detect very subtle differences that cannot be determined by qualitative observations; and data must be derived from measurements made using a variety of scientific equipment. Such observations based on precise measurements of one type or another are **quantitative observations.** Examples of quantitative observations include careful measurements of body or organ dimensions such as mass, size, and volume; measurement of volumes of oxygen consumed during metabolic studies; determination of the concentration of glucose and other chemicals in urine;

and determination of the differences in blood pressure and pulse under conditions of rest and exercise. An apple and an orange could be compared quantitatively by performing chemical measurements of the relative amounts of sugar and water in a given volume of fruit flesh, by analyzing the pigments and vitamins present in the apple skin and orange peel, and so on.

A valuable part of data gathering is the use of experiments to verify or disprove a hypothesis. An **experiment** is a procedure designed to describe the factors in a given situation that affect one another (that is, to discover cause and effect) under certain conditions.

Two general rules govern experimentation. The first of these rules is that the experiment(s) should be conducted in such a manner that every **variable** (any factor that might affect the outcome of the experiment) is under the control of the experimenter. The experimenter manipulates the **independent variables** and observes the effects of this manipulation on the **dependent** (or **response**) **variable.** For example, if the goal is to determine the effect of body temperature on breathing rate, the value measured (breathing rate) is called the dependent variable because it "depends on" the value chosen for the independent variable (body temperature). The ideal way to perform such an experiment is to set up and run a series of tests that are all identical, except for one specific factor that is varied.

One specimen (or group of specimens) is used as the **control** against which all other experimental samples are compared. The importance of the control sample cannot be overemphasized. It is essential to know how the system you are investigating works under normal circumstances before you can be sure that the results obtained from experimentation are due solely to the manipulation of the independent variable(s). Taking our example one step further, if we wanted to investigate the effects of body temperature (the independent variable) on breathing rate (the dependent variable), we could collect data on the breathing rate of individuals with "normal" body temperature (the implicit control group), and compare these data to breathing-rate measurements obtained from groups of individuals with higher and lower body temperatures. The control group would provide the "normal standard" against which all other samples would be compared relative to the dependent variable.

The second rule governing experimentation is that valid results require that testing be done on large numbers of subjects. It is essential to understand that it is nearly impossible to control all possible variables in biological tests. Indeed, there is a bit of scientific wisdom that mirrors this truth—that is, that laboratory animals, even in the most rigidly controlled and carefully designed experiments, "will do as they damn well please." Thus, stating that the testing of a drug for its pain-killing effects was successful after having tested it on only one postoperative patient would be scientific suicide. Large numbers of patients would have to receive the drug and be monitored for a decrease in postoperative pain before such a statement could have any scientific validity. Then, other researchers would have to be able to uphold those conclusions by running similar experiments. *Repeatability* is an important part of the scientific method, and is the primary basis for acceptance or rejection of many hypotheses.

During experimentation and observation, data must be carefully recorded. Usually, such initial, or raw, data are recorded in tabular (table) form. The table should be labeled

---

* *Compare* means to emphasize the similarities between two things, whereas *contrast* means that the differences are to be emphasized.

to show the variables investigated and the results for each sample. At this point, *accurate recording* of observations is the primary concern. Later, these raw data will be reorganized and manipulated to show more explicitly the outcome of the experimentation.

Some of the observations that you will be asked to make in the anatomy and physiology laboratory will require that a drawing be made. Don't panic! The purpose of making drawings (in addition to providing a record) is to force you to observe things very closely. You need not be an artist (most biological drawings are simple outline drawings), but you do need to be neat and as accurate as possible. It is advisable to use a 4H pencil to do your drawings because it is easily erased and doesn't smudge. Before beginning to draw, you should examine your specimen closely, studying it as though you were going to have to draw it from memory. For example, when looking at cells you should ask yourself questions such as "What is their shape—the relationship of length and width? How are they joined together?" Then decide precisely what you are going to show and how large the drawing must be to show the necessary detail. After making the drawing, add labels in the margins and connect them by straight lines (leader lines) to the structures being named.

Step 4: Manipulation and Analysis of Data

The form of the final data varies, depending on the nature of the data collected. Usually, the final data represent information converted from the original measured values (raw data) to some other form. This may mean that averaging or some other statistical treatment must be applied, or it may require conversions from one kind of units to another. In other cases, graphs may be needed to display the data.

Elementary Treatment of Data

Only very elementary statistical treatment of data is required in this manual. For example, you will be expected to understand and/or compute an average (mean), percentages, and a range.

Two of these statistics, the average and the range, are useful in describing the *typical* case among a large number of samples evaluated. Let us use a simple example. We will assume that the following heart rates (in beats/min) were recorded during an experiment: 64, 70, 82, 94, 85, 75, 72, 78. If you put these numbers in numerical order, the **range** is easily computed, because the range is the difference between the highest and lowest numbers obtained (highest number minus lowest number). What is the range of the set of numbers just provided?

1. _____*

The **average,** or **mean,** is obtained by summing the items and dividing the sum by the number of items. Compute the average for the set of numbers just provided:

2. _____

The word *percent* comes from the Latin meaning "for 100"; thus *percent*, indicated by the percent sign, %, means parts per 100 parts. Thus, if we say that 45% of Americans have type O blood, what we are really saying is that among each group of 100 Americans, 45 (45/100) can be expected to have type O blood.

It is very easy to convert any number (including decimals) to a percent. The rule is to move the decimal point two places to the right and add the percent sign. If no decimal point appears, it is *assumed* to be at the end of the number; and zeros are added to fill any empty spaces. Two examples follow:

$$0.25 = 0.2\,5 = 25\%$$
$$5 = 5 = 500\%$$

Change the following numbers to percents:

3. 38.2 = _____   5. 1.6 = _____

4. 402 = _____

Note that although you are being asked here to convert numbers to percents, percents by themselves are meaningless. We always speak in terms of a percentage *of* something.

To change a percent to a whole number (or decimal), remove the percent sign, and move the decimal point two places to the left. Change the following percents to whole numbers or decimals:

6. 36% = _____   8. 25777% = _____

7. 800% = _____   9. 0.05% = _____

Making and Reading Line Graphs

For some laboratory experiments you will be required to show your data (or part of them) graphically. Simple line graphs allow relationships within the data to be shown interestingly and allow trends (or patterns) in the data to be demonstrated. An advantage of properly drawn graphs is that they save the reader's time because the essential meaning of large numbers of statistical data can be visualized at a glance.

To aid in making accurate graphs, graph paper (or a printed grid in the manual) is used. Line graphs have both horizontal and vertical scales. Each scale should have uniform intervals—that is, each unit measured on the scale should require the same distance along the scale as any other. Variations from this rule may be misleading and result in false interpretations of the data. By convention, the condition that is manipulated (the independent variable) in the experimental series is plotted on the X-axis (the horizontal axis); and the value that we then measure (the dependent variable) is plotted on the Y-axis (the vertical axis). To plot the data, a dot or a small x is placed at the precise point where the two variables (measured for each sample) meet; and then a line (this is called the **curve**) is drawn to connect the plotted points.

Sometimes, you will see the curve on a line graph extended beyond the last plotted point. This is (supposedly) done to predict "what comes next." When you see this done, be skeptical. The information provided by such a technique is

* Answers are given on page xviii.

only slightly more accurate than that provided by a crystal ball!

To read a line graph, pick any point on the line, and match it with the information directly below on the horizontal scale and with that directly to the left of it on the vertical scale. Figure G.1 is a graph that illustrates the relationship between breaths per minute (respiratory rate) and body temperature. Answer the following questions about this graph:

10. What was the respiratory rate at a body temperature of

96°F? _____

11. Between 98° and 102°F, the respiratory rate increased

from _____ to _____ breaths per minute.

12. Between which two body temperature readings was the increase in breaths per minute greatest?

_____

13. Are the intervals on each scale uniform?

_____

Step 5: Reporting Conclusions of the Study

Drawings, tables, and graphs alone do not suffice as the final presentation of scientific results. The final step requires that you provide a straightforward description of the conclusions drawn from your results. If possible, your findings should be

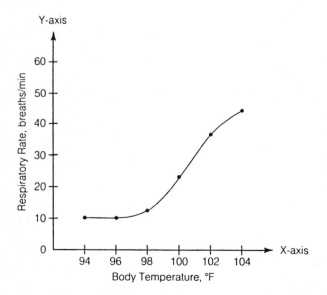

Figure G.1 Example of graphically presented data.

compared to those of other investigators working on the same problem. (For laboratory investigations conducted by students, these comparative figures are provided by classmates.)

It is important to realize that scientific investigations do not always yield the anticipated results. If there are discrepancies between your results and those of others, or what you expected to find based on your class notes or textbook readings, this is the place to try to explain those discrepancies.

Results are often only as good as the observation techniques used. Depending on the type of experiment conducted, several questions may need to be answered. Did you weigh the specimen carefully enough? Did you balance the scale first? Was the subject's blood pressure actually as high as you recorded it, or did you record it hastily (and inaccurately)? If you did record it accurately, is it possible that the subject was emotionally upset about something, which (even though the matter of concern had nothing to do with the experiment) might have given falsely high data for the variable being investigated? Attempting to explain an unexpected result will often teach you more than you would have learned from anticipated results.

When the experiment produces results that are consistent with the hypothesis, then the hypothesis can be said to have reached a higher level of certainty. There is now a greater probability that the hypothesis is correct.

A hypothesis that has been validated by many different investigators is called a **theory.** Theories are useful in two important ways. First, they link sets of data; and second, they make predictions that may lead to additional avenues of investigation. (Okay, we know this with a high degree of certainty; what's next?)

When a theory has been repeatedly verified and appears to have wide applicability in biology, it may assume the status of a **biological principle.** A principle is a statement that applies with a high degree of probability to a range of events. For example, "Living matter is made of cells or cell products" is a principle stated in many biology texts. It is a sound and useful principle, and will continue to be used as such—unless new findings prove it wrong.

We have been through quite a bit of background concerning the scientific method and what its use entails. Because it is important that you remember the phases of the scientific method, they are summarized here:

1. Observation of some phenomenon

2. Statement of a hypothesis (based on the observations)

3. Collection of data (testing the hypothesis with controlled experiments)

4. Manipulation and analysis of the data

5. Reporting of the conclusions of the study

Scientific Notation and Metrics

No matter how highly developed our ability to observe, observations have scientific value only if they can be communicated to others. This necessitates the use of scientific notation and the widely accepted system of metric measurements.

Scientific Notation

Because quantitative measurements often yield very large or very small numbers, you are quite likely to encounter numbers such as 3.5×10^{12} or 10^{-3}. It is important that you understand what this **scientific notation** means.

Scientific notation is dependent on the properties of exponents and on the movement of the decimal point when multiplying or dividing by 10. When you multiply 10 by itself, you get a product that is one followed by zeros. The number of zeros (two, in this case) in the product is equal to the number of times you have used 10 as a factor and is shown as an exponent. Thus, the notation 10^2 has these parts:

$$\text{base} \rightarrow 10^2 \leftarrow \text{exponent}$$

and translates to "the base 10 multiplied by itself (10×10)."

The powers of 10 are represented as follows:

$10^0 = 1$ (Any number followed by a zero exponent is one.)
$10^1 = 10$ ($10 \times 1 = 10$)
$10^2 = 100$ ($10 \times 10 = 100$)
$10^3 = 1000$ ($10 \times 10 \times 10 = 1000$)
$10^4 = 10,000$ ($10 \times 10 \times 10 \times 10 = 10,000$)

As you can see, each time the exponent is increased by one, another zero ($\times 10$) is added to the answer.

When you multiply any number by a power of 10 written with exponents, the decimal point is moved to the right the number of times shown in the exponent. Thus:

$$3.25 \times 10^1 = 3.25 \qquad = 32.5$$
$$3.25 \times 10^3 = 3.25 \qquad = 3250$$
$$3.25 \times 10^5 = 3.25 \qquad = 325,000$$

By using such exponential notation, very large numbers may be written in far simpler form.

Write the following numbers using the proper scientific notation:

14. $140,000 = 1.4 \times$ _____

15. $9,650,000 = 9.65 \times$ _____

16. $852 = 8.52 \times$ _____

17. $10 = 1.0 \times$ _____

Notice that proper scientific notation entails only one number to the left of the decimal point. Thus 1.03×10^3 is correct, but 10.3×10^2 is not.

In the above examples, all of the numbers used were greater than one. Scientific notation can also be used to report numbers less than one. To do this, negative exponents are used. For example, in

$$3.25 \times 10^2$$

the positive exponent means that the decimal point is to be moved two places to the right, and the number designated is 325 ($3.25 \times 10 \times 10$). However, in

$$3.25 \times 10^{-2}$$

the negative exponent means that the number is to be divided by the power of 10 indicated by the exponent and the decimal point is to be moved two places to the left. The number so designated is 0.0325 [$3.25 \div (10 \times 10)$].

Thus, the rule for converting scientific notation (using powers of 10) to decimal notation is to move the decimal point the number of places indicated by the exponent. When the exponent is positive (with or without a plus sign), the decimal point is moved to the right. When the exponent is negative (always indicated with a minus sign), the decimal point is moved to the left.

For a little practice, write the following numbers in scientific notation: (18–23)

$140,000 = 1.4 \times$ _____ $45,000 = 4.5 \times$ _____

$0.0000063 = 6.3 \times$ _____ $0.265 = 2.65 \times$ _____

$0.00054 = 5.4 \times$ _____ $0.10 = 1.0 \times$ _____

Metrics

Without measurement, we would be limited to qualitative description. However, with a system of measurement, quantitative description becomes possible.

Anyone can establish a system of measurement. All that is required is a reference point; and historically, much of our common (the British) system of measurement evolved from units based on objects everyone knew. For example, horses were measured in "hands," and a "fathom" was the distance between outstretched arms. However, the variability in such measurements is immediately apparent—for example, an infant's hand is substantially smaller than that of an adult. Therefore, for precise and repeatable communication of information, the agreed-upon system of measurement used by scientists is the **metric system,** a nonvarying standard of reference.

A major advantage of the metric system is that it is based on units of 10. This allows rapid conversion to workable numbers so that neither very large nor very small figures need be used in calculations. Fractions or multiples of the standard units of length, volume, mass, time, and temperature have been assigned specific names. Table G.1 shows the commonly used units of the metric system, along with the prefixes used to designate fractions and multiples thereof.

To change from smaller units to larger units, you must *divide* by the appropriate factor of 10 (because there are fewer of the larger units). For example, a milliunit (milli = one thousandth), such as a milliliter or millimeter, is one step smaller than a centiunit (centi = one hundredth), such as a centiliter or centimeter. Thus to change milliunits to centiunits, you must divide by 10. On the other hand, when converting from larger units to smaller ones, you must *multiply* by the appropriate factor of 10 (because there will be more of the smaller units). A partial scheme for conversions between the metric units is shown below.

Students studying a science or preparing for a profession in the health-related fields find that certain of the metric units are encountered and dealt with more frequently than others. Thus, the objectives of the sections that follow are to provide a brief overview of these most-used measurements and to help you gain some measure of confidence in dealing with

Table G.1	Commonly Used Units of the Metric System, and Their Fractions and Multiples			

Measurement	Unit	Fraction or multiple	Prefix	Symbol
Length	Meter (m)	10^6 one million	mega	M
Volume	Liter (L; l with prefix)	10^3 one thousand	kilo	k
Mass	Gram (g)	10^{-1} one tenth	deci	d
Time*	Second (s)	10^{-2} one hundredth	centi	c
Temperature	Degree Celsius (°C)	10^{-3} one thousandth	milli	m
		10^{-6} one millionth	micro	μ
		10^{-9} one billionth	nano	n

* The accepted standard for time is the second; and thus hours and minutes are used in scientific, as well as everyday, measurement of time. The prefixes used in the designation of units of length, mass, and volume are also used in specifying units of time. However, because minutes and hours are terms that indicate *multiples* of seconds, the only prefixes generally used are those indicating *fractional portions* of seconds—for example, millisecond and microsecond.

them. (A listing of the most frequently used conversion factors, for conversions between British and metric system units, is provided in Appendix A.)

Length Measurements
The metric unit of length is the **meter (m)**. In addition to measuring things in meters, you will be expected to measure smaller objects in centimeters or millimeters. Subcellular structures are measured in micrometers.

To help you picture these units of length, some equivalents follow:

One meter (m) is slightly longer than one yard (1 m = 39.37 in.).

One centimeter (cm) is approximately the width of a piece of chalk. (Note: there are 2.54 cm in 1 in.)

One millimeter (mm) is approximately the thickness of the wire of a paper clip or of a mark made by a No. 2 pencil lead.

One micrometer (μm) is extremely tiny and can be measured only microscopically.

Make the following conversions between metric units of length: (24–28)

352 cm = _____ mm   12 cm = _____ mm

150 km = _____ m   1 mm = _____ m

2000 μm = _____ mm

Now, circle the answer that would make the most sense in each of the following statements:

29. A match (in a matchbook) is (0.3, 3, 30) cm long.

30. A standard-size American car is about 4 (mm, cm, m, km) long.

31. John pole-vaults a height of 5 meters, whereas Gerry vaults a height of 5 yards. Does John or Gerry make the more difficult vault?

Volume Measurements
The metric unit of volume is the liter. A **liter (l,** or sometimes **L,** especially without a prefix) is slightly more than a quart (1 L = 1.057 quarts). Liquid products measured in liters are becoming more common, and laboratory solutions often come in 1-liter quantities. Liquid volumes measured out for lab experiments are usually measured in milliliters (ml). (The terms *ml* and *cc,* cubic centimeter, are used interchangeably in laboratory and medical settings.)

To help you visualize metric volumes, the equivalents of some common substances follow:

A 12-oz can of soda is just slightly more than 360 ml.

A cup of coffee is approximately 180 ml.

A fluid ounce is 30 ml (cc).

A teaspoon of vanilla is about 5 ml (cc), and many drug injections are given in 5-ml volumes.

$$\text{microunit} \underset{\times 1000}{\overset{\div 1000}{\rightleftarrows}} \text{milliunit} \underset{\times 10}{\overset{\div 10}{\rightleftarrows}} \text{centiunit} \underset{\times 100}{\overset{\div 100}{\rightleftarrows}} \text{unit} \underset{\times 1000}{\overset{\div 1000}{\rightleftarrows}} \text{kilounit}$$

smallest ⇌ largest

Compute the following:

32. How many 5-ml injections can be prepared from 1 liter of a medicine?

_____

33. A 450-ml volume of alcohol is _____ L.

34. The volume of one grape is approximately 0.004 L. What is the volume of the grape in milliliters?

_____

Mass Measurements

Although many people use the terms *mass* and *weight* interchangeably, this usage is inaccurate. **Mass** is the amount of matter in an object; and an object has a constant mass, regardless of where it is—that is, at sea level, on a mountain top, or in outer space. However, weight varies with gravitational pull; the greater the gravitational pull, the greater the weight. Thus, our astronauts are said to be weightless* when in outer space, but they still have the same mass as they do on earth.

The metric unit of mass is the **gram (g),** and most objects weighed in the laboratory will be measured in terms of this unit or fractions thereof. Medical dosages are usually prescribed in milligrams (mg) or micrograms (μg); and in the clinical agency, body weight (particularly of infants) is typically specified in kilograms (kg) (1 kg 5 2.2 lb).

The following examples are provided to help you become familiar with the masses of some common objects:

Two aspirin tablets have a mass of approximately 1 g.

A nickel has a mass of 5 g.

The mass of an average woman (132 lb) is 60 kg.

Make the following conversions:

35. 300 g = _____ mg = _____ μg

36. 4000 μg = _____ mg = _____ g

* Astronauts are not *really* weightless. It is just that they and their surroundings are being pulled toward the earth at the same speed; and so, in reference to their environment, they appear to float.

37. A nurse must administer to her patient, Mrs. Smith, 5 mg of a drug per kg of body mass. Mrs. Smith weighs 140 lb. How many grams of the drug should the nurse administer to her patient?

_____ g

Temperature Measurements

In the laboratory and in the clinical agency, temperature is measured both in metric units (degrees Celsius, °C) and in British units (degrees Fahrenheit, °F). Thus it helps to be familiar with both temperature scales.

The temperatures of boiling and freezing water can be used to compare the two scales:

The boiling point of water is 100°C and 212°F.

The freezing point of water is 0°C and 32°F.

As you can see, the range from the freezing point to the boiling point of water on the Celsius scale is 100 degrees, whereas the comparable range on the Fahrenheit scale is 180 degrees. Hence, one degree on the Celsius scale represents a greater change in temperature. Normal body temperature is approximately 98.6°F and 37°C.

To convert from the Fahrenheit scale to the Celsius scale (or vice versa), the following equation is used:

$$°C = 5(°F - 32)/9$$

For example, to convert 180°C to °F:

$$°C = 5(°F - 32)/9$$
$$180 = 5(°F - 32)/9$$
$$1620 = 5(°F - 32)$$
$$1620 = 5(°F) - 160$$
$$1780 = 5(°F)$$
$$356 = °F$$

and to convert 72°F to °C:

$$°C = 5(°F - 32)/9$$
$$°C = 5(72 - 32)/9$$
$$°C = 5(40)/9$$
$$°C = 200/9$$
$$°C = 22.2$$

Perform the following temperature conversions:

38. Convert 38°C to °F: _____

39. Convert 158°F to °C: _____

Answers

1. range of 94–64; 30 beats/min
2. average 77.5
3. 3820%
4. 40200%
5. 160%
6. 0.36
7. 8
8. 257.77
9. 0.0005
10. 10 breaths/min
11. 12 to 36
12. interval between 100–102° (went from 22 to 36 breaths/min)
13. yes
14. 10^5
15. 10^6
16. 10^2
17. 10^1
18. 1.4×10^5
19. 6.3×10^{-6}
20. 5.4×10^{-4}

21. 4.5×10^4
22. 2.65×10^{-1}
23. 1.0×10^{-1}
24. cm = 3520 mm
25. km = 150,000 m
26. μm = 2 mm
27. cm = 120 mm
28. mm = 0.001 m
29. 3 cm
30. m long
31. John
32. 200
33. 0.45 L
34. 4 ml
35. 300 g = $\underline{3 \times 10^5}$ mg = $\underline{3 \times 10^8}$ μg
36. 4000 μg = $\underline{4}$ mg = $\underline{4 \times 10^{-3}}$ g (0.004)
37. 0.32 g
38. 100.4°F
39. 70°C

The Language of Anatomy

Most of us are naturally curious about our bodies. This fact is amply demonstrated by infants, who are fascinated with their own waving hands or their mother's nose. The study of the gross anatomy of the human body elaborates on this fascination. Unlike the infant, however, the student of anatomy must learn to observe and identify the dissectible body structures formally. The purpose of any gross-anatomy experience is to examine the three-dimensional relationships of body structures—a goal that can never completely be achieved by using illustrations and models, regardless of their excellence.

When beginning the study of any science, the student is often initially overcome by jargon unique to the subject. The study of anatomy is no exception. But without this specialized terminology, confusion is inevitable. For example, what do *over, on top of, superficial to, above,* and *behind* mean in reference to the human body? Anatomists have an accepted set of reference terms that are universally understood. These allow body structures to be located and identified with a minimum of words and a high degree of clarity. Thus it is not surprising that physicians' orders and progress notes, therapists' records, and nurses' notes use *anatomical terminology* to describe body parts, regions, positions, and activities. The ability to understand and use correct anatomical terminology is a skill that distinguishes health care personnel who are successful and comfortable in their chosen profession from those perpetually unsure of just what is expected of them.

This exercise presents some of the most important anatomical terminology used to describe the body and introduces you to basic concepts of **gross anatomy,** the study of body structures visible to the naked eye.

Anatomical Position

When anatomists or doctors refer to specific areas of the human body, they do so in accordance with a universally accepted standard position called the **anatomical position.** It is essential to understand this position because much of the body terminology employed in this book refers to this body positioning, regardless of the position the body happens to be in. In the anatomical position the human body is erect, with feet together, head and toes pointed forward, and arms hanging at the sides with palms facing forward (Figure 1.1).

• Assume the anatomical position, and notice that it is not particularly comfortable. The hands are held unnaturally forward rather than hanging partially cupped toward the thighs.

Objectives

1. To describe the anatomical position verbally or by demonstration.
2. To use proper anatomical terminology to describe body directions, planes, and surfaces.
3. To name the body cavities and indicate the important organs in each.

Materials

❑ Human torso model (dissectible)
❑ Human skeleton
❑ Demonstration: sectioned and labeled kidneys (three separate kidneys uncut or cut so that (a) entire, (b) transverse section, and (c) longitudinal sectional views are visible)

 See Appendix C, Exercise 1 for links to A.D.A.M. Interactive Anatomy.

Surface Anatomy

Body surfaces provide a wealth of visible landmarks for study of the body.

Anterior Body Landmarks

Note the following regions in Figure 1.2a:

Abdominal: Pertaining to the anterior body trunk region inferior to the ribs

Antebrachial: Pertaining to the forearm

Antecubital: Pertaining to the anterior surface of the elbow

Axillary: Pertaining to the armpit

Brachial: Pertaining to the arm

Buccal: Pertaining to the cheek

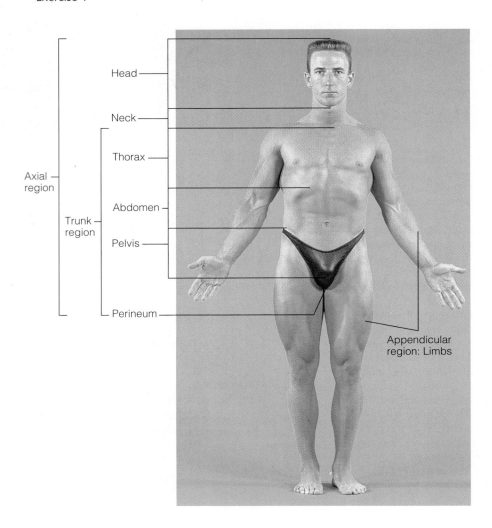

Figure 1.1   Anatomical position.

Carpal: Pertaining to the wrist

Cervical: Pertaining to the neck region

Coxal: Pertaining to the hip

Crural: Pertaining to the leg

Deltoid: Pertaining to the roundness of the shoulder caused by the underlying deltoid muscle

Digital: Pertaining to the fingers or toes

Femoral: Pertaining to the thigh

Frontal: Pertaining to the forehead

Hallux: Pertaining to the great toe

Inguinal: Pertaining to the groin

Mammary: Pertaining to the breast

Manus: Pertaining to the hand

Mental: Pertaining to the chin

Nasal: Pertaining to the nose

Oral: Pertaining to the mouth

Orbital: Pertaining to the bony eye socket (orbit)

Palmar: Pertaining to the palm of the hand

Patellar: Pertaining to the anterior knee (kneecap) region

Pedal: Pertaining to the foot

Pelvic: Pertaining to the pelvis region

Peroneal: Pertaining to the side of the leg

Pollex: Pertaining to the thumb

Pubic: Pertaining to the genital region

Sternal: Pertaining to the region of the breastbone

Tarsal: Pertaining to the ankle

Thoracic: Pertaining to the chest

Umbilical: Pertaining to the navel

Posterior Body Landmarks

Note the following body surface regions in Figure 1.2b:

Acromial: Pertaining to the point of the shoulder

Calcaneal: Pertaining to the heel of the foot

Cephalic: Pertaining to the head

Dorsum: Pertaining to the back

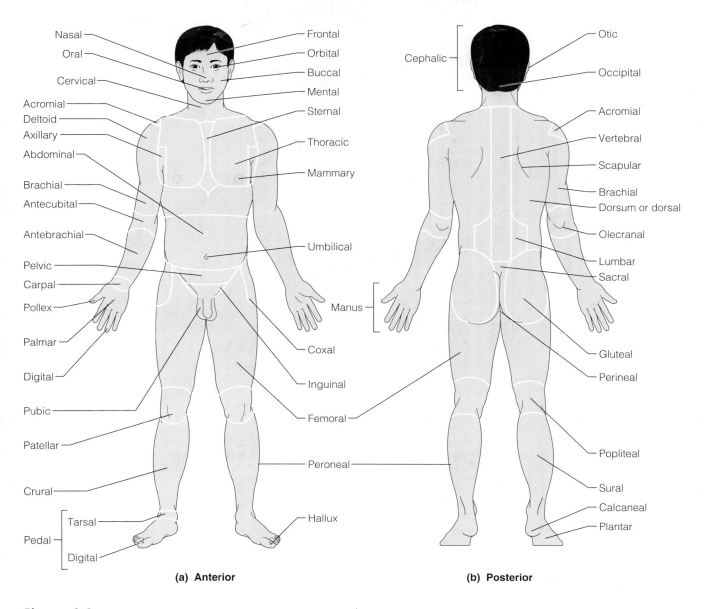

Figure 1.2 Surface anatomy. (a) Anterior body landmarks. **(b)** Posterior body landmarks.

Gluteal: Pertaining to the buttocks or rump

Lumbar: Pertaining to the area of the back between the ribs and hips; the loin

Occipital: Pertaining to the posterior aspect of the head or base of the skull

Olecranal: Pertaining to the posterior aspect of the elbow

Otic: Pertaining to the ear

Perineal: Pertaining to the region between the anus and external genitalia

Plantar: Pertaining to the sole of the foot

Popliteal: Pertaining to the back of the knee

Sacral: Pertaining to the region between the hips (overlying the sacrum)

Scapular: Pertaining to the scapula or shoulder blade area

Sural: Pertaining to the calf or posterior surface of the leg

Vertebral: Pertaining to the area of the spinal column

Activity:
Locating Body Landmarks

Locate the anterior and posterior body landmarks on yourself, your lab partner, and a torso model before continuing.

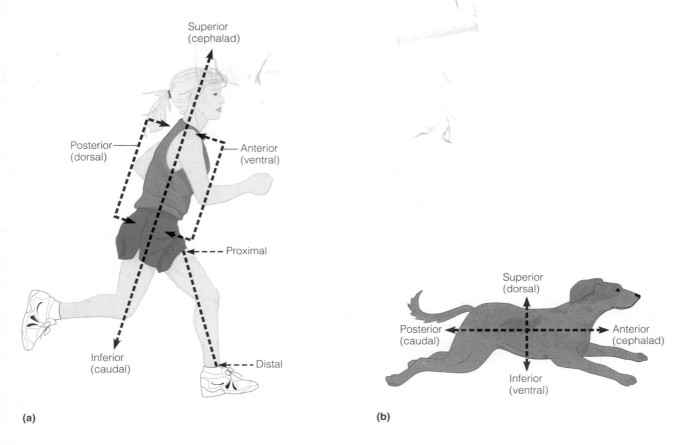

Figure 1.3   Anatomical terminology describing body orientation and direction.
(a) With reference to a human. **(b)** With reference to a four-legged animal.

Body Orientation and Direction

Study the terms below, referring to Figure 1.3. Notice that certain terms have a different meaning for a four-legged animal than they do for a human.

Superior/inferior (*above/below*): These terms refer to placement of a structure along the long axis of the body. Superior structures always appear above other structures, and inferior structures are always below other structures. For example, the nose is superior to the mouth, and the abdomen is inferior to the chest.

Anterior/posterior (*front/back*): In humans the most anterior structures are those that are most forward—the face, chest, and abdomen. Posterior structures are those toward the backside of the body. For instance, the spine is posterior to the heart.

Medial/lateral (*toward the midline/away from the midline or median plane*): The ear is lateral to the nose; the sternum (breastbone) is medial to the ribs.

The terms of position described above depend on an assumption of anatomical position. The next four term pairs are more absolute. Their applicability is not relative to a particular body position, and they consistently have the same meaning in all vertebrate animals.

Cephalad/caudal (*toward the head/toward the tail*): In humans these terms are used interchangeably with *superior* and *inferior.* But in four-legged animals they are synonymous with *anterior* and *posterior* respectively.

Dorsal/ventral (*backside/belly side*): These terms are used chiefly in discussing the comparative anatomy of animals, assuming the animal is standing. *Dorsum* is a Latin word meaning "back." Thus, *dorsal* refers to the animal's back or the *back*side of any other structures; e.g., the posterior surface of the human leg is its dorsal surface. The term *ventral* derives from the Latin term *venter,* meaning "belly," and always refers to the belly side of animals. In humans the terms *ventral* and *dorsal* are used interchangeably with the terms *anterior* and *posterior,* but in four-legged animals *ventral* and *dorsal* are synonymous with *inferior* and *superior* respectively.

Proximal/distal (*nearer the trunk or attached end/farther from the trunk or point of attachment*): These terms are used primarily to locate various areas of the body limbs. For example, the fingers are distal to the elbow; the knee is proximal to the toes.

Superficial/deep (*toward or at the body surface/away from the body surface or more internal*): These terms locate body organs according to their relative closeness to the body surface. For example, the lungs are deep to the rib cage, and the skin is superficial to the skeletal muscles.

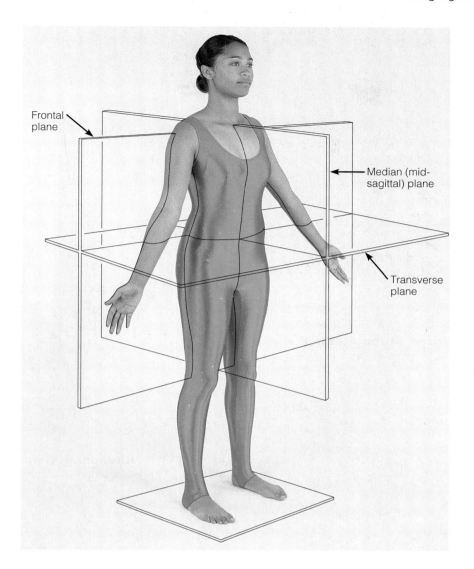

Frontal plane

Median (mid-sagittal) plane

Transverse plane

Figure 1.4   Planes of the body.

A c t i v i t y :
Practicing Using Correct Anatomical Terminology

Before continuing, use a human torso model, a skeleton, or your own body to specify the relationship between the following structures.

1.   The wrist is _____ to the hand.

2.   The trachea (windpipe) is _____ to the spine.

3.   The brain is _____ to the spinal cord.

4.   The kidneys are _____ to the liver.

5.   The nose is _____ to the cheekbones. ■

Body Planes and Sections

The body is three-dimensional, and in order to observe its internal structures, it is often helpful and necessary to make use of a **section,** or cut. When the section is made through the body wall or through an organ, it is made along an imaginary surface or line called a **plane.** Anatomists commonly refer to three planes (Figure 1.4) or sections that lie at right angles to one another.

Sagittal plane: A plane that runs longitudinally, dividing the body into right and left parts, is referred to as a sagittal plane. If it divides the body into equal parts, right down the median plane of the body, it is called a **median,** or **midsagittal, plane.** All other sagittal planes are referred to as **parasagittal planes.**

Frontal plane: Sometimes called a **coronal plane,** the frontal plane is a longitudinal plane that divides the body (or an organ) into anterior and posterior parts.

Transverse plane: A transverse plane runs horizontally, dividing the body into superior and inferior parts. When organs are sectioned along the transverse plane, the sections are commonly called **cross sections.**

As shown in Figure 1.5, a sagittal or frontal plane section of any nonspherical object, be it a banana or a body organ, provides quite a different view than a transverse section.

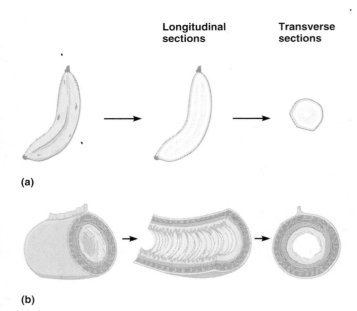

Longitudinal sections

Transverse sections

(a)

(b)

Figure 1.5   Comparison of longitudinal and transverse sections. Sections of **(a)** a banana and **(b)** the small intestine.

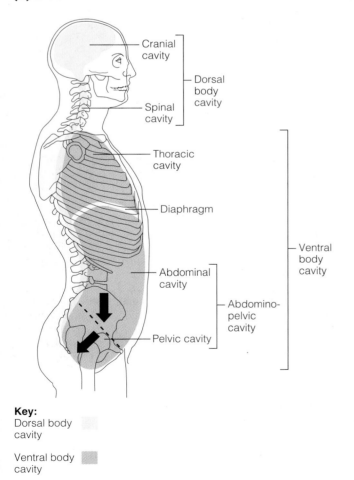

Cranial cavity

Dorsal body cavity

Spinal cavity

Thoracic cavity

Diaphragm

Ventral body cavity

Abdominal cavity

Abdomino-pelvic cavity

Pelvic cavity

Key:
Dorsal body cavity

Ventral body cavity

Figure 1.6   Body cavities. Arrows indicate the angle of the relationship between the abdominal and pelvic cavities.

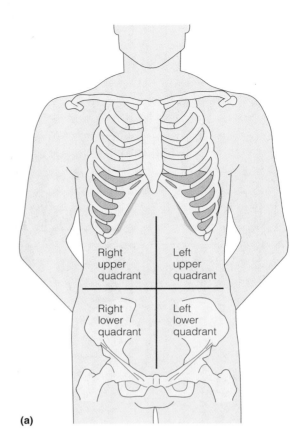

Right upper quadrant

Left upper quadrant

Right lower quadrant

Left lower quadrant

(a)

Figure 1.7   Abdominopelvic surface and cavity. (a) The four quadrants. *(continues on page 7)*

A c t i v i t y :
Observing Organ Specimens

Go to the demonstration area and observe the transversely and longitudinally cut organ specimens. Pay close attention to the different structural details in the samples. ■

Body Cavities

The axial portion of the body has two large cavities that provide different degrees of protection to the organs within them (Figure 1.6).

Dorsal Body Cavity

The dorsal body cavity can be subdivided into the **cranial cavity,** in which the brain is enclosed within the rigid skull, and the **spinal cavity,** within which the delicate spinal cord is protected by the bony vertebral column. Because the spinal cord is a continuation of the brain, these cavities are continuous with each other.

Ventral Body Cavity

Like the dorsal cavity, the ventral body cavity is subdivided. The superior **thoracic cavity** is separated from the rest of the

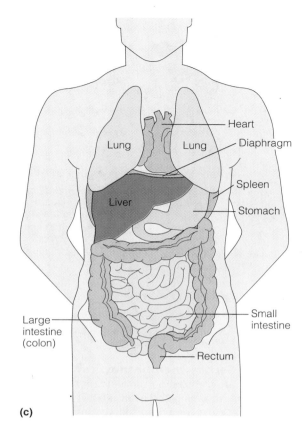

(b)

(c)

Figure 1.7 (continued) **(b)** Nine regions delineated by four planes. The superior horizontal plane is just inferior to the ribs; the inferior horizontal plane is at the superior aspect of the hip bones. The vertical planes are just medial to the nipples. **(c)** Anterior view of the abdominopelvic cavity showing superficial organs.

ventral cavity by the dome-shaped diaphragm. The heart and lungs, located in the thoracic cavity, are afforded some measure of protection by the bony rib cage. The cavity inferior to the diaphragm is often referred to as the **abdominopelvic cavity.** Although there is no further physical separation of the ventral cavity, some prefer to describe the abdominopelvic cavity in terms of a superior **abdominal cavity,** the area that houses the stomach, intestines, liver, and other organs, and an inferior **pelvic cavity,** the region that is partially enclosed by the bony pelvis and contains the reproductive organs, bladder, and rectum. Notice in Figure 1.6 that the abdominal and pelvic cavities are not continuous with each other in a straight plane but that the pelvic cavity is tipped away from the perpendicular.

Abdominopelvic Quadrants and Regions Because the abdominopelvic cavity is quite large and contains many organs, it is helpful to divide it up into smaller areas for discussion or study. The scheme used by most physicians and nurses divides the abdominal surface (and the abdominopelvic cavity deep to it) into four approximately equal regions called **quadrants.** These quadrants are named according to their relative position—that is, *right upper quadrant, right lower quadrant, left upper quadrant,* and *left lower quadrant* (see Figure 1.7a).

Another scheme, commonly used by anatomists, divides the abdominal surface and abdominopelvic cavity into nine separate regions by four planes, as shown in Figure 1.7b. Although the names of these nine regions are unfamiliar to you now, with a little patience and study they will become easier to remember. As you read through the descriptions of these nine regions below and locate them in the figure, notice the organs they contain by referring to Figure 1.7c.

Umbilical region: The centermost region, which includes the umbilicus

Epigastric region: Immediately superior to the umbilical region; overlies most of the stomach

Hypogastric (pubic) region: Immediately inferior to the umbilical region; encompasses the pubic area

Iliac regions: Lateral to the hypogastric region and overlying the superior parts of the hip bones

Lumbar regions: Between the ribs and the flaring portions of the hip bones; lateral to the umbilical region

Hypochondriac regions: Flanking the epigastric region laterally and overlying the lower ribs

A c t i v i t y :

Locating Abdominal Surface Regions

Locate the regions of the abdominal surface on a torso model and on yourself before continuing. ■

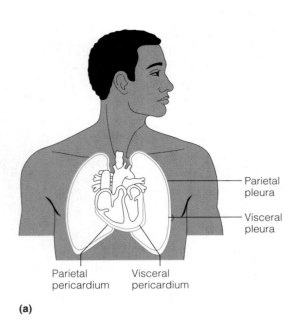

Parietal
pleura

Visceral
pleura

Parietal
pericardium

Visceral
pericardium

(a)

Parietal
peritoneum

Visceral
peritoneum

(b)

Figure 1.8   **Serous membranes. (a)** Serosae in the thoracic cavity are the pleurae surrounding the lungs and the pericardia surrounding the heart. **(b)** The peritoneums surround the digestive viscera in the abdominopelvic cavity.

Serous Membranes of the Ventral Body Cavity   The walls of the ventral body cavity and the outer surfaces of the organs it contains are covered with an exceedingly thin, double-layered membrane called the **serosa,** or **serous membrane.** The part of the membrane lining the cavity walls is referred to as the **parietal serosa,** and it is continuous with a similar membrane, the **visceral serosa,** covering the external surface of the organs within the cavity. These membranes produce a thin lubricating fluid that allows the visceral organs to slide over one another or to rub against the body wall without friction. Serous membranes also compartmentalize the various organs so that infection of one organ is prevented from spreading to others.

The specific names of the serous membranes depend on the structures they envelop. Thus the serosa lining the abdominal cavity and covering its organs is the **peritoneum,** that enclosing the lungs is the **pleura,** and that around the heart is the **pericardium** (Figure 1.8).

Organ Systems Overview

The basic unit or building block of all living things is the **cell.** Cells fall into four different categories according to their structures and functions. Each of these corresponds to one of the four **tissue** types: epithelial, muscular, nervous, and connective. An **organ** is a structure composed of two or more tissue types that performs a specific function for the body. For example, the small intestine, which digests and absorbs nutrients, is composed of all four tissue types.

An **organ system** is a group of organs that act together to perform a particular body function. For example, the organs of the digestive system work together to ensure that food moving through the digestive system is properly broken down and that the end products are absorbed into the bloodstream to provide nutrients and fuel for all the body's cells. In all, there are 11 organ systems, which are described in Table 2.1. The lymphatic system also encompasses a *functional system* called the immune system, which is composed of an army of mobile *cells* that act to protect the body from foreign substances. Read through this summary before beginning your examination of the predissected human cadaver. If a human cadaver is not available, Figures 2.1 through 2.4 will serve as a partial replacement.

Cadaver Observation

Now you will have a chance to observe the general size, shape, location, and distribution of the organs and organ systems.

The following instructions have been written to complement and direct your superficial observations of a previously dissected human cadaver. If a human cadaver is not available, omit the sections concerning external and oral cavity structures and begin your observations of ventral body cavity structures using Figures 2.1–2.4 as references.

Note that four of the organ systems listed in Table 2.1 will not be studied at this time (integumentary, skeletal, muscular, and nervous), as they require microscopic study or more detailed dissection.

Activity:
Observing External Structures

1.   If your instructor has provided a predissected human cadaver, go to the demonstration area to make these observations.

2.   Obtain a probe, forceps, and disposable gloves. Don the gloves before beginning your observations.

3.   Observe the major divisions of the body—head, trunk, and extremities. ■

Objectives

1. To name the human organ systems and indicate the major functions of each.

2. To list two or three organs of each system, and categorize the various organs by organ system.

3. To identify these organs in a human cadaver or on a dissectible human torso model.

4. To identify the correct organ system for each organ when presented with a list of organs (as studied in the laboratory).

Materials

❑ Predissected human cadaver
❑ Probes
❑ Forceps
❑ Disposable plastic gloves
❑ Human torso model (dissectible)

 See Appendix C, Exercise 2 for links to A.D.A.M. Interactive Anatomy.

Activity:
Examining the Oral Cavity

Examine the structures of the oral cavity. Identify the teeth and tongue. Observe the extent of the hard palate (the portion underlain by bone) and the soft palate (immediately posterior to the hard palate, with no bony support). Notice that the posterior end of the oral cavity leads into the throat, or pharynx. The pharynx is a passageway used by both the digestive and respiratory systems. ■

Table 2.1	Overview of Organ Systems of the Body	
Organ system	**Major component organs**	**Function**
Integumentary (Skin)	Epidermal and dermal regions; cutaneous sense organs and glands	• Protects deeper organs from mechanical, chemical, and bacterial injury, and desiccation (drying out) • Excretes salts and urea • Aids in regulation of body temperature • Produces vitamin D
Skeletal	Bones, cartilages, tendons, ligaments, and joints	• Body support and protection of internal organs • Provides levers for muscular action • Cavities provide a site for blood cell formation
Muscular	Muscles attached to the skeleton	• Primary function is to contract or shorten; in doing so, skeletal muscles allow locomotion (running, walking, etc.), grasping and manipulation of the environment, and facial expression • Generates heat
Nervous	Brain, spinal cord, nerves, and sensory receptors	• Allows body to detect changes in its internal and external environment and to respond to such information by activating appropriate muscles or glands • Helps maintain homeostasis of the body via rapid transmission of electrical signals
Endocrine	Pituitary, thyroid, parathyroid, adrenal, and pineal glands; ovaries, testes, and pancreas	• Helps maintain body homeostasis, promotes growth and development; produces chemical "messengers" (hormones) that travel in the blood to exert their effect(s) on various "target organs" of the body
Cardiovascular	Heart, blood vessels, and blood	• Primarily a transport system that carries blood containing oxygen, carbon dioxide, nutrients, wastes, ions, hormones, and other substances to and from the tissue cells where exchanges are made; blood is propelled through the blood vessels by the pumping action of the heart • Antibodies and other protein molecules in the blood act to protect the body
Lymphatic	Lymphatic vessels, lymph nodes, spleen, thymus, tonsils, and scattered collections of lymphoid tissue	• Picks up fluid leaked from the blood vessels and returns it to the blood • Cleanses blood of pathogens and other debris • Houses lymphocytes that act via the immune response to protect the body from foreign substances (antigens)
Respiratory	Nasal passages, pharynx, larynx, trachea, bronchi, and lungs	• Keeps the blood continuously supplied with oxygen while removing carbon dioxide
Digestive	Oral cavity, esophagus, stomach, small and large intestines, and accessory structures (teeth, salivary glands, liver, and pancreas)	• Breaks down ingested foods to minute particles, which can be absorbed into the blood for delivery to the body cells • Undigested residue removed from the body as feces
Urinary	Kidneys, ureters, bladder, and urethra	• Rids the body of nitrogen-containing wastes (urea, uric acid, and ammonia), which result from the breakdown of proteins and nucleic acids by body cells • Maintains water, electrolyte, and acid-base balance of blood
Reproductive	Male: testes, scrotum, penis, and duct system, which carries sperm to the body exterior	• Provides germ cells (sperm) for perpetuation of the species
	Female: ovaries, uterine tubes, uterus, and vagina	• Provides germ cells (eggs) and the female uterus houses the developing fetus until birth

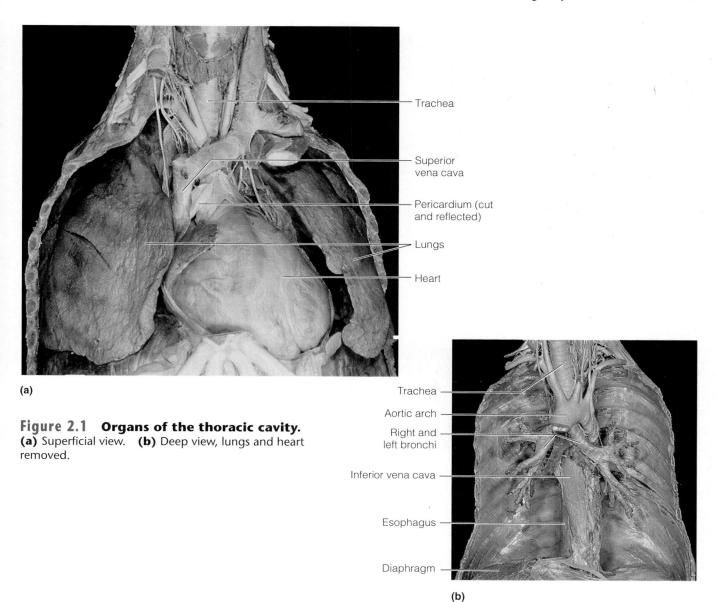

(a)

Figure 2.1   Organs of the thoracic cavity.
(a) Superficial view.   **(b)** Deep view, lungs and heart removed.

(b)

Activity:
Examining the Ventral Body Cavity

1.   Begin your examination with the structures of the thoracic cavity, starting with the most superficial structures and working deeper. As you work, refer to Figure 2.1, which shows the thoracic organs.

Thymus: An irregular mass of glandular tissue overlying the heart (not illustrated).

With the probe, push the thymus to the side to view the heart.

Heart: Medial oval structure enclosed within the pericardium (serous membrane sac).

Lungs: Flanking the heart on either side.

Now observe the throat region to identify the trachea.

Trachea: Tubelike "windpipe" running medially down the throat; part of the respiratory system.

Follow the trachea into the thoracic cavity; notice where it divides into two branches. These are the bronchi.

Bronchi: Two passageways that plunge laterally into the tissue of the two lungs.

To expose the esophagus, push the trachea to one side.

Esophagus: A food chute; the part of the digestive system that transports food from the pharynx (throat) to the stomach.

Diaphragm: A thin muscular membrane attached to the inferior boundary of the rib cage; separates the thoracic and abdominal cavities.

Follow the esophagus through the diaphragm to its junction with the stomach.

Figure 2.2 Abdominal organs. (a) Superficial view. **(b)** Intermediate view. Stomach reflected superiorly to reveal the pancreas, spleen, and duodenum.

Stomach: A C-shaped organ important in food digestion and temporary food storage.

2. Examine the superficial structures of the abdominopelvic cavity. Lift the *greater omentum,* an extension of the peritoneum that covers the abdominal viscera. Continuing from the stomach, trace the rest of the digestive tract (Figure 2.2).

Small intestine: Connected to the stomach and ending just before the saclike cecum.

Cecum: The initial portion of the large intestine.

Large intestine: A large muscular tube coiled within the abdomen.

Follow the course of the large intestine to the rectum, which is partially covered by the urinary bladder.

Rectum: Terminal part of the large intestine; continuous with the anal canal.

Anus: The opening of the digestive tract (anal canal) to the exterior.

Now lift the small intestine with the forceps to view the mesentery.

Mesentery: An apronlike serous membrane; suspends many of the digestive organs in the abdominal cavity. Notice that it is heavily invested with blood vessels and, more likely than not, riddled with large fat deposits.

Locate the remaining abdominal structures.

Adrenal gland

Kidneys

Inferior vena cava

Descending aorta

Ureters

Urinary bladder

Figure 2.3   Deep structures of the abdominopelvic cavity.

Pancreas: A diffuse gland; rests dorsal to and in the mesentery between the first portion of the small intestine and the stomach. You will need to lift the stomach to view the pancreas (Figure 2.2b).

Spleen: A dark red organ curving around the left lateral side of the stomach; considered part of the lymphatic system and often called the red blood cell graveyard.

Liver: Large and brownish red; the most superior organ in the abdominal cavity, directly beneath the diaphragm.

3.   To locate the deeper structures of the abdominopelvic cavity, move the stomach and the intestines to one side with the probe. (Refer to Figure 2.3 as you work.)

Examine the posterior wall of the abdominal cavity to locate the two kidneys.

Kidneys: Bean-shaped organs; retroperitoneal (behind the peritoneum).

Adrenal glands: Large glands that sit astride the superior margin of each kidney; considered part of the endocrine system.

Carefully strip away part of the peritoneum with forceps and attempt to follow the course of one of the ureters to the bladder.

Ureter: Tube running from the indented region of a kidney to the urinary bladder (see also Figure 2.4b).

Urinary bladder: The sac that serves as a reservoir for urine.

4.   In the midline of the body cavity lying between the kidneys are the two principal abdominal blood vessels. Identify each.

Inferior vena cava: The large vein that returns blood to the heart from the lower regions of the body.

Descending aorta: Deep to the inferior vena cava; the largest artery of the body; carries blood away from the heart down the midline of the body.

(a)

(b)

(c)

Figure 2.4 Reproductive organs.
(a) Male external genitalia. **(b)** Sagittal section of the male pelvis. **(c)** Saggital section of the female pelvis.

5. Only a cursory examination of reproductive organs will be done.

Male Cadaver Make a shallow incision into the **scrotum** (Figure 2.4a). Loosen and lift out the oval **testis.** Exert a gentle pull on the testis to identify the slender **ductus (vas) deferens,** or sperm duct, which carries sperm from the testis superiorly into the abdominal cavity and joins with the urethra (Figure 2.4b). The urethra runs through the penis of the male and carries both urine and sperm out of the body. Identify the **penis,** extending from the bladder to the ventral body wall. Figure 2.4b also indicates some glands of the male reproductive system, but they need not be identified at this time.

Female Cadaver Inspect the pelvic cavity to identify the pear-shaped **uterus** lying against the dorsal body wall and beneath the bladder (Figure 2.4c). Follow one of the **uterine tubes** superiorly to identify an **ovary,** a small oval structure at the end of the uterine tube. The inferior part of the uterus is continuous with the **vagina,** which leads to the body exterior. Identify the **vaginal orifice** (external vaginal opening).

6. When you have finished your observations, rewrap or store the cadaver according to your instructor's directions. Wash the dissecting tools with laboratory detergent. Dispose of the gloves. Then wash and dry your hands before continuing with the examination of the torso model. ■

Activity:
Examining the Human Torso Model

Examine a human torso model to identify the organs listed below. (Note: If a torso model is not available, Figure 2.5 may be used for this part of the exercise.)

Dorsal body cavity: Brain, spinal cord

Thoracic cavity: Heart, lungs, bronchi, trachea, esophagus, diaphragm

Abdominopelvic cavity: Liver, stomach, pancreas, spleen, small intestine, large intestine, rectum, kidneys, ureters, bladder, adrenal gland, descending aorta, inferior vena cava

As you observe these structures, locate the nine abdominopelvic areas studied earlier and determine which organs would be found in each area.

1. Umbilical region: _____

2. Epigastric region: _____

3. Hypogastric region: _____

4. Right iliac region: _____

5. Left iliac region: _____

6. Right lumbar region: _____

7. Left lumbar region: _____

8. Right hypochondriac region: _____

9. Left hypochondriac region: _____

Assign each of the organs just identified to one of the organ system categories below.

Digestive: _____

_____

Urinary: _____

_____

Cardiovascular: _____

_____

Endocrine: _____

_____

Reproductive: _____

_____

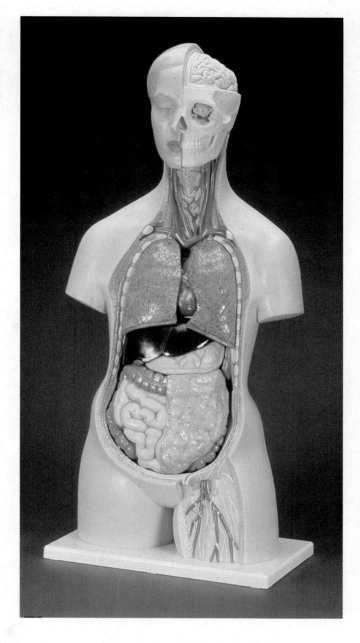

Figure 2.5   Human torso model.

Respiratory: _____

_____

Lymphatic: _____

_____

Nervous: _____

_____

3

The Microscope

Objectives

1. To identify the parts of the microscope and list the function of each.
2. To describe and demonstrate the proper techniques for care of the microscope.
3. To define *total magnification* and *resolution.*
4. To demonstrate proper focusing technique.
5. To define *parfocal, field,* and *depth of field.*
6. To estimate the size of objects in a field.

Materials

- ❏ Compound microscope
- ❏ Millimeter ruler
- ❏ Prepared slides of the letter *e* or newsprint
- ❏ Immersion oil
- ❏ Lens paper
- ❏ Prepared slide of grid ruled in millimeters (grid slide)
- ❏ Prepared slide of 3 crossed colored threads
- ❏ Clean microscope slide and coverslip
- ❏ Toothpicks (flat-tipped)
- ❏ Physiologic saline in a dropper bottle
- ❏ Methylene blue stain (dilute) in a dropper bottle
- ❏ Filter paper
- ❏ Forceps
- ❏ Beaker containing fresh 10% household bleach solution for wet mount disposal
- ❏ Disposable autoclave bag

Note to the Instructor: The slides and coverslips used for viewing cheek cells are to be soaked for 2 hours (or longer) in 10% bleach solution and then drained. The slides, coverslips, and disposable autoclave bag (containing used toothpicks) are to be autoclaved for 15 min at 121°C and 15 pounds pressure to ensure sterility. After autoclaving, the disposable autoclave bag may be discarded in any disposal facility and the glassware washed with laboratory detergent and reprepared for use. These instructions apply as well to any bloodstained glassware or disposable items used in other experimental procedures.

With the invention of the microscope, biologists gained a valuable tool to observe and study structures (like cells) that are too small to be seen by the unaided eye. As a result, many of the theories basic to the understanding of biologic sciences have been established. This exercise will familiarize you with the workhorse of microscopes—the compound microscope—and provide you with the necessary instructions for its proper use.

Care and Structure of the Compound Microscope

The **compound microscope** is a precision instrument and should always be handled with care. At all times you must observe the following rules for its transport, cleaning, use, and storage:

- When transporting the microscope, hold it in an upright position with one hand on its arm and the other supporting its base. Avoid jarring the instrument when setting it down.

- Use only special grit-free lens paper to clean the lenses. Clean all lenses before and after use.

- Always begin the focusing process with the lowest-power objective lens in position, changing to the higher-power lenses as necessary.

- Never use the coarse adjustment knob with the high-power or oil immersion lenses.

- Always use a coverslip with temporary (wet mount) preparations.

- Before putting the microscope in the storage cabinet, remove the slide from the stage, rotate the lowest-power objective lens into position, and replace the dust cover.

- Never remove any parts from the microscope; inform your instructor of any mechanical problems that arise.

Activity:
Identifying the Parts of a Microscope

1. Obtain a microscope and bring it to the laboratory bench. (Use the proper carrying technique!)

- Record the number of your microscope in the summary chart on p. 18.

Compare your microscope with the illustration in Figure 3.1 and identify the following microscope parts:

Head

Arm

Mechanical
stage

Coarse
adjustment
knob

Fine
adjustment
knob

Mechanical
stage
control

Base

Power
switch

Ocular lenses

Rotating
nosepiece

Stage

Objective lenses

Condenser

Iris diaphragm
lever

Substage light

Light control

Figure 3.1   Compound microscope and its parts. (Courtesy of Leica, Inc.)

Base: Supports the microscope. (Note: Some microscopes are provided with an inclination joint, which allows the instrument to be tilted backward for viewing dry preparations.)

Substage light (or *mirror*): Located in the base. In microscopes with a substage light source, the light passes directly upward through the microscope. If a mirror is used, light must be reflected from a separate free-standing lamp.

Stage: The platform the slide rests on while being viewed. The stage has a hole in it to permit light to pass through both it and the specimen. Some microscopes have a stage equipped with *spring clips;* others have a clamp-type *mechanical stage* as shown in Figure 3.1. Both hold the slide in position for viewing; in addition, the mechanical stage permits precise movement of the specimen.

Condenser: Concentrates the light on the specimen. The condenser may have a height-adjustment knob that raises and lowers the condenser to vary light delivery. Generally, the best position for the condenser is close to the inferior surface of the stage.

Iris diaphragm lever: Arm attached to the condenser that regulates the amount of light passing through the condenser. The iris diaphragm permits the best possible contrast when viewing the specimen.

Coarse adjustment knob: Used to focus on the specimen.

Fine adjustment knob: Used for precise focusing once coarse focusing has been completed.

Head or **body tube:** Supports the objective lens system (which is mounted on a movable nosepiece), and the ocular lens or lenses.

Arm: Vertical portion of the microscope connecting the base and head.

Ocular (or *eyepiece*): Depending on the microscope, there will be one or two lenses at the superior end of the head or body tube. Observations are made through the ocular(s). An ocular lens has a magnification of $10\times$ (it increases the apparent size of the object by ten times or ten diameters). If your microscope has a **pointer** (used to indicate a specific area of the viewed specimen), it is attached to the ocular and can be positioned by rotating the ocular lens.

Nosepiece: Generally carries three objective lenses and permits sequential positioning of these lenses over the light beam passing through the hole in the stage.

Objective lenses: Adjustable lens system that permits the use of a **low-power lens,** a **high-power lens,** or an **oil immersion lens.** The objective lenses have different magnifying and resolving powers.

2.   Examine the objectives carefully; note their relative lengths and the numbers inscribed on their sides. On most microscopes, the low-power (l.p.) objective is the shortest and typically has a magnification of $10\times$. The high-power (h.p.) objective is of intermediate length and has a magnification range from $40\times$ to $50\times$, depending on the microscope. The oil immersion objective is usually the longest of the objectives and has a magnifying power of $95\times$ to $100\times$. Note that some microscopes lack the oil immersion lens but have a very low magnification lens called the **scanning lens.** A scanning lens is a very short objective with a magnification of $4\times$ or $5\times$.

•   Record the magnification of each objective lens of your microscope in the first row of the chart below. Also, cross out the column relating to a lens that your microscope does not have.

3.   Rotate the low-power objective until it clicks into position, and turn the coarse adjustment knob about 180 degrees. Notice how far the stage (or objective) travels during this adjustment. Move the fine adjustment knob 180 degrees, noting again the distance that the stage (or the objective) moves. ■

Magnification and Resolution

The microscope is an instrument of magnification. In the compound microscope, magnification is achieved through the interplay of two lenses—the ocular lens and the objective lens. The objective lens magnifies the specimen to produce a **real image** that is projected to the ocular. This real image is magnified by the ocular lens to produce the **virtual image** seen by your eye (Figure 3.2).

The **total magnification** of any specimen being viewed is equal to the power of the ocular lens multiplied by the power of the objective lens used. For example, if the ocular lens magnifies $10\times$ and the objective lens being used magnifies $45\times$, the total magnification is $450\times$ (10×45).

•   Determine the total magnification you may achieve with each of the objectives on your microscope and record the figures on the second row of the chart.

The compound light microscope has certain limitations. Although the level of magnification is almost limitless, the **resolution** (or resolving power), the ability to discriminate two close objects as separate, is not. The human eye can resolve objects about 100 μm apart, but the compound microscope has a resolution of 0.2 μm under ideal conditions. Objects closer than 0.2 μm are seen as a single fused image.

Resolving power (RP) is determined by the amount and physical properties of the visible light that enters the microscope. In general, the more light delivered to the objective lens, the greater the resolution. The size of the objective lens aperture (opening) decreases with increasing magnification, allowing less light to enter the objective. Thus, you will probably find it necessary to increase the light intensity at the higher magnifications.

Summary Chart for Microscope # _____

	Scanning		Low power		High power		Oil immersion	
Magnification of objective lens		$\times$		$\times$		$\times$		$\times$
Total magnification		$\times$		$\times$		$\times$		$\times$
Detail observed								
Field size(diameter)	mm	μm	mm	μm	mm	μm	mm	μm
Working distance		mm		mm		mm		mm

Activity:
Viewing Objects Through the Microscope

1. Obtain a millimeter ruler, a prepared slide of the letter *e* or newsprint, a dropper bottle of immersion oil, and some lens paper. Adjust the condenser to its highest position and switch on the light source of your microscope. (If the light source is not built into the base, use the curved surface of the mirror to reflect the light up into the microscope.)

2. Secure the slide on the stage so that the letter *e* is centered over the light beam passing through the stage. If you are using a microscope with spring clips, make sure the slide is secured at both ends. If your microscope has a mechanical stage, open the jaws of its slide retainer (holder) by using the control lever (typically) located at the rear left corner of the mechanical stage. Insert the slide squarely within the confines of the slide retainer. Check to see that the slide is resting on the stage (and not on the mechanical stage frame) before releasing the control lever.

3. With your lowest power (scanning or low-power) objective in position over the stage, use the coarse adjustment knob to bring the objective and stage as close together as possible.

4. Look through the ocular and adjust the light for comfort. Now use the coarse adjustment knob to focus slowly away from the *e* until it is as clearly focused as possible. Complete the focusing with the fine adjustment knob.

5. Sketch the letter in the circle just as it appears in the **field** (the area you see through the microscope).

What is the total magnification? _____ ×

How far is the bottom of the objective from the specimen? In other words, what is the **working distance**? Use a millimeter ruler to make this measurement and record it here and in the chart on p. 18.

_____ mm

How has the apparent orientation of the *e* changed top to bottom, right to left, and so on?

_____

_____

6. Move the slide slowly away from you on the stage as you view it through the ocular. In what direction does the image move?

_____

Figure 3.2 Image formation in light microscopy. (a) Light passing through the objective lens forms a real image. (b) The real image serves as the object for the ocular lens, which remagnifies the image and forms the virtual image. (c) The virtual image passes through the lens of the eye and is focused on the retina.

Move the slide to the left. In what direction does the image move?

_____

At first this change in orientation will confuse you, but with practice you will learn to move the slide in the desired direction with no problem.

7. Without touching the focusing knobs, increase the magnification by rotating the next higher magnification lens (low-power or high-power) into position over the stage. Make sure it clicks into position. Using the fine adjustment only, sharpen the focus.* What new details become clear?

_____

_____

What is the total magnification now? _____ ×

*Today most good laboratory microscopes are parfocal; that is, the slide should be in focus (or nearly so) at the higher magnifications once you have properly focused in 1.p. If you are unable to swing the objective into position without raising the objective, your microscope is not parfocal. Consult your instructor.

As best you can, measure the distance between the objective and the slide (the working distance) and record it on the chart.

Why should the coarse focusing knob *not* be used when focusing with the higher-powered objective lenses?

_____

_____

Is the image larger or smaller? _____

Approximately how much of the letter *e* is visible now?

_____

Is the field larger or smaller? _____

Why is it necessary to center your object (or the portion of the slide you wish to view) before changing to a higher power?

_____

_____

Move the iris diaphragm lever while observing the field. What happens?

_____

Is it more desirable to increase *or* decrease the light when changing to a higher magnification?

_____ Why? _____

_____

8. If you have just been using the low-power objective, repeat the steps given in direction 7 using the high-power objective lens. Record the total magnification, approximate working distance, and information on detail observed on the chart on p. 18.

9. Without touching the focusing knob, rotate the high-power lens out of position so that the area of the slide over the opening in the stage is unobstructed. Place a drop of immersion oil over the e on the slide and rotate the oil immersion lens into position. Set the condenser at its highest point (closest to the stage), and open the diaphragm fully. Adjust the fine focus and fine-tune the light for the best possible resolution.

Is the field again decreased in size? _____

What is the total magnification with the oil immersion lens?

_____ ×

Is the working distance less *or* greater than it was when the high-power lens was focused?

_____

Figure 3.3 Relative working distances of the 10×, 45×, and 100× objectives.

Compare your observations on the relative working distances of the objective lenses with the illustration in Figure 3.3. Explain why it is desirable to begin the focusing process in low power.

_____

_____

_____

10. Rotate the oil immersion lens slightly to the side and remove the slide. Clean the oil immersion lens carefully with lens paper and then clean the slide in the same manner with a fresh piece of lens paper. ■

Size of the Microscope Field

By this time you should know that the size of the microscope field decreases with increasing magnification. For future microscope work, it will be useful to determine the diameter of each of the microscope fields. This information will allow you to make a fairly accurate estimate of the size of the objects you view in any field. For example, if you have calculated the field diameter to be 4 mm and the object being observed extends across half this diameter, you can estimate the length of the object to be approximately 2 mm.

Microscopic specimens are usually measured in micrometers and millimeters, both units of the metric system. You can get an idea of the relationship and meaning of these units from Table 3.1. A more detailed treatment appears in Appendix A.

| Table 3.1 | Comparison of Metric Units of Length |

Metric unit	Abbreviation	Equivalent
Meter	m	(about 39.3 in.)
Centimeter	cm	10^{-2} m
Millimeter	mm	10^{-3} m
Micrometer (or micron)	μm (μ)	10^{-6} m
Nanometer (or millimicrometer or millimicron)	nm (mμ)	10^{-9} m
Angstrom	Å	10^{-10} m

Activity:
Determining the Size of the Microscope Field

1. Return the letter *e* slide to the supplies area and obtain a grid slide (a slide prepared with graph paper ruled in millimeters). Each of the squares in the grid is 1 mm on each side. Use your lowest power objective to bring the grid lines into focus.

2. Move the slide so that one grid line touches the edge of the field on one side, and then count the number of squares you can see across the diameter of the field. If you can see only part of a square, as in the accompanying diagram, estimate the part of a millimeter that the partial square represents.

~2.5 mm

For future reference, record this figure in the appropriate space marked "field size" on the summary chart on page 18. (If you have been using the scanning lens, repeat the procedure with the low-power objective lens.) Complete the chart by computing the approximate diameter of the high-power and immersion fields. Say the diameter of the low-power field (total magnification of 50×) is 2 mm. You would compute the diameter of a high-power field with a total magnification of 100× as follows:

$$2 \text{ mm} \times 50 = Y \text{ (diameter of h.p. field)} \times 100$$
$$100 \text{ mm} = 100Y$$
$$1 \text{ mm} = Y \text{ (diameter of the h.p. field)}$$

The formula is:
Diameter of the l.p. field (mm) × total magnification of the l.p. field = diameter of field *Y* × total magnification of field *Y*

3. Estimate the length (longest dimension) of the following microscopic objects. *Base your calculations on the field sizes you have determined for your microscope.*

a. Object seen in low-power field:

approximate length:

———— mm

b. Object seen in high-power field:

approximate length:

———— mm

or ———— μm

c. Object seen in oil immersion field:

approximate length:

———— μm

4. If an object viewed with the oil immersion lens looked like the field depicted just below, could you determine its approximate size from this view?

If not, then how could you determine it? ——————

————————————————

————————————————

———————————————— ■

Perceiving Depth

Any microscopic specimen has depth as well as length and width; it is rare indeed to view a tissue slide with just one layer of cells. Normally you can see two or three cell thicknesses. Therefore, it is important to learn how to determine relative depth with your microscope.*

*In microscope work the **depth of field** (the depth of the specimen clearly in focus) is greater at lower magnifications.

Activity:
Perceiving Depth

1.   Return the grid slide and obtain a slide with colored crossed threads. Focusing at low magnification, locate the point where the three threads cross each other.

2.   Use the iris diaphragm lever to greatly reduce the light, thus increasing the contrast. Focus down with the coarse adjustment until the threads are out of focus, then slowly focus upward again, noting which thread comes into clear focus first. This one is the lowest or most inferior thread. (You will see two or even all three threads, so you must be very careful in determining which one comes into clear focus first.) Record your observations:

_____ thread over _____

Continue to focus upward until the uppermost thread is clearly focused. Again record your observation.

_____ thread over _____

Which thread is uppermost? _____

Lowest? _____ ■

Activity:
Preparing and Observing a Wet Mount

1.   Obtain the following: a clean microscope slide and coverslip, a flat-tipped toothpick, a dropper bottle of physiologic saline, a dropper bottle of methylene blue stain, forceps, and filter paper.

2.   Place a drop of physiologic saline in the center of the slide. Using the flat end of the toothpick, *gently* scrape the inner lining of your cheek. Agitate the end of the toothpick containing the cheek scrapings in the drop of saline (Figure 3.4a).

⚠️   *Immediately* discard the used toothpick in the disposable autoclave bag provided at the supplies area.

3.   Add a tiny drop of the methylene blue stain to the preparation. (These epithelial cells are nearly transparent and thus difficult to see without the stain, which colors the nuclei of the cells and makes them look much darker than the cytoplasm.) Stir again and then dispose of the toothpick as described above.

4.   Hold the coverslip with the forceps so that its bottom edge touches one side of the fluid drop (Figure 3.4b), then *carefully* lower the coverslip onto the preparation (Figure

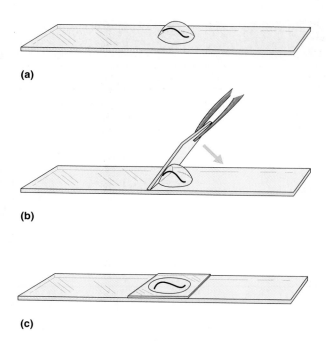

(a)

(b)

(c)

Figure 3.4 Procedure for preparation of a wet mount. (a) The object is placed in a drop of water (or saline) on a clean slide, **(b)** a coverslip is held at a 45° angle with forceps, and **(c)** it is lowered carefully over the water and the object.

3.4c). *Do not just drop the coverslip,* or you will trap large air bubbles under it, which will obscure the cells. *A coverslip should always be used with a wet mount* to prevent soiling the lens if you should misfocus.

5.   Examine your preparation carefully. The coverslip should be closely apposed to the slide. If there is excess fluid around its edges, you will need to remove it. Obtain a piece of filter paper, fold it in half, and use the folded edge to absorb the excess fluid.

⚠️   Before continuing, discard the filter paper in the disposable autoclave bag.

6.   Place the slide on the stage and locate the cells in low power. You will probably want to dim the light with the iris diaphragm to provide more contrast for viewing the lightly stained cells. Furthermore, a wet mount will dry out quickly in bright light, because a bright light source is hot.

7.   Cheek epithelial cells are very thin, six-sided cells. In the cheek, they provide a smooth, tilelike lining, as shown in Figure 3.5.

8.   Make a sketch of the epithelial cells that you observe.

Approximately how wide are the cheek epithelial cells?

_____ mm

Why do *your* cheek cells look different than those illustrated in Figure 3.5? (Hint: what did you have to *do* to your cheek to obtain them?)

_____

_____

⚠ 9.   When you complete your observations, dispose of your wet mount preparation in the beaker of bleach solution.

10. Before leaving the laboratory, make sure all other materials are properly discarded or returned to the appropriate laboratory station. Clean the microscope lenses and put the dust cover on the microscope before you return it to the storage cabinet. ■

Figure 3.5 Epithelial cells of the cheek cavity (surface view, 488×).

The Cell—Anatomy and Division

Objectives

1. To define *cell, organelle,* and *inclusion.*

2. To identify on a cell model or diagram the following cellular regions and to list the major function of each: nucleus, cytoplasm, and plasma membrane.

3. To identify and list the major functions of the various organelles studied.

4. To compare and contrast specialized cells with the concept of the "generalized cell."

5. To define *interphase, mitosis,* and *cytokinesis.*

6. To list the stages of mitosis and describe the events of each stage.

7. To identify the mitotic phases on projected slides or appropriate diagrams.

8. To explain the importance of mitotic cell division and its product.

Materials

❑ Three-dimensional model of the "composite" animal cell or laboratory chart of cell anatomy

❑ Prepared slides of simple squamous epithelium (AgNO$_3$ stain), teased smooth muscle, human blood cell smear, and sperm

❑ Compound microscope

❑ Prepared slides of whitefish blastulae

❑ Three-dimensional models of mitotic stages

Note to the Instructor: See directions for handling wet mount preparations and disposable supplies on p. 16, Exercise 3.

The **cell,** the structural and functional unit of all living things, is a very complex entity. The cells of the human body are highly diverse, and their differences in size, shape, and internal composition reflect their specific roles in the body. Nonetheless, cells do have many common anatomical features, and there are some functions that all must perform to sustain life. For example, all cells have the ability to maintain their boundaries, to metabolize, to digest nutrients and dispose of wastes, to grow and reproduce, to move, and to respond to a stimulus. Most of these functions are considered in detail in later exercises. This exercise fo-

cuses on structural similarities that typify the "composite," or "generalized," cell and considers only the function of cell reproduction (cell division). Transport mechanisms (the means by which substances cross the plasma membrane) are dealt with separately in Exercise 5.

Anatomy of the Composite Cell

In general, all cells have three major regions, or parts, that can readily be identified with a light microscope: the **nucleus,** the **plasma membrane,** and the **cytoplasm.** The nucleus is usually seen as a round or oval structure near the center of the cell. It is surrounded by cytoplasm, which in turn is enclosed by the plasma membrane. Since the advent of the electron microscope, even smaller cell structures—organelles—have been identified. Figure 4.1a is a diagrammatic representation of the fine structure of the composite cell; Figure 4.1b depicts cellular structure as revealed by the electron microscope.

Nucleus

The nucleus is often described as the control center of the cell and is necessary for cell reproduction. A cell that has lost or ejected its nucleus (for whatever reason) is literally programmed to die because the nucleus is the site of the "genes," or genetic material—DNA.

When the cell is not dividing, the genetic material is loosely dispersed throughout the nucleus in a threadlike form called **chromatin.** When the cell is in the process of dividing to form daughter cells, the chromatin coils and condenses to form dense, darkly staining rodlike bodies called **chromosomes**—much in the way a stretched spring becomes shorter and thicker when it is released. (Cell division is discussed later in this exercise.) Notice the appearance of the nucleus carefully—it is somewhat nondescript when a cell is healthy. When the nucleus appears dark and the chromatin becomes clumped, this is an indication that the cell is dying and undergoing degeneration.

The nucleus also contains one or more small round bodies, called **nucleoli,** composed primarily of proteins and ribonucleic acid (RNA). The nucleoli are assembly sites for ribosomal particles (particularly abundant in the cytoplasm), which are the actual protein-synthesizing "factories."

The nucleus is bound by a double-layered porous membrane, the **nuclear membrane** (or nuclear envelope). The nuclear membrane is similar in composition to other cellular membranes, but it is distinguished by its large *nuclear pores,* which permit easy passage of protein and RNA molecules.

(a)

(b)

Figure 4.1 Anatomy of the composite animal cell. (a) Diagrammatic view. **(b)** Transmission electron micrograph (10,000×).

Activity:

Identifying Parts of a Cell

Identify the nuclear membrane, chromatin, nucleoli, and the nuclear pores in Figure 4.1a and b. ■

Plasma Membrane

The **plasma membrane** separates cell contents from the surrounding environment. Its main structural building blocks are phospholipids (fats) and globular protein molecules, but some of the externally facing proteins and lipids have sugar (carbohydrate) side chains attached to them that are important in cellular interactions (Figure 4.2). Described by the fluid-mosaic model, the membrane appears to have a bimolecular lipid core that the protein molecules float in. Occasional cholesterol molecules dispersed in the fluid phospholipid bilayer help stabilize it.

Besides providing a protective barrier for the cell, the plasma membrane plays an active role in determining which substances may enter or leave the cell and in what quantity, maintains a resting potential that is essential to normal functioning of excitable cells, and plays a vital role in cell signaling and cell-to-cell interactions. In some cells the membrane is thrown into minute fingerlike projections or folds called **microvilli,** which greatly increase the surface area of the cell available for absorption or passage of materials, and binding of signaling molecules.

Activity:

Identifying Components of a Plasma Membrane

Identify the phospholipid and protein portions of the plasma membrane in Figure 4.2. Also locate the sugar side chains and cholesterol molecules. Identify the microvilli in Figure 4.1. ■

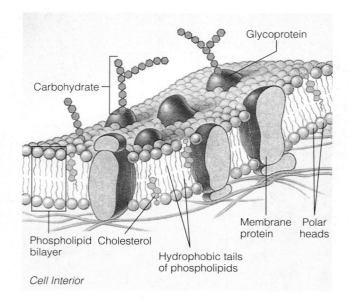

Figure 4.2 Structural details of the plasma membrane.

Cytoplasm and Organelles

The cytoplasm consists of the cell contents outside the nucleus. It is the major site of most activities carried out by the cell. Suspended in the **cytosol,** the fluid cytoplasmic material, are many small structures called **organelles** (literally, "small organs"). The organelles are the metabolic machinery of the cell, and they are highly organized to carry out specific functions for the cell as a whole. The organelles include the ribosomes, endoplasmic reticulum, Golgi apparatus, lysosomes, peroxisomes, mitochondria, cytoskeletal elements, and centrioles.

Activity:
Locating Organelles

Each organelle type is summarized in Table 4.1 and described briefly next. Read through this material and then, as best you can, locate the organelles in both Figure 4.1a and b. ■

- The **ribosomes** are densely staining spherical bodies composed of RNA and protein. They are the actual sites of protein synthesis. They are seen floating free in the cytoplasm or attached to a membranous structure. When they are attached, the whole ribosome-membrane complex is called the *rough endoplasmic reticulum.*

- The **endoplasmic reticulum (ER)** is a highly folded system of membranous tubules and cisternae (sacs) that extends throughout the cytoplasm. The ER is continuous with the Golgi apparatus, nuclear membrane, and plasma membrane. Thus it is assumed that the ER provides a system of channels for the transport of cellular substances (primarily proteins) from one part of the cell to another or to the cell exterior. The ER exists in two forms; a particular cell may have

both or only one, depending on its specific functions. The **rough ER,** as noted earlier, is studded with ribosomes. Its cisternae modify and store the newly formed proteins and dispatch them to other areas of the cell. The external face of the rough ER is involved in phospholipid and cholesterol synthesis. The amount of rough ER is closely correlated with the amount of protein a cell manufactures and is especially abundant in cells that make protein products for export—for example, the pancreas cells that produce digestive enzymes destined for the small intestine. The **smooth ER** has no protein synthesis–related function but is present in conspicuous amounts in cells that produce steroid-based hormones—for example, the interstitial cells of the testes, which produce testosterone. Smooth ER is also abundant in cells that are highly active in lipid metabolism and drug detoxification activities—liver cells, for instance.

- The **Golgi apparatus** is a stack of flattened sacs with bulbous ends that is generally found close to the nucleus. Within its cisterns, the proteins delivered to it from the rough ER are modified (by attachment of sugar groups), segregated, and packaged into membranous vesicles that ultimately (1) are incorporated into the plasma membrane, (2) become secretory vesicles that release their contents from the cell, or (3) become lysosomes.

- The **lysosomes,** which appear in various sizes, are membrane-bound sacs containing an array of powerful digestive enzymes. A product of the packaging activities of the Golgi apparatus, the lysosomes contain *acid hydrolase* enzymes capable of digesting worn-out cell structures and foreign substances that enter the cell through phagocytosis or pinocytosis (see Exercise 5). Lysosomes also bring about some of the changes that occur during menstruation, when the uterine lining is sloughed off. Because they have the capacity of total cell destruction, the lysosomes are often referred to as the "suicide sacs" of the cell.

- **Peroxisomes,** like lysosomes, are enzyme-containing sacs. However, their *oxidase* enzymes have a different task. Using oxygen, they detoxify a number of harmful substances, most importantly free radicals. Peroxisomes are particularly abundant in kidney and liver cells, cells that are actively involved in detoxification.

- The **mitochondria** are generally rod-shaped bodies with a double-membrane wall; the inner membrane is thrown into folds, or *cristae.* Oxidative enzymes on or within the mitochondria catalyze the reactions of the Krebs cycle and the electron transport chain (collectively called oxidative respiration), in which foods are broken down to produce energy. The released energy is captured in the bonds of ATP (adenosine triphosphate) molecules, which are then transported out of the mitochondria to provide a ready energy supply to power the cell. Every living cell requires a constant supply of ATP for its many activities. Because the mitochondria provide the bulk of this ATP, they are referred to as the powerhouses of the cell.

- The **cytoskeletal elements** ramify throughout the cytoplasm forming an internal scaffolding called the *cytoskeleton* that supports and moves substances within the cell. The **microtubules** are basically slender tubules formed of proteins called *tubulins,* which have the ability to aggregate and then disaggregate spontaneously. Microtubules organize the cy-

Table 4.1	Cytoplasmic Organelles

Organelle	Location and function
Ribosomes	Tiny spherical bodies composed of RNA and protein; actual sites of protein synthesis; floating free or attached to a membranous structure (the rough ER) in the cytoplasm
Endoplasmic reticulum (ER)	Membranous system of tubules that extends throughout the cytoplasm; two varieties: rough ER—studded with ribosomes (tubules of the rough ER provide an area for storage and transport of the proteins made on the ribosomes to other cell areas; external face synthesizes phospholipids and cholesterol); smooth ER—no protein synthesis–related function (a site of steroid and lipid synthesis, lipid metabolism, and drug detoxification)
Golgi apparatus	Stack of flattened sacs with bulbous ends and associated small vesicles; found close to the nucleus; role in packaging proteins or other substances for export from the cell or incorporation into the plasma membrane and in packaging lysosomal enzymes
Lysosomes	Various-sized membranous sacs containing digestive enzymes (acid hydrolases); function to digest worn-out cell organelles and foreign substances that enter the cell; have the capacity of total cell destruction if ruptured
Peroxisomes	Small lysosome-like membranous sacs containing oxidase enzymes that detoxify alcohol, hydrogen peroxide, and other harmful chemicals
Mitochondria	Generally rod-shaped bodies with a double-membrane wall; inner membrane is thrown into folds, or cristae; contain enzymes that oxidize foodstuffs to produce cellular energy (ATP); often referred to as "powerhouses of the cell"
Centrioles	Paired, cylindrical bodies lie at right angles to each other, close to the nucleus; direct the formation of the mitotic spindle during cell division; form the bases of cilia and flagella
Cytoskeletal elements: microtubules, intermediate filaments, and microfilaments	Provide cellular support; function in intracellular transport; microtubules form the internal structure of the centrioles and help determine cell shape; intermediate filaments, stable elements composed of a variety of proteins, resist mechanical forces acting on cells; microfilaments are formed largely of actin, a contractile protein, and thus are important in cell mobility (particularly in muscle cells)

toskeleton and direct formation of the spindle formed by the centrioles during cell division. They also act in the transport of substances down the length of elongated cells (such as neurons), suspend organelles, and help maintain cell shape by providing rigidity to the soft cellular substance. **Intermediate filaments** are *stable* proteinaceous cytoskeletal elements that act as internal guy wires to resist mechanical (pulling) forces acting on cells. **Microfilaments,** ribbon or cordlike elements, are formed of contractile proteins. Because of their ability to shorten and then relax to assume a more elongated form, these are important in cell mobility and are very conspicuous in cells that are specialized to contract (such as muscle cells). A cross-linked network of microfilaments braces and strengthens the internal face of the plasma membrane.

The cytoskeletal structures are labile and minute. With the exception of the microtubules of the spindle, which are very obvious during cell division (see pp. 30–31), and the microfilaments of skeletal muscle cells (see p. 123), they are rarely seen, even in electron micrographs, and are not depicted in Figure 4.1b. However, special stains can reveal the plentiful supply of these very important organelles (see Plate 5 in the Histology Atlas).

• The paired **centrioles** lie close to the nucleus in all animal cells capable of reproducing themselves. They are rod-shaped bodies that lie at right angles to each other. Internally each centriole is composed of nine triplets of microtubules. During cell division, the centrioles direct the formation of the mitotic spindle. Centrioles also form the basis for cell projections called cilia and flagella.

The cell cytoplasm contains various other substances and structures, including stored foods (glycogen granules and lipid droplets), pigment granules, crystals of various types, water vacuoles, and ingested foreign materials. But these are not part of the active metabolic machinery of the cell and are therefore called **inclusions.**

Activity:
Examining the Cell Model

Once you have located all of these structures in Figure 4.1a, examine the cell model (or cell chart) to repeat and reinforce your identifications. ∎

Differences and Similarities in Cell Structure

Activity:
Observing Various Cell Structures

1.   Obtain a compound microscope and prepared slides of simple squamous epithelium, sperm, smooth muscle cells (teased), and human blood.

2.   Observe each slide under the microscope, carefully noting similarities and differences in the cells. (The oil immersion lens will be needed to observe blood and sperm.) Distinguish the limits of the individual cells, and notice the shape and position of the nucleus in each case. When you look at the human blood smear, direct your attention to the red blood cells, the pink-stained cells that are most numerous. The color photomicrographs illustrating a blood smear (Plate 57) and sperm (Plate 52) that appear in the Histology Atlas may be helpful in this cell structure study. Sketch your observations in the circles provided.

3.   How do these four cell types differ in shape and size?

_____

_____

_____

_____

How might cell shape affect cell function?

_____

_____

_____

_____

_____

Which cells have visible projections?

_____

How do these projections relate to the function of these cells?

_____

_____

_____

Do any of these cells lack a cell membrane? _____

A nucleus? _____

In the cells with a nucleus, can you discern nucleoli?

_____

Were you able to observe any of the organelles in these cells?

_____ Why or why not? _____

_____

_____

_____

_____ ■

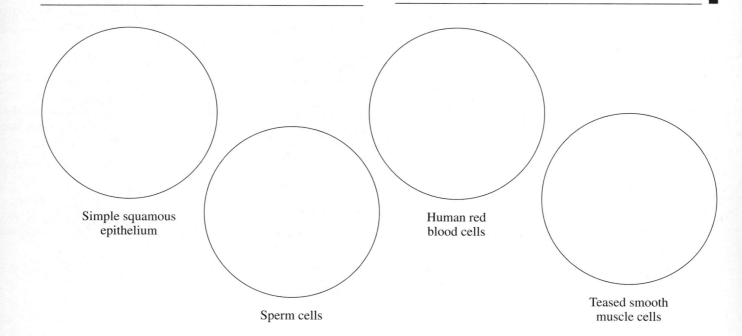

Simple squamous epithelium

Sperm cells

Human red blood cells

Teased smooth muscle cells

Cell Division: Mitosis and Cytokinesis

A cell's *life cycle* is the series of changes it goes through from the time it is formed until it reproduces itself. It encompasses two stages—**interphase,** the longer period during which the cell grows and carries out its usual activities, and **cell division,** when the cell reproduces itself by dividing. In an interphase cell about to divide, the genetic material (DNA) is replicated (duplicated exactly). Once this important event has occurred, cell division ensues.

Cell division in all cells other than bacteria consists of a series of events collectively called mitosis and cytokinesis. **Mitosis** is nuclear division; **cytokinesis** is the division of the cytoplasm, which begins after mitosis is nearly complete. Although mitosis is usually accompanied by cytokinesis, in some instances cytoplasmic division does not occur, leading to the formation of binucleate (or multinucleate) cells. This is relatively common in the human liver and during embryonic development of skeletal muscle cells.

The process of **mitosis** results in the formation of two daughter nuclei that are genetically identical to the mother nucleus. This distinguishes mitosis from **meiosis,** a specialized type of nuclear division that occurs only in the reproductive organs (testes or ovaries). Meiosis, which yields four daughter nuclei that differ genetically in composition from the mother nucleus, is used only for the production of eggs and sperm (gametes) for sexual reproduction. The function of cell division, including mitosis and cytokinesis in the body, is to increase the number of cells for growth and repair while maintaining their genetic heritage.

The stages of mitosis illustrated in Figure 4.3 include the following events:

Prophase (Figure 4.3b and c): At the onset of cell division, the chromatin threads coil and shorten to form densely staining, short, barlike **chromosomes.** By the middle of prophase the chromosomes appear as double-stranded structures (each strand is a **chromatid**) connected by a small median body called a **centromere.** The centrioles separate from one another and act as focal points for the assembly of two systems of microtubules: the **mitotic spindle,** which forms between the centrioles, and the **asters,** which radiate outward from the ends of the spindle and anchor it to the plasma membrane. Some of the spindle microtubules, the *kinetochore microtubules,* attach to special protein complexes on each chromosome's centromere. Spindle fibers that do not attach to the chromosomes are called *polar microtubules.* The spindle acts as a scaffolding for the attachment and movement of the chromosomes during later mitotic stages. Meanwhile, the nuclear membrane and the nucleolus break down and disappear.

Metaphase (Figure 4.3d): A brief stage, during which the chromosomes migrate to the central plane or equator of the spindle and align along that plane in a straight line (the so-called *metaphase plate*) from the superior to the inferior region of the spindle (lateral view). Viewed from the poles of the cell (end view), the chromosomes appear to be arranged in a "rosette," or circle, around the widest dimension of the spindle.

Anaphase (Figure 4.3e): During anaphase, the centromeres split, and the chromatids (now called chromosomes again) separate from one another and then progress slowly toward opposite ends of the cell. The chromosomes are pulled by the kinetochore microtubules attached to their centromeres, their "arms" dangling behind them. Anaphase is complete when poleward movement ceases.

Telophase (Figure 4.3f): During telophase, the events of prophase are essentially reversed. The chromosomes clustered at the poles begin to uncoil and resume the chromatin form, the spindle breaks down and disappears, a nuclear membrane forms around each chromatin mass, and nucleoli appear in each of the daughter nuclei.

Mitosis is essentially the same in all animal cells, but depending on the type of tissue, it takes from 5 minutes to several hours to complete. In most cells, centriole replication is deferred until interphase of the next cell cycle.

Cytokinesis, or the division of the cytoplasmic mass, begins during telophase (Figure 4.3f), and provides a good guideline for where to look for the mitotic figures of telophase. In animal cells, a *cleavage furrow* begins to form approximately over the equator of the spindle, and eventually splits or pinches the original cytoplasmic mass into two portions. Thus at the end of cell division two daughter cells exist, each smaller in cytoplasmic mass than the mother cell but genetically identical to it. The daughter cells grow and carry out the normal spectrum of metabolic processes until it is their turn to divide.

Cell division is extremely important during the body's growth period. Most cells (excluding nerve cells) undergo mitosis until puberty, when normal body size is achieved and overall body growth ceases. After this time in life, only certain cells carry out cell division routinely—for example, cells subjected to abrasion (epithelium of the skin and lining of the gut). Other cell populations—such as liver cells—stop dividing but retain this ability should some of them be removed or damaged. Skeletal muscle, cardiac muscle, and nervous tissue completely lose this ability to divide and thus are severely handicapped by injury. Throughout life, the body retains its ability to repair cuts and wounds and to replace some of its aged cells.

A c t i v i t y :
Identifying the Mitotic Stages

1. Using the three-dimensional models of dividing cells provided, identify each of the mitotic states described above.

2. Obtain a prepared slide of whitefish blastulae to study the stages of mitosis. The cells of each *blastula* (a stage of embryonic development consisting of a hollow ball of cells) are at approximately the same mitotic stage, so it may be necessary to observe more than one blastula to view all the mitotic stages. The exceptionally high rate of mitosis observed in this tissue is typical of embryos, but if occurring in specialized tissues, it can indicate cancerous cells, which also have an extraordinarily high mitotic rate. Examine the slide carefully, identifying the four mitotic stages and the process of cytokinesis. Compare your observations with Figure 4.3, and verify your identifications with your instructor. ■

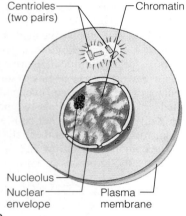

Centrioles (two pairs) — Chromatin

Nucleolus —
Nuclear envelope — Plasma membrane

(a)
Interphase ⟶

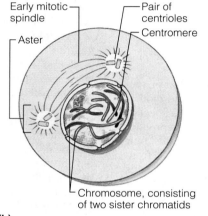

Early mitotic spindle — Pair of centrioles / Centromere

Aster

Chromosome, consisting of two sister chromatids

(b)
Early prophase ⟶

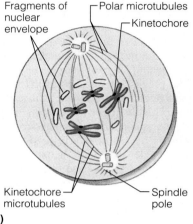

Fragments of nuclear envelope — Polar microtubules / Kinetochore

Kinetochore microtubules — Spindle pole

(c)
Late prophase ⟶

Interphase is the period of a cell's life when it is carrying out its normal metabolic activities and growing. During interphase, the chromosomal material is seen in the form of diffuse (extended) chromatin, and the nuclear membrane and nucleolus are intact and visible. Microtubule arrays (asters) are seen extending from the centrosomes. During various periods of this phase, the centrioles begin replicating (G_1 through G_2), DNA is replicated (S), and the final preparations for mitosis are completed (G_2). The centriole pair finishes replicating into two pairs during G_2.

Prophase, the first and longest phase of mitosis, begins when the chromatin threads start to coil and condense, forming barlike *chromosomes* that are visible with a light microscope. Since DNA replication has occurred during interphase, each chromosome is actually made up of two identical chromatin threads, now called *chromatids*. The chromatids of each chromosome are held together by a small, but-tonlike body called a *centromere*. After the chromatids separate, each will be considered a new chromosome.

As the chromosomes appear, the nucleoli disappear, and the cytoskeletal microtubules disassemble. The centriole pairs separate from one another. The centrioles act as focal points for growth of a new assembly of microtubules called the **mitotic spindle**. As these microtubules lengthen, they push the centrioles farther and farther apart, propelling them toward opposite ends (poles) of the cell.

While the centrioles are still moving away from each other, the nuclear membrane fragments, allowing the spindle to occupy the center of the cell and to interact with the chromosomes. The centrioles also cause the *asters*, microtubules that radiate outward from the ends of the spindle and anchor it to the plasma membrane. Meanwhile, some of the growing spindle microtubules attach to special protein complexes, called *kinetochores* (ki-ne′to-korz), on each chromosome's centromere. Such microtubules are called *kinetochore microtubules*. The remaining spindle microtubules, which do not attach to any chromosomes, are called *polar*, or *nonkinetochore*, *microtubules*. The kinetochore microtubules pull on each chromosome from both poles, resulting in a tug-of-war that ultimately draws the chromosomes to the middle of the cell.

Figure 4.3 The interphase cell and the stages of mitosis. The cells shown are from an early embryo of a whitefish. Photomicrographs are above; corresponding diagrams are below. (Micrographs approximately 600×).

(d)

Metaphase ——————————————►

Metaphase is the second phase of mitosis. The chromosomes cluster at the middle of the cell, with their centromeres precisely aligned at the exact center, or *equator*, of the spindle. This arrangement of the chromosomes along a plane midway between the poles is called the *metaphase plate*.

(e)

Anaphase ——————————————►

Anaphase, the third phase of mitosis, begins abruptly as the centromeres of the chromosomes split, and each chromatid now becomes a chromosome in its own right. The kinetochore fibers shorten (probably due to their disassembly at their kinetochore ends) and, like a recoiling rubberband, gradually pull each chromosome toward the pole it faces. The polar microtubules slide past each other (a process presumed to be driven by kinesin motor molecules), lengthen, and push the two poles of the cell apart, causing the whole cell to elongate. Anaphase is easy to recognize because the moving chromosomes look V-shaped. The centromeres, which are attached to the kinetochore microtubules, lead the way, and the chromosomal "arms" dangle behind them. Anaphase is the shortest stage of mitosis; it typically lasts only a few minutes.

This process of moving and separating the chromosomes is helped by the fact that the chromosomes are short, compact bodies. Diffuse chromatin threads would tangle, trail, and break, which would damage the genetic material and result in its imprecise "parceling out" to the daughter cells.

(f)

Telophase and cytokinesis

Telophase begins as soon as chromosomal movement stops. This final phase is like prophase in reverse. The identical sets of chromosomes at the opposite poles of the cell uncoil and resume their threadlike extended chromatin form. A new nuclear membrane, derived from the rough ER, reforms around each chromatin mass. Nucleoli reappear within the nuclei, and the spindle breaks down and disappears. Mitosis is now ended. The cell, for just a brief period, is a binucleate cell (has two nuclei) and each new nucleus is identical to the original mother nucleus.

As a rule, as mitosis draws to a close, *cytokinesis* occurs, completing the division of the cell into two daughter cells. Cytokinesis occurs as a ring of peripheral microfilaments (not shown) contracts at the *cleavage furrow* and squeezes the cells apart.

Figure 4.3 (*continued*)

The Cell—Transport Mechanisms and Cell Permeability: Wet Lab

Objectives

1. To define *differential permeability; diffusion (simple diffusion* and *osmosis); isotonic, hypotonic,* and *hypertonic solutions; passive transport; active transport; pinocytosis; phagocytosis;* and *solute pump.*

2. To describe the processes that account for the movement of substances across the plasma membrane and to indicate the driving force for each.

3. To determine which way substances will move passively through a differentially permeable membrane (given appropriate information on concentration differences).

Materials

For passive transport experiments:
- ❏ Clean slides and coverslips
- ❏ Forceps
- ❏ Glass stirring rods
- ❏ 15-ml graduated cylinders
- ❏ Compound microscopes
- ❏ Hot plate and large beaker for hot water bath

Diffusion:
- ❏ Petri plate containing 12 ml of 1.5% agar-agar
- ❏ Methylene blue dye crystals
- ❏ Potassium permanganate dye crystals
- ❏ Millimeter rulers

- ❏ Four dialysis sacs or small Hefty "alligator" sandwich bags
- ❏ Beakers (250 ml)
- ❏ Distilled water
- ❏ 40% glucose solution
- ❏ Fine twine or dialysis tubing clamps
- ❏ 10% NaCl solution, boiled starch solution
- ❏ Laboratory balance
- ❏ Benedict's solution in dropper bottle
- ❏ Test tubes in racks, test tube holder
- ❏ Wax marker
- ❏ Small funnel

- ❏ Silver nitrate ($AgNO_3$) in dropper bottle
- ❏ Lugol's iodine solution in dropper bottle

- ❏ Vials of animal (mammalian) blood obtained from a biological supply house or veterinarian—at option of instructor
- ❏ Physiologic (mammalian) saline solution in dropper bottle
- ❏ 1.5% sodium chloride (NaCl) solution in dropper bottle
- ❏ Distilled water in dropper bottle
- ❏ Medicine dropper
- ❏ Basin and wash bottles containing 10% household bleach solution
- ❏ Disposable autoclave bag
- ❏ Disposable plastic gloves

Diffusion demonstrations:

1: Diffusion of a dye through water

Prepared the morning of the laboratory session with setup time noted. Potassium permanganate crystals are placed in a 1000-ml graduated cylinder, and distilled water is added slowly and with as little turbulence as possible to fill to the 1000-ml mark.

2: Osmometer

Just before the laboratory begins, the broad end of a thistle tube is closed with a differentially permeable dialysis membrane, and the tube is secured to a ring stand. Molasses is added to approximately 5 cm above the thistle tube bulb and the bulb is immersed in a beaker of distilled water. At the beginning of the lab session, the level of the molasses in the tube is marked with a wax pencil.

Filtration:
- ❏ Ring stand, ring, clamp
- ❏ Filter paper, funnel
- ❏ Solution containing a mixture of uncooked starch, powdered charcoal, and copper sulfate ($CuSO_4$)

Active transport:
- ❏ Culture of starved amoeba (*Amoeba proteus*)
- ❏ Depression slide
- ❏ Coverslip (glass)

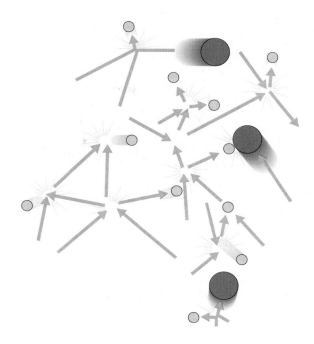

Because of its molecular composition, the plasma membrane is selective about what passes through it. It allows nutrients to enter the cell but keeps out undesirable substances. By the same token, valuable cell proteins and other substances are kept within the cell, and excreta or wastes pass to the exterior. This property is known as **differential,** or **selective, permeability.** Transport through the plasma membrane occurs in two basic ways. In **active transport,** the cell provides energy (ATP) to power the transport process. In the other, **passive transport,** the transport process is driven by concentration or pressure differences.

Passive Transport

The two important passive processes of membrane transport are *diffusion* and *filtration*. Diffusion is an important transport process for every cell in the body. By contrast, filtration usually occurs only across capillary walls. Each of these will be considered in turn.

Diffusion

Recall that all molecules possess *kinetic energy* and are in constant motion. As molecules move about randomly at high speeds, they collide and ricochet off one another, changing direction with each collision (Figure 5A.1). In general, the smaller the particle, the more kinetic energy it has and the faster it moves.

When a **concentration gradient** (difference in concentration) exists, the net effect of this random molecular movement is that the molecules eventually become evenly distributed throughout the environment, i.e., the process called diffusion occurs. Hence, **diffusion** is the movement of molecules from a region of their higher concentration to a region of their lower concentration. Its driving force is the kinetic energy of the molecules themselves.

There are many examples of diffusion in nonliving systems. For example, if a bottle of ether was uncorked at the front of the laboratory, very shortly thereafter you would be nodding as the ether molecules become distributed throughout the room. The ability to smell a friend's cologne shortly after he or she has entered the room is another example.

The diffusion of particles into and out of cells is modified by the plasma membrane, which constitutes a physical barrier. In general, molecules diffuse passively through the plasma membrane if they are small enough to pass through its pores (and are aided by an electrical gradient), or if they can dissolve in the lipid portion of the membrane (as in the case of CO_2 and O_2). The diffusion of solutes (particles dissolved in water) through a semipermeable membrane is called **simple diffusion.** The diffusion of water through a semipermeable membrane is called **osmosis.** Both simple diffusion and

Figure 5A.1   Random movement and numerous collisions cause molecules to become evenly distributed. The small spheres represent water molecules; the large spheres represent glucose molecules.

osmosis involve the movement of a substance from an area of its higher concentration to one of its lower concentration, i.e., down its concentration gradient.

Diffusion of Dye Through Agar Gel and Water   The relationship between molecular weight and the rate of diffusion can be examined easily by observing the diffusion of the molecules of two different types of dye through an agar gel. The dyes used in this experiment are methylene blue, which has a molecular weight of 320 and is deep blue in color, and potassium permanganate, a purple dye with a molecular weight of 158. Although the agar gel appears quite solid, it is primarily (98.5%) water and allows free movement of the diffusing dye molecules through it.

Activity:
Observing Diffusion of Dye Through Agar Gel

1.   Obtain a petri dish containing agar gel, a forceps, a millimeter ruler, and containers of methylene blue crystals and potassium permanganate crystals, and bring them to your bench.

⚠ *Avoid contact between your skin and the dye crystals by using the forceps to pick up the crystals.*

2.   Select a crystal of potassium permanganate dye, and place it gently on the agar gel surface. Place an equal amount of the smaller methylene blue crystals on the agar surface approximately 10 centimeters away from the potassium permanganate crystal (Figure 5A.2). Record the time.

Equal amounts of methylene blue and potassium permanganate crystals are placed on agar medium with forceps.

Diffusion distances measured at 15-minute intervals for 1½ hr. Measure the radius of the diffusion ring.

Figure 5A.2 Setup for comparing the diffusion rates of molecules of methylene blue and potassium permanganate through an agar gel.

Time (min)	Diffusion of methylene blue (mm)	Diffusion of potassium permanganate (mm)
15		
30		
45		
60		
75		
90		

3. At 15-minute intervals, use the millimeter ruler to measure the distance the dye has diffused from each crystal source. These observations should be continued for 1½ hours, and the results recorded in the chart above.

Which dye diffused more rapidly? _____

What is the relationship between molecular weight and rate of molecular movement (diffusion)?

_____

Why did the dye molecules move? _____

_____

Compute the rate of diffusion of the potassium permanganate molecules in millimeters per minute (mm/min) and record.

_____ mm/min

Compute the rate of diffusion of the methylene blue molecules in mm/min and record.

_____ mm/min

Make a mental note to yourself to go to demonstration area 1 at the end of the laboratory session to observe the extent of diffusion of the potassium permanganate dye through water. At that time, follow the directions given next. ■

A c t i v i t y :

Observing the Diffusion of Dye Through Water

1. Measure the number of millimeters from the bottom of the graduated cylinder the dye has diffused and record.

_____ mm

2. Record the time the demonstration was set up and the time of your observation. Then compute the rate of the dye's diffusion through water and record below.

Time of setup _____

Time of observation _____

Rate of diffusion _____ mm/min

3. Does the potassium permanganate dye move (diffuse) more rapidly through water or the agar gel? (Explain your answer.)

_____

_____ ■

Activity:
Observing Diffusion Through Nonliving Membranes

The following experiment provides information on the diffusion of water and solutes through semipermeable membranes, which may be applied to the study of transport mechanisms in living membrane-bound cells.

1. Obtain four dialysis sacs,* a small funnel, a graduated cylinder, a wax marker, fine twine or dialysis tubing clamps, and four beakers (250 ml). Number the beakers 1 to 4 with the wax marker, and half fill all of them with distilled water except beaker 2, to which you should add 40% glucose solution.

2. Prepare the dialysis sacs one at a time. Using the funnel, half fill each with 20 ml of the specified liquid (see below). Press out the air, fold over the open end of the sac, and tie it securely with fine twine or clamp it. Before proceeding to the next sac, quickly and carefully blot the sac dry by rolling it on a paper towel, and weigh it with a laboratory balance. Record the weight, and then drop the sac into the corresponding beaker. Be sure the sac is completely covered by the beaker solution, adding more solution if necessary.

*   Sac 1: 40% glucose solution. Weight: _____ g

*   Sac 2: 40% glucose solution. Weight: _____ g

*   Sac 3: 10% NaCl solution. Weight: _____ g

*   Sac 4: boiled starch solution. Weight: _____ g

Allow sacs to remain undisturbed in the beakers for 1 hour. (Use this time to continue with other experiments.)

3. After an hour, get a beaker of water boiling on the hot plate. Obtain the supplies you will need to determine your experimental results: dropper bottles of Benedict's solution, silver nitrate solution, and Lugol's iodine, a test tube rack, four test tubes, and a test tube holder.

4. Quickly and gently blot sac 1 dry and weigh it. (Note: Do not squeeze the sac during the blotting process.)

Weight of sac 1: _____ g

Has there been any change in weight? _____

*Dialysis sacs are selectively permeable membranes with pores of a particular size. The selectivity of living membranes depends on more than just pore size, but using the dialysis sacs will allow you to examine selectivity due to this factor.

Conclusions? _____

_____

_____

Place 5 ml of Benedict's solution in each of two test tubes. Put 4 ml of the beaker fluid into one test tube and 4 ml of the sac fluid into the other. Mark the tubes for identification and then place them in a beaker containing boiling water. Boil 2 minutes. Cool slowly. If a green, yellow, or rusty red precipitate forms, the test is positive, meaning that glucose is present. If the solution remains the original blue color, the test is negative.

Was glucose still present in the sac? _____

Was glucose present in the beaker? _____

Conclusions? _____

_____

_____

5. Blot gently and weigh sac 2: _____ g

Was there an _increase_ or _decrease_ in weight? _____

With 40% glucose in the sac and 40% glucose in the beaker, would you expect to see any net movements of water (osmosis) or of glucose molecules (simple diffusion)?

_____ Why or why not? _____

_____

6. Blot gently and weigh sac 3: _____ g

Was there any change in weight? _____

Conclusions? _____

_____

Take a 5 ml sample of beaker 3 solution and put it in a clean test tube. Add a drop of silver nitrate. The appearance of a white precipitate or cloudiness indicates the presence of AgCl, which is formed by the reaction of $AgNO_3$ with NaCl (sodium chloride).

Results? _____

Conclusions? _____

7. Blot gently and weigh sac 4: _____ g

Was there any change in weight? _____

Conclusions? _____

_____

Take a 5 ml sample of beaker 4 solution and add a couple of drops of Lugol's iodine solution. The appearance of a black color is a positive test for the presence of starch. Did any starch diffuse from the sac into the beaker?

_____ Explain: _____

_____

8.   In which of the test situations did net osmosis occur?

_____

In which of the test situations did net simple diffusion occur?

_____

What conclusions can you make about the relative size of glucose, starch, NaCl, and water molecules?

_____

_____

_____

With what cell structure can the dialysis sac be compared?

_____

9.   Before leaving the laboratory, observe demonstration 2, the *osmometer demonstration* set up before the laboratory session to follow the movement of water through a membrane (osmosis). Measure the distance the water column has moved during the laboratory period and record below. (The position of the meniscus in the thistle tube at the beginning of the laboratory period is marked with wax pencil.)

Distance the meniscus has moved: _____mm ■

A c t i v i t y :
Investigating Diffusion Through Living Membranes

To examine permeability properties of cell membranes, conduct the following two experiments.

Experiment 1:

1.   The following supplies should be available at your laboratory bench to conduct this experimental series: a clean slide and coverslip, vial of animal blood, medicine dropper, physiologic saline, 1.5% sodium chloride solution, three test tubes, test tube rack, glass stirring rod, 15-ml graduated cylinder, filter paper, and plastic gloves.

2.   Label three test tubes A, B, and C, and prepare them as follows:

*   A: add 2 ml physiologic saline

*   B: add 2 ml 1.5% sodium chloride solution

*   C: add 2 ml distilled water

⚠️   3.   Don the gloves, and use a medicine dropper to add 5 drops of animal blood to each test tube. Stir each test tube with the glass rod, rinsing between each sample.

4.   Hold each test tube in front of this printed page. **Record** the clarity of print seen through the fluid in each tube.

Test tube A _____

Test tube B _____

Test tube C _____

Experiment 2: Now you will conduct a microscopic study of red blood cells suspended in the same three solutions. The objective is to determine if these solutions have any effect on cell shape by promoting net osmosis.

1.   Place a very small drop of physiologic saline on a slide. Using the medicine dropper, add a small drop of animal blood to the saline on the slide. Tilt the slide to mix, cover with a coverslip, and immediately examine the preparation under the high-power lens. Notice that the red blood cells retain their normal smooth disclike shape (see Figure 5A.3a). This is because the physiologic saline is **isotonic** to the cells. That is, it contains a concentration of nonpenetrating solutes equal to that in the cells (same solute-solvent ratio). Consequently, the cells neither gain nor lose water by osmosis.

2.   Prepare another wet mount of animal blood, but this time use 1.5% saline solution as the suspending medium. After 5 minutes, carefully observe the red blood cells under high power. What is happening to the normally smooth disc shape of the red blood cells?

_____

_____

_____

This crinkling-up process, called **crenation,** is due to the fact that the 1.5% sodium chloride solution is slightly hypertonic to the cell sap of the red blood cell. A **hypertonic** solution contains more nonpenetrating solutes (thus less water) than are present in the cell. Under these circumstances, water tends to leave the cells by osmosis. Compare your observations to Figure 5A.3b.

3.   Add a drop of distilled water to the edge of the coverslip. Fold a piece of filter paper in half and place its folded edge at the opposite edge of the coverslip; it will absorb the saline

(a)

(b)

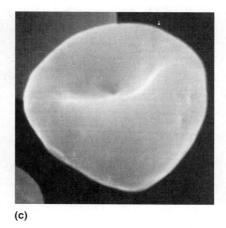

(c)

Figure 5A.3   Influence of hypertonic and hypotonic solutions on red blood cells.   **(a)** Red blood cells suspended in an isotonic solution, where the cells retain their normal size and shape.   **(b)** Red blood cells suspended in hypertonic solution. As the cells lose water to the external environment, they shrink and become prickly, a phenomenon called crenation.   **(c)** Red blood cells suspended in a hypotonic solution. Notice their spherical bloated shape, a result of excessive water intake.

solution and draw the distilled water across the cells. Watch the red blood cells as they float across the field. After about 5 minutes have passed, describe the change in their appearance.

_____

_____

Distilled water contains *no* solutes (it is 100% water). Distilled water and *very* dilute solutions (that is, those containing less than 0.9% nonpenetrating solutes) are **hypotonic** to the cell. In a hypotonic solution, the red blood cells first "plump up" (Figure 5A.3c) but then they suddenly start to disappear. The red blood cells burst as the water floods into them, leaving "ghosts" in their wake—a phenomenon called **hemolysis.**

How do your observations of test tube C in Experiment 1 correlate with what you just observed under the microscope?

_____

_____

_____

⚠️ 4.   Place the blood-soiled slides and test tube in the bleach-containing basin. Put the gloves you used into the disposable autoclave bag. Obtain a wash (squirt) bottle containing 10% bleach solution and squirt the bleach liberally over the bench area where blood was handled. Wipe the bench down with a paper towel wet with the bleach solution and allow it to dry before continuing. ■

Filtration

Filtration is the process by which water and solutes are forced through a membrane from an area of higher hydrostatic (fluid) pressure into an area of lower hydrostatic pressure.

Like diffusion, it is a passive process. For example, fluids and solutes filter out of the capillaries in the kidneys into the kidney tubules because the blood pressure in the capillaries is greater than the fluid pressure in the tubules. Filtration is not a selective process. The amount of filtrate (fluids and solutes) formed depends almost entirely on the pressure gradient (difference in pressure on the two sides of the membrane) and on the size of the membrane pores.

Activity:
Observing the Process of Filtration

1.   Obtain the following equipment: a ring stand, ring, and ring clamp; a funnel; a piece of filter paper; a beaker; a solution containing uncooked starch, powdered charcoal, and copper sulfate; and a dropper bottle of Lugol's iodine. Attach the ring to the ring stand with the clamp.

2.   Fold the filter paper in half twice, open it into a cone, and place it in a funnel. Place the funnel in the ring of the ring stand and place a beaker under the funnel. Shake the starch solution, and fill the funnel with it to just below the top of the filter paper. When the steady stream of filtrate changes to countable filtrate drops, count the number of drops formed in 10 seconds and record.

_____ drops

When the funnel is half empty, again count the number of drops formed in 10 seconds and record the count.

_____ drops

3.   After all the fluid has passed through the filter, check the filtrate and paper to see which materials were retained by the paper. (Note: If the filtrate is blue, the copper sulfate passed. Check both the paper and filtrate for black particles to see if

the charcoal passed. Finally, add Lugol's iodine to a 2-ml filtrate sample in a test tube. If the sample turns blue/black
when iodine is added, starch is present in the filtrate.)

Passed: _____

Retained: _____

What does the filter paper represent? _____

During which counting interval was the filtration rate

greatest? _____

Explain: _____

_____

_____

What characteristic of the three solutes determined whether
or not they passed through the filter paper?

_____ ■

Active Transport

Whenever a cell expends cellular energy (ATP) to move substances across its boundaries, the process is referred to as an
active transport process. Substances moved by active means
are generally unable to pass by diffusion. They may be too
large to pass through the pores; they may not be lipid soluble;
or they may have to move against rather than with a concentration gradient.

In one type of active transport, substances move across
the membrane by combining with a protein carrier molecule;
the process resembles an enzyme-substrate interaction. ATP
provides the driving force, and in many cases the substances
move against concentration or electrochemical gradients or
both. Some of the substances that are moved into the cells by
such carriers, commonly called **solute pumps,** are amino
acids and some sugars. Both solutes are lipid insoluble and
too large to pass through the pores, but are necessary for cell
life. On the other hand, sodium ions (Na^+) are ejected from
cells by active transport. There is more Na^+ outside the cell
than there is inside, so the Na^+ tends to remain in the cell unless actively transported out.

Engulfment processes such as pinocytosis and phagocytosis also require ATP. In **pinocytosis** (cell drinking), the cell
membrane sinks beneath the material to form a small vesicle,
which then pinches off into the cell interior (see Figure
5A.4). Pinocytosis is most common for taking in liquids containing protein or fat.

In **phagocytosis** (cell eating), parts of the plasma membrane and cytoplasm expand and flow around a relatively
large or solid material (for example, bacteria or cell debris)
and engulf it (Figure 5A.4). The membranous sac thus
formed, called a *phagosome,* is then fused with a lysosome
and its contents are digested. In the human body, phagocytic

(a) Phagocytosis

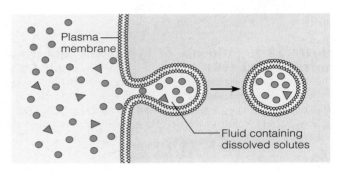

(b) Pinocytosis

Figure 5A.4 Phagocytosis and pinocytosis.
(a) In phagocytosis, cellular extensions (pseudopodia) flow
around the external particle and enclose it within a vacuole. **(b)** In pinocytosis, dissolved proteins gather on the
external surface of the plasma membrane, causing the
membrane to invaginate and to incorporate a droplet of
the fluid.

cells are mainly found among the white blood cells and
macrophages that act as scavengers and help protect the body
from disease-causing microorganisms and cancer cells.

A c t i v i t y :
Observing Phagocytosis in Amoeba

1. Obtain a drop of starved *Amoeba proteus* culture and
place it on a coverslip. Add a drop of *Tetrahymena pyriformis*
culture (an amoeba "meal") to the amoeba-containing drop,
and then quickly but gently invert the coverslip over the well
of a depression slide.

2. Locate an amoeba under low power. Keep the light as
dim as possible; otherwise the amoeba will "ball up" and begin to disintegrate.

3. Watch as the amoeba phagocytizes the *Tetrahymena* by
forming pseudopods that engulf it. In unicellular organisms
like the amoeba, phagocytosis is an important food-getting
mechanism, but in higher organisms, it is more important as
a protective device, as mentioned above.

4. Return all equipment to the appropriate supply areas and
rinse glassware used. ■

The Cell—Transport Mechanisms and Cell Permeability: Computer Simulation

Objectives

1. To define *differential permeability; diffusion (simple diffusion, facilitated diffusion,* and *osmosis); isotonic, hypotonic, and hypertonic solutions; passive transport; active transport; pinocytosis; phagocytosis;* and *solute pump.*
2. To describe the processes that account for the movement of substances across the plasma membrane and to indicate the driving force for each.
3. To determine which way substances will move passively through a differentially permeable membrane (given the appropriate information on concentration differences).

Materials

Part I—Demonstrations
❏ Millimeter rulers

1. Diffusion of a dye through agar
Prepared the morning of the laboratory session with setup time noted. A crystal of potassium permanganate dye is placed gently in a petri plate containing 12 ml of 1.5% agar gel. An equal amount of methylene blue crystals is placed on the agar surface approximately 10 centimeters away from the potassium permanganate crystal.

2. Osmometer
Just before the laboratory begins, the broad end of a thistle tube is closed with a differentially permeable dialysis membrane, and the tube is secured to a ring stand. Molasses is added to approximately 5 cm above the thistle tube bulb and the bulb is immersed in a beaker of distilled water. At the beginning of the lab session, the level of the molasses in the tube is marked with a wax pencil.

Part II—Cell transport simulation
Minimum computer equipment required:
Windows:
❏ 486/66 MHz or better recommended
❏ Windows 95/98 recommended
❏ SVGA display (256 colors at 640 × 480)
❏ Minimum 16 MB available RAM
❏ Double speed CD-ROM drive (quad-speed or higher recommended)
❏ Sound card
❏ Speakers or headphones
❏ Printer
Macintosh:
❏ 68040 processor or Power Macintosh recommended
❏ System 7.1 or higher
❏ Minimum 16 MB available RAM
❏ 13-inch or larger color monitor (640 × 480 resolution)
❏ Double speed CD-ROM drive (quad-speed or higher recommended)
❏ Printer
Software:
❏ Benjamin/Cummings PhysioEx CD-ROM (Cell Transport Mechanisms module)

The molecular composition of the plasma membrane allows it to be selective about what passes through it. It allows nutrients to enter the cell but keeps out undesirable substances. By the same token, valuable cell proteins and other substances are kept within the cell, and excreta or wastes pass to the exterior. This property is known as **differential,** or **selective, permeability.** Transport through the plasma membrane occurs in two basic ways. In **active transport,** the cell provides energy (ATP) to power the transport process. In the other, **passive transport,** the transport process is driven by concentration or pressure differences between the interior and exterior of the cell.

Passive Transport

The two key passive processes of membrane transport are diffusion and filtration. Diffusion is an important transport process for every cell in the body. By contrast, filtration usually occurs only across capillary walls. Each of these will be considered in turn.

Diffusion

Recall that all molecules possess *kinetic energy* and are in constant motion. As molecules move about randomly at high speeds, they collide and ricochet off one another, changing direction with each collision (see Figure 5A.1 on page 33). In general, the smaller the particle, the more kinetic energy it has and the faster it moves.

When a **concentration gradient** (difference in concentration) exists, the net effect of this random molecular movement is that the molecules eventually become evenly distributed throughout the environment, i.e., the process called diffusion occurs. Hence, **diffusion** is the movement of molecules from a region of their higher concentration to a region of their lower concentration. Diffusion's driving force is the kinetic energy of the molecules themselves.

There are many examples of diffusion in nonliving systems. For example, if you uncorked a bottle of ether at the front of the laboratory, very shortly thereafter you would be sleepily nodding off as the ether molecules distributed throughout the room. Similarly, the ability to smell a friend's cologne shortly after he or she has entered the room is an example of diffusion.

The diffusion of particles into and out of cells is modified by the plasma membrane, which constitutes a physical barrier. In general, molecules diffuse passively through the plasma membrane if they are small enough to pass through its pores (and are aided by an electrical gradient), or if they can dissolve in the lipid portion of the membrane as in the case of CO_2 and O_2. The diffusion of solute particles dissolved in water through a semipermeable membrane is called **simple diffusion.** The diffusion of water through a semipermeable membrane is called **osmosis.** Both simple diffusion and osmosis involve movement of a substance from an area of its higher concentration to one of its lower concentration, i.e., down its concentration gradient.

Diffusion of a Dye Through Agar Make a mental note to yourself to go to the demonstration areas near the end of the laboratory session to observe the extent of diffusion of different dyes. At that time, follow the directions for the two activities provided at the end of the diffusion simulations (p. 44).

Solute Transport Through Nonliving Membranes

This computerized simulation provides information on the passage of water and solutes through semipermeable membranes, which may be applied to the study of transport mechanisms in living membrane-bound cells.

Getting Started

Begin by making sure you have the computer equipment and software listed in the Materials section on p. 39.

1. Quit all applications currently running on your computer.

2. Insert the PhysioEx CD-ROM into the CD-ROM drive and keep PhysioEx in the drive during the entire time you use the program.

3. Follow the instructions below for your computer type.

Windows 95/Windows 98
The program will launch automatically after you load the CD-ROM into the drive.
If autorun is disabled on your machine follow these instructions:
1. Double-click the My Computer icon on the Windows desktop.
2. Double-click the PhysioEx CD icon.
3. Look for the icon that looks like a tiny movie projector and double-click it.

Macintosh
Double-click the PhysioEx CD icon that appears in the window on your desktop to launch the program.

A c t i v i t y :
Simulating Dialysis

Choose Cell Transport Mechanisms and Permeability from the main menu. The opening screen will appear in a few seconds (Figure 5B.1). The primary features on the screen when the program starts are a pair of glass beakers perched atop a solutions dispenser, a dialysis membranes cabinet at the right side of the screen, and a data collection unit at the bottom of the display.

The beakers are joined by a membrane holder, which can be equipped with any of the dialysis membranes from the cabinet. Each membrane is represented by a thin colored line suspended in a gray supporting frame. The solute concentration of dispensed solutions is displayed at the side of each beaker. As you work through the experiments, keep in mind that membranes are three-dimensional; thus what appears as a slender line is actually the edge of a membrane sheet.

The solutions you can dispense are listed beneath each beaker. You can choose more than one solution, and the amount to be dispensed is controlled by clicking (+) to increase concentration, or (−) to decrease concentration. The chosen solutions are then delivered to their beaker by clicking the **Dispense** button on the same side. Clicking the **Start** button opens the membrane holder and begins the experiment. To clean the beakers and prepare them for the next run, click **Flush.** Clicking **Pause** and then **Flush** during a run stops the experiment and prepares the beakers for another run. You can adjust the timer for any interval between 5 and 300; the elapsed time is shown in the small window to the right of the timer.

To move dialysis membranes from the cabinet to the membrane holder, click and hold the mouse on the selected membrane, drag it into position between the beakers, and then release the mouse button to drop it into place. Each membrane possesses a different molecular weight cutoff (MWCO), indicated by the number below it. You can think of

Figure 5B.1   Opening screen of the Simple Diffusion experiment.

MWCO in terms of pore size; the larger the MWCO number, the larger the pores in the membrane.

The Run Number window in the data collection unit at the bottom of the screen displays each experimental trial (run). When you click the **Record Data** button, your data is recorded in the computer's memory and is displayed in the data grid at the bottom of the screen. Data displayed in the data grid include the solute (Solute) and membrane (MWCO) used in a run, the starting concentrations in the left and right beakers (Start Conc. L and Start Conc. R), and the average diffusion rate (Avg. Dif. Rate). If you are not satisfied with a run, you can click **Delete Run.**

1.   Click and hold the mouse on the 20 MWCO membrane and drag it to the membrane holder between the beakers. Release the mouse button to lock the membrane into place.

2.   Now increase the NaCl concentration to be dispensed by clicking the (+) button under the left beaker until the display window reads 9.00 m*M*. Click **Dispense** to fill the left beaker with 9.00 m*M* NaCl solution.

3.   Click the **Deionized Water** button under the right beaker and then click **Dispense** to fill the right beaker with deionized water.

4.   Adjust the timer to 60 min (compressed time), then click the **Start** button. When Start is clicked, the barrier between the beakers descends, allowing the solutions in each beaker to have access to the dialysis membrane separating them. Notice that the Start button becomes a Pause button that allows you to momentarily halt the progress of the experiment so you can see instantaneous diffusion or transport rates.

5.   Watch the concentration windows at the side of each beaker for any activity. A level above zero in NaCl concentration in the right beaker indicates that Na^+ and Cl^- ions are diffusing from the left into the right beaker through the semipermeable dialysis membrane. Record your results (+ for diffusion, − for no diffusion) in Chart 1 on p. 42. Click the **Record Data** button to keep your data in the computer's memory.

6.   Click the 20 MWCO membrane (in the membrane holder) again to automatically return it to the membranes cabinet and then click **Flush** beneath each beaker to prepare for the next run.

7.   Drag the next membrane (50 MWCO) to the holder and repeat steps 2 through 6. Continue the runs until you have tested all four membranes. (Remember: click **Flush** beneath each beaker between runs.)

Chart 1 — Dialysis Results

Solute	Membrane (MWCO)			
	20	50	100	200
NaCl				
Urea				
Albumin				
Glucose				

8. Now perform the same experiment for urea, albumin, and glucose by repeating steps 1 through 7 three times. In step 2 you will be dispensing first urea, then albumin, and finally glucose, instead of NaCl.

Which solute(s) were able to diffuse into the right beaker from the left?

_____

Which solute(s) did not diffuse?

_____

If the solution in the left beaker contained both urea and albumin, which membrane(s) could you choose to selectively remove the urea from the solution in the left beaker? How would you carry out this experiment?

_____

_____

_____

Assume that the solution in the left beaker contained NaCl in addition to the urea and albumin. How could you set up an experiment so that you removed the urea, but left the NaCl concentration unchanged?

_____

_____

_____ ∎

Facilitated Diffusion

Some molecules are lipid insoluble or too large to pass through plasma membrane pores; instead, they pass through the membrane by a passive transport process called **facilitated diffusion.** In this form of transport, solutes combine with protein carrier molecules in the membrane and are then transported *along* or *down* their concentration gradient. Because facilitated diffusion relies on carrier proteins, solute transport varies with the number of available membrane transport proteins.

Activity:
Simulating Facilitated Diffusion

Click the **Experiment** menu and then choose **Facilitated Diffusion.** The opening screen will appear in a few seconds (Figure 5B.2). The basic screen layout is similar to that of the previous experiment with only a few modifications to the equipment. You will notice that only NaCl and glucose solutes are available in this experiment, and you will see a Membrane Builder on the right side of the screen.

The (+) and (−) buttons underneath each beaker adjust solute concentration in the solutions to be delivered into each beaker. Similarly, the buttons in the Membrane Builder allow you to control the number of carrier proteins implanted in the membrane when you click the **Build Membrane** button.

In this experiment, you will investigate how glucose transport is affected by the number of available carrier molecules.

1. The Glucose Carriers window in the Membrane Builder should read 500. If not, adjust to 500 by using the (+) or (−) button.

2. Now click **Build Membrane** to insert 500 glucose carrier proteins into the membrane. You should see the membrane appear as a slender line encased in a support structure within the Membrane Builder. Remember that we are looking at the edge of a three-dimensional membrane.

3. Click on the membrane and hold the mouse button down as you drag the membrane to the membrane holder between the beakers. Release the mouse to lock the membrane into place.

4. Adjust the glucose concentration to be delivered to the left beaker to 2.00 m*M* by clicking the (+) button next to the glucose window until it reads 2.00.

5. To fill the left beaker with the glucose solution, click the **Dispense** button just below the left beaker.

6. Click the **Deionized Water** button below the right beaker, and then click the **Dispense** button. The right beaker will fill with deionized water.

7. Set the timer to 60 min and click **Start.** Watch the concentration windows next to the beakers. When the 60 minutes have elapsed, click the **Record Data** button to display glucose transport rate information in the grid at the lower edge of the screen. Record the glucose transport rate in Chart 2 on p. 43.

8. Click the **Flush** button beneath each beaker to remove any residual solution.

9. Click the membrane support to return it to the Membrane Builder. Increase the glucose carriers and repeat steps 2 through 8 using membranes with 700 and then 900 glucose carrier proteins. Record your results in Chart 2 each time.

10. Repeat steps 1 through 9 at 8.00 m*M* glucose concentration. Record your results in Chart 2.

Figure 5B.2   Opening screen of the Facilitated Diffusion experiment.

Chart 2	Facilitated Diffusion Results		
	Number of glucose carrier proteins		
Glucose concentration (mg/ml)	500	700	900
2.00			
8.00			

What happened to the rate of facilitated diffusion as the number of protein carriers increased? Explain your answer.

What do you think would happen to the transport rate if you put the same concentration of glucose into both beakers instead of deionized water in the right beaker?

Should NaCl have an effect on glucose diffusion? Explain your answer. Use the simulation to arrive at an answer.

Activity:
Observing Diffusion of Dye Through Agar Gel

The relationship between molecular weight and the rate of diffusion can be examined easily by observing the diffusion of the molecules of two different types of dye through an agar gel. The dyes used in this experiment are methylene blue, which has a molecular weight of 320 and is deep blue in color, and potassium permanganate, a purple dye with a molecular weight of 158. Although the agar gel appears quite solid, it is primarily (98.5%) water and allows free movement of the diffusing dye molecules through it.

1. Go to demonstration area 1, and use the millimeter ruler to measure the distance each dye has diffused from each crystal source and record.

Potassium permanganate _____ mm

Methylene blue _____ mm

2. Make a note of the time the dyes have been diffusing and then compute the rate (mm/min) of each dye's diffusion through the agar gel and record below.

Rate of diffusion:

Potassium permanganate _____ mm/min

Methylene blue _____ mm/min

Which dye diffused more rapidly? _____

What is the relationship between molecular weight and rate of molecular movement (diffusion)?

_____ ∎

Osmosis

A special form of diffusion, the diffusion of water through a semipermeable membrane, is called **osmosis.** Because water can pass through the pores of most membranes, it can move from one side of a membrane to another relatively unimpeded. Osmosis occurs whenever there is a difference in water concentration on the two sides of a membrane.

If we place distilled water on both sides of a membrane, *net* movement of water will not occur; however, water molecules would still move between the two sides of the membrane. In such a situation, we would say that there is no *net* osmosis. The concentration of water in a solution depends on the number of solutes present. Therefore, increasing the solute concentration coincides with a decrease in water concentration. Because water moves down its concentration gradient, it will always move toward the solution with the highest concentration of solutes. Similarly, solutes also move down their concentration gradient. If we position a *fully* permeable membrane (permeable to solutes and water) between two solutions of differing concentrations, then all substances—solutes and water—will diffuse freely, and an equilibrium will be reached between the two sides of the membrane. However, if we use a semipermeable membrane that is impermeable to the solutes, then we have established a condition where water will move but solutes will not. Consequently, water will move toward the more concentrated solution, resulting in a volume increase. By applying this concept to a closed system where volumes cannot change, we can predict that the pressure in the more concentrated solution would rise.

Activity:
Simulating Osmosis

Click the **Experiment** menu and then select **Osmosis.** The opening screen will appear in a few seconds (Figure 5B.3). The most notable difference in this experiment screen concerns meters atop the beakers that measure pressure changes in the beaker they serve. As before, (+) and (−) buttons control solute concentrations in the dispensed solutions.

1. Drag the 20 MWCO membrane to the holder between the two beakers.

2. Adjust the NaCl concentration to 8.00 mM in the left beaker, and then click the **Dispense** button.

3. Click **Deionized Water** under the right beaker and then click **Dispense.**

4. Set the timer to 60 min and then click **Start** to run the experiment. Now click the **Record Data** button to retain your data in the computer's memory and also record the osmotic pressure in Chart 3 below.

5. Click the membrane to return it to the membrane cabinet.

6. Repeat steps 1 through 5 with the 50, 100, and 200 MWCO membranes.

Do you see any evidence of pressure changes in either beaker, using any of the four membranes? If so, which ones?

_____

Chart 3	Osmosis Results (pressure in mm Hg)			
	Membrane (MWCO)			
Solute	20	50	100	200
Na^+Cl^-				
Albumin				
Glucose				

Figure 5B.3 Opening screen of the Osmosis experiment.

Does NaCl appear in the right beaker? If so, which membrane(s) allowed it to pass?

Will osmotic pressure be generated if solutes are able to diffuse? Explain your answer.

7.   Now perform the same experiment for albumin and glucose by repeating steps 1 through 6. In step 2 you will be dispensing 9.00 m*M* albumin first, and then 10.00 m*M* glucose, instead of NaCl.

Answer the following questions using the results you recorded in Chart 3. Use the simulation if you need help formulating a response.

Explain the relationship between solute concentration and osmotic pressure.

Because the albumin molecule is much too large to pass through a 100 MWCO membrane, you should have noticed the development of osmotic pressure in the left beaker in the albumin run using the 100 MWCO membrane. What do you think would happen to the osmotic pressure if you replaced the deionized water in the right beaker with 9.00 m*M* albumin in that run? (Both beakers would contain 9.00 m*M* albumin.)

What would happen if you doubled the albumin concentration in the left beaker using any membrane?

In the albumin run using the 200 MWCO membrane, what would happen to the osmotic pressure if you put 10 m*M* glucose in the right beaker instead of deionized water? Explain your answer.

_____

_____

_____

What if you used the 100 MWCO membrane in the albumin/glucose run described in the previous question?

_____

_____ ■

A c t i v i t y :
Observing the Osmometer Demonstration

Before continuing on to the filtration simulation, observe the *osmometer demonstration* set up before the laboratory session to follow the movement of water through a membrane (osmosis). Measure the distance the water column has moved during the laboratory period and record it here. The initial position of the molassas meniscus in the thistle tube at the beginning of the laboratory period is marked with wax pencil.

Distance the meniscus has moved: _____ mm. ■

Filtration

Filtration is the process by which water and solutes pass through a membrane from an area of higher hydrostatic (fluid) pressure into an area of lower hydrostatic pressure. Like diffusion, it is a passive process. For example, fluids and solutes filter out of the capillaries in the kidneys into the kid-

Figure 5B.4 Opening screen of the Filtration experiment.

ney tubules because blood pressure in the capillaries is greater than the fluid pressure in the tubules. Filtration is not a selective process. The amount of filtrate—fluids and solutes—formed depends almost entirely on the pressure gradient (the difference in pressure on the two sides of the membrane) and on the size of the membrane pores.

Activity:
Simulating Filtration

Click the **Experiment** menu and then choose **Filtration.** The opening screen will appear in a few seconds (Figure 5B.4). The basic screen elements resemble the other simulations. The top beaker can be pressurized to force fluid through the filtration membrane into the bottom beaker. Any of the filtration membranes can be positioned in the holder between the beakers by drag-and-drop as in the previous experiments. The solutions you can dispense are listed to the right of the top beaker, and are adjusted by clicking the (+) and (−) buttons. The selected solutions are then delivered to the top beaker by clicking **Dispense.** The top beaker is cleaned and prepared for the next run by clicking **Flush.** You can adjust the timer for any interval between 5 and 300; the elapsed time is shown in the window to the right of the timer. When you click the **Record Data** button, your data is recorded in the computer's memory and is displayed in the data grid at the bottom of the screen.

Solute concentrations in the filtrate are automatically monitored by the *Filtrate Analysis Unit* to the right of the bottom beaker. After a run you can detect the presence of any solute remaining on a membrane by using the *Membrane Residue Analysis* unit located above the membrane cabinet.

1.   Click and hold the mouse on the 20 MWCO membrane and drag it to the holder below the top beaker. Release the mouse button to lock the membrane into place.

2.   Now adjust the NaCl, urea, glucose, and powdered charcoal windows to 5.00 mg/ml each, and then click **Dispense.**

3.   If necessary, adjust the pressure unit atop the beaker until its window reads 50 mm Hg.

4.   Set the timer to 60 min and then click **Start.** When the Start button is clicked, the membrane holder below the top beaker retracts, and the solution will flow through the membrane into the beaker below.

5.   Watch the Filtrate Analysis Unit for any activity. A rise in detected solute concentration indicates that the solute particles are moving through the filtration membrane. At the end of the run, record the amount of solute present in the *filtrate* (mg/ml) and the filtration *rate* in Chart 4.

6.   Now drag the 20 MWCO membrane to the holder in the Membrane Residue Analysis unit. Click **Start Analysis** to begin analysis (and cleaning) of the membrane. Record your results for solute *residue* presence on the membrane (+ for present, − for not present) in Chart 4 and click the **Record Data** button to keep your data in the computer's memory.

7.   Click the 20 MWCO membrane again to automatically return it to the membranes cabinet and then click **Flush** to prepare for the next run.

8.   Repeat steps 1 through 7 using 50, 100, and 200 MWCO membranes.

Chart 4		Filtration Results (Filtration Rate, Solute Presence or Absence)			
		Membrane (MWCO)			
Solute		20	50	100	200
	Rate				
NaCl	Filtrate				
	Residue				
Urea	Filtrate				
	Residue				
Glucose	Filtrate				
	Residue				
Powdered charcoal	Filtrate				
	Residue				

Did the membrane's MWCO affect the filtration rate?

_____

_____

Which solute did not appear in the filtrate using any of the membranes?

_____

_____

What would happen if you increased the driving pressure? Use the simulation to arrive at an answer.

_____

_____

Explain how you can increase the filtration rate through living membranes.

_____

_____

By examining the filtration results, we can predict that the

molecular weight of glucose must be greater than _____,

but less than _____. ■

Active Transport

Whenever a cell expends cellular energy (ATP) to move substances across its membrane, the process is referred to as an *active transport process.* Substances moved across cell membranes by active means are generally unable to pass by diffusion. There are several possible reasons why substances may not be able to pass through a membrane by diffusion: they may be too large to pass through the pores, they may not be lipid soluble, or they may have to move against rather than with a concentration gradient.

In one type of active transport, substances move across the membrane by combining with a protein carrier molecule; the process resembles an enzyme-substrate interaction. ATP provides the driving force, and in many cases the substances move against concentration or electrochemical gradients or both. Some of the substances that are moved into the cells by such carriers, commonly called **solute pumps,** are amino acids and some sugars. Both solutes are lipid-insoluble and too large to pass through the pores, but are necessary for cell life. On the other hand, sodium ions (Na^+) are ejected from the cells by active transport. There is more Na^+ outside the cell than inside, so the Na^+ tends to remain in the cell unless actively transported out. In the body, the most common type of solute pump is the coupled Na^+-K^+ pump that moves Na^+ and K^+ in opposite directions across cellular membranes. 3 Na^+ are ejected for every 2 K^+ entering the cell.

Engulfment processes such as pinocytosis and phagocytosis also require ATP. In **pinocytosis** (cell drinking), the cell membrane sinks beneath the material to form a small vesicle, which then pinches off into the cell interior (see Figure 5A.4b on p. 38). Pinocytosis is most common for taking in liquids containing protein or fat.

In **phagocytosis** (cell eating), parts of the plasma membrane and cytoplasm expand and flow around a relatively large or solid material such as bacteria or cell debris and engulf it (see Figure 5A.4a on p. 38), forming a membranous sac called a phagosome. The phagosome is then fused with a lysosome and its contents are digested. In the human body, phagocytic cells are mainly found among the white blood cells and macrophages that act as scavengers and help protect the body from disease-causing microorganisms and cancer cells.

You will examine various factors influencing the function of solute pumps in the following experiment.

Activity:
Simulating Active Transport

Click the **Experiment** menu and then choose **Active Transport.** The opening screen will appear in a few seconds (Figure 5B.5). This experiment screen resembles the osmosis experiment screen, except that an ATP dispenser is substituted for the pressure meters atop the beakers. The (+) and (−) buttons control NaCl, KCl, and glucose concentrations in the dispensed solutions. You will use the Membrane Builder to build membranes containing glucose (facilitated diffusion) carrier proteins and active transport Na^+-K^+ pumps.

In this experiment, we will assume that the left beaker represents the cell's interior and the right beaker represents the extracellular space. The Membrane Builder will insert the Na^+-K^+ pumps into the membrane so Na^+ will be pumped toward the right (out of the cell) while K^+ is simultaneously moved to the left (into the cell).

1.   In the Membrane Builder, adjust the number of glucose carriers and the number of Na^+-K^+ pumps to 500.

2.   Click **Build Membrane,** and then drag the membrane to its position in the membrane holder between the beakers.

3.   Adjust the NaCl concentration to be delivered to the left beaker to 9.00 m*M,* then click the **Dispense** button.

4.   Adjust the KCl concentration to be delivered to the right beaker to 6.00 m*M,* then click **Dispense.**

5.   Adjust the ATP dispenser to 1.00 m*M,* then click **Dispense ATP.** This action delivers the chosen ATP concentration to both sides of the membrane.

6.   Adjust the timer to 60 min, and then click **Start.** Click **Record Data** after each run.

Figure 5B.5   Opening screen of the Active Transport experiment.

Watch the solute concentration windows at the side of each beaker for any changes in Na$^+$ and K$^+$ concentrations. The Na$^+$ transport rate slows and then stops before transport has completed. Why do you think that this happens?

_____

What would happen if you did not dispense any ATP?

_____

7.   Click either Flush button to clean both beakers. Repeat steps 3 through 6, adjusting the ATP concentration to 3.00 mM in step 5.

Has the amount of Na$^+$ transported changed?

_____

Do these results support your ideas in step 6 above? _____

What would happen if you decreased the number of Na$^+$-K$^+$ pumps?

_____

Explain how you could show that this phenomenon is not just simple diffusion. (Hint: adjust the Na$^+$ concentration in the right beaker.)

_____

_____

8.   Now repeat steps 1 through 6 dispensing 9.00 mM NaCl into the left beaker and 10.00 mM NaCl into the right beaker (instead of 6.00 mM KCl). Is Na$^+$ transport affected by this change? Explain your answer.

_____

_____

What would happen to the rate of ion transport if we increased the number of Na^+-K^+ pump proteins?

_____

Would Na^+- and K^+ transport change if we added glucose solution?

_____

Try adjusting various membrane and solute conditions and attempt to predict the outcome of experimental trials. For example, you could dispense 10 mM glucose into the right beaker instead of deionized water. ■

Activity:
Using Your Data

Save Your Data

When you have finished with all five experiments, you can save your data set for later review and additional analysis.

1.   Select **Save Data to File** from the **File** menu.

2.   When the dialog box opens, click a drive and/or directory location.

3.   Type a file name into the **File name** box (your instructor may provide you with a name).

4.   Click **Save** to complete the save. If you want to review previously saved data, select **Open Data File** from the **File** menu, and choose your file.

Print Your Data

If your instructor requires a printed copy of your experiment results, choose **Print Data** from the **File** menu.

1.   Type your name into the space provided.

2.   When you are ready to print, click **OK.** ■

Classification of Tissues

Objectives

1. To name the four major types of tissues in the human body and the major subcategories of each.

2. To identify the tissue subcategories through microscopic inspection or inspection of an appropriate diagram or projected slide.

3. To state the location of the various tissue types in the body.

4. To list the general functions and structural characteristics of each of the four major tissue types.

Materials

☐ Compound microscope

☐ Immersion oil

☐ Prepared slides of simple squamous, simple cuboidal, simple columnar, stratified squamous (nonkeratinized), stratified cuboidal, stratified columnar, pseudostratified ciliated columnar, and transitional epithelium

☐ Prepared slides of mesenchyme; of adipose, areolar, reticular, and dense (both regular [tendon] and irregular [dermis]) connective tissues; of hyaline and elastic cartilage; of fibrocartilage; of bone (cross section); and of blood

☐ Prepared slides of skeletal, cardiac, and smooth muscle (longitudinal sections)

☐ Prepared slide of nervous tissue (spinal cord smear)

Exercise 4 describes cells as the building blocks of life and the all-inclusive functional units of unicellular organisms. But in higher organisms cells do not usually operate as isolated, independent entities. In humans and other multicellular organisms, cells depend on one another and cooperate to maintain homeostasis in the body.

With a few exceptions (parthenogenetic organisms), even the most complex animal starts out as a single cell, the fertilized egg, which divides almost endlessly. The trillions of cells that result become specialized for a particular function; some become supportive bone, others the transparent lens of the eye, still others skin cells, and so on. Thus a division of labor exists, with certain groups of cells highly specialized to perform functions that benefit the organism as a whole. Cell specialization brings about great sophistication of achievement but carries with it certain hazards, because when a small specific group of cells is indispensable, any inability to function on its part can paralyze or destroy the entire body.

Groups of cells that are similar in structure and function are called **tissues.** The four primary tissue types—epithelium, connective tissue, nervous tissue, and muscle—have distinctive structures, patterns, and functions. The four primary tissues are further divided into subcategories, as described shortly.

To perform specific body functions, the tissues are organized into **organs** such as the heart, kidneys, and lungs. Most organs contain several representatives of the primary tissues, and the arrangement of these tissues determines the organ's structure and function. Thus **histology,** the study of tissues, complements a study of gross anatomy and provides the structural basis for a study of organ physiology.

The main objective of this exercise is to familiarize you with the major similarities and dissimilarities of the primary tissues, so that when the tissue makeup of an organ is described, you will be able to more easily understand (and perhaps even predict) the organ's major function. Because epithelium and some types of connective tissue will not be considered again, they are emphasized more than muscle, nervous tissue, and bone (a connective tissue), which are covered in more depth in later exercises.

Epithelial Tissue

Epithelial tissue, or **epithelium,** covers surfaces. For example, epithelium covers the external body surface (as the epidermis), lines its cavities and tubules, and generally marks off our "insides" from our outsides. Since the various endocrine (hormone-producing) and exocrine glands of the body almost invariably develop from epithelial membranes, glands too are logically classed as epithelium.

Epithelial functions include protection, absorption, filtration, excretion, secretion, and sensory reception. For example, the epithelium covering the body protects against bacterial invasion and chemical damage; that lining the respiratory tract is ciliated to sweep dust and other foreign

Squamous

Cuboidal

Columnar

(a)

Simple

Stratified

(b)

Figure 6.1 **Classification of epithelia. (a)** Classification on the basis of cell shape. For each category, a whole cell is shown on the left and a longitudinal section is shown on the right. **(b)** Classification on the basis of arrangement (relative number of layers).

particles away from the lungs. Epithelium specialized to absorb substances lines the stomach and small intestine. In the kidney tubules, the epithelium absorbs, secretes, and filters. Secretion is a specialty of the glands.

Epithelium generally exhibits these characteristics:

• Cells fit closely together to form membranes, or sheets of cells, and are bound together by specialized junctions.

• The membranes always have one free surface, called the *apical surface.*

• The cells are attached to and supported by an adhesive **basement membrane,** an amorphous material secreted partly by the epithelial cells (*basal lamina*) and connective tissue cells (*reticular lamina*) that lie adjacent to each other.

• Epithelial tissues have no blood supply of their own (are avascular), but depend on diffusion of nutrients from the underlying connective tissue.

• If well nourished, epithelial cells can easily regenerate themselves. This is an important characteristic because many epithelia are subjected to a good deal of friction.

The covering and lining epithelia are classified according to two criteria—cell shape and arrangement or relative number of layers (Figure 6.1). **Squamous** (scalelike), **cuboidal** (cubelike), and **columnar** (column-shaped) epithelial cells are the general types based on shape. On the basis of arrangement, there are **simple** epithelia, consisting of one layer of cells attached to the basement membrane, and **stratified** epithelia, consisting of more than one layer of cells. The terms denoting shape and arrangement of the epithelial cells are combined to describe the epithelium fully. *Stratified epithelia are named according to the cells at the apical surface of the epithelial membrane,* not those resting on the basement membrane.

There are, in addition, two less easily categorized types of epithelia. **Pseudostratified epithelium** is actually a simple columnar epithelium (one layer of cells), but because its cells extend varied distances from the basement membrane, it gives the false appearance of being stratified. This epithelium is often ciliated. **Transitional epithelium** is a rather peculiar stratified squamous epithelium formed of rounded, or

(a)

(b)

(c) Exocrine gland

(d) Endocrine gland

Figure 6.2   Formation of endocrine and exocrine glands from epithelial sheets. (a) Epithelial cells grow and push into the underlying tissue. **(b)** A cord of epithelial cells forms. **(c)** In an exocrine gland, a lumen (cavity) forms. The inner cells form the duct, the outer cells produce the secretion. **(d)** In a forming endocrine gland, the connecting duct cells atrophy, leaving the secretory cells with no connection to the epithelial surface. However, they do become heavily invested with blood and lymphatic vessels that receive the secretions.

"plump," cells with the ability to slide over one another to allow the organ to be stretched. Transitional epithelium is found only in urinary system organs subjected to periodic distension, such as the bladder. The superficial cells are flattened (like true squamous cells) when the organ is distended and rounded when the organ is empty.

Epithelial cells forming glands are highly specialized to remove materials from the blood and to manufacture them into new materials, which they then secrete. There are two types of glands, as shown in Figure 6.2. **Endocrine glands** lose their surface connection (duct) as they develop; thus they are referred to as ductless glands. Their secretions (all hormones) are extruded directly into the blood or the lymphatic vessels that weave through the glands. **Exocrine glands** retain their ducts, and their secretions empty through these ducts to an epithelial surface. The exocrine glands—including the sweat and oil glands, liver, and pancreas—are both external and internal; they will be discussed in conjunction with the organ systems to which their products are functionally related.

The most common types of epithelia, their most common locations in the body, and their functions are described in Figure 6.3.

A c t i v i t y :

Examining Epithelial Tissue Under the Microscope

Obtain slides of simple squamous, simple cuboidal, simple columnar, stratified squamous (nonkeratinized), pseudostratified ciliated columnar, stratified cuboidal, stratified columnar, and transitional epithelia. Examine each carefully, and notice how the epithelial cells fit closely together to form intact sheets of cells, a necessity for a tissue that forms linings or covering membranes. Scan each epithelial type for modifications for specific functions, such as cilia (motile cell projections that help to move substances along the cell surface), and microvilli, which increase the surface area for absorption. Also be alert for goblet cells, which secrete lubricating mucus (see Plate 1 of the Histology Atlas). Compare your observations with the photomicrographs in Figure 6.3.

While working, check the questions in the laboratory review section for this exercise. A number of the questions there refer to some of the observations you are asked to make during your microscopic study. ■

Text continues on page 58

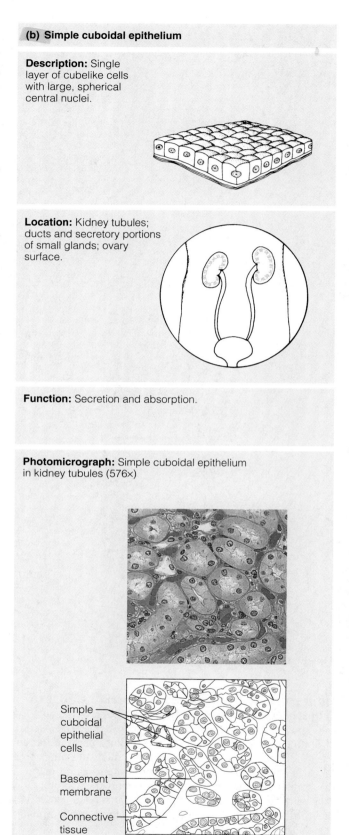

(a) Simple squamous epithelium

Description: Single layer of flattened cells with disc-shaped central nuclei and sparse cytoplasm; the simplest of the epithelia.

Location: Kidney glomeruli; air sacs of lungs; lining of heart, blood vessels, and lymphatic vessels; lining of ventral body cavity (serosae).

Function: Allows passage of materials by diffusion and filtration in sites where protection is not important; secretes lubricating substances in serosae.

Photomicrograph: Simple squamous epithelium forming part of the wall of the glomerular capsules of the kidneys (364×).

Nucleus of simple squamous epithelial cell of the glomerular capsule

Basement membrane

Glomerulus (capillary cluster)

(b) Simple cuboidal epithelium

Description: Single layer of cubelike cells with large, spherical central nuclei.

Location: Kidney tubules; ducts and secretory portions of small glands; ovary surface.

Function: Secretion and absorption.

Photomicrograph: Simple cuboidal epithelium in kidney tubules (576×)

Simple cuboidal epithelial cells

Basement membrane

Connective tissue

Figure 6.3 Epithelial tissues. Simple epithelia **(a** and **b).**

(c) Simple columnar epithelium

Description: Single layer of tall cells with *oval* nuclei; some cells bear cilia; layer may contain mucus-secreting unicellular glands (goblet cells).

Location: Nonciliated type lines most of the digestive tract (stomach to anal canal), gallbladder and excretory ducts of some glands; ciliated variety lines small bronchi, uterine tubes, and some regions of the uterus.

Function: Absorption; secretion of mucus, enzymes, and other substances; ciliated type propels mucus (or reproductive cells) by ciliary action.

Photomicrograph: Simple columnar epithelium of the stomach mucosa (576×).

Simple columnar epithelial cell

Basement membrane

Connective tissue

(d) Pseudostratified columnar epithelium

Description: Single layer of cells of differing heights, some not reaching the free surface; nuclei seen at different levels; may contain goblet cells and bear cilia.

Location: Nonciliated type in ducts of large glands, parts of male urethra; ciliated variety lines the trachea, most of the upper respiratory tract.

Function: Secretion, particularly of mucus; propulsion of mucus by ciliary action.

Photomicrograph: Pseudostratified ciliated columnar epithelium lining the human trachea (612×).

Cilia

Mucous goblet

Pseudo-stratified epithelial layer

Basement membrane

Connective tissue

Figure 6.3 (*continued*) Simple epithelia (**c** and **d**).

(e) Stratified squamous epithelium

Description: Thick membrane composed of several cell layers; basal cells are cuboidal or columnar and metabolically active; surface cells are flattened (squamous); in the keratinized type, the surface cells are full of keratin and dead; basal cells are active in mitosis and produce the cells of the more superficial layers.

Location: Nonkeratinized type forms the moist linings of the esophagus, mouth and vagina; keratinized variety forms the epidermis of the skin, a dry membrane.

Function: Protects underlying tissues in areas subjected to abrasion.

Photomicrograph: Stratified squamous epithelium lining of the esophagus (144×).

(f) Stratified cuboidal epithelium

Description: Generally two layers of cube-like cells.

Location: Largest ducts of sweat glands, mammary glands, and salivary glands.

Function: Protection.

Photomicrograph: Stratified cuboidal epithelium forming a salivary gland duct (117×).

Figure 6.3 (*continued*) **Epithelial tissues.** Stratified epithelia (**e** and **f**).

(g) Stratified columnar epithelium

Description: Several cell layers; basal cells usually cuboidal; superficial cells elongated and columnar.

Location: Rare in the body; small amounts in male urethra and in large ducts of some glands.

Function: Protection; secretion.

Photomicrograph: Stratified columnar epithelium lining of the male urethra (390×).

Stratified columnar epithelium

Underlying connective tissue

Basement membrane

(h) Transitional epithelium

Description: Resembles both stratified squamous and stratified cuboidal; basal cells cuboidal or columnar; surface cells dome-shaped or squamous-like, depending on degree of organ stretch.

Location: Lines the ureters, bladder, and part of the urethra.

Function: Stretches readily and permits distension of urinary organ by contained urine.

Photomicrograph: Transitional epithelium lining of the bladder, relaxed state (252×); note the bulbous, or rounded, appearance of the cells at the surface; these cells flatten and become elongated when the bladder is filled with urine.

Basement membrane

Connective tissue

Transitional epithelium

Figure 6.3 (*continued*) Stratified epithelia (**g** and **h**).

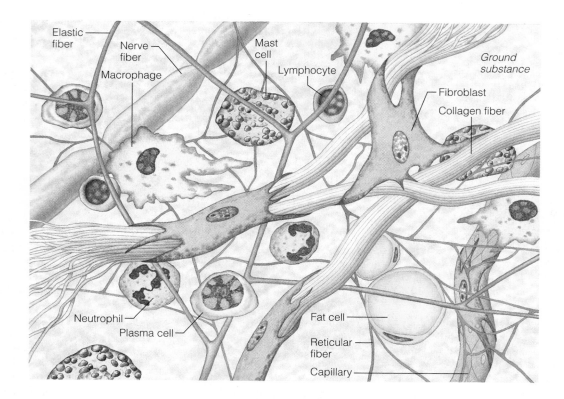

Figure 6.4   Areolar connective tissue: A prototype (model) connective tissue. Note the various cell types and the three classes of fibers (collagen, reticular, elastic) embedded in the ground substance.

Connective Tissue

Connective tissue is found in all parts of the body as discrete structures or as part of various body organs. It is the most abundant and widely distributed of the tissue types.

The connective tissues perform a variety of functions, but they primarily protect, support, and bind together other tissues of the body. For example, bones are composed of connective tissue (**bone** or **osseous tissue),** and they protect and support other body tissues and organs. The ligaments and tendons (**dense connective tissue**) bind the bones together or bind skeletal muscles to bones.

Areolar connective tissue is a soft packaging material that cushions and protects body organs. Fat (**adipose**) tissue provides insulation for the body tissues and a source of stored food. Blood-forming (**hematopoietic**) tissue replenishes the body's supply of red blood cells. In addition, connective tissue also serves a vital function in the repair of all body tissues since many wounds are repaired by connective tissue in the form of scar tissue.

The characteristics of connective tissue include the following:

- With a few exceptions (cartilages, which are avascular, and tendons and ligaments, which are poorly vascularized), connective tissues have a rich supply of blood vessels.

- Connective tissues are composed of many types of cells.

- There is a great deal of noncellular, nonliving material (matrix) between the cells of connective tissue.

The nonliving material between the cells—the **extracellular matrix**—deserves a bit more explanation because it distinguishes connective tissue from all other tissues. It is produced by the cells and then extruded. The matrix is primarily responsible for the strength associated with connective tissue, but there is variation. At one extreme, adipose tissue is composed mostly of cells. At the opposite extreme, bone and cartilage have very few cells and large amounts of matrix.

The matrix has two components—ground substance and fibers. The **ground substance** is composed chiefly of glycoproteins and large, charged polysaccharide molecules. Depending on its specific composition, the ground substance may be liquid, semisolid, gel-like, or very hard. When the matrix is firm, as in cartilage and bone, the connective tissue cells reside in cavities in the matrix called *lacunae*. The fibers, which provide support, include **collagenic** (white) **fibers, elastic** (yellow) **fibers,** and **reticular** (fine collagenic) **fibers.** Of these, the collagen fibers are most abundant.

Generally speaking, the ground substance functions as a molecular sieve, or medium, through which nutrients and other dissolved substances can diffuse between the blood capillaries and the cells. The fibers in the matrix hinder diffusion somewhat and make the ground substance less pliable. The properties of the connective tissue cells and the makeup and arrangement of their matrix elements vary tremendously, accounting for the amazing diversity of this tissue type. Nonetheless, the connective tissues have a common structural plan seen best in *areolar connective tissue* (Figure 6.4), a soft packing tissue that occurs throughout the body. Since all other connective tissues are variations of areolar, it is considered the model or prototype of the connective tissues. Notice in Figure 6.4 that areolar tissue has all three varieties of fibers, but they are sparsely arranged in its transparent gel-like ground substance. The cell type that secretes its matrix is the *fibroblast,* but a wide variety of other cells including phagocytic cells like macrophages and certain white blood cells and mast cells that act in the inflammatory response are present as well. The more durable connective tissues, such as bone, cartilage, and the dense fibrous varieties, characteristically have a firm ground substance and many more fibers.

There are four main types of adult connective tissue, all of which typically have large amounts of matrix. These are connective tissue proper (which includes areolar, adipose, reticular, and dense [fibrous] connective tissues), cartilage, bone, and blood. All of these derive from an embryonic tissue called *mesenchyme.* Figure 6.5 lists the general characteristics, location, and function of some of the connective tissues found in the body.

Activity:

Examining Connective Tissue Under the Microscope

Obtain prepared slides of mesenchyme; of adipose, areolar, reticular, dense regular and irregular connective tissue; of elastic connective tissue; of hyaline and elastic cartilage and fibrocartilage; of osseous connective tissue (bone); and of blood. Compare your observations with the views illustrated in Figure 6.5.

Distinguish between the living cells and the matrix and pay particular attention to the denseness and arrangement of the matrix. For example, notice how the matrix of the dense fibrous connective tissues, making up tendons and the dermis of the skin, is chock-full of collagenic fibers, and that in the *regular* variety (tendon), the fibers are all running in the same direction, whereas in the dermis (a dense *irregular* connective tissue) they appear to be running in many directions.

While examining the areolar connective tissue, a soft "packing tissue," notice how much empty space there appears to be, and distinguish between the collagen fibers and the coiled elastic fibers. Also, try to locate a **mast cell,** which has large, darkly staining granules in its cytoplasm. This cell type releases histamine that makes capillaries quite permeable during inflammatory reactions and allergies and thus is partially responsible for that "runny nose" of allergies.

In adipose tissue, locate a signet ring cell in which the nucleus can be seen pushed to one side by the large, fat-filled vacuole that appears to be a large empty space. Also notice how little matrix there is in fat or adipose tissue. Distinguish between the living cells and the matrix in the dense fibrous, bone, and hyaline cartilage preparations.

Scan the blood slide at low and then high power to examine the general shape of the red blood cells. Then, switch to the oil immersion lens for a closer look at the various types of white blood cells. How does blood differ from all other connective tissues?

Embryonic connective tissue

(a) Mesenchyme

Description: Embryonic connective tissue; gel-like ground substance containing fibers; star-shaped mesenchymal cells.

Location: Primarily in embryo.

Function: Gives rise to all other connective tissue types.

Photomicrograph: Mesenchymal tissue, an embryonic connective tissue (475×); the clear-appearing background is the fluid ground substance of the matrix; notice the fine, sparse fibers.

Mesenchymal cell

Ground substance

Fibers

Connective tissue proper: Loose connective tissue (b to d)

(b) Areolar connective tissue

Description: Gel-like matrix with all three fiber types; cells: fibroblasts, macrophages, mast cells, and some white blood cells.

Location: Widely distributed under epithelia of body, e.g, forms lamina propria of mucous membranes; packages organs; surrounds capillaries.

Epithelium

Lamina propria

Function: Wraps and cushions organs; its macrophages phagocytize bacteria; plays important role in inflammation; holds and conveys tissue fluid.

Photomicrograph: Areolar connective tissue, a soft packaging tissue of the body (182×).

Mast cell

Fibroblast

Fibers of matrix

Figure 6.5 Connective tissues. Embryonic connective tissue **(a)** and connective tissue proper **(b)**.

(c) Adipose tissue

Description: Matrix as in areolar, but very sparse; closely packed adipocytes, or fat cells, have nucleus pushed to the side by large fat droplet.

Location: Under skin; around kidneys and eyeballs; in bones and within abdomen; in breasts.

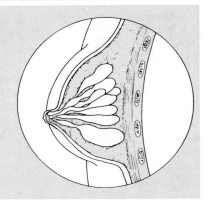

Function: Provides reserve food fuel; insulates against heat loss; supports and protects organs.

Photomicrograph: Adipose tissue from the subcutaneous layer under the skin (839×).

Nuclei of fat cells

Blood vessel

Vacuole containing fat droplet

(d) Reticular connective tissue

Description: Network of reticular fibers in a typical loose ground substance; reticular cells predominate.

Location: Lymphoid organs (lymph nodes, bone marrow, and spleen).

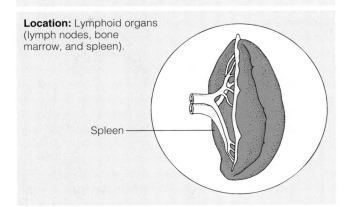

Spleen

Function: Fibers form a soft internal skeleton (stroma) that supports other cell types.

Photomicrograph: Dark-staining network of reticular connective tissue fibers forming the internal skeleton of the spleen (1125×).

Reticular cell

Blood cells (lymphocytes)

Reticular fibers

Figure 6.5 (continued) Connective tissue proper (**c** and **d**).

Connective tissue proper: Dense connective tissue (e and f)

(e) Dense regular connective tissue

Description: Primarily parallel collagen fibers; a few elastin fibers; major cell type is the fibroblast.

Location: Tendons, most ligaments, aponeuroses.

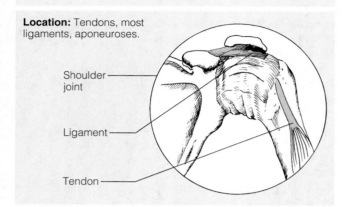

Shoulder joint

Ligament

Tendon

Function: Attaches muscles to bones or to muscles; attaches bones to bones; withstands great tensile stress when pulling force is applied in one direction.

Photomicrograph: Dense regular connective tissue from a tendon (576×).

Collagen fibers

Nuclei of fibroblasts

(f) Dense irregular connective tissue

Description: Primarily irregularly arranged collagen fibers; some elastic fibers; major cell type is the fibroblast.

Location: Dermis of the skin; submucosa of digestive tract; fibrous capsules of organs and of joints.

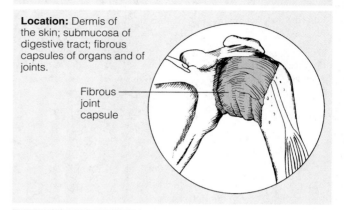

Fibrous joint capsule

Function: Able to withstand tension exerted in many directions; provides structural strength.

Photomicrograph: Dense irregular connective tissue from the dermis of the skin (475×).

Nuclei of fibroblasts

Collagen fibers

Figure 6.5 *(continued)* **Connective tissues.** Connective tissue proper **(e** and **f).**

Cartilage: (g to i)

(g) Hyaline cartilage

Description: Amorphous but firm matrix; collagen fibers form an imperceptible network; chondroblasts produce the matrix and when mature (chondrocytes) lie in lacunae.

Location: Forms most of the embryonic skeleton; covers the ends of long bones in joint cavities; forms costal cartilages of the ribs; cartilages of the nose, trachea, and larynx.

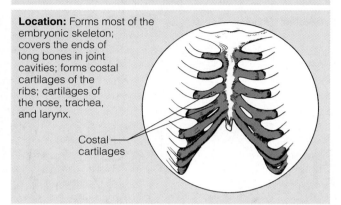

Costal cartilages

Function: Supports and reinforces; has resilient cushioning properties; resists compressive stress.

Photomicrograph: Hyaline cartilage from the trachea (453×).

Chondrocyte in lacuna

Matrix

(h) Elastic cartilage

Description: Similar to hyaline cartilage, but more elastic fibers in matrix.

Location: Supports the external ear (pinna); epiglottis.

Function: Maintains the shape of a structure while allowing great flexibility.

Photomicrograph: Elastic cartilage from the human ear pinna; forms the flexible skeleton of the ear (144×).

Chondrocyte in lacuna

Elastic fibers

Figure 6.5 (*continued*) Cartilage (**g** and **h**).

(i) Fibrocartilage

Description: Matrix similar but less firm than in hyaline cartilage; thick collagen fibers predominate.

Location: Intervertebral discs; pubic symphysis; discs of knee joint.

Function: Tensile strength with the ability to absorb compressive shock.

Photomicrograph: Fibrocartilage of an intervertebral disc (823×).

Lacuna

Chondrocytes

Collagen fiber

Others: (j and k)

(j) Bone (osseous tissue)

Description: Hard, calcified matrix containing many collagen fibers; osteocytes lie in lacunae. Very well vascularized.

Location: Bones

Function: Bone supports and protects (by enclosing); provides levers for the muscles to act on; stores calcium and other minerals and fat; marrow inside bones is the site for blood cell formation (hematopoiesis).

Photomicrograph: Cross-sectional view of bone (100×).

Osteocytes in lacunae

Matrix

Figure 6.5 (continued) **Connective tissues.** Cartilage **(i)** and bone **(j).**

(k) Blood

Description: Red and white blood cells in a fluid matrix (plasma).

Location: Contained within blood vessels.

Function: Transport of respiratory gases, nutrients, wastes, and other substances.

Photomicrograph: Smear of human blood (1000×); two white blood cells (neutrophil in upper left and lymphocyte in lower right) are seen surrounded by red blood cells.

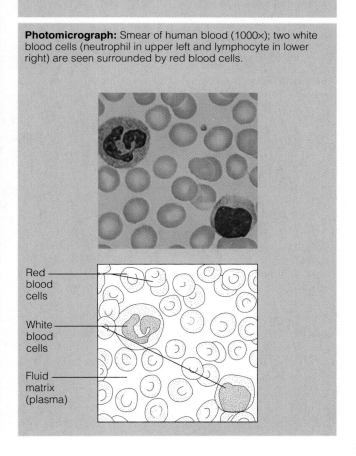

Red blood cells

White blood cells

Fluid matrix (plasma)

Figure 6.5 (*continued*)   Blood **(k).**

Muscle Tissue

Muscle tissue is highly specialized to contract (shorten) in order to produce movement of some body parts. As you might expect, muscle cells tend to be quite elongated, providing a long axis for contraction. The three basic types of muscle tissue are described briefly here. Cardiac and skeletal muscles are treated more completely in later exercises.

   Skeletal muscle, the "meat," or flesh, of the body, is attached to the skeleton. It is under voluntary control (consciously controlled), and its contraction moves the limbs and other external body parts. The cells of skeletal muscles are long, cylindrical, and multinucleate (several nuclei per cell); they have obvious *striations* (stripes).

   Cardiac muscle is found only in the heart. As it contracts, the heart acts as a pump, propelling the blood into the blood vessels. Cardiac muscle, like skeletal muscle, has striations. But cardiac cells are branching uninucleate (or occasionally binucleate) cells that interdigitate (fit together) at junctions called **intercalated discs.** These structural modifications allow the cardiac muscle to act as a unit. Cardiac muscle is under involuntary control, which means that we cannot voluntarily or consciously control the operation of the heart.

   Smooth muscle, or *visceral muscle,* is found mainly in the walls of hollow organs (digestive and urinary tract organs, uterus, blood vessels). Typically there are two layers that run at right angles to each other; consequently its contraction can constrict or dilate the lumen (cavity) of an organ and propel substances along predetermined pathways. Smooth muscle cells are quite different in appearance from those of skeletal or cardiac muscle. No striations are visible, and the uninucleate smooth muscle cells are spindle-shaped.

Activity:
Examining Muscle Tissue Under the Microscope

Obtain and examine prepared slides of skeletal, cardiac, and smooth muscle. Notice their similarities and dissimilarities in your observations and in the illustrations in Figure 6.6. ■

(a) Skeletal muscle

Description: Long, cylindrical, multinucleate cells; obvious striations.

Location: In skeletal muscles attached to bones or occasionally to skin.

Function: Voluntary movement; locomotion; manipulation of the environment; facial expression. Voluntary control.

Photomicrograph: Skeletal muscle (approx. 30×). Notice the obvious banding pattern and the fact that these large cells are multinucleate.

Nuclei

Part of muscle fiber

Figure 6.6   Muscle tissues (a, b, and c).

(b) Cardiac muscle

Description: Branching, striated, generally uninucleate cells that interdigitate at specialized junctions (intercalated discs).

Location: The walls of the heart.

Function: As it contracts, it propels blood into the circulation; involuntary control.

Photomicrograph: Cardiac muscle (250×); notice the striations, branching of fibers, and the intercalated discs.

Intercalated disc

Nucleus

(c) Smooth muscle

Description: Spindle-shaped cells with central nuclei; cells arranged closely to form sheets; no striations.

Location: Mostly in the walls of hollow organs.

Function: Propels substances or objects (foodstuffs, urine, a baby) along internal passageways; involuntary control.

Photomicrograph: Sheet of smooth muscle (approx. 300×).

Smooth muscle cell

Nuclei

Nervous Tissue

Nervous tissue is composed of two major cell populations. The **neuroglia** are special supporting cells that protect, support, and insulate the more delicate neurons. The **neurons** are highly specialized to receive stimuli (irritability) and to conduct waves of excitation, or impulses, to all parts of the body (conductivity). They are the cells that are most often associated with nervous system functioning.

The structure of neurons is markedly different from that of all other body cells. They all have a nucleus-containing cell body, and their cytoplasm is drawn out into long extensions (cell processes)—sometimes as long as 3 feet (about 1 m), which allows a single neuron to conduct an impulse over relatively long distances. More detail about the anatomy of the different classes of neurons and neuroglia appears in Exercise 17.

Activity:
Examining Nervous Tissue Under the Microscope

Obtain a prepared slide of a spinal cord smear. Locate a neuron and compare it to Figure 6.7. Keep the light dim—this will help you see the cellular extensions of the neurons. Also see Plates 4 and 5 in the Histology Atlas. ■

Description: Neurons are branching cells; cell processes that may be quite long extend from the nucleus-containing cell body; also contributing to nervous tissue are nonirritable supporting cells (not illustrated).

Cell processes

Cell body

Location: Brain, spinal cord, and nerves.

Function: Transmit electrical signals from sensory receptors and to effectors (muscles and glands) which control their activity.

Photomicrograph: Neuron (170 ×)

Nuclei of supporting cells

Cell body

Cell processes

Figure 6.7 Nervous tissue.

The Integumentary System

The **skin,** or **integument,** is often considered an organ system because of its extent and complexity. It is much more than an external body covering; architecturally the skin is a marvel. It is tough yet pliable, a characteristic that enables it to withstand constant insult from outside agents.

The skin has many functions, most (but not all) concerned with protection. It insulates and cushions the underlying body tissues and protects the entire body from mechanical damage (bumps and cuts), chemical damage (acids, alkalis, and the like), thermal damage (heat), and bacterial invasion (by virtue of its acid mantle and continuous surface). The hardened uppermost layer of the skin (the cornified layer) prevents water loss from the body surface. The skin's abundant capillary network (under the control of the nervous system) plays an important role in regulating heat loss from the body surface.

The skin has other functions as well. For example, it acts as a mini-excretory system; urea, salts, and water are lost through the skin pores in sweat. The skin also has important metabolic duties. For example, like liver cells, it carries out some chemical conversions that activate or inactivate certain drugs and hormones, and it is the site of vitamin D synthesis for the body. Finally, the cutaneous sense organs are located in the dermis.

Basic Structure of the Skin

The skin has two distinct regions—the superficial *epidermis* composed of epithelium and an underlying connective tissue *dermis*. These layers are firmly "cemented" together along an undulating border. But friction, such as the rubbing of a poorly fitting shoe, may cause them to separate, resulting in a blister. Immediately deep to the dermis is the **hypodermis** or **superficial fascia** (primarily adipose tissue), which is not considered part of the skin. The main skin areas and structures are described below.

Objectives

1. To recount several important functions of the skin, or integumentary system.

2. To recognize and name during observation of an appropriate model, diagram, projected slide, or microscopic specimen the following skin structures: epidermis (and indicate relative positioning of its strata), dermis (papillary and reticular layers), hair follicles and hair, sebaceous glands, and sweat glands.

3. To name the layers of the epidermis and describe the characteristics of each.

4. To compare the properties of the epidermis to those of the dermis.

5. To describe the distribution and function of the skin derivatives—sebaceous glands, sweat glands, and hairs.

6. To differentiate between eccrine and apocrine sweat glands.

7. To enumerate the factors determining skin color.

8. To describe the function of melanin.

9. To identify the major regions of nails.

Materials

- ❑ Skin model (three-dimensional, if available)
- ❑ Compound microscope
- ❑ Prepared slide of human skin with hair follicles
- ❑ Sheet of #20 bond paper ruled to mark off cm^2 areas
- ❑ Scissors
- ❑ Betadine swabs, or Lugol's iodine and cotton swabs
- ❑ Adhesive tape

Hair shaft

Meissner's corpuscle

Pore

Stratum corneum

Dermal papillae

Stratum lucidum

Free nerve ending

Stratum granulosum

Epidermis

Sebaceous (oil) gland

Stratum spinosum

Stratum basale

Arrector pili muscle

Sensory nerve fiber

Papillary layer

Dermis

Reticular layer

Hair follicle

Hair root

Hypodermis (superficial fascia)

Artery

Vein

Eccrine sweat gland

Pacinian corpuscle

Adipose tissue

Root hair plexus

Figure 7.1 Skin structure. Three-dimensional view of the skin and the underlying hypodermis. The epidermis and dermis have been pulled apart at the left corner to reveal the dermal papillae.

Activity:
Locating Structures on a Skin Model

As you read, locate the following structures in Figure 7.1 and on a skin model. ■

Epidermis

Structurally, the avascular epidermis is a keratinized stratified squamous epithelium consisting of four distinct cell types and four or five distinct layers.

Cells of the Epidermis Most epidermal cells are **keratinocytes** (literally, keratin cells), epithelial cells that function mainly to produce keratin fibrils. **Keratin** is a fibrous protein that gives the epidermis its durability and protective capabilities.

Far less abundant are the following types of epidermal cells (Figure 7.2):

• **Melanocytes:** Spidery black cells that produce the brown-to-black pigment called **melanin.** The skin tans be-cause melanin production increases when the skin is exposed to sunlight. The melanin provides a protective pigment umbrella over the nuclei of the cells in the deeper epidermal layers, thus shielding their genetic material (DNA) from the damaging effects of ultraviolet radiation. A concentration of melanin in one spot is called a *freckle.*

• **Langerhans' cells:** Also called *epidermal dendritic cells,* these phagocytic cells (macrophages) play a role in immunity.

• **Merkel cells:** In conjunction with sensory nerve endings, Merkel cells form sensitive touch receptors called *Merkel discs* located at the epidermal-dermal junction.

Layers of the Epidermis From deep to superficial, the layers of the epidermis are the stratum basale, stratum spinosum, stratum granulosum, stratum lucidum, and stratum corneum (Figure 7.1).

The **stratum basale** (basal layer) is a single row of cells immediately adjacent to the dermis. Its cells are constantly undergoing mitotic cell division to produce millions of new cells daily, hence its alternate name *stratum germinativum.*

About a quarter of the cells in this stratum are melanocytes, whose processes thread their way through this and the adjacent layers of keratinocytes (see Figure 7.2).

The **stratum spinosum** (spiny layer) is a stratum consisting of several cell layers immediately superficial to the basal layer. Its cells contain thick weblike bundles of intermediate filaments made of a prekeratin protein. The stratum spinosum cells appear spiky (hence their name) because, as the skin tissue is prepared for histological examination, they shrink but their desmosomes hold tight. Cells divide fairly rapidly in this stratum, but less so than in the stratum basale. Cells in the basal and spiny layers are the only ones to receive adequate nourishment (via diffusion of nutrients from the dermis). So as their daughter cells are pushed upward and away from the source of nutrition, they gradually die.

The **stratum granulosum** (granular layer) is a thin layer named for the abundant granules its cells contain. These granules are of two types: (1) *laminated granules,* which contain a waterproofing glycolipid that is secreted into the extracellular space; and (2) *keratohyalin granules,* which combine with the intermediate filaments in the more superficial layers to form the keratin fibrils. At the upper border of this layer, the cells are beginning to die.

The **stratum lucidum** (clear layer) is a very thin translucent band of flattened dead keratinocytes with indistinct boundaries. It is not present in regions of thin skin.

The outermost epidermal layer, the **stratum corneum** (horny layer), consists of some 20 to 30 cell layers, and accounts for the bulk of the epidermal thickness. Cells in this layer, like those in the stratum lucidum (where it exists), are dead and their flattened scalelike remnants are fully keratinized. They are constantly rubbing off and being replaced by division of the deeper cells.

Dermis

The dense irregular connective tissue making up the dermis consists of two principal regions—the papillary and reticular areas. Like the epidermis, the dermis varies in thickness. For example, the skin is particularly thick on the palms of the hands and soles of the feet and is quite thin on the eyelids.

The **papillary layer** is the more superficial dermal region. It is very uneven and has fingerlike projections from its superior surface, the **dermal papillae,** which attach it to the epidermis above. These projections lie on top of the larger dermal ridges. In the palms of the hands and soles of the feet, they produce the fingerprints, unique patterns of *epidermal ridges* that remain unchanged throughout life. Abundant capillary networks in the papillary layer furnish nutrients for the epidermal layers and allow heat to radiate to the skin surface. The pain and touch receptors (Meissner's corpuscles) are also found here.

The **reticular layer** is the deepest skin layer. It contains many arteries and veins, sweat and sebaceous glands, and pressure receptors.

Both the papillary and reticular layers are heavily invested with collagenic and elastic fibers. The elastic fibers give skin its exceptional elasticity in youth. In old age, the number of elastic fibers decreases and the subcutaneous layer loses fat, which leads to wrinkling and inelasticity of the skin. Fibroblasts, adipose cells, various types of macrophages (which are important in the body's defense), and other cell types are found throughout the dermis.

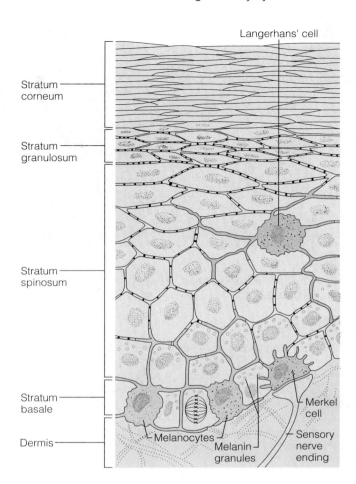

Figure 7.2 Diagram showing the main features—layers and relative numbers of the different cell types—in epidermis of thin skin. The keratinocytes (beige) form the bulk of the epidermis. Less numerous are the melanocytes (gray), which produce the pigment melanin; Langerhans' cells (blue), which function as macrophages; and Merkel cells (purple). A sensory nerve ending (yellow), extending from the dermis, is depicted in association with the Merkel cell forming a Merkel disc (touch receptor). Notice that the keratinocytes, but not the other cell types, are joined by numerous desmosomes.

The abundant dermal blood supply allows the skin to play a role in the regulation of body temperature. When body temperature is high, the arterioles serving the skin dilate, and the capillary network of the dermis becomes engorged with the heated blood. Thus body heat is allowed to radiate from the skin surface. If the environment is cool and body heat must be conserved, the arterioles constrict so that blood bypasses the dermal capillary networks.

Any restriction of the normal blood supply to the skin results in cell death and, if severe enough, skin ulcers (Figure 7.3). **Bedsores (decubitus ulcers)** occur in bedridden patients who are not turned regularly enough. The weight of the body exerts pressure on the skin, especially over bony projections (hips, heels, etc.), which leads to restriction of the blood supply and death of tissue. ■

Figure 7.3   Photograph of a deep (stage III) decubitus ulcer.

The dermis is also richly provided with lymphatic vessels and a nerve supply. Many of the nerve endings bear highly specialized receptor organs that, when stimulated by environmental changes, transmit messages to the central nervous system for interpretation. Some of these receptors—bare nerve endings (pain receptors), a Meissner's corpuscle, a Pacinian corpuscle, and a root hair plexus—are shown in Figure 7.1. (These receptors are discussed in depth in Exercise 23.)

Skin Color

Skin color is a result of three factors—the relative amount of two pigments (melanin and carotene) in skin and the degree of oxygenation of the blood. People who produce large amounts of melanin have brown-toned skin. In light-skinned people, who have less melanin, the dermal blood supply flushes through the rather transparent cell layers above, giving the skin a rosy glow. *Carotene* is a yellow-orange pigment present primarily in the stratum corneum and in the adipose tissue of the hypodermis. Its presence is most noticeable when large amounts of carotene-rich foods (carrots, for instance) are eaten.

Skin color may be an important diagnostic tool. For example, flushed skin may indicate hypertension, fever, or embarrassment, whereas pale skin is typically seen in anemic individuals. When the blood is inadequately oxygenated, as during asphyxiation and serious lung disease, both the blood and the skin take on a bluish or cyanotic cast. **Jaundice,** in which the tissues become yellowed, is almost always diagnostic for liver disease, whereas a bronzing of the skin hints that a person's adrenal cortex is hypoactive (**Addison's disease**). ■

Appendages of the Skin

The appendages of the skin—hair, nails, and cutaneous glands—are all derivatives of the epidermis, but they reside in the dermis. They originate from the stratum basale and grow downward into the deeper skin regions.

Cutaneous Glands

The cutaneous glands fall primarily into two categories: the sebaceous glands and the sweat glands (Figure 7.1). The **sebaceous glands** are found nearly all over the skin, except for the palms of the hands and the soles of the feet. Their ducts usually empty into a hair follicle, but some open directly onto the skin surface.

The product of the sebaceous glands, called **sebum,** is a mixture of oily substances and fragmented cells. The sebum is a lubricant that keeps the skin soft and moist (a natural skin cream) and keeps the hair from becoming brittle. The sebaceous glands become particularly active during puberty when more male hormones (androgens) begin to be produced; thus, the skin tends to become oilier during this period of life. *Blackheads* are accumulations of dried sebum and bacteria; *acne* is due to active infection of the sebaceous glands.

Epithelial openings, called *pores,* are the outlets for the **sweat (sudoriferous) glands.** These exocrine glands are widely distributed in the skin. Sweat glands are subcategorized by the composition of their secretions. The **eccrine glands,** which are distributed all over the body, produce clear perspiration, consisting primarily of water, salts (NaCl), and urea. The **apocrine glands,** found predominantly in the axillary and genital areas, secrete a milky protein- and fat-rich substance (also containing water, salts, and urea) that is an excellent nutrient medium for the microorganisms typically found on the skin.

The sweat glands, under the control of the nervous system, are an important part of the body's heat-regulating apparatus. They secrete perspiration when the external temperature or body temperature is high. When this water-based substance evaporates, it carries excess body heat with it. Thus evaporation of greater amounts of perspiration provides an efficient means of dissipating body heat when the capillary cooling system is not sufficient or is unable to maintain body temperature homeostasis.

Hair

Hairs are found over the entire body surface, except for thick-skinned areas (the palms of the hands, the soles of the feet), parts of the external genitalia, the nipples, and the lips. A hair, enclosed in a hair **follicle,** is also an epithelial structure (Figure 7.4). The portion of the hair enclosed within the follicle is called the **root;** the portion projecting from the scalp surface is called the **shaft.** The hair is formed by mitosis of the well-nourished germinal epithelial cells at the basal end of the follicle (the **hair bulb**). As the daughter cells are pushed farther away from the growing region, they die and become keratinized; thus the bulk of the hair shaft, like the bulk of the epidermis, is dead material.

A hair (Figure 7.4b) consists of a central region (medulla) surrounded first by the cortex and then by a protective cuticle. Abrasion of the cuticle results in "split ends." Hair color is a manifestation of the amount and kind of melanin pigment within the hair cortex.

The hair follicle is structured from both epidermal and dermal cells (Figure 7.4c). Its inner *epithelial root sheath,* with two parts (internal and external), is enclosed by the *connective tissue root sheath,* which is essentially dermal tissue. A small nipple of dermal tissue, the *connective tissue papilla,* protrudes into the hair bulb from the connective tissue sheath

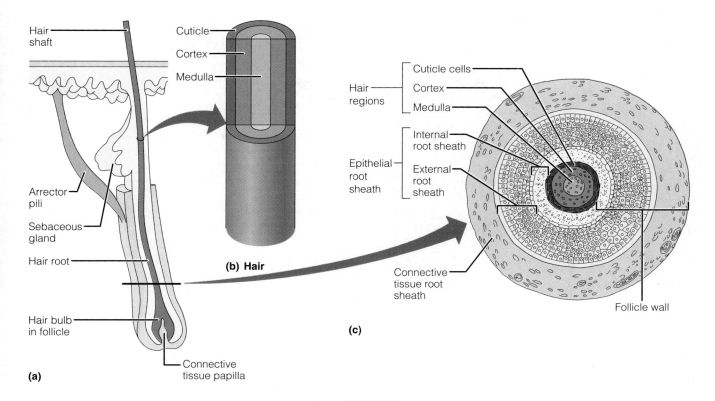

(b) Hair

(c)

Figure 7.4 **Structure of a hair and hair follicle.** **(a)** Longitudinal section of a hair within its follicle. **(b)** Enlarged longitudinal section of a hair. **(c)** Cross section of a hair and hair follicle.

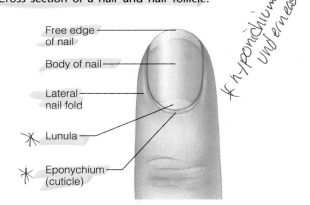

✳ hyponichium underneath

Figure 7.5 **Structure of a nail.** Surface view of the distal part of a finger showing nail parts. The nail matrix that forms the nail lies beneath the lunula; the epidermis of the nail bed underlies the nail.

and provides nutrition to the growing hair. If you look carefully at the structure of the hair follicle (see Figure 7.1), you will see that it generally is in a slanted position. Small bands of smooth muscle cells—**arrector pili**—connect each hair follicle to the papillary layer of the dermis. When these muscles contract (during cold or fright), the hair follicle is pulled upright, dimpling the skin surface with "goose bumps." This phenomenon is especially dramatic in a scared cat, whose fur actually stands on end to increase its apparent size. The activity of the arrector pili muscles also exerts pressure on the sebaceous glands surrounding the follicle, causing a small amount of sebum to be released.

Nails

Nails, the hornlike derivatives of the epidermis, consist of a *free edge,* a *body* (visible attached portion), and a *root* (embedded in the skin and adhering to an epithelial **nail bed**). The borders of the nail are overlapped by skin folds called **nail folds;** the thick proximal nail fold is the **eponychium,** commonly called the cuticle (Figure 7.5).

The germinal cells in the **nail matrix,** the thickened proximal part of the nail bed, are responsible for nail growth. As the nail cells are produced by the matrix, they become heavily keratinized and die. Thus, nails, like hairs, are mostly nonliving material.

Nails are transparent and nearly colorless, but they appear pink because of the blood supply in the underlying dermis. The exception to this is the proximal region of the thickened nail matrix which appears as a white crescent called the *lunula.* When someone is cyanotic due to a lack of oxygen in the blood, the nail beds take on a blue cast.

A c t i v i t y :
Identifying Nail Structures

Identify the nail structures shown in Figure 7.5 on yourself or your lab partner. ■

Epidermis | Sloughing stratum corneum cells | Hair follicles | Hair shaft | Dermis

Figure 7.6 Photomicrograph of skin (34×).

Microscopic Structure of the Skin

Activity:
Examining a Skin Slide

Obtain a prepared slide of human skin, and study it carefully under the microscope. Compare your tissue slide to the view shown in Figure 7.6, and identify as many of the structures diagrammed in Figure 7.1 as possible.

How is this stratified squamous epithelium different from that observed in Exercise 6?

_____

_____

How do these differences relate to the functions of these two similar epithelia?

_____

_____

_____

_____ ■

Activity:
Plotting the Distribution of Sweat Glands

1.   For this simple experiment you will need two squares of bond paper (each 1 cm × 1 cm), adhesive tape, and a Betadine (iodine) swab *or* Lugol's iodine and a cotton-tipped swab. (The bond paper has been preruled in cm²—just cut along the lines to obtain the required squares.)

2.   Paint an area of the medial aspect of your left palm (avoid the crease lines) and a region of your left forearm with the iodine solution, and allow it to dry thoroughly. The painted area in each case should be slightly larger than the paper squares to be used.

3.   Have your lab partner *securely* tape a square of bond paper over each iodine-painted area, and leave them in place for 20 minutes. (If it is very warm in the laboratory while this test is being conducted, good results may be obtained within 10 to 15 minutes.)

4.   After 20 minutes, remove the paper squares, and count the number of blue-black dots on each square. The presence of a blue-black dot on the paper indicates an active sweat gland. (The iodine in the pore is dissolved in the sweat and reacts chemically with the starch in the bond paper to produce the blue-black color.) Thus "sweat maps" have been produced for the two skin areas.

5.   Which skin area tested has the greater density of sweat glands?

_____

_____

_____ ■

Classification of Body Membranes

The body membranes, which cover surfaces, line body cavities, and form protective (and often lubricating) sheets around organs, fall into two major categories. These are the so-called *epithelial membranes* and the *synovial membranes.*

Epithelial Membranes

The term "epithelial membrane" is used in various ways. Here we will define an **epithelial membrane** as a simple organ consisting of an epithelial sheet bound to an underlying layer of connective tissue. Most of the covering and lining epithelia take part in forming one of the three common varieties of epithelial membranes: cutaneous, mucous, or serous.

The **cutaneous membrane** (Figure 8.1a) is the skin, a dry membrane with a keratinizing epithelium (the epidermis). Since the skin is discussed in some detail in Exercise 7, the mucous and serous membranes will receive our attention here.

Mucous Membranes

The **mucous membranes (mucosae)** are composed of epithelial cells resting on a layer of loose connective tissue called the **lamina propria.** They line all body cavities that open to the body exterior—the respiratory (Figure 8.1b), digestive, and urinary tracts. All mucosae are "wet" membranes because they are continuously bathed by secretions (or, in the case of urinary tract mucosae, urine). Although mucous membranes often secrete mucus, this is not a requirement. The mucous membranes of both the digestive and respiratory tracts secrete mucus, but that of the urinary tract does not.

Activity:

Examining the Microscopic Structure of Mucous Membranes

Using Plates 32 and 39 of the Histology Atlas as guides, examine a slide made from a cross section of the trachea and another of the small intestine. Draw the mucosa of each in the appropriate circle, and fully identify each epithelial type. Remember to look for the epithelial cells at the free surface. Also search the epithelial sheets for **goblet cells**—columnar epithelial cells with a large mucus-containing vacuole (goblet) in their apical cytoplasm. Which mucosa type contains goblet cells?

Mucosa of trachea

Mucosa of small intestine

(a) Cutaneous membrane

(b) Mucous membranes

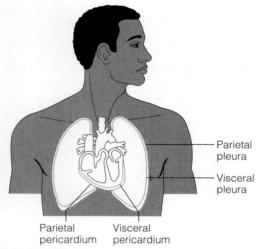

Parietal
pleura

Visceral
pleura

Parietal   Visceral
pericardium   pericardium

(c) Serous membranes

Synovial membrane
lining joint cavity

(d) Synovial membranes

Figure 8.1   Body membranes. The epithelial membranes (a–c) are composite membranes with epithelial and connective tissue elements. **(a)** The cutaneous membrane, or skin, covers and protects the body surface. **(b)** Mucous membranes line body cavities (hollow organs) that open to the exterior. **(c)** Serous membranes line the closed ventral cavity of the body. Two examples, the pleurae and pericardia, are illustrated here. **(d)** Synovial membranes, derived solely from connective tissue, form smooth linings in joint cavities.

Compare and contrast the roles of these two mucous membranes.

_____

_____

_____

_____ ∎

Serous Membranes

The **serous membranes (serosae)** are also epithelial membranes (Figure 8.1c). They are composed of a layer of simple squamous epithelium on a scant amount of loose connective tissue. The serous membranes generally occur in twos. The parietal layer lines a body cavity, and the visceral layer covers the outside of the organs in that cavity (see also Figure 1.8, p. 8). In contrast to the mucous membranes, which line open body cavities, the serous membranes line body cavities that are closed to the exterior (with the exception of the female peritoneal cavity and the dorsal body cavity). The serosae secrete a thin fluid (serous fluid) that lubricates the organs and body walls and thus reduces friction as the organs slide across one another and against the body cavity walls. A serous membrane also lines the interior of blood vessels (endothelium) and the heart (endocardium). In capillaries, the entire wall is composed of serosa that serves as a selectively permeable membrane between the blood and the tissue fluid of the body.

A c t i v i t y :
Examining the Microscopic Structure of a Serous Membrane

Examine a prepared slide of a serous membrane and diagram it in the circle provided here.

What are the specific names of the serous membranes covering the heart and lining the cavity in which it resides (respectively)?

_____

The abdominal viscera and visceral cavity (respectively)?

_____ ■

Synovial Membranes

Synovial membranes, unlike the mucous and serous membranes, are composed entirely of connective tissue; they contain no epithelial cells. These membranes line the cavities surrounding the joints, providing a smooth surface and secreting a lubricating fluid. They also line smaller sacs of connective tissue (bursae and tendon sheaths), which cushion structures moving against each other, as during muscle activity. Figure 8.1d illustrates the positioning of a synovial membrane in the joint cavity.

A c t i v i t y :
Examining the Gross Structure of a Synovial Membrane

If a freshly sawed beef joint is available, visually examine the interior surface of the joint capsule to observe the smooth texture of the synovial membrane. ■

Overview of the Skeleton: Classification and Structure of Bones and Cartilages

Objectives

1. To list three functions of the skeletal system.

2. To identify the four main kinds of bones.

3. To identify surface bone markings and functions.

4. To identify the major anatomical areas on a longitudinally cut long bone (or diagram of one).

5. To identify the major regions and structures of an osteon in a histologic specimen of compact bone (or diagram of one).

6. To explain the role of the inorganic salts and organic matrix in providing flexibility and hardness to bone.

7. To locate and identify the three major types of skeletal cartilages.

Materials

❑ Disarticulated bones (identified by name or number) that demonstrate classic examples of the four bone classifications (long, short, flat, and irregular)

❑ Long bone sawed longitudinally (beef bone from a slaughterhouse, if possible, or prepared laboratory specimen)

❑ Disposable plastic gloves

❑ Long bone soaked in 10% nitric acid (or vinegar) until flexible

❑ Long bone baked at 250°F for more than 2 hours

❑ Compound microscope

❑ Prepared slides of ground bone (cross section), hyaline cartilage, elastic cartilage, and fibrocartilage

❑ 3-D model of microscopic structure of compact bone

❑ Articulated skeleton

 See Appendix C, Exercise 9 for links to A.D.A.M. Interactive Anatomy.

The **skeleton** is constructed of two of the most supportive tissues found in the human body—cartilage and bone. In embryos, the skeleton is predominantly composed of hyaline cartilage, but in the adult, most of the cartilage is replaced by more rigid bone. Cartilage persists only in such isolated areas as the bridge of the nose, the larynx, the trachea, joints, and parts of the rib cage.

Besides supporting and protecting the body as an internal framework, the skeleton provides a system of levers with which the skeletal muscles work to move the body. In addition, the bones store lipids and many minerals (most importantly calcium). Finally, the red marrow cavities of bones provide a site for hematopoiesis (blood cell formation).

The skeleton is made up of bones that are connected at joints, or articulations. The skeleton is subdivided into two divisions: the **axial skeleton** (those bones that lie around the body's center of gravity) and the **appendicular skeleton** (bones of the limbs, or appendages) (Figure 9.1).

Before beginning your study of the skeleton, imagine for a moment that your bones have turned to putty. What if you were running when this metamorphosis took place? Now imagine your bones forming a continuous metal framework within your body, somewhat like a network of plumbing pipes. What problems could you envision with this arrangement? These images should help you understand how well the skeletal system provides support and protection, as well as facilitating movement.

Bone Markings

Even a casual observation of the bones will reveal that bone surfaces are not featureless smooth areas but are scarred with an array of bumps, holes, and ridges. These **bone markings** reveal where bones form joints with other bones, where muscles, tendons, and ligaments were attached, and where blood vessels and nerves passed. Bone markings fall into two categories: projections, or processes that grow out from the bone and serve as sites of muscle attachment or help form joints; and depressions or cavities, indentations or openings in the bone that often serve as conduits for nerves and blood vessels. The bone markings are summarized in Table 9.1.

Classification of Bones

The 206 bones of the adult skeleton are composed of two basic kinds of osseous tissue that differ in their texture. **Compact** bone looks smooth and homogeneous; **spongy** (or *cancellous*) bone is composed of small trabeculae (bars) of bone and lots of open space.

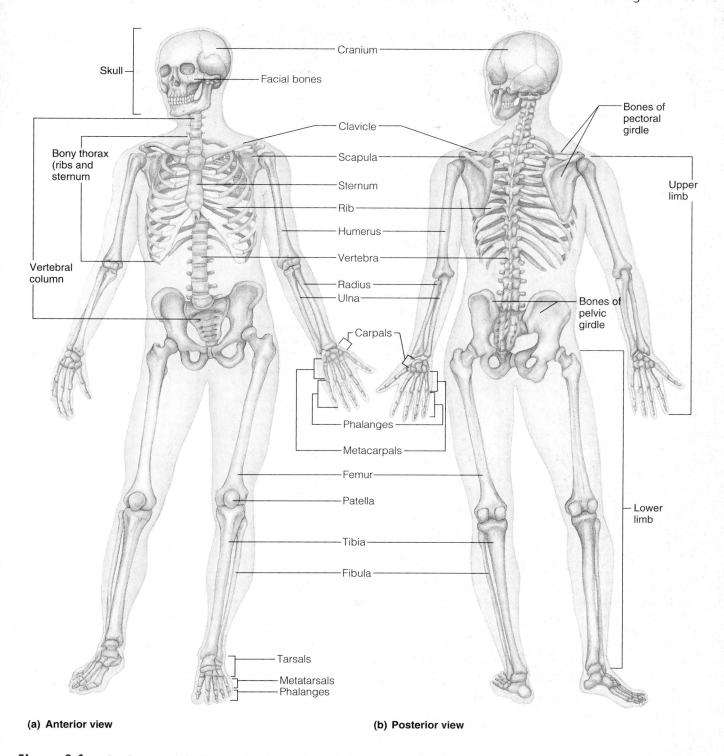

(a) Anterior view

(b) Posterior view

Figure 9.1   The human skeleton. The bones of the axial skeleton are colored green to distinguish them from the bones of the appendicular skeleton.

Bones may be classified further on the basis of their relative gross anatomy into four groups: long, short, flat, and irregular bones.

Long bones, such as the femur (Figure 9.1), are much longer than they are wide, generally consisting of a shaft with heads at either end. Long bones are composed predominantly of compact bone. **Short bones** are typically cube-shaped, and they contain more spongy bone than compact bone. See the tarsals and carpals in Figure 9.1.

Flat bones are generally thin, with two waferlike layers of compact bone sandwiching a layer of spongy bone between them. Although the name "flat bone" implies a structure that is level or horizontal, many flat bones are curved (for example, the bones of the skull). Bones that do not fall

Table 9.1 Bone Markings

Name of bone marking	Description	Illustration
Projections that are sites of muscle and ligament attachment		
Tuberosity (too"bĕ-ros'ĭ-te)	Large rounded projection; may be roughened	
Crest	Narrow ridge of bone; usually prominent	
Trochanter (tro-kan'ter)	Very large, blunt, irregularly shaped process. (The only examples are on the femur.)	
Line	Narrow ridge of bone; less prominent than a crest	
Tubercle (too'ber-kl)	Small rounded projection or process	
Epicondyle (ep"ĭ-kon'dīl)	Raised area on or above a condyle	
Spine	Sharp, slender, often pointed projection	
Projections that help to form joints		
Head	Bony expansion carried on a narrow neck	
Facet	Smooth, nearly flat articular surface	
Condyle (kon'dīl)	Rounded articular projection	
Ramus (ra'mus)	Armlike bar of bone	
Depressions and openings allowing blood vessels and nerves to pass		
Meatus (me-a'tus)	Canal-like passageway	
Sinus	Cavity within a bone, filled with air and lined with mucous membrane	
Fossa (fos'ah)	Shallow, basinlike depression in a bone, often serving as an articular surface	
Groove	Furrow	
Fissure	Narrow, slitlike opening	
Foramen (fo-ra'men)	Round or oval opening through a bone	

into one of the preceding categories are classified as **irregular bones.** The vertebrae are irregular bones (see Figure 9.1).

Some anatomists also recognize two other subcategories of bones. **Sesamoid bones** are small bones formed in tendons. The patellas (kneecaps) are sesamoid bones. **Wormian bones** are tiny bones between cranial bones. Except for the patellas, the sesamoid and Wormian bones are not included in the bone count of 206 because they vary in number and location in different individuals.

Activity:
Examining and Classifying Bones

Examine the isolated (disarticulated) bones on display. See if you can find specific examples of the bone markings described in Table 9.1. Then classify each of the bones into one of the four anatomical groups by recording its name or number in the chart at right. Verify your identifications with your instructor before leaving the laboratory. ∎

Long	Short	Flat	Irregular

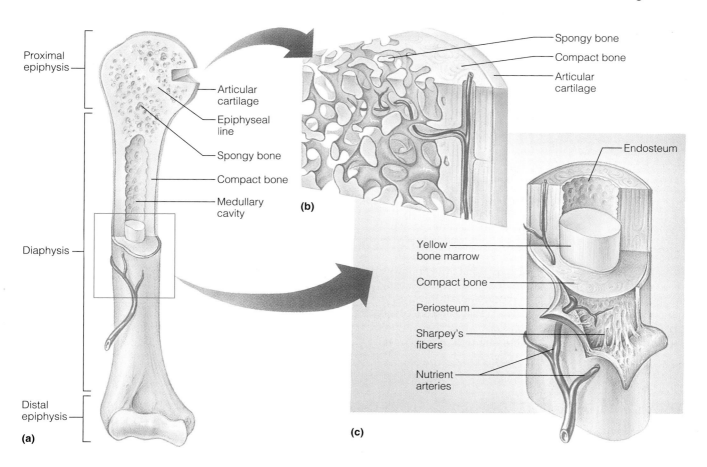

Figure 9.2 The structure of a long bone (humerus of the arm).
(a) Anterior view with longitudinal section cut away at the proximal end.
(b) Pie-shaped, three-dimensional view of spongy bone and compact bone of the epiphysis. **(c)** Cross section of shaft (diaphysis). Note that the external surface of the diaphysis is covered by a periosteum, but the articular surface of the epiphysis is covered with hyaline cartilage.

Gross Anatomy of the Typical Long Bone

A c t i v i t y :
Examining a Long Bone

1. Obtain a long bone that has been sawed along its longitudinal axis. If a cleaned dry bone is provided, no special preparations need be made.

 Note: if the bone supplied is a fresh beef bone, don plastic gloves before beginning your observations.

With the help of Figure 9.2, identify the shaft, or **diaphysis.** Observe its smooth surface, which is composed of compact bone. If you are using a fresh specimen, carefully pull away the **periosteum,** or fibrous membrane covering, to view the bone surface. Notice that many fibers of the periosteum penetrate into the bone. These fibers are called **Sharpey's fibers.** The periosteum is the source of the blood vessels and nerves

that invade the bone. **Osteoblasts** (bone-forming cells) on its inner face secrete the bony matrix that increases the girth of the long bone.

2. Now inspect the **epiphysis,** the end of the long bone. Notice that it is composed of a thin layer of compact bone that encloses spongy bone.

3. Identify the **articular cartilage,** which covers the epiphyseal surface in place of the periosteum. Because it is composed of glassy hyaline cartilage, it provides a smooth surface to prevent friction at joint surfaces.

4. If the animal was still young and growing, you will be able to see the **epiphyseal plate,** a thin area of hyaline cartilage that provides for longitudinal growth of the bone during youth. Once the long bone has stopped growing, these areas are replaced with bone and appear as thin, barely discernible remnants—the **epiphyseal lines.**

5. In an adult animal, the central cavity of the shaft (*medullary cavity*) is essentially a storage region for adipose tissue, or **yellow marrow.** In the infant, this area is involved in forming blood cells, and so **red marrow** is found in the

marrow cavities. In adult bones, the red marrow is confined to the interior of the epiphyses, where it occupies the spaces between the trabeculae of spongy bone.

6. If you are examining a fresh bone, look carefully to see if you can distinguish the delicate **endosteum** lining the shaft. In a living bone, **osteoclasts** (bone-destroying cells) are found on the inner surface of the endosteum, against the compact bone of the diaphysis. As the bone grows in diameter on its external surface, it is constantly being broken down on its inner surface. Thus the thickness of the compact bone layer composing the shaft remains relatively constant.

7. If you have been working with a fresh bone specimen, return it to the appropriate area and properly dispose of your gloves, as designated by your instructor. Wash your hands before continuing on to the microscope study. ■

Longitudinal bone growth at epiphyseal discs follows a predictable sequence and provides a reliable indicator of the age of children exhibiting normal growth. In cases in which problems of long-bone growth are suspected (for example, pituitary dwarfism), X rays are taken to view the width of the growth plates. An abnormally thin epiphyseal plate indicates growth retardation. ■

Chemical Composition of Bone

Bone is one of the hardest materials in the body. Although relatively light, bone has a remarkable ability to resist tension and shear forces that continually act on it. An engineer would tell you that a cylinder (like a long bone) is one of the strongest structures for its mass. Thus nature has given us an extremely strong, exceptionally simple (almost crude), and flexible supporting system without sacrificing mobility.

The hardness of bone is due to the inorganic calcium salts deposited in its ground substance. Its flexibility comes from the organic elements of the matrix, particularly the collagenic fibers.

Activity:
Examining the Effects of Heat and Nitric Acid on Bones

Obtain a bone sample that has been soaked in nitric acid (or vinegar) and one that has been baked. Heating removes the organic part of bone, while acid dissolves out the minerals. Do the treated bones retain the structure of untreated specimens?

_____

Gently apply pressure to each bone sample. What happens to the heated bone?

_____

The bone treated with acid?

_____

_____

What does the acid appear to remove from the bone?

_____

What does baking appear to do to the bone?

_____

_____

In rickets, the bones are not properly calcified. Which of the demonstration specimens would more closely resemble the bones of a child with rickets?

_____ ■

Microscopic Structure of Compact Bone

As you have seen, spongy bone has a spiky, open-work appearance, resulting from the arrangement of the **trabeculae** that compose it, while compact bone appears to be dense and homogeneous. However, microscopic examination of compact bone reveals that it is riddled with passageways carrying blood vessels, nerves, and lymphatic vessels that provide the living bone cells with needed substances and a way to eliminate wastes. Indeed, bone histology is much easier to understand when you recognize that bone tissue is organized around its blood supply.

Activity:
Examining the Microscopic Structure of Compact Bone

1. Obtain a prepared slide of ground bone and examine it under low power. Using Figure 9.3 as a guide, focus on a **central (Haversian) canal.** The central canal runs parallel to the long axis of the bone and carries blood vessels, nerves, and lymph vessels through the bony matrix. Identify the **osteocytes** (mature bone cells) in **lacunae** (chambers), which are arranged in concentric circles (concentric **lamellae**) around the central canal. A central canal and all the concentric lamellae surrounding it are referred to as an **osteon** or **Haversian system.** Also identify **canaliculi,** tiny canals radiating outward from a central canal to the lacunae of the first lamella and then from lamella to lamella. The canaliculi form a dense transportation network through the hard bone matrix, con-

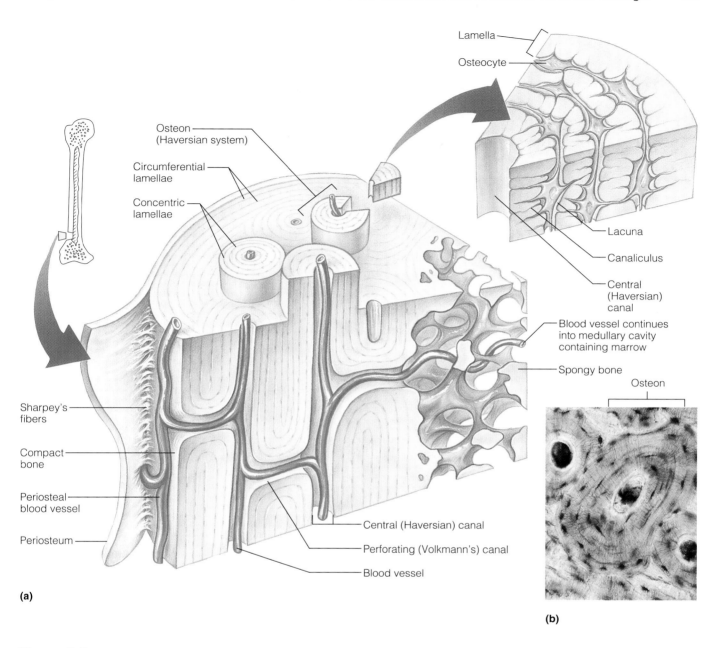

Figure 9.3   Microscopic structure of compact bone. (a) Diagrammatic view of a pie-shaped segment of compact bone, illustrating its structural units (osteons). The inset shows a more highly magnified view of a portion of one osteon. Notice the position of osteocytes in lacunae (cavities of the matrix). **(b)** Photomicrograph of a cross-sectional view of one osteon (90×).

necting all the living cells of the osteon to the nutrient supply. The canaliculi allow each cell to take what it needs for nourishment and to pass along the excess to the next osteocyte. You may need a higher-power magnification to see the fine canaliculi.

2.   Also note the **perforating (Volkmann's) canals** in Figure 9.3. These canals run into the compact bone and marrow

cavity from the periosteum, at right angles to the shaft. With the central canals, the perforating canals complete the communication pathway between the bone interior and its external surface.

3.   If a model of bone histology is available, identify the same structures on the model. ■

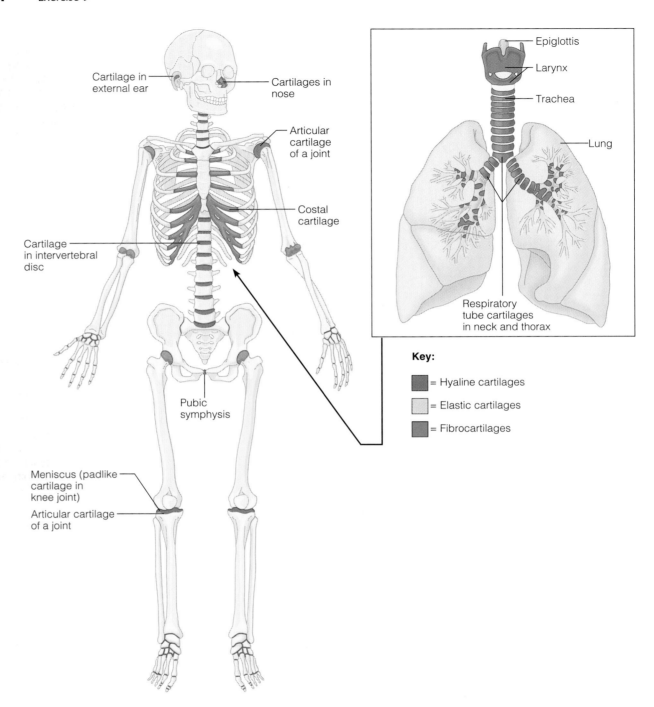

Figure 9.4 Cartilages in the adult skeleton and body. The cartilages that support the respiratory tubes and larynx are shown separately above.

Cartilages of the Skeleton

Location and Basic Structure

As mentioned earlier, cartilaginous regions of the skeleton have a fairly limited distribution in adults (Figure 9.4). The most important of these skeletal cartilages are (1) **articular cartilages,** which cover the bone ends at movable joints; (2) **costal cartilages,** found connecting the ribs to the sternum (breastbone); (3) **laryngeal cartilages,** which largely construct the larynx (voice box); (4) **tracheal** and **bronchial cartilages,** which reinforce other passageways of the respiratory system; (5) **nasal cartilages,** which support the external nose; (6) **intervertebral discs,** which separate and cushion bones of the spine (vertebrae); and (7) the cartilage supporting the external ear.

The skeletal cartilages consist of some variety of *cartilage tissue,* which typically consists primarily of water and is fairly resilient. Additionally, cartilage tissues are distinguished by the fact that they contain no nerves or blood vessels. Like bones, each cartilage is surrounded by a covering of dense connective tissue, called a *perichondrium* (rather than a periosteum). The perichondrium acts like a girdle to resist distortion of the cartilage when the cartilage is subjected to pressure. It also plays a role in cartilage growth and repair.

Classification of Cartilage

The skeletal cartilages have representatives from each of the three cartilage tissue types—hyaline, elastic, and fibrocartilage. Although you have already studied cartilage tissues (Exercise 6), some of that information will be recapped briefly here and you will have a chance to review the microscope structure unique to each cartilage type.

Activity:
Observing the Microscopic Structure of Different Types of Cartilage

Obtain prepared slides of hyaline cartilage, elastic cartilage, and fibrocartilage and bring them to your laboratory bench for viewing.

As you read through the descriptions of these cartilage types, keep in mind that the bulk of cartilage tissue consists of a nonliving *matrix* (containing a jellylike ground substance and fibers) secreted by chondrocytes.

Hyaline Cartilage **Hyaline cartilage** looks like frosted glass when viewed by the unaided eye. As easily seen in Figure 9.4, most skeletal cartilages are composed of hyaline cartilage. Its chondrocytes, snugly housed in lacunae, appear spherical and collagen fibers are the only fiber type in its matrix. Hyaline cartilage provides sturdy support with some resilience or "give." Draw a small section of hyaline cartilage in the circle below. Label the chrondrocytes, lacunae, and the cartilage matrix. Compare your drawing to Figure 6.5g, p. 63.

Elastic Cartilage **Elastic cartilage** can be envisioned as "hyaline cartilage with more elastic fibers." Consequently, it is much more flexible than hyaline cartilage and it tolerates repeated bending better. Essentially, only the cartilages of the external ear and the epiglottis (which flops over and covers the larynx when we swallow) are made of elastic cartilage. Focus on how this cartilage differs from hyaline cartilage as you diagram it in the circle below. Compare your illustration to Figure 6.5h, p. 63.

Fibrocartilage **Fibrocartilage** consists of rows of chondrocytes alternating with rows of thick collagen fibers. This tissue looks like a cartilage-dense regular connective tissue hybrid, and it is always found where hyaline cartilage joins a tendon or ligament. Fibrocartilage has great tensile strength and can withstand heavy compression. Hence, its use to construct the intervertebral discs and the cartilages within the knee joint makes a lot of sense (see Figure 9.4). Sketch a section of this tissue in the circle provided and then compare your sketch to the view seen in Figure 6.5i, p. 64. ■

exercise

10

The Axial Skeleton

Objectives

1. To identify the three bone groups composing the axial skeleton.
2. To identify the bones composing the axial skeleton, either by examining isolated bones or by pointing them out on an articulated skeleton or a skull, and to name the important bone markings on each.
3. To distinguish the different types of vertebrae.
4. To discuss the importance of intervertebral discs and spinal curvatures.
5. To distinguish three abnormal spinal curvatures.

Materials

❑ Intact skull and Beauchene skull
❑ X rays of individuals with scoliosis, lordosis, and kyphosis (if available)
❑ Articulated skeleton, articulated vertebral column
❑ Isolated cervical, thoracic, and lumbar vertebrae, sacrum, and coccyx

 See Appendix C, Exercise 10 for links to A.D.A.M. Interactive Anatomy.

The **axial skeleton** (the green portion of Figure 9.1 on p. 79) can be divided into three parts: the skull, the vertebral column, and the bony thorax.

The Skull

The **skull** is composed of two sets of bones. Those of the **cranium** enclose and protect the fragile brain tissue. The **facial bones** present the eyes in an anterior position and form the base for the facial muscles, which make it possible for us to present our feelings to the world. All but one of the bones of the skull are joined by interlocking joints called *sutures*. The mandible, or lower jawbone, is attached to the rest of the skull by a freely movable joint.

Activity:
Identifying the Bones of the Skull

The bones of the skull, shown in Figures 10.1 through 10.7, are described below. As you read through this material, iden-

tify each bone on an intact (and/or Beauchene*) skull. Note that important bone markings are listed beneath the bones on which they appear and that a color-coded dot before each bone name indicates its color in the figures. ■

The Cranium

The cranium may be divided into two major areas for study—the **cranial vault** or **calvaria,** forming the superior, lateral, and posterior walls of the skull, and the **cranial floor** or **base,** forming the skull bottom. Internally, the cranial floor has three distinct concavities, the **anterior, middle,** and **posterior cranial fossae** (see Figure 10.3). The brain sits in these fossae, completely enclosed by the cranial vault.

Eight large flat bones construct the cranium. *With the exception of two paired bones (the parietals and the temporals), all are single bones.* Sometimes the six ossicles of the middle ear are also considered part of the cranium. Because the ossicles are functionally part of the hearing apparatus, their consideration is deferred to Exercise 25, Special Senses: Hearing and Equilibrium.

○ **Frontal** See Figures 10.1, 10.3, and 10.6. Anterior portion of cranium; forms the forehead, superior part of the orbit, and floor of anterior cranial fossa.

Supraorbital foramen (notch): Opening above each orbit allowing blood vessels and nerves to pass.

Glabella: Smooth area between the eyes.

○ **Parietal** See Figures 10.1 and 10.6. Posterolateral to the frontal bone, forming sides of cranium.

Sagittal suture: Midline articulation point of the two parietal bones.

Coronal suture: Point of articulation of parietals with frontal bone.

○ **Temporal** See Figures 10.1 through 10.3 and 10.6. Inferior to parietal bone on lateral skull. The temporals can be divided into four major parts: the **squamous region** abuts the parietals; the **tympanic region** surrounds the external ear opening; the **mastoid region** is the area posterior to the ear; and the **petrous region** forms the lateral region of the skull base.

Important markings associated with the flaring squamous region (Figures 10.1 and 10.2) include:

Squamous suture: Point of articulation of the temporal bone with the parietal bone.

Zygomatic process: A bridgelike projection joining the zy-

*Two views (Plates A and B) of a Beauchene skull are provided in the Human Anatomy Atlas.

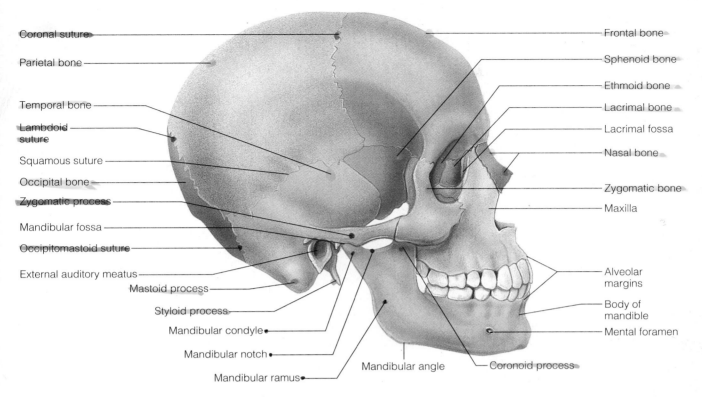

Figure 10.1 **External anatomy of the right lateral aspect of the skull.**

gomatic bone (cheekbone) anteriorly. Together these two bones form the *zygomatic arch.*

Mandibular fossa: Rounded depression on the inferior surface of the zygomatic process (anterior to the ear); forms the socket for the mandibular condyle, the point where the mandible (lower jaw) joins the cranium.

Tympanic region markings (Figures 10.1 and 10.2) include:

External auditory meatus: Canal leading to eardrum and middle ear.

Styloid (*stylo*=stake, pointed object) **process:** Needlelike projection inferior to external auditory meatus; attachment point for muscles and ligaments of the neck. This process is often broken off demonstration skulls.

Prominent structures in the mastoid region (Figures 10.1 and 10.2) are:

Mastoid process: Rough projection inferior and posterior to external auditory meatus; attachment site for muscles.

The mastoid process, full of air cavities and so close to the middle ear—a trouble spot for infections—often becomes infected too, a condition referred to as **mastoiditis.** Because the mastoid area is separated from the brain by only a thin layer of bone, an ear infection that has spread to the mastoid process can inflame the brain coverings or the meninges. The latter condition is known as **meningitis.** ■

Stylomastoid foramen: Tiny opening between the mastoid and styloid processes through which cranial nerve VII leaves the cranium.

The petrous region (Figures 10.2 and 10.3), which helps form the middle and posterior cranial fossae, exhibits several obvi-

ous foramina with important functions:

Jugular foramen: Opening medial to styloid process through which the internal jugular vein and cranial nerves IX, X, and XI pass.

Carotid canal: Opening medial to the styloid process through which the internal carotid artery passes into the cranial cavity.

Internal acoustic meatus: Opening on posterior aspect (petrous portion) of temporal bone allowing passage of cranial nerves VII and VIII (Figure 10.3).

Foramen lacerum: A jagged opening between the petrous temporal bone and the sphenoid providing passage for a number of small nerves, and for the internal carotid artery to enter the middle cranial fossa (after it passes through part of the temporal bone).

● **Occipital** See Figures 10.1, 10.2, 10.3, and 10.6. Most posterior bone of cranium—forms floor and back wall. Joins sphenoid bone anteriorly via its narrow basioccipital region.

Lambdoid suture: Site of articulation of occipital bone and parietal bones.

Foramen magnum: Large opening in base of occipital, which allows the spinal cord to join with the brain.

Occipital condyles: Rounded projections lateral to the foramen magnum that articulate with the first cervical vertebra (atlas).

Hypoglossal canal: Opening medial and superior to the occipital condyle through which the hypoglossal nerve (cranial nerve XII) passes.

External occipital crest and protuberance: Midline prominences posterior to the foramen magnum.

92

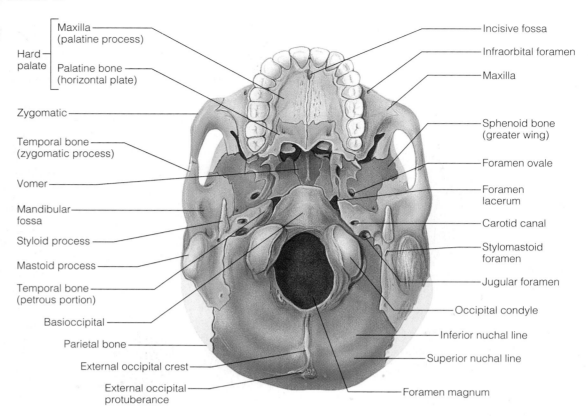

Figure 10.2 Inferior superficial view of the skull, mandible removed.

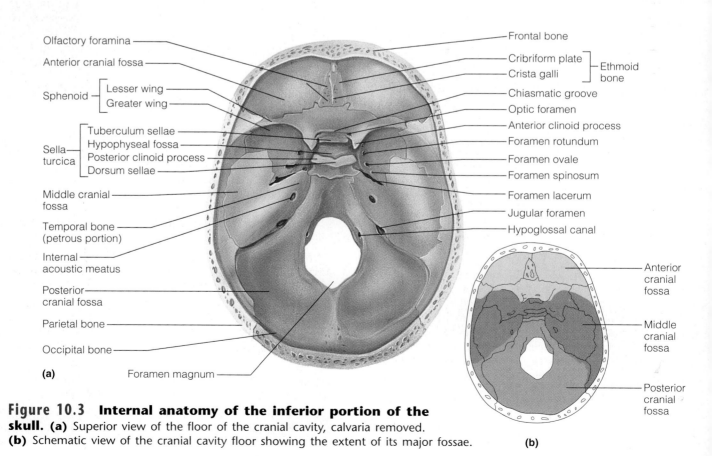

(a)

Figure 10.3 Internal anatomy of the inferior portion of the skull. (a) Superior view of the floor of the cranial cavity, calvaria removed. **(b)** Schematic view of the cranial cavity floor showing the extent of its major fossae.

(b)

Figure 10.4   The sphenoid bone. (a) Superior view. **(b)** Posterior view.

● **Sphenoid** See Figures 10.1 through 10.4 and 10.6. Bat-shaped bone forming the anterior plateau of the middle cranial fossa across the width of the skull.

Greater wings: Portions of the sphenoid seen exteriorly anterior to the temporal and forming a portion of the orbits of the eyes.

Superior orbital fissures: Jagged openings in orbits providing passage for cranial nerves III, IV, V, and VI to enter the orbit where they serve the eye.

The sphenoid bone can be seen in its entire width if the top of the cranium (calvaria) is removed (Figure 10.3).

Sella turcica (Turk's saddle): A saddle-shaped region in the sphenoid midline which nearly encloses the pituitary gland in a living person. The pituitary gland sits in the **hypophyseal fossa** portion of the sella turcica. This fossa is abutted before

and aft respectively by the **tuberculum sellae** and the **dorsum sellae.** The dorsum sellae terminates laterally in the **posterior clinoid processes.**

Lesser wings: Bat-shaped portions of the sphenoid anterior to the sella turcica. Posteromedially these terminate in the pointed **anterior clinoid processes,** which provide an anchoring site for securing the brain within the skull.

Optic foramina: Openings in the bases of the lesser wings through which the optic nerves enter the orbits to serve the eyes; these foramina are connected by the *chiasmatic groove.*

Foramen rotundum: Opening lateral to the sella turcica providing passage for a branch of the fifth cranial nerve. (This foramen is not visible on an inferior view of the skull.)

Foramen ovale: Opening posterior to the sella turcica that allows passage of a branch of the fifth cranial nerve.

-94-

Figure 10.5 The ethmoid bone. Anterior view.

○ **Ethmoid** See Figures 10.1, 10.3, 10.5, and 10.6. Irregularly shaped bone anterior to the sphenoid. Forms the roof of the nasal cavity, upper nasal septum, and part of the medial orbit walls.

Crista galli (cock's comb): Vertical projection providing a point of attachment for the dura mater (outermost membrane covering of the brain).

Cribriform plates: Bony plates lateral to the crista galli through which olfactory fibers pass to the brain from the nasal mucosa. Together the cribriform plates and the midline crista galli form the *horizontal plate* of the ethmoid bone.

Perpendicular plate: Inferior projection of the ethmoid that forms the superior part of the nasal septum.

Lateral masses: Irregularly shaped thin-walled bony regions flanking the perpendicular plate laterally. Their lateral surfaces (*orbital plates*) shape part of the medial orbit wall.

Superior and middle nasal conchae (turbinates): Thin, delicately coiled plates of bone extending medially from the lateral masses of the ethmoid into the nasal cavity. The conchae make air flow through the nasal cavity more efficient and greatly increase the surface area of the mucosa that covers them, thus increasing the mucosa's ability to warm and humidify incoming air.

Facial Bones

Of the 14 bones composing the face, 12 are paired. *Only the mandible and vomer are single bones.* An additional bone, the hyoid bone, although not a facial bone, is considered here because of its location.

○ **Mandible** See Figures 10.1, 10.6, and 10.7. The lower jawbone, which articulates with the temporal bones in the only freely movable joints of the skull.

Body: Horizontal portion; forms the chin.

Ramus: Vertical extension of the body on either side.

Mandibular condyle: Articulation point of the mandible with the mandibular fossa of the temporal bone.

Coronoid process: Jutting anterior portion of the ramus; site of muscle attachment.

Angle: Posterior point at which ramus meets the body.

Mental foramen: Prominent opening on the body (lateral to the midline) that transmits the mental blood vessels and nerve to the lower jaw.

Mandibular foramen: Open the lower jaw of the skull to identify this prominent foramen on the medial aspect of the mandibular ramus. This foramen permits passage of the nerve involved with tooth sensation (mandibular branch of cranial nerve V) and is the site where the dentist injects Novocain to prevent pain while working on the lower teeth.

Alveolar margin: Superior margin of mandible, contains sockets in which the teeth lie.

Mandibular symphysis: Anterior median depression indicating point of mandibular fusion.

○ **Maxillae** See Figures 10.1, 10.2, 10.6, and 10.7. Two bones fused in a median suture; form the upper jawbone and part of the orbits. All facial bones, except the mandible, join the maxillae. Thus they are the main, or keystone, bones of the face.

Alveolar margin: Inferior margin containing sockets (alveoli) in which teeth lie.

Palatine processes: Form the anterior hard palate.

Infraorbital foramen: Opening under the orbit carrying the infraorbital nerves and blood vessels to the nasal region.

Incisive fossa: Large bilateral opening located posterior to the central incisor tooth of the maxilla and piercing the hard palate; transmits the nasopalatine arteries and blood vessels.

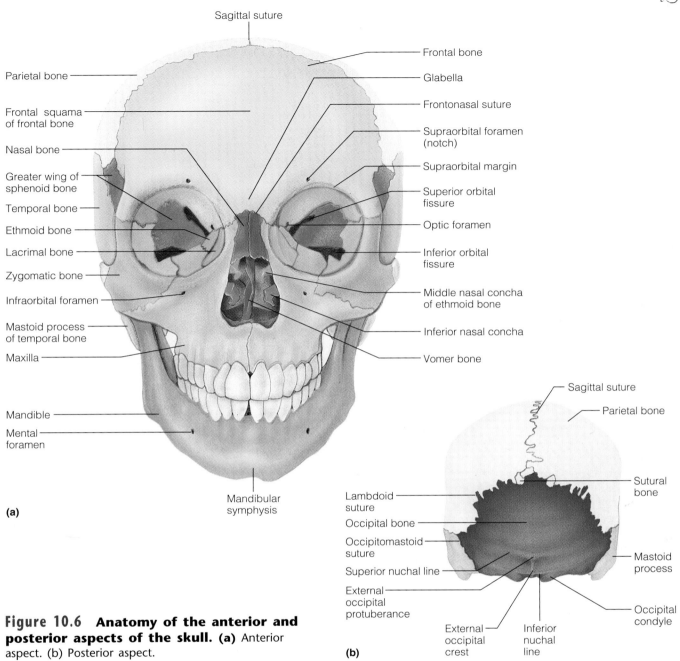

Figure 10.6 Anatomy of the anterior and posterior aspects of the skull. (a) Anterior aspect. **(b)** Posterior aspect.

○ **Palatine** See Figures 10.2 and 10.7. Paired bones posterior to the palatine processes; form posterior hard palate and part of the orbit.

◉ **Zygomatic** See Figures 10.1, 10.2, and 10.6. Lateral to the maxilla; forms the portion of the face commonly called the cheekbone, and forms part of the lateral orbit. Its three processes are named for the bones with which they articulate.

◉ **Lacrimal** See Figures 10.1 and 10.6. Fingernail-sized bones forming a part of the medial orbit walls between the maxilla and the ethmoid. Each lacrimal bone is

pierced by an opening, the **lacrimal fossa,** which serves as a passageway for tears (*lacrima* means "tear").

◉ **Nasal** See Figures 10.1 and 10.6. Small rectangular bones forming the bridge of the nose.

◉ **Vomer** (*vomer* = **plow**) See Figures 10.2 and 10.6. Blade-shaped bone in median plane of nasal cavity that forms the posterior and inferior nasal septum.

◉ **Inferior Nasal Conchae (turbinates)** See Figure 10.6. Thin curved bones protruding medially from the lateral walls of the nasal cavity; serve the same purpose as the turbinate portions of the ethmoid bone (described earlier).

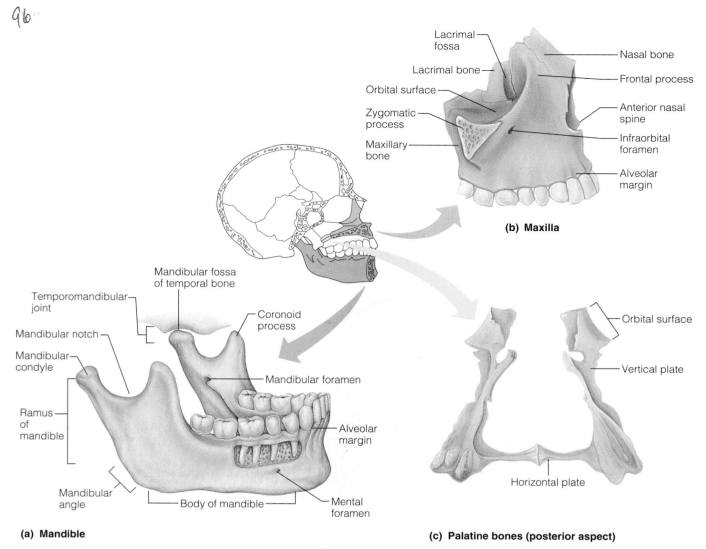

(b) Maxilla

(a) Mandible

(c) Palatine bones (posterior aspect)

Figure 10.7 Detailed anatomy of some isolated facial bones. (Note that the mandible, maxilla, and palatine bones are not drawn in proportion to each other.)

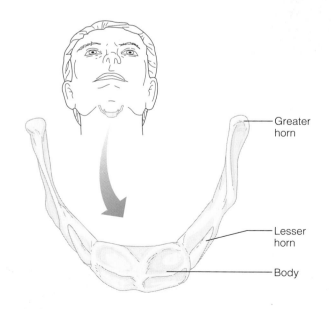

Figure 10.8 Hyoid bone.

Hyoid Bone

Not really considered or counted as a skull bone. Located in the throat above the larynx (Figure 10.8); serves as a point of attachment for many tongue and neck muscles. Does not articulate with any other bone, and is thus unique. Horseshoe-shaped with a body and two pairs of **horns,** or **cornua.**

Paranasal Sinuses

Four skull bones—maxillary, sphenoid, ethmoid, and frontal—contain sinuses (mucosa-lined air cavities), which lead into the nasal passages (see Figure 10.9). These paranasal sinuses lighten the facial bones and may act as resonance chambers for speech. The maxillary sinus is the largest of the sinuses found in the skull.

(a)

(b)

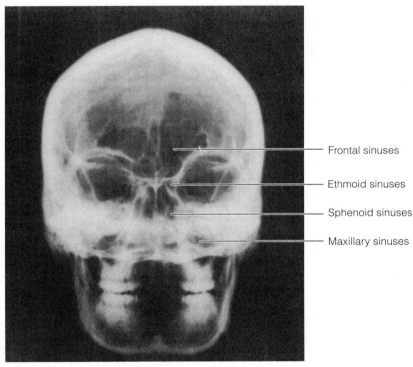

(c)

Figure 10.9   Paranasal sinuses. (a) Anterior "see-through" view.
(b) As seen in a sagittal section of the head. **(c)** Skull X ray showing three of
the paranasal sinuses, anterior view.

 Sinusitis, or inflammation of the sinuses, sometimes occurs as a result of an allergy or bacterial invasion of the sinus cavities. In such cases, some of the connecting passageways between the sinuses and nasal passages may become blocked with thick mucus or infectious material. Then, as the air in the sinus cavities is absorbed, a partial vacuum forms. The result is a sinus headache localized over the inflamed sinus area. Severe sinus infections may require surgical drainage to relieve this painful condition. ■

Activity:
Palpating Skull Markings

Palpate the following areas on yourself:

• Zygomatic bone and arch. (The most prominent part of your cheek is your zygomatic bone. Follow the posterior course of the zygomatic arch to its junction with your temporal bone.)

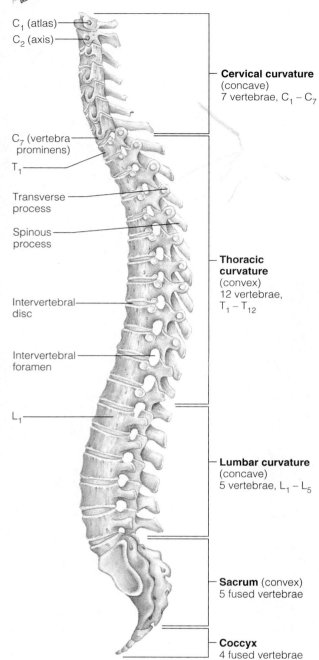

C₁ (atlas)
C₂ (axis)

Cervical curvature
(concave)
7 vertebrae, C₁ – C₇

C₇ (vertebra prominens)
T₁

Transverse process

Spinous process

Thoracic curvature
(convex)
12 vertebrae, T₁ – T₁₂

Intervertebral disc

Intervertebral foramen

L₁

Lumbar curvature
(concave)
5 vertebrae, L₁ – L₅

Sacrum (convex)
5 fused vertebrae

Coccyx
4 fused vertebrae

Figure 10.10 **The vertebral column.** Notice the curvatures in the lateral view. (The terms *convex* and *concave* refer to the curvature of the posterior aspect of the vertebral column.)

- Mastoid process (the rough area behind your ear).

- Temporomandibular joints. (Open and close your jaws to locate these.)

- Greater wing of sphenoid. (Find the indentation posterior to the orbit and superior to the zygomatic arch on your lateral skull.)

- Superior orbital foramen. (Apply firm pressure along the superior orbital margin to find the indentation resulting from this foramen.)

- Inferior orbital foramen. (Apply firm pressure along the inferomedial border of the orbit to locate this large foramen.)

- Mandibular angle (most inferior and posterior aspect of the mandible).

- Mandibular symphysis (midline of chin).

- Nasal bones. (Run your index finger and thumb along opposite sides of the bridge of your nose until they "slip" medially at the inferior end of the nasal bones.)

- External occipital protuberance. (This midline projection is easily felt by running your fingers up the furrow at the back of your neck to the skull.)

- Hyoid bone. (Place a thumb high behind the lateral edge of the mandible and squeeze medially.) ∎

The Vertebral Column

The **vertebral column,** extending from the skull to the pelvis, forms the body's major axial support. Additionally, it surrounds and protects the delicate spinal cord while allowing the spinal nerves to issue from the cord via openings between adjacent vertebrae. The term *vertebral column* might suggest a rather rigid supporting rod, but this is far from the truth. The vertebral column consists of 24 single bones called **vertebrae** and two composite, or fused, bones (the sacrum and coccyx) that are connected in such a way as to provide a flexible curved structure (Figure 10.10). Of the 24 single vertebrae, the seven bones of the neck are called *cervical vertebrae;* the next 12 are *thoracic vertebrae;* and the 5 supporting the lower back are *lumbar vertebrae.* Remembering common mealtimes for breakfast, lunch, and dinner (7 A.M., 12 noon, and 5 P.M.) may help you to remember the number of bones in each region.

The vertebrae are separated by pads of fibrocartilage, **intervertebral discs,** that cushion the vertebrae and absorb shocks. Each disc is composed of two major regions, a central gelatinous *nucleus pulposus* that behaves like a fluid, and an outer ring of encircling collagen fibers called the *annulus fibrosus* that stabilizes the disc and contains the pulposus.

As a person ages, the water content of the discs decreases (as it does in other tissues throughout the body), and the discs become thinner and less compressible. This situation, along with other degenerative changes such as weakening of the ligaments and tendons of the vertebral column, predisposes older people to **ruptured discs.** In a ruptured disc, the nucleus pulposus herniates through the annulus portion and typically compresses adjacent nerves. ∎

The presence of the discs and the S-shaped or springlike construction of the vertebral column prevent shock to the head in walking and running and provide flexibility to the body trunk. The thoracic and sacral curvatures of the spine are referred to as *primary curvatures* because they are present and well developed at birth. Later the *secondary curvatures* are formed. The cervical curvature becomes prominent when the baby begins to hold its head up independently, and the lumbar curvature develops when the baby begins to walk.

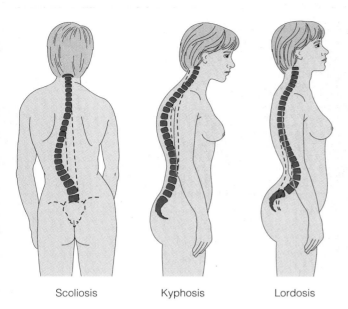

Scoliosis   Kyphosis   Lordosis

Figure 10.11   Abnormal spinal curvatures.

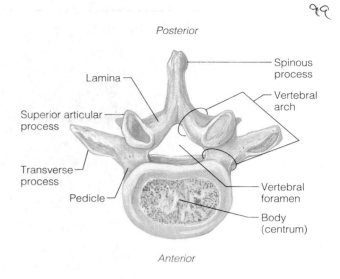

Figure 10.12   A typical vertebra, superior view. Inferior articulating surfaces not shown.

Activity:
Examining Spinal Curvatures

1. Observe the normal curvature of the vertebral column in your laboratory specimen, and compare it to Figure 10.10. Then examine Figure 10.11, which depicts three abnormal spinal curvatures—*scoliosis, kyphosis,* and *lordosis.* These abnormalities may result from disease or poor posture. Also examine X rays, if they are available, showing these same conditions in a living patient.

2. Then using an articulated vertebral column (or an articulated skeleton), examine the freedom of movement between two lumbar vertebrae separated by an intervertebral disc.

When the fibrous disc is properly positioned, are the spinal cord or peripheral nerves impaired in any way?

_____

_____

Remove the disc and put the two vertebrae back together. What happens to the nerve?

_____

_____

What would happen to the spinal nerves in areas of malpositioned or "slipped" discs?

_____

_____

_____ ■

Structure of a Typical Vertebra

Although they differ in size and specific features, all vertebrae have some features in common (Figure 10.12).

Body (or centrum): Rounded central portion of the vertebra, which faces anteriorly in the human vertebral column.

Vertebral arch: composed of pedicles, laminae, and a spinous process, it represents the junction of all posterior extensions from the vertebral body.

Vertebral foramen: Opening enclosed by the body and vertebral arch; a conduit for the spinal cord.

Transverse processes: Two lateral projections from the vertebral arch.

Spinous process: Single medial and posterior projection from the vertebral arch.

Superior and inferior articular processes: Paired projections lateral to the vertebral foramen that enable articulation with adjacent vertebrae. The superior articular processes typically face toward the spinous process, whereas the inferior articular processes face away from the spinous process.

Intervertebral foramina: The right and left pedicles have notches on their inferior and superior surfaces that create openings, the intervertebral foramina, for spinal nerves to leave the spinal cord between adjacent vertebrae.

Figures 10.13 and 10.14 and Table 10.1 show how specific vertebrae differ; refer to them as you read the following sections.

Cervical Vertebrae

The seven cervical vertebrae (referred to as C_1 through C_7) form the neck portion of the vertebral column. The first two cervical vertebrae (atlas and axis) are highly modified to perform special functions (see Figure 10.13). The **atlas** (C_1)

9ª

(a) Superior view of atlas (C₁)

(b) Inferior view of atlas (C₁)

(c) Superior view of axis (C₂)

Figure 10.13 Cervical vertebrae C₁ and C₂. (a) Superior view of the atlas (C₁). **(b)** Inferior view of the atlas (C₁). **(c)** Superior view of the axis (C₂).

lacks a body, and its lateral processes contain large concave depressions on their superior surfaces that receive the occipital condyles of the skull. This joint enables you to nod "yes." The **axis** (C₂) acts as a pivot for the rotation of the atlas (and skull) above. It bears a large vertical process, the **odontoid process,** or **dens,** that serves as the pivot point. The articulation between C₁ and C₂ allows you to rotate your head from side to side to indicate "no."

The more typical cervical vertebrae (C₃ through C₇) are distinguished from the thoracic and lumbar vertebrae by several features (see Table 10.1). They are the smallest, lightest vertebrae and the vertebral foramen is triangular. The spinous process is short and often bifurcated, or divided into two branches. The spinous process of C₇ is not branched, however, and is substantially longer than that of the other cervical vertebrae. Because the spinous process of C₇ is visible through the skin, it is called the *vertebra prominens* and is used as a landmark for counting the vertebrae. Transverse processes of the cervical vertebrae are wide, and they contain foramina through which the vertebral arteries pass superiorly on their way to the brain. Any time you see these foramina in a vertebra, you can be sure that it is a cervical vertebra.

• Palpate your vertebra prominens.

Thoracic Vertebrae

The 12 thoracic vertebrae (referred to as T₁ through T₁₂) may be recognized by the following structural characteristics. As shown in Table 10.1, they have a larger body than the cervical vertebrae. The body is somewhat heart shaped, with two small articulating surfaces, or *costal demifacets,* on each side (one superior, the other inferior) close to the origin of the vertebral arch. These demifacets articulate with the heads of the corresponding ribs. The vertebral foramen is oval or round, and the spinous process is long, with a sharp downward hook. The closer the thoracic vertebra is to the lumbar region, the less sharp and shorter the spinous process. Articular facets on the transverse processes articulate with the tubercles of the ribs. Besides forming the thoracic part of the spine, these vertebrae form the posterior aspect of the bony thoracic cage (rib cage). Indeed, they are the only vertebrae that articulate with the ribs.

Lumbar Vertebrae

The five lumbar vertebrae (L₁ through L₅) have massive blocklike bodies and short, thick, hatchet-shaped spinous processes extending directly backward (see Table 10.1). The superior articular facets face posteromedially; the inferior

Table 10.1	Regional Characteristics of Cervical, Thoracic, and Lumbar Vertebrae		
Characteristic	**(a) Cervical (3–7)**	**(b) Thoracic**	**(c) Lumbar**
Body	Small, wide side to side	Larger than cervical; heart shaped; bears two costal demifacets	Massive; kidney shaped
Spinous process	Short; bifid; projects directly posteriorly	Long; sharp; projects inferiorly	Short; blunt; projects directly posteriorly
Vertebral foramen	Triangular	Circular	Triangular
Transverse processes	Contain foramina	Bear facets for ribs (except T_{11} and T_{12})	No special features
Superior and inferior articulating processes	Superior facets directed superoposteriorly	Superior facets directed posteriorly	Superior facets directed posteromedially (or medially)
	Inferior facets directed inferoanteriorly	Inferior facets directed anteriorly	Inferior facets directed anterolaterally (or laterally)
Movements allowed	Flexion and extension; lateral flexion; rotation; the spine region with the greatest range of movement	Rotation; lateral flexion possible but limited by ribs; flexion and extension prevented	Flexion and extension; some lateral flexion; rotation prevented

Superior view

Right lateral view

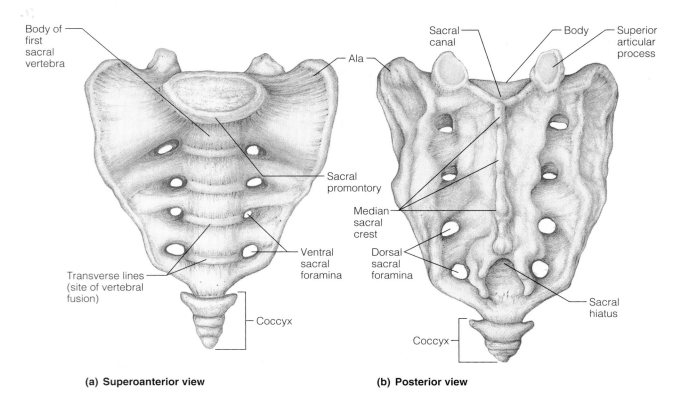

(a) **Superoanterior view**   (b) **Posterior view**

Figure 10.14   Sacrum and coccyx. (a) Superoanterior view. **(b)** Posterior view.

ones are directed anterolaterally. These structural features reduce the mobility of the lumbar region of the spine. Since most stress on the vertebral column occurs in the lumbar region, these are also the sturdiest of the vertebrae.

The spinal cord ends at the superior edge of L$_2$, but the outer covering of the cord, filled with cerebrospinal fluid, extends an appreciable distance beyond. Thus a *lumbar puncture* (for examination of the cerebrospinal fluid) or the administration of "saddle block" anesthesia for childbirth is normally done between L$_3$ and L$_4$ or L$_4$ and L$_5$, where there is little or no chance of injuring the delicate spinal cord.

The Sacrum

The **sacrum** (Figure 10.14) is a composite bone formed from the fusion of five vertebrae. Superiorly it articulates with L$_5$, and inferiorly it connects with the coccyx. The **median sacral crest** is a remnant of the spinous processes of the fused vertebrae. The winglike **alae,** formed by fusion of the transverse processes, articulate laterally with the hip bones. The sacrum is slightly concave anteriorly and forms the posterior border of the pelvis. Four ridges (lines of fusion) cross the anterior part of the sacrum, and **sacral foramina** are located at either end of these ridges. These foramina allow blood vessels and nerves to pass. The vertebral canal continues inside the sacrum as the **sacral canal** and terminates near the coccyx via an enlarged opening called the **sacral hiatus.** The **sacral promontory** (anterior border of the body of S$_1$) is an important anatomical landmark for obstetricians.

•   Attempt to palpate the median sacral crest of your sacrum. (This is more easily done by thin people and [obviously] in privacy.)

The Coccyx

The **coccyx** (see Figure 10.14) is formed from the fusion of three to five small irregularly shaped vertebrae. It is literally the human tailbone, a vestige of the tail that other vertebrates have. The coccyx is attached to the sacrum by ligaments.

A c t i v i t y :
Examining Vertebral Structure

Obtain examples of each type of vertebra and examine them carefully, comparing them to Figures 10.13, 10.14, and Table 10.1, and to each other. ■

The Bony Thorax

The **bony thorax** is composed of the sternum, ribs, and thoracic vertebrae (Figure 10.15). It is also referred to as the **thoracic cage** because of its appearance and because it forms a protective cone-shaped enclosure around the organs of the thoracic cavity (heart and lungs, for example).

The Sternum

The **sternum** (breastbone), a typical flat bone, is a result of the fusion of three bones—the manubrium, body, and xiphoid process. It is attached to the first seven pairs of ribs. The superiormost **manubrium** looks like the knot of a tie; it articulates with the clavicle (collarbone) laterally. The **body (gladiolus)** forms the bulk of the sternum. The **xiphoid**

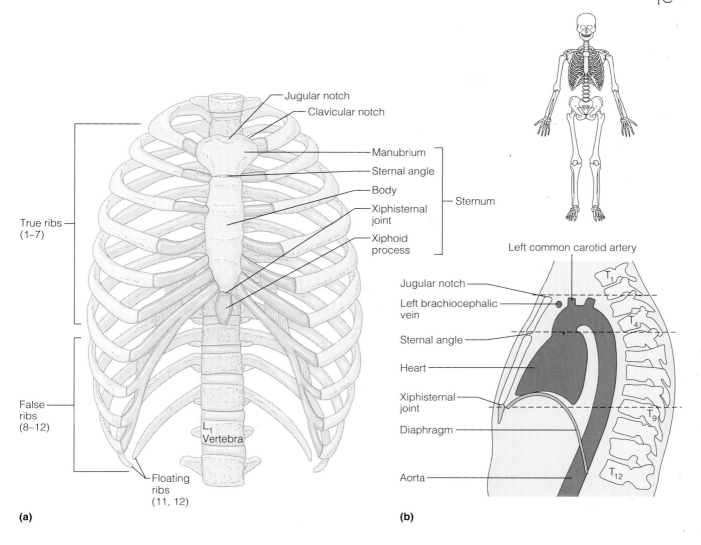

Jugular notch
Clavicular notch
Manubrium
Sternal angle
Body
Xiphisternal joint
Xiphoid process
Sternum
True ribs (1–7)
False ribs (8–12)
Floating ribs (11, 12)
L₁ Vertebra

Left common carotid artery
Jugular notch
Left brachiocephalic vein
Sternal angle
Heart
Xiphisternal joint
Diaphragm
Aorta
T₁
T₄
T₉
T₁₂

(a)   **(b)**

Figure 10.15   The bony thorax. (a) Skeleton of the bony thorax, anterior view (costal cartilages are shown in blue). **(b)** Left lateral view of the thorax, illustrating the relationship of the surface anatomical landmarks of the thorax to the vertebral column (thoracic portion).

process constructs the inferior end of the sternum and lies at the level of the fifth intercostal space. Although it is made of hyaline cartilage in children, it is usually ossified in adults.

In some people, the xiphoid process projects dorsally. This may present a problem because physical trauma to the chest can push such a xiphoid into the heart or liver (both immediately deep to the process), causing massive hemorrhage. ■

The sternum has three important bony landmarks—the jugular notch, the sternal angle, and the xiphisternal joint. The **jugular notch** (concave upper border of the manubrium) can be palpated easily; generally it is at the level of the third thoracic vertebra. The **sternal angle** is a result of the manubrium and body meeting at a slight angle to each other, so that a transverse ridge is formed at the level of the second ribs. It provides a handy reference point for counting ribs to locate the second intercostal space for listening to certain

heart valves, and is an important anatomical landmark for thoracic surgery. The **xiphisternal joint,** the point where the sternal body and xiphoid process fuse, lies at the level of the ninth thoracic vertebra.

•   Palpate your sternal angle and jugular notch.

Because of its accessibility, the sternum is a favored site for obtaining samples of blood-forming (hematopoietic) tissue for the diagnosis of suspected blood diseases. A needle is inserted into the marrow of the sternum and the sample withdrawn (sternal puncture).

(a)

(b)

Figure 10.16 Structure of a "typical" true rib and its articulations. (a) Vertebral and sternal articulations of a typical true rib. **(b)** Superior view of the articulation between a rib and a thoracic vertebra, with costovertebral ligaments shown on left side only.

The Ribs

The 12 pairs of **ribs** form the walls of the thoracic cage (see Figures 10.15 and 10.16). All of the ribs articulate posteriorly with the vertebral column via their heads and tubercles and then curve downward and toward the anterior body surface. The first seven pairs, called the *true,* or *vertebrosternal, ribs,* attach directly to the sternum by their "own" costal cartilages. The next five pairs are called *false ribs;* they attach indirectly to the sternum or entirely lack a sternal attachment. Of these, rib pairs 8–10, which are also called *vertebrochondral ribs,* have indirect cartilage attachments to the sternum. The last two pairs, called *floating,* or *vertebral, ribs,* have no sternal attachment.

Activity:
Examining the Relationship Between Ribs and Vertebrae

First take a deep breath to expand your chest. Notice how your ribs seem to move outward and how your sternum rises. Then examine an articulated skeleton to observe the relationship between the ribs and the vertebrae. ∎

The Appendicular Skeleton

The **appendicular skeleton** (the gold-colored portion of Figure 9.1) is composed of the 126 bones of the appendages and the pectoral and pelvic girdles, which attach the limbs to the axial skeleton. Although the upper and lower limbs differ in their functions and mobility, they have the same fundamental plan, with each limb composed of three major segments connected together by freely movable joints.

Activity:
Examining and Identifying Bones of the Appendicular Skeleton

Carefully examine each of the bones described throughout this exercise and identify the characteristic bone markings of each. The markings aid in determining whether a bone is the right or left member of its pair. *This is a very important instruction because, before completing this laboratory exercise, you will be constructing your own skeleton.* Additionally, when corresponding X rays are available, compare the actual bone specimen to its X-ray image. ■

Bones of the Pectoral Girdle and Upper Extremity
The Pectoral (Shoulder) Girdle

The paired **pectoral,** or **shoulder, girdles** (Figure 11.1) each consist of two bones—the anterior clavicle and the posterior scapula. The shoulder girdles function to attach the upper

Objectives

1. To identify on an articulated skeleton the bones of the pectoral and pelvic girdles and their attached limbs.
2. To arrange unmarked, disarticulated bones in proper relative position to form the entire skeleton.
3. To differentiate between a male and a female pelvis.
4. To discuss the common features of the human appendicular girdles (pectoral and pelvic), and to note how their structure relates to their specialized functions.
5. To identify specific bone markings in the appendicular skeleton.

Materials

❑ Articulated skeletons
❑ Disarticulated skeletons (complete)
❑ Articulated pelves (male and female for comparative study)
❑ X rays of bones of the appendicular skeleton

 See Appendix C, Exercise 11 for links to A.D.A.M. Interactive Anatomy.

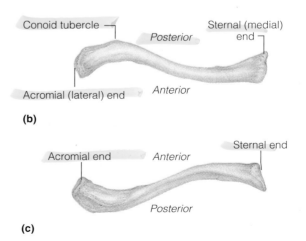

(a)

(b)

(c)

Figure 11.1 Bones of the pectoral (shoulder) girdle. (a) Right pectoral girdle articulated to show the relationship of the girdle to the bones of the thorax and arm. **(b)** Right clavicle, superior view. **(c)** Right clavicle, inferior view.

104

108

Figure 11.2 **Bones of the pectoral (shoulder) girdle.** **(a)** Right pectoral girdle articulated to show the relationship of the girdle to the bones of the thorax and arm. **(b-d)** Right scapula, anterior, lateral, and posterior views. View **(c)** is accompanied by a schematic representation of its orientation.

limbs to the axial skeleton. In addition, the bones of the shoulder girdles serve as attachment points for many trunk and neck muscles.

The **clavicle,** or collarbone, is a slender doubly curved bone—convex forward on its medial two-thirds and concave laterally. Its *sternal* (medial) *end,* which attaches to the sternal manubrium, is rounded or triangular in cross section. The

sternal end projects above the manubrium and can be felt and (usually) seen forming the lateral walls of the *jugular notch* (see Figure 10.15, p. 99). The *acromial* (lateral) *end* of the clavicle is flattened where it articulates with the scapula to form part of the shoulder joint. On its posteroinferior surface is the prominent **conoid tubercle.** This projection anchors a ligament and provides a handy landmark for determining

106
108

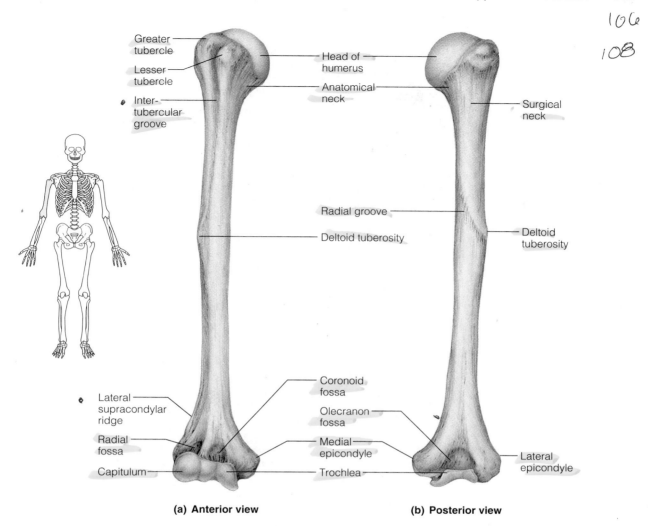

(a) Anterior view   **(b) Posterior view**

Figure 11.3   Bone of the right arm. (a) Humerus, anterior view. **(b)** Humerus, posterior view.

whether a given clavicle is from the right or left side of the body. The clavicle serves as an anterior brace, or strut, to hold the arm away from the top of the thorax.

The **scapulae** (Figure 11.2), or shoulder blades, are generally triangular and are commonly called the "wings" of humans. Each scapula has a flattened body and two important processes—the **acromion** (the enlarged end of the spine of the scapula) and the beaklike **coracoid process** (*corac* = crow, raven). The acromion connects with the clavicle; the coracoid process points anteriorly over the tip of the shoulder joint and serves as an attachment point for some of the upper limb muscles. The **suprascapular notch** at the base of the coracoid process allows nerves to pass. The scapula has no direct attachment to the axial skeleton but is loosely held in place by trunk muscles.

The scapula has three angles: superior, inferior, and lateral. The inferior angle provides a landmark for auscultating (listening to) lung sounds. The scapula also has three named borders: superior, medial (vertebral), and lateral (axillary). Several shallow depressions (fossae) appear on both sides of the scapula and are named according to location; i.e., there are the *anterior subscapular fossa* and the *posterior infraspinous* and *supraspinous fossae*. The **glenoid cavity,** a shal-

low socket that receives the head of the arm bone (humerus), is located in the lateral angle.

The shoulder girdle is exceptionally light and allows the upper limb a degree of mobility not seen anywhere else in the body. This is due to the following factors:

• The sternoclavicular joints are the *only* site of attachment of the shoulder girdles to the axial skeleton.

• The relative looseness of the scapular attachment allows it to slide back and forth against the thorax with muscular activity.

• The glenoid cavity is shallow, and does little to stabilize the shoulder joint.

However, this exceptional flexibility exacts a price: the arm bone (humerus) is very susceptible to dislocation, and fracture of the clavicle disables the entire upper limb.

The Arm

The arm (Figure 11.3) consists of a single bone—the **humerus,** a typical long bone. Proximally its rounded *head* fits into the shallow glenoid cavity of the scapula. The head is

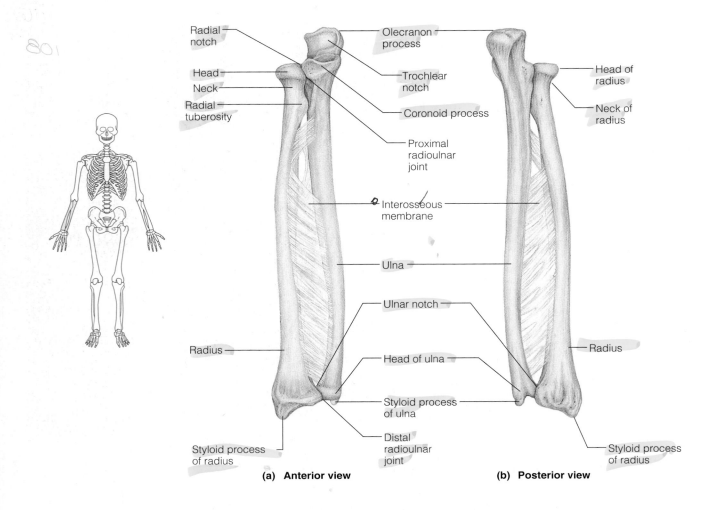

Figure 11.4   Bones of the right forearm. (a) Radius and ulna, anterior view. **(b)** Posterior view.

separated from the shaft by the *anatomical neck* and the more constricted *surgical neck,* which is a common site of fracture. Opposite the head are two prominences, the **greater** and **lesser tubercles** (from lateral to medial aspect), separated by a groove (the **intertubercular** or **bicipital groove**) that guides the tendon of the biceps muscle to its point of attachment (the superior rim of the glenoid cavity). In the midpoint of the shaft is a roughened area, the **deltoid tuberosity,** where the large fleshy shoulder muscle, the deltoid, attaches. Just inferior to the deltoid tuberosity is the **radial groove,** which indicates the pathway of the radial nerve.

At the distal end of the humerus are two condyles—the medial **trochlea** (looking rather like a spool), which articulates with the ulna, and the lateral **capitulum,** which articulates with the radius of the forearm. This condyle pair is flanked medially by the **medial epicondyle** and laterally by the **lateral epicondyle.**

The medial epicondyle is commonly called the "funny bone." The large ulnar nerve runs in a groove beneath the medial epicondyle, and when this region is sharply bumped, we are likely to experience a temporary, but excruciatingly painful, tingling sensation. This event is called "hitting the funny bone," a strange expression, because it is certainly *not* funny!

Above the trochlea on the anterior surface is a depression, the **coronoid fossa;** on the posterior surface is the **olecranon fossa.** These two depressions allow the corresponding processes of the ulna to move freely when the elbow is flexed and extended. The small **radial fossa,** lateral to the coronoid fossa, receives the head of the radius when the elbow is flexed.

The Forearm

Two bones, the radius and the ulna, compose the skeleton of the forearm, or antebrachium (see Figure 11.4). When the body is in the anatomical position, the **radius** is in the lateral position in the forearm and the radius and ulna are parallel. Proximally, the disc-shaped head of the radius articulates with the capitulum of the humerus. Just below the head, on

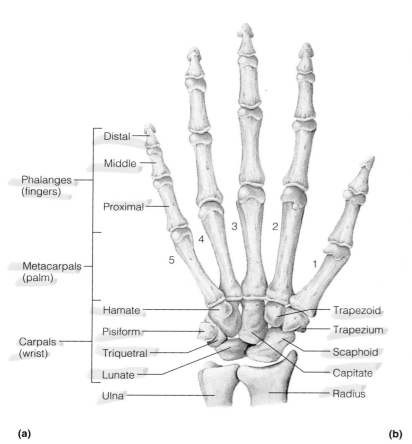

Phalanges
(fingers)

Distal

Middle

Proximal

Metacarpals
(palm)

Carpals
(wrist)

Hamate

Pisiform

Triquetral

Lunate

Ulna

Trapezoid

Trapezium

Scaphoid

Capitate

Radius

3   2

4

5   1

(a)

(b)

Figure 11.5   Bones of the right hand. (a) Anterior view showing the relationships of the carpals, metacarpals, and phalanges. **(b)** X ray. White bar on the proximal phalanx of the ring finger shows the position at which a ring would be worn.

the medial aspect of the shaft, is a prominence called the **radial tuberosity,** the point of attachment for the tendon of the biceps muscle of the arm. Distally, the small **ulnar notch** reveals where it articulates with the end of the ulna.

The **ulna** is the medial bone of the forearm. Its proximal end bears the anterior **coronoid process** and the posterior **olecranon process,** which are separated by the **trochlear notch.** Together these processes grip the trochlea of the humerus in a plierslike joint. The small **radial notch** on the lateral side of the coronoid process articulates with the head of the radius. The slimmer distal end, the ulnar head, bears a small medial **styloid process,** which serves as a point of attachment for the ligaments of the wrist.

The Hand

The skeleton of the hand, or manus (Figure 11.5), includes three groups of bones, those of the carpus (wrist), the metacarpals (bones of the palm), and the phalanges (bones of the fingers).

The wrist is the proximal portion of the hand. It is referred to anatomically as the **carpus;** the eight bones composing it are the **carpals.** (So you actually wear your wristwatch over the distal part of your forearm.) The carpals are arranged in two irregular rows of four bones each, which are illustrated in Figure 11.5. In the proximal row (lateral to me-

dial) are the scaphoid, lunate, triangular, and pisiform bones; the scaphoid and lunate articulate with the distal end of the radius. In the distal row are the trapezium, trapezoid, capitate, and hamate. The carpals are bound closely together by ligaments, which restrict movements between them.

The **metacarpals,** numbered 1 to 5 from the thumb side of the hand toward the little finger, radiate out from the wrist like spokes to form the palm of the hand. The *bases* of the metacarpals articulate with the carpals of the wrist; their more bulbous *heads* articulate with the phalanges of the fingers distally. When the fist is clenched, the heads of the metacarpals become prominent as the knuckles.

Like the bones of the palm, the fingers are numbered from 1 to 5, beginning from the thumb (*pollex*) side of the hand. The 14 bones of the fingers, or digits, are miniature long bones, called **phalanges** as noted above. Each finger contains three phalanges (proximal, middle, and distal) except the thumb which has only two (proximal and distal).

Activity:
Palpating the Surface Anatomy of the Pectoral Girdle and the Upper Limb

Before continuing on to study the bones of the pelvic girdle, take the time to identify the following bone markings on the skin surface of the upper limb. It is usually preferable to observe and palpate the bone markings on your lab partner, particularly since many of these markings can only be seen from the dorsal aspect.

- Clavicle: Palpate the clavicle along its entire length from sternum to shoulder.

- Acromioclavicular joint: The high point of the shoulder, which represents the junction point between the clavicle and the acromion of the scapular spine.

- Spine of the scapula: Extend your arm at the shoulder so that your scapula moves posteriorly. As you do this, your scapular spine will be seen as a winglike protrusion on your dorsal thorax and can be easily palpated by your lab partner.

- Lateral epicondyle of the humerus: The inferiormost projection at the lateral aspect of the distal humerus. After you have located the epicondyle, run your finger posteriorly into the hollow immediately dorsal to the epicondyle. This is the site where the extensor muscles of the hand are attached and is a common site of the excruciating pain of tennis elbow, a condition in which those muscles and their tendons are abused physically.

- Medial epicondyle of the humerus: Feel this medial projection at the distal end of the humerus.

- Olecranon process of the ulna: Work your elbow—flexing and extending—as you palpate its dorsal aspect to feel the olecranon process of the ulna moving into and out of the olecranon fossa on the dorsal aspect of the humerus.

- Styloid process of the ulna: With the hand in the anatomical position, feel out this small inferior projection on the medial aspect of the distal end of the ulna.

- Styloid process of the radius: Find this projection at the distal end of the radius (lateral aspect). It is most easily located by moving the hand medially at the wrist. Once you have palpated the styloid process, move your fingers just medially onto the anterior wrist. Press firmly and then let up slightly on the pressure. You should be able to feel your pulse at this pressure point, which lies over the radial artery (radial pulse).

- Metacarpophalangeal joints (knuckles): Clench your fist and find the first set of flexed-joint protrusions beyond the wrist—these are your metacarpophalangeal joints. ■

Bones of the Pelvic Girdle and Lower Limb

The Pelvic (Hip) Girdle

As with the bones of the pectoral girdle and upper limb, pay particular attention to bone markings needed to identify right and left bones.

The **pelvic girdle,** or **hip girdle** (Figure 11.6), is formed by the two **coxal** (*coxa* = hip) **bones** (also called the **ossa coxae** or hip bones). The two coxal bones together with the sacrum and coccyx form the **bony pelvis.** In contrast to the bones of the shoulder girdle, those of the pelvic girdle are heavy and massive, and they attach securely to the axial skeleton. The sockets for the heads of the femurs (thigh bones) are deep and heavily reinforced by ligaments to en-

sure a stable, strong limb attachment. The ability to bear weight is more important here than mobility and flexibility. The combined weight of the upper body rests on the pelvis (specifically, where the hip bones meet the sacrum).

Each coxal bone is a result of the fusion of three bones—the ilium, ischium, and pubis—which are distinguishable in the young child. The **ilium,** a large flaring bone, forms the major portion of the coxal bone. It connects posteriorly, via its **auricular surface,** with the sacrum at the **sacroiliac joint.** The superior margin of the iliac bone, the **iliac crest,** is rough; when you rest your hands on your hips, you are palpating your iliac crests. The iliac crest terminates anteriorly in the **anterior superior spine** and posteriorly in the **posterior superior spine.** Two inferior spines are located below these. The shallow **iliac fossa** marks its internal surface, and a prominent ridge, the **arcuate line,** outlines the pelvic inlet, or pelvic brim.

The **ischium** is the "sit-down" bone, forming the most inferior and posterior portion of the coxal bone. The most outstanding marking on the ischium is the rough **ischial tuberosity,** which receives the weight of the body when sitting. The **ischial spine,** superior to the ischial tuberosity, is an important anatomical landmark of the pelvic cavity. (See Comparison of the Male and Female Pelves, below.) The obvious **lesser** and **greater sciatic notches** allow nerves and blood vessels to pass to and from the thigh. The sciatic nerve passes through the latter.

The **pubis** is the most anterior portion of the coxal bone. Fusion of the **rami** of the pubic bone anteriorly and the ischium posteriorly forms a bar of bone enclosing the **obturator foramen,** through which blood vessels and nerves run from the pelvic cavity into the thigh. The pubic bones of each hip bone meet anteriorly at the **pubic crest** to form a cartilaginous joint called the **pubic symphysis.** At the lateral end of the pubic crest is the *pubic tubercle* (see Figure 11.6c) to which the important *inguinal ligament* attaches.

The ilium, ischium, and pubis fuse at the deep hemispherical socket called the **acetabulum** (literally, "vinegar cup"), which receives the head of the thigh bone.

Activity:
Observing Pelvic Articulations

Before continuing with the bones of the lower limbs, take the time to examine an articulated pelvis. Notice how each coxal bone articulates with the sacrum posteriorly and how the two coxal bones join at the pubic symphysis. The sacroiliac joint is a common site of lower back problems because of the pressure it must bear. ■

Comparison of the Male and Female Pelves Although bones of males are usually larger, heavier, and have more prominent bone markings, the male and female skeletons are very similar. The outstanding exception to this generalization is pelvic structure.

The female pelvis reflects modifications for childbearing. Generally speaking, the female pelvis is wider, shallower, lighter, and rounder than that of the male. Not only must her pelvis support the increasing size of a fetus, but it must also be large enough to allow the infant's head (its largest dimension) to descend through the birth canal at birth.

To describe pelvic sex differences, we need to introduce

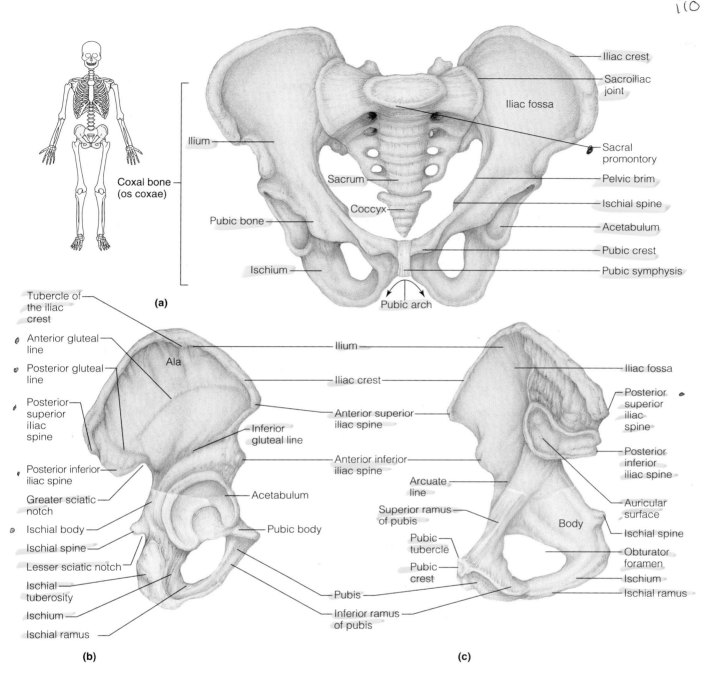

Figure 11.6   Bones of the pelvic girdle. (a) Articulated bony pelvis, showing the two coxal bones, which together comprise the pelvic girdle, and the sacrum. **(b)** Right coxal bone, lateral view, showing the point of fusion of the ilium, ischium, and pubic bones. **(c)** Right coxal bone, medial view.

a few more terms. Anatomically, the pelvis is described in terms of a false pelvis and a true pelvis. The **false pelvis** is that portion superior to the arcuate line; it is bounded by the alae of the ilia laterally and the sacral promontory and lumbar vertebrae posteriorly. Although the false pelvis supports the abdominal viscera, it does not restrict childbirth in any way. The **true pelvis** is the region inferior to the arcuate line that is almost entirely surrounded by bone. Its posterior boundary is formed by the sacrum. The ilia, ischia, and pubic bones define its limits laterally and anteriorly.

The dimensions of the true pelvis, particularly its inlet and outlet, are critical if delivery of a baby is to be uncomplicated; and they are carefully measured by the obstetrician. The **pelvic inlet,** or **pelvic brim,** is the opening delineated by the sacral promontory posteriorly and the arcuate lines of the ilia anterolaterally. It is the superiormost margin of the true pelvis. Its widest dimension is from left to right, that is, along the frontal plane. The **pelvic outlet** is the inferior margin of the true pelvis. It is bounded anteriorly by the pelvic arch, laterally by the ischia, and posteriorly by the sacrum and coccyx.

113

Table 11.1   Comparison of the Male and Female Pelves

Characteristic	Female	Male
General structure and functional modifications	Tilted forward; adapted for childbearing; true pelvis defines the birth canal; cavity of the true pelvis is broad, shallow, and has a greater capacity	Tilted less far forward; adapted for support of a male's heavier build and stronger muscles; cavity of the true pelvis is narrow and deep
Bone thickness	Less; bones lighter, thinner, and smoother	Greater; bones heavier and thicker, and markings are more prominent
Acetabula	Smaller; farther apart	Larger; closer
Pubic angle/arch	Broader (80° to 90°); more rounded	Angle is more acute (50° to 60°)
Anterior view		
	Pelvic brim	
Sacrum	Wider; shorter; sacral curvature is accentuated	Narrow; longer; sacral promontory more ventral
Coccyx	More movable; straighter	Less movable; curves ventrally
Left lateral view		
Pelvic inlet (brim)	Wider; oval from side to side	Narrow; basically heart shaped
Pelvic outlet	Wider; ischial tuberosities shorter, farther apart and everted	Narrower; ischial tuberosities longer, sharper, and point more medially
Posteroinferior view		

Figure 11.7 **Bones of the right thigh and knee. (a)** The patella (kneecap). **(b)** The femur (thigh bone).

(a) Patella

(b) Femur

Since both the coccyx and the ischial spines protrude into the outlet opening, a sharply angled coccyx or large, sharp ischial spines can dramatically narrow the outlet. The largest dimension of the outlet is the anterior-posterior diameter.

The major differences between the male and female pelves are summarized in Table 11.1.

A c t i v i t y :
Comparing Male and Female Pelves

Examine male and female pelves for the following differences:

- The female inlet is larger and more circular.

- The female pelvis as a whole is shallower, and the bones are lighter and thinner.

- The female sacrum is broader and less curved, and the pubic arch is more rounded.

- The female acetabula are smaller and farther apart, and the ilia flare more laterally.

- The female ischial spines are shorter, farther apart, and everted, thus enlarging the pelvic outlet. ■

The Thigh

The **femur,** or thigh bone (Figure 11.7b), is the sole bone of the thigh. It is the heaviest, strongest bone in the body. The ball-like head of the femur articulates with the hip bone via the deep, secure socket of the acetabulum. Obvious in the femur's head is a small central pit called the **fovea capitis** ("pit of the head") from which a small ligament runs to the acetabulum. The head of the femur is carried on a short, constricted *neck,* which angles laterally to join the shaft. The neck is the weakest part of the femur and is a common fracture site (an injury called a broken hip), particularly in the elderly. At the junction of the shaft and neck are the **greater** and **lesser trochanters** (separated posteriorly by the **intertrochanteric crest** and anteriorly by the **intertrochanteric line**). The trochanters and trochanteric crest, as well as the **gluteal tuberosity** and the **linea aspera** located on the shaft, are sites of muscle attachment.

The femur inclines medially as it runs downward to the leg bones; this brings the knees in line with the body's center

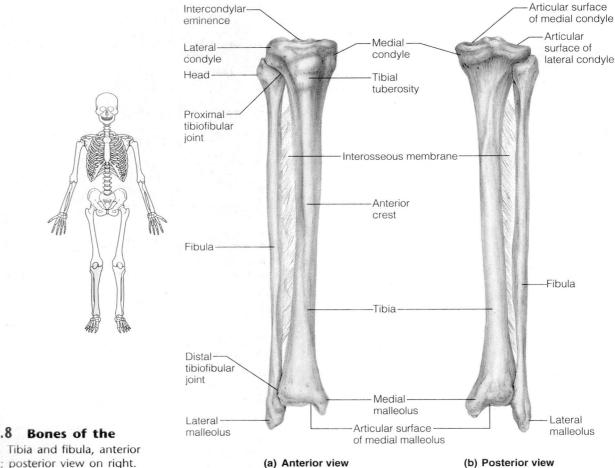

Figure 11.8 Bones of the right leg. Tibia and fibula, anterior view on left; posterior view on right.

(a) Anterior view   **(b) Posterior view**

of gravity, or maximum weight. The medial course of the femur is more noticeable in females because of the wider female pelvis.

Distally, the femur terminates in the **lateral** and **medial condyles,** which articulate with the tibia below, and the **patellar surface,** which forms a joint with the patella (kneecap) anteriorly. The **lateral** and **medial epicondyles,** just superior to the condyles, are separated by the **intercondylar notch.** On the superior part of the medial epicondyle is a bump, the **adductor tubercle,** to which the large adductor magnus muscle attaches.

The **patella** (Figure 11.7a) is a triangular sesamoid bone enclosed in the (quadriceps) tendon that secures the anterior thigh muscles to the tibia. It guards the knee joint anteriorly and improves the leverage of the thigh muscles acting across the knee joint.

The Leg

Two bones, the tibia and the fibula, form the skeleton of the leg (see Figure 11.8). The **tibia,** or *shinbone,* is the larger and more medial of the two leg bones. At the proximal end, the **medial** and **lateral condyles** (separated by the **intercondylar eminence**) receive the distal end of the femur to form the knee joint. The **tibial tuberosity,** a roughened protrusion on the anterior tibial surface (just below the condyles), is the site of attachment of the patellar (kneecap) ligament. Small facets on its superior and inferior lateral surface articulate with the

fibula. Distally, a process called the **medial malleolus** forms the inner (medial) bulge of the ankle, and the smaller distal end articulates with the talus bone of the foot. The anterior surface of the tibia bears a sharpened ridge that is relatively unprotected by muscles. This so-called **anterior crest** is easily felt beneath the skin.

The **fibula,** which lies parallel to the tibia, takes no part in forming the knee joint. Its proximal head articulates with the lateral condyle of the tibia. The fibula is thin and sticklike with a sharp anterior crest. It terminates distally in the **lateral malleolus,** which forms the outer part, or lateral bulge, of the ankle.

The Foot

The bones of the foot include the 7 **tarsal** bones, 5 **metatarsals,** which form the instep, and 14 **phalanges,** which form the toes (see Figure 11.9). Body weight is concentrated on the two largest tarsals which form the posterior aspect of the foot, the *calcaneus* (heel bone) and the *talus,* which lies between the tibia and the calcaneus. The other tarsals are named and identified in Figure 11.9. Like the fingers of the hand, each toe has 3 phalanges except the great toe, which has 2.

The bones in the foot are arranged to produce three strong arches—two longitudinal arches (medial and lateral) and one transverse arch (Figure 11.9b). Ligaments, binding the foot bones together, and tendons of the foot muscles hold

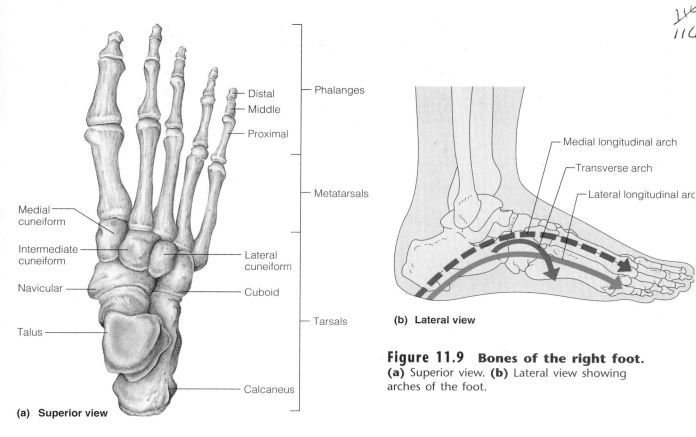

(a) **Superior view**

(b) **Lateral view**

Figure 11.9   Bones of the right foot.
(a) Superior view. **(b)** Lateral view showing
arches of the foot.

the bones firmly in the arched position but still allow a certain degree of give. Weakened arches are referred to as fallen arches or flat feet.

A c t i v i t y :

Palpating the Surface Anatomy of the Pelvic Girdle and Lower Limb

Locate and palpate the following bone markings on yourself and/or your lab partner.

• Iliac crest and anterior superior iliac spine: Rest your hands on your hips—they will be overlying the iliac crests. Trace the crest as far posteriorly as you can and then follow it anteriorly to the anterior superior iliac spine. This latter bone marking is easily felt in almost everyone, and is clearly visible through the skin (and perhaps the clothing) of very slim people. (The posterior superior iliac spine is much less obvious and is usually indicated only by a dimple in the overlying skin. Check it out in the mirror tonight.)

• Greater trochanter of the femur: This is easier to locate in females than in males because of the wider female pelvis; also it is more likely to be clothed by bulky muscles in males. Try to locate it on yourself as the most lateral point of the proximal femur. It typically lies about 6–8 inches below the iliac crest.

• Patella and tibial tuberosity: Feel your kneecap and palpate the ligaments attached to its borders. Follow the inferior patellar ligament to the tibial tuberosity.

• Medial and lateral condyles of the femur and tibia: As you move from the patella inferiorly on the medial (and then the lateral) knee surface, you will feel first the femoral and then the tibial condyle.

• Medial malleolus: Feel the medial protrusion of your ankle, the medial malleolus of the distal tibia.

• Lateral malleolus: Feel the bulge of the lateral aspect of your ankle, the lateral malleolus of the fibula.

• Calcaneus: Attempt to follow the extent of your calcaneus or heel bone. ■

A c t i v i t y :

Constructing a Skeleton

1.   When you finish examining the disarticulated bones of the appendicular skeleton and yourself, work with your lab partner to arrange the disarticulated bones on the laboratory bench in their proper relative positions to form an entire skeleton. Careful observations of the bone markings should help you distinguish between right and left members of bone pairs.

2.   When you believe that you have accomplished this task correctly, ask the instructor to check your arrangement to ensure that it is correct. If it is not, go to the articulated skeleton and check your bone arrangements. Also review the descriptions of the bone markings as necessary to correct your bone arrangement. ■

The Fetal Skeleton

Objectives

1. To define *fontanel* and discuss the function and fate of fontanels in the fetus.

2. To demonstrate important differences between the fetal and adult skeletons.

Materials

❑ Isolated fetal skull
❑ Fetal skeleton
❑ Adult skeleton

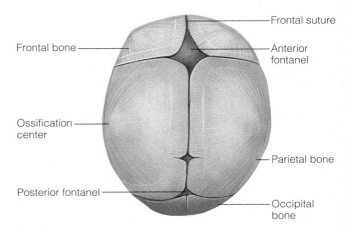

(a) Superior view

A human fetus about to be born has 275 bones, many more than the 206 bones found in the adult skeleton. This is because many of the bones described as single bones in the adult skeleton (for example, the coxal bone, sternum, and sacrum) have not yet fully ossified and fused in the fetus.

Activity:
Examining a Fetal Skull and Skeleton

1. Obtain a fetal skull and study it carefully. Make observations as needed to answer the following questions.

• Does it have the same bones as the adult skull?

• How does the size of the fetal face relate to the cranium?

• How does this compare to what is seen in the adult?

2. Indentations between the bones of the fetal skull, called **fontanels,** are fibrous membranes. These areas will become bony (ossify) as the fetus ages, completing the process by the age of 20 to 22 months. The fontanels allow the fetal skull to be compressed slightly during birth and also allow for brain growth during late fetal life. Locate the following fontanels on the fetal skull with the aid of Figure 12.1: anterior (or frontal) fontanel, mastoid fontanel, sphenoidal fontanel, and posterior (or occipital) fontanel.

3. Notice that some of the cranial bones have conical protrusions. These are growth (ossification) centers. Notice also that the frontal bone is still bipartite, and the temporal bone is incompletely ossified, little more than a ring of bone in the fetus.

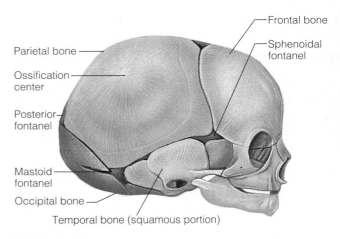

(b) Lateral view

Figure 12.1   The fetal skull. (a) Superior view. **(b)** Lateral view.

4. Obtain a fetal skeleton and examine it carefully, noting differences between it and an adult skeleton. Pay particular attention to the vertebrae, sternum, frontal bone of the cranium, patellae (kneecaps), coxal bones, carpals and tarsals, and rib cage.

5. Check the questions in the review section before completing this study to ensure that you have made all of the necessary observations. ■

Articulations and Body Movements

Objectives

1. To name the three structural categories of joints, and to compare their structure and mobility.
2. To identify the types of synovial joints.
3. To define *origin* and *insertion* of muscles.
4. To demonstrate or identify the various body movements.

Materials

- ❑ Articulated skeleton
- ❑ Skull
- ❑ Diarthrotic beef joint (fresh or preserved), preferably a knee joint
- ❑ Disposable gloves
- ❑ Anatomical chart of joint types (if available)
- ❑ X rays of normal and arthritic joints (if available)

 See Appendix C, Exercise 13 for links to A.D.A.M. Interactive Anatomy. ■

With rare exceptions, every bone in the body is connected to, or forms a joint with, at least one other bone. **Articulations,** or joints, perform two functions for the body. They (1) hold the bones together and (2) allow the rigid skeletal system some flexibility so that gross body movements can occur.

Joints may be classified structurally or functionally. The *structural classification* is based on whether there is connective tissue fiber, cartilage, or a joint cavity between the articulating bones. Structurally, there are *fibrous, cartilaginous,* and *synovial joints.*

The functional classification focuses on the amount of movement allowed at the joint. On this basis, there are **synarthroses,** or immovable joints; **amphiarthroses,** or slightly movable joints; and **diarthroses,** or freely movable joints. Freely movable joints predominate in the limbs, whereas immovable and slightly movable joints are largely restricted to the axial skeleton, where firm bony attachments and protection of enclosed organs are a priority.

As a general rule, fibrous joints are immovable, and synovial joints are freely movable. Cartilaginous joints offer both rigid and slightly movable examples. Since the structural categories are more clear-cut, we will use the structural classification here and indicate functional properties as appropriate.

Fibrous Joints

In **fibrous joints,** the bones are joined by fibrous tissue. No joint cavity is present. The amount of movement allowed depends on the length of the fibers uniting the bones. Although some fibrous joints are slightly movable, most are synarthrotic and permit virtually no movement.

The two major types of fibrous joints are sutures and syndesmoses. In **sutures** (Figure 13.1d) the irregular edges of the bones interlock and are united by very short connective tissue fibers, as in most joints of the skull. In **syndesmoses** the articulating bones are connected by short ligaments of dense fibrous tissue; the bones do not interlock. The joint at the distal end of the tibia and fibula is an example of a syndesmosis (Figure 13.1e). Although this syndesmosis allows some give, it is classed functionally as a synarthrosis.

Activity:
Identifying Fibrous Joints

Examine a human skull again. Notice that adjacent bone surfaces do not actually touch but are separated by fibrous connective tissue. Also examine a skeleton and anatomical chart of joint types for examples of fibrous joints. ■

Cartilaginous Joints

In **cartilaginous joints,** the articulating bone ends are connected by a plate or pad of cartilage. No joint cavity is present. The two major types of cartilaginous joints are synchondroses and symphyses. Although there is variation, most cartilaginous joints are *slightly movable* (amphiarthroses)

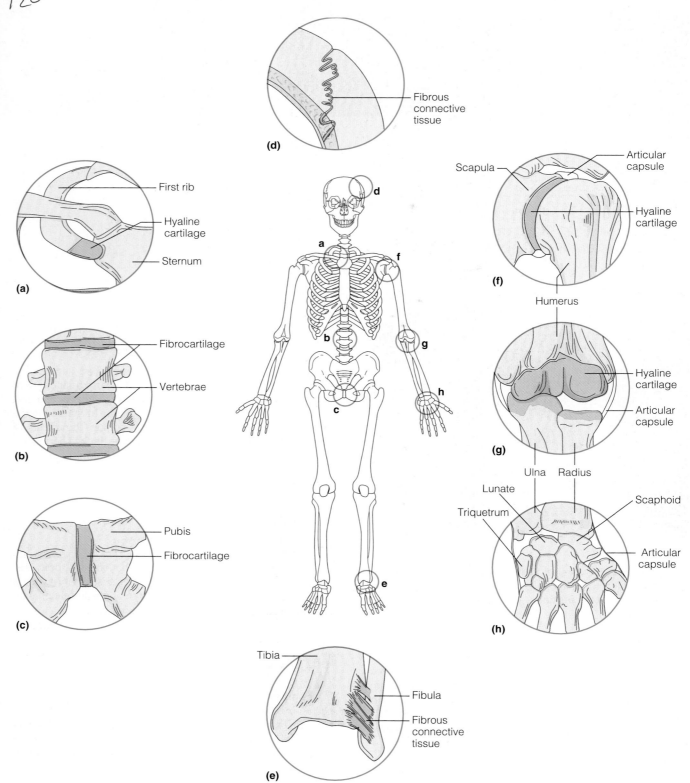

Figure 13.1 Types of joints. Joints to the left of the skeleton are cartilaginous joints; joints above and below the skeleton are fibrous joints; joints to the right of the skeleton are synovial joints. **(a)** Synchondrosis (joint between costal cartilage of rib 1 and the sternum). **(b)** Symphyses (intervertebral discs of fibrocartilage connecting adjacent vertebrae). **(c)** Symphysis (fibrocartilaginous pubic symphysis connecting the pubic bones anteriorly). **(d)** Suture (fibrous connective tissue connecting interlocking skull bones). **(e)** Syndesmosis (fibrous connective tissue connecting the distal ends of the tibia and fibula). **(f)** Synovial joint (multiaxial shoulder joint). **(g)** Synovial joint (uniaxial elbow joint). **(h)** Synovial joints (biaxial intercarpal joints of the hand).

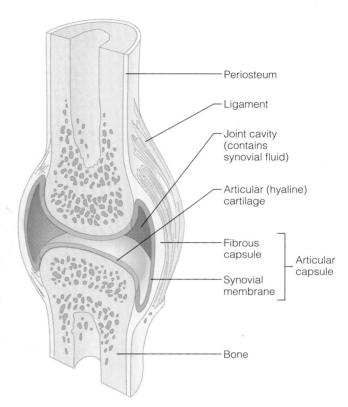

Figure 13.3 Major structural features of a synovial joint.

Figure 13.2 X ray of the hand of a child. Notice the cartilaginous epiphyseal plates, examples of temporary synchondroses.

functionally. In **symphyses** (*symphysis* means "a growth together") the bones are connected by a broad, flat disc of **fibrocartilage.** The intervertebral joints and the pubic symphysis of the pelvis are symphyses (see Figure 13.1b and c). In **synchondroses** the bony portions are united by hyaline cartilage. The articulation of the costal cartilage of the first rib with the sternum (Figure 13.1a) is a synchondrosis, but perhaps the best examples of synchondroses are the epiphyseal plates seen in the long bones of growing children (Figure 13.2). The epiphyseal plates are flexible during childhood but eventually they are totally ossified.

A c t i v i t y :
Identifying Cartilaginous Joints

Identify the cartilaginous joints on a human skeleton and on an anatomical chart of joint types. ■

Synovial Joints

Synovial joints are those in which the articulating bone ends are separated by a joint cavity containing synovial fluid (see Figure 13.1f–h). All synovial joints are diarthroses, or freely movable joints. Their mobility varies, however; some synovial joints can move in only one plane, and others can move in several directions (multiaxial movement). Most joints in the body are synovial joints.

All synovial joints have the following structural characteristics (Figure 13.3):

• The joint surfaces are enclosed by an *articular capsule* (a sleeve of fibrous connective tissue), creating a joint cavity.

• The interior of this capsule is lined with a smooth connective tissue membrane, called *synovial membrane,* which produces a lubricating fluid (synovial fluid) that reduces friction.

• Articulating surfaces of the bones forming the joint are covered with hyaline (*articular*) cartilage.

• The articular capsule is typically reinforced with ligaments and may contain bursae (fluid-filled sacs that reduce friction where tendons cross bone).

• Fibrocartilage pads (articular discs) may be present within the capsule.

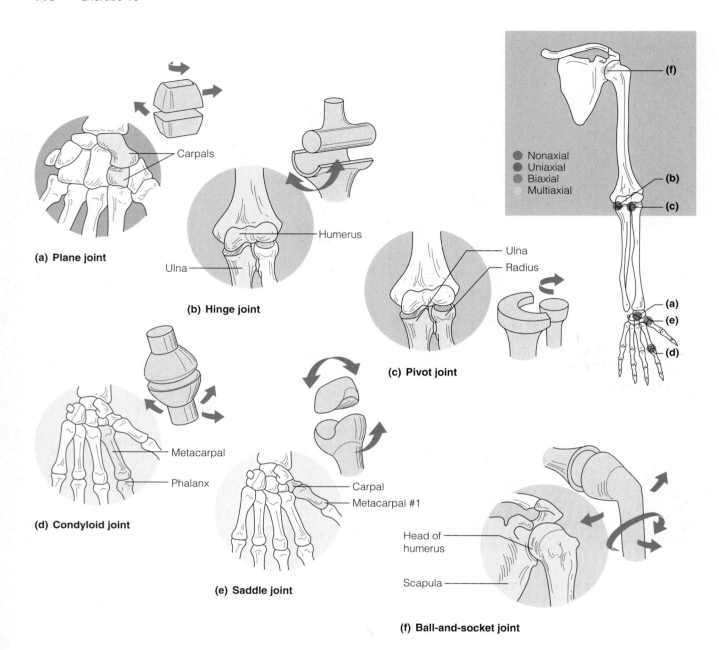

Figure 13.4 Types of synovial joints. Articulating bones being refer-enced are shown in gold; noninvolved bones are left white. **(a)** Plane joint (e.g., intercarpal and intertarsal joints). **(b)** Hinge joint (e.g., elbow joints and interphalangeal joints). **(c)** Pivot joint (e.g., proximal radioulnar joint). **(d)** Condyloid joint (e.g., metacarpophalangeal joints). **(e)** Saddle joint (e.g., carpometacarpal joint of the thumb). **(f)** Ball-and-socket joint (e.g., shoulder joint).

Activity:
Examining Synovial Joint Structure

Examine a beef joint to identify the general structural features of diarthrotic joints.

⚠ If the joint is freshly obtained from the slaughterhouse and you will be handling it, don plastic gloves before beginning your observations. ■

Types of Synovial Joints

Because there are so many types of synovial joints, they have been divided into the following subcategories on the basis of movements allowed (Figure 13.4):

• Plane: Articulating surfaces are flat or slightly curved, allowing sliding movements in one or two planes. Examples are the intercarpal and intertarsal joints and the vertebrocostal joints.

• Hinge: The rounded process of one bone fits into the concave surface of another to allow movement in one plane (uniaxial), usually flexion and extension. Examples are the elbow and interphalangeal joints.

• Pivot: The rounded or conical surface of one bone articulates with a shallow depression or foramen in another bone. Pivot joints allow uniaxial rotation, as in the joint between the atlas and axis (C_1 and C_2).

• Condyloid: The oval condyle of one bone fits into an ellipsoidal depression in another bone, allowing biaxial (two-way) movement. The wrist joint and the metacarpal-phalangeal joints (knuckles) are examples.

• Saddle: Articulating surfaces are saddle shaped; the articulating surface of one bone is convex, and the reciprocal surface is concave. Saddle joints, which are biaxial, include the joint between the thumb metacarpal and the trapezium of the wrist.

• Ball and socket: The ball-shaped head of one bone fits into a cuplike depression of another. These are multiaxial joints, allowing movement in all directions and pivotal rotation. Examples are the shoulder and hip joints.

Activity:
Identifying Types of Synovial Joints

Examine the articulated skeleton, anatomical charts, and yourself to identify the subcategories of synovial joints. Make sure you understand the terms *uniaxial, biaxial,* and *multiaxial.* ■

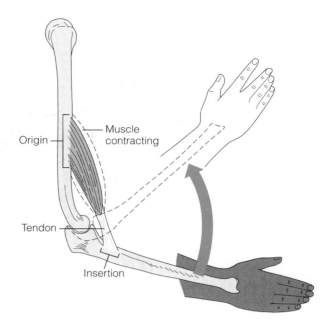

Figure 13.5 Muscle attachments (origin and insertion). When a skeletal muscle contracts, its insertion moves toward its origin.

Movements Allowed by Synovial Joints

Every muscle of the body is attached to bone (or other connective tissue structures) at two points–the **origin** (the stationary, immovable, or less movable attachment) and the **insertion** (the movable attachment). Body movement occurs when muscles contract across diarthrotic synovial joints (Figure 13.5). When the muscle contracts and its fibers shorten, the insertion moves toward the origin. The type of movement depends on the construction of the joint (uniaxial, biaxial, or multiaxial) and on the placement of the muscle relative to the joint. The most common types of body movements are described below and illustrated in Figure 13.6.

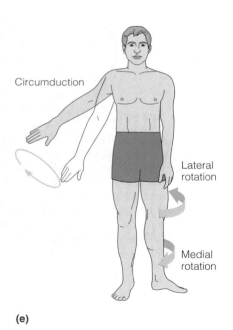

Figure 13.6 Movements occurring at synovial joints of the body. (a) Flexion and extension of the head. (b) Flexion and extension of the knee and shoulder. (c) Abduction and adduction of the arm. (d) Rotation of the head. (e) Circumduction of the arm and lateral and medial rotation of the lower limb around its long axis.

Activity:
Demonstrating Movements of Synovial Joints

Attempt to demonstrate each movement as you read through the following material:

Flexion (Figure 13.6a and b): A movement, generally in the sagittal plane, that decreases the angle of the joint and reduces the distance between the two bones. Flexion is typical of hinge joints (bending the knee or elbow), but is also common at ball-and-socket joints (bending forward at the hip).

Extension (Figure 13.6a and b): A movement that increases the angle of a joint and the distance between two bones or parts of the body (straightening the knee or elbow). Extension is the opposite of flexion. If extension is greater than 180 degrees (bending the trunk backward), it is termed *hyperextension.*

Abduction (Figure 13.6c): Movement of a limb away from the midline or median plane of the body, generally on the frontal plane, or the fanning movement of fingers or toes when they are spread apart.

Adduction (Figure 13.6c): Movement of a limb toward the midline of the body. Adduction is the opposite of abduction.

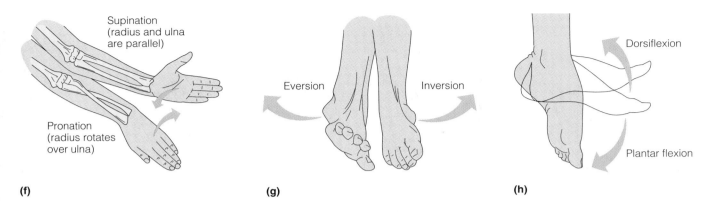

Supination
(radius and ulna
are parallel)

Pronation
(radius rotates
over ulna)

Eversion   Inversion

Dorsiflexion

Plantar flexion

(f)   **(g)**   **(h)**

Figure 13.6 (*continued*)   **(f)** Supination and pronation of the forearm. **(g)** Eversion and inversion of the foot. **(h)** Dorsiflexion and plantar flexion of the foot.

Rotation (Figure 13.6d and e): Movement of a bone around its longitudinal axis without lateral or medial displacement. Rotation, a common movement of ball-and-socket joints, also describes the movement of the atlas around the odontoid process of the axis.

Circumduction (Figure 13.6e): A combination of flexion, extension, abduction, and adduction commonly observed in ball-and-socket joints like the shoulder. The proximal end of the limb remains stationary, and the distal end moves in a circle. The limb as a whole outlines a cone.

Pronation (Figure 13.6f): Movement of the palm of the hand from an anterior or upward-facing position to a posterior or downward-facing position. This action moves the distal end of the radius across the ulna.

Supination (Figure 13.6f): Movement of the palm from a posterior position to an anterior position (the anatomical position). Supination is the opposite of pronation. During supination, the radius and ulna are parallel.

The last four terms refer to movements of the foot:

Inversion (Figure 13.6g): A movement that results in the medial turning of the sole of the foot.

Eversion (Figure 13.6g): A movement that results in the lateral turning of the sole of the foot; the opposite of inversion.

Dorsiflexion (Figure 13.6h): A movement of the ankle joint in a dorsal direction (standing on one's heels).

Plantar flexion (Figure 13.6h): A movement of the ankle joint in which the foot is flexed downward (standing on one's toes or pointing the toes). ■

Activity:
Examining Selected Synovial Joints

Now you will have the opportunity to compare and contrast the structure of the hip and knee joints using Figures 13.7 and 13.8 as your guides. Both of these joints are large weight-bearing joints of the lower limb, but they differ substantially in their security. Read through the brief descriptive material below, and look at the questions in the review section that pertain to this exercise before beginning your comparison.

The Knee Joint   The knee is the largest and most complex joint in the body. Three joints in one (Figure 13.7), it allows extension, flexion, and a little rotation. The tibiofemoral joint, actually a duplex joint between the femoral condyles above and the menisci (semilunar cartilages) of the tibia below, is functionally a hinge joint, a very unstable one made slightly more secure by the menisci. Some rotation occurs when the knee is partly flexed, but during extension, rotation and side-to-side movements are counteracted by the menisci and ligaments. The other joint is the femoropatellar joint, the intermediate joint anteriorly. The knee is unique in that it is only partly enclosed by an articular capsule. The capsule is reinforced by three broad ligaments, the *patellar ligament* and the *medial* and *lateral patellar retinacula,* which merge with the capsule. Extracapsular ligaments including the *fibular* and *tibial collateral ligaments* (which prevent rotation during extension) and the *oblique popliteal* and *arcuate popliteal ligaments* are crucial in reinforcing the knee. The *cruciate ligaments,* which are intracapsular ligaments, prevent anterior-posterior displacement of the joint and overflexion and hyperextension of the joint.

(a)

Tendon of quadriceps femoris
Suprapatellar bursa
Patella
Subcutaneous prepatellar bursa
Synovial cavity
Lateral meniscus
Infrapatellar fat pad
Deep infrapatellar bursa
Patellar ligament

Femur
Articular capsule
Posterior cruciate ligament
Lateral meniscus
Anterior cruciate ligament
Tibia

(b)

Fibular (lateral) collateral ligament
Lateral condyle of femur
Lateral meniscus
Tibia
Fibula

Posterior cruciate ligament
Medial condyle
Tibial (medial) collateral ligament
Anterior cruciate ligament
Medial meniscus (semilunar cartilage)
Patellar ligament
Patella
Quadriceps tendon

(c)

Femoral condyle
Anterior cruciate ligament
Medial tibial condyle
Medial meniscus

(d)

Quadriceps femoris muscle
Tendon of quadriceps femoris muscle
Patella
Lateral patellar retinaculum
Fibular collateral ligament
Fibula

Medial patellar retinaculum
Tibial collateral ligament
Patellar ligament
Tibia

(e)

Medial head of gastrocnemius muscle
Tendon of semimembranosus muscle

Femur
Articular capsule
Oblique popliteal ligament
Lateral head of gastrocnemius muscle
Fibular (lateral) collateral ligament
Arcuate popliteal ligament
Tibia

Figure 13.7   Knee joint relationships. (a) Mid-sagittal section of right knee joint. **(b)** Anterior view of slightly flexed right knee joint showing the cruciate ligaments. Articular capsule has been removed; the quadriceps tendon has been cut and reflected distally. **(c)** Photograph of an opened knee joint corresponds to view in (b). **(d)** Anterior superficial view of the right knee. **(e)** Posterior superficial view of the ligaments clothing the knee joint.

(a)

(c)

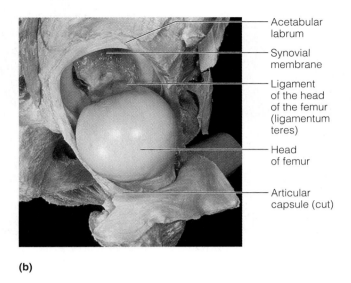

(b)

Figure 13.8   Hip joint relationships. (a) Frontal section through the right hip joint. **(b)** Photograph of the interior of the hip joint, lateral view. **(c)** Anterior superficial view of the right hip joint.

The Hip Joint The hip joint is a ball-and-socket joint, so movements can occur in all possible planes. However, its movements are definitely limited by its deep socket and strong reinforcing ligaments, the two factors that account for its exceptional stability (Figure 13.8).

The deeply cupped acetabulum that receives the head of the femur is enhanced by a circular rim of fibrocartilage called the *acetabular labrum.* Because the diameter of the labrum is smaller than that of the femur's head, dislocations of the hip are rare. A short ligament, *the ligament of the head of the femur* or *ligamentum teres* runs from the pitlike *fovea capitus* on the femur head to the acetabulum where it helps to secure the femur. Several strong ligaments, including the *iliofemoral* and *pubofemoral* anteriorly and the *ischiofemoral* that spirals posteriorly (not shown), are arranged so that they "screw" the femur head into the socket when a person stands upright. ■

Joint Disorders

Most of us don't think about our joints until something goes wrong with them. Joint pains and malfunctions are caused by a variety of things. For example, a hard blow to the knee can cause a painful bursitis, known as "water on the knee," due to damage to, or inflammation of, the patellar bursa. Slippage of a fibrocartilage pad or the tearing of a ligament may result in a painful condition that persists over a long period, since these poorly vascularized structures heal so slowly.

Sprains and dislocations are other types of joint problems. In a **sprain,** the ligaments reinforcing a joint are damaged by excessive stretching or are torn away from the bony attachment. Since both ligaments and tendons are cords of dense connective tissue with a poor blood supply, sprains heal slowly and are quite painful. **Dislocations** occur when bones are forced out of their normal position in the joint cavity. They are normally accompanied by torn or stressed ligaments and considerable inflammation. The process of returning the bone to its proper position, called reduction, should be done only by a physician. Attempts by the untrained person to "snap the bone back into its socket" are often more harmful than helpful.

Advancing years also take their toll on joints. Weight-bearing joints in particular eventually begin to degenerate. *Adhesions* (fibrous bands) may form between the surfaces where bones join, and extraneous bone tissue (*spurs*) may grow along the joint edges. Such degenerative changes lead to the complaint so often heard from the elderly: "My joints are getting so stiff. . . ."

● If possible compare an X ray of an arthritic joint to one of a normal joint. ■

Microscopic Anatomy, Organization, and Classification of Skeletal Muscle

Objectives

1. To describe the structure of skeletal muscle from gross to microscopic levels.

2. To define and explain the role of the following:

   actin   myofilament   tendon
   myosin   perimysium   endomysium
   fiber   aponeurosis   epimysium
   myofibril

3. To describe the structure of a neuromuscular junction and to explain its role in muscle function.

4. To define: agonist (prime mover), antagonist, synergist, fixator, origin, and insertion.

5. To cite criteria used in naming skeletal muscles.

Materials

❑ Three-dimensional model of skeletal muscle cells (if available)

❑ Forceps

❑ Dissecting needles

❑ Microscope slides and coverslips

❑ 0.9% saline solution in dropper bottles

❑ Chicken breast or thigh muscle (freshly obtained from the meat market)

❑ Compound microscope

❑ Histologic slides of skeletal muscle (longitudinal and cross-sectional) and skeletal muscle showing neuromuscular junctions

❑ Three-dimensional model of skeletal muscle showing neuromuscular junction (if available)

The bulk of the body's muscle is called **skeletal muscle** because it is attached to the skeleton (or associated connective tissue structures). Skeletal muscle influences body contours and shape, allows you to grin and frown, provides a means of locomotion, and enables you to manipulate the environment. The balance of the body's muscle—smooth and cardiac muscle—as the major component of the walls of hollow organs and the heart, respectively, is involved with the transport of materials within the body.

Each of the three muscle types has a structure and function uniquely suited to its task in the body. However, because the term *muscular system* applies specifically to skeletal muscle, the primary objective of this unit is to investigate the structure and function of skeletal muscle.

Skeletal muscle is also known as *voluntary muscle* (because it can be consciously controlled) and as *striated muscle* (because it appears to be striped). As you might guess from both of these alternative names, skeletal muscle has some very special characteristics. Thus an investigation of skeletal muscle should begin at the cellular level.

The Cells of Skeletal Muscle

Skeletal muscle is composed of relatively large, long cylindrical cells ranging from 10 to 100 μm in diameter and up to 6 cm in length. However, the cells of large, hard-working muscles like the antigravity muscles of the hip are extremely coarse, ranging up to 25 cm in length, and can be seen with the naked eye.

Skeletal muscle cells (Figure 14.1a) are multinucleate: Multiple oval nuclei can be seen just beneath the plasma membrane (called the *sarcolemma* in these cells). The nuclei are pushed peripherally by the longitudinally arranged **myofibrils,** which nearly fill the sarcoplasm (Figure 14.1b). Alternating light (I) and dark (A) bands along the length of the perfectly aligned myofibrils give the muscle fiber as a whole its striped appearance.

Electron microscope studies have revealed that the myofibrils are made up of even smaller threadlike structures called **myofilaments** (Figure 14.1c and d). The myofilaments are composed largely of two varieties of contractile proteins—**actin** and **myosin**—which slide past each other during muscle activity to bring about shortening or contraction of the muscle cells. It is the highly specific arrangement of the myofilaments within the myofibrils that is responsible for the banding pattern in skeletal muscle. The actual contractile units of muscle, called **sarcomeres,** extend from the middle of one I band (its Z disc) to the middle of the next along the length of the myofibrils. (See Figure 14.1c and d.)

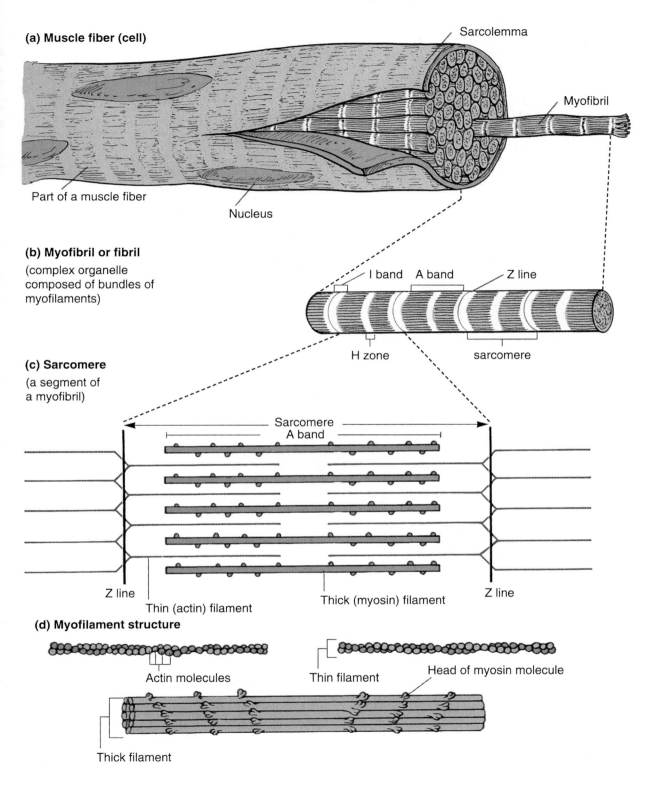

(a) Muscle fiber (cell)

Sarcolemma

Myofibril

Part of a muscle fiber

Nucleus

(b) Myofibril or fibril

(complex organelle composed of bundles of myofilaments)

I band A band Z line

H zone sarcomere

(c) Sarcomere

(a segment of a myofibril)

Sarcomere

A band

Z line Z line

Thin (actin) filament Thick (myosin) filament

(d) Myofilament structure

Actin molecules Thin filament Head of myosin molecule

Thick filament

Figure 14.1 Anatomy of a skeletal muscle cell (fiber). (a) A muscle fiber. One myofibril has been extended. **(b)** Enlarged view of a myofibril showing its banding pattern. **(c)** Enlarged view of one sarcomere (contractile unit) of a myofibril showing its banding pattern. **(d)** Structure of the thick and thin myofilaments found in the sarcomeres.

Activity:
Examining Skeletal Muscle Cell Anatomy

1.   Look at the three-dimensional model of skeletal muscle cells, noting the relative shape and size of the cells. Identify the nuclei, myofibrils, and light and dark bands.

2.   Obtain forceps, two dissecting needles, slide and cover-slip, and a dropper bottle of saline solution. With forceps, re-move a very small piece of muscle from the chicken breast (or thigh). Place the tissue on a clean microscope slide, and add a drop of the saline solution.

3.   Pull the muscle fibers apart with the dissecting needles (tease them) until you have a fluffy-looking mass of tissue. Cover the teased tissue with a coverslip, and observe under the high-power lens of a microscope. Look for the banding pattern. Regulate the light carefully to obtain the highest pos-sible contrast.

4.   Now compare your observations with Figure 14.2 and with what can be seen with professionally prepared muscle tissue. Obtain a slide of skeletal muscle (longitudinal sec-tion), and view it under high power. From your observations, draw a small section of a muscle fiber in the space provided here. Label the nuclei, sarcolemma, and A and I bands.

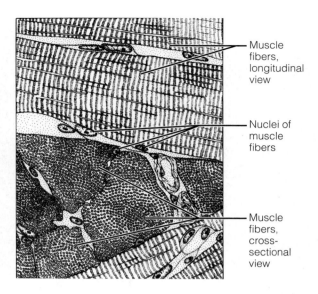

Muscle fibers, longitudinal view

Nuclei of muscle fibers

Muscle fibers, cross-sectional view

Figure 14.2   Muscle fibers, longitudinal and transverse views. (See also Plate 2 in the Histology Atlas.)

What structural details become apparent with the prepared slide?

_____

_____ ∎

Organization of Skeletal Muscle Cells into Muscles

Muscle fibers are soft and surprisingly fragile. Thus thou-sands of muscle fibers are bundled together with connective tissue to form the organs we refer to as skeletal muscles (Fig-ure 14.3). Each muscle fiber is enclosed in a delicate, areolar connective tissue sheath called **endomysium.** Several sheathed muscle fibers are wrapped by a collagenic mem-brane called **perimysium,** forming a bundle of fibers called a **fascicle,** or **fasciculus.** A large number of fascicles are bound together by a substantially coarser "overcoat" of dense con-nective tissue called an **epimysium,** which sheathes the en-tire muscle. These epimysia blend into the **deep fascia,** still coarser sheets of dense connective tissue that bind muscles into functional groups, and into strong cordlike **tendons** or sheetlike **aponeuroses,** which attach muscles to each other or indirectly to bones. As noted in Exercise 13, a muscle's more movable attachment is called its *insertion* whereas its fixed (or immovable) attachment is the *origin.*

Tendons perform several functions, two of the most im-portant being to provide durability and to conserve space. Be-cause tendons are tough collagenic connective tissue, they can span rough bony prominences that would destroy the more delicate muscle tissues. Because of their relatively small size, more tendons than fleshy muscles can pass over a joint.

In addition to supporting and binding the muscle fibers, and providing strength to the muscle as a whole, the connec-tive tissue wrappings provide a route for the entry and exit of nerves and blood vessels that serve the muscle fibers. The larger, more powerful muscles have relatively more connec-tive tissue than muscles involved in fine or delicate move-ments.

As we age, the mass of the muscle fibers decreases, and the amount of connective tissue increases; thus the skeletal muscles gradually become more sinewy, or "stringier." ∎

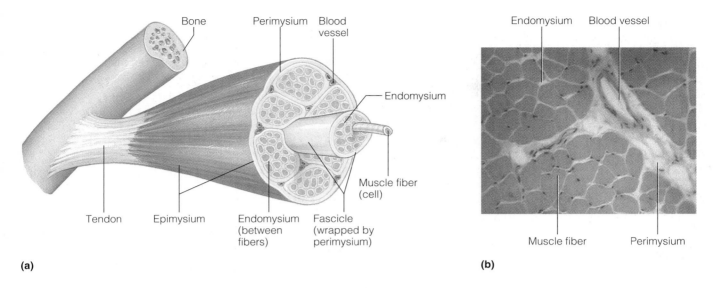

(a)

(b)

Figure 14.3 Connective tissue coverings of skeletal muscle.
(a) Diagrammatic view. **(b)** Photomicrograph of a cross section of skeletal muscle (90×).

A c t i v i t y :
Observing the Structure of Skeletal Muscle Tissue

Obtain a slide showing a cross section of skeletal muscle tissue. Using Figure 14.3 as a reference, identify the muscle fibers, endomysium, perimysium, and epimysium (if visible). ■

The Neuromuscular Junction

The voluntary skeletal muscle cells are always stimulated by motor neurons via nerve impulses. The junction between a nerve fiber (axon) and a muscle cell is called a **neuromuscular**, or **myoneural**, **junction** (Figure 14.4).

Each motor axon breaks up into many branches called *axonal terminals* as it approaches the muscle, and each of these branches participates in forming a neuromuscular junction with a single muscle cell. Thus a single neuron may stimulate many muscle fibers. Together, a neuron and all the muscle cells it stimulates make up the functional structure called the **motor unit.** Part of a motor unit is shown in Figure 14.5 and in Plate 3 of the Histology Atlas.

The neuron and muscle fiber membranes, close as they are, do not actually touch. They are separated by a small fluid-filled gap called the **synaptic cleft** (see Figure 14.4).

Within the axonal terminals are many mitochondria and vesicles containing a neurotransmitter chemical called acetylcholine. When a nerve impulse reaches the axonal endings, some of these vesicles release their contents into the synaptic cleft. The acetylcholine rapidly diffuses across the junction and combines with the receptors on the sarcolemma. If sufficient acetylcholine has been released, a transient change in the permeability of the sarcolemma briefly allows more sodium ions to diffuse into the muscle fiber. The result is depolarization of the sarcolemma and subsequent contraction of the muscle fiber.

A c t i v i t y :
Studying the Structure of a Neuromuscular Junction

1.   If possible, examine a three-dimensional model of skeletal muscle cells that illustrates the neuromuscular junction. Identify the structures just described.

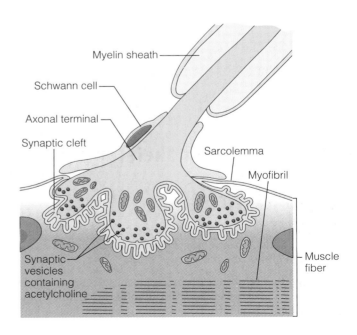

Figure 14.4 The neuromuscular junction.

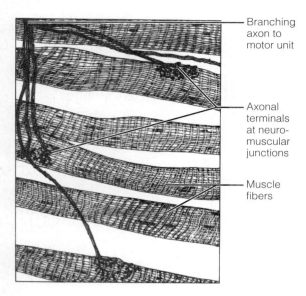

Branching axon to motor unit

Axonal terminals at neuro-muscular junctions

Muscle fibers

Figure 14.5   A portion of a motor unit. (See also Plate 3 in the Histology Atlas.)

2.   Obtain a slide of skeletal muscle stained to show a portion of a motor unit. Examine the slide under high power to identify the axonal fibers extending leashlike to the muscle cells. Follow one of the axonal fibers to its terminus to identify the oval-shaped axonal terminal. Compare your observations to Figure 14.5. Sketch a small section in the space provided, labeling the motor axon, its terminal branches, and muscle fibers. ■

Classification of Skeletal Muscles

Naming Skeletal Muscles

Remembering the names of the skeletal muscles is a monumental task, but certain clues help. Muscles are named on the basis of the following criteria:

• **Direction of muscle fibers:** Some muscles are named in reference to some imaginary line, usually the midline of the body or the longitudinal axis of a limb bone. A muscle with fibers (and fascicles) running parallel to that imaginary line will have the term *rectus* (straight) in its name. For example, the rectus abdominis is the straight muscle of the abdomen. Likewise, the terms *transverse* and *oblique* indicate that the muscle fibers run at right angles and obliquely (respectively) to the imaginary line.

• **Relative size of the muscle:** Terms such as *maximus* (largest), *minimus* (smallest), *longus* (long), and *brevis* (short) are often used in naming muscles—as in gluteus maximus and gluteus minimus.

• **Location of the muscle:** Some muscles are named for the bone with which they are associated. For example, the frontalis muscle overlies the frontal bone.

• **Number of origins:** When the term *biceps, triceps,* or *quadriceps* forms part of a muscle name, you can generally assume that the muscle has two, three, or four origins (respectively). For example, the biceps muscle of the arm has two heads, or origins.

• **Location of the muscle's origin and insertion:** For example, the sternocleidomastoid muscle has its origin on the sternum (*sterno*) and clavicle (*cleido*), and inserts on the mastoid process of the temporal bone.

• **Shape of the muscle:** For example, the deltoid muscle is roughly triangular (*deltoid* = triangle), and the trapezius muscle resembles a trapezoid.

• **Action of the muscle:** For example, all the adductor muscles of the anterior thigh bring about its adduction, and all the extensor muscles of the wrist extend the wrist.

Types of Muscles

Most often, body movements are not a result of the contraction of a single muscle but instead reflect the coordinated action of several muscles acting together. Muscles that are primarily responsible for producing a particular movement are called **prime movers,** or **agonists.**

Muscles that oppose or reverse a movement are called **antagonists.** When a prime mover is active, the fibers of the antagonist are stretched and in the relaxed state. The antagonist can also regulate the prime mover by providing some resistance, to prevent overshoot or to stop its action.

It should be noted that antagonists can be prime movers in their own right. For example, the biceps muscle of the arm (a prime mover of elbow flexion) is antagonized by the triceps (a prime mover of elbow extension).

Synergists aid the action of agonists by reducing undesirable or unnecessary movement. Contraction of a muscle crossing two or more joints would cause movement at all joints spanned if the synergists were not there to stabilize them. For example, you can make a fist without bending your wrist only because synergist muscles stabilize the wrist joint and allow the prime mover to exert its force at the finger joints.

Fixators, or fixation muscles, are specialized synergists. They immobilize the origin of a prime mover so that all the tension is exerted at the insertion. Muscles that help maintain posture are fixators; so too are muscles of the back that stabilize or "fix" the scapula during arm movements.

Gross Anatomy of the Muscular System

Identification of Human Muscles

Muscles of the Head and Neck

The muscles of the head serve many specific functions. For instance, the muscles of facial expression differ from most skeletal muscles because they insert into the skin (or other muscles) rather than into bone. As a result, they move the facial skin, allowing a wide range of emotions to be shown on the face. Other muscles of the head are the muscles of mastication, which manipulate the mandible during chewing, and the six extrinsic eye muscles located within the orbit, which aim the eye. (Orbital muscles are studied in Exercise 24.)

Activity:
Identifying Head and Neck Muscles

Neck muscles are primarily concerned with the movement of the head and shoulder girdle. Figures 15.1 and 15.2 are summary figures illustrating the superficial musculature of the body as a whole. Head and neck muscles are discussed in Tables 15.1 and 15.2 and shown in Figures 15.3 and 15.4.

While reading the tables and identifying the head and neck muscles in the figures, try to visualize what happens when the muscle contracts. Then, use a torso model or an anatomical chart to again identify as many of these muscles as possible. (If a human cadaver is available for observation, specific instructions for muscle examination will be provided by your instructor.) Then carry out the following palpations on yourself:

• To demonstrate the temporalis, clench your teeth. The masseter can also be palpated now at the angle of the jaw. ■

Muscles of the Trunk

The trunk musculature includes muscles that move the vertebral column; anterior thorax muscles that act to move ribs, head, and arms; and muscles of the abdominal wall that play a role in the movement of the vertebral column but more importantly form the "natural girdle," or the major portion of the abdominal body wall.

Activity:
Identifying Muscles of the Trunk

The trunk muscles are described in Tables 15.3 and 15.4 and shown in Figures 15.5 through 15.8. As before, identify the muscles in the figure as you read the tabular descriptions and then identify them on the torso or laboratory chart. ■

Objectives

1. To name and locate the major muscles of the human body (on a torso model, a human cadaver, lab chart, or diagram) and state the action of each.
2. To explain how muscle actions are related to their location.
3. To name muscle origins and insertions as required by the instructor.
4. To identify antagonists of the major prime movers.

Materials

❑ Human torso model or large anatomical chart showing human musculature
❑ Human cadaver for demonstration (if available)
❑ Disposable gloves or protective skin cream
❑ *Human Musculature* videotape*

 See Appendix C, Exercise 15 for links to A.D.A.M. Interactive Anatomy.

 For instructions on animal dissections, see the dissection exercises starting on p. 733 in the cat and fetal pig editions of this manual.

*Available to qualified adopters from Benjamin/Cummings.

Activity:
Demonstrating Operation of Trunk Muscles

Now, work with a partner to demonstrate the operation of the following muscles. One of you can demonstrate the movement (the following steps are addressed to this partner). The other can supply resistance and palpate the muscle being tested.

1. Fully abduct the arm and extend the elbow. Now adduct the arm against resistance. You are using the *latissimus dorsi*.

2. To observe the *deltoid*, attempt to abduct your arm against resistance. Now attempt to elevate your shoulder against resistance; you are contracting the upper portion of the *trapezius*.

3. The *pectoralis major* is used when you press your hands together at chest level with your elbows widely abducted. ■

Text continues on page 141

Figure 15.1 Anterior view of superficial muscles of the body.
The abdominal surface has been partially dissected on the left side of the body to show somewhat deeper muscles.

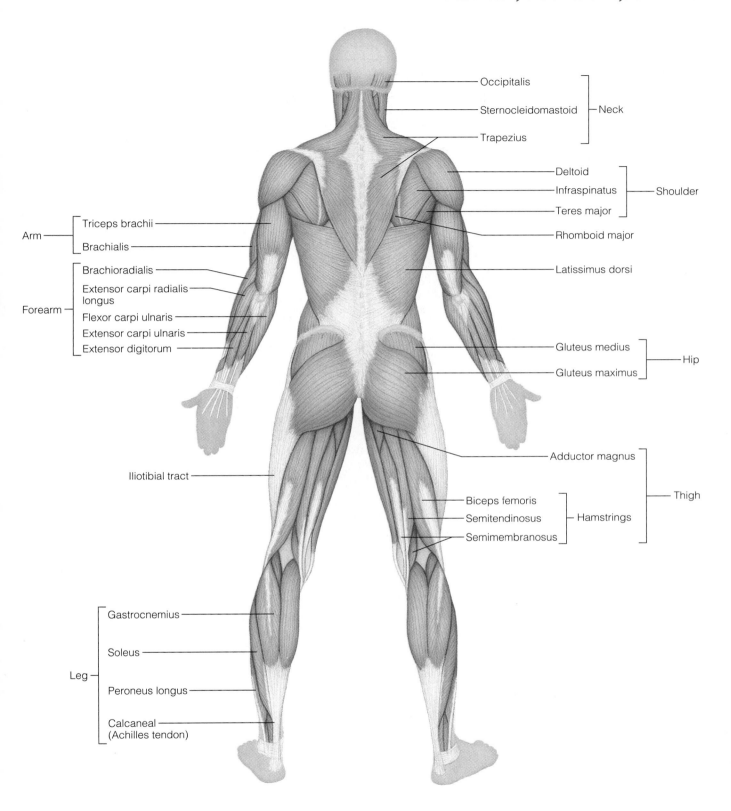

Figure 15.2   Posterior view of superficial muscles of the body.

Table 15.1	Major Muscles of Human Head (see Figure 15.3)			
Muscle	**Comments**	**Origin**	**Insertion**	**Action**
Facial Expression (Figure 15.3a)				
Epicranius—frontalis and occipitalis	Bipartite muscle consisting of frontalis and occipitalis, which covers dome of skull	Frontalis: cranial aponeurosis (galea aponeurotica); occipitalis: occipital bone	Frontalis: skin of eyebrows and root of nose; occipitalis: cranial aponeurosis	With aponeurosis fixed, frontalis raises eyebrows; occipitalis fixes aponeurosis and pulls scalp posteriorly
Orbicularis oculi	Sphincter muscle of eyelids	Frontal and maxillary bones and ligaments around orbit	Encircles orbit and inserts in tissue of eyelid	Various parts can be activated individually; closes eyes, produces blinking, squinting, and draws eyebrows downward
Corrugator supercilii	Small muscle; activity associated with that of orbicularis oculi	Arch of frontal bone above nasal bone	Skin of eyebrow	Draws eyebrows medially; wrinkles skin of forehead vertically
Levator labii superioris	Thin muscle between orbicularis oris and inferior eye margin	Zygomatic bone and infraorbital margin of maxilla	Skin and muscle of upper lip and border of nostril	Raises and furrows upper lip; flares nostril (as in disgust)
Zygomaticus—major and minor	Extends diagonally from corner of mouth to cheekbone	Zygomatic bone	Skin and muscle at corner of mouth	Raises lateral corners of mouth upward (smiling muscle)
Risorius	Slender muscle; runs laterally to zygomaticus	Fascia of masseter muscle	Skin at corner of mouth	Draws corner of lip laterally; tenses lip; zygomaticus synergist
Depressor labii inferioris	Small muscle from lower lip to jawbone	Body of mandible lateral to its midline	Skin and muscle of lower lip	Draws lower lip downward
Depressor anguli oris	Small muscle lateral to depressor labii inferioris	Body of mandible below incisors	Skin and muscle at angle of mouth below insertion of zygomaticus	Zygomaticus antagonist; draws corners of mouth downward and laterally
Orbicularis oris	Multilayered sphincter muscle of lips with fibers that run in many different directions	Arises indirectly from maxilla and mandible; fibers blended with fibers of other muscles associated with lips	Encircles mouth; inserts into muscle and skin at angles of mouth	Closes mouth; purses and protrudes lips (kissing muscle)
Mentalis	One of muscle pair forming V-shaped muscle mass on chin	Mandible below incisors	Skin of chin	Protrudes lower lip; wrinkles chin
Buccinator	Principal muscle of cheek; runs horizontally, deep to the masseter	Molar region of maxilla and mandible	Orbicularis oris	Draws corner of mouth laterally; compresses cheek (as in whistling); holds food between teeth during chewing

(continued on page 132)

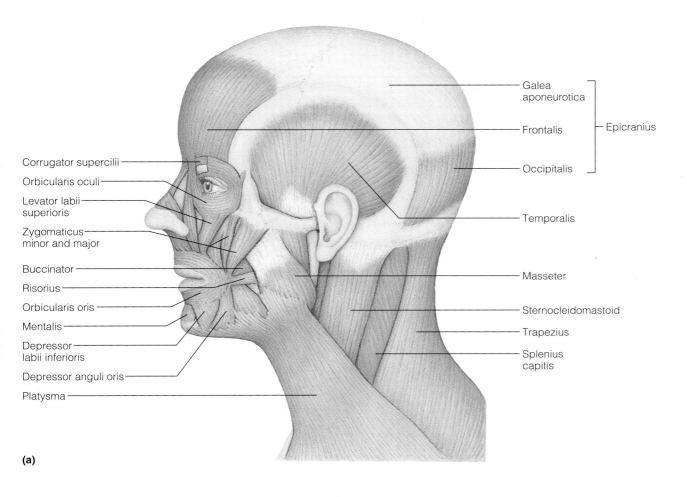

(a)

Galea aponeurotica
Frontalis
⎱ Epicranius
Occipitalis

Corrugator supercilii
Orbicularis oculi
Levator labii superioris
Zygomaticus minor and major
Buccinator
Risorius
Orbicularis oris
Mentalis
Depressor labii inferioris
Depressor anguli oris
Platysma

Temporalis
Masseter
Sternocleidomastoid
Trapezius
Splenius capitis

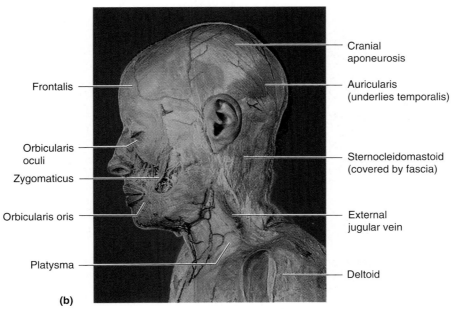

Frontalis

Orbicularis oculi
Zygomaticus
Orbicularis oris
Platysma

(b)

Cranial aponeurosis
Auricularis (underlies temporalis)
Sternocleidomastoid (covered by fascia)
External jugular vein
Deltoid

Figure 15.3   Muscles of the scalp, face, and neck; left lateral view.
(a) Superficial muscles. **(b)** Photo of superficial structures of head and neck.

| Table 15.1 | Major Muscles of Human Head *(continued)* |

Muscle	Comments	Origin	Insertion	Action
Mastication (Figure 15.3c,d)				
Masseter	Extends across jawbone; can be palpated on forcible closure of jaws	Zygomatic process and arch	Angle and ramus of mandible	Closes jaw and elevates mandible
Temporalis	Fan-shaped muscle over temporal bone	Temporal fossa	Coronoid process of mandible	Closes jaw; elevates and retracts mandible
Buccinator	(See muscles of facial expression.)			
Pterygoid—medial	Runs along internal (medial) surface of mandible (thus largely concealed by that bone)	Sphenoid, palatine, and maxillary bones	Medial surface of mandibular ramus and angle	Synergist of temporalis and masseter; closes and elevates mandible; in conjunction with lateral pterygoid, aids in grinding movements
Pterygoid—lateral	Superior to medial pterygoid	Greater wing of sphenoid bone	Mandibular condyle	Protracts jaw (moves it anteriorly); in conjunction with medial pterygoid, aids in grinding movements of teeth

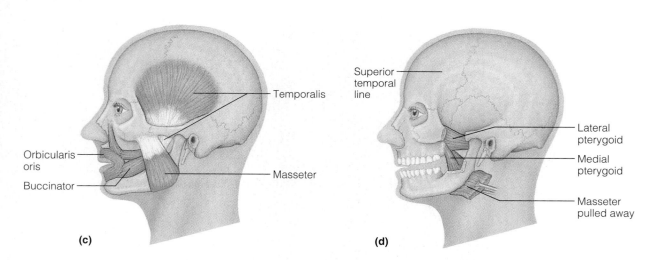

(c) (d)

Figure 15.3 (continued) Muscles promoting mastication.
(c) Lateral view of the temporalis, masseter, and buccinator muscles. **(d)** Lateral view of the deep chewing muscles, the medial and lateral pterygoid muscles.

Table 15.2	Anterolateral Muscles of Human Neck (see Figure 15.4)

Muscle	Comments	Origin	Insertion	Action
Superficial				
Platysma	Unpaired muscle: thin, sheetlike superficial neck muscle, not strictly a head muscle but plays role in facial expression (see also Fig. 15.3a)	Fascia of chest (over pectoral muscles) and deltoid	Lower margin of mandible, skin, and muscle at corner of mouth	Depresses mandible; pulls lower lip back and down; i.e., produces downward sag of the mouth
Sternocleidomastoid	Two-headed muscle located deep to platysma on anterolateral surface of neck; fleshy parts on either side indicate limits of anterior and posterior triangles of neck	Manubrium of sternum and medial portion of clavicle	Mastoid process of temporal bone	Simultaneous contraction of both muscles of pair causes flexion of neck forward, generally against resistance (as when lying on the back); acting independently, rotate head toward shoulder on opposite side
Scalenes—anterior, middle, and posterior	Located more on lateral than anterior neck; deep to platysma (see Fig. 15.4c)	Transverse processes of cervical vertebrae	Anterolaterally on first two ribs	Flex and slightly rotate neck; elevate first two ribs (aid in inspiration)

(continued)

(a)

Platysma

External jugular vein

Sternocleidomastoid

Pectoralis major

Mylohyoid

Digastric

Submandibular gland

Sternocleidomastoid (reflected)

Omohyoid

Sternohyoid

Left clavicle

Sternal manubrium

Figure 15.4 Muscles of the anterolateral neck and throat.
(a) Photo of the anterior and lateral regions of the neck. The fascia has been partially removed (left side of the photo) to expose the sternocleidomastoid muscle. On the right side of the photo, the sternocleidomastoid muscle is reflected to expose the sternohyoid and omohyoid muscles.

Table 15.2	Anterolateral Muscles of Human Neck *(continued)*			
Muscle	**Comments**	**Origin**	**Insertion**	**Action**
Deep (Figure 15.4a,b)				
Digastric	Consists of two bellies united by an intermediate tendon; assumes a V-shaped configuration under chin	Lower margin of mandible (anterior belly) and mastoid process (posterior belly)	By a connective tissue loop to hyoid bone	Acting in concert, elevate hyoid bone; open mouth and depress mandible
Stylohyoid	Slender muscle parallels posterior border of digastric	Styloid process of temporal	Hyoid bone	Elevates and retracts hyoid bone
Mylohyoid	Just deep to digastric; forms floor of mouth	Medial surface of mandible	Hyoid bone	Elevates hyoid bone and base of tongue during swallowing
Sternohyoid	Runs most medially along neck; straplike	Posterior surface of manubrium	Lower margin of body of hyoid bone	Acting with sternothyroid and omohyoid (all inferior to hyoid bone), depresses larynx and hyoid bone if mandible is fixed; may also flex skull
Sternothyroid	Lateral to sternohyoid; straplike	Manubrium and medial end of clavicle	Thyroid cartilage of larynx	(See Sternohyoid above)
Omohyoid	Straplike with two bellies; lateral to sternohyoid	Superior surface of scapula	Hyoid bone	(See Sternohyoid above)
Thyrohyoid	Appears as a slender continuation of sternothyroid muscle	Thyroid cartilage	Hyoid bone	Depresses hyoid bone; elevates thyroid cartilage

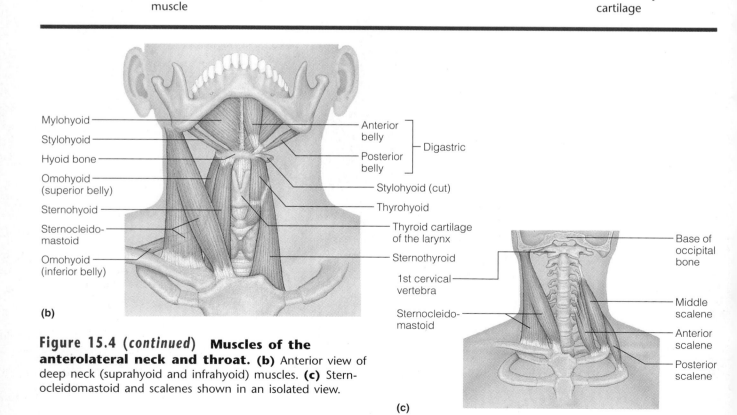

Mylohyoid
Stylohyoid
Hyoid bone
Omohyoid (superior belly)
Sternohyoid
Sternocleidomastoid
Omohyoid (inferior belly)

Anterior belly
Posterior belly
— Digastric
Stylohyoid (cut)
Thyrohyoid
Thyroid cartilage of the larynx
Sternothyroid
1st cervical vertebra
Sternocleidomastoid

Base of occipital bone
Middle scalene
Anterior scalene
Posterior scalene

(b)

(c)

Figure 15.4 (*continued*)   **Muscles of the anterolateral neck and throat. (b)** Anterior view of deep neck (suprahyoid and infrahyoid) muscles. **(c)** Sternocleidomastoid and scalenes shown in an isolated view.

Table 15.3	Anterior Muscles of Human Thorax, Shoulder, and Abdominal Wall (see Figures 15.5, 15.6, and 15.7)			
Muscle	**Comments**	**Origin**	**Insertion**	**Action**
Thorax and Shoulder, Superficial (Figure 15.5)				
Pectoralis major	Large fan-shaped muscle covering upper portion of chest	Clavicle, sternum, cartilage of first six ribs, and aponeurosis of external oblique muscle	Fibers converge to insert by short tendon into greater tubercle of humerus	Prime mover of arm flexion; adducts, medially rotates arm; with arm fixed, pulls chest upward (thus also acts in forced inspiration)
Serratus anterior	Deep and superficial portions; beneath and inferior to pectoral muscles on lateral rib cage	Lateral aspect of first to eighth (or ninth) ribs	Vertebral border of anterior surface of scapula	Moves scapula forward toward chest wall; rotates scapula causing inferior angle to move laterally and upward
Deltoid	Fleshy triangular muscle forming shoulder muscle mass	Lateral third of clavicle; acromion and spine of scapula	Deltoid tuberosity of humerus	Acting as a whole, prime mover of arm abduction; when only specific fibers are active, can aid in flexion, extension, and rotation of humerus
Pectoralis minor	Flat, thin muscle directly beneath and obscured by pectoralis major	Anterior surface of third, fourth, and fifth ribs, near their costal cartilages	Coracoid process of scapula	With ribs fixed, draws scapula forward and inferiorly; with scapula fixed, draws rib cage superiorly
Thorax, Deep: Muscles of Respiration (Figure 15.6)				
Intercostals— external	11 pairs lie between ribs; fibers run obliquely downward and forward toward sternum	Inferior border of rib above (not shown in figure)	Superior border of rib below	Pulls ribs toward one another to elevate rib cage; aids in inspiration

(continued)

Figure 15.5 Superficial muscles of the thorax and shoulder acting on the scapula and arm, anterior view. The superficial muscles, which effect arm movements, are shown on the left. These muscles have been removed on the right side of the figure to show the muscles that stabilize or move the pectoral girdle.

Table 15.3		Anterior Muscles of Human Thorax, Shoulder, and Abdominal Wall *(continued)*		
Muscle	**Comments**	**Origin**	**Insertion**	**Action**
Intercostals—internal	11 pairs lie between ribs; fibers run deep and at right angles to those of external intercostals	Superior border of rib below	Inferior border of rib above (not shown in figure)	Draws ribs together to depress rib cage; aids in forced expiration; antagonistic to external intercostals
Diaphragm	Broad muscle; forms floor of thoracic cavity; dome shaped in relaxed state; fibers converge from margins of thoracic cage toward a central tendon	Inferior border of rib and sternum, costal cartilages of last six ribs and lumbar vertebrae	Central tendon	Prime mover of inspiration flattens on contraction increasing vertical dimensions of thorax; increases intra-abdominal pressure

(continued)

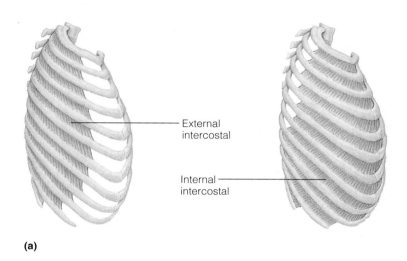

(a)

Figure 15.6 Deep muscles of the thorax: muscles of respiration.
(a) The external intercostals (inspiratory muscles) are shown on the left and the internal intercostals (expiratory muscles) are shown on the right. These two muscle layers run obliquely and at right angles to each other. **(b)** Inferior view of the diaphragm, the prime mover of inspiration. Notice that its muscle fibers converge toward a central tendon, an arrangement that causes the diaphragm to flatten and move inferiorly as it contracts. The diaphragm and its tendon are pierced by the great vessels (aorta and inferior vena cava) and the esophagus.

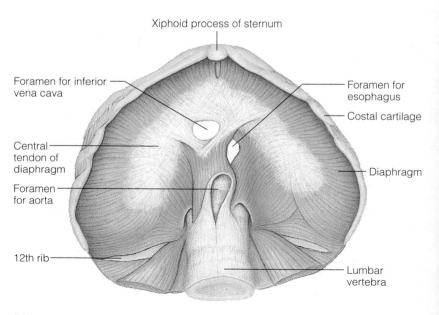

Xiphoid process of sternum

Foramen for inferior vena cava

Foramen for esophagus

Costal cartilage

Central tendon of diaphragm

Diaphragm

Foramen for aorta

12th rib

Lumbar vertebra

(b)

Table 15.3	*(continued)*			
Muscle	**Comments**	**Origin**	**Insertion**	**Action**
Abdominal Wall (Figure 15.7a and b)				
Rectus abdominis	Medial superficial muscle, extends from pubis to rib cage; ensheathed by aponeuroses of oblique muscles; segmented	Pubic crest and symphysis	Xiphoid process and costal cartilages of fifth through seventh ribs	Flexes vertebral column; increases abdominal pressure; fixes and depresses ribs; stabilizes pelvis during walking
External oblique	Most superficial lateral muscle; fibers run downward and medially; ensheathed by an aponeurosis	Anterior surface of last eight ribs	Linea alba,* pubic tubercles, and iliac crest	See Rectus abdominis, above; also aids muscles of back in trunk rotation and lateral flexion
Internal oblique	Fibers run at right angles to those of external oblique, which it underlies	Lumbodorsal fascia, iliac crest, and inguinal ligament	Linea alba, pubic crest, and costal cartilages of last three ribs	As for External oblique
Transversus abdominis	Deepest muscle of abdominal wall; fibers run horizontally	Inguinal ligament, iliac crest, and cartilages of last five or six ribs	Linea alba and pubic crest	Compresses abdominal contents

*The linea alba ("white line") is a narrow, tendinous sheath that runs along the middle of the abdomen from the sternum to the pubic symphysis. It is formed by the fusion of the aponeurosis of the external oblique and transversus muscles.

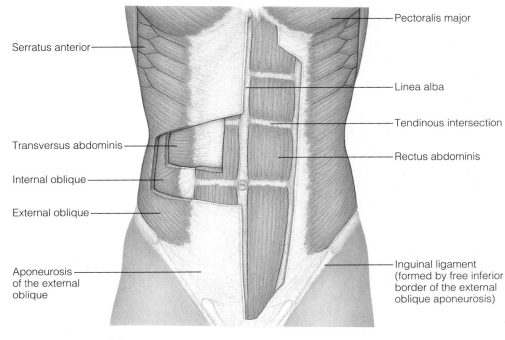

Serratus anterior

Transversus abdominis

Internal oblique

External oblique

Aponeurosis of the external oblique

Pectoralis major

Linea alba

Tendinous intersection

Rectus abdominis

Inguinal ligament (formed by free inferior border of the external oblique aponeurosis)

(a)

Figure 15.7 Anterior view of the muscles forming the anterolateral abdominal wall.
(a) The superficial muscles have been partially cut away on the left side of the diagram to reveal the deeper internal oblique and transversus abdominis muscles.

Table 15.4	Posterior Muscles of Human Trunk (see Figure 15.8)			
Muscle	**Comments**	**Origin**	**Insertion**	**Action**
Muscles of the Neck, Shoulder, and Thorax (Figure 15.8a)				
Trapezius	Most superficial muscle of posterior neck and thorax; very broad origin and insertion	Occipital bone; ligamentum nuchae; spines of C_7 and all thoracic vertebrae	Acromion and spinous process of scapula; lateral third of clavicle	Extends head; retracts (adducts) scapula and stabilizes it; upper fibers elevate scapula; lower fibers depress it
Latissimus dorsi	Broad flat muscle of lower back (lumbar region); extensive superficial origins	Indirect attachment to spinous processes of lower six thoracic vertebrae, lumbar vertebrae, lower 3 to 4 ribs, and iliac crest	Floor of intertubercular groove of humerus	Prime mover of arm extension; adducts and medially rotates arm; depresses scapula; brings arm down in power stroke, as in striking a blow
Infraspinatus	Partially covered by deltoid and trapezius; a rotator cuff muscle	Infraspinous fossa of scapula	Greater tubercle of humerus	Lateral rotation of humerus; helps hold head of humerus in glenoid cavity
Teres minor	Small muscle inferior to infraspinatus; a rotator cuff muscle	Lateral margin of scapula	Greater tuberosity of humerus	As for infraspinatus
Teres major	Located inferiorly to teres minor	Posterior surface at inferior angle of scapula	Crest of lesser tubercle of humerus	Extends, medially rotates, and adducts humerus; synergist of latissimus dorsi

(continued)

(b)

Figure 15.7 (continued) Anterior view of the muscles forming the anterolateral abdominal wall. (b) Photo of the anterolateral abdominal wall.

Table 15.4	(continued)			
Muscle	**Comments**	**Origin**	**Insertion**	**Action**
Supraspinatus	Obscured by trapezius and deltoid; a rotator cuff muscle	Supraspinous fossa of scapula	Greater tubercle of humerus	Assists abduction of humerus; stabilizes shoulder joint
Levator scapulae	Located at back and side of neck, deep to trapezius	Transverse processes of C_1 through C_4	Superior vertebral border of scapula	Raises and adducts scapula; with fixed scapula, flexes neck to the same side
Rhomboids—major and minor	Beneath trapezius and inferior to levator scapulae; run from vertebral column to scapula	Spinous processes of C_7 and T_1 through T_5	Vertebral border of scapula	Pull scapula medially (retraction) and elevate it
Muscles Associated with the Vertebral Column (Figure 15.8b)				
Semispinalis	Deep composite muscle of the back—thoracis, cervicis, and capitis portions	Transverse processes of C_7–T_{12}	Occipital bone and spinous processes of cervical vertebrae and T_1–T_4	Acting together, extend head and vertebral column; acting independently (right vs. left) causes rotation toward the opposite side

(continued)

(a)

Figure 15.8   Muscles of the neck, shoulder, and thorax, posterior view. (a) The superficial muscles of the back are shown for the left side of the body, with a corresponding photograph. The superficial muscles are removed on the right side of the illustration to reveal the deeper muscles acting on the scapula and the rotator cuff muscles that help to stabilize the shoulder joint.

Table 15.4	Posterior Muscles of Human Trunk *(continued)*			
Muscle	**Comments**	**Origin**	**Insertion**	**Action**
Erector spinae	A long tripartite muscle composed of iliocostalis (lateral), longissimus, and spinalis (medial) muscle columns; superficial to semispinalis muscles; extends from pelvis to head	Sacrum, iliac crest, transverse processes of lumbar, thoracic, and cervical vertebrae, and/or ribs 3–12 depending on specific part	Ribs and transverse processes of vertebrae about six segments above origin. Longissimus also inserts into mastoid process	All act to extend and abduct the vertebral column; fibers of the longissimus also extend head

(continued)

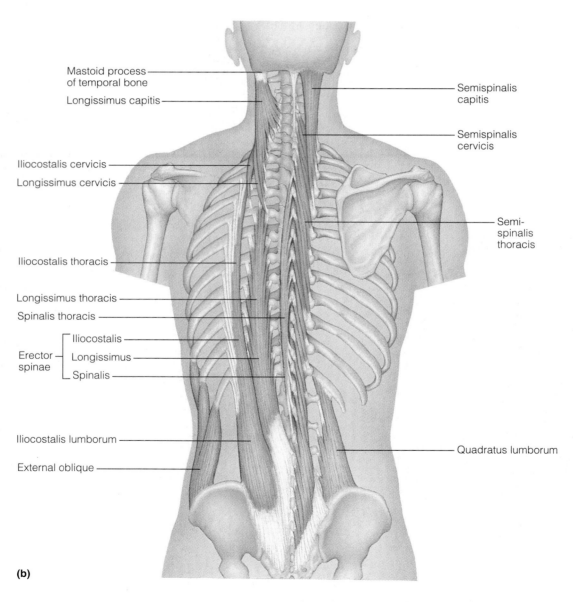

(b)

Figure 15.8 *(continued)* **Muscles of the neck, shoulder, and thorax, posterior view. (b)** The erector spinae and semispinalis muscles which respectively form the intermediate and deep muscle layers of the back associated with the vertebral column.

Table 15.4	(continued)			
Muscle	**Comments**	**Origin**	**Insertion**	**Action**
Splenius (see Figure 15.8c)	Superficial muscle (capitis and cervicis parts) just deep to levator scapulae and superficial to erector spinae	Ligamentum nuchae and spinous processes of C_7–T_6	Mastoid process, occipital bone, and transverse processes of C_2–C_4	As a group, extend or hyperextend head; when only one side is active, head is rotated and bent toward the same side
Quadratus lumborum	Forms greater portion of posterior abdominal wall	Iliac crest and iliolumbar fascia	Inferior border twelfth rib; transverse processes of lumbar vertebrae	Each flexes vertebral column laterally; together extend the lumbar spine and fix the twelfth rib

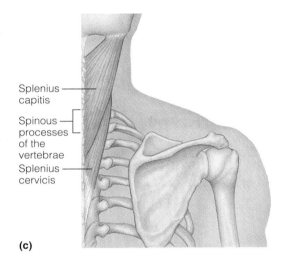

Splenius capitis

Spinous processes of the vertebrae

Splenius cervicis

(c)

Figure 15.8 (continued) **Muscles of the neck, shoulder, and thorax, posterior view. (c)** Deep (splenius) muscles of the posterior neck. Superficial muscles have been removed.

Muscles of the Upper Limb

The muscles that act on the upper limb fall into three groups: those that move the arm, those causing movement at the elbow, and those effecting movements of the wrist and hand.

The muscles that cross the shoulder joint to insert on the humerus and move the arm (subscapularis, supraspinatus and infraspinatus, deltoid, and so on) are primarily trunk muscles that originate on the axial skeleton or shoulder girdle. These muscles are included with the trunk muscles.

The second group of muscles, which cross the elbow joint and move the forearm, consists of muscles forming the musculature of the humerus. These muscles arise primarily from the humerus and insert in forearm bones. They are responsible for flexion, extension, pronation, and supination. The origins, insertions, and actions of these muscles are summarized in Table 15.5 and the muscles are shown in Figure 15.9.

The third group composes the musculature of the forearm. For the most part, these muscles insert on the digits and produce movements at the wrist and fingers. In general, muscles acting on the wrist and hand are more easily identified if their insertion tendons are located first. These muscles are described in Table 15.6 and illustrated in Figure 15.10.

Activity:
Identifying Muscles of the Upper Limb

First study the tables and figures, then see if you can identify these muscles on a torso model, anatomical chart, or cadaver. Complete this portion of the exercise with palpation demonstrations as outlined next.

• To observe the *biceps brachii*, attempt to flex your forearm (hand supinated) against resistance. The insertion tendon of this biceps muscle can also be felt in the lateral aspect of the antecubital fossa (where it runs toward the radius to attach).

• If you acutely flex your elbow and then try to extend it against resistance, you can demonstrate the action of your *triceps brachii*.

• Strongly flex your wrist and make a fist. Palpate your contracting wrist flexor muscles (which originate from the medial epicondyle of the humerus) and their insertion tendons, which can be easily felt at the anterior aspect of the wrist.

• Flare your fingers to identify the tendons of the *extensor digitorum* muscle on the dorsum of your hand. ■

Text continues on page 146

Table 15.5	Muscles of Human Humerus That Act on the Forearm (see Figure 15.9)			
Muscle	**Comments**	**Origin**	**Insertion**	**Action**
Triceps brachii	Sole, large fleshy muscle of posterior humerus; three-headed origin	Long head: inferior margin of glenoid cavity; lateral head: posterior humerus; medial head: distal radial groove on posterior humerus	Olecranon process of ulna	Powerful forearm extensor; antagonist of forearm flexors (brachialis and biceps brachii)
Biceps brachii	Most familiar muscle of anterior humerus because this two-headed muscle bulges when forearm is flexed	Short head: coracoid process; tendon of long head runs in intertubercular groove and within capsule of shoulder joint	Radial tuberosity	Flexion (powerful) of elbow and supination of forearm; "it turns the corkscrew and pulls the cork"; weak arm flexor
Brachioradialis	Superficial muscle of lateral forearm; forms lateral boundary of antecubital fossa	Lateral ridge at distal end of humerus	Base of styloid process of radius	Forearm flexor (weak)
Brachialis	Immediately deep to biceps brachii	Distal portion of anterior humerus	Coronoid process of ulna	A major flexor of forearm

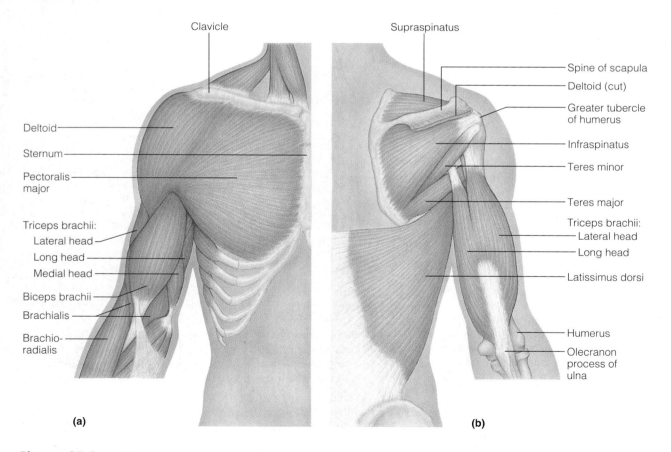

Figure 15.9 Muscles causing movements of the forearm.
(a) Superficial muscles of the anterior thorax, shoulder, and arm, anterior view.
(b) Posterior aspect of the arm showing the lateral and long heads of the triceps brachii muscle.

Table 15.6	Muscles of Human Forearm That Act on Hand and Fingers (see Figure 15.10)			
Muscle	**Comments**	**Origin**	**Insertion**	**Action**
Anterior Compartment (Figure 15.10a, b, c)				
Superficial				
Pronator teres	Seen in a superficial view between proximal margins of brachioradialis and flexor carpi ulnaris	Medial epicondyle of humerus and coronoid process of ulna	Midshaft of radius	Acts synergistically with pronator quadratus to pronate forearm; weak forearm flexor
Flexor carpi radialis	Superficial; runs diagonally across forearm	Medial epicondyle of humerus	Base of second and third metacarpals	Powerful flexor of wrist; abducts hand
Palmaris longus	Small fleshy muscle with a long tendon; medial to flexor carpi radialis	Medial epicondyle of humerus	Palmar aponeurosis	Flexes wrist (weak)
Flexor carpi ulnaris	Superficial; medial to palmaris longus	Medial epicondyle of humerus and olecranon process of ulna	Base of fifth metacarpal	Powerful flexor of wrist; adducts hand

(continued)

Figure 15.10   Muscles of the forearm and wrist. (a) Superficial anterior view of right forearm and hand. **(b)** The brachioradialis, flexors carpi radialis and ulnaris, and palmaris longus muscles have been removed to reveal the position of the somewhat deeper flexor digitorum superficialis. **(c)** Deep muscles of the anterior compartment. Superficial muscles have been removed. Note: The thenar muscles of the thumb and the lumbricals that help move the fingers are illustrated here but are not described in Table 15.6.

Table 15.6	Muscles of Human Forearm That Act on Hand and Fingers *(continued)*			
Muscle	**Comments**	**Origin**	**Insertion**	**Action**
Flexor digitorum superficialis	Deeper muscle; overlain by muscles named above; visible at distal end of forearm	Medial epicondyle of humerus, medial surface of ulna, and anterior border of radius	Middle phalanges of second through fifth fingers	Flexes wrist and middle phalanges of second through fifth fingers
Deep				
Flexor pollicis longus	Deep muscle of anterior forearm; distal to and paralleling lower margin of flexor digitorum superficialis	Anterior surface of radius, and interosseous membrane	Distal phalanx of thumb	Flexes thumb (*pollix* is Latin for "thumb"); weak flexor of wrist
Flexor digitorum profundus	Deep muscle; overlain entirely by flexor digitorum superficialis	Anteromedial surface of ulna and interosseous membrane	Distal phalanges of second through fifth fingers	Sole muscle that flexes distal phalanges; assists in wrist flexion
Pronator quadratus	Deepest muscle of distal forearm	Distal portion of anterior ulnar surface	Anterior surface of radius, distal end	Pronates forearm
Posterior Compartment (Figure 15.10d, e, f)				
Superficial				
Extensor carpi radialis longus	Superficial; parallels brachioradialis on lateral forearm	Lateral supracondylar ridge of humerus	Base of second metacarpal	Extends and abducts wrist
Extensor carpi radialis brevis	Posterior to extensor carpi radialis longus	Lateral epicondyle of humerus	Base of third metacarpal	Extends and abducts wrist; steadies wrist during finger flexion
Extensor digitorum	Superficial; between extensor carpi ulnaris and extensor carpi radialis brevis	Lateral epicondyle of humerus	By four tendons into distal phalanges of second through fifth fingers	Prime mover of finger extension; extends wrist; can flare (abduct) fingers
Extensor carpi ulnaris	Superficial; medial posterior forearm	Lateral epicondyle of humerus	Base of fifth metacarpal	Extends and adducts wrist
Deep				
Extensor pollicis longus and brevis	Deep muscle pair with a common origin and action; overlain by extensor carpi ulnaris	Dorsal shaft of ulna and radius, interosseous membrane	Base of distal phalanx of thumb (longus) and proximal phalanx of thumb (brevis)	Extends thumb
Abductor pollicis longus	Deep muscle; lateral and parallel to extensor pollicis longus	Posterior surface of radius and ulna; interosseous membrane	First metacarpal	Abducts and extends thumb
Supinator	Deep muscle at posterior aspect of elbow	Lateral epicondyle of humerus	Proximal end of radius	Acts with biceps brachii to supinate forearm; antagonist of pronator muscles

(d)

Brachioradialis

Insertion of triceps brachii

Extensor carpi radialis longus

Anconeus

Extensor carpi radialis brevis

Flexor carpi ulnaris

Extensor digitorum

Extensor carpi ulnaris

Abductor pollicis longus

Extensor digit minimi

Extensor pollicis brevis

Extensor indicis

Extensor pollicis longus

Tendons of extensor carpi radialis brevis and longus

Tendons of extensor digitorum

Extensor expansion

(e)

Olecranon process of ulna

Anconeus

Supinator

Abductor pollicis longus

Extensor pollicis longus

Extensor pollicis brevis

Extensor indicis

Interossei

(f)

Extensor carpi radialis longus

Extensor carpi radialis brevis

Abductor pollicis longus

Extensor pollicis brevis

Extensor pollicis longus

Brachioradialis

Anconeus

Olecranon process

Ulna

Extensor indicis

Extensor retinaculum

Figure 15.10 (*continued*)   Muscles of the forearm and wrist.

(d) Superficial muscles, posterior view. **(e)** Deep posterior muscles; superficial muscles have been removed. The interossei, the deepest layer of instrinsic hand muscles, are also illustrated. **(f)** Photo of deep posterior muscles of the right forearm. The superficial muscles have been removed.

Muscles of the Lower Limb

Muscles that act on the lower limb cause movement at the hip, knee, and foot joints. Since the human pelvic girdle is composed of heavy fused bones that allow very little movement, no special group of muscles is necessary to stabilize it. This is unlike the shoulder girdle, where several muscles (mainly trunk muscles) are needed to stabilize the scapulae.

Muscles acting on the thigh (femur) cause various movements at the multiaxial hip joint (flexion, extension, rotation, abduction, and adduction). These include the iliopsoas, the adductor group, and other muscles summarized in Tables 15.7 and 15.8 and illustrated in Figures 15.11 and 15.12.

Muscles acting on the leg form the major musculature of the thigh. (Anatomically the term *leg* refers only to that portion between the knee and the ankle.) The thigh muscles cross

Text continues on page 148

Figure 15.11 Anterior and medial muscles promoting movements of the thigh and leg. (a) Anterior view of the deep muscles of the pelvis and superficial muscles of the right thigh. **(b)** Adductor muscles of the medial compartment of the thigh. **(c)** The vastus muscles (isolated) of the quadriceps group.

Table 15.7	Muscles Acting on Human Thigh and Leg, Anterior and Medial Aspects (see Figure 15.11)			
Muscle	**Comments**	**Origin**	**Insertion**	**Action**
Origin on the Pelvis				
Iliopsoas—iliacus and psoas major	Two closely related muscles; fibers pass under inguinal ligament to insert into femur via a common tendon	Iliacus: iliac fossa; psoas major: transverse processes, bodies, and discs of T_{12} and lumbar vertebrae	Lesser trochanter of femur	Flex trunk on thigh; major flexor of hip (or thigh on pelvis when pelvis is fixed)
Sartorius	Straplike superficial muscle running obliquely across anterior surface of thigh to knee	Anterior superior iliac spine	By an aponeurosis into medial aspect of proximal tibia	Flexes and laterally rotates thigh; flexes knee; known as "tailor's muscle" because it helps bring about cross-legged position in which tailors are often depicted
Medial Compartment				
Adductors—magnus, longus, and brevis	Large muscle mass forming medial aspect of thigh; arise from front of pelvis and insert at various levels on femur	Magnus: ischial and pubic rami; longus: pubis near pubic symphysis; brevis: body and inferior ramus of pubis	Magnus: linea aspera and adductor tubercle of femur; longus and brevis: linea aspera	Adduct and laterally rotate and flex thigh; posterior part of magnus is also a synergist in thigh extension
Pectineus	Overlies adductor brevis on proximal thigh	Pectineal line of pubis	Inferior to lesser trochanter of femur	Adducts, flexes, and laterally rotates thigh
Gracilis	Straplike superficial muscle of medial thigh	Inferior ramus and body of pubis	Medial surface of head of tibia	Adducts thigh; flexes and medially rotates leg, especially during walking
Anterior Compartment				
Quadriceps*				
Rectus femoris	Superficial muscle of thigh; runs straight down thigh; only muscle of group to cross hip joint; arises from two heads	Anterior inferior iliac spine and superior margin of acetabulum	Tibial tuberosity	Extends knee and flexes thigh at hip
Vastus lateralis	Forms lateral aspect of thigh	Greater trochanter and linea aspera	Tibial tuberosity	Extends knee
Vastus medialis	Forms medial aspect of thigh	Linea aspera	Tibial tuberosity	Extends knee
Vastus intermedius	Obscured by rectus femoris; lies between vastus lateralis and vastus medialis on anterior thigh	Anterior and lateral surface of femur (not shown in figure)	Tibial tuberosity	Extends knee
Tensor fasciae latae	Enclosed between fascia layers of thigh	Anterior aspect of iliac crest and anterior superior iliac spine	Iliotibial band of fascia lata	Flexes, abducts, and medially rotates thigh

*The quadriceps form the flesh of the anterior thigh and have a common insertion in the tibial tuberosity via the patellar tendon. They are powerful leg extensors, enabling humans to kick a football, for example.

Table 15.8		Muscles Acting on Human Thigh and Leg, Posterior Aspect (see Figure 15.12)		

Muscle	Comments	Origin	Insertion	Action
Origin on the Pelvis				
Gluteus maximus	Largest and most superficial of gluteal muscles (which form buttock mass)	Dorsal ilium, sacrum, and coccyx	Gluteal tuberosity of femur and iliotibial tract*	Complex, powerful hip extensor (most effective when hip is flexed, as in climbing stairs—but not as in walking); antagonist of iliopsoas; laterally rotates thigh
Gluteus medius	Partially covered by gluteus maximus	Upper lateral surface of ilium	Greater trochanter of femur	Abducts and medially rotates thigh; steadies pelvis during walking
Gluteus minimus (not shown in figure)	Smallest and deepest gluteal muscle	Inferior surface of ilium	Greater trochanter of femur	Abducts and medially rotates thigh
Posterior Compartment				
Hamstrings†				
Biceps femoris	Most lateral muscle of group; arises from two heads	Ischial tuberosity (long head); linea aspera and distal femur (short head)	Tendon passes laterally to insert into head of fibula and lateral condyle of tibia	Extends thigh; laterally rotates leg on thigh; flexes knee
Semitendinosus	Medial to biceps femoris	Ischial tuberosity	Medial aspect of upper tibial shaft	Extends thigh; flexes knee; medially rotates leg
Semimembranosus	Deep to semitendinosus	Ischial tuberosity	Medial condyle of tibia	Extends thigh; flexes knee; medially rotates leg

*The iliotibial tract, a thickened lateral portion of the fascia lata, ensheathes all the muscles of the thigh. It extends as a tendinous band from the iliac crest to the knee.
†The hamstrings are the fleshy muscles of the posterior thigh. The name comes from the butchers' practice of using the tendons of these muscles to hang hams for smoking. As a group, they are strong extensors of the hip; they counteract the powerful quadriceps by stabilizing the knee joint when standing.

the knee to allow its flexion and extension. They include the hamstrings and the quadriceps and, along with the muscles acting on the thigh, are described in Tables 15.7 and 15.8 and illustrated in Figures 15.11 and 15.12. Since some of these muscles also have attachments on the pelvic girdle, they can cause movement at the hip joint.

The muscles originating on the leg and acting on the foot and toes are described in Table 15.9 and shown in Figures 15.13 and 15.14.

Activity:
Identifying Muscles of the Lower Limb

Identify the muscles acting on the thigh, leg, foot, and toes as instructed previously for other muscle groups. ■

Activity:
Palpating Muscles of the Hip and Lower Limb

Complete this exercise by performing the following palpation demonstrations with your lab partner.

• Go into a deep knee bend and palpate your own *gluteus maximus* muscle as you extend your hip to resume the upright posture.

• Demonstrate the contraction of the anterior *quadriceps femoris* by trying to extend your knee against resistance. Do this while seated and note how the patellar tendon reacts. The *biceps femoris* of the posterior thigh comes into play when you flex your knee against resistance.

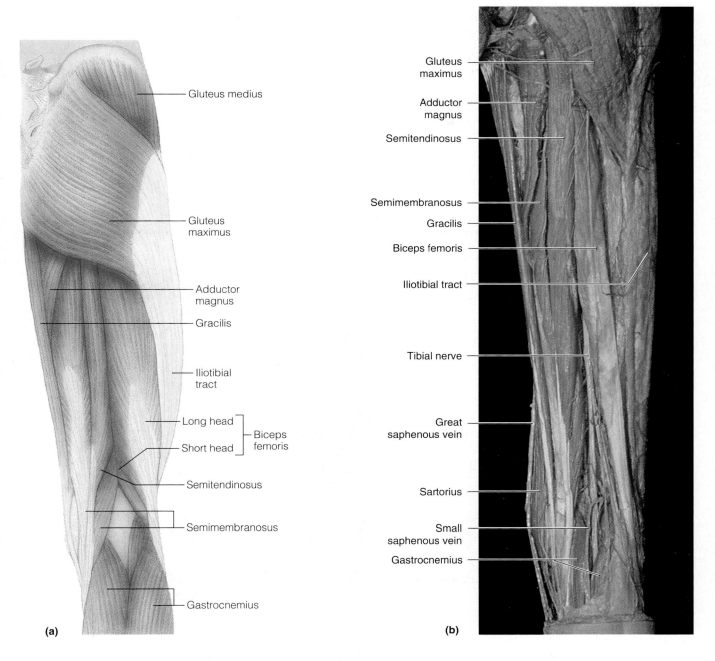

(a)

(b)

Figure 15.12 Muscles of the posterior aspect of the right hip and thigh. (a) Superficial view showing the gluteus muscles of the buttock and hamstring muscles of the thigh. (b) Photo of muscles of the posterior thigh.

• Now stand on your toes. Have your partner palpate the lateral and medial heads of the *gastrocnemius* and follow it to its insertion in the calcaneal tendon.

• Dorsiflex and invert your foot while palpating your *tibialis anterior* muscle (which parallels the sharp anterior crest of the tibia laterally). ■

 For instructions on animal dissections, see the dissection exercises starting on p. 733 in the cat and fetal pig editions of this manual.

Table 15.9	Muscles Acting on Human Foot and Ankle (see Figures 15.13 and 15.14)			
Muscle	**Comments**	**Origin**	**Insertion**	**Action**
Posterior Compartment				
Superficial (Figure 15.13a, b)				
Triceps surae	Muscle pair that shapes posterior calf		Via common tendon (calcaneal or Achilles) into heel	Plantar flex foot
Gastrocnemius	Superficial muscle of pair; two prominent bellies	By two heads from medial and lateral condyles of femur	Calcaneus via calcaneal tendon	Crosses knee joint; thus also can flex knee (when foot is dorsiflexed)

(continued)

Figure 15.13 Muscles of the posterior aspect of the right leg. (a) Superficial view of the posterior leg. **(b)** Photo of posterior aspect of right leg. The gastrocnemius has been transected and its superior part removed.

Table 15.9	(continued)			
Muscle	**Comments**	**Origin**	**Insertion**	**Action**
Soleus	Deep to gastrocnemius	Proximal portion of tibia and fibula	Calcaneus via calcaneal tendon	Plantar flexion; is an important muscle for locomotion
Deep (Figure 15.13c, d)				
Popliteus	Thin muscle at posterior aspect of knee	Lateral condyle of femur	Proximal tibia	Flexes and rotates leg medially to "unlock" extended knee when knee flexion begins

(continued)

O = origin
I = insertion

Plantaris (cut)

Gastrocnemius lateral head (cut)

Gastrocnemius medial head (cut)

Popliteus

Soleus (cut)

Tibialis posterior

Fibula

Tibialis posterior

Peroneus longus

Flexor digitorum longus

Flexor hallucis longus

Peroneus brevis

Tendon of tibialis posterior

Medial malleolus

Calcaneal tendon (cut)

Calcaneus

(c)

(d)

Figure 15.13 (continued) Muscles of the posterior aspect of the right leg. (c) The triceps surae has been removed to show the deep muscles of the posterior compartment. **(d)** Tibialis posterior shown in isolation so that its origin and insertion may be visualized.

Table 15.9	Muscles Acting on Human Foot and Ankle *(continued)*			
Muscle	**Comments**	**Origin**	**Insertion**	**Action**
Tibialis posterior	Thick muscle deep to soleus	Superior portion of tibia and fibula and interosseous membrane	Tendon passes obliquely behind medial malleolus and under arch of foot; inserts into several tarsals and metatarsals 2–4	Prime mover of foot inversion; plantar flexes foot
Flexor digitorum longus	Runs medial to and partially overlies tibialis posterior	Posterior surface of tibia	Distal phalanges of second through fifth toes	Flexes toes; plantar flexes and inverts foot
Flexor hallucis longus (see also Figure 15.14)	Lies lateral to inferior aspect of tibialis posterior	Middle portion of fibula shaft	Tendon runs under foot to insert on distal phalanx of great toe	Flexes great toe; plantar flexes and inverts foot; the "push-off muscle" during walking

Lateral Compartment (Figure 15.13c and Figure 15.14a, b)

Peroneus longus	Superficial lateral muscle; overlies fibula	Head and upper portion of fibula	By long tendon under foot to first metatarsal and medial cuneiform	Plantar flexes and everts foot; helps keep foot flat on ground
Peroneus brevis	Smaller muscle; deep to peroneus longus	Distal portion of fibula shaft	By tendon running behind lateral malleolus to insert on proximal end of fifth metatarsal	Plantar flexes and everts foot, as part of peronei group

Anterior Compartment (Figure 15.14a, b)

Tibialis anterior	Superficial muscle of anterior leg; parallels sharp anterior margin of tibia	Lateral condyle and upper 2/3 of tibia; interosseous membrane	By tendon into inferior surface of first cuneiform and metatarsal 1	Prime mover of dorsiflexion; inverts foot
Extensor digitorum longus	Anterolateral surface of leg; lateral to tibialis anterior	Lateral condyle of tibia; proximal 3/4 of fibula; interosseous membrane	Tendon divides into four parts; insert into middle and distal phalanges of toes 2–5	Prime mover of toe extension; dorsiflexes foot
Peroneus tertius	Small muscle; often fused to distal part of extensor digitorum longus	Distal anterior surface of fibula	Tendon passes anterior to lateral malleolus and inserts on dorsum of fifth metatarsal	Dorsiflexes and everts foot
Extensor hallucis longus	Deep to extensor digitorum longus and tibialis anterior	Anteromedial shaft of fibula and interosseous membrane	Tendon inserts on distal phalanx of great toe	Extends great toe; dorsiflexes foot

- Patella
- Head of fibula
- Gastrocnemius
- Soleus
- Peroneus longus
- Extensor digitorum longus
- Tibialis anterior
- Extensor hallucis longus
- Peroneus tertius
- Superior and inferior extensor retinacula
- Extensor digitorum brevis

Peroneus brevis

Flexor hallucis longus

Peroneal retinaculum

Lateral malleolus

(a)

5th metatarsal

- Peroneus longus
- Gastrocnemius
- Tibia
- Tibialis anterior
- Extensor digitorum longus
- Soleus
- Extensor hallucis longus
- Peroneus tertius
- Superior and inferior extensor retinacula

(b)

Figure 15.14 Muscles of the anterolateral aspect of the right leg. (a) Superficial view of lateral aspect of the leg, illustrating the positioning of the lateral compartment muscles (peroneus longus and brevis) relative to anterior and posterior leg muscles. **(b)** Superficial view of anterior leg muscles.

exercise
16A

Skeletal Muscle Physiology—
Frog Experimentation:
Wet Lab

Objectives

1. To observe muscle contraction on the microscopic level and describe the role of ATP and various ions in muscle contraction.

2. To define and explain the physiologic basis of the following:
   action potential
   subthreshold stimulus
   threshold stimulus
   maximal stimulus
   treppe
   wave summation
   multiple motor unit summation
   tetanus
   muscle fatigue
   absolute and *relative refractory periods*
   depolarization
   repolarization

3. To trace the events that result from the electrical stimulation of a muscle.

4. To explain why the "all-or-none" law is demonstrated by the activity of a single muscle cell but not an intact skeletal muscle.

5. To recognize that a graded response of skeletal muscle is a function of the number of muscle fibers stimulated and the frequency of the stimulus.

6. To name and describe the phases of a muscle twitch.

7. To distinguish between a muscle twitch and a sustained (tetanic) contraction and to describe their importance in normal muscle activity.

8. To demonstrate how the kymograph or polygraph can be used to obtain pertinent and representative recordings of various physiologic events of skeletal muscle activity.

9. To explain the significance of muscle tracings obtained during experimentation.

Some instructors may prefer students to study muscle physiology with a computer simulation exercise. Exercise 16B should satisfy this preference.

Materials

❏ ATP muscle kits (glycerinated rabbit psoas muscle;* ATP and salt solutions obtainable from Carolina Biological Supply, Item #20-3525)
❏ Petri dishes
❏ Microscope slides
❏ Cover glasses
❏ Millimeter ruler
❏ Compound microscope
❏ Dissecting microscope
❏ Pointed glass probes (teasing needles)
❏ Small beaker (50 ml)
❏ Frog Ringer's solution
❏ Scissors
❏ Metal needle probes
❏ Medicine dropper
❏ Cotton thread
❏ Forceps
❏ Disposable gloves
❏ Glass or porcelain plate
❏ Pithed bullfrog†
❏ Apparatus A or B

   A: physiograph (polygraph), physiograph paper and ink, myograph, pin and clip electrodes, stimulator output extension cable, transducer cable, straight pins, frog board, laboratory stand, clamp

   B: kymograph, kymograph paper; writing points (smoking stand, burner, and glazing fluid if using a smoke-writing apparatus), skeletal muscle lever, signal magnet, laboratory stand, clamps, electronic stimulator, 10-g weights

Notes to the Instructor:
*At the beginning of the lab, the muscle bundle should be removed from the test tube and cut into ~ 2-cm lengths. Both the cut muscle segments and the entubed glycerol should be put into a petri dish. One muscle *segment* is sufficient for each two to four students making observations.
†Bullfrogs to be pithed by lab instructor as needed for student experimentation. (If instructor prefers that students pith their own specimens, an instructional sheet on that procedure suitable for copying for student handouts is provided in the Instructor's Guide.)

Muscle Activity

The contraction of skeletal and cardiac muscle fibers can be considered in terms of three events—electrical excitation of the muscle cell, excitation-contraction coupling, and shortening of the muscle cell due to sliding of the myofilaments within it.

At rest, all cells maintain a potential difference, or voltage, across their plasma membrane; the inner face of the membrane is approximately −60 to −90 millivolts (mV) compared with the cell exterior. This potential difference is a result of differences in membrane permeability to cations, most importantly sodium (Na^+) and potassium (K^+) ions. Intracellular potassium concentration is much greater than its extracellular concentration, and intracellular sodium concentration is considerably less than its extracellular concentration. Hence, steep concentration gradients across the membrane exist for both cations. However, because the plasma membrane is slightly more permeable to K^+ than to Na^+, Na^+ influx into the cell is inadequate to balance K^+ outflow. The result of this unequal Na^+-K^+ diffusion across the membrane establishes the **cell's resting membrane potential.** The resting membrane potential is of particular interest in excitable cells, like muscle cells and neurons, because changes in that voltage underlie their ability to do work (to contract or to signal respectively in muscle cells and neurons).

Action Potential

When a muscle cell is stimulated, the sarcolemma becomes temporarily permeable to sodium, which rushes into the cell. This sudden influx of sodium ions alters the membrane potential. That is, the cell interior becomes less negatively charged at that point, an event called **depolarization.** When depolarization reaches a certain level and the sarcolemma momentarily changes its polarity, a depolarization wave travels along the sarcolemma. Even as the influx of sodium ions occurs, the sarcolemma becomes impermeable to sodium and permeable to potassium ions. Consequently, potassium ions leak out of the cell, restoring the resting membrane potential (but not the original ionic conditions), an event called **repolarization.** The repolarization wave follows the depolarization wave across the sarcolemma. This rapid depolarization and repolarization of the membrane that is propagated along the entire membrane from the point of stimulation is called the **action potential.**

While the sodium gates are still open, there is no possibility of another response and the muscle cell is said to be in the **absolute refractory period. The relative refractory period** is the period after the sodium gates have closed when potassium gates are open and repolarization is ongoing. Especially strong stimuli may provoke a contraction during this part of the refractory period. Repolarization restores the muscle cell's irritability. Temporarily, the sodium-potassium pump, which actively transports K^+ into the cell and Na^+ out of the cell, need not be "revved up." But, if the cell is stimulated to contract again and again in rapid-fire order, the loss of potassium and gain of sodium occurring during action potential generation begins to reduce its ability to respond. And so, eventually the sodium-potassium pump must be activated to reestablish the ionic concentrations of the resting state.

Contraction

Propagation of the action potential along the sarcolemma causes the release of calcium ions (Ca^{2+}) from storage depots (tubules of the sarcoplasmic reticulum) within the muscle cell. When the calcium ions bind to regulatory proteins on the actin myofilaments, they act as an ionic trigger that initiates contraction, and the actin and myosin filaments slide past each other. Once the action potential ends, the calcium ions are almost immediately transported back into the tubules of the sarcoplasmic reticulum. Instantly the muscle cell relaxes.

The events of the contraction process can most simply be summarized as follows: muscle cell contraction is initiated by generation and transmission of an action potential along the sarcolemma. This electrical event is coupled with the sliding of the myofilaments—contraction—by the release of Ca^{2+}. Keep in mind this sequence of events as you conduct the experiments.

A c t i v i t y :
Observing Muscle Fiber Contraction

In this simple observational experiment, you will have the opportunity to review your understanding of muscle cell anatomy and to watch fibers contracting (or not contracting) in response to the presence of certain chemicals (ATP, and potassium and magnesium ions).

1.   Obtain the following materials from the supply area: 2 glass teasing needles, 3 glass microscope slides and cover glasses, millimeter ruler, dropper vials containing the following solutions: (a) 0.25% ATP in triply distilled water; (b) 0.25% ATP plus 0.05 M KCl plus 0.001 M $MgCl_2$ in distilled water; and (c) 0.05 M KCl plus 0.001 M $MgCl_2$ in distilled water; a petri dish, and a small portion of a previously cut muscle bundle segment. While you are at the supply area, place the muscle fibers in the petri dish and pour a small amount of glycerol (the fluid in the supply petri dish) over your muscle cells. Also obtain both a compound and a dissecting microscope and bring them to your laboratory bench.

2.   Using the fine glass needles, tease the muscle segment to separate its fibers. The objective is to isolate *single* muscle cells or fibers for observation. Be patient and work carefully so that the fibers do not get torn during this isolation procedure.

3.   Transfer one or more of the fibers (or the thinnest strands you have obtained) onto a clean microscope slide with a glass needle, and cover it with a cover glass. Examine the fiber under low and then high power magnifications to observe the striations and the smoothness of the fibers when they are in the relaxed state.

4.   Transfer three or four fibers to a second clean microscope slide with a glass needle. Using the needle as a prod, carefully position the fibers so that they are parallel to one another and as straight as possible. Place this slide under a dissecting microscope and measure the length of each fiber by holding a millimeter ruler adjacent to it. Alternatively, you can rest the microscope slide *on* the millimeter ruler to make your length determinations. Record the fiber lengths on the chart at the top of page 156.

	Muscle fiber 1	Muscle fiber 2	Muscle fiber 3
Intial length (mm)			
Contracted length (mm)			
% contraction			

5. Flood the fibers (situated under the dissecting microscope) with several drops of the solution containing ATP, potassium ions, and magnesium ions. Watch the reaction of the fibers after adding the solution. After 30 seconds (or slightly longer), remeasure each fiber and record the observed lengths on the chart. Also, observe the fibers to see if any width changes have occurred. Calculate the degree (or percentage) of contraction by using the following simple formula, and record this information on the chart also.

$$\text{initial length (mm)} - \text{contracted length (mm)} = \text{degree of contraction (mm)}$$

then:

$$\frac{\text{degree of contraction (mm)}}{\text{initial length (mm)}} \times 100 = \underline{\hspace{1cm}} \text{ \% contraction}$$

6. Carefully transfer one of the contracted fibers to a clean microscope slide, cover with a cover glass, and observe with the compound microscope. Mentally compare your initial observations with the view you are observing now. What differences do you see? (Be specific.)

_____

_____

_____

What zones (or bands) have disappeared?

_____

7. Repeat steps 3 to 6 twice more, using clean slides and fresh muscle cells. First use the solution of ATP in distilled water (no salts). Then, use the solution containing only salts (no ATP) for the third series.

What degree of contraction was observed when ATP was applied in the absence of potassium and magnesium ions?

_____

What degree of contraction was observed when the muscle fibers were flooded with a K^+- and Mg^{2+}-containing solution that lacked ATP?

_____

What conclusions can you draw about the importance of ATP, and potassium and magnesium ions to the contractile process?

_____

_____ ■

A c t i v i t y :
Induction of Contraction in the Frog Gastrocnemius Muscle

Physiologists have learned a great deal about the way muscles function by isolating muscles from laboratory animals and then stimulating these muscles to observe their responses. Various stimuli—electrical shock, temperature changes, extremes of pH, certain chemicals—elicit muscle activity, but laboratory experiments of this type typically use electrical shock. This is because it is easier to control the onset and cessation of electrical shock, as well as the strength of the stimulus.

Preparing a Muscle for Experimentation

The preparatory work that precedes the recording of muscle activity tends to be quite time consuming. If you work in teams of two or three, the work can be divided. While one of you is setting up the recording apparatus (kymograph or physiograph), one or two students can dissect the frog leg (Figure 16A.1). Experimentation should begin as soon as the dissection is completed.

Various types of apparatus are used to record muscle contraction. All include a way to mark time intervals, a way to indicate exactly when the stimulus was applied, and a way to measure the magnitude of the contractile response. Instructions are provided here for setting up physiograph (Figure 16A.2) and kymograph (Figure 16A.3) apparatus. Specific instructions for use of recording apparatus during recording will be provided by your instructor.

D i s s e c t i o n :
Frog Hind Limb

1. Before beginning the frog dissection, have the following supplies ready at your laboratory bench: a small beaker containing 20 to 30 ml of frog Ringer's solution, scissors, a metal needle probe, a glass probe with a pointed tip, a medicine

Gastrocnemius
muscle

Glass
probe

**Figure 16A.1 Preparation of
the frog gastrocnemius muscle.**
Numbers indicate the sequence of
manipulation.

dropper, cotton thread, forceps, a glass or porcelain plate, and disposable gloves. While these supplies are being accumulated, one member of your team should notify the instructor that you are ready to begin experimentation, so that the frog can be prepared (pithed). Preparation of the frog in this manner renders it unable to feel pain and prevents reflex movements (like hopping) that would interfere with the experiments.

2. All students that will be handling the frog should obtain and don disposable plastic gloves. Obtain a pithed frog and place it ventral surface down on the glass plate. Make an incision into the skin approximately midthigh

(Figure 16A.1), and then continue the cut completely around the thigh. Grasp the skin with the forceps and strip it from the leg and hindfoot. The skin adheres more at the joints, but a careful, persistent pulling motion—somewhat like pulling off a nylon stocking—will enable you to remove it in one piece. From this point on, the exposed muscle tissue should be kept moistened with the Ringer's solution to prevent spontaneous twitches.

3. Identify the gastrocnemius muscle (the fleshy muscle of the posterior calf) and the calcaneal (Achilles) tendon that secures it to the heel.

Myograph tension adjuster

Myograph

Transducer cable

Laboratory transducer stand

Myograph hook

Thread

Clip on stimulator output extension cable

Pin electrodes

Stimulator output extension cable

Pin electrodes

Gastrocnemius muscle

Transducer coupler

RF isolated output

Frog board

Channel amplifier

Stimulator panel

Figure 16A.2 Physiograph setup for frog gastrocnemius experiments.

Materials:
Channel amplifier and stimulator transducer cable
Stimulator panel and stimulator output extension cable
Myograph
Myograph tension adjuster
Transducer stand
Two pin electrodes
Frog board and straight pins
Prepared frog (gastrocnemius muscle freed and calcaneal tendon ligated with thread)
Frog Ringer's solution

1. Connect myograph to transducer stand and attach frog board to stand.

2. Attach transducer cable to myograph and to input connection on channel amplifier.

3. Attach stimulator output extension cable to output on stimulator panel (red to red, black to black).

4. Using clip at opposite end of extension cable, attach cable to bottom of transducer stand adjacent to frog board.

5. Attach two pin electrodes securely to electrodes on clip.

6. Place knee of prepared frog in clip on frog board and secure by inserting a straight pin through tissues of frog. Keep frog muscle moistened with Ringer's solution.

7. Attach thread ligating the calcaneal tendon of frog to myograph leaf-spring hook.

8. Adjust position of myograph on stand to produce a constant tension on thread attached to muscle (taut but not tight). Gastrocnemius muscle should hang vertically directly below myograph hook.

9. Insert free ends of pin electrodes into the muscle, one at proximal end and other at distal end.

4. Slip a glass probe under the gastrocnemius muscle and run it along the entire length and under the calcaneal tendon to free them from the underlying tissues.

5. Cut a piece of thread about 10 in. long and use the glass probe to slide the thread under the calcaneal tendon. Knot the thread firmly around the tendon and then sever the tendon distal to the thread. Alternatively, you can bend a common pin into a Z-shape and insert the pin securely into the tendon. The thread is then attached to the opposite end of the pin. If you are using a physiograph, once the tendon has been tied or pinned, the frog is ready for experimentation (see Figure 16A.2). If you are using a kymograph, the gastrocnemius muscle must be completely isolated, as described in step 6.

Materials:
Recording paper
Kymograph
Kymograph base
Muscle clamp and muscle lever
Writing points
Signal magnet
Ring stand
Right-angle clamps
Eight 10-g weights
Frog Ringer's solution and gauze
Isolated frog gastrocnemius muscle
Electronic stimulator

1. Attach muscle clamp to supporting rod and secure cut femur in clamp.

2. Position and clamp muscle lever to supporting rod beneath muscle tissue. Tie thread ligating the calcaneal tendon to hook on the muscle lever so that muscle hangs vertically. Keep muscle moist with Ringer's solution from this point until completion of preparations and experimentation.

3. Hang a 10-g weight on muscle lever just below thread attachment to provide proper tension on lever. Muscle lever should rest horizontally. Writing tip should lightly touch paper surface on drum. Test its position by pulling lightly and directly upward on thread. Writing tip should remain in contact with drum surface.

4. Clamp signal magnet to supporting rod. Tip of its writing stylus should line up directly beneath that of muscle lever. Precise alignment is important so that a correlation can be seen between the point of stimulus and onset of muscle contractions.

5. Position stimulator to one side so that area immediately in front of recording apparatus is free for manipulations during and after experimentation. Turn all its switches to OFF and all dials to left. Plug the stimulator into the power outlet.

6. Connect two gray wires from signal magnet outlet (back of stimulator) to binding posts on signal magnet. To operate signal magnet, toggle switch (upper left on front of stimulator) must be ON (flipped up). Signal magnet will indicate up to 30 pulses per second when it is operating. If you use frequencies greater than 30 pulses per second during experimentation, turn signal magnet toggle switch to OFF (down) position.

7. Wash platinum tips of stimulator electrode in distilled water and dry. Connect electrode to white and black output terminals of stimulator and clamp electrode to supporting rod so that its tips rest on muscle tissue.

Figure 16A.3 Kymograph setup for frog gastrocnemius experiments.

6. Cut away the fibulotibial bone just distal to the knee. Expose the femur of the thigh and cut it completely through at midthigh. Remove as much of the thigh muscle tissue as possible by carefully cutting it away with the scissors. The isolated gastrocnemius muscle can now be mounted on the muscle bar, and the stimulating electrodes of the kymograph can be attached (see Figure 16A.3). About halfway through the laboratory period, dissect the second leg for use. ■

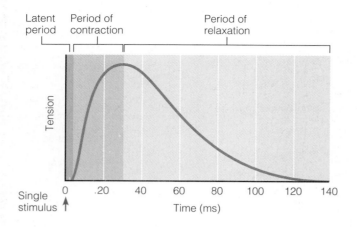

Latent period | Period of contraction | Period of relaxation

Figure 16A.4 Tracing of a muscle twitch.

Recording Muscle Activity

The **"all-or-none"** law of muscle physiology states that a muscle cell will contract maximally when stimulated adequately. Skeletal muscles, however, consist of thousands of muscle cells and react to stimuli with graded responses. Thus muscle contractions can be weak or vigorous, depending on the requirements of the task. Graded responses (different degrees of shortening) of a skeletal muscle depend on the number of muscle cells being stimulated. In the intact organism, the number of motor units firing at any one time determines how many muscle cells will be stimulated. In this laboratory, the frequency and strength of an electrical current determines the response.

A single contraction of skeletal muscle is called a **muscle twitch.** A tracing of a muscle twitch (Figure 16A.4) shows three distinct phases: latent, contraction, and relaxation. The **latent phase** is the interval from stimulus application until the muscle begins to shorten. Although no activity is indicated on the tracing during this phase, important electrical and chemical changes are occurring within the muscle. During the **contraction phase,** the muscle fibers shorten; the tracing shows an increasingly higher needle deflection and the tracing peaks. During the **relaxation phase,** represented by a downward curve of the tracing, the muscle fibers relax and lengthen. On a slowly moving recording surface, the single muscle twitch appears as a spike (rather than a bell-shaped curve, as in Figure 16A.4), but on a rapidly moving recording surface, the three distinct phases just described become recognizable.

Determining the Threshold Stimulus

1. Assuming that you have already set up the recording apparatus, set the time marker to deliver one pulse per second and set the paper speed at a slow rate, approximately 0.1 cm per second.

2. Set the duration control on the stimulator between 7 and 15 msec, multiplier × one and the voltage control at zero V, multiplier × one. Turn the sensitivity control knob of the stimulator fully clockwise (lowest value, greatest sensitivity).

3. Administer single stimuli to the muscle at intervals of 1 to 2 sec, beginning with 0.1 V and increasing each successive

stimulus by 0.1 V until a contraction is obtained (shown by a spike on the paper).

At what voltage did contraction occur? _____ V

The voltage at which the first perceptible contractile response is obtained is called the **threshold stimulus.** All stimuli applied prior to this point are termed **subthreshold stimuli,** because at those voltages no response was elicited.

4. Stop the recording and mark the record to indicate the threshold stimulus, voltage, and time. Do not remove the record from the recording surface; continue with the next experiment. Remember: keep the muscle preparation moistened with Ringer's solution at all times.

Observing Graded Muscle Response to Increased Stimulus Intensity

1. Follow the previous setup instructions, but set the voltage control at the threshold voltage (as determined in the first experiment).

2. Deliver single stimuli at 1- or 2-sec intervals. Initially increase the voltage between shocks by 0.5 V; then increase the voltage by 1 to 2 V between shocks as the experiment continues, until contraction height increases no further. Stop the recording apparatus.

What voltage produced the highest spike (thus the maximal strength of contraction)?

_____ V

This voltage, called the **maximal stimulus** (for *your* muscle specimen), is the weakest stimulus at which all muscle cells are being stimulated. As the voltage was increased to this point, more and more muscle cells (motor units) were activated, resulting in a stronger and stronger contraction. Past this point, an increase in the intensity of the stimulus will not produce any greater contractile response. This phenomenon, called **multiple motor unit summation,** or **recruitment,** is the process by which the increased contractile strength reflects the relative number of muscle cells stimulated.

3. Mark the record *multiple motor unit summation.* Record the maximal stimulus voltage and the time you completed the experiment. Continue on to the next experiment.

Timing the Muscle Twitch

1. Follow the previous setup directions, but set the voltage for the maximal stimulus (as determined in the preceding experiment) and set the paper advance or recording speed at maximum. Record the paper speed setting:

_____ mm/sec

2. Determine the time required for the paper to advance 1 mm by using the formula

$$\frac{1 \text{ mm}}{\text{mm/sec (paper speed)}}$$

(Thus, if your paper speed is 25 mm/sec, each mm on the chart equals 0.04 sec.) Record the computed value:

1 mm = _____ sec

3.   Deliver single stimuli at 2- to 3-sec intervals to obtain several "twitch" curves. Stop the recording.

4.   Determine the duration of the latent, contraction, and relaxation phases of the twitches and record here:

Duration of latent period: _____ sec

Duration of contraction period: _____ sec

Duration of relaxation period: _____ sec

5.   Label the record to indicate the point of stimulus, the beginning of contraction, the end of contraction, and the end of relaxation.

6.   Allow the muscle to rest (but keep it moistened) before continuing with the next experiment.

Inducing Treppe, or the Staircase Phenomenon As a muscle is stimulated to contract, a curious phenomenon is observed in the tracing pattern of the first few twitches. Even though the stimulus intensity is unchanged, the height of the individual spikes increases in a stepwise manner—producing a sort of staircase pattern called **treppe** (Figure 16A.5). This phenomenon is not well understood, but the following explanation has been offered: in the muscle cell's resting state, there is much less Ca^{2+} in the sarcoplasm and the enzyme systems in the muscle cell are less efficient than after it has contracted a few times. As the muscle cell begins to contract, the intracellular concentration of Ca^{2+} rises dramatically, and the heat generated by muscle activity increases the efficiency of the enzyme systems. As a result, the muscle becomes more efficient and contracts more vigorously. This is the physiologic basis of the warm-up period prior to competition in sports events.

1.   Set up the apparatus as in the previous experiment, again setting the voltage to the maximal stimulus.

2.   Deliver single stimuli at 1-sec intervals until the strength of contraction does not increase further.

3.   Stop the recording apparatus and mark the record *treppe*. Note also the number of contractions (and seconds) required to reach the constant contraction magnitude. Record the voltage used and the time when you completed this experiment. Continue on to the next experiment.

Observing Graded Muscle Response to Increased Stimulus Frequency Muscles subjected to frequent stimulation, without a chance to relax, exhibit two kinds of responses—wave summation and tetanus—depending on the level of stimulus frequency (Figure 16A.5).

Wave Summation: If a muscle is stimulated with a rapid series of stimuli of the same intensity before it has had a chance to relax completely, the response to the second and subsequent stimuli will be greater than to the first stimulus. This phenomenon, called **wave,** or **temporal, summation,** occurs because the muscle is already in a partially contracted state when subsequent stimuli are delivered.

1.   With the recorder running at slow speed and the stimulus intensity set to the maximal stimulus, stimulate the muscle at a rate of 15 to 25 stimuli per second.

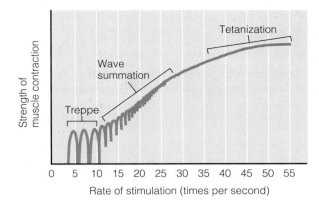

Figure 16A.5 Treppe, wave summation, and tetanization. Progressive summation of successive contractions occurs as the rate of stimulation is increased. Tetanization occurs when the rate of stimulation reaches approximately 35 per second, and maximum contraction force occurs at a stimulation rate of approximately 50 per second.

2.   Shut off the recorder and label the record as wave *summation*. Note also the time, the voltage, and the frequency.

Tetanus: Stimulation of a muscle at an even higher frequency will produce a "fusion" (tetanization) of the summated twitches. In effect, a single sustained contraction is achieved in which no evidence of relaxation can be seen (Figure 16A.5). **Tetanus** is a feature of normal skeletal muscle functioning; the single muscle twitch is primarily a laboratory phenomenon.

1.   To demonstrate tetanus, maintain the conditions used for wave summation except for the frequency of stimulation. Set the stimulator to deliver 60 stimuli per second.

2.   As soon as you obtain a single smooth, sustained contraction (with no evidence of relaxation), discontinue stimulation and shut off the recorder.

3.   Label the tracing with the conditions of experimentation, the time, and the area of tetanus.

Inducing Muscle Fatigue **Muscle fatigue,** the loss of the ability to contract, is believed to be a result of the oxygen debt that occurs in the tissue after prolonged activity (through the accumulation of such waste products as lactic acid as well as the depletion of ATP). True muscle fatigue rarely occurs in the body because it is most often preceded by a subjective feeling of fatigue. Furthermore, fatigue of the neuromuscular junctions typically precedes fatigue of the muscle.

1.   To demonstrate muscle fatigue, set up an experiment like the tetanus experiment, but continue stimulation until the muscle completely relaxes and the contraction curve returns to the base line.

2.   Measure the time interval between the beginning of complete tetanus and the beginning of fatigue (when the tracing begins its downward curve). Mark the record appropriately.

3. Determine the time required for complete fatigue to occur (the time interval from the beginning of fatigue until the return of the curve to the base line). Mark the record appropriately.

4. Allow the muscle to rest (keeping it moistened with Ringer's solution) for 10 min, and then repeat the experiment.

What was the effect of the rest period on the fatigued muscle?

_____

_____

What is the physiologic basis for this reaction?

_____

_____

Determining the Effect of Load on Skeletal Muscle When the fibers of a skeletal muscle are slightly stretched by a weight or tension, the muscle responds by contracting more forcibly and thus is capable of doing more work. If the load is increased beyond the optimum, the latent period becomes longer, contractile force decreases, and relaxation (fatigue) occurs more quickly. With excessive stretching, the muscle is unable to develop any tension and no contraction occurs. Since the filaments no longer overlap at all with this degree of stretching, the sliding force cannot be generated.

If your equipment allows you to add more weights to the muscle specimen or to increase the tension on the muscle, perform the following experiment to determine the effect of loading on skeletal muscle, and to develop a work curve for the frog's gastrocnemius muscle.

1. Set the stimulator to deliver the maximal voltage as previously determined.

2. Stimulate the unweighted muscle with single shocks at 1- to 2-sec intervals to achieve three or four muscle twitches.

3. Stop the recording apparatus and add 10 g of weight or tension to the muscle. Restart and advance the recording about 1 cm, and then stimulate again to obtain three or four spikes.

4. Repeat the previous step seven more times, increasing the weight by 10 g each time until the total load on the muscle is 80 g or the muscle fails to respond. If the calcaneal tendon tears, the weight will drop, thus ending the trial. In such cases, you will need to prepare another frog's leg to continue the experiments and the maximal stimulus will have to be determined for the new muscle preparation.

5. When these "loading" experiments are completed, discontinue recording and remove the tracing. Mark the curves on the record to indicate the load (in grams).

6. Measure the height of contraction (in millimeters) for each sequence of twitches obtained with each load, and insert this information into the chart to the right.

7. Compute the work done by the muscle for each twitch (load) sequence.

Weight of load (g) × distance load lifted (mm) = work done

Load (g)	Distance load lifted (mm)		Work done	
	Trial 1	Trial 2	Trial 1	Trial 2
0				
10				
20				
30				
40				
50				
60				
70				
80				

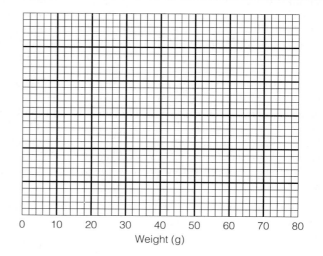

Weight (g)

Enter these calculations into the chart in the columns labeled Trial 1.

8. Allow the muscle to rest for 5 minutes. Then conduct a second trial in the same manner (i.e., repeat steps 2 through 7). Record this second set of calculations in the columns labeled Trial 2. Be sure to keep the muscle well moistened with Ringer's solution during the resting interval.

9. Using two different colors, plot a line graph of work done against the weight on the accompanying grid for each trial. Label each plot appropriately.

10. Dismantle all apparatus and prepare the equipment for storage. Dispose of the frog remains in the appropriate container. Discard the gloves as instructed and wash and dry your hands.

11. Inspect your records of the experiments and make sure each is fully labeled with the experimental conditions, the date, and the names of those who conducted the experiments. (If you used a smoked-drum kymograph recording system, "fix" the records before leaving the laboratory.) For future reference, attach a tracing (or a photocopy of the tracing) for each experiment to this page. ∎

Skeletal Muscle Physiology: Computer Simulation

Objectives

1. To define these terms used in describing muscle physiology: *maximal stimulus, treppe, wave summation, multiple motor unit summation, tetanus*

2. To identify two ways that the mode of stimulation can affect muscle force production.

3. To plot a graph relating stimulus strength and twitch force to illustrate graded muscle response.

4. To explain how slow, smooth, sustained contraction is possible in a skeletal muscle.

5. To graphically understand the relationships between passive, active, and total forces.

6. To identify the conditions under which muscle contraction is isometric or isotonic.

7. To describe in terms of length and force the transitions between isometric and isotonic conditions during a single muscle twitch.

8. To describe the effects of resistance and starting length on the initial velocity of shortening.

9. To explain why muscle force remains constant during isotonic shortening.

10. To explain experimental results in terms of muscle structure.

Materials

Minimum equipment required:

Windows:

- ❑ 486/66 MHz or better recommended
- ❑ Windows 95/98 recommended
- ❑ SVGA display (256 colors at 640 X 480)
- ❑ Minimum 16 MB available RAM
- ❑ Double-speed CD-ROM drive (quad-speed or higher recommended)
- ❑ Sound card
- ❑ Speakers or headphones
- ❑ Printer

Macintosh:

- ❑ 68040 processor or Power Macintosh recommended

- ❑ System 7.1 or higher
- ❑ Minimum 16 MB available RAM
- ❑ 13-inch or larger color monitor (640 X 480 resolution)
- ❑ Double-speed CD-ROM drive (quad-speed or higher recommended)
- ❑ Printer

Software:

- ❑ Benjamin/Cummings PhysioEx CD-ROM—Skeletal Muscle Physiology module

Skeletal muscles are composed of hundreds to thousands of individual cells, each doing their share of work in the production of force. As their name suggests, skeletal muscles move the skeleton. Skeletal muscles are remarkable machines; while allowing us the manual dexterity to create magnificent works of art, they are also capable of generating the brute force needed to lift a 100-lb sack of concrete. When a skeletal muscle from an experimental animal is electrically stimulated, it behaves in the same way as a stimulated muscle in the intact body, that is, *in vivo*. Hence, such an experiment gives us valuable insight into muscle behavior.

This set of computer simulations demonstrates many important physiological concepts of skeletal muscle contraction. The program graphically provides all the equipment and materials necessary for you, the investigator, to set up experimental conditions and observe the results. In student-conducted laboratory investigations there are many ways to approach a problem, and the same is true of these simulations. The instructions will guide you in your investigation, but you should also try out alternate approaches to gain insight into the logical methods used in scientific experimentation.

Try this approach: As you work through the simulations for the first time, follow the instructions closely and answer the questions posed as you go. Then try asking "What if . . . ?" questions to test the validity of your theories. The major advantages of these computer simulations are that the muscle cannot be accidentally damaged, lab equipment will not break down at the worst possible time, and you will have ample time to think critically about the processes being investigated.

Because you will be working with a simulated muscle and oscilloscope display, you need to watch both carefully during the experiments. Think about what is happening in each situation. You need to understand how you are experimentally manipulating the muscle in order to understand your results.

Getting Started

Begin by making sure you have the computer equipment and software listed in the Materials section on p. 163.

1. Quit all applications currently running on your computer.

2. Insert the PhysioEx CD-ROM into the CD-ROM drive and keep PhysioEx in the drive the entire time you use the program.

3. Follow the instructions below for your computer type.

Windows 95/Windows 98
The program will launch automatically after you load the CD-ROM into the drive.
If autorun is disabled on your machine follow these instructions:
1. Double-click the My Computer icon on the Windows desktop.
2. Double-click the PhysioEx CD icon.
3. Look for the icon that looks like a tiny movie projector and double-click it.

Macintosh
Double-click the PhysioEx CD icon that appears in the window on your desktop to launch the program.

Electrical Stimulation

A contracting skeletal muscle will produce force and/or shortening when nervous or electrical stimulation is applied. Unlike single cells or motor units, which follow the all-or-none law of muscle physiology, a whole muscle responds to stimuli with a graded response. A motor unit consists of a motor neuron and all the muscle cells it innervates. Hence, activation of the neuron innervating a single motor unit will cause all muscle cells in that unit to fire simultaneously in an all-or-none fashion. The graded contractile response of a whole muscle reflects the number of motor units firing at a given time. Strong muscle contraction implies that many motor units are activated and each unit has maximally contracted. Weak contraction means that few motor units are active; however, the activated units are maximally contracted. By increasing the number of motor units firing, we can produce a steady increase in muscle force, a process called recruitment or motor unit summation.

Regardless of the number of motor units activated, a single contraction of skeletal muscle is called a muscle twitch. A tracing of a muscle twitch is divided into three phases: latent, contraction, and relaxation (see Figure 16A.4, p. 160). The latent phase is a short period between the time of stimulation and the beginning of contraction. Although no force is generated during this interval, chemical changes occur intracellularly in preparation for contraction. During contraction, the myofilaments are sliding past each other and the muscle shortens. Relaxation takes place when contraction has ended and the muscle returns to its normal resting state and length.

The first activity you will conduct simulates an isometric, or fixed length, contraction of an isolated skeletal muscle. This activity allows you to investigate how the strength and frequency of an electrical stimulus affect whole muscle function. Note that these simulations involve indirect stimulation by an electrode placed on the surface of the muscle. This differs from the situation *in vivo* where each fiber in the muscle receives direct stimulation via a nerve ending. In other words, increasing the intensity of the electrical stimulation mimics how the nervous system increases the number of motor units activated.

Single Stimulus

Choose **Skeletal Muscle Physiology** from the main menu. The opening screen will appear in a few seconds (Figure 16B.1). The oscilloscope display, the grid at the top of the screen, is the most important part of the screen because it graphically displays the contraction data for analysis. Time is displayed on the horizontal axis. A full sweep is initially set at 200 msec. However, you can adjust the sweep time from 200 msec to 1000 msec by clicking and dragging the **200** msec button at the lower right corner of the oscilloscope display to the left to a new position on the time axis. The force (in grams) produced by muscle contraction is displayed on the vertical axis. Clicking the **Clear Tracings** button erases all muscle twitch tracings from the oscilloscope display.

The *electrical stimulator* is the equipment seen just beneath the oscilloscope display. Clicking **Stimulate** delivers the electrical shock to the muscle through the electrodes lying on the surface of the muscle. Stimulus voltage is set by clicking the (+) or (−) buttons next to the voltage window. Three small windows to the right of the Stimulate button display the force measurements. *Active force* is produced during muscle contraction, while *passive force* results from the muscle being stretched (much like a rubber band). The *total force* is the sum of active and passive forces. The red arrow at the left of the oscilloscope display is an indicator of passive force. After the muscle is stimulated, the **Measure** button at the right edge of the electrical stimulator becomes active. When the Measure button is clicked, a vertical orange line will be displayed at the left edge of the oscilloscope display. Clicking the arrow buttons below the Measure button moves the orange line horizontally across the screen. The Time window displays the difference in time between the zero point on the X-axis and the intersection between the orange measure line and the muscle twitch tracing.

The muscle is suspended in the support stand to the left of the oscilloscope display. The hook through the upper tendon of the muscle is part of the force transducer, which measures the force produced by the muscle. The hook through the lower tendon secures the muscle in place. The weight cabinet just below the muscle support stand is not active in this experiment; it contains weights you will use in the isotonic contraction part of the simulation. You can adjust the starting length of the muscle by clicking the (+) or (−) buttons located next to the Muscle Length display window.

When you click the **Record Data** button in the data collection unit below the electrical stimulator, your data is recorded in the computer's memory and is displayed in the

Figure 16B.1 Opening screen of the Single Stimulus experiment.

data grid at the bottom of the screen. Data displayed in the data grid include the voltage, muscle length, and active, passive, and total force measurements. If you are not satisfied with a single run, you can click **Delete Line** to erase a single line of data. Clicking the **Clear Table** button will remove all accumulated data in the experiment and allow you to start over.

Activity:
Practicing Generating a Tracing

1.   Click the **Stimulate** button once. Because the beginning voltage is set to zero, no muscle activity should result. You will see a blue line moving across the bottom of the oscilloscope display. This blue line will indicate muscle force in the experiments. If you are using Windows 3.1 or an older Macintosh and the tracings move too slowly across the screen, click and hold the **200** button at the lower right corner of the oscilloscope and drag it to the left to the 40 msec mark and

release it. This action resets the total sweep time to 1000 msec to speed up the display time.

2.   Click and hold the (+) button beneath the Stimulate button until the voltage window reads 3.0 volts. Click **Stimulate** once. You will see the muscle react, and a contraction tracing will appear on the screen. Notice that the muscle tracing color alternates between blue and yellow each time the Stimulate button is clicked to enhance the visual difference between twitch tracings. You can click the **Clear Tracings** button as needed to clean up the oscilloscope display. To retain your data, click the **Record Data** button at the end of each stimulus.

3.   Change the voltage to 5.0 volts and click **Stimulate** again. Notice how the force of contraction also changes. Identify the latent, contraction, and relaxation phases in the tracings.

Feel free to experiment with anything that comes to mind to get a sense of how whole muscle responds to an electrical stimulus. ■

Activity:
Determining the Latent Period

1. Click **Clear Tracings** to erase the oscilloscope display. The voltage should be set to 5.0 volts.

2. Drag the **200** msec button to the right edge of the oscilloscope.

3. Click the **Stimulate** button once and allow the tracing to complete (this may be quite slow on Windows 3.1 or an older Macintosh).

4. When the tracing is completed, click the **Measure** button. Click the right arrow button next to the Time window repeatedly until you notice the first increase in the active force window. Click the left arrow button next to the Time window until the active force window reads zero. The time now displayed in the Time window represents the difference between when the muscle was stimulated and the beginning of force production. This time represents the latent period of muscle contraction.

How long is the latent period? _____ msec

What occurs in the muscle during this apparent lack of activity?

_____

_____ ∎

Activity:
Determining the Threshold (Minimal) Stimulus

1. Click **Clear Tracings** to erase the oscilloscope display.

2. Set the voltage control to 0.0 volts by clicking the (−) button next to the voltage window.

3. Increase the voltage by 0.1 volt and click **Stimulate.**

4. Repeat step 3 until you see a number greater than 0.0 appear in the Active Force window. This number indicates that a contraction has occurred. The number in the voltage window represents the threshold voltage, below which no contraction occurs.

What is the threshold voltage? _____ V ∎

Activity:
Investigating Graded Muscle Response to Increased Stimulus Intensity

1. Examine the tracings just recorded in the previous activity.

2. Now stimulate the muscle several times using increasingly higher voltages. Start at 6.0 volts and increase the voltage in approximately 1.0-volt increments. If the display is slow, drag the **200** msec button back to the 40 msec mark.

Notice that as voltage is increased, the amount of force produced by the muscle also increases. As more voltage is delivered to the whole muscle, greater numbers of muscle fibers are activated, thereby increasing the total force produced by the muscle. This result mimics the muscle activity occurring *in vivo* where the recruitment of additional motor units increases the total force produced. This phenomenon is called *multiple motor unit summation.*

Is there a stimulus voltage above which there appears to be no further increase in muscle contraction?

_____

Why is this so?

_____

_____

3. Click **Clear Tracings** to prepare for the next experiment. You will perform a systematic study of force produced over the entire range of stimulus voltages in 0.5-volt increments up to 10.0 volts.

4. Set the voltage to 0.0 and click **Stimulate.**

5. Click **Record Data**. If you decide to redo a single stimulus, choose the data line in the grid and click **Delete Line** to erase that single line of data. If you want to repeat the entire experiment, click the **Clear Table** button to erase all data recorded to that point.

6. Repeat steps 4 and 5, increasing the voltage by 0.5 each time until you reach the maximum voltage of 10.0.

Based on the twitch tracings you see in the display, is a muscle twitch an all-or-none response?

_____

Explain your answer.

_____

_____

_____

_____

7. Observe the twitch tracings. Click the **Tools** menu and then choose **Plot Data.** Note that the oscilloscope tracings will be automatically erased from the oscilloscope display, but your data will be retained in the grid.

8. Use the slider bars to display Active Force on the Y-axis and Voltage on the X-axis. Click **Draw It** to graph your collected data. Estimate the maximal stimulus by looking at your graph.

What is the maximal stimulus? _____ V

This voltage represents the lowest stimulus intensity or strength necessary to activate all cells within the muscle and is called the **maximal stimulus.** The maximum contraction thus produced is called the maximal response.

9.   When finished, click the close box at the top of the plot window. ■

Multiple Stimulus

Choose **Multiple Stimulus** from the **Experiment** menu. The opening screen will appear in a few seconds (Figure 16B.2).

The only significant change to the on-screen equipment is found in the electrical stimulator. The measuring equipment has been removed and other controls have been added: The **Multiple Stimulus** button is a toggle that allows you to alternately start and stop the electrical stimulator. When Multiple Stimulus is first clicked, its name changes to Stop Stimulus and electrical stimuli are delivered to the muscle at the rate specified in the Stimuli/sec window until the muscle completely fatigues or the stimulator is turned off. The stimulator is turned off by clicking the **Stop Stimulus** button. The stimulus rate is adjusted by clicking the (+) or (−) buttons next to the stimuli/sec window.

A c t i v i t y :
Investigating Treppe

When a muscle first contracts, the force it is able to produce is less than the force it is able to produce in subsequent contractions within a relatively narrow time span. A myogram, a recording of a muscle twitch, reveals this phenomenon as the **treppe,** or staircase, effect. For the first few twitches, each

Figure 16B.2   Opening screen of the Multiple Stimulus experiment.

successive stimulation produces slightly more force than the previous contraction as long as the muscle is allowed to fully relax between stimuli, and the stimuli are delivered relatively close together. Treppe is thought to be caused by increased efficiency of the enzyme systems within the cell and increased availability of intracellular calcium.

1. The voltage should be set to 8.2 volts and the muscle length should be 75 mm.

2. Drag the **200** msec button to the center of the X-axis time range.

3. Be sure that you fully understand the following three steps before you proceed.

• Click **Single Stimulus.** Watch the twitch tracing carefully.

• After the tracing shows that the muscle has completely relaxed, immediately click **Single Stimulus** again.

• When the second twitch completes, click **Single Stimulus** once more and allow the tracing to complete.

What happens to force production with each subsequent stimulus?

_____

_____ ■

A c t i v i t y :
Investigating Wave Summation

As demonstrated in the single-stimulus activity earlier, multiple motor unit summation is one way to increase the amount of force produced by muscle. **Multiple motor unit summation** relies on increased stimulus *intensity* in this simulation. Another way to increase force is by wave, or temporal, summation. **Wave summation** is achieved by increasing the stimulus *frequency,* or rate of stimulus delivery to the muscle. Wave summation occurs because the muscle is already in a partially contracted state when subsequent stimuli are delivered.

Tetanus can be considered an extreme form of wave summation that results in a steady, sustained contraction. In effect, the muscle does not have any chance to relax because it is being stimulated at such a high frequency. This "fuses" the force peaks so that we observe a smooth tracing.

1. Click **Clear Tracings** to erase the oscilloscope display.

2. Set and keep the voltage at the maximal stimulus (8.2 volts) and the muscle length at 75 mm.

3. Drag the **200** msec button to the right edge of the oscilloscope display unless you are using a slower computer.

4. Click **Single Stimulus,** and then click **Single Stimulus** again when the muscle has relaxed about halfway.

Is the peak force produced in the second contraction greater than that produced by the first stimulus?

_____

5. Try stimulating again at greater frequencies by clicking the Single Stimulus button several times in rapid succession.

Is the total force production even greater? _____

6. To see if you can produce smooth, sustained contraction at Active Force = 2 gms, try rapidly clicking **Single Stimulus** several times.

Is it possible to produce a smooth contraction (fused force peaks), or does the force rise and fall periodically?

_____

7. Using the same method as in step 6, try to produce a smooth contraction at 3 gms.

Is the trace smoother (less height between peaks and valleys) this time?

_____

Because there is a limit to how fast you can manually click the Single Stimulus button, what do you think would happen to the smoothness of the tracing if you could click even faster?

_____

_____

_____

8. So far you have been using the maximal stimulus. What do you think would happen if you used a lower voltage?

_____

_____

_____

9. Use the concepts of motor unit summation and stimulus frequency to explain how human skeletal muscles work to achieve smooth, steady contractions at all desired levels of force.

_____

_____

_____

_____ ■

Activity:
Investigating Fusion Frequency

1. Click **Clear Tracings** to erase the oscilloscope display.

2. The voltage should be set to 8.2 volts and the muscle length should be 75 mm.

3. Adjust the stimulus rate to 30 stimuli/sec.

4. The following steps constitute a single "run." Become familiar with the procedure for completing a run before continuing.

• Click **Multiple Stimulus.**

• When the tracing is close to the right side of the screen, click **Stop Stimulus** to turn off the stimulator.

• Click **Record Data** to retain your data in the grid in the bottom of the screen and the computer's memory.

   If you decide to redo a single run, choose the data line in the grid and click **Delete Line** to erase that single line of data. If you want to repeat the entire experiment, click the **Clear Table** button to erase all data recorded thus far.

Describe the appearance of the tracing.

_____

_____

5. Repeat step 4, increasing the stimulation rate by 10 stimuli/sec each time up to 150 stimuli/sec.

How do the tracings change as the stimulus rate is increased?

_____

6. When you have finished observing the twitch tracings, click the **Tools** menu and then choose **Plot Data.** Notice that the oscilloscope tracings will be automatically erased from the oscilloscope display, but your data will be retained in the grid.

7. Set the Y-axis slider to display Active Force and the X-axis slider to display Stimuli/sec. Click **Draw It** to display a graph of your collected data.

From your graph, estimate the stimulus rate above which there appears to be no significant increase in force.

_____ stimuli/sec

This rate is the fusion frequency, also called tetanus.

8. When finished, click the close box at the top of the plot window.

9. Reset the stimulus rate to the fusion frequency.

10. Try to produce a smooth contraction at Force = 2 gms and Force = 3 gms by adjusting only the stimulus intensity, or voltage, using the following procedure.

• Decrease the voltage to a starting point of 1.0 volt and click **Multiple Stimulus.**

• Click **Stop Stimulus** to turn off the stimulator when the tracing is near the right side of the oscilloscope display.

• If the force produced is not smooth and continuous at the desired level of force, increase the voltage in 0.1-volt increments and stimulate as above until you achieve a smooth force at 2 gms and again at 3 gms.

What stimulus intensity produced smooth force at Force = 2 gms?

_____ V

Which intensity produced smooth contraction at Force = 3 gms?

_____ V

Explain what must be happening in the muscle to achieve smooth contraction at different force levels.

_____

_____

_____

_____

_____ ∎

Activity:
Investigating Muscle Fatigue

A prolonged period of sustained contraction will result in muscle fatigue, a condition in which the tissue has lost its ability to contract. Fatigue results when a muscle cell's ATP consumption is faster than its production. Consequently, increasingly fewer ATP molecules are available for the contractile parts within the muscle cell.

   If you are using Windows 3.1 or an older Macintosh and the tracings move too slowly across the screen, click and drag the **200** button at the lower right corner of the oscilloscope to the 40 msec mark to speed up the display time.

1. Click **Clear Tracings** to erase the oscilloscope display.

2. The voltage should be set to 8.2 volts and the muscle length should be 75 mm.

3. Adjust the stimulus rate to 120 stimuli/sec.

4. Click **Multiple Stimulus** and allow the tracing to sweep through three screens and then click **Stop Stimulus** to stop the stimulator.

Why does the force begin to fall with time? Note that a fall in force indicates muscle fatigue.

_____

_____

5. Click **Clear Tracings** to erase the oscilloscope display. Keep the same settings as before.

6. You will be clicking **Multiple Stimulus** on and off three times to demonstrate fatigue with recovery. Read the steps below before proceeding.

- Click **Multiple Stimulus.**

- When the tracing reaches the middle of the screen, briefly turn off the stimulator by clicking **Stop Stimulus,** then immediately click **Multiple Stimulus** again.

- You will see a dip in the force tracing where you turned the stimulator off and then on again. The force tracing will continue to drop as the muscle fatigues.

- Before the muscle fatigues completely, repeat the on/off cycle twice more without clearing the screen.

   Turning the stimulator off allows a small measure of recovery. The muscle will produce force for a longer period if the stimulator is briefly turned off than if the stimulations were allowed to continue without interruption. Explain why.

_____

_____

_____

7. To see the difference between continuous multiple stimulation and multiple stimulation with recovery, click **Multiple Stimulus** and let the tracing fall without interruption to zero force. This tracing will follow the original myogram exactly until the first "dip" is encountered, after which you will notice a difference in the amount of force produced between the two runs.

Describe the difference between the current tracing and the myogram generated in step 6.

_____

_____

_____

_____

_____

_____ ■

Isometric Contraction

Isometric contraction is the condition in which muscle length does not change regardless of the amount of force generated by the muscle (_iso_ = same, _metric_ = length). This is accomplished experimentally by keeping both ends of the muscle in a fixed position while stimulating it electrically. Resting length (length of the muscle before contraction) is an important factor in determining the amount of force that a muscle can develop. Passive force is generated by stretching the muscle, and is due to the elastic properties of the tissue itself. Active force is generated by the physiological contraction of the muscle. Think of the muscle as having two force properties: it exerts passive force when it is stretched (like a rubber band exerts passive force), and active force when it contracts. Total force is the sum of passive and active forces, and it is what we experimentally measure.

   This simulation allows you to set the resting length of the experimental muscle and stimulate it with individual maximal stimulus shocks. A graph relating the forces generated to the length of the muscle will be automatically plotted as you stimulate the muscle. The results of this simulation can then be applied to human muscles in order to understand how optimum resting length will result in maximum force production. In order to understand why muscle tissue behaves as it does, it is necessary to comprehend contraction at the cellular level. Hint: If you have difficulty understanding the results of this exercise, review the sliding filament model of muscle contraction. Then think in terms of sarcomeres that are too short, too long, or just the right length.

   Choose **Isometric Contraction** from the **Experiment** menu. The opening screen will appear in a few seconds (Figure 16B.3). Notice that the oscilloscope is now divided into two parts. The left side of the scope displays the muscle twitch tracing. The active, passive, and total force data points are plotted on the right side of the screen.

A c t i v i t y :
Investigating Isometric Contraction

1. The voltage should be set to the maximal stimulus (8.2 volts) and the muscle length should be 75 mm.

2. To see how the equipment works, stimulate once by clicking **Stimulate.** You should see a single muscle twitch tracing on the left oscilloscope display, and three data points representing active, passive, and total force on the right display. The yellow box represents the total force and the red dot it contains symbolizes the superimposed active force. The green square represents the passive force data point.

3. Try adjusting the muscle length by clicking the (+) or (−) buttons located next to the Muscle Length window and watch the effect on the muscle.

4. When you feel comfortable with the equipment, click **Clear Tracings** and **Clear Plot.**

5. Now stimulate at different muscle lengths using the following procedure.

- Shorten the muscle to a length of 50 mm by clicking the (−) button next to the Muscle Length window.

Figure 16B.3   Opening screen of the Isometric Contraction experiment.

• Click **Stimulate** and when the tracing is complete, click **Record Data.**

• Repeat the **Stimulate** and **Record Data** sequence, increasing the muscle length by 2 mm each time until you reach the maximum muscle length of 100 mm.

6.   Carefully examine the active, passive, and total force plots in the right oscilloscope display.

What happens to the passive and active forces as the muscle length is increased from 50 mm to 100 mm?

Passive force:

_____

_____

Active force:

_____

_____

Total force:

_____

_____

Explain the dip in the total force curve. (Hint: keep in mind you are measuring the sum of active and passive forces.)

_____

_____

_____

_____

_____ ■

Isotonic Contraction

During isotonic contraction, muscle length changes, but the force produced stays the same (*iso* = same, *tonic* = force). Unlike the isometric exercise in which both ends of the muscle are held in a fixed position, one end of the muscle remains free in the isotonic contraction exercise. Different weights can then be attached to the free end while the other end is fixed in position on the force transducer. If the weight is not too great, the muscle will be able to lift it with a certain velocity. You can think of lifting an object from the floor as an example: if the object is light it can be lifted quickly (high velocity), whereas a heavier weight will be lifted with a slower velocity. Try to transfer the idea of what is happening in the simulation to the muscles of your arm when you lift a weight. The two important variables in this exercise are starting length of the muscle and resistance (weight) applied. You

have already examined the effect of starting length on muscle force production in the previous exercise. Now you will change both muscle length and resistance to investigate how such changes affect the speed of skeletal muscle shortening. Both variables can be independently altered and the results are graphically presented on the screen.

Choose **Isotonic Contraction** from the **Experiment** menu. The opening screen will appear in a few seconds (Figure 16B.4). The general operation of the equipment is the same as in the previous experiments. In this simulation, the weight cabinet doors are open. You will attach weights to the lower tendon of the muscle by clicking and holding the mouse on any weight in the cabinet and then dragging-and-dropping the weight's hook onto the lower tendon. The length window displays the muscle length achieved when the muscle is stretched by hanging a weight from it's lower tendon. You can click the (+) and (−) buttons next to the Platform Height window to change the position of the platform

Figure 16B.4 **Opening screen of the Isotonic Contraction experiment.**

on which the weight rests. Click on the weight again to automatically return it to the weight cabinet. The electrical stimulator displays the initial velocity of muscle shortening in the small window to the right of the Voltage control.

If you are using Windows 3.1 or an older Macintosh and the tracings move too slowly across the screen, click and hold the **200** button at the lower right corner of the oscilloscope and drag it to the left to the 40 msec mark and release it. This action resets the total sweep time to 1000 msec to speed up the display time.

A c t i v i t y :
Investigating the Effect of Load on Skeletal Muscle

1. Set the voltage to the maximal stimulus (8.2 volts).

2. Drag-and-drop the .5-g weight onto the muscle's lower tendon.

3. Platform height should be 75 mm.

4. Click **Stimulate** and simultaneously watch the muscle action and the oscilloscope tracing.

5. Click the **Record Data** button to retain and display the data in the grid.

What do you see happening to the muscle during the flat part of the tracing? Click **Stimulate** to repeat if you wish to see the muscle action again.

_____

_____

Does the force the muscle produces change during the flat part of the tracing (increases, decreases, or stays the same)?

_____

Describe the muscle activity during the flat part of the tracing in terms of isotonic contraction and relaxation.

_____

_____

_____

_____

Circle the correct terms in parentheses in the following sequence: The force rises during the first part of the muscle tracing due to (isometric, isotonic) contraction while the fall in force on the right side of the tracing corresponds to (isometric, isotonic) relaxation.

6. Return the .5-g weight to the cabinet. Drag the 1.5-g weight to the muscle. Click **Stimulate** and then click **Record Data.**

Which of the two weights used so far results in the highest initial velocity of shortening?

_____

7. Repeat step 6 for the remaining two weights.

8. Choose **Plot Data** from the **Tools** menu.

9. Set Weight as the X-axis and Force Produced as the Y-axis by dragging the slider bars. Click **Draw It.**

What does the plot reveal about the resistance and the initial velocity of shortening?

_____

_____

_____

10. Close the plot window and then click **Clear Table** in the data control unit at the bottom of the screen. Click **Yes** when you are asked if you want to erase all data in the table.

11. Return the current weight to the weight cabinet.

12. Attach the 1.5-g weight to the muscle and run through the range of starting lengths from 60–90 mm in 5-mm increments. Be sure to click **Record Data** after each stimulus.

13. After all runs have been completed, choose **Plot Data** from the **Tools** menu.

14. Set the Length as the X-axis and the Velocity as the Y-axis by dragging the slider bars.

Describe the relationship between starting length and initial velocity of shortening.

_____

_____

_____

_____

Do these results support your ideas in the isometric contraction part of this exercise?

_____

15. Close the plot window.

Can you set up a contraction that is completely isometric?

_____ One that is completely isotonic? _____

Explain your answers.

_____

_____

_____

_____ ■

Activity:
Using Your Data

Save Your Data

When you have finished with all experiments, you can save your data set for later review and additional analysis.

1.   Select **Save Data to File** from the **File** menu.

2.   When the dialog box opens, click a drive and/or directory location.

3.   Type a file name into the File name box (your instructor may provide you with a name).

4.   Click **Save** to complete the save. If you want to review previously saved data, select **Open Data File** from the **File** menu, and choose your file.

Print Your Data

If your instructor requires a hard copy of your experiment results, choose **Print Data** from the **File** menu.

1.   Type your name into the provided space.

2.   When you are ready to print, click **OK.**   ■

Histology of Nervous Tissue

The nervous system is the master integrating and coordinating system, continuously monitoring and processing sensory information both from the external environment and from within the body. Every thought, action, and sensation is a reflection of its activity. Like a computer, it processes and integrates new "inputs" with information previously fed into it ("programmed") to produce an appropriate response ("readout"). However, no computer can possibly compare in complexity and scope to the human nervous system.

Despite its complexity, nervous tissue is made up of just two principal cell populations: **neurons** and their **supporting cells.** The supporting cells in the CNS (central nervous system: brain and spinal cord) are usually referred to as **neuroglia** or **glial cells**. The *neuroglia,* literally "nerve glue," include *astrocytes, oligodendrocytes, microglia,* and *ependymal cells* (Figure 17.1). The most important supporting cells in the PNS (peripheral nervous system), that is, in the neural structures outside the CNS, are *Schwann cells* and *satellite cells.* Supporting cells serve the needs of the neurons by acting as phagocytes and by bracing, protecting, and myelinating the delicate neurons. In addition, they probably serve some nutritive function by acting as a selective barrier between the capillary blood supply and the neurons. Although neuroglia resemble neurons in some ways (they have fibrous cellular extensions), they are not capable of generating and transmitting nerve impulses, a capability that is highly developed in neurons. Our focus in this exercise is the highly irritable neurons.

Neuron Anatomy

Neurons are the structural units of nervous tissue. They are highly specialized to transmit messages (nerve impulses) from one part of the body to another. Although neurons differ structurally, they have many identifiable features in common (Figure 17.2a). All have a **cell body** from which slender processes or fibers extend. Although neuron cell bodies are typically found in the CNS in clusters called **nuclei,** occasionally they reside in **ganglia** (collections of neuron cell bodies outside the CNS). They make up the gray matter of the nervous system. Neuron processes running through the CNS form **tracts** of white matter; in the PNS they form the peripheral **nerves.**

Objectives

1. To differentiate between the functions of neurons and neuroglia.
2. To list four types of neuroglia cells.
3. To identify the important anatomical characteristics of a neuron on an appropriate diagram or projected slide.
4. To state the functions of axons, dendrites, axonal terminals, neurofibrils, and myelin sheaths.
5. To explain how a nerve impulse is transmitted from one neuron to another.
6. To explain the role of Schwann cells in the formation of the myelin sheath.
7. To classify neurons according to structure and function.
8. To distinguish between a nerve and a tract and between a ganglion and a nucleus.
9. To describe the structure of a nerve, identifying the connective tissue coverings (endoneurium, perineurium, and epineurium) and citing their functions.

Materials

❑ Model of a "typical" neuron (if available)
❑ Compound microscope
❑ Histologic slides of an ox spinal cord smear and teased myelinated nerve fibers
❑ Prepared slides of Purkinje cells (cerebellum), pyramidal cells (cerebrum), and a dorsal root ganglion
❑ Prepared slide of a nerve (cross section)

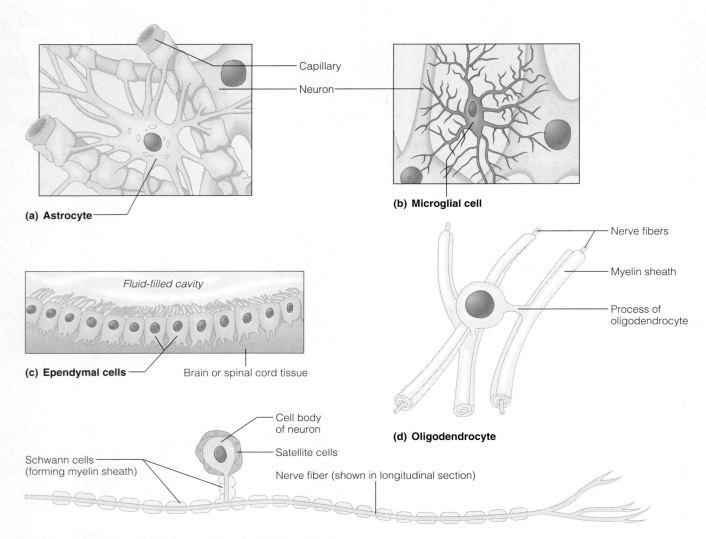

(a) **Astrocyte**

Capillary
Neuron

(b) **Microglial cell**

Nerve fibers

Myelin sheath

Process of
oligodendrocyte

(d) **Oligodendrocyte**

Fluid-filled cavity

(c) **Ependymal cells** — Brain or spinal cord tissue

Cell body
of neuron

Satellite cells

Schwann cells
(forming myelin sheath)

Nerve fiber (shown in longitudinal section)

(e) **Sensory neuron with Schwann cells and satellite cells**

Figure 17.1 Supporting cells of nervous tissue. (a) Astrocyte.
(b) Microglial cell. **(c)** Ependymal cells. **(d)** Oligodendrocyte. **(e)** Neuron with
Schwann cells and satellite cells.

The neuron cell body contains a large round nucleus surrounded by cytoplasm (*neuroplasm*). The cytoplasm is riddled with neurofibrils and with darkly staining structures called Nissl bodies. **Neurofibrils,** the cytoskeletal elements of the neuron, have a support and intracellular transport function. **Nissl bodies,** an elaborate type of rough endoplasmic reticulum, are involved in the metabolic activities of the cell.

According to the older, traditional scheme, neuron processes that conduct electrical currents *toward* the cell body are called **dendrites;** and those that carry impulses away *from* the nerve cell body are called **axons.** When it was discovered that this functional scheme had pitfalls (some axons carry impulses *both* toward and away from the cell body), a newer functional definition of neuron processes was

adopted. According to this scheme, dendrites are *receptive regions* (they bear receptors for neurotransmitters released by other neurons), whereas axons are *nerve impulse generators* and *transmitters*. Neurons have only one axon (which may branch into **collaterals**) but may have many dendrites, depending on the neuron type. Notice that the term *nerve fiber* is a synonym for axon and is, thus, quite specific.

In general, a neuron is excited by other neurons when their axons release neurotransmitters close to its dendrites or cell body. The electrical current produced travels across the cell body and (given a threshold stimulus) down the axon. As Figure 17.2a shows, the axon (in motor neurons) begins at a slightly enlarged cell body structure called the **axon hillock** and ends in many small structures called **axonal terminals,**

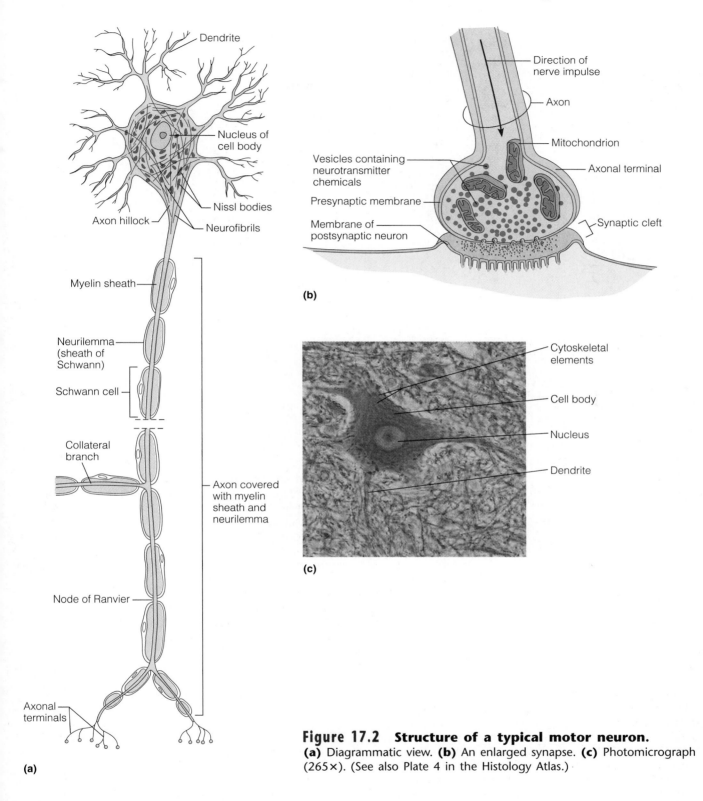

(a)

(b)

(c)

Figure 17.2 **Structure of a typical motor neuron.**
(a) Diagrammatic view. **(b)** An enlarged synapse. **(c)** Photomicrograph
(265×). (See also Plate 4 in the Histology Atlas.)

or synaptic knobs. These terminals store the neurotransmitter chemical in tiny vesicles. Each axonal terminal is separated from the cell body or dendrites of the next (postsynaptic) neuron by a tiny gap called the **synaptic cleft** (Figure 17.2b). Thus, although they are close, there is no actual physical contact between neurons. When an impulse reaches the axonal terminals, some of the synaptic vesicles rupture and release

neurotransmitter into the synaptic cleft. The neurotransmitter then diffuses across the synaptic cleft to bind to membrane receptors on the next neuron, initiating the action potential.*

---

* Specialized synapses in skeletal muscle are called neuromuscular junctions. They are discussed in Exercise 14.

Figure 17.3 Myelination of neuron processes by individual Schwann cells. (a–d). A Schwann cell becomes apposed to an axon and envelops it in a trough. It then begins to rotate around the axon, wrapping it loosely in successive layers of its plasma membrane. Eventually, the Schwann cell cytoplasm is forced from between the membranes and comes to lie peripherally just beneath the exposed portion of the Schwann cell membrane. The tight membrane wrappings surrounding the axon form the myelin sheath. The area of Schwann cell cytoplasm and its exposed membrane are referred to as the neurilemma or sheath of Schwann. **(e)** Longitudinal view of myelinated axon showing portions of adjacent Schwann cells and the node of Ranvier between them. **(f)** Unmyelinated fibers. Schwann cells may associate loosely with several axons, which they partially invest. In such cases, Schwann cell coiling around the axons does not occur.

Most long nerve fibers are covered with a fatty material called *myelin,* and such fibers are referred to as **myelinated fibers.** Axons in the peripheral nervous system are typically heavily myelinated by special supporting cells called **Schwann cells,** which wrap themselves tightly around the axon jelly-roll fashion (Figure 17.3). During the wrapping process, the cytoplasm is squeezed from between adjacent layers of the Schwann cell membranes, so that when the process is completed a tight core of plasma membrane material (protein-lipoid material) encompasses the axon. This wrapping is the **myelin sheath.** The Schwann cell nucleus and the bulk of its cytoplasm ends up just beneath the outer-most portion of its plasma membrane. This peripheral part of the Schwann cell and its plasma membrane is referred to as the **neurilemma.** Since the myelin sheath is formed by many individual Schwann cells, it is a discontinuous sheath. The gaps or indentations in the sheath are called **nodes of Ranvier** (see Figure 17.2).

Within the CNS, myelination is accomplished by glial cells called **oligodendrocytes** (see Figure 17.1d). These CNS sheaths do not exhibit the neurilemma seen in fibers myelinated by Schwann cells. Because of its chemical composition, myelin insulates the fibers and greatly increases the speed of neurotransmission by neuron fibers.

Activity:
Identifying Parts of a Neuron

1.   Study the typical motor neuron shown in Figure 17.2, noting the structural details described above, and then identify these structures on a neuron model.

2.   Obtain a prepared slide of the ox spinal cord smear, which has large, easily identifiable neurons. Study one representative neuron under oil immersion and identify the cell body; the nucleus; the large, prominent "owl's eye" nucleolus; and the granular Nissl bodies. If possible, distinguish the axon from the many dendrites. Sketch the cell in the space provided here, and label the important anatomical details you have observed. Compare your sketch to Plate 4 of the Histology Atlas. Also reexamine Figure 17.2a, which differentiates the neuronal processes more clearly.

3.   Obtain a prepared slide of teased myelinated nerve fibers. Using Figure 17.4 as a guide, identify the following: nodes of Ranvier, neurilemma, axis cylinder (the axon itself), Schwann cell nuclei, and myelin sheath.

Sketch a portion of a myelinated nerve fiber in the space provided here, illustrating two or three nodes of Ranvier. Label the axon, myelin sheath, nodes, and neurilemma.

Fibroblast nucleus

Myelin sheath

Schwann cell nucleus

Nodes of Ranvier

Figure 17.4   Photomicrograph of a small portion of a peripheral nerve in longitudinal section.

Explain the significance of this finding: _____

_____

_____

_____

_____

_____

_____   ■

Do the nodes seem to occur at consistent intervals, or are they irregularly distributed?

_____

Neuron Classification

Neurons may be classified on the basis of structure or of function.

Basis of Structure

Structurally, neurons may be differentiated according to the number of processes attached to the cell body (Figure 17.5a). In **unipolar neurons,** one very short process, which divides

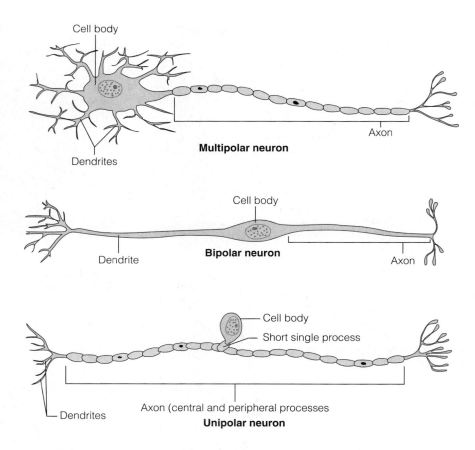

(a) **Classification of neurons based on structure (number of processes extending from the cell body).**

(b) **Structure variations within the classes**

Figure 17.5 Classification of neurons according to structure.

(a) Classification of neurons based on structure (number of processes extending from the cell body). (b) Structural variations within the classes.

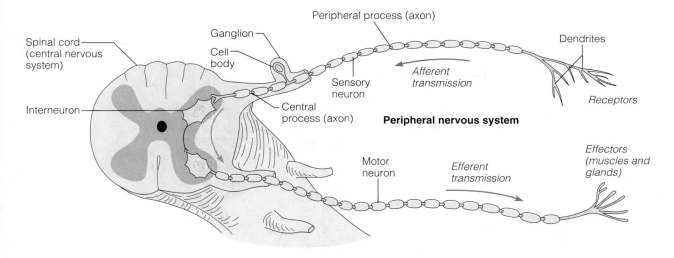

Figure 17.6 Classification of neurons on the basis of function.
Sensory (afferent) neurons conduct impulses from the body's sensory receptors to the central nervous system; most are unipolar neurons with their nerve cell bodies in ganglia in the peripheral nervous system (PNS). Motor (efferent) neurons transmit impulses from the CNS to effectors such as muscles and glands. Association neurons (interneurons) complete the communication line between sensory and motor neurons. They are typically multipolar and their cell bodies reside in the CNS.

into *peripheral* and *central processes,* extends from the cell body. Functionally, only the most distal portions of the peripheral process act as dendrites; the rest acts as an axon along with the central process. Nearly all neurons that conduct impulses toward the CNS are unipolar.

Bipolar neurons have two processes—one axon and one dendrite—attached to the cell body. This neuron type is quite rare, typically found only as part of the receptor apparatus of the eye, ear, and olfactory mucosa.

Many processes issue from the cell body of **multipolar neurons,** all classified as dendrites except for a single axon. Most neurons in the brain and spinal cord (CNS neurons) and those whose axons carry impulses away from the CNS fall into this last category.

Activity:
Studying the Microscopic Structure of Selected Neurons

Obtain prepared slides of Purkinje cells of the cerebellar cortex, pyramidal cells of the cerebral cortex, and a dorsal root ganglion. As you observe them under the microscope, try to pick out the anatomical details depicted in Figure 17.5b, and in Plates 5 and 6 of the Histology Atlas. Notice that the neurons of the cerebral and cerebellar tissues (both brain tissues) are extensively branched; in contrast, the neurons of the dorsal root ganglion are more rounded. You may also be able to identify astrocytes (a type of neuroglia) in the brain tissue slides if you examine them closely, and satellite cells can be seen surrounding the neurons in the dorsal root ganglion.

Which of these neuron types would be classified as multipolar neurons?

_____

Which as unipolar?

_____ ∎

Basis of Function

In general, neurons carrying impulses from the sensory receptors in the internal organs (viscera) or in the skin are termed **sensory,** or **afferent, neurons** (see Figure 17.6). The dendritic endings of sensory neurons are often equipped with specialized receptors that are stimulated by specific changes in their immediate environment. The structure and function of these receptors is considered separately in Exercise 23 (General Sensation). The cell bodies of sensory neurons are always found in a ganglion outside the CNS, and these neurons are typically unipolar.

Neurons carrying activating impulses from the CNS to the viscera and/or body muscles and glands are termed **motor,** or **efferent, neurons.** Motor neurons are most often multipolar and their cell bodies are almost always located in the CNS.

The third functional category of neurons is the **association neurons,** or **interneurons,** which are situated between and contribute to pathways that connect sensory and motor neurons. Their cell bodies are always located within the CNS and they are multipolar neurons structurally.

(a)

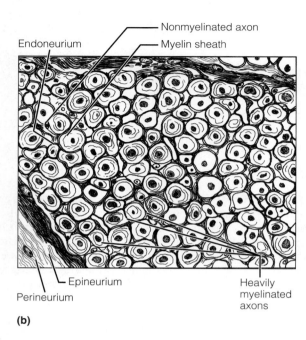

(b)

Figure 17.7 **Structure of a nerve showing connective tissue wrappings.** **(a)** Three-dimensional view of a portion of a nerve. **(b)** Cross-sectional view. (See also Plate 9 in the Histology Atlas.)

Structure of a Nerve

A nerve is a bundle of neuron fibers or processes wrapped in connective tissue coverings that extends to and/or from the CNS and visceral organs or structures of the body periphery (such as skeletal muscles, glands, and skin).

Within a nerve, each fiber is surrounded by a delicate connective tissue sheath called an **endoneurium,** which insulates it from the other neuron processes adjacent to it. The endoneurium is often mistaken for the myelin sheath; it is instead an additional sheath that surrounds the myelin sheath. Groups of fibers are bound by a coarser connective tissue, called the **perineurium,** to form bundles of fibers called **fascicles.** Finally, all the fascicles are bound together by a tough, white, fibrous connective tissue sheath called the **epineurium,** forming the cordlike nerve (Figure 17.7). In addition to the connective tissue wrappings, blood vessels and lymphatic vessels serving the fibers also travel within a nerve.

Like neurons, nerves are classified according to the direction in which they transmit impulses. Nerves carrying both sensory (afferent) and motor (efferent) fibers are called **mixed nerves;** all spinal nerves are mixed nerves. Nerves that carry only sensory processes and conduct impulses only toward the CNS are referred to as **sensory,** or **afferent, nerves.** A few of the cranial nerves are pure sensory nerves, but the majority are mixed nerves. The ventral roots of the spinal cord, which carry only motor fibers, can be considered **motor,** or **efferent, nerves.**

Activity:
Examining the Microscopic Structure of a Nerve

Examine under the compound microscope a prepared cross section of a peripheral nerve. Identify nerve fibers, myelin sheaths, fascicles, and endoneurium, perineurium, and epineurium sheaths. If desired, sketch the nerve in the space below. ■

Neurophysiology of Nerve Impulses

Objectives

1. To list the two major physiological properties of neurons.

2. To describe the polarized and depolarized states of the nerve cell membrane and to describe the events that lead to generation and conduction of a nerve impulse.

3. To explain how a nerve impulse is transmitted from one neuron to another.

4. To define *action potential, depolarization, repolarization, relative refractory period,* and *absolute refractory period.*

5. To list various substances and factors that can stimulate neurons.

6. To recognize that neurotransmitters may be either stimulatory or inhibitory in nature.

7. To state the site of action of the blocking agents ether and curare.

Materials

❑ *Rana pipiens*[*]
❑ Dissecting instruments and tray
❑ Ringer's solution (frog) in dropper bottles, some at room temperature and some in an ice bath
❑ Thread
❑ Glass rods or probes
❑ Glass plates or slides
❑ Ring stand and clamp
❑ Stimulator; platinum electrodes
❑ Oscilloscope
❑ Nerve chamber
❑ Filter paper
❑ 0.01% hydrochloric acid (HCl) solution
❑ Sodium chloride (NaCl) crystals
❑ Heat-resistant mitts
❑ Bunsen burner
❑ Absorbent cotton
❑ Ether
❑ Pipettes
❑ 1-cc syringe with small-gauge needle

❑ 0.5% tubocurarine solution
❑ Frog board
❑ Disposable plastic gloves
❑ Safety goggles

*Instructor to provide freshly pithed frogs (*Rana pipiens*) for student experimentation.

The Nerve Impulse

Neurons have two major physiologic properties: **irritability,** or the ability to respond to stimuli and convert them into nerve impulses, and **conductivity,** the ability to transmit the impulse to other neurons, muscles, or glands. In a resting neuron (as in resting muscle cells), the exterior surface of the membrane is slightly more positively charged than the inner surface, as shown in Figure 18.1a. This difference in electrical charge on the two sides of the membrane results in a voltage across the plasma membrane referred to as the **resting membrane potential,** and a neuron in this state is said to be **polarized.** In the resting state, the predominant intracellular ion is potassium (K^+), and sodium ions (Na^+) are found in greater concentration in the extracellular fluids. The resting potential is maintained by a very active sodium-potassium pump, which transports Na^+ out of the cell and K^+ into the cell.

When the neuron is activated by a stimulus of adequate intensity—a **threshold stimulus**—the membrane at its *trigger zone,* typically the axon hillock (or the most peripheral part of a sensory neuron's axon), briefly becomes more permeable to sodium (sodium gates are opened). Sodium ions rush into the cell, increasing the number of positive ions inside the cell and reversing the polarity (Figure 18.1b). Thus the interior of the membrane becomes less negative at that point and the exterior surface becomes less positive—a phenomenon called **depolarization.** When depolarization reaches a certain point such that the local membrane polarity changes (momentarily the external face becomes negative and the internal face becomes positive), it initiates an **action potential**[†] (Figure 18.1c).

[†] If the stimulus is of less than threshold intensity, depolarization is limited to a small area of the membrane, and no action potential is generated.

Figure 18.1 **The nerve impulse.** **(a)** Resting membrane potential (~85 mV). There is an excess of positive ions outside the cell, with Na$^+$ the predominant extracellular fluid ion and K$^+$ the predominant intracellular ion. The plasma membrane has a low permeability to Na$^+$. **(b)** Depolarization—reversal of the resting potential. Application of a stimulus changes the membrane permeability, and Na$^+$ ions are allowed to diffuse rapidly into the cell. **(c)** Generation of the action potential or nerve impulse. If the stimulus is of adequate intensity, the depolarization wave spreads rapidly along the entire length of the membrane. **(d)** Repolarization—reestablishment of the resting potential. The negative charge on the internal plasma membrane surface and the positive charge on its external surface are reestablished by diffusion of K$^+$ ions out of the cell, proceeding in the same direction as in depolarization. **(e)** The original ionic concentrations of the resting state are restored by the sodium-potassium pump. **(f)** A tracing of an action potential.

Within a millisecond after the inward influx of sodium, the membrane permeability is again altered. As a result, Na$^+$ permeability decreases, K$^+$ permeability increases, and K$^+$ rushes out of the cell. Since K$^+$ ions are positively charged, their movement out of the cell reverses the membrane poten- tial again, so that the external membrane surface is again pos- itive relative to the internal membrane face (Figure 18.1d). This event, called **repolarization,** reestablishes the resting membrane potential. When the sodium gates are open, the neuron is totally insensitive to additional stimuli and is said to be in an **absolute refractory period.** During the time of repolarization, the neuron is nearly insensitive to further stimulation. A very strong stimulus may reactivate it, however, thus this period is referred to as the **relative refractory period.**

Once generated, the action potential is a self-propagating phenomenon that spreads rapidly along the entire length of the neuron. It is never partially transmitted; that is, it is an all-or-none response. This propagation of the action potential in neurons is also called the **nerve impulse.** When the nerve impulse reaches the axonal terminals, they release a neurotransmitter that acts either to stimulate or to inhibit the next neuron in the transmission chain. (Note that only stimulatory transmitters are considered here.)

Because only minute amounts of sodium and potassium ions change places, once repolarization has been completed, the neuron can quickly respond again to a stimulus. In fact, thousands of impulses can be generated before ionic imbalances prevent the neuron from transmitting impulses. Eventually, however, it is necessary to restore the original ionic concentrations on the two sides of the membrane; this is accomplished by enhanced activity of the Na^+-K^+ pump (Figure 18.1e).

Physiology of Nerve Fibers

In this laboratory session, you will investigate the functioning of nerve fibers by subjecting the sciatic nerve of a frog to various types of stimuli and blocking agents. Work in groups of two to four to lighten the work load.

Dissection:
Isolating the Gastrocnemius Muscle and Sciatic Nerve

⚠️ 1. Don gloves to protect yourself from any parasites the frogs might have. Request and obtain a pithed frog from your instructor and bring it to your laboratory bench. Also obtain dissecting instruments and a tray from the supply area.

2. Prepare the sciatic nerve as illustrated in Figure 18.2. Place the pithed frog on the dissecting tray, dorsal side down. Make a cut through the skin around the circumference of the frog approximately halfway down the trunk, and then pull the skin down over the muscles of the legs. Open the abdominal cavity and push the abdominal organs to one side to expose the origin of the glistening white sciatic nerve, which arises from the last three spinal nerves. *Once the sciatic nerve has been exposed, it should be kept continually moist with room temperature Ringer's solution.*

3. Using a glass probe, slip a piece of thread moistened with Ringer's solution under the sciatic nerve close to its origin at the vertebral column. Make a single ligature (tie it

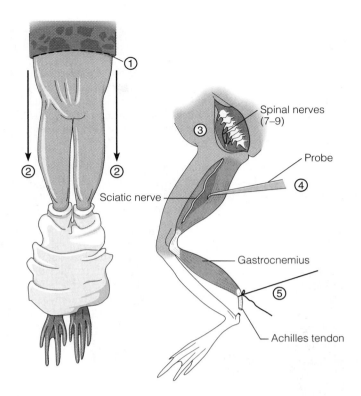

Figure 18.2 Removal of the sciatic nerve and gastrocnemius muscle. (1) Cut through the frog's skin around the circumference of the trunk. **(2)** Pull the skin down over the trunk and legs. **(3)** Make a longitudinal cut through the abdominal musculature and expose the roots of the sciatic nerve (arising from spinal nerves 7–9). Ligate the nerve and cut the roots proximal to the ligature. **(4)** Use a glass probe to expose the sciatic nerve beneath the posterior thigh muscles. **(5)** Ligate the Achilles tendon and cut it free distal to the ligature. Release the gastrocnemius muscle from the connective tissue of the knee region.

firmly with the thread), and then cut through the nerve roots to free the proximal end of the sciatic nerve from its attachments. Using a glass rod or probe, carefully separate the posterior thigh muscles to locate and then free the sciatic nerve, which runs down the posterior aspect of the thigh.

4. Tie a piece of thread around the Achilles tendon of the gastrocnemius muscle, and then cut through the tendon distal to the ligature to free the gastrocnemius muscle from the heel. Using a scalpel, carefully release the gastrocnemius muscle from the connective tissue in the knee region. At this point you should have completely freed both the gastrocnemius muscle and the sciatic nerve, which innervates it. ■

Activity:
Eliciting a Nerve Impulse

In this first set of experiments, stimulation of the nerve and generation of the action potential will be indicated by contraction of the gastrocnemius muscle. Because you will make no mechanical recording (unless your instructor asks you to), you must keep complete and accurate records of all experimental procedures and results.

1. With glass rods, transfer the isolated muscle-nerve preparation to a glass plate or slide, and then attach the slide to a ring stand with a clamp. Allow the end of the sciatic nerve to hang over the free edge of the glass slide, so that it is easily accessible for stimulation. Remember to keep the nerve moist at all times.

2. You are now ready to investigate the response of the sciatic nerve to various stimuli, beginning with electrical stimulation. Using the stimulator and platinum electrodes, stimulate the sciatic nerve with single shocks, gradually increasing the intensity of the stimulus until the threshold stimulus is determined.

(The muscle as a whole will just barely contract at the threshold stimulus.) Record the voltage of this stimulus:

Threshold stimulus: _____ V

Continue to increase the voltage until you find the point beyond which no further increase occurs in the strength of muscle contraction—that is, the point at which the maximal contraction of the muscle is obtained. Record this voltage below.

Maximal stimulus: _____ V

Delivering multiple or repeated shocks to the sciatic nerve causes volleys of impulses in the nerve. Shock the nerve with multiple stimuli. Observe the response of the muscle. How does this response compare with the response to the single electrical shocks?

_____

_____

_____

3. To investigate mechanical stimulation, pinch the free end of the nerve by firmly pressing it between two glass rods or by pinching it with forceps. What is the result?

_____

4. Chemical stimulation can be tested by applying a small piece of filter paper saturated with 0.01% hydrochloric acid (HCl) solution to the free end of the nerve. What is the result?

_____

Drop a few grains of salt (NaCl) on the free end of the nerve. What is the result?

_____

5. Now test thermal stimulation. Wearing the heat-resistant mitts, heat a glass rod for a few moments over a Bunsen burner. Then touch the rod to the free end of the nerve. What is the result?

_____

What do these muscle reactions say about the irritability and conductivity of neurons?

_____

_____ ■

Although most neurons within the body are stimulated to the greatest degree by a particular stimulus (in many cases, a chemical neurotransmitter), a variety of other stimuli may trigger nerve impulses, as illustrated by the experimental series just conducted. Generally, no matter what type of stimulus is present, if the affected part responds by becoming activated, it will always react in the same way. Familiar examples are the well-known phenomenon of "seeing stars" when you receive a blow to the head or press on your eyeball (try it), both of which trigger impulses in your optic nerves.

Activity:
Inhibiting the Nerve Impulse

Numerous physical factors and chemical agents can impair the ability of nerve fibers to function. For example, deep pressure and cold temperature both block nerve impulse transmission by preventing the local blood supply from reaching the nerve fibers. Local anesthetics, alcohol, and numerous other chemicals are also very effective at blocking nerve transmission. Ether, one such chemical blocking agent, will be investigated first.

⚠ Since ether is extremely volatile and explosive, be sure that laboratory fans are on and that *all Bunsen burners are off* during this procedure. *Don safety glasses before beginning this procedure.*

1. Clamp a second glass slide to the ring stand slightly below the first slide of the apparatus setup for the previous experiment. With glass rods, gently position the sciatic nerve on this second slide, allowing a small portion of the nerve's distal end to extend over the edge. Place a piece of absorbent cotton soaked with ether under the midsection of the nerve on the slide, prodding it into position with a glass rod. Using a voltage slightly above the threshold stimulus, stimulate the distal end of the nerve at 2-min intervals until the muscle fails to respond. (If the cotton dries before this, rewet it with ether using a pipette.) How long did it take for anesthesia to occur?

_____ sec

2.   Once anesthesia has occurred, stimulate the nerve beyond the anesthetized area, between the ether-soaked pad and the muscle. What is the result?

_____

3.   Remove the ether-soaked pad and flush the nerve fibers with saline. Again stimulate the nerve at its distal end at 2-min intervals. How long does it take for recovery?

_____

Does ether exert its blocking effect on the nerve fibers *or* on the muscle cells?

_____ Explain your reasoning.

_____

_____

If sufficient frogs are available and time allows, you may do the following classic experiment. In the 1800s Claude Bernard described an investigation into the effect of curare on nerve-muscle interaction. *Curare* was used by some South American Indian tribes to tip their arrows. Victims struck with these arrows were paralyzed, but the paralysis was not accompanied by loss of sensation.

1.   Prepare another frog as described in steps 1 through 3 of the dissection instructions. However, in this case position the frog ventral side down on a frog board. In exposing the sciatic nerve, take care not to damage the blood vessels in the thigh region, as the success of the experiment depends on maintaining the blood supply to the muscles of the leg.

2.   Expose and gently tie the left sciatic nerve so that it can be lifted away from the muscles of the leg for stimulation. Slip another length of thread under the nerve, and then tie the thread tightly around the thigh muscles to cut off circulation to the leg. The sciatic nerve should be above the thread and *not in* the ligatured tissue. Expose and ligature the sciatic nerve of the *right* leg in the same manner, but this time do *not* ligate the thigh muscles.

⚠ 3. Using a syringe and needle, slowly and carefully inject 1 cc of 0.5% tubocurarine into the dorsal lymph sac of the frog.* The dorsal lymph sacs are located dorsally at the level of the scapulae, so introduce the needle of the syringe just beneath the skin between the scapulae and toward one side of the spinal column. *Handle the tubocurarine very carefully, because it is extremely poisonous.* Do not get any on your skin.

4.   Wait 15 min after injection of the tubocurarine to allow it to be distributed throughout the body in the blood and lymphatic stream. Then stimulate electrically the left sciatic nerve. Be careful not to touch any of the other tissues with the electrode. Gradually increasing the voltage, deliver single shocks until the threshold stimulus is determined for this specimen.

Threshold stimulus: _____ V

Now stimulate the right sciatic nerve with the same voltage intensity. Is there any difference in the reaction of the two muscles?

_____ If so, explain. _____

_____

_____

If you did not find any difference, wait an additional 10 to 15 min and restimulate both sciatic nerves.

What is the result? _____

_____

5.   To determine the site at which tubocurarine acts, directly stimulate each gastrocnemius muscle. What is the result?

_____

_____

_____

Explain the difference between the responses of the right and left sciatic nerves.

_____

_____

_____

Explain the results when the muscles were stimulated directly.

_____

_____

_____

At what site does tubocurarine (or curare) act?

_____ ■

* To obtain 1 cc of the tubocurarine, inject 1 cc of air into the vial through the rubber membrane, and then draw up 1 cc of the chemical into the syringe.

The Oscilloscope: An Experimental Tool

The *oscilloscope* is an instrument that visually displays the rapid but extremely minute changes in voltage that occur during an action potential. The oscilloscope is similar to a TV set in that the screen display is produced by a stream of electrons generated by an electron gun (cathode) at the rear of a tube. The electrons pass through the tube and between two sets of plates that lie alongside the beam pathway. When the electrons reach the fluorescent screen, they create a tiny glowing spot. The plates determine the placement of the glowing spot by controlling the vertical or horizontal sweep of the beam. Vertical movement represents the voltage of the input signal, and horizontal movement indicates the time base. When there is no electrical output signal to the oscilloscope, the electron beam sweeps horizontally (left to right) across the screen, but when the plates are electrically stimulated, the path of electrons is deflected vertically.

In this exercise, a frog's sciatic nerve will be electrically stimulated, and the action potentials generated will be observed on the oscilloscope. The dissected nerve will be placed in contact with two pairs of electrodes—*stimulating* and *recording*. The stimulating electrodes will be used to deliver a pulse of electricity to a point on the sciatic nerve. At another point on the nerve, a pair of recording electrodes connected to the oscilloscope will deliver the current to the plates inside the tube, and the electrical pulse will be recorded on the screen as a vertical deflection, or a *stimulus artifact* (Figure 18.3a). As the nerve is stimulated with increasingly higher voltage, the stimulus artifact increases in amplitude as well. When the stimulus voltage reaches a high enough level (threshold), an action potential will be generated by the nerve, and a *second* vertical deflection will appear on the screen, approximately 2 milliseconds after the stimulus artifact (Figure 18.3b–d). This second deflection reports the potential difference between the two recording electrodes—that is, between the first recording electrode which has already depolarized (and is in the process of repolarizing as the action potential travels along the nerve) and the second recording electrode.

Activity:
Visualizing the Action Potential with an Oscilloscope

1. Obtain a nerve chamber, an oscilloscope, a stimulator, frog Ringer's solution (room temperature), a dissecting needle, and glass probes. Set up the experimental apparatus as illustrated in Figure 18.4. Connect the two stimulating electrodes to the output terminals of the stimulator and the two recording electrodes to the preamplifier of the oscilloscope.

2. Obtain another pithed frog, and prepare one of its sciatic nerves for experimentation as indicated on p. 185 in steps 1 through 3 in the dissection instructions. While working, be careful not to touch the nerve with your fingers, and do not allow the nerve to touch the frog's skin.

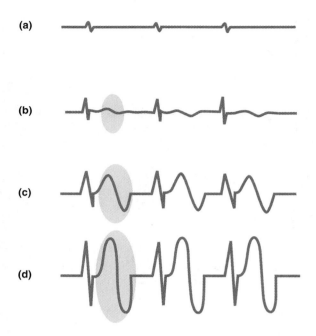

Figure 18.3 Oscilloscope scans of nerve stimulation using stimuli with increasing intensities. The first action potential in each scan is circled. **(a)** Stimulus artifacts only; no action potential produced. Subthreshold stimulation. **(b)** Threshold stimulation. **(c)** Submaximal stimulation. **(d)** Maximal stimulus.

3. When you have freed the sciatic nerve to the knee region with the glass probe, slip another thread length beneath that end of the nerve and make a ligature. Cut the nerve distal to this tied thread and then carefully lift the cut nerve away from the thigh of the frog by holding the threads at the nerve's proximal and distal ends. Place the nerve in the nerve chamber so that it rests across all four electrodes (the two stimulating and two recording electrodes) as shown in Figure 18.4. Flush the nerve with room temperature frog Ringer's solution.

4. Adjust the horizontal sweep according to the instructions given in the manual or by your instructor, and set the stimulator duration, frequency, and amplitude to their lowest settings.

5. Begin to stimulate the nerve with single stimuli, slowly increasing the voltage until a threshold stimulus is achieved. The action potential will appear as a small rounded "hump" immediately following the stimulus artifact. Record the voltage of the threshold stimulus:

Threshold stimulus: _____ V

6. Flush the nerve with the Ringer's solution and continue to increase the voltage, watching as the vertical deflections produced by the action potentials become diphasic (show

Stimulator

Oscilloscope

Nerve chamber

Output terminals

Input terminals

Stimulating electrodes

Recording electrodes

Nerve to be positioned across all four electrodes

Figure 18.4 **Setup for oscilloscope visualization of action potentials in a nerve.**

both upward and downward vertical deflections). Record the voltage at which the action potential reaches its maximal amplitude; this is the maximal stimulus:

Maximal stimulus: _____ V

7. Set the stimulus voltage at a level just slightly lower than the maximal stimulus and gradually increase the frequency of stimulation. What is the effect on the size (amplitude) of the action potential?

_____

8. Flush the nerve with saline once again, and allow it to sit for a few minutes while you obtain a bottle of Ringer's solution from the ice bath. Repeat steps 5 and 6 while your partner continues to flush the nerve preparation with the cold saline. Record the threshold and maximal stimuli and watch the oscilloscope pattern carefully to detect any differences in

the velocity or speed of conduction from what was seen previously.

Threshold stimulus: _____ V

Maximal stimulus: _____ V

9. Flush the nerve preparation with room temperature Ringer's solution again and then gently lift the nerve by its attached threads and turn it around so that the end formerly resting on the stimulating electrodes now rests on the recording electrodes and vice versa. Stimulate the nerve. Is the impulse conducted in the opposite direction?

_____

10. Dispose of the frog remains and gloves in the appropriate containers, and return your equipment to the proper supply area. ∎

Gross Anatomy of the Brain and Cranial Nerves

Objectives

1. To identify the following brain structures on a dis-sected specimen, human brain model (or slices), or appropriate diagram, and to state their functions:

   - *Cerebral hemisphere structures:* lobes, impor-tant fissures, lateral ventricles, basal nuclei, corpus callosum, fornix, septum pellucidum

   - *Diencephalon structures:* thalamus, intermedi-ate mass, hypothalamus, optic chiasma, pitu-itary gland, mammillary bodies, pineal body, choroid plexus of the third ventricle, interven-tricular foramen

   - *Brain stem structures:* corpora quadrigemina, cerebral aqueduct, cerebral peduncles of the midbrain, pons, medulla, fourth ventricle

   - *Cerebellum structures:* cerebellar hemispheres, vermis, arbor vitae

2. To describe the composition of gray and white matter.

3. To locate the well-recognized functional areas of the human cerebral hemispheres.

4. To define *gyri* and *fissures* (*sulci*).

5. To identify the three meningeal layers and state their function, and to locate the falx cerebri, falx cerebelli, and tentorium cerebelli.

6. To state the function of the arachnoid villi and dural sinuses.

7. To discuss the formation, circulation, and drainage of cerebrospinal fluid.

8. To identify at least four pertinent anatomical dif-ferences between the human brain and that of the sheep (or other mammal).

9. To identify the cranial nerves by number and name on an appropriate model or diagram, stating the origin and function of each.

Materials

- ☐ Human brain model (dissectible)
- ☐ 3-D model of ventricles
- ☐ Preserved human brain (if available)
- ☐ Coronally sectioned human brain slice (if avail-able)
- ☐ Preserved sheep brain (meninges and cran nerves intact)
- ☐ Dissecting tray and instruments
- ☐ Protective skin cream or disposable gloves
- ☐ Materials as needed for cranial nerve testing (s Table 19.1)
- ☐ *The Human Nervous System: The Brain and Cran Nerves* videotape*

 See Appendix C, Exercise 19 for links to A.D.A. Interactive Anatomy.

*Available to qualified adopters from Benjamin/Cummings.

When viewed alongside all nature's animals, huma are indeed unique, and the key to their uniqueness found in the brain. Only in humans has the brain gion called the cerebrum become so elaborated a grown out of proportion that it overshadows other brain eas. Other animals are primarily concerned with inform tional input and response for the sake of survival and prese vation of the species, but human beings devote considerab time to nonsurvival ends. They are the only animals who n nipulate abstract ideas and search for knowledge for its ov sake, who are capable of emotional response and artistic cr ativity, or who can anticipate the future and guide their liv according to ethical and moral values. For all this, huma can thank their overgrown cerebrum (cerebral hemispheres

Each of us is a composite reflection of our brain's exp rience. If all past sensory input could mysteriously and su denly be "erased," we would be unable to walk, talk, or co municate in any manner. Spontaneous movement wou occur, as in a fetus, but no voluntary integrated function any type would be possible. Clearly we would cease to be t same individuals.

Because of the complexity of the nervous system, anatomical structures are usually considered in terms of tv principal divisions: the **central nervous system (CNS)** a the **peripheral nervous system (PNS).** The central nervo system consists of the brain and spinal cord, which primari interpret incoming sensory information and issue instructio based on past experience. The peripheral nervous syste consists of the cranial and spinal nerves, ganglia, and senso receptors. These structures serve as communication lines they carry impulses—from the sensory receptors to the CN and from the CNS to the appropriate glands or muscles.

In this exercise both CNS (brain) and PNS (crani nerves) structures will be studied because of their clo anatomical relationship.

The Human Brain

During embryonic development of all vertebrates, the CNS first makes its appearance as a simple tubelike structure, the **neural tube,** that extends down the dorsal median plane. By the fourth week, the human brain begins to form as an expansion of the anterior or rostral end of the neural tube (the end toward the head). Shortly thereafter, constrictions appear, dividing the developing brain into three major regions—**forebrain, midbrain,** and **hindbrain** (Figure 19.1). The remainder of the neural tube becomes the spinal cord.

During fetal development, two anterior outpocketings extend from the forebrain and grow rapidly to form the cerebral hemispheres. Because of space restrictions imposed by the skull, the cerebral hemispheres are forced to grow posteriorly and inferiorly, and finally end up enveloping and obscuring the rest of the forebrain and most midbrain structures. Somewhat later in development, the dorsal part of the hindbrain also enlarges to produce the cerebellum. The central canal of the neural tube, which remains continuous throughout the brain and cord, becomes enlarged in four regions of the brain, forming chambers called **ventricles** (see Figure 19.8a and b, p. 199).

Activity:
Identifying External Brain Structures

Identify external brain structures using the figures cited. Also use a model of the human brain and other learning aids as they are mentioned.

Generally, the brain is considered in terms of four major regions: the cerebral hemispheres, diencephalon, brain stem, and cerebellum. The relationship between these four anatomical regions and the structures of the forebrain, midbrain, and hindbrain is also outlined in Figure 19.1.

Cerebral Hemispheres

The **cerebral hemispheres** are the most superior portion of the brain (Figure 19.2). Their entire surface is thrown into elevated ridges of tissue called **gyri** that are separated by depressed areas called **fissures** or **sulci.** Of the two types of depressions, the fissures are deeper. Many of the fissures and gyri are important anatomical landmarks.

The cerebral hemispheres are divided by a single deep fissure, the **longitudinal fissure.** The **central sulcus** divides the **frontal lobe** from the **parietal lobe,** and the **lateral sulcus** separates the **temporal lobe** from the parietal lobe. The **parieto-occipital sulcus,** which divides the **occipital lobe** from the parietal lobe, is not visible externally. Notice that the cerebral hemisphere lobes are named for the cranial bones that lie over them.

Some important functional areas of the cerebral hemispheres have also been located (Figure 19.2d). The **primary somatosensory area** is located in the **postcentral gyrus** of the parietal lobe. Impulses traveling from the body's sensory receptors (such as those for pressure, pain, and temperature) are localized in this area of the brain. ("This information is from my big toe.") Immediately posterior to the primary somatosensory area is the **somatosensory association area,** in which the meaning of incoming stimuli is analyzed. ("Ouch! I have a *pain* there.") Thus, the somatosensory association

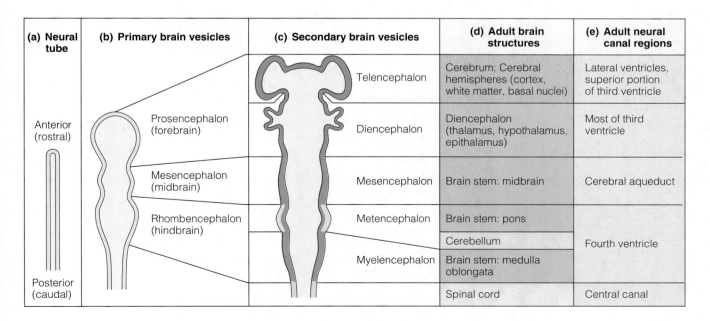

Figure 19.1   Embryonic development of the human brain. (a) The neural tube becomes subdivided into **(b)** the primary brain vesicles, which subsequently form **(c)** the secondary brain vesicles, which differentiate into **(d)** the adult brain structures. **(e)** The adult structures derived from the neural canal.

(a) Neural tube	(b) Primary brain vesicles	(c) Secondary brain vesicles	(d) Adult brain structures	(e) Adult neural canal regions
		Telencephalon	Cerebrum: Cerebral hemispheres (cortex, white matter, basal nuclei)	Lateral ventricles, superior portion of third ventricle
Anterior (rostral)	Prosencephalon (forebrain)	Diencephalon	Diencephalon (thalamus, hypothalamus, epithalamus)	Most of third ventricle
	Mesencephalon (midbrain)	Mesencephalon	Brain stem: midbrain	Cerebral aqueduct
	Rhombencephalon (hindbrain)	Metencephalon	Brain stem: pons	Fourth ventricle
			Cerebellum	
		Myelencephalon	Brain stem: medulla oblongata	
Posterior (caudal)			Spinal cord	Central canal

Figure 19.2 External structure (lobes and fissures) of the cerebral hemispheres. (a) Left lateral view of the brain. **(b)** Superior view. **(c)** Photograph of the superior aspect of the human brain. **(d)** Functional areas of the left cerebral cortex. The olfactory area, which is deep to the temporal lobe on the medial hemispheric surface, is not identified. Numbers indicate brain regions plotted by the Brodman system.

(a)

(b)   (c)

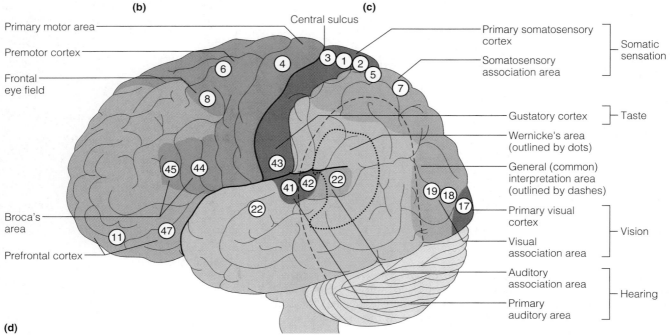

(d)

area allows you to become aware of pain, coldness, a light touch, and the like.

Impulses from the special sense organs are interpreted in other specific areas also noted in Figure 19.2d. For example, the visual areas are in the posterior portion of the occipital lobe and the auditory area is located in the temporal lobe in the gyrus bordering the lateral sulcus. The olfactory area is deep within the temporal lobe along its medial surface, in a region called the **uncus** (see Figure 19.4a, p. 194).

The **primary motor area,** which is responsible for conscious or voluntary movement of the skeletal muscles, is located in the **precentral gyrus** of the frontal lobe. A specialized motor speech area called **Broca's area** is found at the base of the precentral gyrus just above the lateral sulcus. Damage to this area (which is located only in one cerebral hemisphere, usually the left) reduces or eliminates the ability to articulate words. Areas involved in intellect, complex reasoning, and personality lie in the anterior portions of the frontal lobes, in a region called the **prefrontal cortex.**

A rather poorly defined region at the junction of the parietal and temporal lobes is **Wernicke's area,** an area in which unfamiliar words are sounded out. Like Broca's area, Wernicke's area is located in one cerebral hemisphere only, typically the left.

Although there are many similar functional areas in both cerebral hemispheres, such as motor and sensory areas, each hemisphere is also a "specialist" in certain ways. For example, the left hemisphere is the "language brain" in most of us, because it houses centers associated with language skills and speech. The right hemisphere is more specifically concerned with abstract, conceptual, or spatial processes—skills associated with artistic or creative pursuits.

The cell bodies of cerebral neurons involved in these functions are found only in the outermost gray matter of the cerebrum, the area called the **cerebral cortex.** Most of the balance of cerebral tissue—the deeper **cerebral white matter**—is composed of fiber tracts carrying impulses to or from the cortex.

Using a model of the human brain (and a preserved human brain, if available), identify the areas and structures of the cerebral hemispheres described above.

Then continue using the model and preserved brain along with the figures as you read about other structures.

Diencephalon

The **diencephalon,** sometimes considered the most superior portion of the brain stem, is embryologically part of the forebrain, along with the cerebral hemispheres.

Turn the brain model so the ventral surface of the brain can be viewed. Using Figure 19.3 as a guide, start superiorly and identify the externally visible structures that mark the

Frontal lobe of cerebral hemisphere

Temporal lobe

Pituitary gland

Cerebral peduncle of midbrain

Pyramid of medulla oblongata

Decussation of pyramids

Spinal cord

Olfactory bulb (synapse point of cranial nerve I)

Frontal lobe

Optic nerve (II)

Optic chiasma

Optic tract

Mammillary body

Pons

Temporal lobe

Medulla

Cerebellum

Spinal cord

Figure 19.3 Ventral aspect of the human brain, showing the three regions of the brain stem. Only a small portion of the midbrain can be seen; the rest is surrounded by other brain regions. (See also Plate C in the Human Anatomy Atlas.)

Frontal lobe
of cerebral
hemisphere

Septum
pellucidum

Intermediate
mass of
thalamus

Optic
chiasma

Uncus

Pons

Medulla
oblongata

(a)

Parietal lobe
of cerebral
hemisphere

Corpus callosum

Fornix

Choroid plexus
of third ventricle

Pineal body

Corpora
quadrigemina

Cerebral
aqueduct

Midbrain

Arbor vitae

Fourth
ventricle

Cerebellum

Figure 19.4 Diencephalon and brain stem structures as seen in a midsagittal section of the brain. (a) Photograph.

position of the floor of the diencephalon. These are the **olfactory bulbs** and **tracts, optic nerves, optic chiasma** (where the fibers of the optic nerves partially cross over), **optic tracts, pituitary gland,** and **mammillary bodies.**

Brain Stem

Continue inferiorly to identify the **brain stem** structures—the **cerebral peduncles** (fiber tracts in the **midbrain** connecting the pons below with cerebrum above), the pons, and the medulla oblongata. *Pons* means "bridge," and the **pons** consists primarily of motor and sensory fiber tracts connecting the brain with lower CNS centers. The lowest brain stem region, the **medulla,** is also composed primarily of fiber tracts. You can see the **decussation of pyramids,** a crossover point for the major motor tract (pyramidal tract) descending from the motor areas of the cerebrum to the cord, on the medulla's anterior surface. The medulla also houses many vital autonomic centers involved in the control of heart rate, respiratory rhythm, and blood pressure as well as involuntary centers involved in the initiation of vomiting, swallowing, and so on.

Cerebellum

1. Turn the brain model so you can see the dorsal aspect. Identify the large cauliflowerlike **cerebellum,** which projects dorsally from under the occipital lobe of the cerebrum. Notice that, like the cerebrum, the cerebellum has two major hemispheres and a convoluted surface (see Figure 19.6). It also has an outer cortex made up of gray matter with an inner region of white matter.

2. Remove the cerebellum to view the **corpora quadrigemina,** located on the posterior aspect of the midbrain, a

brain stem structure. The two superior prominences are the **superior colliculi** (visual reflex centers); the two smaller inferior prominences are the **inferior colliculi** (auditory reflex centers). ■

Activity:
Identifying Internal Brain Structures

The deeper structures of the brain have also been well mapped. Like the external structures, these can be studied in terms of the four major regions. As the internal brain areas are described, identify them on the figures cited. Also, use the brain model as indicated to help you in this study.

Cerebral Hemispheres

1. Take the brain model apart so you can see a median sagittal view of the internal brain structures (Figure 19.4). Observe the model closely to see the extent of the outer cortex (gray matter), which contains the cell bodies of cerebral neurons. (The pyramidal cells of the cerebral motor cortex [studied in Exercise 17, pp. 180 and 181] are representative of the neurons seen in the precentral gyrus.)

2. Observe the deeper area of white matter, which is composed of fiber tracts. The fiber tracts found in the cerebral hemisphere white matter are called *association tracts* if they connect two portions of the same hemisphere, *projection tracts* if they run between the cerebral cortex and the lower brain or spinal cord, and *commissures* if they run from one hemisphere to another. Observe the large **corpus callosum,** the major commissure connecting the cerebral hemispheres. The corpus callosum arches above the structures of the diencephalon and roofs over the lateral ventricles. Note also the

Third ventricle
Septum pellucidum
Intermediate mass of thalamus
Frontal lobe of cerebral hemisphere
Interventricular foramen (foramen of Monro)
Anterior commissure
Hypothalamus
Optic chiasma
Hypophysis cerebri (pituitary gland)
Temporal lobe of cerebral hemisphere
Mammillary body
Pons
Medulla oblongata
Spinal cord

Parietal lobe of cerebral hemisphere
Corpus callosum
Fornix
Choroid plexus
Occipital lobe of cerebral hemisphere
Thalamus (encloses third ventricle)
Pineal body (part of epithalamus)
Corpora quadrigemina
Cerebral aqueduct
Midbrain
Arbor vitae
Fourth ventricle
Choroid plexus
Cerebellum

(b)

Figure 19.4 (*continued*) Diencephalon and brain stem structures as seen in a midsagittal section of the brain. (b) Diagrammatic view.

fornix, a bandlike fiber tract concerned with olfaction as well as limbic system functions, and the membranous **septum pellucidum,** which separates the lateral ventricles of the cerebral hemispheres.

3. In addition to the gray matter of the cerebral cortex, there are several "islands" of gray matter (clusters of neuron cell bodies) called **nuclei** buried deep within the white matter of the cerebral hemispheres. One important group of cerebral nuclei, called the **basal nuclei,*** flank the lateral and third ventricles. You can see the basal nuclei if you have an appropriate dissectible model or a coronally or cross-sectioned human brain slice. Otherwise, Figure 19.5 will suffice.

The basal nuclei, which are important subcortical motor nuclei (and part of the so-called *extrapyramidal system*), are involved in regulating voluntary motor activities. The most important of them are the arching, comma-shaped **caudate nucleus,** the **claustrum,** the **amygdaloid nucleus** (located at the tip of the caudate nucleus), and the **lentiform nucleus,** which is composed of the **putamen** and **globus pallidus nuclei.** The **corona radiata,** a spray of projection fibers coursing down from the precentral (motor) gyrus, combines with sensory fibers traveling to the sensory cortex to form a broad band of fibrous material called the **internal capsule.** The internal capsule passes between the diencephalon and the basal

nuclei, and gives these basal nuclei a striped appearance. This is why the caudate nucleus and the lentiform nucleus are sometimes referred to collectively as the **corpus striatum,** or "striped body."

4. Examine the relationship of the lateral ventricles and corpus callosum to the diencephalon structures; that is, hypothalamus, thalamus, and third ventricle—from the cross-sectional viewpoint (see Figure 19.5b).

Diencephalon

1. The major internal structures of the diencephalon are the thalamus, hypothalamus, and epithalamus (Figure 19.4). The **thalamus** consists of two large lobes of gray matter that laterally enclose the shallow third ventricle of the brain. A slender stalk of thalamic tissue, the **intermediate mass,** or **massa intermedia,** connects the two thalamic lobes and bridges the ventricle. The thalamus is a major integrating and relay station for sensory impulses passing upward to the cortical sensory areas for localization and interpretation. Locate also the **interventricular foramen** (*foramen of Monro*), a tiny orifice connecting the third ventricle with the lateral ventricle on the same side.

2. The **hypothalamus** makes up the floor and the inferolateral walls of the third ventricle. It is an important autonomic center involved in regulation of body temperature, water balance, and fat and carbohydrate metabolism as well as in many other activities and drives (sex, hunger, thirst). Locate again

*The historical term for these nuclei is *basal ganglia,* a misleading term because ganglia are PNS structures.

Figure 19.5 Basal nuclei. **(a)** Three-dimensional view of the basal nuclei showing their positions within the cerebrum. **(b)** A transverse section of the cerebrum and diencephalon showing the relationship of the basal nuclei to the thalamus and the lateral and third ventricles.

the pituitary gland, or **hypophysis,** which hangs from the anterior floor of the hypothalamus by a slender stalk, the **infundibulum.** (The pituitary gland is usually not present in preserved brain specimens.) In life, the pituitary rests in the hypophyseal fossa of the sella turcica of the sphenoid bone. Its function is discussed in Exercise 27.

Anterior to the pituitary, identify the optic chiasma portion of the optic pathway to the brain. The **mammillary bodies,** relay stations for olfaction, bulge exteriorly from the floor of the hypothalamus just posterior to the pituitary gland.

3.   The **epithalamus** forms the roof of the third ventricle and is the most dorsal portion of the diencephalon. Important structures in the epithalamus are the **pineal body,** or **gland** (a neuroendocrine structure), and the **choroid plexus** of the third ventricle. The choroid plexuses, knotlike collections of capillaries within each ventricle, form the cerebrospinal fluid.

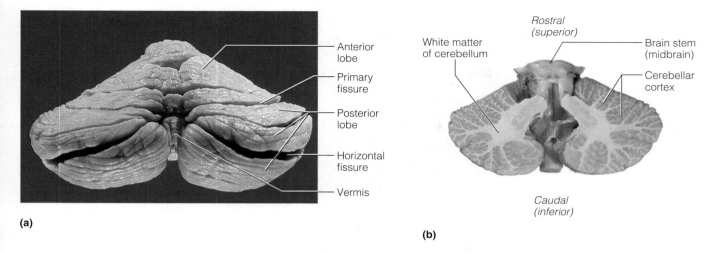

(a)

(b)

Figure 19.6   Cerebellum. (a) Posterior (dorsal) view. **(b)** The cerebellum, sectioned to reveal its cortex and medullary regions. (Note that the cerebellum is sectioned frontally and the brain stem is sectioned horizontally in this posterior view.)

Brain Stem

1.   Now trace the short midbrain from the mammillary bodies to the rounded pons below. Continue to refer to Figure 19.4. The **cerebral aqueduct** is a slender canal traveling through the midbrain; it connects the third ventricle to the fourth ventricle in the hindbrain below. The cerebral peduncles and the rounded corpora quadrigemina make up the midbrain tissue anterior and posterior (respectively) to the cerebral aqueduct.

2.   Locate the hindbrain structures. Trace the rounded pons to the medulla oblongata below, and identify the fourth ventricle posterior to these structures. Attempt to identify the single median aperture and the two lateral apertures, three orifices found in the walls of the fourth ventricle. These apertures serve as conduits for cerebrospinal fluid to circulate into the subarachnoid space from the fourth ventricle.

Cerebellum

Examine the cerebellum. Notice that it is composed of two lateral hemispheres each with three lobes (*anterior, posterior,* and a deep *flocculonodular*) connected by a midline lobe called the **vermis** (Figure 19.6). As in the cerebral hemispheres, the cerebellum has an outer cortical area of gray matter and an inner area of white matter. The treelike branching of the cerebellar white matter is referred to as the **arbor vitae,** or "tree of life." The cerebellum is concerned with unconscious coordination of skeletal muscle activity and con-

trol of balance and equilibrium. Fibers converge on the cerebellum from the equilibrium apparatus of the inner ear, visual pathways, proprioceptors of the tendons and skeletal muscles, and from many other areas. Thus the cerebellum remains constantly aware of the position and state of tension of the various body parts. ∎

Meninges of the Brain

The brain (and spinal cord) are covered and protected by three connective tissue membranes called **meninges** (Figure 19.7). The outermost meninx is the leathery **dura mater,** a double-layered membrane. One of its layers (the *periosteal layer*) is attached to the inner surface of the skull, forming the periosteum. The other (the *meningeal layer*) forms the outermost brain covering and is continuous with the dura mater of the spinal cord.

The dural layers are fused together except in three places where the inner membrane extends inward to form a septum that secures the brain to structures inside the cranial cavity. One such extension, the **falx cerebri,** dips into the longitudinal fissure between the cerebral hemispheres to attach to the crista galli of the ethmoid bone of the skull. The cavity created at this point is the large **superior sagittal sinus,** which collects blood draining from the brain tissue. The **falx cerebelli,** separating the two cerebellar hemispheres, and the **tentorium cerebelli,** separating the cerebrum from the cerebellum below, are two other important inward folds of the inner dural membrane.

(a)

Skin of scalp
Periosteum
Bone of skull
Periosteal ⎤ Dura
Meningeal ⎦ mater
Arachnoid membrane
Pia mater
Arachnoid villus
Blood vessel
Falx cerebri (in longitudinal fissure only)

Superior sagittal sinus
Subdural space
Subarachnoid space

(b)

Falx cerebri   Tentorium cerebelli
Superior sagittal sinus
Crista galli of the ethmoid bone
Falx cerebelli

(c)

Skull
Scalp
Superior sagittal sinus
Dura mater
Tranverse sinus
Temporal bone

Occipital lobe
Tentorium cerebelli
Cerebellum
Arachnoid mater over medulla oblongata

Figure 19.7   Meninges of the brain. (a) Three-dimensional frontal section showing the relationship of the dura mater, arachnoid mater, and pia mater. The meningeal dura forms the falx cerebri fold, which extends into the longitudinal fissure and attaches the brain to the ethmoid bone of the skull. A dural sinus, the superior sagittal sinus, is enclosed by the dural membranes superiorly. Arachnoid villi, which return cerebrospinal fluid to the dural sinus, are also shown. **(b)** Position of the dural folds, the falx cerebri, tentorium cerebelli, and falx cerebelli. **(c)** Posterior view of the brain in place, surrounded by the dura mater.

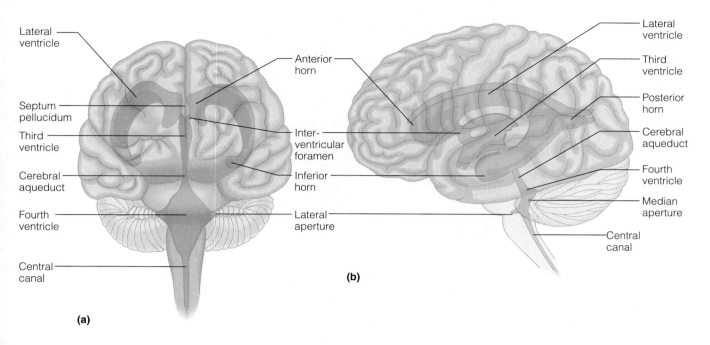

Figure 19.8   Location and circulatory pattern of cerebro-spinal fluid. (a) Anterior view; note that different regions of the large lateral ventricles are indicated by the terms *anterior horn, posterior horn,* and *inferior horn.* **(b)** Lateral view.

The middle meninx, the weblike **arachnoid mater,** or simply the **arachnoid,** underlies the dura mater and is partially separated from it by the **subdural space.** Threadlike projections bridge the **subarachnoid space** to attach the arachnoid to the innermost meninx, the **pia mater.** The delicate pia mater is highly vascular and clings tenaciously to the surface of the brain, following its convolutions.

In life, the subarachnoid space is filled with cerebrospinal fluid. Specialized projections of the arachnoid tissue called **arachnoid villi** protrude through the dura mater to allow the cerebrospinal fluid to drain back into the venous circulation via the superior sagittal sinus and other dural sinuses.

Meningitis, inflammation of the meninges, is a serious threat to the brain because of the intimate association between the brain and meninges. Should infection spread to the neural tissue of the brain itself, life-threatening **encephalitis** may occur. Meningitis is often diagnosed by taking a sample of cerebrospinal fluid from the subarachnoid space. ■

Cerebrospinal Fluid

The cerebrospinal fluid, much like plasma in composition, is continually formed by the **choroid plexuses,** small capillary knots hanging from the roof of the ventricles of the brain. The cerebrospinal fluid in and around the brain forms a watery cushion that protects the delicate brain tissue against blows to the head.

Within the brain, the cerebrospinal fluid circulates from the two lateral ventricles (in the cerebral hemispheres) into the third ventricle via the **interventricular foramina,** and then through the cerebral aqueduct of the midbrain into the fourth ventricle in the hindbrain (Figure 19.8). Some of the fluid reaching the fourth ventricle continues down the central canal of the spinal cord, but the bulk of it circulates into the subarachnoid space, exiting through the three foramina in the walls of the fourth ventricle (the two lateral and the single median apertures). The fluid returns to the blood in the dural sinuses via the arachnoid villi.

Superior sagittal sinus
Superior cerebral vein
Choroid plexus
Cerebrum covered with pia mater
Septum pellucidum
Corpus callosum
Interventricular foramen
Third ventricle
Pituitary gland

(c)

Arachnoid villus
Subarachnoid space
Arachnoid
Meningeal dura mater
Periosteal dura mater
Great cerebral vein
Tentorium cerebelli
Straight sinus
Confluence of sinuses
Cerebellum
Choroid plexus
Cerebral vessels that supply choroid plexus
Central canal of spinal cord
Spinal dura mater (dural sheath)
Inferior end of spinal cord
Filum terminale (inferior end of pia mater)

Cerebral aqueduct
Lateral aperture
Fourth ventricle
Median aperture

Figure 19.8 (continued) Location and circulatory pattern of cerebrospinal fluid. (c) The cerebrospinal fluid flows from the lateral ventricles, through the interventricular foramina, into the third ventricle, and then into the fourth ventricle via the cerebral aqueduct. (The relative position of the right lateral ventricle is indicated by the pale blue area deep to the corpus callosum and septum pellucidum.)

Ordinarily, cerebrospinal fluid forms and drains at a constant rate. However, under certain conditions—for example, obstructed drainage or circulation resulting from tumors or anatomical deviations—the cerebrospinal fluid accumulates and exerts increasing pressure on the brain which, uncorrected, causes neurological damage in adults. In infants, **hydrocephalus** (literally, "water on the brain") is indicated by a gradually enlarging head. Since the infant's skull is still flexible and contains fontanels, it can expand to accommodate the increasing size of the brain. ■

Cranial Nerves

The **cranial nerves** are part of the peripheral nervous system and not part of the brain proper, but they are most appropriately identified in conjunction with the study of brain anatomy. The 12 pairs of cranial nerves primarily serve the head and neck. Only one pair, the vagus nerves, extends into the thoracic and abdominal cavities. All but the first two pairs (olfactory and optic nerves) arise from the brain stem and pass through foramina in the base of the skull to reach their destination.

The cranial nerves are numbered consecutively, and in most cases their names reflect the major structures they control. The cranial nerves are described by name, number (Roman numeral), origin, course, and function in Table 19.1. This information should be committed to memory. A mnemonic device that might be helpful for remembering the cranial nerves in order is "*On occasion, our trusty truck acts funny—very good vehicle anyhow.*" The first letter of each word and the "a" and "h" of the final word "anyhow" will remind you of the first letter of the cranial nerve name.

Most cranial nerves are mixed nerves (containing both motor and sensory fibers). But close scrutiny of Table 19.1 will reveal that three pairs of cranial nerves (optic, olfactory, and vestibulocochlear) are purely sensory in function.

You may recall that the cell bodies of neurons are always located within the central nervous system (cortex or nuclei) or in specialized collections of cell bodies (ganglia) outside the CNS. Neuron cell bodies of the sensory cranial nerves are located in ganglia; those of the mixed cranial nerves are found both within the brain and in peripheral ganglia.

Text continues on page 203

Table 19.1	The Cranial Nerves (see Figure 19.9)

Number and name	Origin and course	Function	Testing
I. Olfactory	Fibers arise from olfactory mucosa and run through cribriform plate of ethmoid bone to synapse with olfactory bulbs.	Purely sensory–carries impulses associated with sense of smell.	Person is asked to sniff aromatic substances, such as oil of cloves and vanilla, and to identify each.
II. Optic	Fibers arise from retina of eye and pass through optic foramen in sphenoid bone. Fibers of the two optic nerves then take part in forming optic chiasma (with partial crossover of fibers) after which they continue on to thalamus as the optic tracts. Final fibers of this pathway travel from the thalamus to the optic cortex as the optic radiation.	Purely sensory—carries impulses associated with vision.	Vision and visual field are determined with eye chart and by testing the point at which the person first sees an object (finger) moving into the visual field. Fundus of eye viewed with ophthalmoscope to detect papilledema (swelling of optic disc, or point at which optic nerve leaves the eye) and to observe blood vessels.
III. Oculomotor	Fibers emerge from midbrain and exit from skull via superior orbital fissure to run to eye.	Mixed—somatic motor fibers to inferior oblique and superior, inferior, and medial rectus muscles, which direct eyeball, and to levator palpebrae muscles of eyelid; parasympathetic fibers to iris and smooth muscle controlling lens shape (reflex responses to varying light intensity and focusing of eye for near vision); contains proprioceptive sensory fibers carrying impulses from extrinsic eye muscles.	Pupils are examined for size, shape, and equality. Pupillary reflex is tested with penlight (pupils should constrict when illuminated). Convergence for near vision is tested, as is subject's ability to follow objects up, down, side to side, and diagonally.
IV. Trochlear	Fibers emerge from midbrain and exit from skull via superior orbital fissure to run to eye.	Mixed—provides somatic motor fibers to superior oblique muscle (an extrinsic eye muscle); conveys proprioceptive impulses from same muscle to brain.	Tested in common with cranial nerve III.
V. Trigeminal	Fibers emerge from pons and form three divisions, which exit separately from skull: mandibular division through foramen ovale in sphenoid bone, maxillary division via foramen rotundum in sphenoid bone, and ophthalmic division through superior orbital fissure of eye socket.	Mixed—major sensory nerve of face; conducts sensory impulses from skin of face and anterior scalp, from mucosae of mouth and nose, and from surface of eyes; mandibular division also contains motor fibers that innervate muscles of mastication and muscles of floor of mouth.	Sensations of pain, touch, and temperature are tested with safety pin and hot and cold objects. Corneal reflex tested with wisp of cotton. Motor branch assessed by asking person to clench his teeth, open mouth against resistance, and move jaw side to side.
VI. Abducens	Fibers leave inferior region of pons and exit from skull via superior orbital fissure to run to eye.	Carries motor fibers to lateral rectus muscle of eye and proprioceptive fibers from same muscle to brain.	Tested in common with cranial nerve III.

Table 19.1	The Cranial Nerves (see Figure 19.9) (continued)		
Number and name	**Origin and course**	**Function**	**Testing**
VII. Facial	Fibers leave pons and travel through temporal bone via internal acoustic meatus, exiting via stylomastoid foramen to reach the face.	Mixed—supplies somatic motor fibers to muscles of facial expression and parasympathetic motor fibers to lacrimal and salivary glands; carries sensory fibers from taste receptors of anterior portion of tongue.	Anterior two-thirds of tongue is tested for ability to taste sweet (sugar), salty, sour (vinegar), and bitter (quinine) substances. Symmetry of face is checked. Subject is asked to close eyes, smile, whistle, and so on. Tearing is assessed with ammonia fumes.
VIII. Vestibulo-cochlear	Fibers run from inner-ear equilibrium and hearing apparatus, housed in temporal bone, through internal acoustic meatus to enter pons.	Purely sensory—vestibular branch transmits impulses associated with sense of equilibrium from vestibular apparatus and semicircular canals; cochlear branch transmits impulses associated with hearing from cochlea.	Hearing is checked by air and bone conduction using tuning fork.
IX. Glosso-pharyngeal	Fibers emerge from medulla and leave skull via jugular foramen to run to throat.	Mixed—somatic motor fibers serve pharyngeal muscles, and parasympathetic motor fibers serve salivary glands; sensory fibers carry impulses from pharynx, tonsils, posterior tongue (taste buds), and pressure receptors of carotid artery.	Position of the uvula is checked. Gag and swallowing reflexes are checked. Subject is asked to speak and cough. Posterior third of tongue may be tested for taste.
X. Vagus	Fibers emerge from medulla and pass through jugular foramen and descend through neck region into thorax and abdomen.	Mixed—fibers carry somatic motor impulses to pharynx and larynx and sensory fibers from same structures; very large portion is composed of parasympathetic motor fibers, which supply heart and smooth muscles of abdominal visceral organs; transmits sensory impulses from viscera.	As for cranial nerve IX (IX and X are tested in common, since they both innervate muscles of throat and mouth).
XI. Accessory	Fibers arise from medulla and superior aspect of spinal cord and travel through jugular foramen to reach muscles of neck and back.	Mixed—provides somatic motor fibers to sternocleido-mastoid and trapezius muscles and to muscles of soft palate, pharynx, and larynx (spinal and medullary fibers respectively); proprioceptive impulses are conducted from these muscles to brain.	Sternocleidomastoid and trapezius muscles are checked for strength by asking person to rotate head and shoulders against resistance.
XII. Hypoglossal	Fibers arise from medulla and exit from skull via hypoglossal canal to travel to tongue.	Mixed—carries somatic motor fibers to muscles of tongue and proprioceptive impulses from tongue to brain.	Person is asked to protrude and retract tongue. Any deviations in position are noted.

Activity:
Identifying and Testing the Cranial Nerves

1.   Observe the anterior surface of the brain model to identify the cranial nerves. Figure 19.9 may also aid you in this study. Notice that the first (olfactory) cranial nerves are not visible on the model because they consist only of short axons that run from the nasal mucosa through the cribriform plate of the ethmoid bone. (However, the synapse points of the first cranial nerves, the *olfactory bulbs,* are visible on the model.)

2.   The last column of Table 19.1 describes techniques for testing cranial nerves, which is an important part of any neurologic examination. This information may help you understand cranial nerve function, especially as it pertains to some aspects of brain function. Conduct tests of cranial nerve function following directions given in the "testing" column of the table.

3.   Several cranial nerve ganglia are named here. *Using your textbook or an appropriate reference,* name the cranial nerve the ganglion is associated with and state its location.

Cranial nerve ganglion	Cranial nerve	Site of ganglion
trigeminal		
geniculate		
inferior		
superior		
spiral		
vestibular		

Figure 19.9   Ventral aspect of the human brain, showing the cranial nerves. (See also Plate C in the Human Anatomy Atlas.)

Dissection: The Sheep Brain

The brain of any mammal is enough like the human brain to warrant comparison. Obtain a sheep brain, protective skin cream or disposable gloves, dissecting pan, and instruments, and bring them to your laboratory bench.

1. Before beginning the dissection, turn your sheep brain so that you are viewing its left lateral aspect. Compare the various areas of the sheep brain (cerebrum, brain stem, cerebellum) to the photo of the human brain in Figure 19.10. Relatively speaking, which of these structures is obviously much larger in the human brain?

_____

2. Place the intact sheep brain ventral surface down on the dissecting pan and observe the dura mater. Feel its consistency and note its toughness. Cut through the dura mater along the line of the longitudinal fissure (which separates the cerebral hemispheres) to enter the superior sagittal sinus. Gently force the cerebral hemispheres apart laterally to expose the corpus callosum deep to the longitudinal fissure.

3. Carefully remove the dura mater and examine the superior surface of the brain. Notice that, like the human brain, its surface is thrown into convolutions (fissures and gyri). Locate the arachnoid mater, which appears on the brain surface as a delicate "cottony" material spanning the fissures. In contrast, the innermost meninx, the pia mater, closely follows the cerebral contours.

Ventral Structures

Figure 19.11a and c shows the important features of the ventral surface of the brain. Turn the brain so that its ventral surface is uppermost.

Figure 19.10 Photo of lateral aspect of the human brain.

Labels: Cerebrum, Brain stem, Cerebellum

1. Look for the clublike olfactory bulbs anteriorly, on the inferior surface of the frontal lobes of the cerebral hemispheres. Axons of olfactory neurons run from the nasal mucosa through the perforated cribriform plate of the ethmoid bone to synapse with the olfactory bulbs.

How does the size of these olfactory bulbs compare with those of humans?

_____

Is the sense of smell more important as a protective and a food-getting sense in sheep or in humans?

_____

2. The optic nerve (II) carries sensory impulses from the retina of the eye. Thus this cranial nerve is involved in the sense of vision. Identify the optic nerves, optic chiasma, and optic tracts.

3. Posterior to the optic chiasma, two structures protrude from the ventral aspect of the hypothalamus—the infundibulum (stalk of the pituitary gland) immediately posterior to the optic chiasma and the mammillary body. Notice that the sheep's mammillary body is a single rounded eminence. In humans it is a double structure.

4. Identify the cerebral peduncles on the ventral aspect of the midbrain, just posterior to the mammillary body of the hypothalamus. The cerebral peduncles are fiber tracts connecting the cerebrum and medulla. Identify the large oculomotor nerves (III), which arise from the ventral midbrain surface, and the tiny trochlear nerves (IV), which can be seen at the junction of the midbrain and pons. Both of these cranial nerves provide motor fibers to extrinsic muscles of the eyeball.

5. Move posteriorly from the midbrain to identify first the pons and then the medulla oblongata, both hindbrain structures composed primarily of ascending and descending fiber tracts.

6. Return to the junction of the pons and midbrain and proceed posteriorly to identify the following cranial nerves, all arising from the pons:

• Trigeminal nerves (V), which are involved in chewing and sensations of the head and face

• Abducens nerves (VI), which abduct the eye (and thus work in conjunction with cranial nerves III and IV)

• Facial nerves (VII), large nerves are involved in taste sensation, gland function (salivary and lacrimal glands), and facial expression

7. Continue posteriorly to identify:

• Vestibulocochlear nerves (VIII), purely sensory nerves that are involved with hearing and equilibrium

• Glossopharyngeal nerves (IX), which contain motor fibers innervating throat structures and sensory fibers transmitting taste stimuli (in conjunction with cranial nerve VII)

(a)

(b)

Figure 19.11   Intact sheep brain. (a) Diagrammatic ventral view. **(b)** Photographs showing ventral and dorsal views.

Olfactory bulb

Cerebrum

Cerebellum

Medulla oblongata

Spinal cord

(c)

Figure 19.11 (*continued*) Intact sheep brain.
(c) Diagrammatic dorsal view.

- Vagus nerves (X), often called "wanderers," which serve many organs of the head, thorax, and abdominal cavity

- Accessory nerves (XI), which serve muscles of the neck, larynx, and shoulder; notice that the accessory nerves arise from both the medulla and the spinal cord

- Hypoglossal nerves (XII), which stimulate tongue and neck muscles

Dorsal Structures

1. Refer to Figure 19.11b and c as a guide in identifying the following structures. Reidentify the now exposed cerebral hemispheres. How does the depth of the fissures in the sheep's cerebral hemispheres compare to that in the human brain?

_____

_____

2. Carefully examine the cerebellum. Notice that, in contrast to the human cerebellum, it is not divided longitudinally, and that its fissures are oriented differently. What dural falx is missing that is present in humans?

_____

3. Locate the three pairs of cerebellar peduncles, fiber tracts that connect the cerebellum to other brain structures, by lifting the cerebellum dorsally away from the brain stem. The most posterior pair, the inferior cerebellar peduncles, connect

Occipital lobe of cerebral hemisphere

Pineal body

Superior colliculi of corpora quadrigemina

Inferior colliculi of corpora quadrigemina

Cerebellum

Figure 19.12 Means of exposing the dorsal midbrain structures of the sheep brain.

(a)

(b)

Figure 19.13 **Sagittal section of the sheep brain showing internal structures. (a)** Diagrammatic view. **(b)** Photograph.

the cerebellum to the medulla. The middle cerebellar peduncles attach the cerebellum to the pons, and the superior cerebellar peduncles run from the cerebellum to the midbrain.

4. To expose the dorsal surface of the midbrain, gently spread the cerebrum and cerebellum apart, as shown in Figure 19.12. Identify the corpora quadrigemina, which appear as four rounded prominences on the dorsal midbrain surface. What is the function of the corpora quadrigemina?

_____

_____

Also locate the pineal body, which appears as a small oval protrusion in the midline just anterior to the corpora quadrigemina.

Internal Structures

1. The internal structure of the brain can only be examined after further dissection. Place the brain ventral side down on the dissecting pan and make a cut completely through it in a superior to inferior direction. Cut through the longitudinal fissure, corpus callosum, and midline of the cerebellum. Refer to Figure 19.13 as you work.

2. The thin nervous tissue membrane immediately ventral to the corpus callosum that separates the lateral ventricles is the septum pellucidum. Pierce this membrane and probe the lateral ventricle cavity. The fiber tract ventral to the septum pellucidum and anterior to the third ventricle is the fornix.

Figure 19.14 Frontal section of a sheep brain. Major structures revealed are the location of major basal nuclei in the interior, the thalamus, hypothalamus, and lateral and third ventricles.

How does the relative size of the fornix in this brain compare with the human fornix?

_____

_____

Why do you suppose this is so? (Hint: What is the function of this band of fibers?)

_____

_____

_____

3. Identify the thalamus, which forms the walls of the third ventricle and is located posterior and ventral to the fornix. The intermediate mass spanning the ventricular cavity appears as an oval protrusion of the thalamic wall. Anterior to the intermediate mass, locate the interventricular foramen, a canal connecting the lateral ventricle on the same side with the third ventricle.

4. The hypothalamus forms the floor of the third ventricle. Identify the optic chiasma, infundibulum, and mammillary body on its exterior surface. You can see the pineal body at the superoposterior end of the third ventricle, just beneath the junction of the corpus callosum and fornix.

5. Locate the midbrain by identifying the corpora quadrigemina that form its dorsal roof. Follow the cerebral aqueduct (the narrow canal connecting the third and fourth ventricles) through the midbrain tissue to the fourth ventricle. Identify the cerebral peduncles, which form its anterior walls.

6. Identify the pons and medulla, which lie anterior to the fourth ventricle. The medulla continues into the spinal cord without any obvious anatomical change, but the point at which the fourth ventricle narrows to a small canal is generally accepted as the beginning of the spinal cord.

7. Identify the cerebellum posterior to the fourth ventricle. Notice its internal treelike arrangement of white matter, the arbor vitae.

8. If time allows, obtain another sheep brain and section it along the frontal plane so that the cut passes through the infundibulum. Compare your specimen to the photograph in Figure 19.14, and attempt to identify all the structures shown in the figure.

9. Check with your instructor to determine if cow spinal cord sections (preserved) are available for the spinal cord studies in Exercise 21. If not, save the small portion of the spinal cord from your brain specimen. Otherwise, dispose of all the organic debris in the appropriate laboratory containers and clean the dissecting instruments and tray before leaving the laboratory. ■

Electroencephalography

Brain Wave Patterns and the Electroencephalogram

Any physiologic investigation of the brain can emphasize and expose only a very minute portion of its activity. Higher brain functions, such as consciousness and logical reasoning, are extremely difficult to investigate. It is obviously much easier to do experiments on the brain's input-output functions, some of which can be detected with appropriate recording equipment. Still, the ability to record brain activity does not necessarily guarantee an understanding of the brain.

The **electroencephalogram (EEG),** a record of the electrical activity of the brain, can be obtained through electrodes placed at various points on the skin or scalp of the head. This electrical activity, which is recorded as waves (Figure 20.1), is not completely understood at present but may be regarded as action potentials generated by brain neurons.

Certain characteristics of brain waves are known. They have a frequency of 1 to 30 hertz (Hz) or cycles per second, a dominant rhythm of 10 Hz, and an average amplitude (voltage) of 20 to 100 microvolts (μV). They vary in frequency in different brain areas, occipital waves having a lower frequency than those associated with the frontal and parietal lobes.

The first of the brain waves to be described by scientists were the **alpha waves** (or alpha rhythm). Alpha waves have an average frequency range of 8 to 13 Hz and are produced when the individual is in a relaxed state with the eyes closed. **Alpha block,** suppression of the alpha rhythm, occurs if the eyes are opened or if the individual begins to concentrate on some mental problem or visual stimulus. Under these conditions, the waves decrease in amplitude but increase in frequency. Under conditions of fright or excitement, the frequency increases still more.

Beta waves, closely related to alpha waves, are faster (14 to 25 Hz) and have a lower amplitude. They are typical of the attentive or alert state.

Very large (high-amplitude) waves with a frequency of 4 Hz or less that are seen in deep sleep are **delta waves. Theta waves** are large, abnormally contoured waves with a frequency of 4 to 7 Hz. Although theta waves are normal in children, they represent emotional problems or some sort of neural imbalance in adults.

Brain waves change with age, sensory stimuli, brain pathology or disease, and the chemical state of the body. (Glucose deprivation, oxygen poisoning, and sedatives all interfere with the rhythmic activity of brain output by disturbing the metabolism of the neurons.) Sleeping individuals and patients in a coma have EEGs that are slower (or lower frequency) than the alpha rhythm of normal adults. Fright, epileptic seizures, and various types of drug intoxication are associated with comparatively faster cortical activity. Thus impairment of cortical function is indicated by neuronal ac-

tivity that is either too fast or too slow; unconsciousness occurs at both extremes of the frequency range.

Since spontaneous brain waves are always present, even during unconsciousness and coma, the absence of brain waves (a "flat" EEG) is taken as clinical evidence of death. The EEG is used clinically to diagnose and localize many types of brain lesions, including epileptic lesions, infections, abscesses, and tumors. ■

Objectives

1. To define *electroencephalogram* and to discuss its clinical significance.
2. To describe or recognize typical tracings of the most common brain wave patterns (alpha, beta, theta, and delta waves) and to indicate the conditions under which each is most likely to be predominant.
3. To state the source of brain waves.
4. To define *alpha block.*
5. To monitor electroencephalography and recognize alpha rhythm.
6. To describe the effect of a sudden sound, mental concentration, and alkalosis on brain wave patterns.

Materials

- ❏ Oscilloscope and EEG lead-selector box or polygraph and high-gain preamplifier
- ❏ Cot
- ❏ Electrode gel
- ❏ EEG electrodes and leads
- ❏ Collodion gel or long elastic EEG straps

Activity:
Observing Brain Wave Patterns

If one electrode (the *active electrode*) is placed over a particular cortical area and another (the *indifferent electrode*) is placed over an inactive part of the head, such as the earlobe, all of the activity of the cortex underlying the active electrode will, theoretically, be recorded. The inactive area provides a zero reference point or a baseline, and the EEG represents the difference between "activities" occurring under the two electrodes.

(a)

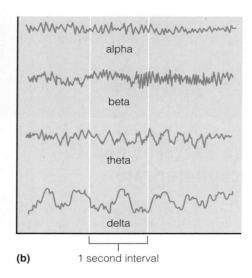

(b)

1 second interval

Figure 20.1 Electroencephalography and brain waves. (a) To obtain a recording of brain wave activity (an EEG), electrodes are positioned on the patient's scalp and attached to a recording device called an electroencephalograph. **(b)** Typical EEGs. Alpha waves are typical of the awake but relaxed state; beta waves occur in the awake, alert state; theta waves are common in children but not in normal awake adults; delta waves occur during deep sleep.

1. Connect the EEG selector box to the oscilloscope pre-amplifier, or connect the high-gain preamplifier to the polygraph channel amplifier. Adjust the horizontal sweep and sensitivity according to the directions given in the instrument manual or by your instructor.

2. Prepare the subject. The subject should lie undisturbed with eyes closed in a quiet, dimly lit area. (Someone who is able to relax easily makes a good subject.) Apply a small amount of electrode gel to the subject's forehead above the left eye and on the left earlobe. Press an electrode to each prepared area and secure them (1) by applying a film of collodion gel to the electrode surface and the adjacent skin, or (2) with a long elastic EEG strap (knot tied at the back of the head). If collodion gel is used, allow it to dry before you continue.

3. Connect the active frontal lead (forehead) to the EEG selector box outlet marked "L Frontal." Connect the lead from the indifferent electrode (earlobe) to the ground outlet (or to the appropriate input terminal on the high-gain preamplifier).

4. Turn the oscilloscope or polygraph on, and observe the EEG pattern of the relaxed subject for a period of 5 min. If the subject is truly relaxed, you should see a typical alpha-wave pattern. (If the subject is unable to relax and the alpha-wave pattern does not appear in this time interval, test another subject.) Discourage all muscle movement during the monitoring period.*

5. Abruptly and loudly clap your hands. The subject's eyes should open and alpha block should occur. Observe the immediate brain wave pattern. How do the frequency and amplitude of the brain waves change?

* Note that 60-cycle "noise" (appearing as fast, regular, low amplitude waves superimposed on the more irregular brain waves) may interfere with the tracings being made, particularly if the laboratory has a lot of electrical equipment.

_____

_____

Would you characterize this as beta rhythm? _____

Why? _____

6. Allow the subject about 5 min to achieve complete relaxation once again, then ask him or her to compute a number problem that requires concentration (for example, add 3 and 36, subtract 7, multiply by 2, add 50 etc.). Observe the brain wave pattern during the period of mental computation.

7. Once again allow the subject to relax until alpha rhythm resumes. Then, instruct him or her to hyperventilate for 3 min. *Be sure to tell the subject when to stop hyperventilating.* Hyperventilation rapidly flushes carbon dioxide out of the lungs, decreasing carbon dioxide levels in the blood and producing respiratory alkalosis.

Observe the changes in the rhythm and amplitude of the brain waves occurring during the period of hyperventilation. Record your observations:

_____

_____

_____

_____

_____

_____

Spinal Cord, Spinal Nerves, and the Autonomic Nervous System

Objectives

1. To identify important anatomical areas on a spinal cord model or appropriate diagram of the spinal cord, and to indicate the neuron type found in these areas (where applicable).

2. To indicate two major areas where the spinal cord is enlarged, and to explain the reasons for this anatomical characteristic.

3. To define *conus medullaris, cauda equina,* and *filum terminale.*

4. To locate on a diagram the fiber tracts in the spinal cord and to state their functional importance.

5. To list two major functions of the spinal cord.

6. To name the meningeal coverings of the spinal cord and state their function.

7. To describe the origin, fiber composition, and distribution of the spinal nerves, differentiating between roots, the spinal nerve proper, and rami, and to discuss the result of transecting these structures.

8. To discuss the distribution of the dorsal rami and ventral rami of the spinal nerves.

9. To identify the four major nerve plexuses, the major nerves of each, and their distribution.

10. To identify the site of origin and the function of the sympathetic and parasympathetic divisions of the autonomic nervous system, and to state how the autonomic nervous system differs from the somatic nervous system.

Materials

❑ Spinal cord model (cross section)

❑ Laboratory charts of the spinal cord and spinal nerves and sympathetic chain

❑ Red and blue pencils

❑ Preserved cow spinal cord sections with meninges and nerve roots intact (or spinal cord segment saved from the brain dissection in Exercise 19)

❑ Dissecting tray and instruments

❑ Dissecting microscope

❑ Histologic slide of spinal cord (cross section)

❑ Protective skin cream or disposable gloves

❑ Compound microscope

❑ *The Human Nervous System: The Spinal Cord and Spinal Nerves* videotape*

 See Appendix C, Exercise 21 for links to A.D.A.M. Interactive Anatomy.

 For instructions on animal dissections, see the dissection exercises starting on p. 733 in the cat and fetal pig editions of this manual.

*Available to qualified adopters from Benjamin/Cummings

Anatomy of the Spinal Cord

The cylindrical **spinal cord,** a continuation of the brain stem, is an association and communication center. It plays a major role in spinal reflex activity and provides neural pathways to and from higher nervous centers. Enclosed within the vertebral canal of the spinal column, the spinal cord extends from the foramen magnum of the skull to the first or second lumbar vertebra, where it terminates in the cone-shaped **conus medullaris** (Figure 21.1). Like the brain, it is cushioned and protected by meninges. The dura mater and arachnoid meningeal coverings extend beyond the conus medullaris, approximately to the level of S_2, and the **filum terminale,** a fibrous extension of the pia mater, extends even farther (into the coccygeal canal) to attach to the posterior coccyx. *Denticulate ligaments,* saw-toothed shelves of pia mater, secure the spinal cord to the bony wall of the vertebral column all along its length (See Figure 21.1c).

The fact that the meninges, filled with cerebrospinal fluid, extend well beyond the end of the spinal cord provides an excellent site for removing cerebrospinal fluid for analysis (as when bacterial or viral infections of the spinal cord or its meningeal coverings are suspected) without endangering the delicate spinal cord. This procedure, called a *lumbar tap,* is usually performed below L_3. Additionally, "saddle block" or caudal anesthesia for childbirth is normally administered (injected) between L_3 and L_5.

In humans, 31 pairs of spinal nerves arise from the spinal cord and pass through intervertebral foramina to serve the body area at their approximate level of emergence. The cord is about the size of a thumb in circumference for most of its length, but there are obvious enlargements in the cervical and lumbar areas where the nerves serving the upper and lower limbs issue from the cord.

Figure 21.1 Gross structure of the spinal cord, posterior view. (a) The bony vertebral arches have been removed to reveal the spinal cord and its nerve roots. The dura and arachnoid mater are cut open and reflected laterally. **(b)** Photograph of the cervical part of the spinal cord. **(c)** Enlargement of the spinal cord, showing the denticulate ligaments. **(d)** Inferior end of the spinal cord, showing the conus medullaris, cauda equina, and filum terminale.

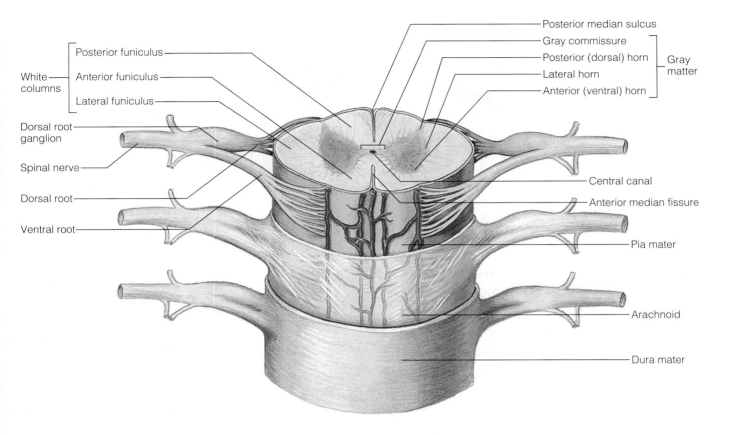

Figure 21.2 **Anatomy of the human spinal cord (three-dimensional view).**

Because the spinal cord does not extend to the end of the vertebral column, the spinal nerves emerging from the inferior end of the cord must travel through the vertebral canal for some distance before exiting at the appropriate intervertebral foramina. This collection of spinal nerves traversing the inferior end of the vertebral canal is called the **cauda equina** (Figure 21.1d) because of its similarity to a horse's tail (the literal translation of *cauda equina*).

Activity:
Identifying Structures of the Spinal Cord

Obtain a model of a cross section of a spinal cord and identify its structures as they are described next. ■

Gray Matter

In cross section, the **gray matter** of the spinal cord looks like a butterfly or the letter H (Figure 21.2). The two posterior projections are called the **posterior,** or **dorsal, horns.** The two anterior projections are the **anterior,** or **ventral, horns.** The tips of the anterior horns are broader and less tapered than those of the posterior horns. In the thoracic and lumbar regions of the cord, there is also a lateral outpocketing of gray

matter on each side referred to as the **lateral horn.** The central area of gray matter connecting the two vertical regions is the **gray commissure.** The gray commissure surrounds the **central canal** of the cord, which contains cerebrospinal fluid.

Neurons with specific functions can be localized in the gray matter. The posterior horns, for instance, contain association neurons and sensory fibers that enter the cord from the body periphery via the **dorsal root.** The cell bodies of these sensory neurons are found in an enlarged area of the dorsal root called the **dorsal root ganglion.** The anterior horns contain cell bodies of motor neurons of the somatic nervous system (voluntary system), which send their axons out via the **ventral root** of the cord to enter the adjacent spinal nerve. The **spinal nerves** are formed from the fusion of the dorsal and ventral roots. The lateral horns, where present, contain nerve cell bodies of motor neurons of the autonomic nervous system (sympathetic division). Their axons also leave the cord via the ventral roots, along with those of the motor neurons of the anterior horns.

White Matter

The **white matter** of the spinal cord is nearly bisected by fissures (see Figure 21.2). The more open anterior fissure is the **anterior median fissure,** and the posterior one is the **posterior median sulcus.** The white matter is composed of myelinated fibers—some running to higher centers, some traveling

Ascending tracts

Descending tracts

Figure 21.3 Cross section of the spinal cord showing the relative positioning of its major tracts.

from the brain to the cord, and some conducting impulses from one side of the cord to the other.

Because of the irregular shape of the gray matter, the white matter on each side of the cord can be divided into three primary regions or *white columns:* the **posterior, lateral,** and **anterior funiculi.** Each funiculus contains a number of fiber **tracts** composed of axons with the same origin, terminus, and function. Tracts conducting sensory impulses to the brain are called *ascending,* or *sensory, tracts;* those carrying impulses from the brain to the skeletal muscles are *descending,* or *motor, tracts.*

Because it serves as the transmission pathway between the brain and the body periphery, the spinal cord is an extremely important functional area. Even though it is protected by meninges and cerebrospinal fluid in the vertebral canal, it is highly vulnerable to traumatic injuries, such as might occur in an automobile accident.

When the cord is transected (or severely traumatized), both motor and sensory functions are lost in body areas normally served by that (and lower) regions of the spinal cord. Injury to certain spinal cord areas may even result in a permanent flaccid paralysis of both legs (paraplegia) or of all four limbs (quadriplegia). ■

Activity:

Identifying Spinal Cord Tracts

With the help of your textbook or a laboratory chart showing the tracts of the spinal cord, label Figure 21.3 with the tract names that follow. Since each tract is represented on both sides of the cord, for clarity you can label the motor tracts on the right side of the diagram and the sensory tracts on the left side of the diagram. *Color ascending tracts blue and de-*

scending tracts red. Then fill in the functional importance of each tract beside its name below. As you work, try to be aware of how the naming of the tracts is related to their anatomical distribution.

Fasciculus gracilis _____

Fasciculus cuneatus _____

Posterior spinocerebellar _____

Anterior spinocerebellar _____

Lateral spinothalamic _____

Anterior spinothalamic _____

Lateral corticospinal _____

Anterior corticospinal _____

Rubrospinal _____

Tectospinal _____

Vestibulospinal _____

Medial reticulospinal _____

Lateral reticulospinal _____

Olivospinal _____ ■

Dissection:
Spinal Cord

1.   Obtain a dissecting tray and instruments, protective skin cream or disposable gloves, and a segment of preserved spinal cord (from a cow or saved from the brain specimen used in Exercise 19). Identify the tough outer meninx (dura mater) and the weblike arachnoid mater.

What name is given to the third meninx, and where is it found?

_____

_____

Peel back the dura mater and observe the fibers making up the dorsal and ventral roots. If possible, identify a dorsal root ganglion.

2.   Cut a thin cross section of the cord and identify the anterior and posterior horns of the gray matter with the naked eye or with the aid of a dissecting microscope.

How can you be certain that you are correctly identifying the anterior and posterior horns?

_____

_____

Also identify the central canal, white matter, anterior median fissure, posterior median sulcus, and posterior, anterior, and lateral funiculi.

3.   Obtain a prepared slide of the spinal cord (cross section) and a compound microscope. Refer to Figure 21.4 as you examine the slide carefully under low power. Observe the shape of the central canal.

Is it basically circular or oval? _____

Name the glial cell type that lines this canal. _____

What would you expect to find in this canal in the living animal?

_____

_____

Can any neuron cell bodies be seen? _____

Where? _____

_____

What type of neurons would these most likely be—motor, association, or sensory?

_____

_____ ■

Posterior median sulcus

Posterior (dorsal) horn

White matter

Anterior (ventral) horn

Anterior median fissure

Figure 21.4 Cross section of the spinal cord. (See also Plate 8 in the Histology Atlas.)

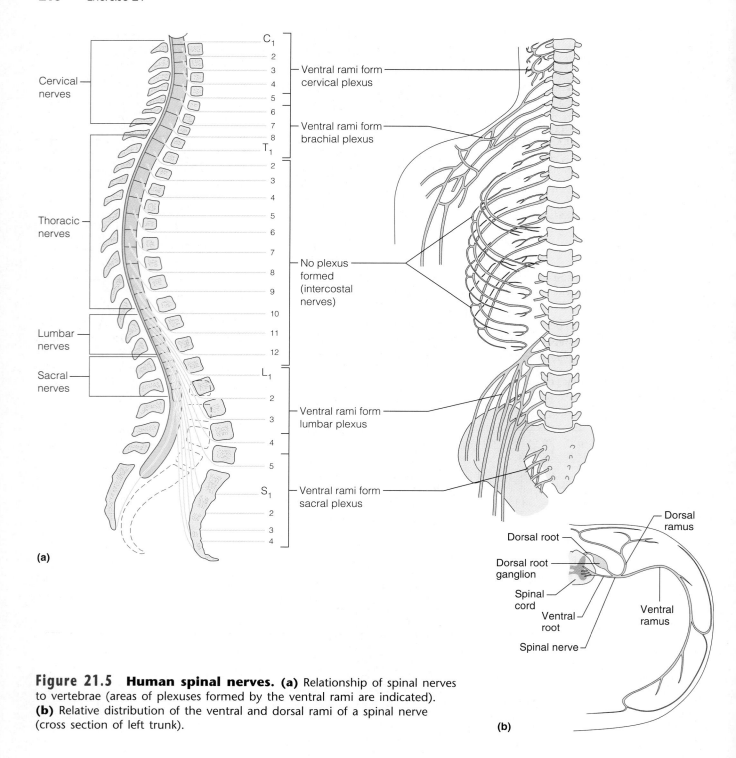

Figure 21.5 Human spinal nerves. (a) Relationship of spinal nerves to vertebrae (areas of plexuses formed by the ventral rami are indicated). **(b)** Relative distribution of the ventral and dorsal rami of a spinal nerve (cross section of left trunk).

Spinal Nerves and Nerve Plexuses

The 31 pairs of human spinal nerves arise from the fusions of the ventral and dorsal roots of the spinal cord. Figure 21.5 shows how the nerves are named according to their point of issue. Because the ventral roots contain myelinated axons of motor neurons located in the cord and the dorsal roots carry sensory fibers entering the cord, all spinal nerves are **mixed nerves.** The first pair of spinal nerves leaves the vertebral canal between the base of the occiput and the atlas, but all the rest exit via the intervertebral foramina. The first through seventh pairs of cervical nerves emerge *above* the vertebra for which they are named. C_8 emerges between C_7 and T_1. (Notice that there are 7 cervical vertebrae, but 8 pairs of cervical nerves.) The remaining spinal nerve pairs emerge from the spinal cord *below* the same-numbered vertebra.

Almost immediately after emerging, each nerve divides into **dorsal** and **ventral rami.** (Thus each spinal nerve is only about 1 or 2 cm long.) The rami, like the spinal nerves, contain both motor and sensory fibers. The smaller dorsal rami serve the skin and musculature of the posterior body trunk at their approximate level of emergence. The ventral rami of spinal nerves T_2-T_{12} pass anteriorly as the **intercostal nerves** to supply the muscles of intercostal spaces, and the skin and muscles of the anterior and lateral trunk. The ventral rami of all other spinal nerves form complex networks of nerves called **plexuses.** These plexuses serve the motor and sensory needs of the muscles and skin of the limbs. The fibers of the ventral rami unite in the plexuses (with a few rami supplying fibers to more than one plexus). From the plexuses the fibers diverge again to form peripheral nerves, each of which contains fibers from more than one spinal nerve. The four major nerve plexuses and their chief peripheral nerves are described below and in Tables 21.1–21.3 and illustrated in Figures 21.6–21.8. Their names and site of origin should be committed to memory.

Cervical Plexus and the Neck

The **cervical plexus** (Figure 21.6 and Table 21.1) arises from the ventral rami of C_1 through C_5 to supply muscles of the shoulder and neck. The major motor branch of this plexus is the **phrenic nerve,** which arises from C_3-C_4 (plus some fibers from C_5) and passes into the thoracic cavity in front of the first rib to innervate the diaphragm. The primary danger of a broken neck is that the phrenic nerve may be severed, leading to paralysis of the diaphragm and cessation of breathing. A jingle to help you remember the rami (roots) forming the phrenic nerves is "C_3, C_4, C_5 keep the diaphragm alive."

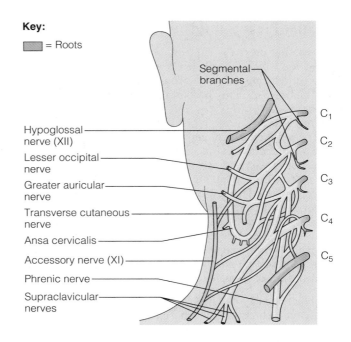

Key:
■ = Roots

Segmental branches

Hypoglossal nerve (XII)
Lesser occipital nerve
Greater auricular nerve
Transverse cutaneous nerve
Ansa cervicalis
Accessory nerve (XI)
Phrenic nerve
Supraclavicular nerves

C_1
C_2
C_3
C_4
C_5

Figure 21.6 The cervical plexus. The nerves colored gray connect to the plexus but do not belong to it. (See Table 21.1.)

Table 21.1	Branches of the Cervical Plexus (See Figure 21.6)

Nerves	Spinal Roots (Ventral Rami)	Structures Served
Cutaneous branches (superfical)		
Lesser occipital	C_2 (C_3)	Skin on posterolateral neck
Greater auricular	C_2, C_3	Skin of ear, skin over parotid gland
Transverse cutaneous	C_2, C_3	Skin on anterior and lateral aspect of neck
Supraclavicular (anterior, middle, and posterior)	C_3, C_4	Skin of shoulder and anterior chest
Motor branches (deep)		
Ansa cervicalis (superior and inferior roots)	C_1-C_3	Infrahyoid muscles of neck
Segmental and other muscular branches	C_1-C_5	Deep muscles of neck and portions of scalenes, levator scapulae, trapezius, and sternocleidomastoid muscles
Phrenic	C_3-C_5	Diaphragm (sole motor nerve supply)

Brachial Plexus and the Upper Limb

The **brachial plexus** is large and complex, arising from the ventral rami of C_5 through C_8 and T_1 (Table 21.2). The plexus, after being rearranged consecutively into *trunks, divisions,* and *cords,* finally becomes subdivided into five major *peripheral nerves* (Figure 21.7 and Plate F of the Human Anatomy Atlas).

The **axillary nerve,** which serves the muscles and skin of the shoulder, has the most limited distribution. The large **radial nerve** passes down the posterolateral surface of the arm and forearm, supplying all the extensor muscles of the arm, forearm, and hand and the skin along its course. The radial nerve is often injured in the axillary region by the pressure of a crutch or by hanging one's arm over the back of a chair. The **median nerve** passes down the anteromedial surface of the arm to supply most of the flexor muscles in the forearm and several muscles in the hand (plus the skin of the lateral surface of the palm of the hand).

- Hyperextend your wrist to identify the long, obvious tendon of your palmaris longus muscle, which crosses the exact midline of the anterior wrist. Your median nerve lies immediately deep to that tendon, and the radial nerve lies just *lateral* to it.

The **musculocutaneous nerve** supplies the arm muscles that flex the forearm and the skin of the lateral surface of the forearm. The **ulnar nerve** travels down the posteromedial surface of the arm. It courses around the medial epicondyle of the humerus to supply the flexor carpi ulnaris, the ulnar head of the flexor digitorum profundus of the forearm, and all intrinsic muscles of the hand not served by the median nerve. It supplies the skin of the medial third of the hand, both the anterior and posterior surfaces. Trauma to the ulnar nerve, which often occurs when the elbow is hit, produces a smarting sensation commonly referred to as "hitting the funny bone."

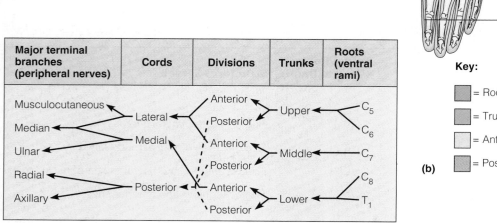

(a)

Figure 21.7   Brachial plexus. (a) Flowchart of consecutive branches formed in the brachial plexus from the spinal roots (ventral rami) to the major nerves formed from the cords. **(b)** Distribution of the major peripheral nerves of the upper limb. (See Table 21.2. See also Plate E in the Human Anatomy Atlas.)

Table 21.2	Branches of the Brachial Plexus (See Figure 21.7)

Nerves	Cord and Spinal Roots (Ventral Rami)	Structures Served
Musculocutaneous	Lateral cord (C_5–C_7)	Muscular branches: flexor muscles in anterior arm (biceps brachii, brachialis, coracobrachialis) Cutaneous branches: skin on anterolateral forearm (extremely variable)
Median	By two branches, one from medial cord (C_8, T_1) and one from the lateral cord (C_5–C_7)	Muscular branches to flexor group of anterior forearm (palmaris longus, flexor carpi radialis, flexor digitorum superficialis, flexor pollicis longus, lateral half of flexor digitorum profundus, and pronator muscles); intrinsic muscles of lateral palm and first two fingers Cutaneous branches: skin of lateral two-thirds of hand, palm side and dorsum of fingers 2 and 3
Ulnar	Medial cord (C_8, T_1)	Muscular branches: flexor muscles in anterior forearm (flexor carpi ulnaris and medial half of flexor digitorum profundus); most intrinsic muscles of hand Cutaneous branches: skin of medial third of hand, both anterior and posterior aspects
Radial	Posterior cord (C_5–C_8, T_1)	Muscular branches: posterior muscles of arm, forearm, and hand (triceps brachii, anconeus, supinator, brachioradialis, extensors carpi radialis longus and brevis, extensor carpi ulnaris, and several muscles that extend the fingers) Cutaneous branches: skin of posterolateral surface of entire limb (except dorsum of fingers 2 and 3)
Axillary	Posterior cord (C_5, C_6)	Muscular branches: deltoid and teres minor muscles Cutaneous branches: some skin of shoulder region
Dorsal scapular	Branches of C_5 rami	Rhomboid muscles and levator scapulae
Long thoracic*	Branches of C_5–C_7 rami	Serratus anterior muscle
Subscapular*	Posterior cord; branches of C_5 and C_6 rami	Teres major and subscapular muscles
Suprascapular	Upper trunk (C_5, C_6)	Shoulder joint; supraspinatus and infraspinatus muscles
Pectoral (lateral and medial)*	Branches of lateral and medial cords (C_5–T_1)	Pectoralis major and minor muscles

* The long thoracic, subscapular, and pectoral nerves are not illustrated.

Severe injuries to the brachial plexus cause weakness or paralysis of the entire upper limb. Such injuries may occur when the upper limb is pulled hard and the plexus is stretched (as when a football tackler yanks the arm of the halfback), and by blows to the shoulder that force the humerus inferiorly (as when a cyclist is pitched headfirst off his motorcycle and grinds his shoulder into the pavement). ■

Lumbosacral Plexus and the Lower Limb

The **lumbosacral plexus,** which serves the pelvic region of the trunk and the lower limbs, is actually a complex of two plexuses, the lumbar plexus and the sacral plexus (see Figure 21.8). These plexuses interweave considerably and many fibers of the lumbar plexus contribute to the sacral plexus.

The Lumbar Plexus   The **lumbar plexus** arises from ventral rami of L_1 through L_4 (and sometimes T_{12}). Its nerves serve the lower abdominopelvic region and the anterior thigh (Table 21.3). The largest nerve of this plexus is the **femoral nerve,** which passes beneath the inguinal ligament to innervate the anterior thigh muscles. The cutaneous branches of the femoral nerve (median and anterior femoral cutaneous and the saphenous nerves) supply the skin of the anteromedial surface of the entire lower limb.

(a)

(b)

Figure 21.8 Distribution of the major peripheral nerves of the lower limb.
(a) Lumbar plexus. **(b)** Sacral plexus. See also Plates F, G, and H in the Human Anatomy Atlas.

The Sacral Plexus Arising from L_4 through S_4, the nerves of the **sacral plexus** supply the buttock, the posterior surface of the thigh, and virtually all sensory and motor fibers of the leg and foot (Table 21.3). The major peripheral nerve of this plexus is the **sciatic nerve,** the largest nerve in the body. The sciatic nerve leaves the pelvis through the greater sciatic notch and travels down the posterior thigh, serving its flexor muscles and skin. In the popliteal region, the sciatic nerve divides into the **common peroneal nerve** and the **tibial nerve,** which together supply the balance of the leg muscles and skin, both directly and via several branches.

Injury to the proximal part of the sciatic nerve, as might follow a fall or disc herniation, results in a number of lower limb impairments. *Sciatica* (si-at' ĭ-kah), characterized by stabbing pain radiating over the course of the sciatic nerve, is common. When the sciatic nerve is completely sev-

ered, the leg is nearly useless. The leg cannot be flexed and the foot drops into plantar flexion (dangles), a condition called *footdrop*. ■

Activity:
Identifying the Major Nerve Plexuses and Peripheral Nerves

Identify each of the four major nerve plexuses (and its major nerves) shown in Figures 21.6 to 21.8 on a large laboratory chart or model. Trace the courses of the nerves and relate those observations to the information provided in Tables 21.1 to 21.3. ■

Table 21.3	Branches of the Lumbosacral Plexus (See Figure 21.8)	

Nerves	Spinal Roots (Ventral Rami)	Structures Served
The Lumbar Plexus		
Femoral	L_2–L_4	Skin of anterior and medial thigh via *anterior femoral cutaneous* branch; skin of medial leg and foot, hip and knee joints via *saphenous* branch; motor to anterior muscles (quadriceps and sartorius) of thigh; pectineus, iliacus
Obturator	L_2–L_4	Motor to adductor magnus (part), longus and brevis muscles, gracilis muscle of medial thigh, obturator externus; sensory for skin of medial thigh and for hip and knee joints
Lateral femoral cutaneous	L_2, L_3	Skin of lateral thigh; some sensory branches to peritoneum
Iliohypogastric	L_1	Skin of lower abdomen, lower back, hip; muscles of anterolateral abdominal wall (obliques and transversus) and pubic region
Ilioinguinal	L_1	Skin of external genitalia and proximal medial aspect of the thigh; inferior abdominal muscles
Genitofemoral	L_1, L_2	Skin of scrotum in males, of labia majora in females, and of anterior thigh inferior to middle portion of inguinal region; cremaster muscle in males
The Sacral Plexus		
Sciatic nerve	L_4, L_5, S_1–S_3	Composed of two nerves (tibial and common peroneal) in a common sheath that diverge just proximal to the knee
• Tibial (including sural branch and medial and lateral plantar branches)	L_4–S_3	Cutaneous branches: to skin of posterior surface of leg and sole of foot Motor branches: to muscles of back of thigh, leg, and foot (hamstrings [except short head of biceps femoris], posterior part of adductor magnus, triceps surae, tibialis posterior, popliteus, flexor digitorum longus, flexor hallucis longus, and intrinsic muscles of foot)
• Common peroneal (superficial and deep branches)	L_4–S_2	Cutaneous branches: to skin of anterior surface of leg and dorsum of foot Motor branches: to short head of biceps femoris of thigh, peroneal muscles of lateral compartment of leg, tibialis anterior, and extensor muscles of toes (extensor hallucis longus, extensors digitorum longus and brevis)
Superior gluteal	L_4, L_5, S_1	Motor branches: to gluteus medius and minimus and tensor fasciae latae
Inferior gluteal	L_5–S_2	Motor branches: to gluteus maximus
Posterior femoral cutaneous	S_1–S_3	Skin of buttock, posterior thigh, and popliteal region; length variable; may also innervate part of skin of calf and heel
Pudendal	S_2–S_4	Supplies most of skin and muscles of perineum (region encompassing external genitalia and anus); external anal sphincter

The Autonomic Nervous System

The **autonomic nervous system (ANS)** is the subdivision of the PNS that regulates body activities that are generally not under conscious control. It is composed of a special group of motor neurons serving cardiac muscle (the heart), smooth muscle (found in the walls of the visceral organs and blood vessels), and internal glands. Because these structures typically function without conscious control, this system is often referred to as the *involuntary nervous system.*

There is a basic anatomical difference between the motor pathways of the **somatic** (voluntary) **nervous system,** which

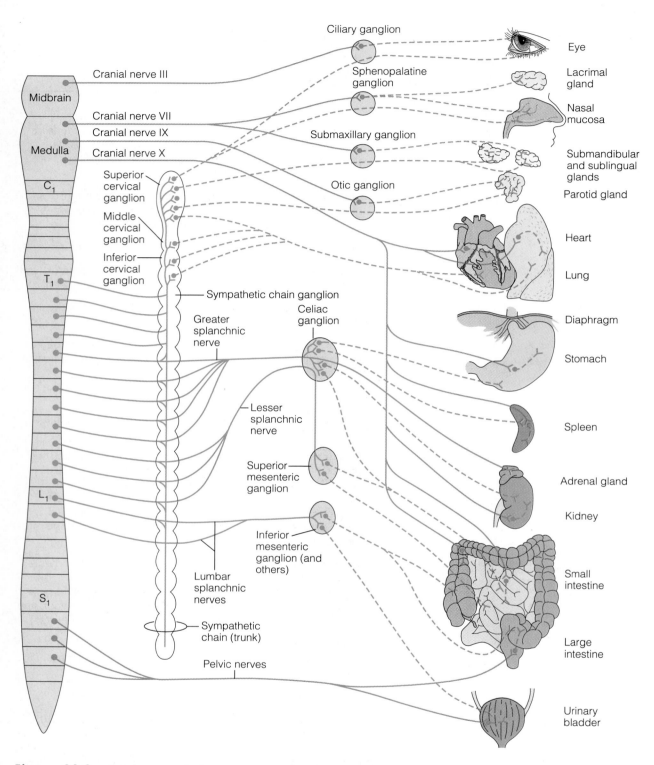

Figure 21.9 Anatomy of the autonomic nervous system. Parasympathetic fibers (craniosacral division) are shown in purple, sympathetic fibers (thoracolumbar division) in green. Solid lines represent preganglionic fibers; dashed lines indicate postganglionic fibers.

innervates the skeletal muscles, and those of the autonomic nervous system. In the somatic division, the cell bodies of the motor neurons reside in the CNS (spinal cord or brain), and their axons, sheathed in spinal nerves, extend all the way to the skeletal muscles they serve. However, the autonomic ner-

vous system consists of chains of two motor neurons. The first motor neuron of each pair, called the *preganglionic neuron,* resides in the brain or cord. Its axon leaves the CNS to synapse with the second motor neuron *(postganglionic neuron),* whose cell body is located in a ganglion outside the

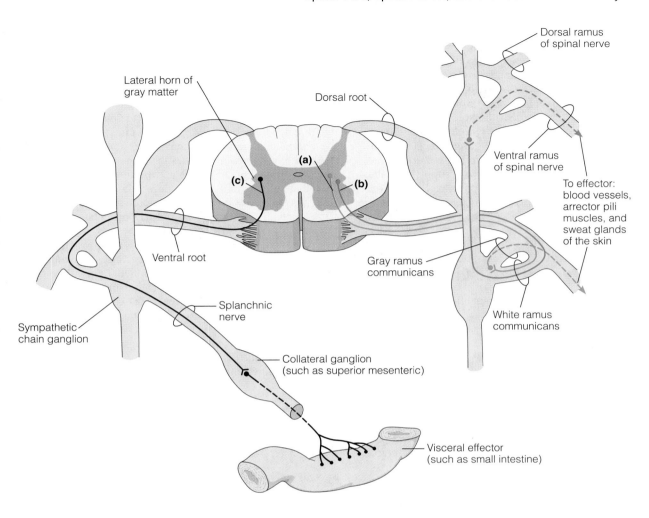

Figure 21.10   Sympathetic pathways. (a) Synapse in a sympathetic chain (paravertebral) ganglion at the same level. **(b)** Synapse in a sympathetic chain ganglion at a different level. **(c)** Synapse in a collateral (prevertebral) ganglion. Preganglionic neurons are shown with solid lines; postganglionic neurons are dashed lines.

CNS. The axon of the postganglionic neuron then extends to the organ it serves.

The autonomic nervous system has two major functional subdivisions (Figure 21.9). These, the sympathetic and parasympathetic divisions, serve most of the same organs, but generally cause opposing or antagonistic effects.

Parasympathetic Division

The preganglionic neurons of the **parasympathetic,** or **craniosacral,** division are located in brain nuclei of cranial nerves III, VII, IX, X and in the S_2 through S_4 level of the spinal cord. The axons of the preganglionic neurons of the cranial region travel in their respective cranial nerves to the *immediate area* of the head and neck organs to be stimulated. There they synapse with the postganglionic neuron in a **terminal,** or **intramural** (literally, "within the walls"), **ganglion.** The postganglionic neuron then sends out a very short axon to the organ it serves. In the sacral region, the preganglionic axons leave the ventral roots of the spinal cord and collectively form the **pelvic splanchnic nerves,** which travel to the pelvic cavity. In the pelvic cavity, the preganglionic ax-

ons synapse with the postganglionic neurons in ganglia located on or close to the organs served.

Sympathetic Division

The preganglionic neurons of the **sympathetic,** or **thoracolumbar,** division are in the lateral horns of the gray matter of the spinal cord from T_1 through L_2. The preganglionic axons leave the cord via the ventral root (in conjunction with the axons of the somatic motor neurons), enter the spinal nerve, and then travel briefly in the ventral ramus (Figure 21.10). From the ventral ramus, they pass through a small branch called the **white ramus communicans** to enter a **paravertebral ganglion** in the **sympathetic chain,** or **trunk,** which lies alongside the vertebral column (the literal meaning of *paravertebral*).

Having reached the ganglion, a preganglionic axon may take one of three main courses (see Figure 21.10). First, it may synapse with a postganglionic neuron in the sympathetic chain at that level. Second, the axon may travel upward or downward through the sympathetic chain to synapse with a postganglionic neuron in a paravertebral ganglion at another level. In either of these two instances, the postganglionic axons then reenter the ventral or dorsal ramus of a spinal nerve via a **gray ramus communicans** and travel in the ramus to innervate skin structures (sweat glands, arrector pili muscles attached to hair follicles, and the smooth muscles of blood vessel walls). Third, the axon may pass through the ganglion without synapsing and form part of the **splanchnic nerves,** which travel to the viscera to synapse with a postganglionic neuron in a **prevertebral** or **collateral ganglion.** The major prevertebral ganglia—the *celiac, superior mesenteric, inferior mesenteric,* and *hypogastric ganglia*—supply the abdominal and pelvic visceral organs. The postganglionic axon then leaves the ganglion and travels to a nearby visceral organ which it innervates.

Activity:
Locating the Sympathetic Chain

Locate the sympathetic chain on the spinal nerve chart. ■

Autonomic Functioning

As noted earlier, most body organs served by the autonomic nervous system receive fibers from both the sympathetic and parasympathetic divisions. The only exceptions are the structures of the skin (sweat glands and arrector pili muscles attached to the hair follicles), the pancreas and liver, the adrenal medulla, and essentially all blood vessels except those of the external genitalia, all of which receive sympathetic innervation only. When both divisions serve an organ, they have antagonistic effects. This is because their postganglionic axons release different neurotransmitters. The parasympathetic fibers, called **cholinergic fibers,** release acetylcholine; the sympathetic postganglionic fibers, called **adrenergic fibers,** release norepinephrine. (However, there are isolated examples of postganglionic sympathetic fibers, such as those serving blood vessels in the skeletal muscles, that release acetylcholine.) The preganglionic fibers of both divisions release acetylcholine.

The parasympathetic division is often referred to as the housekeeping, or "resting and digesting," system because it maintains the visceral organs in a state most suitable for normal functions and internal homeostasis; that is, it promotes normal digestion and elimination. In contrast, activation of the sympathetic division is referred to as the "fight or flight" response because it readies the body to cope with situations that threaten homeostasis. Under such emergency conditions, the sympathetic nervous system induces an increase in heart rate and blood pressure, dilates the bronchioles of the lungs, increases blood sugar levels, and promotes many other effects that help the individual cope with a stressor.

As we grow older, our sympathetic nervous system gradually becomes less and less efficient, particularly in causing vasoconstriction of blood vessels. When elderly people stand up quickly after sitting or lying down, they often become light-headed or faint. This is because the sympathetic nervous system is not able to react quickly enough to counteract the pull of gravity by activating the vasoconstrictor fibers. So, blood pools in the feet. This condition, **orthostatic hypotension,** is a type of low blood pressure resulting from changes in body position as described. Orthostatic hypotension can be prevented to some degree if *slow* changes in position are made. This gives the sympathetic nervous system a little more time to react and adjust. ■

Activity:
Comparing Sympathetic and Parasympathetic Effects

Several body organs are listed in the chart below. *Using your textbook as a reference,* list the effect of the sympathetic and parasympathetic divisions on each. ■

For instructions on animal dissections, see the dissection exercises starting on p. 733 in the cat and fetal pig editions of this manual.

Organ or function	Parasympathetic effect	Sympathetic effect
Heart		
Bronchioles of lungs		
Digestive tract activity		
Urinary bladder		
Iris of the eye		
Blood vessels (most)		
Penis/clitoris		
Sweat glands		
Adrenal medulla		
Pancreas		

Human Reflex Physiology

The Reflex Arc

Reflexes are rapid, predictable, involuntary motor responses to stimuli; they are mediated over neural pathways called **reflex arcs.**

Reflexes can be categorized into one of two large groups: autonomic reflexes and somatic reflexes. **Autonomic (or visceral) reflexes** are mediated through the autonomic nervous system and are not subject to conscious control. These reflexes activate smooth muscles, cardiac muscle, and the glands of the body and they regulate body functions such as digestion, elimination, blood pressure, salivation, and sweating. **Somatic reflexes** include all those reflexes that involve stimulation of skeletal muscles by the somatic division of the nervous system. An example of such a reflex is the rapid withdrawal of a hand from a hot object.

Reflex testing is an important diagnostic tool for assessing the condition of the nervous system. Distorted, exaggerated, or absent reflex responses may indicate degeneration or pathology of portions of the nervous system, often before other signs are apparent.

If the spinal cord is damaged, the easily performed reflex tests can help pinpoint the area (level) of spinal cord injury. Motor nerves above the injured area may be unaffected, whereas those at or below the lesion site may be unable to participate in normal reflex activity. ■

Components of a Reflex Arc

All reflex arcs have five essential components (Figure 22.1):

1. The *receptor* reacts to a stimulus.

2. The *sensory neuron* conducts the afferent impulses to the CNS.

3. The *integration center* consists of one or more synapses in the CNS.

4. The *motor neuron* conducts the efferent impulses from the integration center to an effector.

5. The *effector,* muscle fibers or glands, responds to the efferent impulses by contracting or secreting a product, respectively.

Objectives

1. To define *reflex* and *reflex arc.*

2. To name, identify, and describe the function of each element of a reflex arc.

3. To indicate why reflex testing is an important part of every physical examination.

4. To describe and discuss several types of reflex activities as observed in the laboratory; to indicate the functional or clinical importance of each; and to categorize each as a somatic or autonomic reflex action.

5. To explain why cord-mediated reflexes are generally much faster than those involving input from the higher brain centers.

6. To investigate differences in reaction time of reflexes and unlearned responses.

Materials

❑ Reflex hammer

❑ Sharp pencils

❑ Cot (if available)

❑ Absorbent cotton (sterile)

❑ Tongue depressor

❑ Metric and 12-in. ruler

❑ Flashlight

❑ 100- or 250-ml beaker

❑ 10- or 25-ml graduated cylinder

❑ Lemon juice in dropper bottle

❑ Wide-range pH paper

❑ Large laboratory bucket containing freshly prepared 10% household bleach solution (for saliva-soiled glassware)

❑ Disposable autoclave bag

❑ Wash bottle containing 10% bleach solution

Figure 22.1 Simple reflex arcs. Components of all human reflex arcs: receptor, sensory neuron, integration center (one or more synapses in the CNS), motor neuron, and effector.

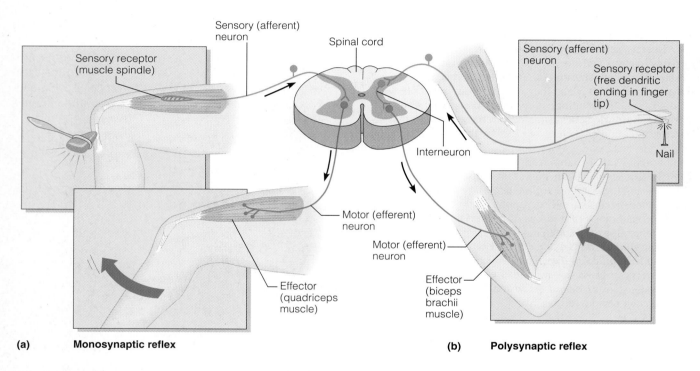

(a) Monosynaptic reflex

(b) Polysynaptic reflex

Figure 22.2 Monosynaptic and polysynaptic reflex arcs. The integration center is in the spinal cord, and in each example the receptor and effector are in the same limb. **(a)** The patellar reflex, a two-neuron monosynoptic reflex. **(b)** A flexor reflex, an example of a polysynoptic reflex.

The simple patellar or knee-jerk reflex shown in Figure 22.2a is an example of a simple, two-neuron, *monosynaptic* (literally, "one synapse") reflex arc. It will be demonstrated in the laboratory. However, most reflexes are more complex and *polysynaptic*, involving the participation of one or more association neurons in the reflex arc pathway. A three-neuron reflex arc (flexor reflex) is diagrammed in Figure 22.2b. Since delay or inhibition of the reflex may occur at the synapses, the more synapses encountered in a reflex pathway, the more time is required to effect the reflex.

Reflexes of many types may be considered programmed into the neural anatomy. Many *spinal reflexes* (reflexes that are initiated and completed at the spinal cord level such as the flexor reflex) occur without the involvement of higher brain centers. These reflexes work equally well in decerebrate animals (those in which the brain has been destroyed), as long as the spinal cord is functional. Conversely, other reflexes require the involvement of functional brain tissue, since many different inputs must be evaluated before the appropriate reflex is determined. Superficial cord reflexes and pupillary responses to light are in this category. In addition, although many spinal reflexes do not require the involvement of higher centers, the brain is "advised" of spinal cord reflex activity and may alter it by facilitating or inhibiting the reflexes.

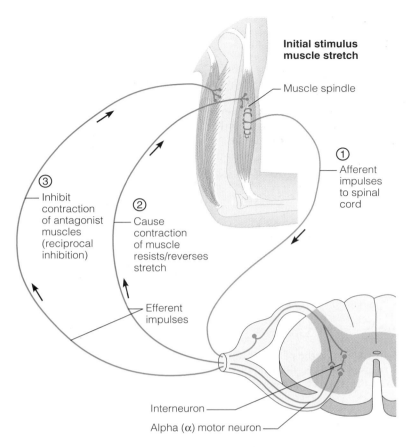

Initial stimulus muscle stretch

Muscle spindle

③ Inhibit contraction of antagonist muscles (reciprocal inhibition)

② Cause contraction of muscle resists/reverses stretch

① Afferent impulses to spinal cord

Efferent impulses

Interneuron

Alpha (α) motor neuron

Figure 22.3 Events of the stretch reflex by which muscle stretch is damped. The events are shown in circular fashion. **(1)** Stretching of the muscle activates a muscle spindle. **(2)** Impulses transmitted by afferent fibers from muscle spindle to alpha motor neurons in the spinal cord result in activation of the stretched muscle, causing it to contract. **(3)** Impulses transmitted by afferent fibers from muscle spindle to interneurons in the spinal cord result in reciprocal inhibition of the antagonist muscle.

Somatic Reflexes

There are several types of somatic reflexes, including several that you will be eliciting during this laboratory session—the stretch, crossed extensor, superficial cord, corneal, and gag reflexes. Some require only spinal cord activity; others require brain involvement as well.

Spinal Reflexes

Stretch Reflexes **Stretch reflexes** are important postural reflexes, normally acting to maintain posture, balance, and locomotion. Stretch reflexes are initiated by tapping a tendon, which stretches the muscle the tendon is attached to. This stimulates the muscle spindles and causes reflex contraction of the stretched muscle or muscles, which resists further

stretching. Even as the primary stretch reflex is occurring, impulses are being sent to other destinations as well. For example, branches of the afferent fibers (from the muscle spindles) also synapse with association neurons (interneurons) controlling the antagonist muscles (Figure 22.3). The inhibition of the antagonist muscles that follows, called *reciprocal inhibition,* causes them to relax and prevents them from resisting (or reversing) the contraction of the stretched muscle caused by the main reflex arc. Additionally, impulses are relayed to higher brain centers (largely via the dorsal white columns) to advise of muscle length, speed of shortening, and the like—information needed to maintain muscle tone and posture. Stretch reflexes tend to be hypoactive or absent in cases of peripheral nerve damage or ventral horn disease, and hyperactive in corticospinal tract lesions. They are absent in deep sedation and coma.

Activity:
Initiating Stretch Reflexes

1. Test the **patellar,** or knee-jerk, **reflex** by seating a subject on the laboratory bench with legs hanging free (or with knees crossed). Tap the patellar ligament sharply with the reflex hammer just below the knee to elicit the knee-jerk response, which assesses the L_2–L_4 level of the spinal cord (Figure 22.4). Test both knees and record your observations.

_____

_____

Which muscles contracted? _____

_____

What nerve is carrying the afferent and efferent impulses?

_____

2. Test the effect of mental distraction on the patellar reflex by having the subject add a column of three-digit numbers while you test the reflex again. Is the response greater than *or* less than the first response?

_____

What are your conclusions about the effect of mental distraction on reflex activity?

_____

_____

3. Now test the effect of muscular activity occurring simultaneously in other areas of the body. Have the subject clasp the edge of the laboratory bench and vigorously attempt to pull it upward with both hands. At the same time, test the patellar reflex again. Is the response more or less vigorous than the first response?

_____

4. Fatigue also influences the reflex response. The subject should jog in position until she or he is very fatigued (**really fatigued**—no slackers). Test the patellar reflex again and record whether it is more or less vigorous than the first response.

_____

Would you say that nervous system activity *or* muscle function is responsible for the changes you have just observed?

_____

Figure 22.4   Testing the patellar reflex. The examiner supports the subject's knee so that the subject's muscles are relaxed, and then strikes the patellar ligament with the reflex hammer. The proper location may be ascertained by palpation of the patella.

Explain your reasoning. _____

_____

_____

_____

5. The **Achilles,** or ankle-jerk, **reflex** assesses the first two sacral segments of the spinal cord. With your shoe removed and your foot dorsiflexed slightly to increase the tension of the gastrocnemius muscle, have your partner sharply tap your calcaneal (Achilles) tendon with the reflex hammer (Figure 22.5).

What is the result? _____

_____

Does the contraction of the gastrocnemius normally result in the activity you have observed?

_____ ■

Crossed Extensor Reflex   The **crossed extensor reflex** is more complex than the stretch reflex. It consists of a flexor, or withdrawal, reflex followed by extension of the opposite limb.

This reflex is quite obvious when, for example, a stranger suddenly and strongly grips one's arm. The immediate response is to withdraw the clutched arm and push the in-

Figure 22.5 Testing the Achilles reflex. The examiner slightly dorsiflexes the subject's ankle by supporting the foot lightly in the hand, and then taps the Achilles tendon just above the ankle.

truder away with the other arm. The reflex is more difficult to demonstrate in a laboratory because it is anticipated, and under these conditions the extensor part of the reflex may be inhibited.

Activity:
Initiating the Crossed Extensor Reflex

The subject should sit with eyes closed and with the dorsum of one hand resting on the laboratory bench. Obtain a sharp pencil and suddenly prick the subject's index finger. What are the results?

_____

_____

_____

Did the extensor part of this reflex seem to be slow compared to the other reflexes you have observed?

_____

What are the reasons for this? _____

_____

_____ ∎

The reflexes that have been demonstrated so far—the stretch and crossed extensor reflexes—are examples of reflexes in which the reflex pathway is initiated and completed at the spinal cord level.

Superficial Cord Reflexes The **superficial cord reflexes** (abdominal, cremaster, and plantar reflexes) result from pain and temperature changes. They are initiated by stimulation of receptors in the skin and mucosae. The superficial cord reflexes depend *both* on functional upper-motor pathways and on the cord-level reflex arc. Since only the plantar reflex can be tested conveniently in a laboratory setting, we will use this as our example.

The **plantar reflex,** an important neurological test, is elicited by stimulating the cutaneous receptors in the sole of the foot. In adults, stimulation of these receptors causes the toes to flex and move closer together. Damage to the pyramidal (or corticospinal) tract, however, produces *Babinski's sign,* an abnormal response in which the toes flare and the great toe moves in an upward direction. (In newborn infants, Babinski's sign is seen due to incomplete myelination of the nervous system.)

Activity:
Initiating the Plantar Reflex

Have the subject remove a shoe and lie on the cot or laboratory bench with knees slightly bent and thighs rotated so that the lateral side of the foot rests on the cot. Alternatively, the subject may sit up and rest the lateral surface of the foot on a chair. Draw the handle of the reflex hammer firmly down the lateral side of the exposed sole from the heel to the base of the great toe (Figure 22.6).

Figure 22.6 Testing the plantar reflex. Using a moderately sharp object, the examiner strokes the lateral border of the subject's sole, starting at the heel and continuing toward the big toe across the ball of the foot.

What is the response? _____

_____

Is this a normal plantar reflex or Babinski's sign?

_____ ■

Cranial Nerve Reflex Tests

In these experiments, you will be working with your lab partner to illustrate two somatic reflexes mediated by cranial nerves.

Corneal Reflex The **corneal reflex** is mediated through the trigeminal nerve (cranial nerve V). The absence of this reflex is an ominous sign because it often indicates damage to the brain stem resulting from compression of the brain or other trauma.

Activity:
Initiating the Corneal Reflex

Stand to one side of the subject; the subject should look away from you toward the opposite wall. Wait a few seconds and then quickly, *but gently,* touch the subject's cornea (on the side toward you) with a wisp of absorbent cotton. What is the reaction?

_____

What is the function of this reflex?

_____

_____

_____

Was the sensation that of touch *or* of pain?

_____

Why? _____

_____ ■

Gag Reflex The **gag reflex** tests the somatic motor responses of cranial nerves IX and X. When the oral mucosa on the side of the uvula is stroked, each side of the mucosa should rise, and the amount of elevation should be equal.*

* The uvula is the fleshy tab hanging from the roof of the mouth just above the root of the tongue.

Activity:
Initiating the Gag Reflex

For this experiment, select a subject who does not have a queasy stomach, because regurgitation is a possibility. Stroke the oral mucosa on each side of the subject's uvula with a tongue depressor. What happens?

_____

_____

⚠ Discard the used tongue depressor in the disposable autoclave bag before continuing. Do *not* lay it on the laboratory bench at any time. ■

Autonomic Reflexes

The autonomic reflexes include the pupillary, ciliospinal, and salivary reflexes, as well as a multitude of other reflexes. Work with your partner to demonstrate the four autonomic reflexes described next.

Pupillary Reflexes

There are several types of pupillary reflexes. The **pupillary light reflex** and the **consensual reflex** will be examined here. In both of these pupillary reflexes, the retina of the eye is the receptor, the optic nerve (cranial nerve II) contains the afferent fibers, the oculomotor nerve (cranial nerve III) is responsible for conducting efferent impulses to the eye, and the smooth muscle of the iris is the effector. Many central nervous system centers are involved in the integration of these responses. Absence of the normal pupillary reflexes is generally a late indication of severe trauma or deterioration of the vital brain stem tissue due to metabolic imbalance.

Activity:
Initiating Pupillary Reflexes

1. Conduct the reflex testing in an area where the lighting is relatively dim. Before beginning, obtain a metric ruler to measure and record the size of the subject's pupils as best you can.

Right pupil: _____ mm   Left pupil: _____ mm

2. Stand to the left of the subject to conduct the testing. The subject should shield his or her right eye by holding a hand vertically between the eye and the right side of the nose.

3. Shine a flashlight into the subject's left eye. What is the pupillary response?

_____

Measure the size of the left pupil: _____ mm

4.   Observe the right pupil. Has the same type of change (called a *consensual response*) occurred in the right eye?

_____

_____

Measure the size of the right pupil: _____ mm

The consensual response, or any reflex observed on one side of the body when the other side has been stimulated, is called a **contralateral response.** The pupillary light response, or any reflex occurring on the same side stimulated, is referred to as an **ipsilateral response.**

When a contralateral response occurs, what does this indicate about the pathways involved?

_____

_____

_____

Was the sympathetic *or* the parasympathetic division of the autonomic nervous system active during the testing of these reflexes?

_____

_____

What is the function of these pupillary responses?

_____

_____

_____ ■

Ciliospinal Reflex

The **ciliospinal reflex** is another example of reflex activity in which pupillary responses can be observed. This response may initially seem a little bizarre, especially in view of the consensual reflex just demonstrated.

Activity:
Initiating the Ciliospinal Reflex

1.   While observing the subject's eyes, gently stroke the skin (or just the hairs) on the left side of the back of the subject's neck, close to the hairline.

What is the reaction of the left pupil? _____

The reaction of the right pupil? _____

2.   If you see no reaction, repeat the test using a gentle pinch in the same area.
    The response you should have noted—pupillary dilation—is consistent with the pupillary changes occurring when the sympathetic nervous system is stimulated. Such a response may also be elicited in a single pupil when more impulses from the sympathetic nervous system reach it for any

reason. For example, when the left side of the subject's neck was stimulated, sympathetic impulses to the left iris increased, resulting in the ipsilateral reaction of the left pupil.
    On the basis of your observations, would you say that the sympathetic innervation of the two irises is closely integrated?

_____Why or why not? _____

_____

_____ ■

Salivary Reflex

Unlike the other reflexes, in which the effectors were smooth or skeletal muscles, the effectors of the **salivary reflex** are glands. The salivary glands secrete varying amounts of saliva in response to reflex activation.

Activity:
Initiating the Salivary Reflex

1.   Obtain a small beaker, a graduated cylinder, lemon juice, and wide-range pH paper. After refraining from swallowing for 2 minutes, the subject is to expectorate (spit) the accumulated saliva into a small beaker. Using the graduated cylinder, measure the volume of the expectorated saliva and determine its pH.

Volume: _____ cc   pH:_____

2.   Now place 2 or 3 drops of lemon juice on the subject's tongue. Allow the lemon juice to mix with the saliva for 5 to 10 seconds, and then determine the pH of the subject's saliva by touching a piece of pH paper to the tip of the tongue.

pH: _____

As before, the subject is to refrain from swallowing for 2 minutes. After the 2 minutes is up, again collect and measure the volume of the saliva and determine its pH.

Volume: _____ cc   pH:_____

3.   How does the volume of saliva collected after the application of the lemon juice compare with the volume of the first saliva sample?

_____

How does the final saliva pH reading compare to the initial reading?

_____

To that obtained 10 seconds after the application of lemon juice?

_____

What division of the autonomic nervous system mediates the reflex release of saliva?

_____
_____

⚠ Dispose of the saliva-containing beakers and the graduated cylinders in the laboratory bucket that contains bleach and put the used pH paper into the disposable autoclave bag. Wash the bench down with 10% bleach solution before continuing. ■

Reaction Time of Unlearned Responses

The time required for reaction to a stimulus depends on many factors—sensitivity of the receptors, velocity of nerve conduction, the number of neurons and synapses involved, and the speed of effector activation, to name just a few. The type of response to be elicited is also important. If the response involves a reflex arc, the synapses are facilitated and response time will be short. If, on the other hand, the response can be categorized as an unlearned response, then a far larger number of neural pathways and many types of higher intellectual activities—including choice and decision making—will be involved, and the time for response will be considerably lengthened.

There are various ways of testing reaction time of unlearned responses. The tests range in difficulty from simple to ultrasophisticated. Since the objective here is to demonstrate the major time difference between reflexes and unlearned responses, the simple approach will suffice.

Activity:
Testing Reaction Time for Unlearned Responses

1. Using a reflex hammer, elicit the patellar reflex in your partner. Note the relative reaction time needed for this reflex to occur.

2. Now test the reaction time for unlearned responses. The subject should hold a hand out, with the thumb and index finger extended. Hold a 12-in. ruler so that its end is exactly 1 in. above the subject's outstretched hand. The ruler should be in the vertical position with the numbers reading from the bottom up. When the ruler is dropped, the subject should be able to grasp it between thumb and index finger as it passes, without having to change position. Have the subject catch the ruler five times, varying the time between trials. The relative speed of reaction can be determined by reading the number on the ruler at the point of the subject's fingertips. (Thus if the number at the fingertips is 6 in., the subject was unable to catch the ruler until 7 in. of length had passed through his or her fingers; 6 in. of ruler length plus 1 in. to account for the distance of the ruler above the hand.) Record the number of inches that pass through the subject's fingertips for each trial:

Trial 1:_____ in.   Trial 4:_____ in.

Trial 2:_____ in.   Trial 5:_____ in.

Trial 3:_____ in.

3. Perform the test again, but this time say a simple word each time you release the ruler. Designate a specific word as a signal for the subject to catch the ruler. On all other words, the subject is to allow the ruler to pass through his fingers. Trials in which the subject erroneously catches the ruler are to be disregarded. Record the distance the ruler travels for five *successful* trials:

Trial 1:_____ in.   Trial 4:_____ in.

Trial 2:_____ in.   Trial 5:_____ in.

Trial 3:_____ in.

Did the addition of a specific word to the stimulus increase or decrease the reaction time?

_____

4. Perform the testing once again to investigate the subject's reaction to word association. As you drop the ruler, say a word—for example, *hot*. The subject is to respond with a word he or she associates with the stimulus word—for example, *cold*—catching the ruler while responding. If unable to make a word association, the subject must allow the ruler to pass through his or her fingers. Record the distance the ruler travels for five successful trials, as well as the number of times the ruler is not caught by the subject.

Trial 1:_____ in.   Trial 4:_____ in.

Trial 2:_____ in.   Trial 5:_____ in.

Trial 3:_____ in.

The number of times the subject was unable to catch the ruler:

_____

You should have noticed quite a large variation in reaction time in this series of trials. Why is this so?

_____
_____
_____
_____ ■

General Sensation

It can be said without reservation that people are irritable creatures. Hold a sizzling steak before them and their mouths water. Flash your high beams in their eyes on the highway and they cuss. Tickle them and they giggle. These "irritants" (the steak, the light, and the tickle) and many others are stimuli that continually assault us.

The body's **sensory receptors** react to stimuli or changes within the body and in the external environment. The tiny sensory receptors of the *general senses* react to touch, pressure, pain, heat, cold, and changes in position and are distributed throughout the body. In contrast to these widely distributed *general sensory receptors,* the receptors of the special senses are large, complex *sense organs* or small, localized groups of receptors. The *special senses* include sight, hearing, equilibrium, smell, and taste.

Sensory receptors may be classified according to the source of their stimulus. **Exteroceptors** react to stimuli in the external environment, and typically they are found close to the body surface. Exteroceptors include the simple cutaneous receptors in the skin and the highly specialized receptor structures of the special senses (the vision apparatus of the eye, and the hearing and equilibrium receptors of the ear, for example). **Interoceptors** respond to stimuli arising within the body. Interoceptors are found in the internal visceral organs, and include stretch receptors (in walls of hollow organs), chemoreceptors, and others. **Proprioceptors,** a special subdivision of interoceptors, are located in joint capsules and in skeletal muscles and their tendons. They provide information on the position and degree of stretch of those structures.

The receptors of the special sense organs are complex and deserve considerable study. Thus the special senses (vision, hearing, equilibrium, taste, and smell) are covered separately in Exercises 24 through 26. Only the anatomically simpler **general sensory receptors**—cutaneous receptors and proprioceptors—will be studied in this exercise.

Structure of General Sensory Receptors

You cannot become aware of changes in the environment unless your sensory neurons and their receptors are operating properly. Sensory receptors are modified dendritic endings (or specialized cells associated with the dendrites) that are sensitive to certain environmental stimuli. They react to such stimuli by initiating a nerve impulse. Several histologically distinct types of general sensory receptors have been identified in the skin. Their structures are depicted in Figure 23.1.

Many references link each type of receptor to specific stimuli, but there is still considerable controversy about the precise qualitative function of each receptor. It may be that

Objectives

1. To recognize various types of general sensory receptors as studied in the laboratory, and to describe the function and location of each type.
2. To define *exteroceptor, interoceptor,* and *proprioceptor.*
3. To demonstrate and relate differences in relative density and distribution of tactile and thermoreceptors in the skin.
4. To define *tactile localization* and describe how this ability varies in different areas of the body.
5. To explain the tactile two-point discrimination test, and to state its anatomical basis.
6. To define *referred pain.*
7. To define *adaptation, negative afterimage,* and *projection.*

Materials

❑ Compound microscope
❑ Histologic slides of Pacinian corpuscles (longitudinal section), Meissner's corpuscles, Golgi tendon organs, and muscle spindles
❑ Mall probes (obtainable from Carolina Biological Supply, code #62-7400), or pointed metal rods of approximately 1-in. diameter and 2- to 3-in. length, with insulated blunt end
❑ Large beaker of ice water; chipped ice
❑ Hot water bath set at 45°C
❑ Laboratory thermometer
❑ Fine-point, felt-tipped markers (black, red, and blue)
❑ Small metric ruler
❑ Von Frey's hairs (or 1-in. lengths of horsehair glued to a matchstick) or sharp pencil
❑ Towel
❑ Caliper or esthesiometer
❑ Four coins (nickels or quarters)
❑ Three large finger bowls or 1000-ml beakers

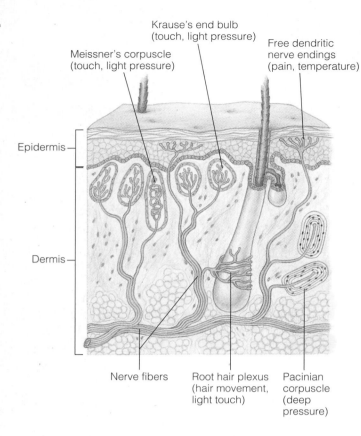

Krause's end bulb
(touch, light pressure)

Meissner's corpuscle
(touch, light pressure)

Free dendritic
nerve endings
(pain, temperature)

Epidermis

Dermis

Nerve fibers

Root hair plexus
(hair movement,
light touch)

Pacinian
corpuscle
(deep
pressure)

Figure 23.1 Cutaneous receptors. Free dendritic nerve endings, root hair plexus, Meissner's corpuscle, Pacinian corpuscle, and Krause's end bulb. (See also Plates 10, 11, and 12 in the Histology Atlas.)

Activity:
Studying the Structure of Selected Sensory Receptors

1. Obtain histologic slides of Pacinian and Meissner's corpuscles. Locate, under low power, a Meissner's corpuscle in the dermal layer of the skin. As mentioned above, these are usually found in the dermal papillae. Then switch to the oil immersion lens for a detailed study. Notice that the naked dendritic fibers within the capsule are aligned parallel to the skin surface. Compare your observations to Figure 23.1 and Plate 10 of the Histology Atlas.

2. Next observe a Pacinian corpuscle located much deeper in the dermis. Try to identify the slender naked dendrite ending in the center of the receptor and the heavy capsule of connective tissue surrounding it (which looks rather like an onion cut lengthwise). Also, notice how much larger the Pacinian corpuscles are than the Meissner's corpuscles. Compare your observations to the views shown in Figure 23.1 and Plate 12 of the Histology Atlas.

3. Obtain slides of muscle spindles and Golgi tendon organs, the two major types of proprioceptors (see Figure 23.2). In the slide of **muscle spindles,** note that minute extensions of the dendrites of the sensory neurons coil around specialized slender skeletal muscle cells called **intrafusal cells,** or **fibers.** The **Golgi tendon organs** are composed of dendrites that ramify through the tendon tissue close to the muscle tendon attachment. Stretching of muscles or tendons excites these receptors, which then transmit impulses that ultimately reach the cerebellum for interpretation. Compare your observations to Figure 23.2. ■

Receptor Physiology

Sensory receptors act as *transducers,* changing environmental stimuli into afferent nerve impulses. Since the action potential generated in all nerve fibers is essentially identical, the stimulus is identified entirely by the area of the brain's sensory cortex that is stimulated (which, of course, differs for the various afferent nerves).

Four qualities of cutaneous sensations have traditionally been recognized: tactile (touch), heat, cold, and pain. Mapping these sensations on the skin has revealed that the sensory receptors for these qualities are not distributed uniformly. Instead, they have discrete locations and are characterized by clustering at certain points—**punctate distribution.**

The simple pain receptors, extremely important in protecting the body, are the most numerous. Touch receptors cluster where greater sensitivity is desirable, as on the hands and face. It is surprising to learn that rather large areas of the skin are quite insensitive to touch because of a relative lack of touch receptors.

There are several simple experiments you can conduct to investigate the location and physiology of cutaneous receptors. In each of the following activities, work in pairs with one person as the subject and the other as the experimenter. After you have completed an experiment, switch roles and go through the procedures again so that all class members obtain

the responses of all these receptors overlap considerably. Certainly, intense stimulation of any of them is always interpreted as pain.

The least specialized of the cutaneous receptors are the **free dendritic endings** of sensory neurons (Figure 23.1 and Plate 11 of the Histology Atlas), which respond chiefly to pain and temperature. (The pain receptors are widespread in the skin and make up a sizable portion of the visceral interoceptors.) Certain free dendritic endings associate with specific epidermal cells to form **Merkel discs,** or entwine in hair follicles to form **root hair plexuses.** Both Merkel discs and root hair plexuses function as light touch receptors.

The other cutaneous receptors are a bit more complex, with the dendritic endings *encapsulated* by connective tissue cells. **Meissner's corpuscles,** commonly referred to as *tactile receptors* because they respond to light touch, are located in the dermal papillae of the skin. Although **Krause's end bulbs** and the anatomically similar **Ruffini's corpuscles** were presumed to be thermoreceptors, now most authorities classify Krause's end bulbs with Meissner's corpuscles as a special kind of touch, or light pressure, receptor. Ruffini's corpuscles appear to respond to deep pressure and stretch stimuli. As you inspect Figure 23.1, notice that all of the encapsulated receptors are quite similar, with the possible exception of the **Pacinian corpuscles** (deep pressure receptors), which are anatomically more distinctive and lie deepest in the dermis.

Figure 23.2 Proprioceptors. (a) Diagrammatic view of a muscle spindle and Golgi tendon organ. **(b)** Drawing of a photomicrograph of a muscle spindle. (See also Plate 13 in the Histology Atlas.)

individual results. Keep an accurate account of each test that you perform.

Density and Location of Touch and Temperature Receptors

A c t i v i t y :

Plotting the Relative Density and Location of Touch and Temperature Receptors

1. Obtain 2 Mall probes (or temperature rods), Von Frey hairs (or a sharp pencil), and black, red, and blue felt-tipped markers and bring them to your laboratory bench.

2. Place one Mall probe (or temperature rod) in a beaker of ice water and the other in a water bath controlled at 45°C. With a felt marker, draw a square (2 cm on each side) on the ventral surface of the subject's forearm. During the following tests, the subject's eyes are to remain closed. The subject should tell the examiner when a stimulus is detected.

3. Working in a systematic manner from one side of the marked square to the other, gently touch the Von Frey's hairs or a sharp pencil tip to different points within the square. The *hairs* should be applied with a pressure that just causes them to bend; use the same pressure for each contact. Do not apply deep pressure. The goal is to stimulate only the more superficially located Meissner's corpuscles (as opposed to the Pacinian corpuscles located in the subcutaneous tissue). Mark with a *black dot* all points at which the touch is perceived.

4. Remove the Mall probe (or temperature rod) from the ice water and quickly wipe it dry. Repeat the procedure outlined above, noting all points of cold perception with a *blue dot.** Perception should be for temperature, not simply touch.

5. Remove the second Mall probe from the 45°C water bath and repeat the procedure once again, marking all points of heat perception with a *red dot.**

*The Mall probes will have to be returned to the water baths approximately every 2 minutes to maintain the desired testing temperatures.

6. After each student has acted as the subject, produce a "map" of your own receptor areas in the squares provided here.

Student 1 _____

2 cm ▢

Student 2 _____

2 cm ▢

How does the density of the heat receptors correspond to that of the touch receptors?

_____

_____

To that of the cold receptors? _____

_____

On the basis of your observations, which of these three types of receptors appears to be most abundant (at least in the area tested)?

_____ ■

Two-Point Discrimination Test

The density of the touch receptors varies significantly in different areas of the body. In general, areas that have the greatest density of tactile receptors have a heightened ability to "feel." These areas correspond to areas that receive the greatest motor innervation; thus they are also typically areas of fine motor control.

On the basis of this information, which areas of the body do you *predict* will have the greatest density of touch receptors?

_____

_____

A c t i v i t y :
Determining the Two-Point Threshold

1. Using a caliper or esthesiometer and a metric ruler, test the ability of the subject to differentiate two distinct sensations when the skin is touched simultaneously at two points. Beginning with the face, start with the caliper arms completely together. Gradually increase the distance between the arms, testing the subject's skin after each adjustment. Continue with this testing procedure until the subject reports that *two points* of contact can be felt. This measurement, the

smallest distance at which two points of contact can be felt, is the **two-point threshold.**

2. Repeat this procedure on the back and palm of the hand, fingertips, lips, back of the neck, and back of the calf. Record your results in Chart 1 below. ■

Chart 1	Determining Two-point Threshold
Body area tested	**Two-point threshold (millimeters)**
Face	
Back of hand	
Palm of hand	
Fingertips	
Lips	
Back of neck	
Back of calf	

Tactile Localization

Tactile localization is the ability to determine which portion of the skin has been touched. The tactile receptor field of the body periphery has a corresponding "touch" field in the brain's somatosensory association area. Some body areas are well represented with touch receptors, allowing tactile stimuli to be localized with great accuracy, but density of touch receptors in other body areas allows only crude discrimination.

A c t i v i t y :
Testing Tactile Localization

1. The subject's eyes should be closed during the testing. The experimenter touches the palm of the subject's hand with a pointed black felt-tipped marker. The subject should then try to touch the exact point with his or her own marker, which should be of a different color. Measure the error of localization in millimeters.

2. Repeat the test in the same spot twice more, recording the error of localization for each test. Average the results of the three determinations and record it in Chart 2.

Does the ability to localize the stimulus improve the second time?

_____ The third time? _____ Explain. _____

_____

_____

3. Repeat the above procedure on a fingertip, the ventral forearm, the ventral surface of the upper arm, and the upper back between the shoulder blades. Record the averaged results in Chart 2 below. ■

Chart 2	Testing Tactile Location
Body area tested	**Average error of localization (millimeters)**
Palm of hand	
Fingertip	
Ventral forearm	
Ventral surface of arm	
Upper back	

Adaptation of Sensory Receptors

The number of impulses transmitted by sensory receptors often changes both with the intensity of the stimulus and with the length of time the stimulus is applied. In many cases, when a stimulus is applied for a prolonged period, the rate of receptor discharge slows and conscious awareness of the stimulus declines or is lost until some type of stimulus change occurs. This phenomenon is referred to as **adaptation.** The touch receptors adapt particularly rapidly, which is highly desirable. Who, for instance, would want to be continually aware of the pressure of clothing on their skin? The simple experiments to be conducted next allow you to investigate the phenomenon of adaptation.

A c t i v i t y :
Demonstrating Adaptation of Touch Receptors

1. The subject's eyes should be closed. Place a coin on the anterior surface of the subject's forearm, and determine how long the sensation persists for the subject. Duration of the sensation:

_____ sec

2. Repeat the test, placing the coin at a different forearm location. How long does the sensation persist at the second location?

_____ sec

3. After awareness of the sensation has been lost at the second site, stack three more coins atop the first one.

Does the pressure sensation return? _____

If so, for how long is the subject aware of the pressure in this instance?

_____ sec

Are the same receptors being stimulated when the four coins, rather than the one coin, are used?

_____ Explain. _____

_____

4. To further illustrate the adaptation of touch receptors—in this case, the root hair plexuses of the hair follicles—gently and slowly bend one hair shaft with a pen or pencil until it springs back (away from the pencil) to its original position. Is the tactile sensation greater when the hair is being slowly bent or when it springs back?

_____

Why is the adaptation of the touch receptors in the hair follicles particularly important to a woman who wears her hair in a ponytail? If the answer is not immediately apparent, consider the opposite phenomenon: what would happen, in terms of sensory input from her hair follicles, if these receptors did not exhibit adaptation?

_____

_____ ■

A c t i v i t y :
Demonstrating Adaptation of Temperature Receptors

Adaptation of the temperature receptors can be tested using some very unsophisticated methods.

1. Obtain three large finger bowls or 1000-ml beakers and fill the first with 45°C water. Have the subject immerse her or his left hand in the water and report the sensation. Keep the left hand immersed for 1 min and then also immerse the right hand in the same bowl.

What is the sensation of the left hand when it is first immersed?

_____

What is the sensation of the left hand after 1 min as compared to the sensation in the right hand just immersed?

_____

Had adaptation occurred in the left hand? _____

2. Rinse both hands in tap water, dry them, and wait 5 min before conducting the next test. Just before beginning the test, refill the finger bowl with fresh 45°C water, fill a second with ice water, and fill a third with water at room temperature.

3. Place the *left* hand in the ice water and the *right* hand in the 45°C water. What is the sensation in each hand after 2 min as compared to the sensation perceived when the hands were first immersed?

_____

_____

_____

Which hand seemed to adapt more quickly?

_____

_____

4. After reporting these observations, the subject should then place both hands simultaneously into the finger bowl containing the water at room temperature. Record the sensation in the left hand:

_____

_____

The right hand: _____

The sensations that the subject experiences when both hands were put into room-temperature water are called **negative afterimages.** They are explained by the fact that sensations of heat and cold depend on the rapidity of heat loss or gain by the skin and differences in the temperature gradient. ■

Referred Pain

Experiments on pain receptor localization and adaptation are commonly conducted in the laboratory. However, there are certain problems with such experiments. Pain receptors are densely distributed in the skin, and they adapt very little, if at all. (This lack of adaptability is due to the protective function of the receptors. The sensation of pain often indicates tissue damage or trauma to body structures.) Thus no attempt will be made in this exercise to localize the pain receptors or to prove their nonadaptability, since both would cause needless discomfort to those of you acting as subjects and would not add any additional insight.

However, the phenomenon of referred pain is easily demonstrated in the laboratory, and such experiments provide information that may be useful in explaining common examples of this phenomenon. **Referred pain** is a sensory experience in which pain is perceived as arising in one area of the body when in fact another, often quite remote area is receiving the painful stimulus. Thus the pain is said to be "referred" to a different area. The phenomenon of **projection,** the process by which the brain refers sensations to their *usual* point of stimulation, provides the most simple explanation of such experiences. Many of us have experienced referred pain as a radiating pain in the forehead after quickly swallowing an ice-cold drink. Referred pain is important in many types of clinical diagnosis because damage to many visceral organs results in this phenomenon. For example, inadequate oxygenation of the heart muscle often results in pain being referred to the chest wall and left shoulder (*angina pectoris*), and the reflux of gastric juice into the esophagus causes a sensation of intense discomfort in the thorax referred to as *heartburn*. In addition, amputees often report *phantom limb pain*—feelings of pain that appear to be coming from a part of the body that is no longer there.

Activity:

Demonstrating the Phenomenon of Referred Pain

Immerse the subject's elbow in a finger bowl containing ice water. In Chart 3 below, record the quality (such as discomfort, tingling, or pain) and the quality progression of the sensations he or she reports for 2 min. Also record the location of the perceived sensations. The ulnar nerve, which serves the medial third of the hand, is involved in the phenomenon of referred pain experienced during this test. How does the localization of this referred pain correspond to the areas served by the ulnar nerve?

_____

_____

_____

_____ ■

Chart 3	Demonstrating Referred Pain	
Time of observation	**Quality of sensation**	**Localization of sensation**
On immersion		
After 1 min		
After 2 min		

Special Senses: Vision

Objectives

1. To describe the structure and function of the accessory visual structures.
2. To identify the structural components of the eye when provided with a model, an appropriate diagram, or a preserved sheep or cow eye, and list the function(s) of each.
3. To describe the cellular makeup of the retina.
4. To discuss the mechanism of image formation on the retina.
5. To trace the visual pathway to the optic cortex and indicate the effects of damage to various parts of this pathway.
6. To define the following terms:

   refraction   myopia
   accommodation   hyperopia
   convergence   cataract
   astigmatism   glaucoma
   emmetropia   conjunctivitis

7. To discuss the importance of the pupillary and convergence reflexes.
8. To explain the difference between rods and cones with respect to visual perception and retinal localization.
9. To state the importance of an ophthalmoscopic examination.

Materials

❏ Dissectible eye model
❏ Chart of eye anatomy
❏ Preserved cow or sheep eye
❏ Dissecting pan and instruments
❏ Protective skin cream or disposable gloves

❏ Histologic section of an eye showing retinal layers
❏ Compound microscope

❏ Common straight pins
❏ Snellen eye chart (floor marked with chalk to indicate 20-ft distance from posted Snellen chart)
❏ Ishihara's color-blindness plates

❏ 1-inch-diameter discs of colored paper (white, red, blue, green)
❏ White, red, blue, and green chalk
❏ Metric ruler; meter stick
❏ Test tubes
❏ Laboratory lamp or penlight
❏ Ophthalmoscope

 See Appendix C, Exercise 24 for links to A.D.A.M. Interactive Anatomy.

Anatomy of the Eye

External Anatomy and Accessory Structures

The adult human eye is a sphere measuring about 2.5 cm (1 inch) in diameter. Only about one-sixth of the eye's anterior surface is observable (Figure 24.1); the remainder is enclosed and protected by a cushion of fat and the walls of the bony orbit.

The **lacrimal apparatus** consists of the **lacrimal gland, lacrimal canals, lacrimal sac,** and the **nasolacrimal duct.** The lacrimal glands are situated superior to the lateral aspect of each eye. They continually liberate a dilute salt solution (tears) that flows onto the anterior surface of the eyeball through several small ducts. The tears flush across the eyeball into the lacrimal canals medially, then into the lacrimal sac, and finally into the nasolacrimal duct, which empties into the nasal cavity. The lacrimal secretion also contains **lysozyme,** an antibacterial enzyme. Because it constantly flushes the eyeball, the lacrimal fluid cleanses and protects the eye surface as it moistens and lubricates it. As we age, our eyes tend to become dry due to decreased lacrimation, and thus are more vulnerable to bacterial invasion and irritation.

The anterior surface of each eye is protected by the **eyelids,** or **palpebrae.** (See Figure 24.1.) The medial and lateral junctions of the upper and lower eyelids are referred to as the **medial** and **lateral canthus** (respectively). The **caruncle,** a fleshy elevation at the medial canthus, produces a whitish oily secretion. A mucous membrane, the **conjunctiva,** lines the internal surface of the eyelids (as the *palpebral conjunctiva*) and continues over the anterior surface of the eyeball to its junction with the corneal epithelium (as the *ocular,* or *bulbar, conjunctiva*). The conjunctiva secretes mucus, which aids in lubricating the eyeball. Inflammation of the conjunctiva, often accompanied by redness of the eye, is called **conjunctivitis.**

(a)

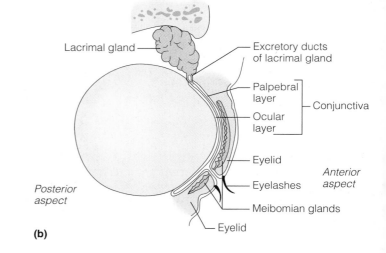

(b)

Figure 24.1 External anatomy of the eye and accessory structures. (a) Anterior view. (b) Sagittal section.

Projecting from the border of each eyelid is a row of short hairs, the **eyelashes.** The **ciliary glands,** modified sweat glands, lie between the eyelash hair follicles and help lubricate the eyeball. An inflammation of one of these glands is called a **sty.** Small sebaceous glands associated with the hair follicles and the larger **meibomian glands,** located posterior to the eyelashes, secrete an oily substance.

Six **extrinsic eye muscles** attached to the exterior surface of each eyeball control eye movement and make it possible for the eye to follow a moving object. The names and positioning of these extrinsic muscles are noted in Figure 24.2. Their actions are given in the chart accompanying that figure.

Activity:
Identifying Accessory Eye Structures

Observe the eyes of another student and identify as many of the accessory structures as possible. Ask the student to look to the left. What extrinsic eye muscles are responsible for this action?

Right eye _____

Left eye _____ ■

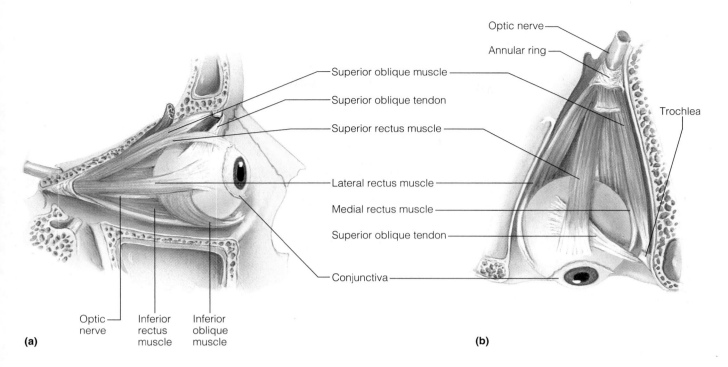

Figure 24.2 Extrinsic muscles of the eye. (a) Lateral view of the right eye. **(b)** Superior view of right eye. **(c)** Summary of actions of the extrinsic eye muscles, and cranial nerves that control them.

Internal Anatomy of the Eye

Anatomically, the wall of the eye is constructed of three tunics, or coats. The outermost **fibrous tunic** is a protective layer composed of dense avascular connective tissue. It has two obviously different regions: The opaque white **sclera** forms the bulk of the fibrous tunic and is observable anteriorly as the "white of the eye." Its anteriormost portion is modified structurally to form the transparent **cornea,** through which light enters the eye. (See Figure 24.3.)

The middle tunic, called the **uvea,** is the **vascular tunic.** Its posteriormost part, the **choroid,** is a richly vascular nutritive layer containing a dark pigment that prevents light scattering within the eye. Anteriorly, the choroid is modified to form the **ciliary body,** to which the lens is attached, and then

the pigmented **iris.** The iris is incomplete, resulting in a rounded opening, the **pupil,** through which light passes.

The iris is composed of circularly and radially arranged smooth muscle fibers and acts as a reflexively activated diaphragm to regulate the amount of light entering the eye. In close vision and bright light, the circular muscles of the iris contract, and the pupil constricts. In distant vision and in dim light, the radial fibers contract, enlarging (dilating) the pupil and allowing more light to enter the eye.

The innermost **sensory tunic** of the eye is the delicate, two-layered **retina.** The outer **pigmented epithelial layer** abuts and lines the entire uvea. The transparent inner **neural (nervous) layer** extends anteriorly only to the ciliary body. It contains the photoreceptors, **rods** and **cones,** which begin the chain of electrical events that ultimately result in the

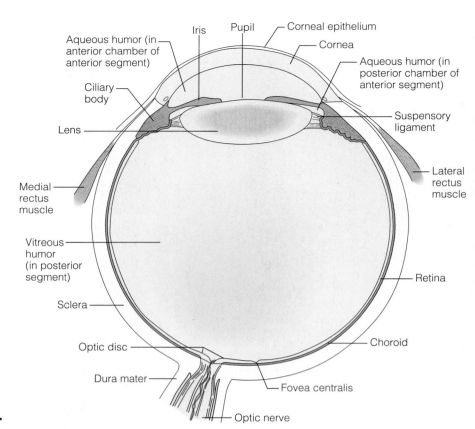

Figure 24.3 Internal anatomy of the eye (transverse section).

transduction of light energy into nerve impulses that are transmitted to the optic cortex of the brain. Vision is the result. The photoreceptor cells are distributed over the entire neural retina, except where the optic nerve leaves the eyeball. This site is called the **optic disc,** or blind spot. Lateral to each blind spot, and directly posterior to the lens, is an area called the **macula lutea** (yellow spot), an area of high cone density. In its center is the **fovea centralis,** a minute pit about ½ mm in diameter, which contains only cones and is the area of greatest visual acuity. Focusing for discriminative vision occurs in the fovea centralis.

Light entering the eye is focused on the retina by the **lens,** a flexible crystalline structure held vertically in the eye's interior by the **suspensory ligament** attached to the ciliary body. Activity of the ciliary muscle, which accounts for the bulk of ciliary body tissue, changes lens thickness to allow light to be properly focused on the retina.

In the elderly the lens becomes increasingly hard and opaque. **Cataracts,** which often result from this process, cause vision to become hazy or entirely obstructed. ∎

The lens divides the eye into two segments: the **anterior segment** anterior to the lens, which contains a clear watery fluid called the **aqueous humor,** and the **posterior segment** behind the lens, filled with a gel-like substance, the **vitreous humor,** or **vitreous body.** The anterior segment is further di-

vided into anterior and **posterior chambers,** located before and after the iris, respectively. The aqueous humor is continually formed by the capillaries of the **ciliary processes** of the ciliary body. It helps to maintain the intraocular pressure of the eye and provides nutrients for the avascular lens and cornea. The aqueous humor is reabsorbed into the **scleral venous sinus (canal of Schlemm).** The vitreous humor provides the major internal reinforcement of the posterior part of the eyeball, and helps to keep the neural layer of the retina pressed firmly against the wall of the eyeball. It is formed *only* before birth.

Anything that interferes with drainage of the aqueous fluid increases intraocular pressure. When intraocular pressure reaches dangerously high levels, the retina and optic nerve are compressed, resulting in pain and possible blindness, a condition called **glaucoma.** ∎

Activity:
Identifying Internal Structures of the Eye

Obtain a dissectible eye model and identify its internal structures described above. As you work, also refer to Figure 24.3. ∎

Figure 24.4   Microscopic anatomy of the cellular layers of the retina.
(a) Diagrammatic view. **(b)** Photomicrograph of the retina (1092X).

Microscopic Anatomy of the Retina

As described above, the retina consists of two main types of cells: a pigmented *epithelial* layer, which abuts the choroid, and an inner cell layer composed of *neurons,* which is in contact with the vitreous humor (Figure 24.4). The inner nervous layer is composed of three major neuronal populations. These are, from outer to inner aspect, the **photoreceptors** (rods and cones), the **bipolar cells,** and the **ganglion cells.**

The **rods** are the specialized receptors for dim light. Visual interpretation of their activity is in gray tones. The **cones** are color receptors that permit high levels of visual acuity, but they function only under conditions of high light intensity; thus, for example, no color vision is possible in moonlight. Only cones are found in the fovea centralis, and their number decreases as the retinal periphery is approached. By contrast, rods are most numerous in the periphery, and their density decreases as the macula is approached.

Light must pass through the ganglion cell layer and the bipolar neuron layer to reach and excite the rods and cones. As a result of a light stimulus, the photoreceptors undergo changes in their membrane potential that ultimately influence the bipolar neurons. These in turn stimulate the ganglion cells, whose axons leave the retina in the tight bundle of fibers known as the optic nerve. The retinal layer is thickest where the optic nerve attaches to the eyeball because an in-

creasing number of ganglion cell axons converge at this point. It thins as it approaches the ciliary body.

Activity:
Studying the Microscopic Anatomy of the Retina

Obtain a histologic slide of a longitudinal section of the eye. Identify the retinal layers by comparing your view to Figure 24.4. ■

Visual Pathways to the Brain

The axons of the ganglion cells of the retina converge at the posterior aspect of the eyeball and exit from the eye as the optic nerve. At the **optic chiasma,** the fibers from the medial side of each eye cross over to the opposite side. The fiber tracts thus formed are called the **optic tracts.** Each optic tract contains fibers from the lateral side of the eye on the same side and from the medial side of the opposite eye.

The optic tract fibers synapse with neurons in the **lateral geniculate nucleus** of the thalamus, whose axons form the **optic radiation,** terminating in the **optic,** or **visual, cortex** in the occipital lobe of the brain. Here they synapse with the cortical cells, and visual interpretation occurs.

Figure 24.5 Visual pathway to the brain. (a) Diagram. (Note that fibers from the lateral portion of each retinal field do not cross at the optic chiasma.) **(b)** Photograph.

Activity:
Predicting the Effects of Visual Pathway Lesions

After examining Figure 24.5, determine what effects lesions in the following areas would have on vision:

In the right optic nerve _____

_____

Through the optic chiasma _____

_____

In the left optic tract _____

_____

In the right cerebral cortex (visual area) _____

_____ ■

Dissection:
The Cow (Sheep) Eye

1.   Obtain a preserved cow or sheep eye, dissecting instruments, and a dissecting pan. Apply protective skin cream or don disposable gloves if desired.

2.   Examine the external surface of the eye, noting the thick cushion of adipose tissue. Identify the optic nerve (cranial nerve II) as it leaves the eyeball, the remnants of the extrinsic eye muscles, the conjunctiva, the sclera, and the cornea. The normally transparent cornea is opalescent or opaque if the eye has been preserved. Refer to Figure 24.6 as you work.

3.   Trim away most of the fat and connective tissue, but leave the optic nerve intact. Holding the eye with the cornea facing downward, carefully make an incision with a sharp scalpel into the sclera about ¼ inch above the cornea. (The sclera of the preserved eyeball is *very* tough so you will have to apply substantial pressure to penetrate it.) Using scissors, complete the incision around the circumference of the eyeball paralleling the corneal edge.

4.   Carefully lift the anterior part of the eyeball away from the posterior portion. Conditions being proper, the vitreous body should remain with the posterior part of the eyeball.

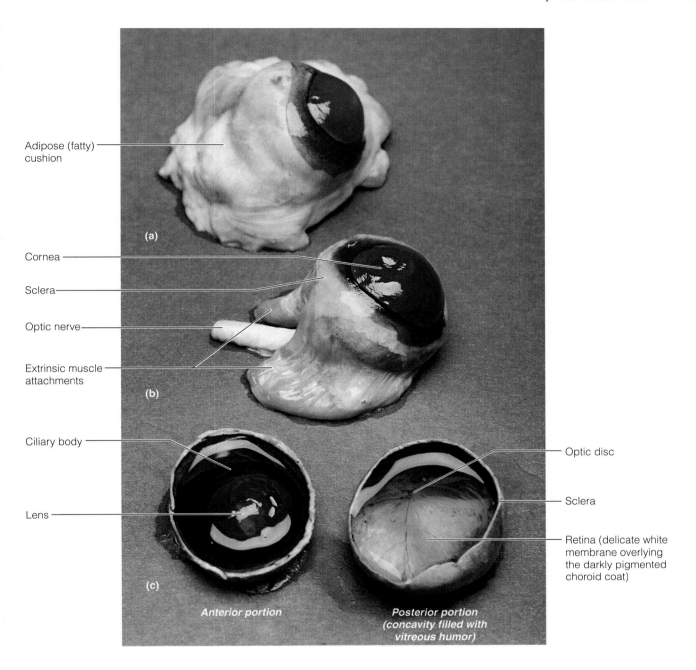

Adipose (fatty) cushion

(a)

Cornea

Sclera

Optic nerve

Extrinsic muscle attachments

(b)

Ciliary body

Lens

(c)

Optic disc

Sclera

Retina (delicate white membrane overlying the darkly pigmented choroid coat)

Anterior portion

Posterior portion (concavity filled with vitreous humor)

Figure 24.6 Anatomy of the cow eye. (a) Cow eye (entire) removed from orbit (notice the large amount of fat cushioning the eyeball). **(b)** Cow eye (entire) with fat removed to show the extrinsic muscle attachments and optic nerve. **(c)** Cow eye cut along the coronal plane to reveal internal structures.

5.   Examine the anterior part of the eye and identify the following structures:

Ciliary body: Black pigmented body that appears to be a halo encircling the lens.

Lens: Biconvex structure that is opaque in preserved specimens.

Suspensory ligament: A halo of delicate fibers attaching the lens to the ciliary body.

Carefully remove the lens and identify the adjacent structures:

Iris: Anterior continuation of the ciliary body penetrated by the pupil.

Cornea: More convex anteriormost portion of the sclera; normally transparent but cloudy in preserved specimens.

6.   Examine the posterior portion of the eyeball. Remove the vitreous humor, and identify the following structures:

Retina: The neural layer of the retina appears as a delicate white, probably crumpled membrane that separates easily from the pigmented choroid.

Figure 24.7   Blind spot test figure.

Note its point of attachment. What is this point called?

_____

Pigmented choroid coat: Appears iridescent in the cow or sheep eye owing to a special reflecting surface called the **tapetum lucidum.** This specialized surface reflects the light within the eye and is found in the eyes of animals that live under conditions of low-intensity light. It is not found in humans. ■

Visual Tests and Experiments
The Blind Spot

A c t i v i t y :
Demonstrating the Blind Spot

1.   Hold Figure 24.7 about 18 inches from your eyes. Close your left eye, and focus your right eye on the X, which should be positioned so that it is directly in line with your right eye. Move the figure slowly toward your face, keeping your right eye focused on the X. When the dot focuses on the blind spot, which lacks photoreceptors, it will disappear.

2.   Have your laboratory partner record in metric units the distance at which this occurs. The dot will reappear as the figure is moved closer. Distance at which the dot disappears:

Right eye _____

Repeat the test for the left eye, this time closing the right eye and focusing the left eye on the dot. Record the distance at which the X disappears:

Left eye _____ ■

Afterimages

When light from an object strikes **rhodopsin,** the purple pigment contained in the rods of the retina, it triggers a photo-

chemical reaction that splits rhodopsin into its colorless precursor molecules (vitamin A and a protein called opsin). This event, called *bleaching of the pigment,* initiates a chain of events leading to impulse transmission along fibers of the optic nerve. Once bleaching has occurred in a rod, the photoreceptor pigment must be resynthesized before the rod can be restimulated. This takes a certain period of time. Both phenomena—that is, the stimulation of the photoreceptor cells and their subsequent inactive period—can be demonstrated indirectly in terms of positive and negative afterimages.

A c t i v i t y :
Demonstrating Afterimages

1.   Stare at a bright lightbulb for a few seconds, and then gently close your eyes for approximately 1 minute.

2.   Record, in sequence of occurrence, what you "saw" after closing your eyes:

_____

_____

_____

The bright image of the lightbulb initially seen was a **positive afterimage** caused by the continued firing of the rods. The dark image of the lightbulb that subsequently appeared against a lighter background was the **negative afterimage,** an indication that the rhodopsin in the affected photoreceptor cells had been bleached. ■

Refraction, Visual Acuity, and Astigmatism

When light rays pass from one medium to another, their velocity, or speed of transmission, changes, and the rays are bent or **refracted.** Thus the light rays in the visual field are refracted as they encounter the cornea, lens, and vitreous humor of the eye.

The refractive index (bending power) of the cornea and vitreous humor are constant. But the lens's refractive index, or strength, can be varied by changing the lens's shape—that is, by making it more or less convex so that the light is properly converged and focused on the retina. The greater the lens convexity, or bulge, the more the light will be bent and the stronger the lens. Conversely, the less the lens convexity (the flatter it is), the less it bends the light.

In general, light from a distant source (over 20 feet) approaches the eye as parallel rays, and no change in lens convexity is necessary for it to focus properly on the retina. However, light from a close source tends to diverge, and the convexity of the lens must increase to make close vision possible. To achieve this, the ciliary muscle contracts, decreasing the tension on the suspensory ligament attached to the lens and allowing the elastic lens to "round up." Thus, a lens capable of bringing a *close* object into sharp focus is stronger (more convex) than a lens focusing on a more distant object. The ability of the eye to focus differentially for objects of near vision (less than 20 feet) is called **accommodation.** It should be noted that the image formed on the retina as a re-

(a) Myopic eye (nearsighted)

Figure 24.8 Refraction of light in the eye, resulting in the production of a real image on the retina.

sult of the refractory activity of the lens (see Figure 24.8) is a **real image** (reversed from left to right, inverted, and smaller than the object).

The normal or **emmetropic eye** is able to accommodate properly. However, visual problems may result from (1) lenses that are too strong or too "lazy" (overconverging and underconverging, respectively), (2) from structural problems such as an eyeball that is too long or too short to provide for proper focusing by the lens, or (3) a cornea or lens with improper curvatures.

Individuals in whom the image normally focuses in front of the retina are said to have **myopia**, or "nearsightedness" (Figure 24.9a); they can see close objects without difficulty, but distant objects are blurred or seen indistinctly. Correction requires a concave lens, which causes the light reaching the eye to diverge (Figure 24.9b).

If the image focuses behind the retina, the individual is said to have **hyperopia** or farsightedness. Such persons have no problems with distant vision but need glasses with convex lenses to augment the converging power of the lens for close vision (Figure 24.9c and d).

Irregularities in the curvatures of the lens and/or the cornea lead to a blurred vision problem called **astigmatism.** Cylindrically ground lenses, which compensate for inequalities in the curvatures of the refracting surfaces, are prescribed to correct the condition. ■

Near-Point Accommodation The elasticity of the lens decreases dramatically with age, resulting in difficulty in focusing for near or close vision. This condition is called **presbyopia**—literally, old vision. Lens elasticity can be tested by measuring the **near point of accommodation.** The near point of vision is about 10 cm from the eye in young adults. It is closer in children and farther in old age.

(b) Concave lens

(c) Hyperopic eye (farsighted)

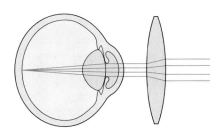

(d) Convex lens

Figure 24.9 Common refraction problems and their correction. In myopia: **(a)** light from a distant object focuses in front of the retina, and **(b)** correction involves the use of a concave lens that diverges the light before it enters the eye. In hyperopia: **(c)** light from a distant object focuses behind the retina, and **(d)** correction requires a convex lens that converges the light rays before they enter the eye.

A c t i v i t y :
Determining Near Point of Accommodation

To determine your near point of accommodation, hold a common straight pin at arm's length in front of one eye. Slowly move the pin toward that eye until the pin image becomes distorted. Have your lab partner measure the distance from your eye to the pin at this point, and record the distance below. Repeat the procedure for the other eye.

Near point for right eye _____

Near point for left eye _____ ■

Visual Acuity Visual acuity, or sharpness of vision, is generally tested with a Snellen eye chart, which consists of letters of various sizes printed on a white card. This test is based on the fact that letters of a certain size can be seen clearly by

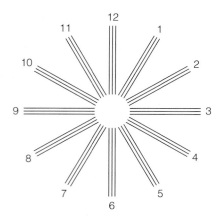

Figure 24.10 Astigmatism testing chart.

eyes with normal vision at a specific distance. The distance at which the normal, or emmetropic, eye can read a line of letters is printed at the end of that line.

Activity:
Testing Visual Acuity

1. Have your partner stand 20 feet from the posted Snellen eye chart and cover one eye with a card or hand. As your partner reads each consecutive line aloud, check for accuracy. If this individual wears glasses, give the test twice—first with glasses off and then with glasses on.

2. Record the number of the line with the smallest-sized letters read. If it is 20/20, the person's vision for that eye is normal. If it is 20/40, or any ratio with a value less than one, he or she has less than the normal visual acuity. (Such an individual is myopic.) If the visual acuity is 20/15, vision is better than normal, because this person can stand at 20 feet from the chart and read letters that are only discernible by the normal eye at 15 feet. Give your partner the number of the line corresponding to the smallest letters read, to record in step 4.

3. Repeat the process for the other eye.

4. Have your partner test and record your visual acuity. If you wear glasses, the test results *without* glasses should be recorded first.

Visual acuity, right eye _____

Visual acuity, left eye _____ ■

Activity:
Testing for Astigmatism

The astigmatism chart (Figure 24.10) is designed to test for defects in the refracting surface of the lens and/or cornea.

View the chart first with one eye and then with the other, focusing on the center of the chart. If all the radiating lines appear equally dark and distinct, there is no distortion of your refracting surfaces. If some of the lines are blurred or appear less dark than others, at least some degree of astigmatism is present.

Is astigmatism present in your left eye? _____

Right eye? _____ ■

Color Blindness

Ishihara's color plates are designed to test for deficiencies in the cones or color photoreceptor cells. There are three cone types, each containing a different light-absorbing pigment. One type primarily absorbs the red wavelengths of the visible light spectrum, another the blue wavelengths, and a third the green wavelengths. Nerve impulses reaching the brain from these different photoreceptor types are then interpreted (seen) as red, blue, and green, respectively. Interpretation of the intermediate colors of the visible light spectrum is a result of overlapping input from more than one cone type.

Activity:
Testing for Color Blindness

1. View the color plates in bright light or sunlight while holding them about 30 inches away and at right angles to your line of vision. Report to your laboratory partner what you see in each plate. Take no more than 3 seconds for each decision.

2. Your partner is to write down your responses and then check their accuracy with the correct answers provided in the color plate book. Is there any indication that you have some degree of color blindness?

_____ If so, what type? _____

_____

Repeat the procedure to test your partner's color vision. ■

Relative Positioning of Rods and Cones on the Retina

The test subject for this demonstration should have shown no color vision problems during the test for color blindness. Students may work in pairs, or two students may perform the test for the entire class. White, red, blue, and green paper discs and chalk will be needed for this demonstration.

Activity:
Mapping the Rods and Cones

1. Position the subject about 1 foot away from the blackboard.

2. Make a small white chalk circle on the board immediately in front of the subject's right eye. Have the subject close the left eye and stare fixedly at the circle with the right eye throughout the test.

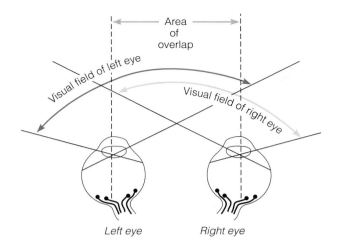

Figure 24.11 Overlapping of the visual fields.

3.   To map the extent of the rod field, begin to move a white paper disc into the field of vision from various sides of the visual field (beginning at least 2 feet away from the white chalk circle) and plot with white chalk dots the points at which the disc first becomes visible to the test subject.

4.   Repeat the procedure, using red, green, and blue paper discs and like-colored chalk to map the cone fields. Color (not object) identification is required.

5.   After the test has been completed for all four discs, connect all dots of the same color. It should become apparent that each of the color fields has a different radius and that the rod and cone distribution on the retina is not uniform. (Normal fields have white outermost, followed by blue, red, and green as the fovea is approached.)

6.   Record the rod and cone color field distribution observed in the laboratory review section using appropriately colored pencils. ∎

Binocular Vision

Humans, cats, predatory birds, and most primates are endowed with **binocular** (two-eyed) **vision.** Although both eyes look in approximately the same direction, they see slightly different views. Their visual fields, each about 170 degrees, overlap to a considerable extent; thus there is two-eyed vision at the overlap area (Figure 24.11).

In contrast, the eyes of many animals (rabbits, pigeons, and others) are more on the sides of their head. Such animals see in two different directions and thus have a panoramic field of view and **panoramic vision.**

Although both types of vision have their good points, binocular vision provides three-dimensional vision and an accurate means of locating objects in space. The slight differ-ences between the views seen by the two eyes are fused by the higher centers of the visual cortex to give us *depth perception.* Because of the manner in which the visual cortex resolves these two different views into a single image, it is sometimes referred to as the "cyclopean eye of the binocular animal."

Activity:
Tests for Binocular Vision

1.   To demonstrate that a slightly different view is seen by each eye, perform the following simple experiment.

Close your left eye. Hold a pencil at arm's length directly in front of your right eye. Position another pencil directly beneath it and then move the lower pencil about half the distance toward you. As you move the lower pencil, make sure it remains in the *same plane* as the stationary pencil, so that the two pencils continually form a straight line. Then, without moving the pencils, close your right eye and open your left eye. Notice that with only the right eye open, the moving pencil stays in the same plane as the fixed pencil, but that when viewed with the left eye, the moving pencil is displaced laterally away from the plane of the fixed pencil.

2.   To demonstrate the importance of two-eyed binocular vision for depth perception, perform this second simple experiment.

Have your laboratory partner hold a test tube erect about arm's length in front of you. With both eyes open, quickly insert a pencil into the test tube. Remove the pencil, bring it back close to your body, close one eye, and quickly and without hesitation insert the pencil into the test tube. (Do not feel for the test tube with the pencil!) Repeat with the other eye closed.

Was it as easy to dunk the pencil with one eye closed as with both eyes open?

_____ ∎

Eye Reflexes

Both intrinsic (internal) and extrinsic (external) muscles are necessary for proper eye functioning. The *intrinsic muscles,* controlled by the autonomic nervous system, are those of the ciliary body (which alters the lens curvature in focusing) and the radial and circular muscles of the iris (which control pupillary size and thus regulate the amount of light entering the eye). The *extrinsic muscles* are the rectus and oblique muscles, which are attached to the eyeball exterior (see Figure 24.1). These muscles control eye movement and make it possible to keep moving objects focused on the fovea centralis. They are also responsible for **convergence,** or medial eye movements, which is essential for near vision. When convergence occurs, both eyes are directed toward the near object viewed. The extrinsic eye muscles are controlled by the somatic nervous system.

Activity:
Demonstrating Reflex Activity of Intrinsic and Extrinsic Eye Muscles

Involuntary activity of both the intrinsic and extrinsic muscle types is brought about by reflex actions that can be observed in the following experiments.

Photopupillary Reflex

Sudden illumination of the retina by a bright light causes the pupil to constrict reflexively in direct proportion to the light intensity. This protective response prevents damage to the delicate photoreceptor cells.

Obtain a laboratory lamp or penlight. Have your laboratory partner sit with eyes closed and hands over his or her eyes. Turn on the light and position it so that it shines on the subject's right hand. After 1 minute, ask your partner to uncover and open the right eye. Quickly observe the pupil of that eye. What happens to the pupil?

_____

Shut off the light and ask your partner to uncover and open the opposite eye. What are your observations?

_____

_____

Accommodation Pupillary Reflex

Have your partner gaze for approximately 1 minute at a distant object in the lab—*not* toward the windows or another light source. Observe your partner's pupils. Then hold some printed material 6 to 10 inches from his or her face, and direct him or her to focus on it.

How does pupil size change as your partner focuses on the printed material?

_____

Explain the value of this reflex. _____

_____

_____

_____

Convergence Reflex

Repeat the previous experiment, this time using a pen or pencil as the close object to be focused on. Note the position of your partner's eyeballs both while he or she is gazing at the distant and at the close object. Do they change position as the object of focus is changed?

_____ In what way? _____

Viewing window
(look at eye interior
through this)

Diopter window
(indicates lens in use)

Lens
selection disk
(rotate to change lens in use)

Aperture
selection dial

Rheostat control
(rotate to control light intensity)

Rheostat lock
(On-Off switch)

Handle

(a)

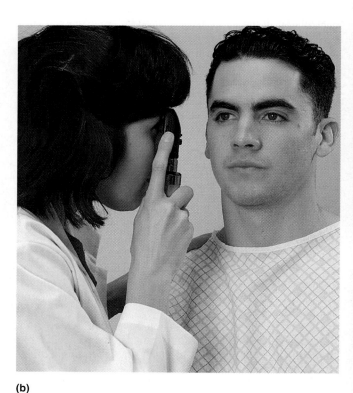

(b)

Figure 24.12 Structure and use of an ophthalmoscope. (a) Structure of an ophthalmoscope. **(b)** Proper position for beginning to examine the right eye with an ophthalmoscope.

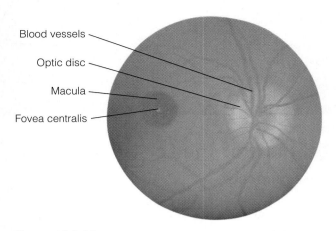

Blood vessels

Optic disc

Macula

Fovea centralis

Figure 24.13   Posterior portion of right retina.

_____

_____

_____

Explain the importance of the convergence reflex.

_____

_____

_____ ■

Ophthalmoscopic Examination of the Eye (Optional)

The ophthalmoscope is an instrument used to examine the *fundus,* or eyeball interior, to determine visually the condition of the retina, optic disc, and internal blood vessels. Certain pathologic conditions such as diabetes mellitus, arteriosclerosis, and degenerative changes of the optic nerve and retina can be detected by such an examination. The ophthalmoscope consists of a set of lenses mounted on a rotating disc (the **lens selection disc**), a light source regulated by a **rheostat control,** and a mirror that reflects the light so that the eye interior can be illuminated (Figure 24.12a).

The lens selection disc is positioned in a small slit in the mirror, and the examiner views the eye interior through this slit, appropriately called the **viewing window.** The focal length of each lens is indicated in diopters preceded by a + sign if the lens is convex and by a − sign if the lens is concave. When the zero (0) is seen in the **diopter window,** there is no lens positioned in the slit. The depth of focus for viewing the eye interior is changed by changing the lens.

The light is turned on by depressing the red **rheostat lock button** and then rotating the rheostat control in the clockwise direction. The aperture selection disc on the front of the instrument allows the nature of the light beam to be al-

tered. Generally, green light allows for clearest viewing of the blood vessels in the eye interior and is most comfortable for the subject.

Once you have examined the ophthalmoscope and have become familiar with it, you are ready to conduct an eye examination.

Activity:
Conducting an Ophthalmoscopic Examination

1.   Conduct the examination in a dimly lit or darkened room with the subject comfortably seated and gazing straight ahead. To examine the right eye, sit face-to-face with the subject, hold the instrument in your right hand, and use your right eye to view the eye interior (Figure 24.12b). You may want to steady yourself by resting your left hand on the subject's head. To view the left eye, use your left eye, hold the instrument in your left hand, and steady yourself with your right hand.

2.   Begin the examination with the 0 (no lens) in position. Grasp the instrument so that the lens disc may be rotated with the index finger. Holding the ophthalmoscope about 6 inches from the subject's eye, direct the light into the pupil at a slight angle—through the pupil edge rather than directly through its center. You will see a red circular area that is the illuminated eye interior.

3.   Move in as close as possible to the subject's cornea (to within 2 in.) as you continue to observe the area. Steady your instrument-holding hand on the subject's cheek if necessary. If both your eye and that of the subject are normal, the fundus can be viewed clearly without further adjustment of the ophthalmoscope. If the fundus cannot be focused, slowly rotate the lens disc counterclockwise until the fundus can be clearly seen. When the ophthalmoscope is correctly set, the fundus of the right eye should appear as shown in Figure 24.13. (Note: If a positive [convex] lens is required and your eyes are normal, the subject has hyperopia. If a negative [concave] lens is necessary to view the fundus and your eyes are normal, the subject is myopic.)

When the examination is proceeding correctly, the subject can often see images of retinal vessels in his own eye that appear rather like cracked glass. If you are unable to achieve a sharp focus or to see the optic disc, move medially or laterally and begin again.

4.   Examine the optic disc for color, elevation, and sharpness of outline, and observe the blood vessels radiating from near its center. Locate the macula, lateral to the optic disc. It is a darker area in which blood vessels are absent, and the fovea appears to be a slightly lighter area in its center. The macula is most easily seen when the subject looks directly into the light of the ophthalmoscope.

⚠   Do not examine the macula for longer than 1 second at a time.

5.   When you have finished examining your partner's retina, shut off the ophthalmoscope. Change places with your partner (become the subject) and repeat steps 1–4. ■

Special Senses: Hearing and Equilibrium

Objectives

1. To identify the anatomical structures of the outer, middle, and inner ear by appropriately labeling a diagram.

2. To describe the anatomy of the organ of hearing (organ of Corti in the cochlea) and explain its function in sound reception.

3. To describe the anatomy of the equilibrium organs of the inner ear (cristae ampullares and maculae), and to explain their relative function in maintaining equilibrium.

4. To define or explain *central deafness, conduction deafness,* and *nystagmus.*

5. To state the purpose of the Weber, Rinne, Barany, and Romberg tests.

6. To explain how one is able to localize the source of sounds.

7. To describe the effects of acceleration on the semicircular canals.

8. To explain the role of vision in maintaining equilibrium.

Materials

❑ Three-dimensional dissectible ear model and/or chart of ear anatomy
❑ Histologic slides of the cochlea of the ear
❑ Compound microscope
❑ *Demonstration:* Microscope focused on a crista ampullaris receptor of a semicircular canal
❑ Otoscope (if available)
❑ Alcohol swabs
❑ Absorbent cotton
❑ Pocket watch or clock that ticks
❑ Tuning forks (range of frequencies)
❑ Rubber mallet
❑ 12-inch ruler
❑ Audiometer
❑ Red and blue pencils

Anatomy of the Ear

Gross Anatomy

The ear is a complex structure containing sensory receptors for hearing and equilibrium. The ear is divided into three major areas: the *outer ear,* the *middle ear,* and the *inner ear* (Figure 25.1). The outer and middle ear structures serve the needs of the sense of hearing *only,* while inner ear structures function both in equilibrium and hearing reception.

Activity:
Identifying Structures of the Ear

Obtain a dissectible ear model and identify the structures described below. Refer to Figure 25.1 as you work. ■

The **outer,** or **external, ear** is composed primarily of the **pinna,** or **auricle,** and the **external auditory canal.** The pinna is the skin-covered cartilaginous structure encircling the auditory canal opening. In many animals, it collects and directs sound waves into the external auditory canal. In humans this function of the pinna is largely lost. The portion of the pinna lying inferior to the external auditory canal is the **lobule.**

The external auditory canal is a short, narrow (about 1 inch long by ¼ inch wide) chamber carved into the temporal bone. In its skin-lined walls are wax-secreting glands called **ceruminous glands.** The sound waves that enter the external auditory canal eventually encounter the **tympanic membrane,** or **eardrum,** which vibrates at exactly the same frequency as the sound wave(s) hitting it. The membranous eardrum separates the outer from the middle ear.

The **middle ear** is essentially a small chamber—the **tympanic cavity**—found within the temporal bone. The cavity is spanned by three small bones, collectively called the **ossicles** (hammer, anvil, and stirrup),* which articulate to form a lever system that transmits the vibratory motion of the eardrum to the fluids of the inner ear via the **oval window.**

Connecting the middle ear chamber with the nasopharynx is the **pharyngotympanic,** or **auditory, tube.** Normally this tube is flattened and closed, but swallowing or yawning can cause it to open temporarily to equalize the pressure of the middle ear cavity with external air pressure. This is an important function. The eardrum does not vibrate properly unless the pressure on both of its surfaces is the same.

Because the mucosal membranes of the middle ear cavity and nasopharynx are continuous through the

---

* The ossicles are often referred to by their Latin names, that is, **malleus, incus,** and **stapes,** respectively.

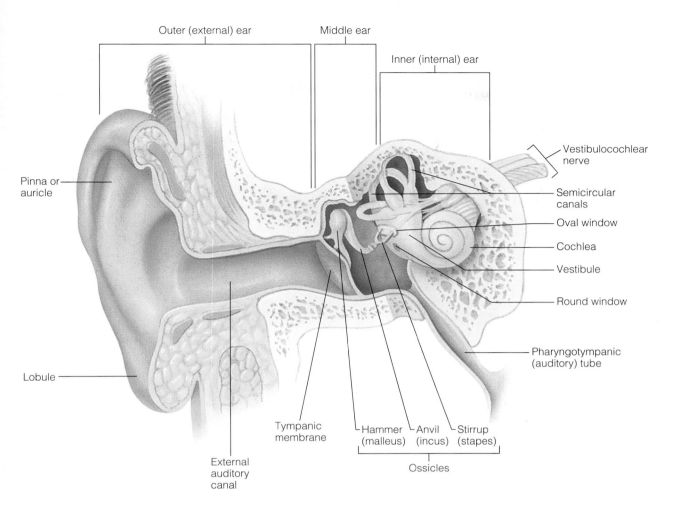

Figure 25.1 Anatomy of the ear.

pharyngotympanic tube, **otitis media,** or inflammation of the middle ear, is a fairly common condition, especially among youngsters prone to sore throats. In cases where large amounts of fluid or pus accumulate in the middle ear cavity, an emergency myringotomy (lancing of the eardrum) may be necessary to relieve the pressure. Frequently, tiny ventilating tubes are put in during the procedure. ■

The **inner,** or **internal, ear** consists of a system of bony and rather tortuous chambers called the **osseous,** or **bony, labyrinth,** which is filled with an aqueous fluid called **perilymph** (Figure 25.2). Suspended in the perilymph is the **membranous labyrinth,** a system that mostly follows the contours of the osseous labyrinth. The membranous labyrinth is filled with a more viscous fluid called **endolymph.** The three subdivisions of the bony labyrinth are the **cochlea,** the **vestibule,** and the **semicircular canals,** with the vestibule situated between the cochlea and semicircular canals.

The snail-like cochlea (see Figures 25.2 and 25.3) contains the sensory receptors for hearing. The cochlear membranous labyrinth, the **cochlear duct,** is a soft wormlike tube about 1½ inches long. It winds through the full two and three-quarter turns of the cochlea and separates the perilymph-containing cochlear cavity into upper and lower

chambers, the **scala vestibuli** and **scala tympani,** respectively. The scala vestibuli terminates at the oval window, which "seats" the foot plate of the stirrup located laterally in the tympanic cavity. The scala tympani is bounded by a membranous area called the **round window.** The cochlear duct, itself filled with endolymph, supports the **organ of Corti,** which contains the receptors for hearing—the sensory hair cells and nerve endings of the cochlear division of the vestibulocochlear nerve (VIII).

Activity:
Examining the Ear with an Otoscope (Optional)

1. Obtain an otoscope and two alcohol swabs. Inspect your partner's ear canal and then select the largest-*diameter* (not length!) speculum that will fit comfortably into his or her ear to permit full visibility. Clean the speculum thoroughly with an alcohol swab, and then attach the speculum to the battery-containing otoscope handle. Before beginning, check that the otoscope light beam is strong. (If not, obtain another otoscope or new batteries.)

Figure 25.2 Inner ear. Right membranous labyrinth shown within the bony labyrinth.

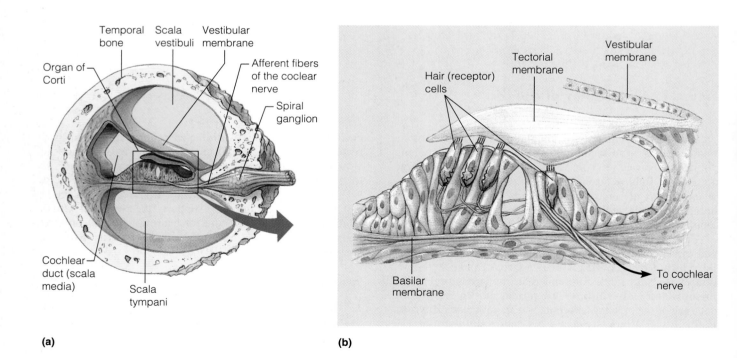

(a)

(b)

Figure 25.3 Anatomy of the cochlea. (a) A cross-sectional view of one turn of the cochlea, showing the relationship of the three scalae. The scalae vestibuli and tympani contain perilymph. The cochlear duct (scala media), which houses the organ of Corti, contains endolymph. **(b)** Detailed structure of the organ of Corti. The receptor cells (hair cells) rest on the basilar membrane.

2. When you are ready to begin the examination, hold the lighted otoscope securely between your thumb and forefinger (like a pencil), and rest the little finger of the otoscope-holding hand against your partner's head. This maneuver forms a brace that allows the speculum to move as your partner moves and prevents the speculum from penetrating too deeply into the ear canal during unexpected movements.

3. Grasp the ear pinna firmly and pull it up, back, and slightly laterally. If your partner experiences pain or discomfort when the pinna is manipulated, an inflammation or infection of the external ear may be present. If this occurs, do not attempt to examine the ear canal.

4. Carefully insert the speculum of the otoscope into the external auditory canal in a downward and forward direction only far enough to permit examination of the tympanic membrane or eardrum. Note its shape, color, and vascular network. The healthy tympanic membrane is pearly white. During the examination, notice if there is any discharge or redness in the canal and identify earwax.

5. After the examination, thoroughly clean the speculum with the second alcohol swab before returning the otoscope to the supply area. ■

Microscopic Anatomy of the Organ of Corti and the Mechanism of Hearing

The anatomical details of the organ of Corti are shown in Figure 25.3. The hair (auditory receptor) cells rest on the **basilar membrane**, which forms the floor of the cochlear duct, and their "hairs" (stereocilia) project into a gelatinous membrane, the **tectorial membrane,** that overlies them. The roof of the cochlear duct is called the **vestibular membrane.** The endolymph-filled chamber of the cochlear duct is the **scala media.**

A c t i v i t y :
Examining the Microscopic Structure of the Cochlea

Obtain a compound microscope and a prepared microscope slide of the cochlea and identify the areas shown in Figure 25.4. ■

The mechanism of hearing begins as sound waves pass through the external auditory canal and through the middle ear into the inner ear, where the vibration eventually reaches the organ of Corti, which contains the receptors for hearing. Many theories have attempted to explain how the organ of Corti actually responds to sound.

The popular "traveling wave" hypothesis of Von Békésy suggests that vibration of the stirrup at the oval window initiates traveling waves that cause maximal displacements of the basilar membrane where they peak and stimulate the hair cells of the organ of Corti in that region. Since the area at which the traveling waves peak is a high-pressure area, the vestibular membrane is compressed at this point and, in turn, compresses the endolymph and the basilar membrane of the

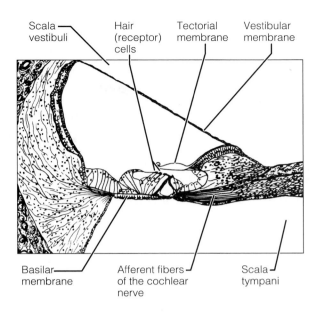

Scala vestibuli   Hair (receptor) cells   Tectorial membrane   Vestibular membrane

Basilar membrane   Afferent fibers of the cochlear nerve   Scala tympani

Figure 25.4 Microscopic view of the organ of Corti. (See also Plate 15 in the Histology Atlas.)

cochlear duct. The resulting pressure on the perilymph in the scala tympani causes the membrane of the round window to bulge outward into the middle ear chamber, thus acting as a relief valve for the compressional wave (Figure 25.5). Von Békésy found that high-frequency waves (high-pitched sounds) peaked close to the oval window and that low-frequency waves (low-pitched sounds) peaked farther up the basilar membrane near the apex of the cochlea. Although the mechanism of sound reception by the organ of Corti is not completely understood, we do know that hair cells on the basilar membrane are uniquely stimulated by sounds of various frequencies and amplitude and that once stimulated they depolarize and begin the chain of nervous impulses to the auditory centers of the temporal lobe cortex. This series of events results in the phenomenon we call hearing.

By the time most people are in their 60s, a gradual deterioration and atrophy of the organ of Corti begins, and leads to a loss in the ability to hear high tones and speech sounds. This condition, **presbycusis,** is a type of sensorineural deafness. Because many elderly people refuse to accept their hearing loss and resist using hearing aids, they begin to rely more and more on their vision for clues as to what is going on around them, and may be accused of ignoring people.

Although presbycusis is considered to be a disability of old age, it is becoming much more common in younger people as our world grows noisier. The damage (breakage of the "hairs" of the hair cells) caused by excessively loud sounds is progressive and cumulative. Each assault causes a bit more damage. Rock music played and listened to at deafening levels is definitely a contributing factor to the deterioration of hearing receptors. ■

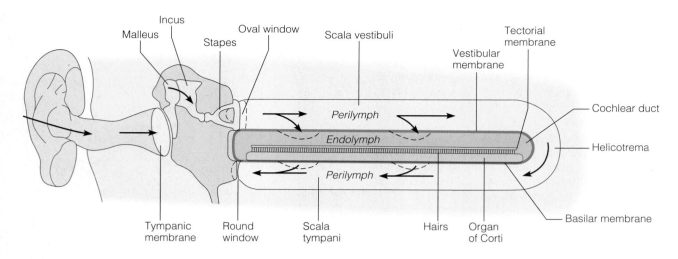

Figure 25.5 Fluid movement in the cochlea following the stirrup thrust at the oval window. Compressional wave created causes the round window to bulge into the middle ear.

Activity:
Conducting Laboratory Tests of Hearing

Perform the following hearing tests in a quiet area.

Acuity Test

Have your lab partner pack one ear with cotton and sit quietly with eyes closed. Obtain a ticking clock or pocket watch and hold it very close to his or her *unpacked* ear. Then slowly move it away from the ear until your partner signals that the ticking is no longer audible. Record the distance in inches at which ticking is inaudible.

Right ear _____ Left ear _____

Is the threshold of audibility sharp or indefinite?

_____

Sound Localization

Ask your partner to close both eyes. Hold the pocket watch at an audible distance (about 6 inches) from his or her ear, and move it to various locations (front, back, sides, and above his or her head). Have your partner locate the position by pointing in each instance. Can the sound be localized equally well at all positions?

_____ If not, at what position(s) was the sound less easily located?

_____

_____

The ability to localize the source of a sound depends on two factors—the difference in the loudness of the sound reaching each ear and the time of arrival of the sound at each ear. How does this information help to explain your findings?

_____

_____

Frequency Range of Hearing

Obtain three tuning forks: one with a low frequency (75 to 100 Hz [cps]), one with a frequency of approximately 1000 Hz, and one with a frequency of 4000 to 5000 Hz. Strike the lowest-frequency fork on the heel of your hand or with a rubber mallet, and hold it close to your partner's ear. Repeat with the other two forks.

Which fork was heard most clearly and comfortably?

_____ Hz

Which was heard least well?_____ Hz

Weber Test to Determine Conductive and Sensorineural Deafness

Strike a tuning fork and place the handle of the tuning fork medially on your partner's head (see Figure 25.6a). Is the tone equally loud in both ears, or is it louder in one ear?

_____

If it is equally loud in both ears, you have equal hearing or equal loss of hearing in both ears. If sensorineural deafness is present in one ear, the tone will be heard in the unaffected ear,

(a)

(b)

(c)

Figure 25.6   The Weber and Rinne tuning fork tests. (a) The Weber test to evaluate whether the sound remains centralized (normal) or lateralizes to one side or the other (indicative of some degree of conductive or sensorineural deafness). **(b** and **c)** The Rinne test to compare bone conduction and air conduction.

but not in the ear with sensorineural deafness. If conduction deafness is present, the sound will be heard more strongly in the ear in which there is a hearing loss. Conduction deafness can be simulated by plugging one ear with cotton to interfere with the conduction of sound to the inner ear.

Rinne Test for Comparing Bone- and Air-Conduction Hearing

1.   Strike the tuning fork, and place its handle on your partner's mastoid process (Figure 25.6b).

2.   When your partner indicates that the sound is no longer audible, hold the still-vibrating prongs close to his auditory canal (Figure 25.6c). If your partner hears the fork again (by air conduction) when it is moved to that position, hearing is not impaired and the test result is to be recorded as positive (+). (Record below.)

3.   Repeat the test, but this time test air-conduction hearing first.

4.   After the tone is no longer heard by air conduction, hold the handle of the tuning fork on the bony mastoid process. If the subject hears the tone again by bone conduction after hearing by air conduction is lost, there is some conductive deafness and the result is recorded as negative (−).

5.   Repeat the sequence for the opposite ear.

Right ear _____ Left ear _____

Does the subject hear better by bone or by air conduction?

_____ ■

Audiometry

When the simple tuning fork tests reveal a problem in hearing, audiometer testing is usually prescribed to determine the precise nature of the hearing deficit. An *audiometer* is an instrument (specifically, an electronic oscillator with earphones) used to determine hearing acuity by exposing each ear to sound stimuli of differing *frequencies* and *intensities.* The hearing range of human beings during youth is from 30 to 22,000 Hz, but hearing acuity declines with age, with reception for the high-frequency sounds lost first. Though this loss represents a major problem for some people, such as musicians, most of us tend to be fairly unconcerned until we begin to have problems hearing sounds in the range of 125 to 8000 Hz, the normal frequency range of speech.

The basic procedure of audiometry is to initially deliver tones of different frequencies to one ear of the subject at an intensity of 0 decibels (dB). (Zero decibels is not the complete absence of sound, but rather the softest sound intensity that can be heard by a person of normal hearing at each frequency.) If the subject cannot hear a particular frequency stimulus of 0 dB, the hearing threshold level control is adjusted until the subject reports that he or she can hear the tone. The number of decibels of intensity required above 0 dB is recorded as the hearing loss. For example, if the subject cannot hear a particular frequency tone until it is delivered at 30 dB intensity, then he or she has a hearing loss of 30 dB for that frequency.

Activity:
Audiometry Testing

1.   Before beginning the tests, examine the audiometer to identify the two tone controls: one to regulate frequency and a second to regulate the intensity (loudness) of the sound stimulus. Also identify the two output control switches that regulate the delivery of sound to one ear or the other (*red* to the right ear, *blue* to the left ear). Also find the *hearing threshold level control,* which is calibrated to deliver a basal tone of 0 dB to the subject's ears.

2.   Place the earphones on the subject's head so that the red cord or ear-cushion is over the right ear and the blue cord or ear-cushion is over the left ear. Instruct the subject to raise one hand when he or she hears a tone.

3.   Set the frequency control at 125 Hz and the intensity control at 0 dB. Press the red output switch to deliver a tone to the subject's right ear. If the subject does not respond, raise the sound intensity slowly by rotating the hearing level control counterclockwise until the subject reports (by raising a hand) that a tone is heard. Repeat this procedure for frequencies of 250, 500, 1000, 2000, 4000, and 8000.

4.   Record the results in the chart below by marking a small red circle on the chart at each frequency-dB junction indicated. Then connect the circles with a red line to produce a hearing acuity graph for the right ear.

5.   Repeat steps 3 and 4 for the left (blue) ear and record the results with blue circles and connecting lines on the chart. ■

Microscopic Anatomy of the Equilibrium Apparatus and Mechanisms of Equilibrium

The equilibrium apparatus of the inner ear is in the vestibular and semicircular canal portions of the bony labyrinth. Their chambers are filled with perilymph, in which membranous labyrinth structures are suspended. The vestibule contains the saclike **utricle** and **saccule,** and the semicircular chambers contain **membranous semicircular ducts.** Like the cochlear duct, these membranes are filled with endolymph and contain receptor cells that are activated by the disturbance of their cilia.

The semicircular canals are centrally involved in the **mechanism of dynamic equilibrium.** They are about ½ inch in circumference and are oriented in three planes—horizontal, frontal, and sagittal. At the base of each semicircular duct is an enlarged region, the **ampulla,** which communicates with the utricle of the vestibule. Within each ampulla is a receptor region called a **crista ampullaris,** which consists of a tuft of hair cells covered with a gelatinous cap, or **cupula** (Figure 25.7). When your head position changes in an angular direction, as when twirling on the dance floor or when taking a rough boat ride, the endolymph in the canal lags behind, pushing the cupula—like a swinging door—in a direction opposite to that of the angular motion (Figure 25.7c). This movement depolarizes the hair cells, resulting in enhanced impulse transmission up the vestibular division of the eighth cranial nerve to the brain. Likewise, when the angular motion stops suddenly, the inertia of the endolymph causes it to continue to move, pushing the cupula in the same direction as the previous body motion. This movement again initiates electrical changes in the hair cells (in this case, it hyperpolarizes them). (This phenomenon accounts for the reversed motion sensation you feel when you stop suddenly after twirling.) If you begin to move at a constant rate of motion, the endolymph comes to rest and the cupula gradually returns to its original position. The hair cells, no longer bent, send no new signals, and you lose the sensation of spinning. Thus the response of these dynamic equilibrium receptors is a reaction to *changes* in angular motion rather than to motion itself.

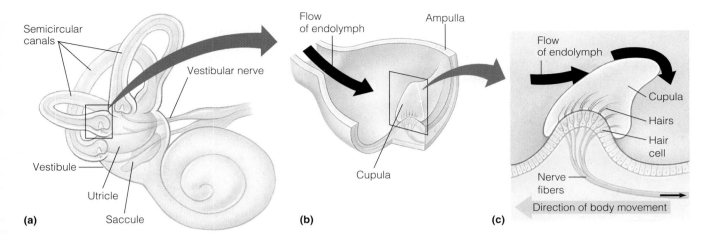

Figure 25.7 **Structure and function of the crista ampullaris.** **(a)** Arranged in the three spatial planes, the semicircular ducts in the semicircular canals each have a swelling called an ampulla at their base. **(b)** Each ampulla contains a crista ampullaris, a receptor that is essentially a cluster of hair cells with hairs projecting into a gelatinous cap called the cupula. **(c)** When the rate of rotation changes, inertia prevents the endolymph in the semicircular canals from moving with the head, so the fluid presses against the cupula, bending the hair cells in the opposite direction. The bending increases the frequency of action potentials in the sensory neurons in direct proportion to the amount of rotational acceleration. The mechanism adjusts quickly if rotation continues at a constant speed.

Activity:

Examining the Microscopic Structure of the Crista Ampullaris

Go to the demonstration area and examine the slide of a crista ampullaris. Identify the areas depicted in Figure 25.7b and c. ∎

The vestibule contains **maculae,** receptors that are essential to the **mechanism of static equilibrium** (Figure 25.8). The maculae respond to gravitational pull, thus providing information on which way is up or down, and to linear or straightforward changes in speed. They are located on the walls of the saccule and utricle. The **otolithic membrane,** a gelatinous material containing small grains of calcium carbonate (**otoliths**), overrides the hair cells in each macula. As the head moves, the otoliths move in response to variations in gravitational pull (Figure 25.8b). As they deflect different hair cells, they trigger hyperpolarization or depolarization of the hair cells and modify the rate of impulse transmission along the vestibular nerve.

Although the receptors of the semicircular canals and the vestibule are responsible for dynamic and static equilibrium respectively, they rarely act independently. Complex interaction of many of the receptors is the rule. Also, the information these equilibrium, or balance, senses provide is significantly augmented by the proprioceptors and sight, as some of the following laboratory experiments demonstrate.

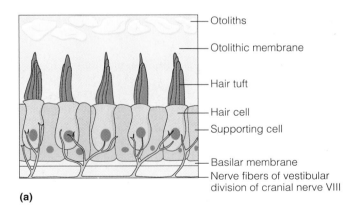

Figure 25.8 **Structure and function of static equilibrium receptors (maculae).** **(a)** Diagrammatic view of part of a macula. Figure continues on p. 260.

(b)

Head upright

Head inverted

Head horizontal

Figure 25.8 *(continued)* **Structure and function of static equilibrium receptors (maculae).**
(b) Stimulation of the maculae by movement of otoliths in the gelatinous otolithic membrane creates a pull on the hair cells. Arrows indicate direction of gravitational pull.

Activity:
Conducting Laboratory Tests on Equilibrium

The function of the semicircular canals and vestibule are not routinely tested in the laboratory, but the following simple tests should serve to illustrate normal equilibrium apparatus functioning.

Balance Test

Have your partner walk a straight line, placing one foot directly in front of the other.

Is he or she able to walk without undue wobbling from side to side?

_____

Did he or she experience any dizziness? _____

The ability to walk with balance and without dizziness, unless subject to rotational forces, indicates normal function of the equilibrium apparatus.

Was nystagmus* present? _____

* **Nystagmus** is the involuntary rolling of the eyes in any direction or the trailing of the eyes slowly in one direction, followed by their rapid movement in the opposite direction. It is normal after rotation; abnormal otherwise. The direction of nystagmus is that of its quick phase on acceleration.

Barany Test (Induction of Nystagmus and Vertigo†)

This experiment evaluates the semicircular canals and should be conducted as a group effort to protect test subject(s) from possible injury.

⚠ The following precautionary notes should be read before beginning:

• The subject(s) chosen should not be easily inclined to dizziness during rotational or turning movements.

• Rotation should be stopped immediately if the subject complains of feeling nauseous.

• Because the subject(s) will experience vertigo and loss of balance as a result of the rotation, several classmates should be prepared to catch, hold, or support the subject(s) as necessary until the symptoms pass.

1. Instruct the subject to sit on a rotating chair or stool, and to hold on to the arms or seat of the chair, feet on stool rungs. The subject's head should be tilted forward approximately 30 degrees (almost touching the chest). The horizontal (lateral) semicircular canal will be stimulated when the head is in this position. The subject's eyes are to remain *open* during the test.

2. Four classmates should position themselves so that the subject is surrounded on all sides. The classmate posterior to the subject will rotate the chair.

† **Vertigo** is a sensation of dizziness and rotational movement when such movement is not occurring or has ceased.

3.   Rotate the chair to the subject's right approximately 10 revolutions in 10 seconds, then suddenly stop the rotation.

4.   Immediately note the direction of the subject's resultant nystagmus; and ask him or her to describe the feelings of movement, indicating speed and direction sensation. Record below.

_____

_____

_____

_____

If the semicircular canals are operating normally, the subject will experience a sensation that the stool is still rotating immediately after it has stopped and *will* demonstrate nystagmus.

When the subject is rotated to the right, the cupula will be bent to the left, causing nystagmus during rotation in which the eyes initially move slowly to the left and then quickly to the right. Nystagmus will continue until the cupula has returned to its initial position. Then, when rotation is stopped abruptly, the cupula will be bent to the right, producing nystagmus with its slow phase to the right and its rapid phase to the left. In many subjects, this will be accompanied by a feeling of vertigo and a tendency to fall to the right.

Romberg Test

The Romberg test determines the integrity of the dorsal white column of the spinal cord, which transmits impulses to the brain from the proprioceptors involved with posture.

1.   Have your partner stand with his or her back to the blackboard.

2.   Draw one line parallel to each side of your partner's body. He or she should stand erect, with eyes open and staring straight ahead for 2 minutes while you observe any movements. Did you see any gross swaying movements?

_____

3.   Repeat the test. This time the subject's eyes should be closed. Note and record the degree of side-to-side movement.

_____

4.   Repeat the test with the subject's eyes first open and then closed. This time, however, the subject should be positioned with his or her left shoulder toward, but not touching, the board so that you may observe and record the degree of front-to-back swaying.

_____

Do you think the equilibrium apparatus of the inner ear was operating equally well in all these tests?

_____

The proprioceptors? _____

Why was the observed degree of swaying greater when the eyes were closed?

_____

_____

What conclusions can you draw regarding the factors necessary for maintaining body equilibrium and balance?

_____

_____

_____

Role of Vision in Maintaining Equilibrium

To further demonstrate the role of vision in maintaining equilibrium, perform the following experiment. (Ask your lab partner to record observations and act as a "spotter.") Stand erect, with your eyes open. Raise your left foot approximately 1 foot off the floor, and hold it there for 1 minute.

Record the observations: _____

_____

_____

Rest for 1 or 2 minutes; and then repeat the experiment with the same foot raised, but with your eyes closed. Record the observations:

_____

_____   ■

Special Senses: Olfaction and Taste

Objectives

1. To describe the location and cellular composition of the olfactory epithelium.

2. To describe the structure and function of the taste receptors.

3. To name the four basic qualities of taste sensation and list the chemical substances that elicit them.

4. To point out on a diagram of the tongue the predominant location of the basic types of taste receptors (salty, sweet, sour, bitter).

5. To explain the interdependence between the senses of smell and taste.

6. To name two factors other than olfaction that influence taste appreciation of foods.

7. To define *olfactory adaptation*.

Materials

❑ Prepared histologic slides: the tongue showing taste buds; nasal olfactory epithelium (longitudinal section)

❑ Compound microscope

❑ Small mirror

❑ Paper towels

❑ Granulated sugar

❑ Cotton-tipped swabs

❑ Disposable autoclave bag

❑ Paper cups; paper plates

❑ Prepared vials of 10% NaCl, 0.1% quinine or Epsom salt solution, 5% sucrose solution, and 1% acetic acid

❑ Beaker containing 10% bleach solution

❑ Prepared dropper bottles of oil of cloves, oil of peppermint, and oil of wintergreen or corresponding flavors found in the condiment section of a supermarket

❑ Equal-size food cubes of cheese, apple, raw potato, dried prunes, banana, raw carrot, and hard-cooked egg white (These prepared foods should be in an opaque container; a foil-lined egg carton would work well.)

❑ Chipped ice

❑ Absorbent cotton

The receptors for olfaction and taste are classified as **chemoreceptors** because they respond to chemicals or volatile substances in solution. Although four relatively specific types of taste receptors have been identified, the olfactory receptors are considered sensitive to a much wider range of chemical sensations. The sense of smell is the least understood of the special senses.

Localization and Anatomy of the Olfactory Receptors

The **olfactory epithelium** (organ of smell) occupies an area of about 2.5 cm in the roof of each nasal cavity (Figure 26.1a). Since the air entering the human nasal cavity must make a hairpin turn to enter the respiratory passages below, the nasal epithelium is in a rather poor position for performing its function. This is why sniffing, which brings more air into contact with the receptors, intensifies the sense of smell.

The specialized receptor cells in the olfactory epithelium are surrounded by **supporting cells,** nonsensory epithelial cells. The **olfactory receptor cells** are bipolar neurons whose **olfactory hairs** (actually cilia) extend outward from the epithelium. Axonal nerve fibers emerging from their basal ends penetrate the cribriform plate of the ethmoid bone and proceed as the *olfactory nerves* to synapse in the olfactory bulbs lying on either side of the crista galli of the ethmoid bone. Impulses from neurons of the olfactory bulbs are then conveyed to the olfactory portion of the cortex (uncus).

Activity:
Microscopic Examination of the Olfactory Epithelium

Obtain a longitudinal section of olfactory epithelium. Examine it closely, comparing it to Figure 26.1b. ■

(a)

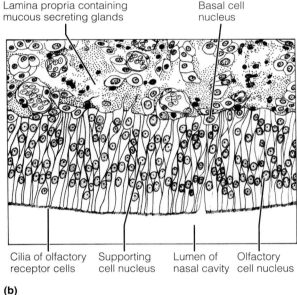

(b)

Figure 26.1 Location and cellular composition of olfactory epithelium. (a) Diagrammatic representation. **(b)** Microscopic view. (See also Plate 17 in the Histology Atlas.)

Localization and Anatomy of Taste Buds

The **taste buds,** specific receptors for the sense of taste, are widely but not uniformly distributed in the oral cavity. Most are located on the dorsal surface of the tongue (as described next). A few are found on the soft palate, epiglottis, and inner surface of the cheeks.

The dorsal tongue surface is covered with small projections, or **papillae,** of three major types: sharp *filiform papillae* and the rounded *fungiform* and *circumvallate papillae.* The taste buds are located primarily on the sides of the circumvallate papillae (arranged in a V-formation on the posterior surface of the tongue) and on the more numerous fungiform papillae. The latter look rather like minute mushrooms and are widely distributed on the tongue. (See Figure 26.2.)

• Use a mirror to examine your tongue. Can you pick out the various papillae types?

_____ If so, which? _____

_____

Each taste bud consists largely of a globular arrangement of two types of modified epithelial cells: the **gustatory,** or **taste cells,** which are the actual receptor cells, and **supporting cells.** Several nerve fibers enter each taste bud and supply sensory nerve endings to each of the taste cells. The long microvilli of the receptor cells penetrate the epithelial surface through an opening called the **taste pore.** When these microvilli, called **gustatory hairs,** contact specific chemicals in the solution, the taste cells depolarize. The afferent fibers from the taste buds to the sensory cortex in the postcentral gyrus of the brain are carried in three cranial nerves: the *facial nerve (VII)* serves the anterior two-thirds of the tongue; the *glossopharyngeal nerve (IX)* serves the posterior third of the tongue; and the *vagus nerve (X)* carries a few fibers from the pharyngeal region.

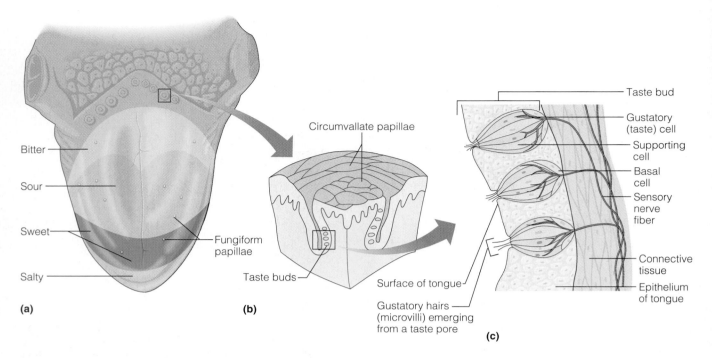

Figure 26.2 **Location and structure of taste buds. (a)** Taste buds on the tongue are associated with papillae, projections of the tongue mucosa. Major location of each type of taste bud is indicated by different color bands. **(b)** A sectioned circumvallate papilla shows the position of the taste buds in its lateral walls. **(c)** An enlarged view of three taste buds. See also Plate 16 in the Histology Atlas.

Activity:
Microscopic Examination of Taste Buds

Obtain a microscope and a prepared slide of a tongue cross section. Use Figure 26.2b and Plate 16 in the Histology Atlas as guides to aid you in locating the taste buds on the tongue papillae. Make a detailed study of one taste bud. Identify the taste pore and gustatory hairs if observed. ■

Laboratory Experiments

Activity:
Stimulating Taste Buds

1.   Obtain several paper towels and a disposable autoclave bag and bring them to your bench.

2.   With a paper towel, dry the dorsal surface of your tongue.

⚠   Immediately dispose of the paper towel in the autoclave bag.

3.   Place a few sugar crystals on your dried tongue. Do *not* close your mouth. Time how long it takes to taste the sugar.

_____ sec

Why couldn't you taste the sugar immediately?

_____ ■

Activity:
Plotting Taste Bud Distribution

When taste is tested with pure chemical compounds, most taste sensations can be grouped into one of four basic qualities—sweet, sour, bitter, or salty. Although all taste buds are believed to respond in some degree to all four classes of chemical stimuli, each type responds optimally to only one. This characteristic makes it possible to map the tongue to show the relative density of each type of taste bud.

The *sweet receptors* respond to a number of seemingly unrelated compounds such as sugars (fructose, sucrose, glucose), saccharine, and some amino acids. Some believe the common factor is the hydroxyl (OH^-) group. *Sour receptors* respond to hydrogen ions (H^+) or the acidity of the solution, *bitter receptors* to alkaloids, and *salty receptors* to metallic ions in solution.

1.   Prepare to make a taste sensation map of your lab partner's tongue by obtaining the following: cotton-tipped swabs; one vial each of NaCl, quinine or Epsom salt solution, sucrose solution, and acetic acid; paper cups; and a flask of distilled or tap water.

2.   Before each test, the subject should rinse his or her mouth thoroughly with water and lightly dry his or her tongue with a paper towel.

⚠   Dispose of used paper towels in the autoclave bag.

3.   Generously moisten a swab with 5% sucrose solution and touch it to the center, back, tip, and sides of the dorsal surface of the subject's tongue.

4.   Map, with O's on the tongue outline below, the location of the sweet receptors.

 Put the used swab in the autoclave bag. Do not redip the swab into the sucrose solution.

5.   Repeat the procedure with quinine (or Epsom salt solution) to map the location of the bitter receptors (use the symbol B), with NaCl to map the salt receptors (symbol +), and with acetic acid to map the sour receptors (symbol −).

![warning] *Use a fresh swab for each test,* and properly dispose of the swabs immediately after use.

What area of the tongue dorsum seems to lack taste receptors?

_____

_____

How closely does your localization of the different taste receptors coincide with the information in your textbook?

_____

_____ ▪

Activity:

Examining the Combined Effects of Smell, Texture, and Temperature on Taste

Effects of Smell and Texture

1.   Ask the subject to sit with eyes closed and to pinch his or her nostrils shut.

2.   Using a paper plate, obtain samples of the food items listed in the chart below. At no time should the subject be allowed to see the foods being tested.

3.   Use an out-of-sequence order of food testing. For each test, place a cube of food in the subject's mouth and ask him or her to identify the food by using the following sequence of activities:

- First, manipulate the food with the tongue.

- Second, chew the food.

- Third, if a positive identification is not made with the first two techniques and the taste sense, ask the subject to release the pinched nostrils and to continue chewing with the nostrils open to determine if a positive identification can be made.

Record the results on the chart by checking the appropriate column.

Was the sense of smell equally important in all cases?

_____

_____

Where did it seem to be important and why?

_____

_____

| Chart 1 | Method of Identification | | | |
Food	Texture only	Chewing with nostrils pinched	Chewing with nostrils open	Identification not made
Cheese	_____	_____	_____	_____
Apple	_____	_____	_____	_____
Raw potato	_____	_____	_____	_____
Banana	_____	_____	_____	_____
Dried prunes	_____	_____	_____	_____
Raw carrot	_____	_____	_____	_____
Hard-cooked egg white	_____	_____	_____	_____

Effect of Olfactory Stimulation

There is no question that what is commonly referred to as taste depends heavily on stimulation of the olfactory receptors, particularly in the case of strongly odoriferous substances. The following experiment should illustrate this fact.

1. Obtain vials of oil of wintergreen, peppermint, and cloves and some fresh cotton-tipped swabs. Ask the subject to sit so that he or she cannot see which vial is being used, and to dry the tongue and close the nostrils.

2. Apply a drop of one of the oils to the subject's tongue. Can he or she distinguish the flavor?

_____

3. Have the subject open the nostrils, and record the change in sensation he or she reports.

_____

4. Have the subject rinse the mouth well and dry the tongue.

5. Prepare two swabs, each with one of the two remaining oils.

6. Hold one swab under the subject's open nostrils, while touching the second swab to the tongue.

Record the reported sensations. _____

_____

_____

7. ⚠ Dispose of the used swabs and paper towels in the autoclave bag before continuing.

Which sense, taste or smell, appears to be more important in the proper identification of a strongly flavored volatile substance?

_____

Effect of Temperature

In addition to the effect that olfaction and food texture play in determining our taste sensations, the temperature of foods also helps determine if the food is appreciated or even tasted. To illustrate this, have your partner hold some chipped ice on the tongue for approximately a minute and then close his or her eyes. Immediately place any of the foods previously identified in his or her mouth and ask for an identification.

Results? _____

_____

_____ ■

A c t i v i t y :
Demonstrating Olfactory Adaptation

Obtain some absorbent cotton and two of the following oils (oil of wintergreen, peppermint, or cloves). Press one nostril shut. Hold the bottle of oil under the open nostril and exhale through the mouth. Record the time required for the odor to disappear (for olfactory adaptation to occur).

_____ sec

Repeat the procedure with the other nostril.

_____ sec

Immediately test another oil with the nostril that has just experienced olfactory adaptation. What are the results?

_____

_____

What conclusions can you draw? _____

_____

_____ ■

Anatomy and Basic Function of the Endocrine Glands

The **endocrine system** is the second major controlling system of the body. Acting with the nervous system, it helps coordinate and integrate the activity of the body's cells. However, the nervous system employs electrochemical impulses to bring about rapid control, while the more slowly acting endocrine system employs chemical "messengers," or **hormones,** which are released into the blood to be transported throughout the body.

The term *hormone* comes from a Greek word meaning "to arouse." The body's hormones, which are steroids or amino acid–based molecules, arouse the body's tissues and cells by stimulating changes in their metabolic activity. These changes lead to growth and development and to the physiologic homeostasis of many body systems. Although all hormones are bloodborne, a given hormone affects only the biochemical activity of a specific organ or organs. Organs that respond to a particular hormone are referred to as the **target organs** of that hormone. The ability of the target tissue to respond seems to depend on the ability of the hormone to bind with specific receptors (proteins) occurring on the cells' plasma membrane or within the cells.

Although the function of some hormone-producing glands (the anterior pituitary, thyroid, adrenals, parathyroids) is purely endocrine, the function of others (the pancreas and gonads) is mixed—both endocrine and exocrine. Both types of glands are derived from epithelium, but the endocrine, or ductless, glands release their product (always hormonal) directly into the blood or lymph. The exocrine glands release their products at the body's surface or outside an epithelial membrane via ducts. In addition, there are varied numbers of hormone-producing cells within the intestine, stomach, kidney, and placenta, organs whose functions are primarily nonendocrine. Only the major endocrine organs are considered here.

Gross Anatomy and Basic Function of the Endocrine Glands

Pituitary Gland (Hypophysis)

The *pituitary gland,* or *hypophysis,* is located in the concavity of the sella turcica of the sphenoid bone. It consists largely of two functional areas, the **adenohypophysis,** or **anterior pituitary,** and the **neurohypophysis,** consisting mainly of the **posterior pituitary.** The pituitary gland is attached to the hypothalamus by a stalk called the **infundibulum.**

Anterior Pituitary Hormones The anterior pituitary, or adenohypophysis, secretes a number of hormones. Four of these are **tropic** hormones. In each case, a tropic hormone released by

Objectives

1. To identify and name the major endocrine glands and tissues of the body when provided with an appropriate diagram.
2. To list the hormones produced by the endocrine glands and discuss the general function of each.
3. To indicate the means by which hormones contribute to body homeostasis by giving appropriate examples of hormonal actions.
4. To cite the mechanism by which the endocrine glands are stimulated to release their hormones.
5. To describe the structural and functional relationship between the hypothalamus and the pituitary.
6. To describe a major pathologic consequence of hypersecretion and hyposecretion of each hormone considered.
7. To correctly identify the histologic structure of the anterior and posterior pituitary, thyroid, parathyroid, adrenal cortex and medulla, pancreas, testis, and ovary by microscopic inspection or when presented with an appropriate photomicrograph or diagram. (Optional)
8. To name and point out the specialized hormone-secreting cells in the above tissues as studied in the laboratory. (Optional)

Materials

- ❑ Human torso model
- ❑ Anatomical chart of the human endocrine system
- ❑ Compound microscopes
- ❑ Colored pencils
- ❑ Histologic slides of the anterior pituitary,* posterior pituitary, thyroid gland, parathyroid glands, adrenal gland, and pancreas,* ovary, and testis tissue

 See Appendix C, Exercise 27 for links to A.D.A.M. Interactive Anatomy.

 For instructions on animal dissections, see the dissection exercises starting on p. 733 in the cat and fetal pig editions of this manual.

*Differential-staining if possible

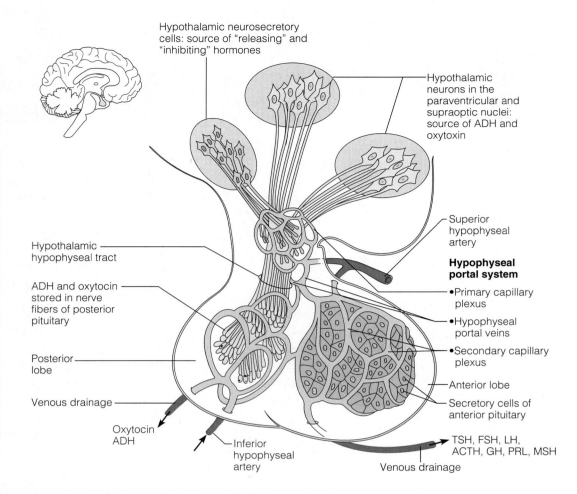

Hypothalamic neurosecretory cells: source of "releasing" and "inhibiting" hormones

Hypothalamic neurons in the paraventricular and supraoptic nuclei: source of ADH and oxytoxin

Hypothalamic hypophyseal tract

Superior hypophyseal artery

Hypophyseal portal system

ADH and oxytocin stored in nerve fibers of posterior pituitary

•Primary capillary plexus

•Hypophyseal portal veins

•Secondary capillary plexus

Posterior lobe

Anterior lobe

Venous drainage

Secretory cells of anterior pituitary

Oxytocin ADH

Inferior hypophyseal artery

TSH, FSH, LH, ACTH, GH, PRL, MSH

Venous drainage

Figure 27.1   Neural and vascular relationships between the hypothalamus and the anterior and posterior lobes of the pituitary.

the anterior pituitary stimulates its target organ, which is also an endocrine gland, to secrete its hormones. Target organ hormones then exert their effects on other body organs and tissues. The tropic hormones include:

•   The **gonadotropins—follicle-stimulating hormone (FSH)** and **luteinizing hormone (LH)—regulate** gamete production and hormonal activity of the gonads (ovaries and testes). The precise roles of the gonadotropins are described in Exercise 44 along with other considerations of reproductive system physiology.

•   **Adrenocorticotropic hormone (ACTH)** regulates the endocrine activity of the cortex portion of the adrenal gland.

•   **Thyrotropic hormone (TSH)** influences the growth and activity of the thyroid gland.

The three other hormones produced by the anterior pituitary are not directly involved in the regulation of other endocrine glands of the body.

•   **Growth hormone (GH)** is a general metabolic hormone that plays an important role in determining body size. It affects many tissues of the body; however, its major effects are exerted on the growth of muscle and the long bones of the body.

•   **Prolactin (PRL)** stimulates breast development and promotes and maintains lactation by the mammary glands after childbirth. Its function in males is unknown.

•   **Melanocyte-stimulating hormone (MSH),** which stimulates melanocytes to increase their synthesis of melanin pigment, does not appear to be of major significance in humans.

The anterior pituitary controls the activity of so many other endocrine glands that it has often been called the *master endocrine gland.* However, the anterior pituitary is not autonomous in its control because release of the anterior pituitary hormones is controlled by neurosecretions, *releasing or inhibiting hormones,* produced by the hypothalamus. These hypothalamic hormones are liberated into the **hypophyseal portal system,** which serves the circulatory needs of the anterior pituitary (Figure 27.1).

Posterior Pituitary Hormones   The posterior pituitary is not an endocrine gland in a strict sense because it does not synthesize the hormones it releases. (This relationship is also indicated in Figure 27.1.) Instead, it acts as a storage area for two hormones transported to it from the paraventricular and supraoptic nuclei of the hypothalamus. The first of these hormones is **oxytocin,** which stimulates powerful uterine contractions during birth and coitus and also causes milk ejection in the lactating mother. The second, **antidiuretic hormone**

(ADH), causes the distal and collecting tubules of the kidneys to reabsorb more water from the urinary filtrate, thereby reducing urine output and conserving body water. It also plays a minor role in increasing blood pressure because of its vasoconstrictor effect on the arterioles.

Hyposecretion of ADH results in dehydration from excessive urine output, a condition called **diabetes insipidus.** Individuals with this condition experience an insatiable thirst. ■

Thyroid Gland

The *thyroid gland* is composed of two lobes joined by a central mass, or isthmus. It is located in the throat, just inferior to the larynx. It produces two major hormones, thyroid hormone and calcitonin.

Thyroid hormone (TH) is actually two physiologically active hormones known as T_4 (thyroxine) and T_3 (triiodothyronine). Because its primary function is to control the rate of body metabolism and cellular oxidation, TH affects virtually every cell in the body.

Hyposecretion of thyroxine leads to a condition of mental and physical sluggishness, which is called **myxedema** in the adult. ■

Calcitonin (also called **thyrocalcitonin**) decreases blood calcium levels by stimulating calcium deposit in the bones. It acts antagonistically to parathyroid hormone, the hormonal product of the parathyroid glands.

• Try to palpate your thyroid gland by placing your fingers against your windpipe. As you swallow, the thyroid gland will move up and down on the sides and front of the windpipe.

Parathyroid Glands

The *parathyroid glands* are found embedded in the posterior surface of the thyroid gland. Typically, there are two small oval glands on each lobe, but there may be more and some may be located in other regions of the neck. They secrete **parathyroid hormone (PTH),** the most important regulator of calcium-phosphate ion homeostasis of the blood. When blood calcium levels decrease below a certain critical level, the parathyroids release PTH, which causes release of calcium from bone matrix and prods the kidney to reabsorb more calcium and less phosphate from the filtrate. If blood calcium levels fall too low, **tetany** results and may be fatal.

Adrenal Glands

The two bean-shaped *adrenal,* or *suprarenal, glands* are located atop or close to the kidneys. Anatomically, the **adrenal medulla** develops from neural crest tissue and it is directly controlled by sympathetic nervous system neurons. The medullary cells respond to this stimulation by releasing **epinephrine** (80%) or **norepinephrine** (20%), which act in conjunction with the sympathetic nervous system to elicit the "flight or fight" response to stressors.

The **adrenal cortex** produces three major groups of steroid hormones, collectively called the **corticosteroids.** The **mineralocorticoids,** chiefly **aldosterone,** regulate water and electrolyte balance in the extracellular fluids, mainly by regulating sodium ion reabsorption by kidney tubules. The **glucocorticoids** (cortisone, hydrocortisone, and corticosterone) enable the body to resist long-term stressors, primar-

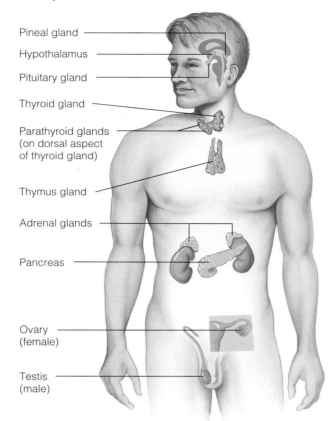

Figure 27.2 Human endocrine organs.

Pineal gland
Hypothalamus
Pituitary gland
Thyroid gland
Parathyroid glands
(on dorsal aspect
of thyroid gland)
Thymus gland
Adrenal glands
Pancreas
Ovary
(female)
Testis
(male)

ily by increasing blood glucose levels. The **gonadocorticoids,** or **sex hormones,** produced by the adrenal cortex are chiefly androgens (male sex hormones), but some estrogens (female sex hormones) are formed. The gonadocorticoids are produced throughout life in relatively insignificant amounts; however, hypersecretion of these hormones produces abnormal hairiness (**hirsutism**), and masculinization occurs.

Pancreas

The *pancreas,* located partially behind the stomach in the abdomen, functions as both an endocrine and exocrine gland. It produces digestive enzymes as well as insulin and glucagon, important hormones concerned with the regulation of blood sugar levels.

Elevated blood glucose levels stimulate release of **insulin,** which decreases blood sugar levels, primarily by accelerating the transport of glucose into the body cells, where it is oxidized for energy or converted to glycogen or fat for storage.

Hyposecretion of insulin or some deficiency in the insulin receptors leads to **diabetes mellitus,** which is characterized by the inability of body cells to utilize glucose and the subsequent loss of glucose in the urine. Alterations of protein and fat metabolism also occur secondary to derangements in carbohydrate metabolism. ■

Glucagon acts antagonistically to insulin. Its release is stimulated by low blood glucose levels, and its action is basically hyperglycemic. It stimulates the liver, its primary target organ, to break down its glycogen stores to glucose and subsequently to release the glucose to the blood.

The Gonads

The female *gonads,* or *ovaries,* are paired, almond-sized organs located in the pelvic cavity. In addition to producing the female sex cells (ova), the ovaries produce two steroid hormone groups, the estrogens and progesterone. The endocrine and exocrine functions of the ovaries do not begin until the onset of puberty, when the anterior pituitary gonadotropic hormones prod the ovary into action that produces rhythmic ovarian cycles in which ova develop and hormonal levels rise and fall. The **estrogens** are responsible for the development of the secondary sex characteristics of the female at puberty (primarily maturation of the reproductive organs and development of the breasts) and act with progesterone to bring about cyclic changes of the uterine lining that occur during the menstrual cycle. The estrogens also help prepare the mammary glands for lactation.

Progesterone, as already noted, acts with estrogen to bring about the menstrual cycle. During pregnancy it maintains the uterine musculature in a quiescent state and helps to prepare the breast tissue for lactation.

The paired oval *testes* of the male are suspended in a pouchlike sac, the scrotum, outside the pelvic cavity. In addition to the male sex cells, sperm, the testes produce the male sex hormone, **testosterone.** Testosterone promotes the maturation of the reproductive system accessory structures, brings about the development of the secondary sex characteristics, and is responsible for the male sexual drive, or libido. Both the endocrine and exocrine functions of the testes begin at puberty under the influence of the anterior pituitary gonadotropins.

Two glands not mentioned earlier as major endocrine glands should also be briefly considered here, the thymus and the pineal gland.

Thymus

The *thymus* is a bilobed gland situated in the superior thorax, posterior to the sternum and anterior to the heart and lungs. Conspicuous in the infant, it begins to atrophy at puberty, and by old age it is relatively inconspicuous. The thymus produces a hormone called **thymosin** which helps direct the maturation and specialization of a unique population of white blood cells called T lymphocytes or T cells. T lymphocytes are responsible for the cellular immunity aspect of body defense; that is, rejection of foreign grafts, tumors, or virus-infected cells.

Pineal Body

The *pineal body,* or *epiphysis cerebri,* is a small cone-shaped gland located in the roof of the third ventricle of the brain. Its major endocrine product is **melatonin.**

The endocrine role of the pineal body in humans is still controversial, but it is known to play a role in the biological rhythms (particularly mating and migratory behavior) of other animals. In humans, melatonin appears to exert some inhibitory effect on the reproductive system that prevents precocious sexual maturation.

Identifying the Endocrine Organs

Locate the endocrine organs on Figure 27.2. Also locate these organs on the anatomical charts or torso. ■

 For instructions on animal dissections, see the dissection exercises starting on p. 733 in the cat and fetal pig editions of this manual.

Microscopic Anatomy of Selected Endocrine Glands (Optional)

Examining the Microscopic Structure of Endocrine Glands

To prepare for the histologic study of the endocrine glands, obtain a microscope, one each of the slides listed in the list of materials, and colored pencils. We will study only organs in which it is possible to identify the endocrine-producing cells. Compare your observations with the line drawings in Figure 27.3a–f of the endocrine tissue photomicrographs.

Thyroid Gland

1. Scan the thyroid under low power, noting the **follicles,** spherical sacs containing a pink-stained material (*colloid*). Stored T_3 and T_4 are attached to the protein colloidal material stored in the follicles as **thyroglobulin** and are released gradually to the blood. Compare the tissue viewed to Figure 27.3a and Plate 18 in the Histology Atlas.

2. Observe the tissue under high power. Note that the walls of the follicles are formed by simple cuboidal or squamous epithelial cells that synthesize the follicular products. The **parafollicular,** or **C, cells** you see between the follicles are responsible for calcitonin production.

3. Color appropriately two or three follicles in Figure 27.3a. Label the colloid and parafollicular cells.

When the thyroid gland is actively secreting, the follicles appear small, and the colloidal material has a ruffled border. When the thyroid is hypoactive or inactive, the follicles are large and plump and the follicular epithelium appears to be squamouslike. What is the physiologic state of the tissue you have been viewing?

_____

_____

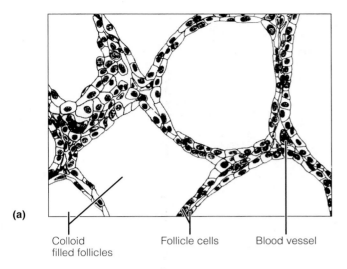

(a)

Colloid
filled follicles   Follicle cells   Blood vessel

(b)

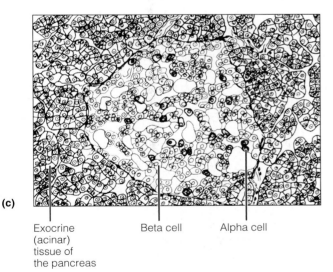

(c)

Exocrine
(acinar)
tissue of
the pancreas   Beta cell   Alpha cell

Figure 27.3 Microscopic anatomy of selected endocrine organs. (See also Plates 18–23 in the Histology Atlas.) **(a)** Thyroid; **(b)** parathyroid; **(c)** pancreas showing a pancreatic islet.

Parathyroid Glands

1.   Observe the parathyroid tissue under low power to view its two major cell types, the **chief cells** and the **oxyphil cells.** Compare your observations to the view in Plate 19 of the Histology Atlas. The chief cells, which synthesize parathyroid hormone (PTH), are small and abundant, and arranged in thick branching cords. The function of the scattered, much larger oxyphil cells is unknown.

2.   Color a small portion of the parathyroid tissue in Figure 27.3b. Label the chief cells and the oxyphil cells.

Pancreas

1.   Observe pancreas tissue under low power to identify the roughly circular **pancreatic islets (islets of Langerhans),** the endocrine portions of the pancreas. The islets are scattered amid the more numerous acinar cells and stain differently (usually lighter), which makes their identification possible. (See Figure 39.14 [p. 407] and Plate 43 in the Histology Atlas.)

2.   Focus on an islet and examine its cells under high power. Notice that the islet cells are densely packed and have no definite arrangement. In contrast, the cuboidal acinar cells are arranged around secretory ducts. Unless special stains are used, it will not be possible to distinguish the **alpha cells,** which tend to cluster at the periphery of the islets and produce glucagon, from the **beta cells,** which synthesize insulin. With these specific stains, the beta cells are larger and stain gray-blue, and the alpha cells are smaller and appear bright pink, as hinted at in Figure 27.3c and shown in Plate 20 in the Histology Atlas. What is the product of the acinar cells?

_____

3.   Draw a section of the pancreas in the space below. Label the islets and the acinar cells. If possible, differentiate the alpha and beta cells of the islets by color.

Pituitary Gland

1. Observe the general structure of the pituitary gland under low power to differentiate between the glandular anterior pituitary and the neural posterior pituitary. Figure 27.3d and e should help you get started.

2. Using the high-power lens, focus on the nests of cells of the anterior pituitary. It is possible to identify the specialized cell types that secrete the specific hormones when differential stains are used. Using Plate 21 in the Histology Atlas as a guide, locate the reddish-brown-stained **acidophil cells,** which produce growth hormone and prolactin, and the **basophil cells,** whose deep-blue granules are responsible for the production of the tropic hormones (TSH, ACTH, FSH, and LH). **Chromophobes,** the third cellular population, do not take up the stain and appear rather dull and colorless. The role of the chromophobes is controversial, but they apparently are not directly involved in hormone production.

3. Use appropriately colored pencils to identify acidophils, basophils, and chromophobes in Figure 27.3d.

4. Switch your focus to the posterior pituitary. Observe the nerve fibers (axons of hypophyseal neurons) that compose most of this portion of the pituitary. Also note the **pituicytes,** glial cells which are randomly distributed among the nerve fibers. Refer to Plate 22 in the Histology Atlas as you scan the slide.

What two hormones are stored here?

_____

What is their source?

_____

Adrenal Gland

1. Hold the slide of the adrenal gland up to the light to distinguish the outer cortex and inner medulla areas. Then scan the cortex under low power to distinguish the differences in cell appearance and arrangement in the three cortical areas. Refer to Figure 27.3f, Plate 23 in the Histology Atlas, and the descriptions below to identify the following cortical areas:

*   Connective tissue capsule of the adrenal gland.

*   The outermost **zona glomerulosa,** where most mineralocorticoid production occurs and where the tightly packed cells are arranged in spherical clusters.

*   The deeper intermediate **zona fasciculata,** which produces glucocorticoids. This is the thickest part of the cortex. Its cells are arranged in parallel cords.

(d)

Basophils   Chromophobes   Acidophils

(e)

Fenestrated   Nuclei of pituicytes
capillaries

(f)

Capsule
Zona glomerulosa
Zona fasciculata
Zona reticularis
Adrenal medulla

Figure 27.3 (continued) Microscopic anatomy of selected endocrine organs. (d) anterior pituitary; **(e)** posterior pituitary; **(f)** adrenal gland.

• The innermost cortical zone, the **zona reticularis** abutting the medulla, which produces sex hormones and some glucocorticoids. The cells here stain intensely and form a branching network.

2. Switch focus to view the large, lightly stained cells of the adrenal medulla under high power. Notice their clumped arrangement.

What hormones are produced by the medulla?

_____ and _____

3. Draw a representative area of each of the adrenal regions, indicating in your sketch the differences in relative cell size and arrangement.

Zona glomerulosa   Zona fasciculata

Zona reticularis   Adrenal medulla

Ovary

Because you will consider the *ovary* in greater histologic detail when you study the reproductive system, the objective in this laboratory exercise is just to identify the endocrine-producing parts of the ovary.

1. Scan an ovary slide under low power, and look for a **vesicular (Graafian) follicle,** a circular arrangement of cells enclosing a central cavity. See Figure 44.5 (p. 452) and Plate 24 in the Histology Atlas. This structure synthesizes estrogens.

2. Examine the vesicular follicle under high power, identifying the follicular cells that produce estrogens, the antrum (fluid-filled cavity), and developing ovum (if present). The ovum will be the largest cell in the follicle.

3. Draw and color a vesicular follicle below, labeling the antrum, follicle cells, and developing ovum.

4. Switch to low power, and scan the slide to find a **corpus luteum,** a large amorphous-looking area that produces progesterone (and some estrogens). A corpus luteum is shown in Plate 25 in the Histology Atlas.

Testis

1. Examine a section of a *testis* under low power. Identify the seminiferous tubules, which produce sperm, and the **interstitial cells,** which produce testosterone. The interstitial cells are scattered between the seminiferous tubules in the connective tissue matrix. The photomicrograph of seminiferous tubules, Plate 51 in the Histology Atlas, will be helpful here.

2. Draw a representative area of the testis in the space below. Label the seminiferous tubules and area of the interstitial cells. ■

Experiments on Hormonal Action

Objectives

1. To describe and explain the effect of pituitary hormones on the ovary.
2. To describe and explain the effects of hyperinsulinism.
3. To describe and explain the effect of epinephrine on the heart.
4. To understand the physiologic (and clinical) importance of metabolic rate measurement.
5. To investigate the effect of hypo-, hyper-, and euthyroid conditions on oxygen consumption and metabolic rate.
6. To assemble the necessary apparatus and properly use a manometer to obtain experimental results.
7. To calculate metabolic rate in terms of ml O_2/kg/hr.

Materials

Experiment 1:

❏ Glass desiccator; manometer, 20-ml glass syringe, two-hole rubber stopper, and T-valve (1 for every 3 to 4 students)
❏ Soda lime (desiccant)
❏ Hardware cloth squares
❏ Petrolatum
❏ Rubber tubing; tubing clamps; scissors
❏ 3-in. pieces of glass tubing
❏ Animal balances
❏ Heavy animal-handling gloves
❏ Chart set up on chalkboard so that each student group can record its computed metabolic rate figures under the appropriate headings
❏ Young rats of the same sex, obtained 2 weeks prior to the laboratory session and treated as follows for 14 days:

Group 1: control group—fed normal rat chow and water

Group 2: experimental group A—fed normal rat chow, and drinking water containing 0.02% 6-n-propylthiouracil*

Group 3: experimental group B—fed rat chow containing desiccated thyroid (2% by weight), and normal drinking water

Experiment 2:[†]

❏ Female frogs (*Rana pipiens*)
❏ Syringe (2-ml capacity)
❏ 20- to 25-gauge needle
❏ Frog pituitary extract
❏ Physiological saline
❏ Battery jar
❏ Spring or pond water
❏ Wax marking pencils

Experiment 3:[†]

❏ 500- or 600-ml beakers
❏ 20% glucose solution
❏ Commercial insulin solution (400 IU per 100 ml H_2O)
❏ Finger bowls
❏ Small ($1\frac{1}{2}$–2 in.) freshwater fish (guppy, bluegill, or sunfish—listed in order of preference)
❏ Wax marking pencils

Experiment 4:[†]

❏ Frog (*Rana pipiens*)
❏ 1:1000 epinephrine (Adrenalin) solution in dropper bottles
❏ Dissecting pan, instruments, and pins
❏ Frog Ringer's solution in dropper bottle
❏ Disposable gloves

*Note to the Instructor: 6-n-propylthiouracil (PTU) is degraded by light and should be stored in light-resistant containers or in the dark.

[†]*The Selected Actions of Hormones and Other Chemical Messengers* videotape (available to qualified adopters from Benjamin/Cummings) may be used in lieu of student participation in Experiments 2–4 of Exercise 28.

The endocrine system exerts many complex and interrelated effects on the body as a whole, as well as on specific organs and tissues. Most scientific knowledge about this system is contemporary, and new information is constantly being presented. Many experiments on the endocrine system require relatively large laboratory animals; are time-consuming (requiring days to weeks of observation); and often involve technically difficult surgical procedures to remove the glands or parts of them, all of which makes it difficult to conduct more general types of laboratory experiments. Nevertheless, the four technically unsophisticated experiments presented here should illustrate how dramatically hormones affect body functioning.

To conserve laboratory specimens, experiments can be conducted by groups of four students. The use of larger working groups should not detract from benefits gained, since the major value of these experiments lies in observation.

Activity:
Experiment 1: Determining the Effect of Thyroid Hormone on Metabolic Rate

Metabolism is a broad term referring to all chemical reactions that are necessary to maintain life. It involves both *catabolism,* enzymatically controlled processes in which substances are broken down to simpler substances, and *anabolism,* processes in which larger molecules or structures are built from smaller ones. During catabolic reactions, energy is released as chemical bonds are broken. Some of the liberated energy is captured to make ATP, the energy-rich molecule used by body cells to energize all their activities; the balance is lost in the form of thermal energy or heat. Maintaining body temperature is critically related to the heat-liberating aspects of metabolism.

Various foodstuffs make different contributions to the process of metabolism. For example, carbohydrates, particularly glucose, are generally broken down or oxidized to make ATP, whereas fats are utilized to form cell membranes, myelin sheaths, and to insulate the body with a fatty cushion. (Fats are used secondarily for producing ATP, particularly when there are inadequate carbohydrates in the diet.) Proteins and amino acids tend to be carefully conserved by body cells, and understandably so, since most structural elements of the body are proteinaceous in nature.

Thyroid hormone (collectively T_3 and T_4), produced by the thyroid gland, is the single most important hormone influencing the rate of cellular metabolism and body heat production.

Under conditions of excess thyroid hormone production (hyperthyroidism), an individual's basal metabolic rate (BMR), heat production, and oxygen consumption increase, and the individual tends to lose weight and become heat-intolerant and irritable. Conversely, hypothyroid individuals become mentally and physically sluggish, obese, and are cold-intolerant because of their low BMR. ■

Many factors other than thyroid hormone levels contribute to metabolic rate (for example, body size and weight, age, and activity level), but the focus of the following experiment is to investigate how differences in thyroid hormone concentration impact metabolism.

Three groups of laboratory rats will be used. The *control group* animals are assumed to be euthyroid and to have normal metabolic rates for their relative body weights. *Experimental group A* animals have received water containing the chemical 6-n-propylthiouracil, which counteracts or antagonizes the effects of thyroid hormone in the body. *Experimental group B* animals have been fed rat chow containing dried thyroid tissue, which contains thyroid hormone. The rates of oxygen consumption (an indirect means of determining metabolic rate) in the animals of the three groups will be measured and compared to investigate the effects of hyperthyroid, hypothyroid, and euthyroid conditions.

Oxygen consumption will be measured with a simple respirometer-manometer apparatus. Each animal will be placed in a closed chamber, containing soda lime. As carbon dioxide is evolved and expired, it will be absorbed by the soda lime; therefore, the pressure changes observed will indicate the volume of oxygen consumed by the animal during the testing interval. Students will work in groups of 3 to 4 to assemble the apparatus, make preliminary weight measurements on the animals, and record the data.

Preparing the Respirometer-Manometer Apparatus

1.   Obtain a desiccator, a two-hole rubber stopper, a 20-ml glass syringe, a hardware cloth square, a T-valve, scissors, rubber tubing, two short pieces of glass tubing, soda lime, a manometer, a clamp, and petrolatum, and bring them to your laboratory bench. The apparatus will be assembled as illustrated in Figure 28.1.

2.   Shake soda lime into the bottom of the desiccator to thoroughly cover the glass bottom. Then place the hardware cloth on the ledge of the desiccator over the soda lime. The hardware cloth should be well above the soda lime, so that the animal will not be able to touch it. Soda lime is quite caustic and can cause chemical burns.

3.   Lubricate the ends of the two pieces of glass tubing with petrolatum, and twist them into the holes in the rubber stopper until their distal ends protrude from the opposite side. *Do not plug the tubing with petrolatum.* Place the stopper into the desiccator cover, and set the cover on the desiccator temporarily.

4.   Cut off a short (3-in.) piece of rubber tubing, and attach it to the top of one piece of glass tubing extending from the stopper. Cut and attach a 12- to 14-in. piece of rubber tubing to the other glass tubing. Insert the T-valve stem into the distal end of the longer-length tubing.

5.   Cut another short piece of rubber tubing; attach one end to the T-valve and the other to the nib of the 20-ml syringe. Remove the plunger of the syringe and grease its sides generously with petrolatum. Insert the plunger back into the syringe barrel and work it up and down to evenly disperse the petrolatum on the inner walls of the syringe, then pull the plunger out to the 20-ml marking.

6.   Cut a piece of rubber tubing long enough to reach from the third arm of the T-valve to one arm of the manometer.

Figure 28.1 **Respirometer-manometer apparatus.**

(The manometer should be partially filled with water so that a U-shaped water column is seen.) Attach the tubing to the T-valve and the manometer arm.

7. Remove the desiccator cover and generously grease the cover's bottom edge with petrolatum. Place the cover back on the desiccator and firmly move it from side to side to spread the lubricant evenly.

8. Test the system for leaks as follows: Firmly clamp the *short* length of rubber tubing extending from the stopper. Now gently push in on the plunger of the syringe. If the system is properly sealed, the fluid in the manometer will move away from the rubber tubing attached to its arm. If there is an air leak, the manometer fluid level will not change, or it will change and then quickly return to its original level. If either of these events occurs, check all glass-to-glass or glass-to-rubber tubing connections. Smear additional petrolatum on suspect areas and test again. The apparatus must be airtight before experimentation can begin.

9. After ensuring that there are no leaks in the system, unclamp the short rubber tubing, and remove the desiccator cover.

Preparing the Animal

1. Put on the heavy animal-handling gloves, and obtain one of the animals as directed by your instructor. Handling it gently, weigh the animal to the nearest 0.1 g on the animal balance.

2. Carefully place the animal on the hardware cloth in the desiccator. The objective is to measure oxygen usage at basal

levels, so you do not want to prod the rat into high levels of activity, which would produce errors in your measurements.

3. Record the animal's group (control or experimental group A or B) and its weight in kilograms (that is, weight in grams/1000) on the data sheet on p. 277.

Equilibrating the Chamber

1. Place the lid on the desiccator, and move it slightly from side to side to seal it firmly.

2. Leave the short tubing unclamped for 7 to 10 minutes to allow for temperature equilibration in the chamber. (Since the animal's body heat will warm the air in the container, the air will expand initially. This must be allowed to occur before any measurements of oxygen consumption are taken. Otherwise, it would appear that the animal is evolving oxygen rather than consuming it.)

Determining Oxygen Consumption of the Animal

1. Once again clamp the short rubber tubing extending from the desiccator lid stopper.

2. Check the manometer to make sure that the fluid level is the same in both arms. (If not, manipulate the syringe plunger to make the fluid levels even.) Record the time and the position of the bottom of the plunger (use ml marking) in the syringe.

3. Observe the manometer fluid levels at 1-minute intervals. Each time make the necessary adjustment to bring the fluid levels even in the manometer by carefully pushing the syringe plunger further into the barrel. Determine the amount of oxygen used per minute intervals by computing the difference in air volumes within the syringe. For example, if after the first minute, you push the plunger from the 20- to the 17-ml marking, the oxygen consumption is 3 ml/min. Then, if the plunger is pushed from 17 ml to 15 ml at the second minute reading, the oxygen usage during minute 2 would be 2 ml, and so on.

4. Continue taking readings (and recording oxygen consumption per minute interval on the data sheet) for 10 con-

secutive minutes or until the syringe plunger has been pushed nearly to the 0-ml mark. Then unclamp the short rubber tubing, remove the desiccator cover, and allow the apparatus to stand open for 2 to 3 minutes to flush out the stale air.

5. Repeat the recording procedures for another 10-minute interval. *Make sure that you equilibrate the temperature within the chamber before beginning this second recording series.*

6. After you have recorded the animal's oxygen consumption for two 10-minute intervals, unclamp the short rubber tubing, remove the desiccator lid, and carefully return the rat to its cage.

Metabolic Rate Data Sheet

Animal used from group: _____

14-day prior treatment of animal: _____

Body weight in grams: _____ /1000 = body weight in kg: _____

O_2 consumption/min: Test 1	**O_2 consumption/min: Test 2**
Beginning syringe reading _____ ml	Beginning syringe reading _____ ml

Test 1: min 1 _____ min 6 _____ ; min 2 _____ min 7 _____ ; min 3 _____ min 8 _____ ; min 4 _____ min 9 _____ ; min 5 _____ min 10 _____ ; _____ Total O_2/10 min

Test 2: min 1 _____ min 6 _____ ; min 2 _____ min 7 _____ ; min 3 _____ min 8 _____ ; min 4 _____ min 9 _____ ; min 5 _____ min 10 _____ ; _____ Total O_2/10 min

Average ml O_2 consumed/10 min: _____

Milliliters O_2 consumed/hr: _____

Metabolic rate: _____ ml O_2/kg/hr

Averaged class results:

Metabolic rate of control animals: _____ ml O_2/kg/hr

Metabolic rate of experimental group A animals (PTU-treated): _____ ml O_2/kg/hr

Metabolic rate of experimental group B animals (desiccated thyroid–treated): _____ ml O_2/kg/hr

Computing Metabolic Rate

Metabolic rate calculations are generally reported in terms of $kcal/m^2/hr$ and require that corrections be made to present the data in terms of standardized pressure and temperature conditions. These more complex calculations will not be used here, since the object is simply to arrive at some generalized conclusions concerning the effect of thyroid hormone on metabolic rate.

1.   Obtain the average figure for milliliters of oxygen consumed per 10-minute interval by adding up the minute-interval consumption figures for each 10-minute testing series and dividing the total by 2.

$$\frac{\text{Total ml } O_2 \text{ test series 1 + total ml } O_2 \text{ test series 2}}{2}$$

2.   Determine oxygen consumption per hour using the following formula, and record the figure on the data sheet:

$$\frac{\text{Average ml } O_2 \text{ consumed} \times 60 \text{ min}}{10 \text{ min}} \quad \frac{}{\text{hr}} = \text{ml } O_2/hr$$

3. To determine the metabolic rate in milliliters of oxygen consumed per kilogram of body weight per hour so that the results of all experiments can be compared, divide the figure just obtained in procedure 2 by the animal's weight in kilograms (kg = wt in lb ÷ 2.2).

$$\text{Metabolic rate} = \frac{\text{ml } O_2/hr}{\text{wt in kg}} = \text{ml} \underline{\hspace{1cm}} O_2/kg/hr$$

Record the metabolic rate on the data sheet and also in the appropriate space on the chart on the chalkboard.

4.   Once all groups have recorded their final metabolic rate figures on the chalkboard, average the results of each animal grouping to obtain the mean for each experimental group. Also record this information on your data sheet. ■

Activity:
Experiment 2: Determining the Effect of Pituitary Hormones on the Ovary

As indicated in Exercise 27, anterior pituitary hormones called *gonadotropins*, specifically follicle-stimulating hormone (FSH) and luteinizing hormone (LH), regulate the ovarian cycles of the female. Although amphibians normally ovulate seasonally, many can be stimulated to ovulate "on demand" by injecting an extract of pituitary hormones.

⚠ 1.   Don plastic gloves, and obtain two frogs. Place them in separate battery jars to bring them to your laboratory bench. Also bring back a syringe and needle, a wax marking pencil, pond or spring water, and containers of pituitary extract and physiological saline.

2.   Before beginning, examine each frog for the presence of eggs. Hold the frog firmly with one hand and exert pressure on its abdomen toward the cloaca (in the direction of the legs). If ovulation has occurred, any eggs present in the oviduct will be forced out and will appear at the cloacal opening. If no eggs are present, continue with step 3. If eggs are

expressed, return the animal to your instructor and obtain another frog for experimentation. Repeat the procedure for determining if eggs are present until two frogs that lack eggs have been obtained.

3.   Aspirate 1 to 2 ml of the pituitary extract into a syringe. Inject the extract subcutaneously into the anterior abdominal (peritoneal) cavity of the frog you have selected to be the experimental animal. To inject into the peritoneal cavity, hold the frog with its ventral surface superiorly. Insert the needle through the skin and muscles of the abdominal wall in the lower quarter of the abdomen. Do not insert the needle far enough to damage any of the vital organs. With a wax marker, label its large battery jar "experimental," and place the frog in it. Add a small amount of pond water to the battery jar before continuing.

4.   Aspirate 1 to 2 ml of normal saline into a syringe and inject it into the peritoneal cavity of the second frog—this will be the control animal. (Make sure you inject the same volume of fluid into both frogs.) Place this frog into the second battery jar, marked "control." Allow the animals to remain undisturbed for 24 hours.

5.   After 24 hours,* again check each frog for the presence of eggs in the cloaca. (See step 2.) If no eggs are present, make arrangements with your laboratory instructor to return to the lab on the next day (at 48 hours after injection) to check your frogs for the presence of eggs.

6.   Return the frogs to the terrarium before leaving or continuing with the lab.

In which of the prepared frogs was ovulation induced?

_____

Specifically, what hormone in the pituitary extract causes ovulation to occur?

_____ ■

Activity:
Experiment 3: Observing the Effects of Hyperinsulinism

Many people with diabetes mellitus need injections of insulin to maintain blood sugar (glucose) homeostasis. Adequate levels of blood glucose are essential for proper functioning of the nervous system; thus, the administration of insulin must be carefully controlled. If blood glucose levels fall precipitously, the patient will go into insulin shock.

A small fish will be used to demonstrate the effects of hyperinsulinism. Since the action of insulin on the fish parallels that in the human, this experiment should provide valid information concerning its administration to humans.

---

* The student will need to inject the frog the day before the lab session or return to check results the day after the scheduled lab session.

1.   Prepare two finger bowls. Using a wax marker, mark one A and the other B. To finger bowl A, add 100 ml of the commercial insulin solution. To finger bowl B, add 200 ml of 20% glucose solution.

2.   Place a small fish in finger bowl A and observe its actions carefully as the insulin diffuses into its bloodstream through the capillary circulation of its gills.

Approximately how long did it take for the fish to become comatose?

_____

What types of activity did you observe in the fish before it became comatose?

_____

_____

_____

3.   When the fish is comatose, carefully transfer it to finger bowl B and observe its actions. What happens to the fish after it is transferred?

_____

_____

_____

Approximately how long did it take for this recovery?

_____

4.   After all observations have been made and recorded, carefully return the fish to the aquarium. ■

Activity:
Experiment 4: Testing the Effect of Epinephrine on the Heart

As noted in Exercise 27, the adrenal medulla and the sympathetic nervous system are closely interrelated, specifically because the cells of the adrenal medulla and the postgan-

glionic axons of the sympathetic nervous system both release catecholamines. This experiment demonstrates the effects of epinephrine on the frog heart.

⚠ 1. Obtain a frog, dissecting instruments and pan, disposable gloves, frog Ringer's solution, dropper bottle of 1:1000 epinephrine solution, and bring them to your laboratory bench. Don the gloves before beginning step 2.

2.   Destroy the nervous system of the frog. (A frog used in the experiment on pituitary hormone effects may be used if your test results have already been obtained and are positive. Otherwise, obtain another frog.) Insert one blade of a scissors into its mouth as far as possible and quickly cut off the top of its head, posterior to the eyes. Then identify the spinal cavity and insert a dissecting needle into it to destroy the spinal cord.

3.   Place the frog dorsal side down on a dissecting pan, and carefully open its ventral body cavity by making a vertical incision with the scissors.

4.   Identify the beating heart, and carefully cut through the saclike pericardium to expose the heart tissue.

5.   Visually count the heart rate for 1 minute, and record below. Keep the heart moistened with frog Ringer's solution during this interval.

Beats per minute: _____

6.   Flush the heart with epinephrine solution. Record the heartbeat rate per minute for 5 consecutive minutes.

minute 1 _____   minute 4 _____

minute 2 _____   minute 5 _____

minute 3 _____

What was the effect of epinephrine on the heart rate?

_____

Was the effect long-lived? _____

7.   Dispose of the frog in an appropriate container, and clean the dissecting pan and instruments before returning them to the supply area. ■

Objectives

1. To name the two major components of blood and state their average percentages in whole blood.

2. To describe the composition and functional importance of plasma.

3. To define *formed elements* and list the cell types composing them, cite their relative percentages, and describe their major functions.

4. To identify red blood cells, basophils, eosinophils, monocytes, lymphocytes, and neutrophils when provided with a microscopic preparation or appropriate diagram.

5. To conduct the following blood test determinations in the laboratory, and to state their norms and the importance of each.

   hematocrit
   hemoglobin determination
   clotting time
   sedimentation rate
   differential white blood cell count
   total white blood cell count
   total red blood cell count
   ABO and Rh blood typing
   plasma cholesterol concentration

6. To discuss the reason for transfusion reactions resulting from the administration of mismatched blood.

7. To define *anemia, polycythemia, leukopenia, leukocytosis,* and *leukemia* and to cite a possible reason for each condition.

Materials

❑ Compound microscope
❑ Immersion oil
❑ Models and charts of blood cells
❑ Safety glasses (student provided)

Demonstration station:

❑ Microscopes set up with prepared slides demonstrating the following blood (or bone marrow) conditions:

   Macrocytic hypochromic anemia

   Microcytic hypochromic anemia

   Sickle-cell anemia

   Lymphocytic leukemia (chronic)

   Eosinophilia

General supply area:

❑ Plasma (obtained from an animal hospital or prepared by centrifuging animal (e.g., cattle or sheep) blood obtained from a biological supply house

❑ Wide-range pH paper

❑ *Stained smears of human blood.* If desired by the instructor, heparinized animal blood obtained from a biological supply house, or an animal hospital (e.g., dog blood), or EDTA-treated red cells (Reference cells*) with blood type labels obscured (available from Immunocor, Inc.; catalogue #2338)

❑ Clean microscope slides
❑ Sterile lancets
❑ Glass stirring rods
❑ Alcohol swabs (wipes)
❑ Absorbent cotton balls
❑ Wright's stain in dropper bottle
❑ Distilled water in dropper bottle
❑ Test tubes
❑ Test tube racks
❑ Disposable gloves

❑ Pipette cleaning solutions—(1) 10% household bleach solution, (2) distilled water, (3) 70% ethyl alcohol, (4) acetone

❑ Bucket or large beaker containing 10% household bleach solution for slide and glassware disposal

❑ Spray bottles containing 10% bleach solution

Because many blood tests are to be conducted in this exercise, it is advisable to set up a number of appropriately labeled supply areas for the various tests, as designated below.

---

*The blood in these kits (each containing 4 blood cell types—A1, A2, B, and 0—individually supplied in 10-ml vials) is used to calibrate cell counters and other automated clinical laboratory equipment. This blood has been carefully screened and can be safely used by students for blood typing and determining hematocrits. It is not usable for hemoglobin determinations or coagulation studies.

Blood cell count supply area:
- ❑ Hemacytometer
- ❑ Unopette reservoir system apparatus for conducting blood counts (available from Becton-Dickinson [Test #5851 for RBCs and #5855 for WBCs])
- ❑ Mechanical hand counters

Hematocrit supply area:
- ❑ Heparinized capillary tubes
- ❑ Microhematocrit centrifuge and reading gauge (if the reading gauge is not available, millimeter ruler may be used)
- ❑ Seal-ease (Clay Adams Co.) or modeling clay

Hemoglobin determination supply area:
- ❑ Tallquist hemoglobin scales and test paper or a hemoglobinometer, hemolysis applicator, and lens paper

Sedimentation rate supply area:
- ❑ Landau Sed-rate pipettes with tubing and rack*
- ❑ Wide-mouthed bottle of 5% sodium citrate
- ❑ Mechanical suction device
- ❑ Millimeter ruler

Coagulation time supply area:
- ❑ Capillary tubes (nonheparinized)
- ❑ Fine triangular file

Note to the Instructor: See directions for handling of soiled glassware and disposable items on p. 16.

Blood typing supply area:
- ❑ Blood typing sera (anti-A, anti-B, and anti-Rh [D])
- ❑ Rh typing box
- ❑ Wax marker
- ❑ Toothpicks
- ❑ Clean microscope slides
- ❑ Medicine dropper

Cholesterol-measurement supply area:
- ❑ Mechanical pipettor
- ❑ 0.1-ml pipettes (3)
- ❑ 10- to 15-ml graduated cylinder
- ❑ Test tubes (3) and rack
- ❑ Cholesterol standard solution (200 mg/100 ml)
- ❑ Cholesterol reagent (Harleco)†
- ❑ Water bath set at 37°C
- ❑ Spectrophotometer
- ❑ Cuvettes for the spectrophotometer
- ❑ Wax marker

*An alternative method is the Westergren ESR method. See the Instructor's Guide for ordering information.

†Instructions for preparing Harleco reagent are provided in the Instructor's Guide.

I n this exercise you will study plasma and formed elements of blood and conduct various hematological tests. These tests are extremely useful diagnostic tools for the physician because blood composition (number and types of blood cells, and chemical composition) reflects the status of many body functions and malfunctions.

⚠ ALERT: The decision to use animal blood for testing or to have students test their own blood will be made by the instructor in accordance with the educational goals of the student group. For example, for students in the nursing or laboratory technician curricula, learning how to safely handle human blood or other human wastes is essential. If blood samples are provided and they are human blood samples, gloves should be worn while conducting the blood tests. If human blood is being tested, yours or that obtained from a clinical agency, precautions provided in the text for disposal of human waste *must be observed*. All soiled glassware is to be immersed in household bleach solution immediately after use, and disposable items (lancets, cotton balls, alcohol swabs, etc.) are to be placed in a disposable autoclave bag so that they can be sterilized before disposal. Safety glasses are to be worn throughout the laboratory session.

Composition of Blood

The blood circulating to and from the body cells within the blood vessels is a rather viscous substance that varies from bright scarlet to a dull brick red, depending on the amount of oxygen it is carrying. The circulatory system of the average adult contains about 5.5 liters of blood.

Blood is classified as a type of connective tissue because it consists of a nonliving fluid matrix (the **plasma**) in which living cells (**formed elements**) are suspended. The fibers typical of a connective tissue matrix become visible in blood only when clotting occurs. They then appear as fibrin threads, which form the structural basis for clot formation.

More than 100 different substances are dissolved or suspended in plasma (Figure 29.1), which is over 90% water. These include nutrients, gases, hormones, various wastes and metabolites, many types of proteins, and mineral salts. The composition of plasma varies continuously as cells remove or add substances to the blood.

Three types of formed elements are present in blood. The most numerous are the **erythrocytes,** or **red blood cells (RBCs),** which are literally sacs of hemoglobin molecules

Figure 29.1 **The composition of blood.**

that transport the bulk of the oxygen carried in the blood (and a small percentage of the carbon dioxide). **Leukocytes,** or **white blood cells (WBCs),** are part of the body's nonspecific defenses and the immune system, and **platelets** function in hemostasis (blood clot formation). Formed elements normally constitute 45% of whole blood; plasma accounts for the remaining 55%.

Activity:
Determining the Physical Characteristics of Plasma

Go to the general supply area and carefully pour a few milliliters of plasma into a test tube. Also obtain some wide-range pH paper and then return to your laboratory bench to make the following simple observations.

pH of Plasma

Test the pH of the plasma with wide-range pH paper. Record the pH observed.

_____

Color and Clarity of Plasma

Hold the test tube up to a source of natural light. Note and record its color and degree of transparency. Is it clear, translucent, or opaque?

Color_____

Degree of transparency_____

Consistency

Dip your finger and thumb into the plasma and then press them firmly together for a few seconds. Gently pull them apart. How would you describe the consistency of plasma? Slippery, watery, sticky, or granular? Record your observations.

———————————————————————— ■

Activity:
Examining the Formed Elements of Blood Microscopically

In this section, you will conduct your observations of blood cells on an already prepared (purchased) blood slide, or on a slide prepared from your own blood or blood provided by your instructor. Those using the purchased blood slide are to obtain a slide and begin their observations at step 6. Those testing blood provided by a biological supply source or an animal hospital are to obtain a tube of the supplied blood, disposable gloves, and the supplies listed in step 1, except for the lancets and alcohol swabs. After donning gloves, those students will jump to step 3b to begin their observations. If you are examining your own blood, you will perform all the steps described below *except* step 3b.

1. Obtain two glass slides, a glass stirring rod, dropper bottles of Wright's stain and distilled water, two or three lancets, cotton balls, and alcohol swabs. Bring this equipment to the laboratory bench. Clean the slides thoroughly and dry them.

2. Open the alcohol swab packet and scrub your third or fourth finger with the swab. (Because the pricked finger may be a little sore later, it is better to prepare a finger on the hand used less often.) Circumduct your hand (swing it in a cone-shaped path) for 10 to 15 seconds. This will dry the alcohol and cause your fingers to become engorged with blood. Then, open the lancet packet and grasp the lancet by its blunt end. Quickly jab the pointed end into the prepared finger to produce a free flow of blood. It is *not* a good idea to squeeze or "milk" the finger, as this forces out tissue fluid as well as blood. If the blood is not flowing freely, another puncture should be made.

⚠ *Under no circumstances is a lancet to be used for <u>more than one puncture</u>.* Dispose of the lancets in the disposable autoclave bag immediately after use.

3a. With a cotton ball, wipe away the first drop of blood; then allow another large drop of blood to form. Touch the blood to one of the cleaned slides approximately ½ inch from the end. Then quickly (to prevent clotting) use the second slide to form a blood smear as shown in Figure 29.2. When properly prepared, the blood smear is uniformly thin. If the blood smear appears streaked, the blood probably began to clot or coagulate before the smear was made, and another slide should be prepared. Continue at step 4.

3b. Dip a glass rod in the blood provided, and transfer a generous drop of blood to the end of a cleaned microscope slide. For the time being, lay the glass rod on a paper towel on the bench. Then, as described just above, use the second slide to make your blood smear.

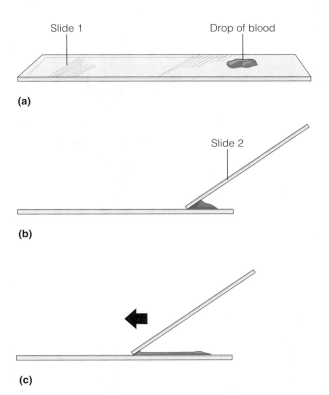

(a)

(b)

(c)

Figure 29.2 Procedure for making a blood smear. (a) Place a drop of blood on slide 1 approximately ½ inch from one end. **(b)** Hold slide 2 at a 30° to 40° angle to slide 1 (it should touch the drop of blood) and allow blood to spread along entire bottom edge of angled slide. **(c)** Smoothly advance slide 2 to end of slide 1 (blood should run out before reaching the end of slide 1).

4. Dry the slide by waving it in the air. When it is completely dry, it will look dull. Place it on a paper towel, and flood it with Wright's stain. Count the number of drops of stain used. Allow the stain to remain on the slide for 3 to 4 minutes and then flood the slide with an equal number of drops of distilled water. Allow the water and Wright's stain mixture to remain on the slide for 4 or 5 minutes or until a metallic green film or scum is apparent on the fluid surface. Blow on the slide gently every minute or so to keep the water and stain mixed during this interval.

5. Rinse the slide with a stream of distilled water. Then flood it with distilled water, and allow it to lie flat until the slide becomes translucent and takes on a pink cast. Then stand the slide on its long edge on the paper towel, and allow it to dry completely. Once the slide is dry, you can begin your observations.

6. Obtain a microscope and scan the slide under low power to find the area where the blood smear is the thinnest. After scanning the slide in low power to find the areas with the largest numbers of nucleated WBCs, read the following descriptions of cell types, and find each one on Figure 29.1. (The formed elements are also shown in Plates 57 through 62 in the Histology Atlas.) Then, switch to the oil immersion lens and observe the slide carefully to identify each cell type.

Erythrocytes

Erythrocytes, or red blood cells, which average 7.5 μm in diameter, vary in color from a salmon red color to pale pink, depending on the effectiveness of the stain. They have a distinctive biconcave disk shape and appear paler in the center than at the edge (see Plate 57 in the Histology Atlas).

As you observe the slide, notice that the red blood cells are by far the most numerous blood cells seen in the field. Their number averages 4.5 million to 5.0 million cells per cubic millimeter of blood (for women and men, respectively).

Red blood cells differ from the other blood cells because they are anucleate when mature and circulating in the blood. As a result, they are unable to reproduce and have a limited life span of 100 to 120 days, after which they begin to fragment and are destroyed in the spleen and other reticuloendothelial tissues of the body.

In various anemias, the red blood cells may appear pale (an indication of decreased hemoglobin content) or may be nucleated (an indication that the bone marrow is turning out cells prematurely). ■

Leukocytes

Leukocytes, or white blood cells, are nucleated cells that are formed in the bone marrow from the same stem cell (*hemocytoblast*) as red blood cells. They are much less numerous than the red blood cells, averaging from 4000 to 11,000 cells per cubic millimeter. Basically, white blood cells are protective, pathogen-destroying cells that are transported to all parts of the body in the blood or lymph. Important to their protective function is their ability to move in and out of blood vessels, a process called **diapedesis,** and to wander through body tissues by **amoeboid motion** to reach sites of inflammation or tissue destruction. They are classified into two major groups, depending on whether or not they contain conspicuous granules in their cytoplasm.

Granulocytes make up the first group. The granules in their cytoplasm stain differentially with Wright's stain, and they have peculiarly lobed nuclei, which often consist of expanded nuclear regions connected by thin strands of nucleoplasm. There are three types of granulocytes:

Neutrophil: The most abundant of the white blood cells (40% to 70% of the leukocyte population); nucleus consists of 3 to 7 lobes and the pale lilac cytoplasm contains fine cytoplasmic granules, which are generally indistinguishable and take up both the acidic (red) and basic (blue) dyes (*neutrophil* = neutral loving); functions as an active phagocyte. The number of neutrophils increases exponentially during acute infections. (See Plates 57 and 58.)

Eosinophil: Represents 1% to 4% of the leukocyte population; nucleus is generally figure 8 or bilobed in shape; contains large cytoplasmic granules (elaborate lysosomes) that stain red-orange with the acid dyes in Wright's stain (see Plate 61). Precise function is unknown, but they increase in number during allergies and parasite infections and may selectively phagocytize antigen-antibody complexes.

Basophil: Least abundant leukocyte type representing less than 1% of the population; large U- or S-shaped nucleus with two or more indentations. Cytoplasm contains coarse, sparse granules that are stained deep purple by the basic dyes in Wright's stain (see Plate 62). The granules contain several

chemicals including histamine, a vasodilator which is discharged on exposure to antigens and helps mediate the inflammatory response.

The second group, **agranulocytes,** or **agranular leukocytes,** contains no observable cytoplasmic granules. Although found in the bloodstream, they are much more abundant in lymphoid tissues. Their nuclei tend to be closer to the norm, that is, spherical, oval, or kidney-shaped. Specific characteristics of the two types of agranulocytes are listed below.

Lymphocyte: The smallest of the leukocytes, approximately the size of a red blood cell (see Plates 57 and 59). The nucleus stains dark blue to purple, is generally spherical or slightly indented, and accounts for most of the cell mass. Sparse cytoplasm appears as a thin blue rim around the nucleus. Concerned with immunologic responses in body; one population, the B lymphocytes, oversees the production of antibodies that are released to blood. The second population, T lymphocytes, plays a regulatory role and destroys grafts, tumors, and virus-infected cells. Represents 20% to 45% of the WBC population.

Monocyte: The largest of the leukocytes; approximately twice the size of red blood cells (see Plate 60). Represents 4% to 8% of the leukocyte population. Dark blue nucleus is generally kidney-shaped; abundant cytoplasm stains gray-blue. Functions as an active phagocyte (the "long-term cleanup team"), increasing dramatically in number during chronic infections such as tuberculosis.

Students are often asked to list the leukocytes in order from the most abundant to the least abundant. The following silly phrase may help you with this task: *Never let monkeys eat bananas* (neutrophils, lymphocytes, monocytes, eosinophils, basophils).

Platelets

Platelets are cell fragments of large multinucleate cells (**megakaryocytes**) formed in the bone marrow. They appear as darkly staining, irregularly shaped bodies interspersed among the blood cells (see Plate 57). The normal platelet count in blood ranges from 250,000 to 500,000 per cubic millimeter. Platelets are instrumental in the clotting process that occurs in plasma when blood vessels are ruptured.

After you have identified these cell types on your slide, observe three-dimensional models of blood cells if these are available. *Do not dispose of your slide,* as it will be used later for the differential white blood cell count. ■

Hematologic Tests

When someone enters a hospital as a patient, several hematologic tests are routinely done to determine general level of health as well as the presence of pathologic conditions. You will be conducting the most common of these tests in this exercise.

Materials such as cotton balls, lancets, and alcohol swabs are used in nearly all of the following diagnostic tests. These supplies are at the general supply area and should be properly disposed of (glassware to the "bleach bucket" and disposable items to the autoclave bag) immediately after use.

Other necessary supplies and equipment are at specific supply areas marked according to the test with which they are used. Since nearly all of the tests require a finger stab, if you will be using your own blood it might be wise to quickly read through the tests to determine in which instances more than one preparation can be done from the same finger stab. For example, the hematocrit capillary tubes and sedimentation rate samples might be prepared at the same time the chamber is being prepared for the total blood cell counts. A little preplanning will save you the discomfort of a multiply punctured finger.

An alternative to using blood obtained from the finger stab technique is using heparinized blood samples supplied by your instructor. The purpose of using heparinized tubes is to prevent the blood from clotting. Thus blood collected and stored in such tubes will be suitable for all tests except coagulation time testing.

Total White and Red Blood Cell Counts

To conduct a **total WBC** or **RBC count,** you will dilute a known volume of blood with a fluid that prevents blood coagulation and place the mixture into a counting chamber of known volume—a **hemacytometer.** The cells are then counted by microscopic inspection, and corrections for dilution and chamber volume are made to obtain the result in cells per cubic millimeter. The white and red blood cell diluents differ. They are isotonic for the cell type to be counted and cause dissolution of the cell types not being counted. In other words, if the WBC diluent is used, the RBCs are hemolyzed and thus do not interfere with the WBC counting process. The Unopette apparatus for conducting blood counts is shown in Figure 29.3. The reservoir contains the required amount of the appropriate diluting fluid (acetic acid for WBC counts and Hayem's solution for RBC counts). Note that this hand counting technique is rather outdated, since most clinical agencies now have computerized equipment for performing blood counts.

Activity:
Conducting a Total White Blood Cell Count

Since white blood cells are an important part of the body's defense system, it is essential to note any abnormalities in them. **Leukocytosis,** an abnormally high WBC count, may indicate bacterial or viral infection, metabolic disease, hemorrhage, or poisoning by drugs or chemicals. A decrease in the white cell number below $4000/mm^3$ (**leukopenia**) may indicate typhoid fever, measles, infectious hepatitis or cirrhosis, tuberculosis, or excessive antibiotic or X-ray therapy. A person with leukopenia lacks the usual protective mechanisms. **Leukemia,** a malignant disorder of the lymphoid tissues characterized by uncontrolled proliferation of abnormal WBCs accompanied by a reduction in the number of RBCs and platelets, is detectable not only by a total WBC count but also by a differential WBC count (see below).

1. Obtain a hemacytometer and cover glass, cotton balls, lancets, alcohol swabs, and a hand counter. Also obtain the

Pipette shield	Pipette (capillary tube)	Reservoir containing diluent

Figure 29.3 The Unopette reservoir system for blood counts. The system consists of a reservoir containing a premeasured amount of RBC or WBC diluent and a plastic capillary tube pipette with a shield that is used for puncturing the reservoir top.

Unopette system for conducting WBC counts (the reservoir has a *white* bottom).

2. Puncture the top (diaphragm) of the reservoir container with the pointed end of the protective pipette shield. Then transfer a drop of animal blood to a clean slide with the glass rod. If you will be using your own blood, cleanse your finger thoroughly, and obtain a second drop of blood as described on p. 283 (step 2).

3. Now you are ready to charge the Unopette pipette with blood. Twist off the protective shield and, holding the pipette horizontally, touch the tip of the capillary tube pipette to the blood drop. When the pipette has filled by capillary action, wipe its tip with a cotton ball.

Squeeze the reservoir slightly to expel some air. While the reservoir is still compressed, insert the blood-charged pipette into the reservoir and then release the pressure on the reservoir. As you do so, the blood will be drawn into the diluent in the reservoir.

4. Squeeze the reservoir and invert it gently several times to thoroughly mix the contents. Remove the pipette from the reservoir, and then reverse it and insert its opposite end into the reservoir to convert the capillary tube to a dropper.

5. Place a coverslip on the hemacytometer, and prepare to charge it. Discard the first 2 to 3 drops—which is only diluent—from the Unopette onto a paper towel. Squeeze the reservoir to apply one drop of the diluted blood at each end of the coverslip at the junction of the hemacytometer and coverslip (Figure 29.4a), and then immediately lift the pipette

(a)

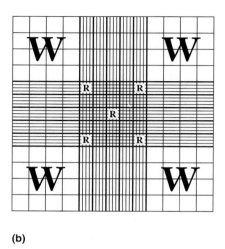

(b)

Figure 29.4 **Apparatus and preparation for blood cell counts.** (a) Procedure for introducing diluted blood to the hemacytometer. (b) Areas of the hemacytometer used for red and white blood cell counts.

away from the hemacytometer surface. *Do this quickly;* otherwise the hemacytometer will overfill, and the mixture will flow under the coverslip by capillary action. You now have two charged chambers. If the chambers are properly charged, no fluid will appear in the moat.

6. Place the hemacytometer on the microscope stage and allow it to remain undisturbed for 2 to 3 minutes to allow the cells to become evenly distributed.

7. Bring the grid lines into focus under low power (10× objective lens) and check for uniformity of white blood cell distribution. If it is uneven, switch to the other chamber for observation.

8. Move the slide into position so that one of the W areas can be seen (see Figure 29.4b), and proceed to count all WBCs observed in the four corner W areas. Use a hand counter to facilitate the counting process. *To prevent overcounts of cells at boundary lines, count only those that touch the left and upper boundary lines of the W area but not those touching the right and lower boundaries.* Record the number of WBCs counted and multiply this number by 50 to obtain the number of WBCs per cubic millimeter.*

No. WBCs: _____ WBC/mm^3: _____

*The factor of 50 is obtained by multiplying the volume dilution factor (20) by the volume correction factor (2.5): 20 × 2.5 = 50. The 2.5 volume correction factor is obtained in the following manner: Each W area on the grid is exactly 1 mm^2 × 0.1 mm deep; therefore, the volume of each W section is 0.1 mm^3. Since WBCs in four W areas are counted (a total of 0.4 mm^3), this volume must be multiplied by 2.5 (correction factor) to obtain the number of cells in 1 mm^3.

Record this information on the data sheet on p. 287. How does your result compare with the norms for a white blood cell count?

9. Discard the used Unopette equipment in the disposable autoclave bag. Clean the hemacytometer and coverslip in preparation for the red blood cell count. ∎

Activity:
Conducting a Total Red Blood Cell Count

The red blood cell count, like the white blood cell count, determines the total number of this cell type per unit volume of blood. Since RBCs are absolutely necessary for oxygen transport, a doctor typically investigates any excessive change in their number immediately.

An increase in the number of RBCs (**polycythemia**) may result from bone marrow cancer or from living at high altitudes where less oxygen is available. A decrease in the number of RBCs results in anemia. (The term **anemia** simply indicates a decreased oxygen-carrying capacity of blood that may result from a decrease in RBC number or size or a decreased hemoglobin content of the RBCs.) A decrease in RBCs may result suddenly from hemorrhage or more gradually from conditions that destroy RBCs or hinder RBC production. ∎

1. Obtain a Unopette apparatus for performing a RBC count (*red* bottom on the reservoir). Also obtain other sup-

Hematologic Test Data Sheet

Differential WBC count:

_____ % granulocytes   _____ % agranulocytes

_____ % neutrophils   _____ % lymphocytes

_____ % eosinophils   _____ % monocytes

_____ % basophils

Total WBC count _____ WBCs/mm^3

Total RBC count _____ RBCs/mm^3

Hematocrit (PCV):

RBC _____ % of blood volume

WBC _____ % of blood volume } not generally reported

Plasma _____ % of blood

Hemoglobin content:

Tallquist method:

_____ g/100 ml blood; _____ %

Hemoglobinometer (type: _____

_____ g/100 ml blood; _____ %

Ratio (PCV/grams Hb per 100 ml blood): _____

Sedimentation rate _____ mm/hr

Coagulation time _____

Blood typing:

ABO group _____   Rh factor _____

Cholesterol concentration _____ mg/100 ml plasma

---

plies as required for the WBC count. Follow the same procedure for obtaining blood and charging the chamber as for the WBC count, except be very careful when reinserting the pipette into the reservoir (step 4) and when mixing the blood with the diluent (Hayem's solution, in this case).

⚠ Do not squeeze the reservoir hard because this might force Hayem's solution out the overflow opening in the reservoir. Hayem's solution contains sodium azide, an antibacterial agent that is very toxic. Hence, it is best to prevent it from splashing on your skin.

2. Allow the hemacytometer to rest on the microscope stage as before. Using the high-power lens, count all RBCs in the R-marked areas (see Figure 29.4b). Again, at the boundary lines, count only those cells touching the left and upper lines.

3. Record the number of cells counted, and multiply that number by 10,000 to compute the RBC/mm^3. (Note: the dilution factor (200) × volume correction factor (50) = 10,000.)

No. RBCs: _____ RBCs/mm^3: _____

How does your result compare with normal RBC values?

_____

Record the RBCs/mm^3 on the data sheet at the end of this exercise.

⚠ 4. Place all used glassware (including the hemacytometer) in the beaker containing bleach. Put disposable items, including the Unopette apparatus, in the disposable autoclave bag. ■

Differential White Blood Cell Count

To make a **differential white blood cell count,** 100 WBCs are counted and classified according to type. Such a count is routine in a physical examination and in diagnosing illness, since any abnormality or significant elevation in percentages of WBC types may indicate a problem or the source of pathology. Use the slide prepared for the identification of the blood cells (p. 283) for the count.

Activity:
Conducting a Differential WBC Count

1. Begin at the edge of the smear and move the slide in a systematic manner on the microscope stage—either up and down or from side to side as indicated in Figure 29.5.

2. Record each type of white blood cell you observe by making a count on the chart on page 288 (for example, ⫫⫲ = 7 cells) until you have observed and recorded a total of 100 WBCs. Using the following equation, compute the percentage of each WBC type counted, and record the percentages on the data sheet at the top of this page.

$$\text{Percent (\%)} = \frac{\text{\# observed}}{\text{Total \# counted (100)}} \times 100$$

How does your differential white blood cell count correlate with the percentages given for each type on p. 284? ■

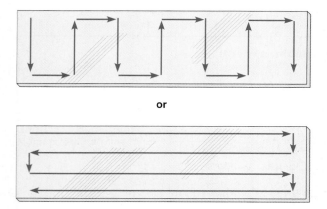

Figure 29.5 Alternative methods of moving the slide for a differential WBC count.

Cell type	Number observed
Neutrophils	
Eosinophils	
Basophils	
Lymphocytes	
Monocytes	

Hematocrit

The **hematocrit,** or **packed cell volume (PCV),** is routinely determined when anemia is suspected. Centrifuging whole blood spins the formed elements to the bottom of the tube, with plasma forming the top layer (see Figure 29.1). Since the blood cell population is primarily RBCs, the PCV is generally considered equivalent to the RBC volume, and this is the only value reported. However, the relative percentage of WBCs can be differentiated, and both WBC and plasma volume will be reported here. Normal hematocrit values for the male and female, respectively, are 47.0 ± 7 and 42.0 ± 5.

Activity:
Determining the Hematocrit

The hematocrit is determined by the micromethod, so only a drop of blood is needed. If possible (and the centrifuge allows), all members of the class should prepare their capillary tubes at the same time so the centrifuge can be properly balanced and run only once.

1. Obtain two heparinized capillary tubes, Seal-ease or modeling clay, a lancet, alcohol swabs, and some cotton balls.

2. If you are using your own blood, cleanse a finger, and allow the blood to flow freely. Wipe away the first few drops and, holding the red-line-marked end of the capillary tube to the blood drop, allow the tube to fill at least three-fourths full by capillary action (Figure 29.6a). If the blood is not flowing freely, the end of the capillary tube will not be completely submerged in the blood during filling, air will enter, and you will have to prepare another sample.

(a)

(b)

(c)

Figure 29.6 Steps in a hematocrit determination. (a) Load a heparinized capillary tube with blood. **(b)** Plug the blood-containing end of the tube with clay. **(c)** Place the tube in a microhematocrit centrifuge. (Centrifuge must be balanced.)

If you are using instructor-provided blood, simply immerse the red-marked end of the capillary tube in the blood sample and fill it three-quarters full as just described.

3.   Plug the blood-containing end by pressing it into the Seal-ease or clay (Figure 29.6b). Prepare a second tube in the same manner.

4.   Place the prepared tubes opposite one another in the radial grooves of the microhematocrit centrifuge with the sealed ends abutting the rubber gasket at the centrifuge periphery (Figure 29.6c). This loading procedure balances the centrifuge and prevents blood from spraying everywhere by centrifugal force. *Make a note of the numbers of the grooves your tubes are in.* When all the tubes have been loaded, make sure the centrifuge is properly balanced, and secure the centrifuge cover. Turn the centrifuge on, and set the timer for 4 or 5 minutes.

5.   Determine the percentage of RBCs, WBCs, and plasma by using the microhematocrit reader. The RBCs are the bottom layer, the plasma is the top layer, and the WBCs are the buff-colored layer between the two. If the reader is not available, use a millimeter ruler to measure the length of the filled capillary tube occupied by each element, and compute its percentage by using the following formula:

$$\frac{\text{Height of the column composed of the element (mm)}}{\text{Height of the original column of whole blood (mm)}} \times 100$$

Record your calculations below and on the data sheet.

% RBC _____   % WBC _____% plasma _____

Usually WBCs constitute 1% of the total blood volume. How do your blood values compare to this figure and to the normal percentages for RBCs and plasma? (See p. 282.)

As a rule, a hematocrit is considered a more accurate test for determining the RBC composition of the blood than the total RBC count. A hematocrit within the normal range generally indicates a normal RBC number, whereas an abnormally high or low hematocrit is cause for concern. ■

Hemoglobin Concentration

As noted earlier, a person can be anemic even with a normal RBC count. Since hemoglobin is the RBC protein responsible for oxygen transport, perhaps the most accurate way of measuring the oxygen-carrying capacity of the blood is to determine its hemoglobin content. Oxygen, which combines reversibly with the heme (iron-containing portion) of the hemoglobin molecule, is picked up by the blood cells in the lungs and unloaded in the tissues. Thus, the more hemoglobin molecules the RBCs contain, the more oxygen they will be able to transport. Normal blood contains 12 to 16 g hemoglobin per 100 ml blood. Hemoglobin content in men is slightly higher (13 to 18 g) than in women (12 to 16 g).

Determining Hemoglobin Concentration

Several techniques have been developed to estimate the hemoglobin content of blood, ranging from the old, rather inaccurate Tallquist method to expensive colorimeters, which are precisely calibrated and yield highly accurate results. Directions for both the Tallquist method and a hemoglobinometer are provided here.

Using the Talliquist Method

1.   Obtain a Tallquist hemoglobin scale, lancets, alcohol swabs, and cotton balls.

2.   Use instructor-provided *animal* blood or prepare the finger as previously described. (For best results, make sure the alcohol evaporates before puncturing your finger.) Place one good-sized drop of blood on the special absorbent paper provided with the color chart. The blood stain should be larger than the holes on the color chart.

3.   As soon as the blood has dried and lost its glossy appearance, match its color, under natural light, with the color standards by moving the specimen under the comparison chart so that the blood stain appears at all the various apertures. (The blood should not be allowed to dry to a brown color, as this will result in an inaccurate reading.) Because the colors on the chart represent 1% variations in hemoglobin content, it may be necessary to estimate the percentage if the color of your blood sample is intermediate between two color standards.

4.   On the data sheet on p. 287, record your results as the percentage of hemoglobin concentration and as grams per 100 ml of blood.

Using a Hemoglobinometer

1.   Obtain a hemoglobinometer, hemolysis applicator stick, alcohol swab, and lens paper and bring them to your bench. Test the hemoglobinometer light source to make sure it is working; if not, request new batteries before proceeding and test it again.

2.   Remove the blood chamber from the slot in the side of the hemoglobinometer and disassemble the blood chamber by separating the glass plates from the metal clip. Notice as you do this that the larger glass plate has an H-shaped depression cut into it that acts as a moat to hold the blood, whereas the smaller glass piece is flat and serves as a coverslip.

3.   Clean the glass plates with an alcohol swab and then wipe dry with lens paper. Hold the plates by their sides to prevent smearing during the wiping process.

4.   Reassemble the blood chamber (remember: larger glass piece on the bottom with the moat up), but leave the moat plate about halfway out to provide adequate exposed surface to charge it with blood.

(a) A drop of blood is added to the moat plate of the blood chamber. The blood must flow freely.

(b) The blood sample is hemolyzed with a wooden hemolysis applicator. Complete hemolysis requires 35 to 45 seconds.

(c) The charged blood chamber is inserted into the slot on the side of the hemoglobinometer.

(d) The colors of the green split screen are found by moving the slide with the right index finger. When the two colors match in density, the grams/100 ml and %Hb are read on the scale.

Figure 29.7   Hemoglobin determination using a hemoglobinometer.

5.   Obtain a drop of blood (from the provided sample or from your fingertip as before), and place it on the depressed area of the moat plate that is closest to you (Figure 29.7a).

6.   Using the wood hemolysis applicator, stir or agitate the blood to rupture (lyse) the RBCs (Figure 29.7b). This usually takes 35 to 45 seconds. Hemolysis is complete when the blood appears transparent rather than cloudy.

7.   Push the blood-containing glass plate all the way into the metal clip and then firmly insert the charged blood chamber back into the slot on the side of the instrument (Figure 29.7c).

8.   Hold the hemoglobinometer in your left hand with your left thumb resting on the light switch located on the underside of the instrument. Look into the eyepiece and notice that there is a green area divided into two halves (a split field).

9.   With the index finger of your right hand, slowly move the slide on the right side of the hemoglobinometer back and forth until the two halves of the green field match (Figure 29.7d).

10.   Note and record on the data sheet on p. 287 the grams Hb (hemoglobin)/100 ml blood indicated on the uppermost scale by the index mark on the slide. Also record % Hb, indicated by one of the lower scales.

11.   Disassemble the blood chamber once again, and carefully place its parts (glass plates and clip) into a bleach-containing beaker.

Generally speaking, the relationship between the PCV and grams of hemoglobin per 100 ml blood is 3:1. How do your values compare?

_____

Record on the data sheet the value obtained from your data. ■

Sedimentation Rate

The speed at which red blood cells settle to the bottom of a vertical tube when allowed to stand is called the **sedimentation rate.** The normal rate for adults is 0 to 6 mm/hr (averaging 3 mm/hr) and for children 0 to 8 mm/hr (averaging 4 mm/hr). Sedimentation of RBCs apparently proceeds in three stages: rouleaux formation, rapid settling, and final packing. _Rouleaux formation_ (alignment of RBCs like a stack of pennies) does not occur with abnormally shaped red blood cells (as in sickle-cell anemia); therefore, the sedimentation rate is decreased. The size and number of RBCs affect the packing phase. In anemia the sedimentation rate increases; in polycythemia the rate decreases. The sedimentation rate is greater than normal during menses and pregnancy, and very high sedimentation rates may indicate infectious conditions or tissue destruction occurring somewhere in the body. Although this test is nonspecific, it alerts the diagnostician to the need for further tests to pinpoint the site of pathology. The Landau micromethod, which uses just one drop of blood, is used here. An alternative method is the Westergren ESR method. See the Instructor's Guide for ordering information.

Activity:
Determining Sedimentation Rate

1. Obtain lancets, cotton balls, alcohol swabs, and Landau Sed-rate pipette and tubing, the Landau rack, a wide-mouthed bottle of 5% sodium citrate, a mechanical suction device, and a millimeter ruler.

2. Use the mechanical suction device to draw up the sodium citrate to the first (most distal) marking encircling the pipette. Prepare the finger for puncture, and produce a free flow of blood.

3. Wipe off the first drop, and then draw the blood into the pipette until the mixture reaches the second encircling line. Keep the pipette tip immersed in the blood to avoid air bubbles.

4. Thoroughly mix the blood with the citrate (an anticoagulant) by drawing the mixture into the bulb and then forcing it back down into the lumen. Repeat this mixing procedure six times, and then adjust the top level of the mixture as close to the zero marking as possible. _If any air bubbles are introduced during the mixing process, discard the sample and begin again._

5. Seal the tip of the pipette by holding it tightly against the tip of your index finger, and then carefully remove the suction device from the upper end of the pipette.

6. Stand the pipette in an exactly vertical position on the Sed-rack with its lower end resting on the base of the rack. Record the time, and allow it to stand for exactly 1 hour.

Time _____

7. After 1 hour, measure the number of millimeters of visible clear plasma (which indicates the amount of settling of the RBCs), and record this figure here and on the data sheet.

Sedimentation rate _____ mm/hr

⚠️ 8. Clean the pipette by using the mechanical suction device to draw up each of the cleaning solutions (available at the general supply area) in succession—bleach, distilled water, alcohol, and acetone. Discharge the contents each time before drawing up the next solution. Place the cleaned pipette in the bleach-containing beaker at the general supply area, after first drawing bleach up into the pipette. ■

Coagulation Time

Blood clotting, or **coagulation,** is a protective mechanism that minimizes blood loss when blood vessels are ruptured. This process requires the interaction of many substances normally present in the plasma (clotting factors, or procoagulants) as well as some released by platelets and injured tissues. Basically hemostasis proceeds as follows (Figure 29.8a): The injured tissues and platelets release **tissue factor (TF)** and PF_3 respectively, which trigger the clotting mechanism, or cascade. Tissue factor and PF_3 interact with other blood protein clotting factors and calcium ions to form **prothrombin activator,** which in turn converts **prothrombin** (present in plasma) to **thrombin.** Thrombin then acts enzymatically to polymerize the soluble **fibrinogen** proteins (present in plasma) into insoluble **fibrin,** which forms a meshwork of strands that traps the RBCs and forms the basis of the clot (Figure 29.8b). Normally, blood removed from the body clots within 2 to 6 minutes.

Activity:
Determining Coagulation Time

1. Obtain a _nonheparinized_ capillary tube, a lancet, cotton balls, a triangular file, and alcohol swabs.

2. Clean and prick the finger to produce a free flow of blood.

3. Place one end of the capillary tube in the blood drop, and hold the opposite end at a lower level to collect the sample.

4. Lay the capillary tube on a paper towel.

Record the time. _____

5. At 30-sec intervals, make a small nick on the tube close to one end with the triangular file, and then carefully break the tube. Slowly separate the ends to see if a gel-like thread of fibrin spans the gap. When this occurs, record below and on the data sheet the time for coagulation to occur. Are your results within the normal time range?

_____

6. Dispose of the capillary tube and used supplies in the disposable autoclave bag. ■

(b)

Figure 29.8 Events of hemostasis and blood clotting. (a) Simple schematic of events. Steps numbered 1–3 represent the major events of coagulation. **(b)** Photomicrograph of RBCs trapped in a fibrin mesh (1600×).

Blood Typing

Blood typing is a system of blood classification based on the presence of specific glycoproteins on the outer surface of the RBC plasma membrane. Such proteins are called **antigens,** or **agglutinogens,** and are genetically determined. In many cases, these antigens are accompanied by plasma proteins, **antibodies** or **agglutinins,** that react with RBCs bearing different antigens, causing them to be clumped, agglutinated, and eventually hemolyzed. It is because of this phenomenon that a person's blood must be carefully typed before a whole blood or packed cell transfusion.

Several blood typing systems exist, based on the various possible antigens, but the factors routinely typed for are antigens of the ABO and Rh blood groups which are most commonly involved in transfusion reactions. Other blood factors, such as Kell, Lewis, M, and N, are not routinely typed for unless the individual will require multiple transfusions. The basis of the ABO typing is shown in the chart at the top of p. 293.

Individuals whose red blood cells carry the Rh antigen are Rh positive (approximately 85% of the U.S. population); those lacking the antigen are Rh negative. Unlike ABO blood groups neither the blood of the Rh-positive (Rh^+) nor Rh-negative (Rh^-) individuals carries preformed anti-Rh antibodies. This is understandable in the case of the Rh-positive individual. However, Rh-negative persons who receive transfusions of Rh-positive blood become sensitized by the Rh antigens of the donor RBCs and their systems begin to produce anti-Rh antibodies. On subsequent exposures to Rh-positive blood, typical transfusion reactions occur, resulting in the clumping and hemolysis of the donor blood cells.

ALERT: Although the blood of dogs and other mammals does react with some of the human agglutinins (present in the antisera), the reaction is not as pronounced and varies with the animal blood used. Hence, the most accurate and predictable blood typing results are obtained with human blood.

A c t i v i t y :
Typing for ABO and Rh Blood Groups

1. Obtain two clean microscope slides, a wax marking pencil, anti-A, anti-B, and anti-Rh typing sera, toothpicks, lancets, alcohol swabs, medicine dropper, and the Rh typing box.

2. Divide slide 1 into two equal halves with the wax marking pencil. Label the lower left-hand corner "anti-A" and the lower right-hand corner "anti-B." Mark the bottom of slide 2 "anti-Rh."

3. Place one drop of anti-A serum on the *left* side of slide 1. Place one drop of anti-B serum on the *right* side of slide 1. Place one drop of anti-Rh serum in the center of slide 2.

4. If you are using your own blood, cleanse your finger with an alcohol swab, pierce the finger with a lancet, and wipe away the first drop of blood. Obtain 3 drops of freely flowing blood, placing one drop on each side of slide 1 and a drop on slide 2.

If using instructor-provided animal blood or EDTA-treated red cells, use a medicine dropper to place one drop of blood on each side of slide 1 and a drop of blood on slide 2.

ABO blood type	Antigens present on RBC membranes	Antibodies present in plasma	% of U.S. population		
			White	Black	Asian
A	A	Anti-B	40	27	28
B	B	Anti-A	11	20	27
AB	A and B	None	4	4	5
O	Neither	Anti-A and anti-B	45	49	40

Result	Observed (+)	Not observed (−)
Presence of clumping with anti-A		
Presence of clumping with anti-B		
Presence of clumping with anti-Rh		

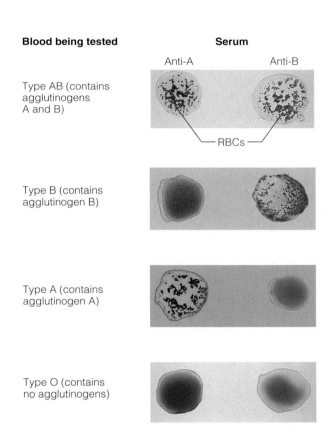

5. Quickly mix each blood-antiserum sample with a *fresh* toothpick. Then dispose of the toothpicks, lancet, and used alcohol swab in the autoclave bag.

6. Place slide 2 on the Rh typing box and rock gently back and forth. (A slightly higher temperature is required for precise Rh typing than for ABO typing.)

7. After 2 minutes, observe all three blood samples for evidence of clumping. The agglutination that occurs in the positive test for the Rh factor is very fine and difficult to perceive; thus if there is any question, observe the slide under the microscope. Record your observations in the chart just above.

8. Interpret your ABO results in light of the information in Figure 29.9. If clumping was observed on slide 2, you are Rh positive. If not, you are Rh negative.

9. Record your blood type on the data sheet.

10. Put the used slides in the bleach-containing bucket at the general supply area; put disposable supplies in the autoclave bag. ■

Activity:
Observing Demonstration Slides

Before continuing on to the cholesterol determination, take the time to look at the five slides that have been put on demonstration by your instructor. Record your observations in the appropriate section of the Review Sheets for Exercise 29. You can refer to your notes later to respond to questions about the blood pathologies represented on the slides. ■

Cholesterol Concentration in Plasma

Atherosclerosis is the disease process in which the body's blood vessels become increasingly occluded by plaques. Because the plaques narrow the arteries, they can contribute to hypertensive heart disease. They also serve as focal points for

Figure 29.9 Blood typing of ABO blood types. When serum containing anti-A or anti-B agglutinins is added to a blood sample, agglutination will occur between the agglutinin and the corresponding agglutinogen (A or B). As illustrated, agglutination occurs with both sera in blood group AB, with anti-B serum in blood group B, with anti-A serum in blood group A, and with neither serum in blood group O.

the formation of blood clots (thrombi) which may break away and block smaller vessels farther downstream in the circulatory pathway, causing heart attacks or strokes.

Ever since medical clinicians discovered that cholesterol is a major component of the smooth muscle plaques formed during atherosclerosis, it has had a bad press. Today, virtually no physical examination of an adult is considered complete until cholesterol levels are assessed along with other lifestyle risk factors. A normal value for plasma cholesterol in adults ranges from 130 to 200 mg per 100 ml plasma; you will be making such a determination on the animal plasma provided by the instructor.

Although the total plasma cholesterol concentration is valuable information, it may be misleading, particularly if a person's high-density lipoprotein (HDL) level is high and low-density lipoprotein (LDL) level is relatively low. Cholesterol, being water insoluble, is transported in the blood complexed to lipoproteins. In general, cholesterol bound into HDLs is destined to be degraded by the liver and then eliminated from the body, whereas that forming part of the LDLs is "traveling" to the body's tissue cells. When LDL levels are excessive, cholesterol is deposited in the blood vessel walls; hence, LDLs are considered to carry the "bad" cholesterol.

Activity:
Measuring Plasma Cholesterol Concentration

1.   Go to the appropriate supply area, put three test tubes in a test tube rack, and add 5 ml of cholesterol reagent into each of the three test tubes. Mark the test tubes 1–3 with a wax marker.

2.   Using the mechanical pipettor and a fresh 0.1-ml pipette each time, add 0.1-ml quantities of the following to the numbered test tubes as designated next.

*   To test tube 1: Add 0.1 ml of cholesterol standard (200 mg/100 ml).

*   To test tube 2: Add 0.1 ml of plasma (available at the general supply area).

*   To test tube 3: Add 0.1 ml of distilled water.

3.   Mix each tube by shaking it gently from side to side and then place the tubes (in the rack) into the water bath set at 37°C. (Note: if there is any question about your ability to identify your test tubes, put your initials on them with the wax marker *before* placing them into the water bath.)

Record the time: _____

4.   Obtain three spectrophotometer cuvettes (containers for the solutions to be "read" by the spectrophotometer). Clean them thoroughly with cuvette brushes and laboratory detergent, rinsing several times with distilled water. Invert them on a paper towel to drain.

5.   After the test tubes have been in the water bath for 10 minutes, remove them and transfer their contents to similarly numbered cuvettes. Standardize the spectrophotometer at 625 nm using sample 3 (distilled water) as the *blank*.

6.   Determine and record the absorbance of samples 1 and 2.

Absorbance of 1: _____   Absorbance of 2: _____

7.   Calculate the cholesterol concentration (*C*) in sample 2 (the "unknown" plasma sample) using the following equation:

$$C_{\text{plasma}} = \frac{\text{absorbance}_{\text{plasma}}}{\text{absorbance}_{\text{standard}}} \times C_{\text{standard}}$$

Record the computed cholesterol concentration here and on the data sheet.

Cholesterol concentration: _____ mg/100 ml plasma

8.   To properly dispose of cuvette solutions, liberally run cold water in the sink while carefully pouring the solutions down the drain.

9.   Clean the pipettes using the mechanical pipettor to draw up each of the pipette cleaning solutions (bleach, distilled water, ethyl alcohol, and acetone) in turn, and then return them to the general supply area.

10.   Before leaving the laboratory, use a paper towel saturated with bleach solution to wash down your laboratory bench. ■

Anatomy of the Heart

The major function of the **cardiovascular system** is transportation. Using blood as the transport vehicle, the system carries oxygen, digested foods, cell wastes, electrolytes, and many other substances vital to the body's homeostasis to and from the body cells. The system's propulsive force is the contracting heart, which can be compared to a muscular pump equipped with one-way valves. As the heart contracts, it forces blood into a closed system of large and small plumbing tubes (blood vessels) within which the blood is confined and circulated. This exercise deals with the structure of the heart or circulatory pump. The anatomy of the blood vessels is considered separately in Exercise 32.

Gross Anatomy of the Human Heart

The **heart,** a cone-shaped organ approximately the size of a fist, is located within the mediastinum, or medial cavity, of the thorax. It is flanked laterally by the lungs, posteriorly by the vertebral column, and anteriorly by the sternum (Figure 30.1). Its more pointed **apex** extends slightly to the left and rests on the diaphragm, approximately at the level of the fifth intercostal space. Its broader **base,** from which the great vessels emerge, lies beneath the second rib and points toward the right shoulder. *In situ,* the right ventricle of the heart forms most of its anterior surface.

If an X ray of a human thorax is available, verify the relationships described above; else Figure 30.1 will suffice.

The heart is enclosed within a double-walled fibroserous sac called the pericardium. The thin **visceral pericardium,**

Objectives

1. To describe the location of the heart.
2. To name and locate the major anatomical areas and structures of the heart when provided with an appropriate model, diagram, or dissected sheep heart, and to explain the function of each.
3. To trace the pathway of blood through the heart.
4. To explain why the heart is called a double pump, and to compare the pulmonary and systemic circuits.
5. To explain the operation of the atrioventricular and semilunar valves.
6. To name and follow the functional blood supply of the heart.
7. To describe the histologic structure of cardiac muscle, and to note the importance of its intercalated discs and the spiral arrangement of its fibers in the heart.

Materials

❑ Torso model or laboratory chart showing heart anatomy
❑ Red and blue pencils
❑ Three-dimensional models of cardiac and skeletal muscle
❑ Heart model (three-dimensional)
❑ X ray of the human thorax for observation of the position of the heart *in situ;* X ray viewing box
❑ Preserved sheep heart, pericardial sacs intact (if possible)
❑ Dissecting pan and instruments
❑ Pointed glass rods for probes
❑ Protective skin cream or disposable gloves
❑ Compound microscope
❑ Histologic slides of cardiac muscle (longitudinal section)

Human Cardiovascular System: The Heart videotape*

 See Appendix C, Exercise 30 for links to A.D.A.M. Interactive Anatomy.

*Available to qualified adopters from Benjamin/Cummings.

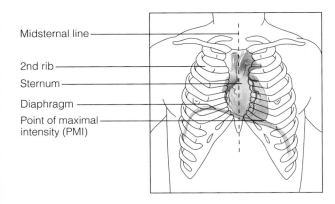

Midsternal line

2nd rib

Sternum

Diaphragm

Point of maximal intensity (PMI)

Figure 30.1   Location of the heart in the thorax.

Brachiocephalic artery

Superior vena cava

Right pulmonary artery

Ascending aorta

Pulmonary trunk

Right pulmonary veins

Right atrium

Right coronary artery (in right atrioventricular groove)

Anterior cardiac vein

Right ventricle

Marginal artery

Small cardiac vein

Inferior vena cava

Left common carotid artery

Left subclavian artery

Aortic arch

Ligamentum arteriosum

Left pulmonary artery

Left pulmonary veins

Left atrium

Auricle

Circumflex artery

Left coronary artery (in left atrioventricular groove)

Left ventricle

Great cardiac vein

Anterior interventricular artery (in anterior interventricular sulcus)

Apex

(a)

Superior vena cava

Right pulmonary artery

Pulmonary trunk

Right atrium

Right pulmonary veins

Fossa ovalis

Pectinate muscles

Tricuspid valve

Right ventricle

Chordae tendineae

Trabeculae carneae

Inferior vena cava

Aorta

Left pulmonary artery

Left atrium

Left pulmonary veins

Bicuspid (mitral) valve

Aortic semilunar valve

Pulmonary semilunar valve

Left ventricle

Papillary muscle

Interventricular septum

Myocardium

Visceral pericardium

(b)

Figure 30.2 Anatomy of the human heart.
(a) External anterior view. (b) Frontal section.

(a)

Anterior

Pulmonary valve
Aortic valve
Area of cutaway
Bicuspid valve
Tricuspid valve

Myocardium
Tricuspid valve (right atrioventricular)
Bicuspid (mitral) valve (left atrioventricular)
Aortic semilunar valve
Pulmonary semilunar valve
Fibrous skeleton

(b)

Figure 30.3   Heart valves. (a) Superior view of the two sets of heart valves (atria removed). **(b)** Photograph of the heart valves, superior view.

or **epicardium,** is closely applied to the heart muscle. It reflects downward at the base of the heart to form its companion serous membrane, the outer, loosely applied **parietal pericardium,** which is attached at the heart apex to the diaphragm. Serous fluid produced by these membranes allows the heart to beat in a relatively frictionless environment. The serous parietal pericardium, in turn, lines the loosely fitting superficial **fibrous pericardium** composed of dense connective tissue.

Inflammation of the pericardium, **pericarditis,** causes painful adhesions between the serous pericardial layers. These adhesions interfere with heart movements. ∎

The walls of the heart are composed primarily of cardiac muscle—the **myocardium**—which is reinforced internally by a dense fibrous connective tissue network. This network—the *fibrous skeleton of the heart*—is more elaborate and thicker in certain areas, for example, around the valves and at the base of the great vessels leaving the heart.

Figure 30.2 shows two views of the heart—an external anterior view and a frontal section. As its anatomical areas are described in the text, consult the figure.

Heart Chambers

The heart is divided into four chambers: two superior **atria** and two inferior **ventricles,** each lined by a thin serous membrane called the **endocardium.** The septum that divides the heart longitudinally is referred to as the **interatrial** or **interventricular septum,** depending on which chambers it partitions. Functionally, the atria are receiving chambers and are

relatively ineffective as pumps. Blood flows into the atria under low pressure from the veins of the body. The right atrium receives relatively oxygen-poor blood from the body via the **superior** and **inferior venae cavae.** Four **pulmonary veins** deliver oxygen-rich blood from the lungs to the left atrium.

The inferior thick-walled ventricles, which form the bulk of the heart, are the discharging chambers. They force blood out of the heart into the large arteries that emerge from its base. The right ventricle pumps blood into the **pulmonary trunk,** which routes blood to the lungs to be oxygenated. The left ventricle discharges blood into the **aorta,** from which all systemic arteries of the body diverge to supply the body tissues. Discussions of the heart's pumping action usually refer to ventricular activity.

Heart Valves

Four valves enforce a one-way blood flow through the heart chambers. The **atrioventricular (AV) valves,** located between the atrial and ventricular chambers on each side, prevent backflow into the atria when the ventricles are contracting. The left atrioventricular valve, also called the **mitral** or **bicuspid valve,** consists of two cusps, or flaps, of endocardium. The right atrioventricular valve, the **tricuspid valve,** has three cusps (Figure 30.3). Tiny white collagenic cords called the **chordae tendineae** (literally, "heart strings") anchor the cusps to the ventricular walls. The chordae tendineae originate from small bundles of cardiac muscle, called **papillary muscles,** that project from the myocardial wall (see Figure 30.2).

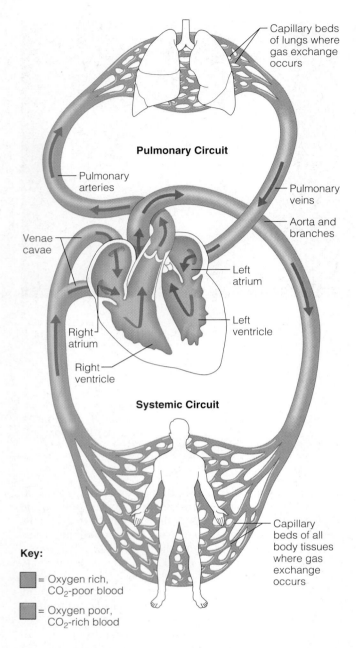

Pulmonary Circuit

Capillary beds of lungs where gas exchange occurs

Pulmonary arteries

Pulmonary veins

Aorta and branches

Venae cavae

Left atrium

Left ventricle

Right atrium

Right ventricle

Systemic Circuit

Capillary beds of all body tissues where gas exchange occurs

Key:

■ = Oxygen rich, CO₂-poor blood

■ = Oxygen poor, CO₂-rich blood

Figure 30.4   The systemic and pulmonary circuits. The heart is a double pump that serves two circulations. The right side of the heart pumps blood through the pulmonary circuit to the lungs and back to the left heart. (For simplicity, the actual number of two pulmonary arteries and four pulmonary veins has been reduced to one each.) The left heart pumps blood via the systemic circuit to all body tissues and back to the right heart. Notice that blood flowing through the pulmonary circuit gains oxygen and loses carbon dioxide as depicted by the color change from blue to red. Blood flowing through the systemic circuit loses oxygen and picks up carbon dioxide (red to blue color change).

When blood is flowing passively into the atria and then into the ventricles during **diastole** (the period of ventricular relaxation), the AV valve flaps hang limply into the ventricular chambers and then are carried passively toward the atria by the accumulating blood. When the ventricles contract (**systole**) and compress the blood in their chambers, the intraventricular blood pressure rises, causing the valve flaps to be reflected superiorly, which closes the AV valves. The chordae tendineae, pulled taut by the contracting papillary muscles, anchor the flaps in a closed position that prevents backflow into the atria during ventricular contraction. If unanchored, the flaps would blow upward into the atria rather like an umbrella being turned inside out by a strong wind.

The second set of valves, the **pulmonary** and **aortic semilunar valves,** each composed of three pocketlike cusps, guards the bases of the two large arteries leaving the ventricular chambers. The valve cusps are forced open and flatten against the walls of the artery as the ventricles discharge their blood into the large arteries during systole. However, when the ventricles relax, blood flows backward toward the heart and the cusps fill with blood, closing the semilunar valves and preventing arterial blood from reentering the heart.

A c t i v i t y :
Using the Heart Model to Study Heart Anatomy

When you have pinpointed in Figure 30.2 all the structures described above, observe the human heart model and reidentify the same structures without reference to the figure. ■

Pulmonary, Systemic, and Cardiac Circulations
Pulmonary and Systemic Circulations

The heart functions as a double pump. The right side serves as the **pulmonary circulation** pump, shunting the carbon dioxide-rich blood entering its chambers to the lungs to unload carbon dioxide and pick up oxygen, and then back to the left side of the heart (Figure 30.4). The function of this circuit is strictly to provide for gas exchange. The second circuit, which carries oxygen-rich blood from the left heart through the body tissues and back to the right heart, is called the **systemic circulation.** It supplies the functional blood supply to all body tissues.

A c t i v i t y :
Tracing the Path of Blood Through the Heart

Trace the pathway of blood through the heart by adding arrows to the frontal section diagram (Figure 30.2b). Use red arrows for the oxygen-rich blood and blue arrows for the less oxygen-rich blood. ■

(a)

Figure 30.6 Longitudinal view of the heart chambers showing the spiral arrangement of the cardiac muscle fibers.

largely drained by the **great, middle,** and **small cardiac veins,** which empty into the **coronary sinus.** The coronary sinus, in turn, empties into the right atrium. In addition, several **anterior cardiac veins** empty directly into the right atrium (Figure 30.5).

(b)

Figure 30.5 Cardiac circulation.

Cardiac Circulation

Even though the heart chambers are almost continually bathed with blood, this contained blood does not nourish the myocardium. The functional blood supply of the heart is provided by the right and left coronary arteries (see Figures 30.2 and 30.5). The **coronary arteries** issue from the base of the aorta just above the aortic semilunar valve and encircle the heart in the **atrioventricular groove** at the junction of the atria and ventricles. They then ramify over the heart's surface, the right coronary artery supplying the posterior surface of the ventricles and the lateral aspect of the right side of the heart, largely through its **posterior interventricular** and **marginal artery** branches. The left coronary artery supplies the anterior ventricular walls and the laterodorsal part of the left side of the heart via its two major branches, the **anterior interventricular artery** and the **circumflex artery.** The coronary arteries and their branches are compressed during systole and fill when the heart is relaxed. The myocardium is

Microscopic Anatomy of Cardiac Muscle

Cardiac muscle is found in only one place—the heart. The heart acts as a vascular pump, propelling blood to all tissues of the body; cardiac muscle is thus very important to life. Cardiac muscle is involuntary, ensuring a constant blood supply.

The cardiac cells, only sparingly invested in connective tissue, are arranged in spiral or figure-8-shaped bundles (Figure 30.6). When the heart contracts, its internal chambers become smaller (or are temporarily obliterated), forcing the blood into the large arteries leaving the heart.

A c t i v i t y :
Examining Heart Anatomy

1. Observe the three-dimensional model of cardiac muscle, examining its branching cells and the areas where the cells interdigitate, the **intercalated discs.** These two structural features provide a continuity to cardiac muscle not seen in other muscle tissues and allow close coordination of heart activity.

2. Compare the model of cardiac muscle to the model of skeletal muscle. Note the similarities and differences between the two kinds of muscle tissue.

Striations   Intercalated   Nucleus of cardiac
                 discs   muscle fiber

Figure 30.7 Photomicrograph of cardiac muscle (688×).

3. Obtain and observe a longitudinal section of cardiac muscle under high power. Identify the nucleus, striations, intercalated discs, and sarcolemma of the individual cells and then compare your observations to the view seen in Figure 30.7. ∎

Dissection:
The Sheep Heart

Dissection of a sheep heart is valuable because it is similar in size and structure to the human heart. Also, a dissection experience allows you to view structures in a way not possible with models and diagrams. Refer to Figure 30.8 as you proceed with the dissection.

1. Obtain a preserved sheep heart, a dissection tray, dissecting instruments, a glass probe and, if desired, protective skin cream or gloves. Rinse the sheep heart in cold water to remove excessive preservatives and to flush out any trapped blood clots. Now you are ready to make your observations.

2. Observe the texture of the pericardium. Also, note its point of attachment to the heart. Where is it attached?

_____

_____

3. If the serous pericardial sac is still intact, slit open the parietal pericardium and cut it from its attachments. Observe the visceral pericardium (epicardium). Using a sharp scalpel, carefully pull a little of this serous membrane away from the myocardium. How do its position, thickness, and apposition to the heart differ from those of the parietal pericardium?

_____

_____

_____

4. Examine the external surface of the heart. Notice the accumulation of adipose tissue, which in many cases marks the separation of the chambers and the location of the coronary arteries that nourish the myocardium. Carefully scrape away some of the fat with a scalpel to expose the coronary blood vessels.

5. Identify the base and apex of the heart, and then identify the two wrinkled **auricles,** earlike flaps of tissue projecting from the atrial chambers. The balance of the heart muscle is ventricular tissue. To identify the left ventricle, compress the ventricular chambers on each side of the longitudinal fissures carrying the coronary blood vessels. The side that feels thicker and more solid is the left ventricle. The right ventricle feels much thinner and somewhat flabby when compressed. This difference reflects the greater demand placed on the left ventricle, which must pump blood through the much longer systemic circulation, a pathway with much higher resistance than the pulmonary circulation served by the right ventricle. Hold the heart in its anatomical position (Figure 30.8a), with the anterior surface uppermost. In this position the left ventricle composes the entire apex and the left side of the heart.

6. Identify the pulmonary trunk and the aorta extending from the superior aspect of the heart. The pulmonary trunk is more anterior, and you may see its division into the right and left pulmonary arteries if it has not been cut too closely to the heart. The thicker-walled aorta, which branches almost immediately, is located just beneath the pulmonary trunk. The first branch of the sheep aorta, the **brachiocephalic artery,** is identifiable unless the aorta has been cut immediately as it leaves the heart. The brachiocephalic artery splits to form the right carotid and subclavian arteries, which supply the right side of the head and right forelimb, respectively.

Carefully clear away some of the fat between the pulmonary trunk and the aorta to expose the **ligamentum arteriosum,** a cordlike remnant of the **ductus arteriosus.** (In the fetus, the ductus arteriosus allows blood to pass directly from the pulmonary trunk to the aorta, thus bypassing the nonfunctional fetal lungs.)

7. Cut through the wall of the aorta until you see the aortic semilunar valve. Identify the two openings to the coronary arteries just above the valve. Insert a probe into one of these holes to see if you can follow the course of a coronary artery across the heart.

8. Turn the heart to view its posterior surface. The heart will appear as shown in Figure 30.8b. Notice that the right and left ventricles appear equal-sized in this view. Try to identify the four thin-walled pulmonary veins entering the left atrium. (It may or may not be possible to locate the pulmonary veins from this vantage point, depending on how they were cut as the heart was removed.) Identify the superior and inferior venae cavae entering the right atrium. Compare the approximate diameter of the superior vena cava with the diameter of the aorta.

Which is larger? _____

Which has thicker walls?_____

Base

Aorta — Ligamentum arteriosum

— Pulmonary trunk

Right auricle — — Left auricle

Right atrium (beneath fat) — Left atrium (beneath fat)

Right ventricle —

Left ventricle — Coronary vessels in anterior interventricular sulcus

Aorta

Pulmonary artery — Opening of superior vena cava

Pulmonary veins — Right atrium

Left atrium (beneath fat) — Opening of inferior vena cava

Left ventricle — Right ventricle

Apex —

Right pulmonary artery   Pulmonary veins   Left pulmonary artery

...gamentum ...teriosum

...orta

...ulmonary ...unk

...uperior ...na cava

...rachio-...ephalic ...tery ...men)

...uricle ...f left ...trium)

...ght ...uricle

...ght ...ntricle

Anterior interventricular-sulcus (fat-filled proximally)   Apex of heart   Left ventricle

(a)

Pulmonary veins   Right pulmonary artery   Aorta

Superior vena cava

Brachio-cephalic artery

Left atrium (fat-covered)

Auricle of right atrium

Inferior vena cava (peg in lumen)

Right ventricle

Left ventricle

Apex of heart (left ventricle)

(b)

Figure 30.8 **Anatomy of the sheep heart.** **(a)** Anterior view.
(b) Posterior view. Diagrammatic views at top; photographs at bottom.

Figure 30.9 **Right side of the sheep heart opened and reflected to reveal internal structures.**

Why do you suppose these differences exist?

_____

_____

_____

9.   Insert a probe into the superior vena cava and use scissors to cut through its wall so that you can view the interior of the right atrium. Do not extend your cut entirely through the right atrium or into the ventricle. Observe the right atrioventricular valve.

How many flaps does it have?_____

Pour some water into the right atrium and allow it to flow into the ventricle. _Slowly and gently_ squeeze the right ventricle to watch the closing action of this valve. (If you squeeze too vigorously, you'll get a face full of water!) Drain the water from the heart before continuing.

10.   Return to the pulmonary trunk and cut through its anterior wall until you can see the pulmonary semilunar valve. Pour some water into the base of the pulmonary trunk to observe the closing action of this valve. How does its action differ from that of the atrioventricular valve?

_____

_____

_____

After observing semilunar valve action, drain the heart once again. Return to the superior vena cava, and continue the cut made in its wall through the right atrium and right atrioventricular valve into the right ventricle. Parallel the anterior border of the interventricular septum until you "round the corner" to the dorsal aspect of the heart (Figure 30.9).

11.   Reflect the cut edges of the superior vena cava, right atrium, and right ventricle to obtain the view seen in Figure 30.9. Observe the comblike ridges of muscle throughout most of the right atrium. This is called **pectinate muscle** (_pectin_ means "comb"). Identify, on the ventral atrial wall, the large opening of the inferior vena cava and follow it to its external opening with a probe. Notice that the atrial walls in the vicinity of the venae cavae are smooth and lack the roughened appearance (pectinate musculature) of the other regions of the atrial walls. Just below the inferior vena caval opening, identify the opening of the **coronary sinus,** which returns venous blood of the coronary circulation to the right atrium. Nearby, locate an oval depression, the **fossa ovalis,** in the interatrial septum. This depression marks the site of an opening in the fetal heart, the **foramen ovale,** which allows blood to pass from the right to the left atrium, thus bypassing the fetal lungs.

12.   Identify the papillary muscles in the right ventricle, and follow their attached chordae tendineae to the flaps of the tricuspid valve. Notice the pitted and ridged appearance (**trabeculae carneae**) of the inner ventricular muscle.

13.   Make a longitudinal incision through the aorta and continue it into the left ventricle. Notice how much thicker the myocardium of the left ventricle is than that of the right ventricle. Compare the _shape_ of the left ventricular cavity to the shape of the right ventricular cavity.

_____

_____

_____

Are the papillary muscles and chordae tendineae observed in the right ventricle also present in the left ventricle?

_____

Count the number of cusps in the left atrioventricular valve. How does this compare with the number seen in the right atrioventricular valve?

_____

_____

How do the sheep valves compare with their human counterparts?

_____

14. Continue your incision from the left ventricle superiorly into the left atrium. Reflect the cut edges of the atrial wall, and attempt to locate the entry points of the pulmonary veins into the left atrium. Follow the pulmonary veins to the heart exterior with a probe. Notice how thin-walled these vessels are.

15. Properly dispose of the organic debris, and clean the dissecting tray and instruments before leaving the laboratory. ◼

Conduction System of the Heart and Electrocardiography

Objectives

1. To list and localize the elements of the intrinsic conduction, or nodal, system of the heart; and to describe how impulses are initiated and conducted through this system and the myocardium.

2. To interpret the ECG in terms of depolarization and repolarization events occurring in the myocardium; and to identify the P, QRS, and T waves on an ECG recording.

3. To calculate the heart rate, QRS interval, P-R interval, and Q-T interval from an ECG obtained during the laboratory period.

4. To define *tachycardia*, *bradycardia*, and *fibrillation*.

Materials

❑ ECG recording apparatus

❑ Electrode paste; alcohol swabs

❑ Cot

The Intrinsic Conduction System

Heart contraction results from a series of electrical potential changes (depolarization waves) that travel through the heart preliminary to each beat. The ability of cardiac muscle to beat is intrinsic—it does not depend on impulses from the nervous system to initiate its contraction and will continue to contract rhythmically even if all nerve connections are severed. However, two types of controlling systems exert their effects on heart activity. One of these involves nerves of the autonomic nervous system, which accelerate or decrease the heartbeat rate depending on which division is activated. The second system is the **intrinsic conduction system,** or **nodal system,** of the heart, consisting of specialized noncontractile myocardial tissue. The intrinsic conduction system ensures that heart muscle depolarizes in an orderly and sequential manner (from atria to ventricles) and that the heart beats as a coordinated unit.

The components of the intrinsic conduction system include the **SA (sinoatrial) node,** located in the right atrium

just inferior to the entrance to the superior vena cava; the **AV (atrioventricular) node** in the lower atrial septum at the junction of the atria and ventricles; the **AV bundle (bundle of His)** and right and left **bundle branches,** located in the interventricular septum; and the **Purkinje fibers,** which ramify within the muscle bundles of the ventricular walls. The Purkinje fiber network is much denser and more elaborate in the left ventricle because of the larger size of this chamber (Figure 31.1).

The SA node, which has the highest rate of discharge, provides the stimulus for contraction. Because it sets the rate of depolarization for the heart as a whole, the SA node is often referred to as the *pacemaker.* From the SA node, the impulse spreads throughout the atria and to the AV node. This electrical wave is immediately followed by atrial contraction. At the AV node, the impulse is momentarily delayed (approximately 0.1 sec), allowing the atria to complete their contraction. It then passes through the AV bundle, the right and left bundle branches, and the Purkinje fibers, finally resulting in ventricular contraction. Note that the atria and ventricles are separated from one another by a region of electrically inert connective tissue; so the depolarization wave can be transmitted to the ventricles only via the tract between the AV node and AV bundle. Thus, any damage to the AV node-bundle pathway partially or totally insulates the ventricles from the influence of the SA node. Although autorhythmic cells are found throughout the heart, their rates of spontaneous depolarization differ. The nodal system increases the rate of heart depolarization and synchronizes heart activity.

Electrocardiography

The conduction of impulses through the heart generates electrical currents that eventually spread throughout the body. These impulses can be detected on the body's surface and recorded with an instrument called an *electrocardiograph.* The graphic recording of the electrical changes (depolarization followed by repolarization) occurring during the cardiac cycle is called an **electrocardiogram (ECG)** (Figure 31.2). The typical ECG consists of a series of three recognizable waves called *deflection waves.* The first wave, the **P wave,** is a small wave that indicates depolarization of the atria immediately before atrial contraction. The large **QRS complex,** resulting from ventricular depolarization, has a complicated shape (primarily because of the variability in size of the two ventricles and the time differences required for these chambers to depolarize). It precedes ventricular contraction. The **T wave** results from currents propagated during ventricular repolarization. The repolarization of the atria, which occurs

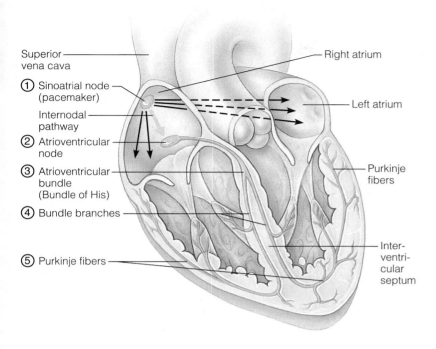

Superior vena cava

Right atrium

① Sinoatrial node (pacemaker)

Left atrium

Internodal pathway

② Atrioventricular node

③ Atrioventricular bundle (Bundle of His)

Purkinje fibers

④ Bundle branches

⑤ Purkinje fibers

Interventricular septum

Figure 31.1 The intrinsic conduction system of the heart. Dashed-line arrows indicate transmission of the impulse from the SA node through the atria. Solid arrow indicates transmission of the impulse from the SA node to the AV node via the internodal pathway.

(a)

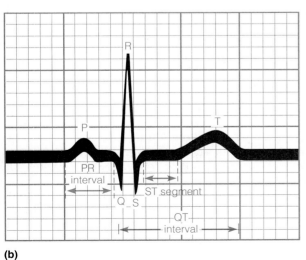

(b)

Time: small squares = 0.04 sec
     1 large square = 0.20 sec
     5 large squares = 1.00 sec

Figure 31.2 The normal electrocardiogram. **(a)** Regular sinus rhythm. **(b)** Waves, segments, and intervals of a normal ECG.

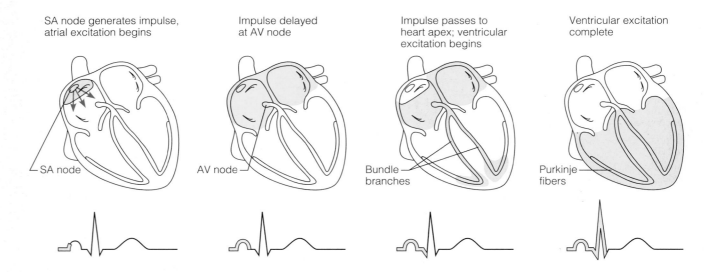

SA node generates impulse, atrial excitation begins

Impulse delayed at AV node

Impulse passes to heart apex; ventricular excitation begins

Ventricular excitation complete

SA node

AV node

Bundle branches

Purkinje fibers

Figure 31.3 **The sequence of excitation of the heart related to the deflection waves of an ECG tracing.**

during the QRS interval, is generally obscured by the large QRS complex. The relationship between the deflection waves of an ECG and sequential excitation of the heart is shown in Figure 31.3.

It is important to understand what an ECG does and does not show: First, an ECG is a record of voltage and time—nothing else. Although we can and do infer that muscle contraction follows its excitation, sometimes it does not. Secondly, an ECG records electrical events occurring in relatively large amounts of muscle tissue (i.e., the bulk of the heart muscle), *not* the activity of nodal tissue which, like muscle contraction, can only be inferred. Nonetheless, abnormalities of the deflection waves and changes in the time intervals of the ECG are useful in detecting myocardial infarcts or problems with the conduction system of the heart. The P-R (P-Q) interval represents the time between the beginning of atrial depolarization and ventricular depolarization. Thus, it typically includes the period during which the depolarization wave passes to the AV node, atrial systole, and the passage of the excitation wave to the balance of the conducting system. Generally, the P-R interval is about 0.16 to 0.18 sec. A longer interval may suggest a partial AV heart block caused by damage to the AV node. In total heart block, no impulses are transmitted through the AV node, and the atria and ventricles beat independently of one another—the atria at the SA node rate and the ventricles at their intrinsic rate, which is considerably slower.

If the QRS interval (normally 0.08 sec) is prolonged, it may indicate a right or left bundle branch block in which one ventricle is contracting later than the other. The Q-T interval is the period from the beginning of ventricular depolarization through repolarization and includes the time of ventricular contraction (the S-T segment). With a heart rate of 70 beats/min, this interval is normally 0.31 to 0.41 sec. As the rate increases, this interval becomes shorter; conversely, when the heart rate drops, the interval is longer.

A heart rate over 100 beats/min is referred to as **tachycardia;** a rate below 60 beats/min is **bradycardia.** Although neither condition is pathological, prolonged tachycardia may

progress to **fibrillation,** a condition of rapid uncoordinated heart contractions which makes the heart useless as a pump. Bradycardia in athletes is a positive finding; that is, it indicates an increased efficiency of cardiac functioning. Because *stroke volume* (the amount of blood ejected by a ventricle with each contraction) increases with physical conditioning, the heart can contract more slowly and still meet circulatory demands.

Twelve standard leads are used to record an ECG for diagnostic purposes. Three of these are bipolar leads that measure the voltage difference between the arms, or an arm and a leg, and nine are unipolar leads. Together the 12 leads provide a fairly comprehensive picture of the electrical activity of the heart.

For this investigation, four electrodes are used (Figure 31.4), and results are obtained from the three *standard limb leads* (also shown in Figure 31.4). Several types of polygraphs or ECG recorders are available. Your instructor will provide specific directions on how to set up and use the available apparatus.

Understanding the Standard Limb Leads

As you might expect, electrical activity recorded by any lead depends on the location and orientation of the recording electrodes. Clinically, it is assumed that the heart lies in the center of a triangle with sides of equal lengths (*Einthoven's triangle*) and that the recording connections are made at the vertices (corners) of that triangle. But in practice, the electrodes connected to each arm and to the left leg are considered to connect to the triangle vertices. The standard limb leads record the voltages generated in the extracellular fluids surrounding the heart by the ion flows occurring simultaneously in many cells between any two of the connections. A recording using lead I (RA-LA), which connects the right arm (RA) and the left arm (LA), is most sensitive to electrical activity spreading horizontally across the heart. Lead II (RA-

LL) and lead III (LA-LL) record activity along the vertical axis (from the base of the heart to its apex), but from different orientations. The significance of Einthoven's triangle is that the sum of the voltages of leads I and III equals that in lead II (Einthoven's law). Hence, if the voltages of two of the standard leads are recorded, that of the third lead can be determined mathematically.

A c t i v i t y :
Recording ECGs

Once the subject is prepared, the ECG will be recorded first under baseline (resting) conditions and then under conditions of fairly strenuous activity. Finally, recordings will be made while the subject holds his or her breath. The activity and breath-holding recordings will be compared to the baseline recordings, and you will be asked to determine the reasons for the observed differences in the recordings.

Preparing the Subject

1.   Place electrode paste on four electrode plates and position each electrode as follows after scrubbing the skin at that attachment site with an alcohol swab. Attach an electrode to the anterior surface of each forearm, about 2 to 3 in. above the wrist, and secure them with rubber straps. In the same manner, attach an electrode to each leg, approximately 2 to 3 in. above the medial malleolus (inner aspect of the ankle).

2.   Attach the appropriate tips of the patient cable to the electrodes. The cable leads are marked RA (right arm), LA (left arm), LL (left leg), and RL (right leg, the ground).

Baseline Recordings

1.   Position the subject comfortably in a supine position on a cot (if available), or sitting relaxed on a laboratory chair.

2.   Turn on the power switch and adjust the sensitivity knob to 1. Set the paper speed to 25 mm/sec and the lead selector to the position corresponding to recording from lead I (RA-LA).

3.   Set the control knob at the RUN position and record the subject's at-rest ECG from lead I for 2 to 3 minutes or until the recording stabilizes. (You will need a tracing long enough to provide each student with a representative segment.) The subject should try to relax and not move unnecessarily, because the skeletal muscle action potentials will also be picked up and recorded.

4.   Stop the recording and mark it "lead I."

5.   Repeat the recording procedure for leads II (RA-LL) and III (LA-LL).

6.   Each student should take a representative segment of one of the lead recordings and label the record with the name of the subject and the lead used. Identify and label the P, QRS, and T waves. The calculations you perform for your recording should be based on the following information: Because

Figure 31.4   ECG recording positions for the standard limb leads.

the paper speed was 25 mm/sec, each millimeter of paper corresponds to a time interval of 0.04 sec. Thus, if an interval requires 4 mm of paper, its duration is 4 mm $\times$ 0.04 sec/mm = 0.16 sec.

7.   Compute the heart rate. Measure the distance (mm) from the beginning of one QRS complex to the beginning of the next QRS complex, and plug this value into the following equation to find the time for one heartbeat.

_____ mm $\times$ 0.04 sec/min = _____ sec/beat

Now find the beats per minute, or heart rate, by using the figure just computed for seconds per beat in the following equation:

Beats/min = $\dfrac{1}{\text{_____ sec/beat}} \times 60$ sec/min

Beats/min = _____

Is the obtained value within normal limits? _____

Measure the QRS interval and compute its duration.

_____

Measure the Q-T interval and compute its duration.

_____

Measure the P-R interval and compute its duration.

_____

Are the computed values within normal limits?

_____

8.   At the bottom of this page, attach segments of the ECG recordings from leads I through III. Make sure you indicate the paper speed, lead, and subject's name on each tracing. To the recording on which you based your previous computations, add your calculations for the duration of the QRS, P-R intervals, and Q-T intervals above the respective area of tracing. Also record the heart rate on that tracing.

"Running in Place" Recording

1.   Make sure the electrodes are securely attached to prevent electrode movement while recording the ECG.

2.   Set the paper speed to 25 mm/sec, and prepare to make the recording using lead I.

3.   Record the ECG while the subject is running in place for 3 min. Then have the subject sit down, but continue to record the ECG for an additional 4 minutes. *Mark the recording* at the end of the 3 minutes of running and at 1 minute after cessation of activity.

4.   Stop the recording. Compute the beats/min during the third minute of running, at 1 minute after exercise, and at 4 minutes after exercise. Record below:

_____ beats/min while running in place

_____ beats/min at 1 minute after exercise

_____ beats/min at 4 minutes after exercise

5.   Compare this recording with the previous recording from lead I. Which intervals are shorter in the "running" recording?

_____

Does the subject's heart rate return to baseline levels by 4 minutes after exercise?

_____

"Breath-Holding" Recording

1.   Position the subject comfortably in the sitting position.

2.   Using lead I and a paper speed of 25 mm/sec, *begin* the recording. After approximately 10 seconds, instruct the subject to begin breath holding and mark the record to indicate the onset of the 1-min breath-holding interval.

3.   Stop the recording after 1 minute and remind the subject to breathe. Compute the beats/minute during the 1-minute experimental (breath-holding) period.

Beats/min during breath holding _____

4.   Compare this recording with the lead I recording obtained under baseline conditions.

What differences are seen? _____

_____

Attempt to *explain* the physiologic reason for the differences you have seen. (Hint: a good place to start might be to check "hypoventilation" or the role of the *respiratory* system in acid-base balance of the blood.)

_____

_____

_____

_____ ■

Anatomy of Blood Vessels

The blood vessels constitute a closed transport system. As the heart contracts, blood is propelled into the large arteries leaving the heart. It moves into successively smaller arteries and then to the arterioles, which feed the capillary beds in the tissues. Capillary beds are drained by the venules, which in turn empty into veins that ultimately converge on the great veins entering the heart.

Arteries, carrying blood away from the heart, and veins, which drain the tissues and return blood to the heart, function simply as conducting vessels or conduits. Only the tiny capillaries that connect the arterioles and venules and ramify throughout the tissues directly serve the needs of the body's cells. It is through the capillary walls that exchanges are made between tissue cells and blood. Respiratory gases, nutrients, and wastes move along diffusion gradients. Thus, oxygen and nutrients diffuse from the blood to the tissue cells, and carbon dioxide and metabolic wastes move from the cells to the blood.

In this exercise you will examine the microscopic structure of blood vessels and identify the major arteries and veins of the systemic circulation and other special circulations.

Microscopic Structure of the Blood Vessels

Except for the microscopic capillaries, the walls of blood vessels are constructed of three coats, or *tunics* (Figure 32.1). The **tunica intima,** or **interna,** which lines the lumen of a vessel, is a single thin layer of *endothelium* (squamous cells underlain by a scant basal lamina) that is continuous with the endocardium of the heart. Its cells fit closely together, forming an extremely smooth blood vessel lining that helps decrease resistance to blood flow.

The **tunica media** is the more bulky middle coat and is composed primarily of smooth muscle and elastic tissue. The smooth muscle, under the control of the sympathetic nervous system, plays an active role in regulating the diameter of blood vessels, which in turn alters peripheral resistance and blood pressure.

The **tunica externa,** or **adventitia,** the outermost tunic, is composed of areolar or fibrous connective tissue. Its function is basically supportive and protective.

In general, the walls of arteries are thicker than those of veins. The tunica media in particular tends to be much heavier and contains substantially more smooth muscle and elastic tissue. This anatomical difference reflects a functional difference in the two types of vessels. Arteries, which are closer to the pumping action of the heart, must be able to expand as an increased volume of blood is propelled into them and then

Objectives

1. To describe the tunics of blood vessel walls and state the function of each layer.
2. To correlate differences in artery, vein, and capillary structure with the functions of these vessels.
3. To recognize a cross-sectional view of an artery and vein when provided with a microscopic view or appropriate diagram.
4. To list and/or identify the major arteries arising from the aorta, and to indicate the body region supplied by each.
5. To list and/or identify the major veins draining into the superior and inferior venae cavae, and to indicate the body regions drained.
6. To point out and/or discuss the unique features of special circulations (hepatic portal system, circle of Willis, pulmonary circulation, fetal circulation) in the body.

Materials

- ❏ Anatomical charts of human arteries and veins (or a three-dimensional model of the human circulatory system)
- ❏ Anatomical charts of the following specialized circulations: pulmonary circulation, hepatic portal circulation, arterial supply and circle of Willis of the brain (or a brain model showing this circulation), fetal circulation
- ❏ Compound microscope
- ❏ Prepared microscope slides showing cross sections of an artery and vein
- ❏ *Human Cardiovascular System: The Blood Vessels* videotape*

*Available to qualified adopters from Benjamin/Cummings.

 For instructions on animal dissections, see the dissection exercises starting on page 733 in the cat and fetal pig editions of this manual.

 See Appendix C, Exercise 32 for links to A.D.A.M. Interactive Anatomy.

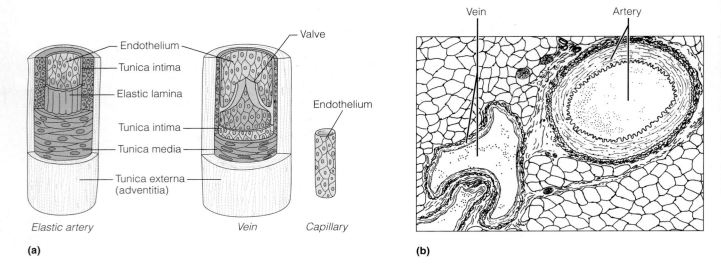

Figure 32.1 Structure of arteries, veins, and capillaries.
(a) Diagrammatic view. **(b)** Line drawing of a small artery and vein, cross-sectional view. (See also Plate 26 in the Histology Atlas.)

recoil passively as the blood flows off into the circulation during diastole. Their walls must be sufficiently strong and resilient to withstand such pressure fluctuations. Since these larger arteries have such large amounts of elastic tissue in their media, they are often referred to as *elastic arteries.* Smaller arteries, further along in the circulatory pathway, are exposed to less extreme pressure fluctuations. They have less elastic tissue but still have substantial amounts of smooth muscle in their media. For this reason, they are called *muscular arteries.*

By contrast, veins, which are far removed from the heart in the circulatory pathway, are not subjected to such pressure fluctuations and are essentially low-pressure vessels. Thus, veins may be thinner-walled without jeopardy. However, the low-pressure condition itself and the fact that blood returning to the heart often flows against gravity require structural modifications to ensure that venous return equals cardiac output. Thus, the lumens of veins tend to be substantially larger than those of corresponding arteries, and valves in larger veins act to prevent backflow of blood in much the same manner as the semilunar valves of the heart. The skeletal muscle "pump" also promotes venous return; as the skeletal muscles surrounding the veins contract and relax, the blood is milked through the veins toward the heart. (Anyone who has been standing relatively still for an extended time will be happy to show you their swollen ankles, caused by blood pooling in their feet during the period of muscle inactivity!) Pressure changes that occur in the thorax during breathing also aid the return of blood to the heart.

• To demonstrate how efficiently venous valves prevent backflow of blood, perform the following simple experiment.

Allow one hand to hang by your side until the blood vessels on the dorsal aspect become distended. Place two fingertips against one of the distended veins and, pressing firmly, move the superior finger proximally along the vein and then release this finger. The vein will remain flattened and collapsed despite gravity. Then remove the distal fingertip and observe the rapid filling of the vein.

The transparent walls of the tiny capillaries are only one cell layer thick, consisting of just the endothelium underlain by a basal lamina, that is, the tunica intima. Because of this exceptional thinness, exchanges are easily made between the blood and tissue cells.

Activity:
Examining the Microscopic Structure of Arteries and Veins

1. Obtain a slide showing a cross-sectional view of blood vessels and a microscope.

2. Using Figure 32.1b as a guide, scan the section to identify a thick-walled artery. Very often, but not always, its lumen will appear scalloped due to the constriction of its walls by the elastic tissue of the media.

3. Identify a vein. Its lumen may be elongated or irregularly shaped and collapsed, and its walls will be considerably thinner. Notice the difference in the relative amount of elastic fibers in the media of the two vessels. Also, note the thinness of the intima layer, which is composed of flat squamous-type cells. ∎

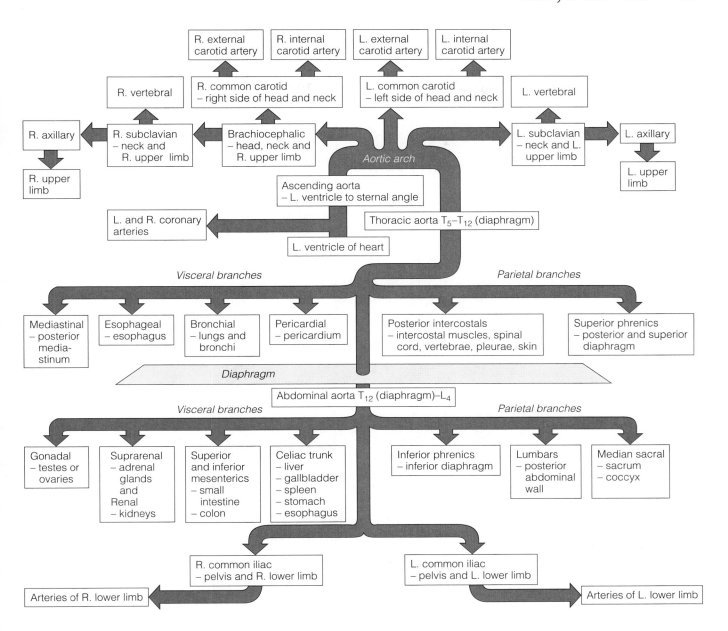

Figure 32.2 Schematic of the systemic arterial circulation.

Major Systemic Arteries of the Body

The **aorta** is the largest artery of the body. Extending upward as the *ascending aorta* from the left ventricle, it arches posteriorly and to the left (*aortic arch*) and then courses downward as the *descending aorta* through the thoracic cavity. It penetrates the diaphragm to enter the abdominal cavity just anterior to the vertebral column.

Figure 32.2 depicts the relationship of the aorta and its major branches. As you locate the arteries on the figure, be aware of ways in which you can make your memorization task easier. In many cases the name of the artery reflects the body region traversed (axillary, subclavian, brachial, popliteal), the organ served (renal, hepatic), or the bone followed (tibial, femoral, radial, ulnar).

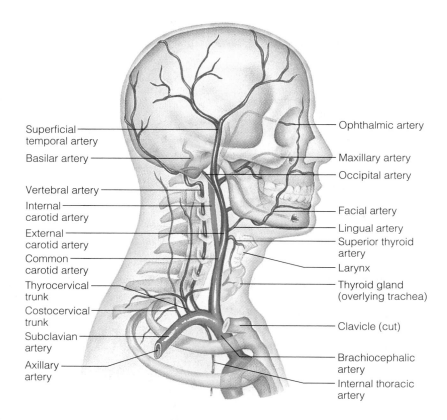

Superficial temporal artery

Basilar artery

Vertebral artery

Internal carotid artery

External carotid artery

Common carotid artery

Thyrocervical trunk

Costocervical trunk

Subclavian artery

Axillary artery

Ophthalmic artery

Maxillary artery

Occipital artery

Facial artery

Lingual artery

Superior thyroid artery

Larynx

Thyroid gland (overlying trachea)

Clavicle (cut)

Brachiocephalic artery

Internal thoracic artery

Figure 32.3 Arteries of the head and neck, right aspect.

Ascending Aorta

The only branches of the ascending aorta are the **right** and the **left coronary arteries,** which supply the myocardium. The coronary arteries are described in Exercise 30 in conjunction with heart anatomy.

Aortic Arch and Thoracic Aorta

The **brachiocephalic** (literally, "arm-head") **artery** is the first branch of the aortic arch (Figure 32.3). The other two major arteries branching off the aortic arch are the **left common carotid artery** and the **left subclavian artery.** The brachiocephalic artery persists briefly before dividing into the **right common carotid artery** and the **right subclavian artery.**

Arteries Serving the Head and Neck The common carotid artery on each side divides to form an **internal carotid artery,** which serves the brain, and an **external carotid artery.** The external carotid artery supplies the extracranial tissues of the neck and head, largely via its **superficial temporal, ophthalmic, maxillary, facial,** and **occipital** arterial branches.

(Notice that several arteries are shown in the figure that are not described here. Ask your instructor which arteries you are required to identify.)

The right and left subclavian arteries each give off several branches to the head and neck. The first of these is the **vertebral artery,** which runs up the posterior neck to supply the cerebellum, part of the brain stem, and the posterior cerebral hemispheres. Issuing just lateral to the vertebral artery are the **thyrocervical trunk,** which mainly serves the thyroid gland and some scapular muscles, and the **costocervical trunk,** which supplies deep neck muscles and some of the upper intercostal muscles. In the armpit, the subclavian artery becomes the axillary artery, which serves the upper limb.

Arteries Serving the Thorax and Upper Limbs As the **axillary artery** runs through the axilla, it gives off several branches to the chest wall and shoulder girdle (Figure 32.4). These include the **thoracoacromial trunk** (to shoulder and pectoral region), the **lateral thoracic artery** (lateral chest wall), the **subscapular artery** (to scapula and dorsal thorax), and the **anterior** and **posterior humeral circumflex arteries** (to the shoulder and the deltoid muscle). The axillary artery becomes the **brachial artery** as it enters the arm. This gives

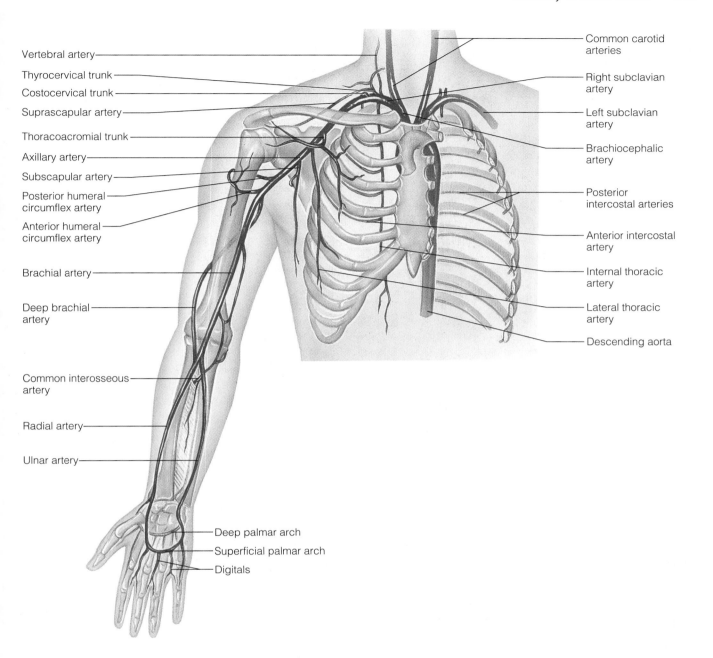

Figure 32.4 Arteries of the right upper limb and thorax.

off a deep branch, and at the elbow, the brachial artery divides into the **radial** and **ulnar arteries,** which follow the same-named bones to supply the forearm and hand.

The **internal thoracic (mammary) arteries** that arise from the subclavian arteries supply the mammary glands, most of the thorax wall, and anterior intercostal structures via their **anterior intercostal artery** branches. The first two

pairs of **posterior intercostal arteries** arise from the costocervical trunk, noted above. (The more inferior pairs arise from the thoracic aorta.) Not shown in Figure 32.4 are the small arteries that serve the diaphragm (*phrenic arteries*), esophagus (*esophageal arteries*), and bronchi (*bronchial arteries*).

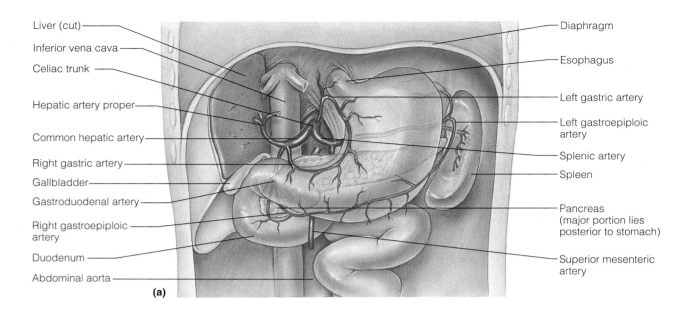

Liver (cut)
Inferior vena cava
Celiac trunk
Hepatic artery proper
Common hepatic artery
Right gastric artery
Gallbladder
Gastroduodenal artery
Right gastroepiploic artery
Duodenum
Abdominal aorta

Diaphragm
Esophagus
Left gastric artery
Left gastroepiploic artery
Splenic artery
Spleen
Pancreas (major portion lies posterior to stomach)
Superior mesenteric artery

(a)

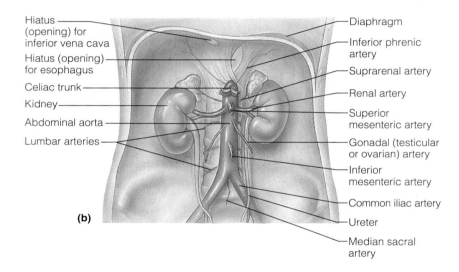

Hiatus (opening) for inferior vena cava
Hiatus (opening) for esophagus
Celiac trunk
Kidney
Abdominal aorta
Lumbar arteries

Diaphragm
Inferior phrenic artery
Suprarenal artery
Renal artery
Superior mesenteric artery
Gonadal (testicular or ovarian) artery
Inferior mesenteric artery
Common iliac artery
Ureter
Median sacral artery

(b)

Figure 32.5 Arteries of the abdomen. (a) The celiac trunk and its major branches. **(b)** Major branches of the abdominal aorta.

Abdominal Aorta

Although several small branches of the descending aorta serve the thorax (see above), the more major branches of the descending aorta are those serving the abdominal organs and ultimately the lower limbs (Figure 32.5).

Arteries Serving Abdominal Organs The **celiac trunk** is an unpaired artery that subdivides almost immediately into three branches: the **left gastric artery** supplying the stomach, the **splenic artery** supplying the spleen, and the **common hepatic artery,** which runs superiorly and gives off branches to the stomach (**right gastric artery**), duodenum, and pancreas. Where the **gastroduodenal artery** branches off, the common hepatic artery becomes the **hepatic artery proper,** which serves the liver. The **right** and **left gastroepiploic arteries,** branches of the gastroduodenal and splenic arteries respectively, serve the left (greater) curvature of the stomach.

The largest branch of the abdominal aorta, the **superior mesenteric artery,** supplies most of the small intestine (via the *intestinal arteries*) and the first half of the large intestine (via the *ileocolic* and *colic arteries*). (These are not shown in Figure 32.5.)

The paired **renal arteries** supply the kidneys, and the **gonadal arteries,** arising from the ventral aortic surface just below the renal arteries, run inferiorly to serve the gonads. They are called **ovarian arteries** in the female and **testicular arteries** in the male. Since these vessels must travel through the inguinal canal to supply the testes, they are considerably longer in the male than the female.

The final major branch of the abdominal aorta is the **inferior mesenteric artery,** which supplies the distal half of the large intestine via several branches. Just below this, four pairs of **lumbar arteries** arise from the posterolateral surface of the aorta to supply the posterior abdominal wall (lumbar region).

In the pelvic region, the descending aorta divides into the two large **common iliac arteries,** which serve the pelvis, lower abdominal wall, and the lower limbs.

Arteries Serving the Lower Limbs Each of the common iliac arteries extends for about 2 inches into the pelvis before it divides into the internal and external iliac arteries (Figure 32.6). The **internal iliac artery** supplies the gluteal muscles via the **superior** and **inferior gluteal arteries** and the adductor muscles of the medial thigh via the **obturator artery,** as well as the external genitalia and perineum (via the *pudendal artery,* not illustrated).

The **external iliac artery** supplies the anterior abdominal wall and the lower limb. As it continues into the thigh, its name changes to **femoral artery.** Proximal branches of the femoral artery, the **femoral circumflex arteries,** supply the head and neck of the femur and the hamstring muscles. Slightly lower, the femoral artery gives off a deep branch, the **deep femoral artery,** which supplies the posterior thigh (knee flexor muscles). In the knee region, the femoral artery briefly becomes the **popliteal artery;** its subdivisions—the **anterior** and **posterior tibial arteries**—supply the leg, ankle, and foot. The posterior tibial, which supplies flexor muscles, gives off one main branch, the **peroneal artery,** that serves the lateral calf (peroneal muscles), and then divides into the **lateral** and **medial plantar arteries** that supply blood to the sole of the foot. The anterior tibial artery supplies the extensor muscles and terminates with the **dorsalis pedis artery.** The dorsalis pedis supplies the dorsum of the foot and continues on as the **arcuate artery.** The dorsalis pedis is often palpated in patients with circulation problems of the leg to determine the circulatory efficiency to the limb as a whole.

- Palpate your own dorsalis pedis artery.

A c t i v i t y :
Locating Arteries on an Anatomical Chart or Model

Now that you have identified the arteries on Figures 32.2–32.6, attempt to locate and name them (without a reference) on a large anatomical chart or three-dimensional model of the vascular system. ∎

Figure 32.6 Arteries of the right pelvis and lower limb.

Common iliac artery
Internal iliac artery
Superior gluteal artery
External iliac artery
Deep femoral artery
Lateral femoral circumflex artery
Medial femoral circumflex artery
Obturator artery
Femoral artery
Adductor hiatus
Popliteal artery
Anterior tibial artery
Posterior tibial artery
Peroneal artery
Dorsalis pedis artery
Arcuate artery
Metatarsal arteries
Plantar arch

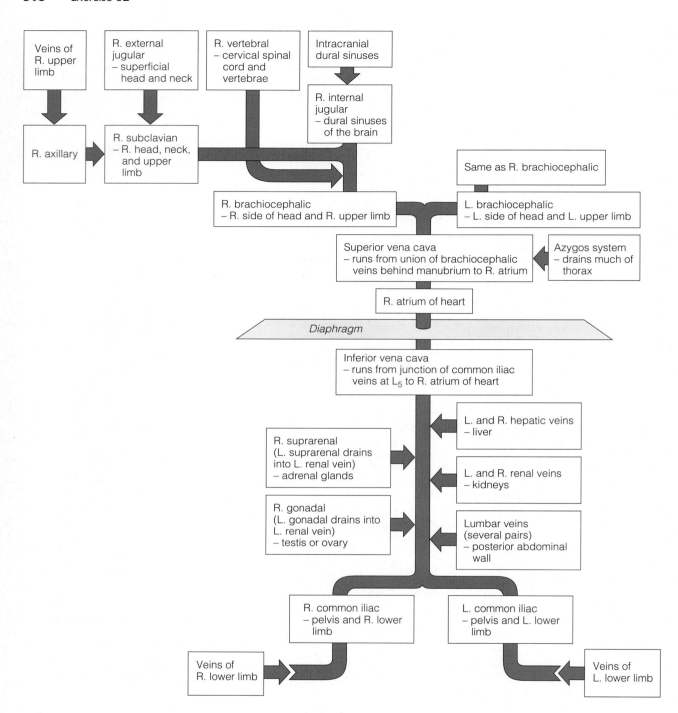

Figure 32.7 Schematic of systemic venous circulation.

Major Systemic Veins of the Body

Arteries are generally located in deep, well-protected body areas. However, many veins follow a more superficial course and are often easily seen and palpated on the body surface. Most deep veins parallel the course of the major arteries, and in many cases the naming of the veins and arteries is identical except for the designation of the vessels as veins. Whereas the major systemic arteries branch off the aorta, the veins tend to converge on the venae cavae, which enter the right atrium of the heart. Veins draining the head and upper extremities empty into the **superior vena cava,** and those draining the lower body empty into the **inferior vena cava.** Figure 32.7 is a schematic of the systemic veins and their relationship to the venae cavae to get you started.

Veins Draining into the Inferior Vena Cava

The inferior vena cava, a much longer vessel than the superior vena cava, returns blood to the heart from all body regions below the diaphragm (see Figure 32.7). It begins in the lower abdominal region with the union of the paired **common iliac veins,** which drain venous blood from the legs and pelvis.

Veins of the Lower Limbs Each common iliac vein is formed by the union of the **internal iliac vein,** draining the pelvis, and the **external iliac vein,** which receives venous blood from the lower limb (Figure 32.8). Veins of the leg include the **anterior** and **posterior tibial veins,** which serve the calf and foot. The anterior tibial vein is a superior continuation of the **dorsalis pedis vein** of the foot. The posterior tibial vein is formed by the union of the **medial** and **lateral plantar veins,** and ascends deep in the calf muscles. It joins with the anterior tibial vein at the knee to produce the **popliteal vein,** which crosses the back of the knee. The popliteal vein becomes the **femoral vein** in the thigh; the femoral vein in turn becomes the external iliac vein in the inguinal region.

The **great saphenous vein,** a superficial vein, is the longest vein in the body. Beginning in common with the **small saphenous vein** from the **dorsal venous arch,** it extends up the medial side of the leg, knee, and thigh to empty into the femoral vein. The small saphenous vein runs along the lateral aspect of the foot and through the calf muscle, which it drains, and then empties into the popliteal vein at the knee (Figure 32.8b).

Veins of the Abdomen Moving superiorly in the abdominal cavity (Figure 32.9), the inferior vena cava receives blood from the posterior abdominal wall via several pairs of **lumbar veins,** and from the right ovary or testis via the **right gonadal vein.** (The **left gonadal [ovarian or testicular] vein** drains into the left renal vein superiorly.) The paired **renal veins** drain the kidneys. Just above the right renal vein, the **right suprarenal vein** (receiving blood from the adrenal gland on the same side) drains into the inferior vena cava, but its partner, the **left suprarenal vein,** empties into the left renal vein

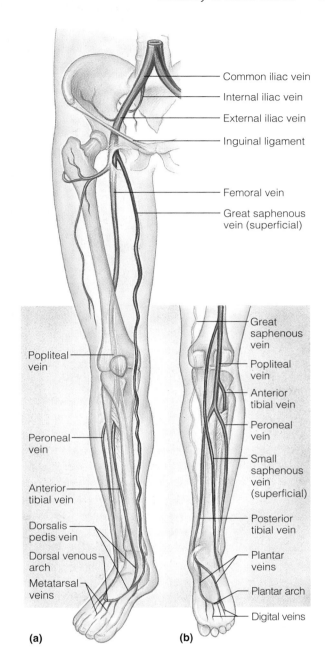

(a)   **(b)**

Figure 32.8 Veins of the right lower limb.
(a) Anterior view. **(b)** Posterior view.

inferiorly. The **right** and **left hepatic veins** drain the liver. The unpaired veins draining the digestive tract organs empty into a special vessel, the **hepatic portal vein,** which carries blood to the liver to be processed before it enters the systemic venous system. (The hepatic portal system is discussed separately on p. 321.)

Figure 32.9 Venous drainage of abdominal organs not drained by the hepatic portal vein.

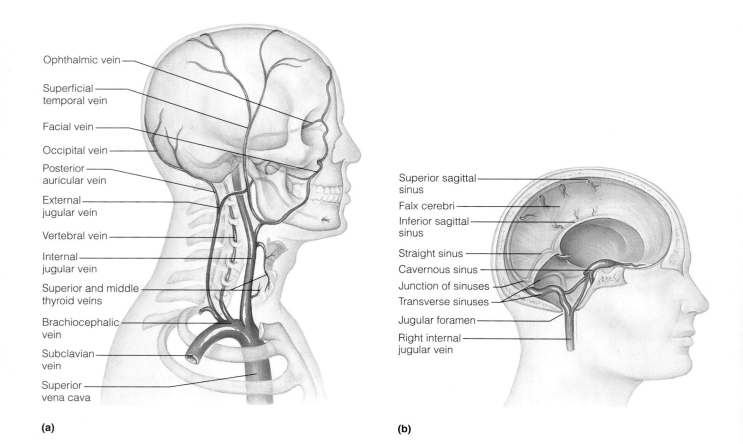

(a)

(b)

Figure 32.10   Venous drainage of the head, neck, and brain.
(a) Veins of the head and neck, right superficial aspect. **(b)** Dural sinuses of the brain, right aspect.

Right subclavian vein

Axillary vein

Brachial vein

Cephalic vein

Basilic vein

Median cubital vein

Median vein of the forearm

Cephalic vein

Radial vein

Internal jugular vein

External jugular vein

Brachiocephalic veins

Left subclavian vein

Superior vena cava

Azygos vein

Accessory hemiazygos vein

Hemiazygos vein

Posterior intercostals

Inferior vena cava

Ascending lumbar vein

Basilic vein

Ulnar vein

Deep palmar venous arch

Superficial palmar venous arch

Digital veins

Figure 32.11 Veins of the right upper limb and shoulder. For clarity, the abundant branching and anastomoses of these vessels are not shown.

Veins Draining into the Superior Vena Cava

Veins draining into the superior vena cava are named from the superior vena cava distally, *but remember that the flow of blood is in the opposite direction.*

Veins of the Head and Neck The **right** and **left brachiocephalic veins** drain the head, neck, and upper extremities and unite to form the superior vena cava (Figure 32.10). Notice that although there is only one brachiocephalic artery, there are two brachiocephalic veins.

Branches of the brachiocephalic veins include the internal jugular, vertebral, and subclavian veins. The **internal jugular veins** are large veins that drain the superior sagittal sinus and other dural sinuses of the brain. As they move inferiorly, they receive blood from the head and neck via the **superficial temporal** and **facial** veins. The **vertebral veins**

drain the posterior aspect of the head including the cervical vertebrae and spinal cord. The **subclavian veins** receive venous blood from the upper extremity. The **external jugular vein** joins the subclavian vein near its origin to return the venous drainage of the extracranial (superficial) tissues of the head and neck.

Veins of the Upper Limb and Thorax As the subclavian vein traverses the axilla, it becomes the **axillary vein** and then the **brachial vein** as it courses along the posterior aspect of the humerus (Figure 32.11). The brachial vein is formed by the union of the deep **radial** and **ulnar veins** of the forearm. The superficial venous drainage of the arm includes the **cephalic vein,** which courses along the lateral aspect of the arm and empties into the axillary vein; the **basilic vein,** found on the medial aspect of the arm and entering the brachial vein; and the **median cubital vein,** which runs between the cephalic and basilic veins in the anterior aspect of the elbow (this vein is often the site of choice for removing blood for testing purposes). The **median vein of the forearm** lies between the radial and ulnar veins, and terminates variably by entering the cephalic or basilic vein at the elbow.

The **azygos system** (Figure 32.11) drains the intercostal muscles of the thorax and provides an accessory venous system to drain the abdominal wall. The **azygos vein,** which drains the right side of the thorax, enters the dorsal aspect of the superior vena cava immediately before that vessel enters the right atrium. Also part of the azygos system are the **hemiazygos** (a continuation of the **left ascending lumbar vein** of the abdomen) and the **accessory hemiazygos veins,** which together drain the left side of the thorax and empty into the azygos vein.

Activity:
Identifying the Systemic Veins

Identify the important veins of the systemic circulation on the large anatomical chart or model without referring to the figures. ■

Special Circulations
Pulmonary Circulation

The pulmonary circulation (discussed previously in relation to heart anatomy on p. 298) differs in many ways from systemic circulation because it does not serve the metabolic needs of the body tissues with which it is associated (in this case, lung tissue). It functions instead to bring the blood into close contact with the alveoli of the lungs to permit gas exchanges that rid the blood of excess carbon dioxide and replenish its supply of vital oxygen. The arteries of the pulmonary circulation are structurally much like veins, and they create a low-pressure bed in the lungs. (If the arterial pressure in the systemic circulation is 120/80, the pressure in the pulmonary artery is likely to be approximately 25/10.) The functional blood supply of the lungs is provided by the **bronchial arteries,** which diverge from the thoracic portion of the descending aorta.

Pulmonary circulation begins with the large **pulmonary trunk,** which leaves the right ventricle and divides into the **right** and **left pulmonary arteries** about 2 inches above its origin (Figure 32.12). The right and left pulmonary arteries plunge into the lungs, where they subdivide into **lobar arteries** (three on the right and two on the left), which accompany the main bronchi into the lobes of the lungs. The lobar arteries branch extensively within the lungs to form arterioles, which finally terminate in the capillary networks surrounding the alveolar sacs of the lungs. Diffusion of the respiratory gases occurs across the walls of the alveoli and **pulmonary capillaries.** The pulmonary capillary beds are drained by venules, which converge to form sequentially larger veins and finally the four **pulmonary veins** (two leaving each lung), which return the blood to the left atrium of the heart.

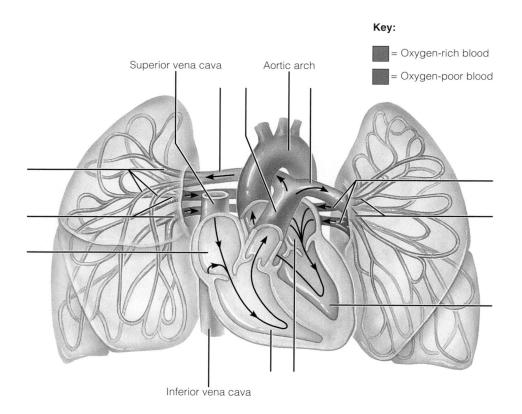

Key:

█ = Oxygen-rich blood

█ = Oxygen-poor blood

Superior vena cava

Aortic arch

Inferior vena cava

a. Left atrium

b. Pulmonary trunk

c. Right pulmonary artery

d. Left pulmonary artery

e. Left pulmonary veins

f. Lobar arteries (left lung)

g. Lobar arteries (right lung)

h. Right atrium

i. Right ventricle

j. Right pulmonary veins

k. Left ventricle

Figure 32.12 The pulmonary circulation.

Activity:
Identifying Vessels of the Pulmonary Circulation

Using the terms provided in Figure 32.12, *label* all structures provided with leader lines. ■

Hepatic Portal Circulation

Blood vessels of the hepatic portal circulation drain the digestive viscera, spleen, and pancreas and deliver this blood to the liver for processing via the **hepatic portal vein.** If a meal has recently been eaten, the hepatic portal blood will be nutrient rich. The liver is the key body organ involved in maintaining proper sugar, fatty acid, and amino acid concentrations in the blood, and this system ensures that these substances pass through the liver before entering the systemic circulation. As blood percolates through the liver sinusoids, some of the nutrients are removed to be stored or processed in various ways for release to the general circulation. At the same time, the hepatocytes are detoxifying alcohol and other possibly harmful chemicals present in the blood, and the liver's macrophages are removing bacteria and other debris from the passing blood. The liver in turn is drained by the hepatic veins that enter the inferior vena cava.

The **inferior mesenteric vein,** draining the distal portions of the large intestine, joins the **splenic vein,** which drains the spleen and part of the pancreas and stomach. The splenic vein and the **superior mesenteric vein,** which receives blood from the small intestine and the ascending and transverse colon, unite to form the hepatic portal vein. The **left gastric vein,** which drains the lesser curvature of the stomach, drains directly into the hepatic portal vein.

Activity:
Tracing the Hepatic Portal Circulation

Locate the vessels named above on Figure 32.13. ■

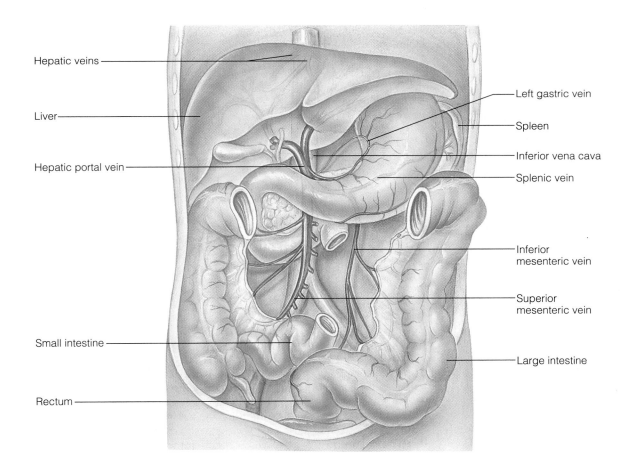

Figure 32.13 **Hepatic portal circulation of the human.**

Fetal Circulation

In a developing fetus, the lungs and digestive system are not yet functional, and all nutrient, excretory, and gaseous exchanges occur through the placenta (see Figure 32.14a). Nutrients and oxygen move across placental barriers from the mother's blood into fetal blood, and carbon dioxide and other metabolic wastes move from the fetal blood supply to the mother's blood.

Fetal blood travels through the umbilical cord, which contains three blood vessels: two smaller umbilical arteries and one large umbilical vein. The **umbilical vein** carries blood rich in nutrients and oxygen to the fetus; the **umbilical arteries** carry carbon dioxide and waste-laden blood from the fetus to the placenta. The umbilical arteries, which transport blood away from the fetal heart, meet the umbilical vein at the *umbilicus* (navel, or belly button) and wrap around the vein within the cord en route to their placental attachments. Newly oxygenated blood flows in the umbilical vein superiorly toward the fetal heart. En route, some of this blood perfuses the liver, but the larger proportion is ducted through the relatively nonfunctional liver to the inferior vena cava via a shunt vessel called the **ductus venosus,** which carries the blood to the right atrium of the heart.

Because fetal lungs are nonfunctional and collapsed, two shunting mechanisms ensure that blood almost entirely bypasses the lungs. Much of the blood entering the right atrium is shunted into the left atrium through the **foramen ovale,** a flaplike opening in the interatrial septum. The left ventricle then pumps the blood out the aorta to the systemic circulation. Blood that does enter the right ventricle and is pumped out of the pulmonary trunk encounters a second shunt, the **ductus arteriosus,** a short vessel connecting the pulmonary trunk and the aorta. Because the collapsed lungs present an extremely high-resistance pathway, blood more readily enters the systemic circulation through the ductus arteriosus.

The aorta carries blood to the tissues of the body; this blood ultimately finds its way back to the placenta via the umbilical arteries. The only fetal vessel that carries highly oxygenated blood is the umbilical vein. All other vessels contain varying degrees of oxygenated and deoxygenated blood.

At birth, or shortly after, the foramen ovale closes and becomes the **fossa ovalis,** and the ductus arteriosus collapses and is converted to the fibrous **ligamentum arteriosum** (Figure 32.14b). Lack of blood flow through the umbilical vessels leads to their eventual obliteration, and the circulatory pattern becomes that of the adult. Remnants of the umbilical arteries persist as the **medial umbilical ligaments** on the inner surface of the anterior abdominal wall, of the umbilical vein as the **ligamentum teres** or **round ligament** of the liver, and of the ductus venosus as a fibrous band called the **ligamentum venosum** on the inferior surface of the liver.

Activity:
Tracing the Pathway of Fetal Blood Flow

The pathway of fetal blood flow is indicated with arrows on Figure 32.14a. Appropriately *label* all specialized fetal circulatory structures provided with leader lines. ■

Arterial Supply of the Brain and the Circle of Willis

A continuous blood supply to the brain is crucial because oxygen deprivation for even a few minutes causes irreparable damage to the delicate brain tissue. The brain is supplied by two pairs of arteries arising from the region of the aortic arch—the *internal carotid arteries* and the *vertebral arteries.* Figure 32.15 is a diagram of the brain's arterial supply.

Figure 32.14 Circulation in fetus and newborn. Arrows indicate direction of blood flow. **(a)** Special adaptations for embryonic and fetal life. The umbilical vein carries oxygen- and nutrient-rich blood from the placenta to the fetus. The umbilical arteries carry waste-laden blood from the fetus to the placenta; the ductus arteriosus and foraman ovale bypass the nonfunctional lungs; and the ductus venosus allows blood to partially bypass the liver. **(b)** Changes in the cardiovascular system at birth. The umbilical vessels are occluded, as are the liver and lung bypasses (ductus venosus and arteriosus, and the foramen ovale).

Fetus

Newborn

Aortic arch

Superior vena cava

Lung

Ligamentum arteriosum

Pulmonary artery

Pulmonary veins

Heart

Fossa ovalis

Liver

Ligamentum venosum

Hepatic portal vein

Ligamentum teres

Inferior vena cava

Umbilicus

Abdominal aorta

Common iliac artery

Medial umbilical ligaments

Urinary bladder

Umbilical cord

(b)

Placenta

(a)

Key:

= High oxygenation

= Moderate oxygenation

= Low oxygenation

= Very low oxygenation

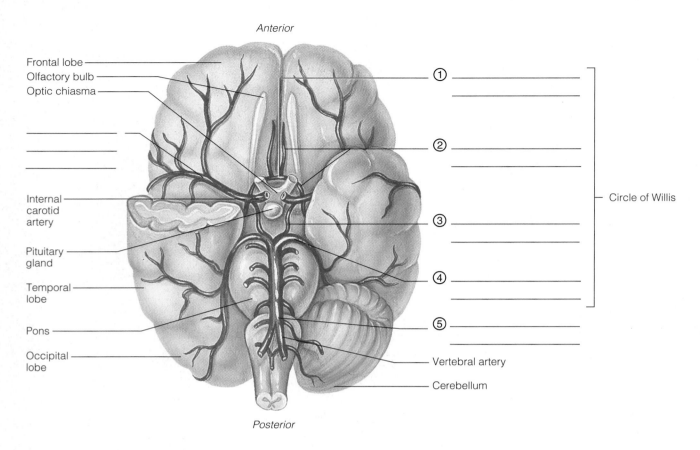

Anterior

Frontal lobe

Olfactory bulb

Optic chiasma

① _____ _____

② _____

Circle of Willis

Internal carotid artery

③ _____ _____

Pituitary gland

④ _____ _____

Temporal lobe

Pons

⑤ _____ _____

Occipital lobe

Vertebral artery

Cerebellum

Posterior

Figure 32.15 Arterial supply of the brain. Cerebellum is not shown on the left side of the figure. (See also Plate D in the Human Anatomy Atlas.)

A c t i v i t y :
Tracing the Arterial Supply of the Brain

The internal carotid and vertebral arteries are labeled. As you read the description of the blood supply below, *complete the labeling of this diagram.* ■

The **internal carotid arteries,** branches of the common carotid arteries, follow a deep course through the neck and along the pharynx, entering the skull through the carotid canals of the temporal bone. Within the cranium, each divides into **anterior** and **middle cerebral arteries,** which supply the bulk of the cerebrum. The internal carotid arteries also contribute to the formation of the **circle of Willis,** an arterial anastomosis at the base of the brain surrounding the pituitary gland and the optic chiasma, by contributing to a **posterior communicating artery** on each side. The circle is completed by the **anterior communicating artery,** a short shunt connecting the right and left anterior cerebral arteries.

The paired **vertebral arteries** diverge from the subclavian arteries and pass superiorly through the foramina of the transverse process of the cervical vertebrae to enter the skull through the foramen magnum. Within the skull, the vertebral arteries unite to form a single **basilar artery,** which continues superiorly along the ventral aspect of the brain stem, giving off branches to the pons, cerebellum, and inner ear. At the base of the cerebrum, the basilar artery divides to form the **posterior cerebral arteries.** These supply portions of the temporal and occipital lobes of the cerebrum and also become part of the circle of Willis by joining with the posterior communicating arteries.

The uniting of the blood supply of the internal carotid arteries and the vertebral arteries via the circle of Willis is a protective device that theoretically provides an alternate set of pathways for blood to reach the brain tissue in the case of arterial occlusion or impaired blood flow anywhere in the system. In actuality, the communicating arteries are tiny, and in many cases the communicating system is defective.

 For instructions on animals dissections, see the dissection exercises starting on p. 733 in the cat and pig editions of this manual.

Cardiovascular Dynamics: Computer Simulation

The physiology of human blood circulation can be divided into two distinct but remarkably harmonized processes: (1) the pumping of blood by the heart, and (2) the transport of blood to all body tissues via the vasculature, or blood vessels. Blood supplies all body tissues with the substances needed for survival, so it is vital that blood delivery is ample for tissue demands.

The Mechanics of Circulation

To understand how blood is transported throughout the body, let's examine three important factors influencing how blood circulates through the cardiovascular system: blood flow, blood pressure, and peripheral resistance.

Blood flow is the amount of blood moving through a body area or the entire cardiovascular system in a given amount of time. While total blood flow is determined by cardiac output (the amount of blood the heart is able to pump per minute), blood flow to specific body areas can vary dramatically in a given time period. Organs differ in their requirements from moment to moment, and blood vessels constrict or dilate to regulate local blood flow to various areas in response to the tissue's immediate needs. Consequently, blood flow can increase to some regions and decrease to other areas at the same time.

Blood pressure is the force blood exerts against the wall of a blood vessel. Owing to cardiac activity, pressure is highest at the heart end of any artery. Because of the effect of peripheral resistance, which will be discussed shortly, pressure within the arteries (or any blood vessel) drops as the distance (vessel length) from the heart increases. This pressure gradient causes blood to move from and then back to the heart, always moving from high- to low-pressure areas.

Peripheral resistance is the opposition to blood flow resulting from the friction developed as blood streams through blood vessels. Three factors affect vessel resistance: blood viscosity, vessel radius, and vessel length.

Blood viscosity is a measure of the "thickness" of the blood, and is caused by the presence of proteins and formed elements in the plasma (the fluid part of the blood). As the viscosity of a fluid increases, its flow rate through a tube decreases. Blood viscosity in healthy persons normally does not change, but certain conditions such as too many or too few blood cells may modify it.

Objectives

1. To define the following: blood flow; viscosity; peripheral resistance; systole; diastole; end diastolic volume; end systolic volume; stroke volume; cardiac output.

2. To explore cardiovascular dynamics using an experimental setup to simulate a human body function.

3. To understand that heart and blood vessel functions are highly coordinated.

4. To comprehend that pressure differences provide the driving force that moves blood through the blood vessels.

5. To recognize that body tissues may differ in their blood demands at a given time.

6. To identify the most important factors in control of blood flow.

7. To comprehend that changing blood vessel diameter can alter the pumping ability of the heart.

8. To examine the effect of stroke volume on blood flow.

Materials

Minimum computer hardware required:

Windows:

❏ 486/66MHz or better recommended

❏ Windows 95/98 recommended

❏ SVGA display (256 colors at 640 × 480)

❏ Minimum 16 MB available RAM

❏ Double-speed CD-ROM drive (quad speed or higher recommended)

❏ Sound card

❏ Speakers or headphones

❏ Printer

Macintosh:

❏ 68040 processor or Power Macintosh recommended

❏ System 7.1 or higher

❏ Minimum 16 MB available RAM

❑ 13-inch or larger color monitor (640 × 480 resolution)
❑ Double-speed CD-ROM drive (quad speed or higher recommended)
❑ Printer
❑ **Software:** Benjamin/Cummings *PhysioEx* CD-ROM—Cardiovascular Dynamics module

Controlling *blood vessel radius* (one-half of the diameter) is the principal method of blood flow control. This is accomplished by contracting or relaxing the smooth muscle within the blood vessel walls. To see why radius has such a pronounced effect on blood flow, we need to explore the physical relationship between blood and the vessel wall. Blood in direct contact with the vessel wall flows relatively slowly because of the friction, or drag, between the blood and the lining of the vessel. In contrast, fluid in the center of the vessel flows more freely because it is not "rubbing" against the vessel wall. When we contrast large- and small-radius vessels, we see that proportionately more blood is in contact with the wall of small vessels, hence blood flow is notably impeded in small-radius vessels.

Although *vessel length* does not ordinarily change in a healthy person, any increase in vessel length causes a corresponding flow decrease. This effect is principally caused by friction between blood and the vessel wall. Consequently, given two blood vessels of the same diameter, the longer vessel will have more resistance, and thus a reduced blood flow.

The Effect of Blood Pressure and Vessel Resistance on Blood Flow

Poiseuille's equation describes the relationship between pressure, vessel radius, viscosity, and vessel length on blood flow:

$$\text{Blood flow } (\Delta Q) = \frac{\pi \Delta P \, r^4}{8\eta l}$$

In the equation, ΔP is the pressure difference between the two ends of the vessel and represents the driving force behind blood flow. Viscosity (η) and blood vessel length (l) are not commonly altered in a healthy adult. We can also see from the equation that blood flow is directly proportional to the fourth power of vessel radius (r^4), which means that small variations in vessel radius translate into large changes in blood flow. In the human body, changing blood vessel radius provides an extremely effective and sensitive method of blood flow control. Peripheral resistance is the most important factor in blood flow control, because circulation to individual organs can be independently regulated even though systemic pressure may be changing.

Getting Started

Begin by making sure you have the computer equipment and software listed in the Materials section on pp. 325–326.

1. Quit all applications currently running on your computer.

2. Insert the PhysioEx CD-ROM into the CD-ROM drive and keep PhysioEx in the drive during the entire time you use the program.

3. Follow the instructions below for your computer type.

Windows 95/98
The program will launch automatically by loading the CD-ROM into the drive.
If autorun is disabled on your machine follow these instructions:
1. Double-click the My Computer icon on the Windows desktop.
2. Double-click the PhysioEx CD icon.
3. Look for the icon that looks like a tiny movie projector and double-click it.

Macintosh
Double-click the PhysioEx CD icon that appears in the window on your desktop to launch the program.

Vessel Resistance

Imagine for a moment that you are one of the first cardiovascular researchers interested in the physics of blood flow. Your first task as the principal investigator for this project is to plan an effective experimental design simulating a simple fluid pumping system that can be related to the mechanics of the cardiovascular system. The initial phenomenon you study is how fluids, including blood, flow through tubes or blood vessels. Questions you might ask include:

1. What role does pressure play in the flow of fluid?

2. How does peripheral resistance affect fluid flow?

The equipment required to solve these and other questions has already been designed for you in the form of a computerized simulation, which frees you to focus on the logic of the experiment. The first part of the computer simulation indirectly investigates the effects of pressure, vessel radius, viscosity, and vessel length on fluid flow. The second part of the experiment will explore the effects of several variables on the output of a single-chamber pump. Follow the specific guidelines in the exercise for collecting data. As you do so, also try to imagine alternate methods of achieving the same experimental goal.

Figure 33.1   Opening screen of the Vessel Resistance experiment.

Choose **Cardiovascular Dynamics** from the main menu. The opening screen will appear in a few seconds (Figure 33.1).

The primary features on the screen when the program starts are a pair of glass beakers perched atop a simulated electronic device called the *equipment control unit,* which is used to set experiment parameters and to operate the equipment. When the **Start** button (beneath the left beaker) is clicked, the simulated blood flows from the left beaker (source) to the right beaker (destination) through the connecting tube. Clicking the **Refill** button refills the source beaker after an experimental trial. Experimental parameters can be adjusted by clicking the plus (**+**) or minus (**−**) buttons to the right of each display window.

The equipment in the lower part of the screen is called the *data collection unit*. This equipment records and displays data you accumulate during the experiments. The data set for the first experiment (Radius) is highlighted in the **Data Sets** window. You can add or delete a data set by clicking the appropriate button to the right of the **Data Sets** window. The **Record Data** button at the lower right part of the screen activates automatically after an experimental trial. Clicking the **Delete Line** or **Clear Data Set** buttons erases any data you want to delete.

You will record the data you accumulate in the experimental values grid in the lower middle part of the screen.

A c t i v i t y :
Studying the Effect of Flow Tube Radius on Fluid Flow

Our initial study will examine the effect of flow tube radius on fluid flow.[1]

1.   Conduct the initial equipment setup.

The **Radius** line in the data collection unit should be highlighted in bright blue. If it is not, choose it by clicking the **Radius** line. The data collection unit will now record flow variations due to changing flow tube radius.

If the data grid is not empty, click **Clear Data Set** to discard all previous values.

If the left beaker is not full, click **Refill.**

2.   Adjust the flow tube radius to 1.5 mm and the viscosity to 1.0 by clicking the appropriate plus or minus button. During the course of this part of the experiment, maintain the other experiment conditions at:

100 mm Hg driving pressure (top left)

50 mm flow tube length (middle right)

3.   Click **Start** and watch the fluid move into the right beaker. (Fluid moves slowly under some conditions—be patient!) Pressure (currently set to 100 mm Hg) propels fluid from the left beaker to the right beaker through the flow tube. The flow rate is displayed in the Flow window when the left beaker has finished draining. Now click **Record Data** to display the flow rate and experiment parameters in the experimental values grid (and retain the data in the computer's memory for printing and saving). Click **Refill** to replenish the left beaker.

4.   Increase the radius in 0.5 mm increments and repeat step 3 until the maximum radius (6.0 mm) is achieved. Be sure to click **Record Data** after each fluid transfer. If you make an error and want to delete a data value, click the data line in the experimental values grid and then click **Delete Line.**

[1] If you need help identifying any piece of equipment, choose Balloons On from the Help menu and move the mouse pointer onto any piece of equipment visible on the computer's screen. As the pointer touches the object, a pop-up window appears identifying the equipment. To close the pop-up window, move the mouse pointer away from the equipment. Repeat the process for all equipment on the screen until you feel confident with it. When finished, choose Balloons On again to turn off this help feature.

5.   View your data graphically by choosing **Plot Data** from the **Tools** menu. Choose **Radius** as the data set to be graphed, and then use the slider bars to select the radius data to be plotted on the X-axis and the flow data to be plotted on the Y-axis. Click **Draw It** to generate the graph. You can highlight individual data points by clicking a line in the data grid. When you are finished, click the close box at the top of the plot window.

What happened to fluid flow as the radius of the flow tube was increased?

_____

_____

_____

Because fluid flow is proportional to the fourth power of the

radius, _____ changes in tube radius cause

_____ changes in fluid flow.

Is the relationship between fluid flow and flow tube radius

linear or exponential? _____

In this experiment, a simulated motor changes the diameter of the flow tube. Explain how our blood vessels alter blood flow.

_____

_____

_____

_____

After a heavy meal when we are relatively inactive, we might expect blood vessels in the skeletal muscles to be somewhat

_____, whereas blood vessels in the digestive

organs are probably _____. ■

A c t i v i t y :
Studying the Effect of Viscosity on Fluid Flow

With a viscosity of 3 to 4, blood is much more viscous than water (1.0 viscosity). Although viscosity is altered by factors such as dehydration and altered blood cell numbers, a body in homeostatic balance has a relatively stable blood consistency. Nonetheless it is useful to examine the effects of viscosity changes on fluid flow, because we can then predict what might transpire in the human cardiovascular system under certain homeostatic imbalances.

1.   Set the starting conditions as follows:

*   100 mm Hg driving pressure
*   5.0 mm flow tube radius
*   1.0 viscosity
*   50 mm flow tube length

2.   Click the **Viscosity** data set in the data collection unit. (This action prepares the experimental values grid to record the viscosity data.)

3.   **Refill** the left beaker if you have not already done so.

4.   Click **Start** to begin the experiment. After all the fluid has drained into the right beaker, click **Record Data** to record this data point, and then click **Refill** to replenish the left beaker.

5.   In 1.0 unit increments, increase the fluid viscosity and repeat step 4 until the maximum viscosity (10.0) is reached.

6.   View your data graphically by choosing **Plot Data** from the **Tools** menu. Choose **Viscosity** as the data set to be graphed, and then use the slider bars to select the viscosity data to be plotted on the X-axis and the flow data to be plotted on the Y-axis. Click **Draw It** to generate the graph. You can highlight individual data points by clicking a line in the data grid. When finished, click the close box at the top of the plot window.

How does fluid flow change as viscosity is modified?

_____

_____

_____

Is fluid flow versus viscosity an inverse or direct relationship?

_____

How does the effect of viscosity compare with the effect of radius on fluid flow?

_____

_____

Predict the effect of anemia (e.g., fewer red blood cells than normal) on blood flow.

_____

_____

What might happen to blood flow if we increased the number of blood cells?

_____

_____

_____

Explain why changing blood viscosity would or would not be a reasonable method for the body to control blood flow.

_____

_____

_____

_____

Blood viscosity would _____ in conditions of

dehydration, resulting in _____ blood flow. ■

A c t i v i t y :
Studying the Effect of Flow Tube Length on Fluid Flow

With the exception of the normal growth that occurs until the body reaches full maturity, blood vessel length does not significantly change. By conducting this part of the exercise, we will be able to understand the physical relationship between vessel length and blood movement by observing how blood flow changes in flow tubes (vessels) of constant radius, but of different lengths.

1.   Set the starting conditions as follows:

*   100 mm Hg driving pressure
*   5.0 mm flow tube radius
*   3.5 viscosity
*   10 mm flow tube length

2.   Click the **Length** data set in the data collection unit. (This action prepares the experimental values grid to display the length data.)

3. **Refill** the left beaker if you have not already done so.

4. Click **Start** to begin the experiment. After all the fluid has drained into the right beaker, click **Record Data** to record this data point, and then click **Refill** to refill the left beaker.

5. In 5 mm increments, increase the flow tube length by clicking the plus button next to the length window and repeat step 4 until the maximum length (50 mm) has been reached.

6. View your data graphically by choosing **Plot Data** from the **Tools** menu. Choose **Length** as the data set to be graphed, and then use the slider bars to select the length data to be plotted on the X-axis and the flow data to be plotted on the Y-axis. Click **Draw It** to generate the graph. You can highlight individual data points by clicking a line in the grid. When finished, click the close box at the top of the plot window.

How does flow tube length affect fluid flow?

_____

_____

_____

Explain why altering blood vessel length would or would not be a good method of controlling blood flow in the body.

_____

_____

_____

_____ ∎

Activity:
Studying the Effect of Pressure on Fluid Flow

The pressure difference between the two ends of a blood vessel is the driving force behind blood flow. In comparison, our experimental setup pressurizes the left beaker, thereby providing the driving force that propels fluid through the flow tube to the right beaker. You will examine the effect of pressure on fluid flow in this part of the experiment.

1. Set the starting conditions as follows:

- 25 mm Hg driving pressure
- 5.0 mm flow tube radius
- 3.5 viscosity
- 50 mm flow tube length

2. Click the **Pressure** data set in the data collection unit. (This action prepares the experimental values grid for the pressure data.)

3. **Refill** the left beaker if you have not already done so.

4. Click **Start** to begin the experiment. After all the fluid has moved into the right beaker, click **Record Data** to record this data point. Click **Refill** to refill the left beaker.

5. In 25 mm Hg increments, increase the driving pressure by clicking the plus button next to the pressure window and repeat step 4 until the maximum pressure (225 mm Hg) has been reached.

6. View your data graphically by choosing **Plot Data** from the **Tools** menu. Choose **Pressure** as the data set to be graphed, and then use the slider bars to select the pressure data to be plotted on the X-axis and the flow data to be plotted on the Y-axis. Click **Draw It** to generate the graph. You can highlight individual data points by clicking a line in the data grid. When finished, click the close box at the top of the plot window.

How does driving pressure affect fluid flow?

_____

_____

_____

How does this plot differ from the others?

_____

_____

_____

Although changing pressure could be used as a means of blood flow control, explain why this approach would not be as effective as altering blood vessel radius.

_____

_____

_____

_____ ∎

Pump Mechanics

In the human body, the heart beats approximately 70 strokes each minute. Each heart beat consists of a filling interval during which blood moves into the chambers of the heart, and an ejection period when blood is actively pumped into the great arteries. The pumping activity of the heart can be described in terms of the phases of the cardiac cycle. Heart chambers fill during **diastol**e (relaxation of the heart) and pump blood out during **systole** (contraction of the heart). As you can imagine, the length of time the heart is relaxed is one factor that determines the amount of blood within the heart at the end of the filling interval. Up to a point, increasing ventricular filling time results in a corresponding increase in ventricular volume. The volume in the ventricles at the end of diastole, just before cardiac contraction, is called the **end diastolic volume,** or **EDV.**

Blood moves from the heart into the arterial system when systolic pressure increases above the residual pressure (from the previous systole) in the great arteries leaving the heart. Although ventricular contraction causes blood ejection, the heart does not empty completely. As a result, a small quantity of blood (**end systolic volume, or ESV**) remains in the ventricles at the end of systole.

Because the oxygen requirements of body tissue change depending on activity levels, we would expect the **cardiac output** (amount of blood pumped by each ventricle per minute) to vary correspondingly. We can calculate the stroke volume (amount of blood pumped per contraction of each ventricle) by subtracting the end systolic volume from the end diastolic volume (SV = EDV − ESV). It is possible to then compute cardiac output by multiplying the stroke volume by heart rate.

The human heart is a complex four-chambered organ, consisting of two individual pumps (the right and left sides) connected together in series. The right heart pumps blood through the lungs into the left heart, which in turn delivers blood to the systems of the body. Blood then returns to the right heart to complete the circuit.

Using the Pump Mechanics part of the program you will explore the operation of a simple one-chambered pump, and apply the physical concepts gained in the simulation to the operation of either of the two pumps comprising the human heart.

In this experiment you can vary the starting and ending volumes of the pump (analogous to EDV and ESV, respectively), driving and resistance pressures, and the diameters of the flow tubes leading to and from the pump chamber. As you proceed through the exercise, try to apply the ideas of ESV, EDV, cardiac output, stroke volume, and blood flow to the on-screen simulated pump system. For example, imagine that the flow tube leading to the pump from the left represents the pulmonary veins, while the flow tube exiting the pump to the right represents the aorta. The pump would then represent the left side of the heart.

Select **Pump Mechanics** from the **Experiment** menu. The equipment for the Pump Mechanics part of the experiment will become visible within a moment (Figure 33.2).

As you discovered in the previous experiment, there are two simulated electronic control units on the computer's screen. The upper apparatus is the *equipment control unit,* which is used to adjust experiment parameters and to operate the equipment. The lower apparatus is the *data collection* unit, in which you will record the data you accumulate during the course of the experiment.

This equipment is slightly different than that used in the Vessel Resistance experiment. There are two beakers: the *source* beaker on the left and the *destination* beaker on the right. The pressure in each beaker is individually controlled by the small pressure units on top of the beakers. Between the two beakers is a simple pump, which can be thought of as one

side of the heart, or even as a single ventricle (e.g., left ventricle). The left beaker and flow tube are analogous to the venous side of human blood flow, while arterial circulation is simulated by the right flow tube and beaker. One-way valves in the flow tubes supplying the pump ensure fluid movement in one direction—from the left beaker into the pump, and then into the right beaker. If you imagine that the pump represents the left ventricle, then think of the valve to the left of the pump as the bicuspid valve and the valve to the right of the pump as the aortic semilunar valve. The pump is driven by a pressure unit mounted on its cap. An important distinction between the pump's pressure unit and the pressure units atop the beakers is that the pump only delivers pressure during its downward stroke. Upward pump strokes are driven by pressure from the left beaker (the pump does not exert any resistance to flow from the left beaker during pump filling). In contrast, pressure in the right beaker works against the pump pressure, which means that the net pressure driving the fluid into the right beaker is calculated (automatically) by subtracting the right beaker pressure from the pump pressure. The resulting pressure difference between the pump and the right beaker is displayed in the experimental values grid in the data collection unit as **Pres.Dif.R.**

Clicking the **Auto Pump** button will cycle the pump through the number of strokes indicated in the Max. strokes window. Clicking the **Single** button cycles the pump through one stroke. During the experiment, the pump and flow rates are automatically displayed when the number of pump strokes is 5 or greater. The radius of each flow tube is individually controlled by clicking the appropriate button. Click the plus button to increase flow tube radius or the minus button to decrease flow tube radius.

The pump's stroke volume (the amount of fluid ejected in one stroke) is automatically computed by subtracting its ending volume from the starting volume. You can adjust starting and ending volumes, and thereby stroke volume, by clicking the appropriate plus or minus button in the equipment control unit.

The data collection unit records and displays data you accumulate during the experiments. The data set for the first experiment (**Rad.R.,** which represents right flow tube radius) is highlighted in the **Data Sets** window. You can add or delete a data set by clicking the appropriate button to the right of the **Data Sets** window. Clicking **Delete Data Set** will erase the data set itself, including all the data it contained. The **Record Data** button at the right edge of the screen activates automatically after an experimental trial. When clicked, the **Record Data** button displays the flow rate data in the experimental values grid and saves it in the computer's memory. Clicking the **Delete Line** or **Clear Data Set** button erases any data you want to delete.

You will record the data you accumulate in the experimental values grid in the lower middle part of the screen.

Figure 33.2 Opening screen of the Pump Mechanics experiment.

Activity:
Studying the Effect of Radius on Pump Activity

Although you will only be manipulating the radius of the right flow tube in this part of the exercise, try to predict the consequence of altering the left flow tube radius as you collect the experimental data. (Remember that the left flow tube simulates the pulmonary veins and the right flow tube simulates the aorta.)

1. Conduct the initial equipment setup.

Click the **Rad.R.** data set to activate it. The data collection unit is now ready to record flow variations due to changing flow tube radius.

If the experimental values grid is not empty, click **Clear Data Set** to discard all previous values.

If the left beaker is not full, click **Refill.**

2. Adjust the **right** flow tube radius to 2.5 mm, and the **left** flow tube radius to 3.0 mm by clicking and holding the ap-

propriate button. During the entire part of the radius experiment, maintain the other experiment conditions at:

- 40 mm Hg left beaker pressure (this pressure drives fluid into the pump, which offers no resistance to filling)
- 120 mm Hg pump pressure (pump pressure is the driving force that propels fluid into the right beaker)
- 80 mm Hg right beaker pressure (this pressure is the resistance to the pump's pressure)
- 120 ml starting volume (analogous to EDV)
- 50 ml ending volume (analogous to ESV)
- 10 strokes in the Max. strokes window

Notice that the displayed 70 ml stroke volume is automatically calculated. Before starting, click the **Single** button one or two times and watch the pump action.

To be sure you understand how this simple mechanical pump can be thought of as a simulation of the human heart, complete the following statements by circling the correct term within the parentheses:

a. When the piston is at the bottom of its travel, the volume remaining in the pump is analogous to (EDV, ESV) of the heart.

b. The amount of fluid ejected into the right beaker by a single pump cycle is analogous to (stroke volume, cardiac output) of the heart.

c. The volume of blood in the heart just before systole is called (EDV, ESV), and is analogous to the volume of fluid present in the simulated pump when it is at the (top, bottom) of its stroke.

3. Click the **Auto Pump** button in the equipment control unit to start the pump. After the 10 stroke volumes have been delivered, the Flow and Rate windows will automatically display the experiment results. Now click **Record Data** to display the figures you just collected in the experimental values grid. Click **Refill** to replenish the left beaker.

4. Increase the *right* flow tube radius in 0.5 mm increments and repeat step 3 above until the maximum radius (6.0 mm) is achieved. Be sure to click **Record Data** after each trial.

5. When you have completed the experiment, view your data graphically by choosing **Plot Data** from the **Tools** menu. Choose **Rad.R.** as the data set to be graphed, and then use the slider bars to select the Rad.R. data to be plotted on the X-axis and the flow data to be plotted on the Y-axis. Click **Draw It** to generate the graph. You can highlight individual data points by clicking a line in the data grid.

The total flow rates you just determined depend on the flow rate into the pump from the left and on the flow rate out of the pump toward the right. Consequently, the shape of the plot is different than what you might predict after viewing the vessel resistance radius graph.

Try to explain why this graph differs from the radius plot in the Vessel Resistance experiment. Remember that the flow rate into the pump did not change, whereas the flow rate out of the pump varied according to your radius manipulations. When you have finished, click the close box at the top of the plot window.

_____

_____

_____

_____

Complete the following statements by circling the correct term within the parentheses.

a. As the right flow tube radius is increased, fluid flow rate (increases, decreases). This is analogous to (dilation, constriction) of blood vessels in the human body.

b. Even though the pump pressure remains constant, the pump rate (increases, decreases) as the radius of the right flow tube is increased. This happens because the resistance to fluid flow is (increased, decreased).

Apply your observations of the simulated mechanical pump to complete the following statements about human heart function. If you are not sure how to formulate your response, use the simulation to arrive at an answer. Circle the correct term within the parentheses.

c. The heart must contract (more, less) forcefully to maintain cardiac output if the resistance to blood flow in the vessels exiting the heart is increased.

d. Increasing the resistance (e.g., a constriction) of the blood vessels entering the heart would (increase, decrease) the time needed to fill the heart chambers.

What do you think would happen to the flow rate and the pump rate if the left flow tube radius is changed (either increased or decreased)?

_____

_____

_____ ■

Activity:
Studying the Effect of Stroke Volume on Pump Activity

Whereas the heart of a person at rest pumps about 60% of the blood in its chambers, 40% of the total amount of blood remains in the chambers after systole. The 60% of blood ejected by the heart is called the stroke volume, and is the difference between EDV and ESV. Even though our simple pump in this experiment does not work exactly like the human heart, you can apply the concepts to basic cardiac function. In this experiment, you will examine how the activity of the simple pump is affected by changing the pump's starting and ending volumes.

1. Click the **Str.V.** (stroke volume) data set to activate it. If the experimental values grid is not empty, click **Clear Data Set** to discard all previous values. If the left beaker is not full, click **Refill.**

2. Adjust the stroke volume to 10 ml by setting the Start volume (EDV) to 120 ml and the End volume (ESV) to 110 ml (stroke volume = start volume – end volume). During the entire stroke volume part of the experiment, keep the other experimental conditions at:

- 40 mm Hg left beaker pressure
- 120 mm Hg pump pressure
- 80 mm Hg right beaker pressure
- 3.0 mm left and right flow tube radius
- 10 strokes in the Max. strokes window

3. Click the **Auto Pump** button to start the experiment. After 10 stroke volumes have been delivered, the Flow and Rate windows will display the experiment results given the current parameters. Click **Record Data** to display the figures you just collected in the experimental values grid. Click **Refill** to replenish the left beaker.

4. Increase the stroke volume in 10 ml increments (by decreasing the End volume) and repeat step 3 until the maximum stroke volume (120 ml) is achieved. Be sure to click the **Record Data** button after each trial. Watch the pump action during each stroke to see how you can apply the concepts of starting and ending pump volumes to EDV and ESV of the heart.

5. View your data graphically by choosing **Plot Data** from the **Tools** menu. Choose **Str.V.** as the data set to be graphed, and then use the slider bars to select the Str.V. data to be plotted on the X-axis and the pump rate data to be plotted on the Y-axis. Click **Draw It** to generate the graph and then answer the following questions. When you have finished, click the close box at the top of the plot window.

What happened to the pump's rate as its stroke volume was increased?

_____

_____

_____

Using your simulation results as a basis for your answer, explain why an athlete's resting heart rate might be lower than that of the average person.

_____

_____

_____

Applying the simulation outcomes to the human heart, predict the effect of increasing the stroke volume on cardiac output (at any given rate).

_____

_____

_____

When heart rate is increased, the time of ventricular filling is (circle one: increased, decreased), which in turn (increases, decreases) the stroke volume.

What do you think might happen to the pressure in the pump during filling if the valve in the right flow tube became leaky? (Remember that the pump offers no resistance to filling.)

_____

_____

_____

Applying this concept to the human heart, what might occur in the left heart and pulmonary blood vessels if the aortic valve became leaky?

_____

_____

What might occur if the aortic valve became slightly constricted?

_____

_____

In your simulation, increasing the pressure in the right beaker is analogous to the aortic valve becoming (circle one: leaky, constricted). ■

A c t i v i t y :
Studying Combined Effects

In this section, you will set up your own experimental conditions to answer the following questions. Carefully examine each question and decide how to set experiment parameters to arrive at an answer. You should try to set up your own experimental conditions to formulate a response, but you can examine your previously collected data at any time if you need additional information. As a rule of thumb, record several data points for each question as evidence for your answer (unless the question calls for a single pump stroke).

Click the **Add Data Set** button in the data collection unit. Next, create a new data set called Combined. Your newly created data set will be displayed beneath Str.V. Now click the **Combined** line to activate the data set. As you collect the supporting data for the following questions, be sure to click **Record Data** each time you have a data point you wish to keep for your records.

How is the flow rate affected when the right flow tube radius is kept constant (at 3.0 mm) and the left flow tube radius is modified (either up or down)?

_____

_____

_____

_____

How does decreasing left flow tube radius affect pump chamber filling time? Does it affect pump chamber emptying?

_____

_____

_____

_____

You have already examined the effect of changing the pump's end volume as a way of manipulating stroke volume. What happens to flow and pump rate when you keep the end volume constant and alter the start volume to manipulate stroke volume?

_____

_____

_____

_____

Try manipulating the pressure delivered to the left beaker. How does changing the left beaker pressure affect flow rate? (This change would be similar to changing pulmonary vein pressure.)

_____

_____

If the left beaker pressure is decreased to 10 mm Hg, how is pump-filling time affected?

_____

_____

_____

What happens to the pump rate if the filling time is shortened?

_____

_____

What happens to fluid flow when the right beaker pressure equals the pump pressure?

_____

_____

_____ ■

Activity:
Studying Compensation

Click the **Add Data Set** button in the data collection unit. Next, create a new data set called **Comp.** Your newly created data set will be displayed beneath Combined. Now click the **Comp. Line** to activate the data set. As you collect the supporting data for the following questions, be sure to click **Record Data** each time you have a data point you wish to keep for your records.

Adjust the experimental conditions to the following:

- 40 mm Hg left beaker pressure
- 120 mm Hg pump pressure
- 80 mm Hg right beaker pressure
- 3.0 mm left and right flow tube radius
- 10 strokes in the Max. strokes window
- 120 ml Starting volume
- 50 ml Ending volume

Click **Auto Pump** and then record your flow rate data. Let's declare the value you just obtained to be the "normal" flow rate for the purpose of this exercise.

Now decrease the right flow tube radius to 2.5 mm, and run another trial. How does this flow rate compare with "normal"?

_____

_____

Leave the right flow tube radius at 2.5 mm radius and try to adjust one or more other conditions to return flow to "normal." Think logically about what condition(s) might compensate for a decrease in flow tube radius. How were you able to accomplish this? (Hint: There are several ways.)

_____

_____

_____

Decreasing the right flow tube radius is similar to a partial (circle one: leakage, blockage) of the aortic valve or (increased, decreased) resistance in the arterial system.

Explain how the human heart might compensate for such a condition.

_____

_____

_____

_____

If we wanted to increase (or decrease) blood flow to a particular body system (e.g., digestive), would it be better to adjust heart rate or blood vessel diameter? Explain.

_____

_____

_____

_____

Complete the following statements by circling the correct response. (If necessary, use the pump simulation to help you with your answers.)

a.   If we decreased overall peripheral resistance in the human body (as in an athlete), the heart would need to generate (more, less) pressure to deliver an adequate amount of blood flow, and arterial pressure would be (higher, lower).

b.   If the diameter of the arteries of the body were partly filled with fatty deposits, the heart would need to generate (more, less) force to maintain blood flow, and pressure in the arterial system would be (higher, lower) than normal. ■

Activity:
Using Your Data

Save Your Data

When you have finished all parts of the exercise, you can save your data set for later review and additional analysis.

1.   Select **Save Data to File** from the **File** menu.

2.   When the dialog box opens, click a drive and/or directory location.

3.   Type a file name into the **File name** box (your instructor may provide you with a name).

4.   Click **Save** to complete the save. If you want to review previously saved data, select **Open Data File** from the File menu and choose your file.

Printing Your Data

If your instructor requires a printed copy of your experiment results, choose **Print** from the **File** menu.

1.   Type your name into the space provided.

2.   When you are ready to print, click OK. ■

Human Cardiovascular Physiology—Blood Pressure and Pulse Determinations

Objectives

1. To define *systole*, *diastole*, and *cardiac cycle*.
2. To indicate the normal length of the cardiac cycle, the relative pressure changes occurring within the atria and ventricles during the cycle, and the timing of valve closure.
3. To use the stethoscope to auscultate heart sounds and to relate heart sounds to cardiac cycle events.
4. To describe the clinical significance of heart sounds and heart murmurs.
5. To demonstrate the thoracic locations where the first and second heart sounds are most accurately auscultated.
6. To define *pulse*, *pulse deficit*, *blood pressure*, and *sounds of Korotkoff*.
7. To accurately determine a subject's apical and radial pulse.
8. To accurately determine a subject's blood pressure with a sphygmomanometer, and to relate systolic and diastolic pressures to events of the cardiac cycle.
9. To investigate the effects of exercise on blood pressure, pulse, and cardiovascular fitness.
10. To note factors affecting and/or determining blood flow and skin color.

Materials

- ❑ Stethoscope
- ❑ Sphygmomanometer
- ❑ Watch (or clock) with a second hand
- ❑ Step stools (16 in. and 20 in. in height)
- ❑ Cot (if available)
- ❑ Alcohol swabs
- ❑ Millimeter ruler
- ❑ Ice
- ❑ Small basin suitable for the immersion of one hand
- ❑ Laboratory thermometer (°C)
- ❑ Record or audiotape: "Interpreting Heart Sounds" (available on free loan from the local chapters of the American Heart Association)
- ❑ Phonograph or tape deck
- ❑ Felt marker

 See Appendix C, Exercise 34 for links to A.D.A.M. Interactive Anatomy.

Any comprehensive study of human cardiovascular physiology takes much more time than a single laboratory period. However, it is possible to conduct investigations of a few phenomena such as pulse, heart sounds, and blood pressure, all of which reflect the heart in action and the function of blood vessels. (The electrocardiogram is studied separately in Exercise 31.) A discussion of the cardiac cycle will provide a basis for understanding and interpreting the various physiologic measurements taken.

Cardiac Cycle

In a healthy heart, the two atria contract simultaneously. As they begin to relax, simultaneous contraction of the ventricles occurs. However, according to general usage, the terms **systole** and **diastole** refer to events of ventricular contraction and relaxation, respectively. The **cardiac cycle** is equivalent to one complete heartbeat—during which both atria and ventricles contract and then relax. It is marked by a succession of changes in blood volume and pressure within the heart. Figure 34.1 is a graphic representation of the events of the cardiac cycle for the left side of the heart. Although pressure changes in the right side are less dramatic than those in the left, the same relationships apply.

We will begin the discussion of the cardiac cycle with the heart in complete relaxation (diastole). At this point, pressure in the heart is very low, blood is flowing passively from the pulmonary and systemic circulations into the atria and on through to the ventricles; the semilunar valves are closed, and the AV valves are open. Shortly, atrial contraction occurs and atrial pressure increases, forcing residual blood into the ventricles. Then ventricular systole begins and intraventricular pressure increases rapidly, closing the AV valves. When ventricular pressure exceeds that of the large arteries leaving the

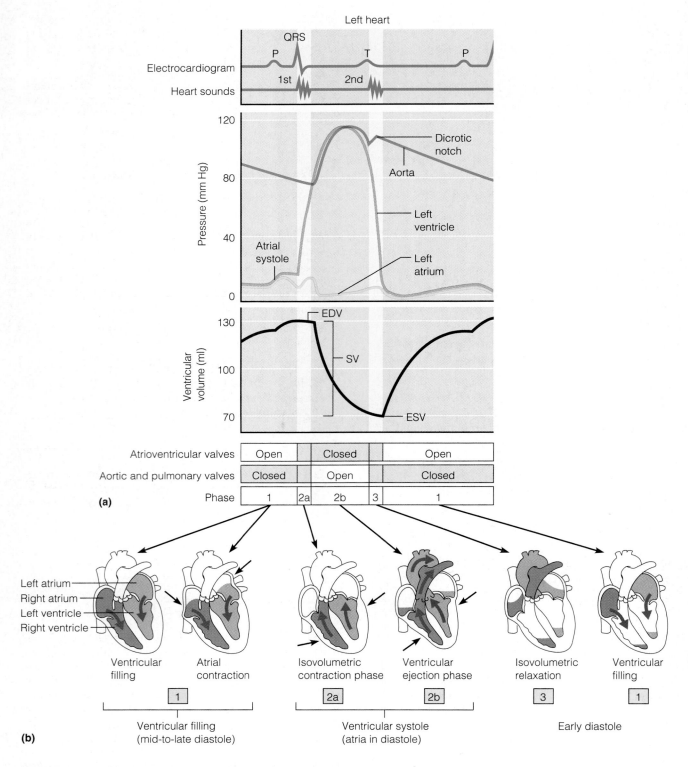

Left heart

Figure 34.1 Summary of events occurring in the heart during the cardiac cycle. (a) Events in the left side of the heart. An ECG tracing is superimposed on the graph (top) so that pressure and volume changes can be related to electrical events occurring at any point. Time occurrence of heart sounds is also indicated. **(b)** Events of phases 1 through 3 of the cardiac cycle are depicted in diagrammatic views of the heart.

heart, the semilunar valves are forced open; and the blood in the ventricular chambers is expelled through the valves. During this phase, the aortic pressure reaches approximately 120 mm Hg. During ventricular systole, the atria relax and their chambers fill with blood, which results in gradually increasing atrial pressure. At the end of ventricular systole, the ventricles relax; the semilunar valves snap shut, preventing backflow, and momentarily, the ventricles are closed chambers. When the aortic semilunar valve snaps shut, a momentary increase in the aortic pressure results from the elastic recoil of the aorta after valve closure. This event results in the pressure fluctuation called the *dicrotic notch* (see Figure 34.1a). As the ventricles relax, the pressure within them begins to drop. When intraventricular pressure is again less than atrial pressure, the AV valves are forced open, and the ventricles again begin to fill with blood. Atrial and aortic pressures decrease, and the ventricles rapidly refill, completing the cycle.

The average heart beats approximately 72 beats per minute, and so the length of the cardiac cycle is about 0.8 second. Of this time period, atrial contraction occupies the first 0.1 second, which is followed by atrial relaxation and ventricular contraction for the next 0.3 second. The remaining 0.4 second is the quiescent, or ventricular relaxation, period. When the heart beats at a more rapid pace than normal, this last period decreases.

Notice that two different types of phenomena control the movement of blood through the heart: the alternate contraction and relaxation of the myocardium, and the opening and closing of valves (which is entirely dependent on the pressure changes within the heart chambers).

Study Figure 34.1 carefully to make sure you understand what has been discussed before continuing with the next portion of the exercise.

Heart Sounds

Two distinct sounds can be heard during each cardiac cycle. These heart sounds are commonly described by the monosyllables "lub" and "dup"; and the sequence is designated lub-dup, pause, lub-dup, pause, and so on. The first heart sound (lub) is associated with closure of the AV valves at the beginning of systole. The second heart sound (dup) occurs as the semilunar valves close and corresponds with the end of systole. Figure 34.1a indicates the timing of heart sounds in the cardiac cycle.

Listen to the recording "Interpreting Heart Sounds" so that you may hear both normal and abnormal heart sounds. Abnormal heart sounds are called **murmur**s and often indicate valvular problems. In valves that do not close tightly, closure is followed by a swishing sound due to the backflow of blood (regurgitation). Distinct sounds, often described as high-pitch "screeching," are associated with the tortuous flow of blood through constricted, or stenosed, valves. ■

A c t i v i t y :
Auscultating Heart Sounds

In the following procedure, you will auscultate your partner's heart sounds with an ordinary stethoscope. A number of more

sophisticated heart-sound amplification systems are on the market, and your instructor may prefer to use one such if it is available. If so, directions for the use of this apparatus will be provided by the instructor.

1. Obtain a stethoscope and some alcohol swabs. Heart sounds are best auscultated (listened to) if the subject's outer clothing is removed, so a male subject is preferable.

2. With an alcohol swab, clean the earpieces of the stethoscope. Allow the alcohol to dry. Notice that the earpieces are angled. For comfort and best auscultation, the earpieces should be angled in a *forward* direction when placed into the ears.

3. Don the stethoscope. Place the diaphragm of the stethoscope on your partner's thorax, just to the sternal side of the left nipple at the fifth intercostal space, and listen carefully for heart sounds. The first sound will be a longer, louder (more booming) sound than the second, which is short and sharp. After listening for a couple of minutes, try to time the pause between the second sound of one heartbeat and the first sound of the subsequent heartbeat.

How long is this interval? _____ sec

How does it compare to the interval between the first and second sounds of a single heartbeat?

_____

_____

4. To differentiate individual valve sounds somewhat more precisely, auscultate the heart sounds over specific thoracic regions. Refer to Figure 34.2 for the positioning of the stethoscope.

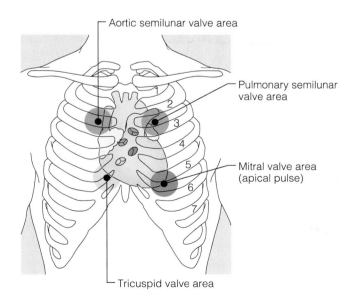

Figure 34.2 Areas of the thorax where valvular sounds can best be detected.

Auscultation of AV Valves

As a rule, the mitral valve closes slightly before the tricuspid valve. You can hear the mitral valve more clearly if you place the stethoscope over the apex of the heart, which is at the fifth intercostal space, approximately in line with the middle region of the left clavicle. Listen to the heart sounds at this region; then move the stethoscope medially to the right margin of the sternum to auscultate the tricuspid valve. Can you detect the slight lag between the closure of the mitral and tricuspid valves?

There are normal variations in the site for "best" auscultation of the tricuspid valve. These range from the site depicted in Figure 34.2 (right sternal margin over fifth intercostal space) to over the sternal body in the same plane, to the left sternal margin over the fifth intercostal space. If you have difficulty hearing closure of the tricuspid valve, try one of these other locations.

Auscultation of Semilunar Valves

Again there is a slight dissynchrony of valve closure; the aortic semilunar valve normally snaps shut just ahead of the pulmonary semilunar valve. If the subject inhales deeply but gently, filling of the right ventricle will be delayed slightly (due to the compression of the thoracic blood vessels by the increased intrapulmonary pressure); and the two sounds can be heard more distinctly. Position the stethoscope over the second intercostal space, just to the *right* of the sternum. The aortic valve is best heard at this position. As you listen, have your partner take a deep breath. Then move the stethoscope to the *left* side of the sternum in the same line; and auscultate the pulmonary valve. Listen carefully; try to hear the "split" between the closure of these two valves in the second heart sound.

Although at first it may seem a bit odd that the pulmonary valve issuing from the *right* heart is heard most clearly to the *left* of the sternum and the aortic valve of the left heart is best heard at the right sternal border, this is easily explained by reviewing heart anatomy. Because the heart is twisted, with the right ventricle forming most of the anterior ventricular surface, the pulmonary trunk actually crosses to the right as it issues from the right ventricle. Similarly, the aorta issues from the left ventricle at the left side of the pulmonary trunk before arching up and over that vessel. ■

The Pulse

The term **pulse** refers to the alternating surges of pressure (expansion and then recoil) in an artery that occur with each contraction and relaxation of the left ventricle. Normally the pulse rate (pressure surges per minute) equals the heart rate (beats per minute), and the pulse averages 70 to 76 beats per minute in the resting state.

Parameters other than pulse rate are also useful clinically. You may also assess the regularity (or rhythmicity) of the pulse, and its amplitude and/or tension—does the blood vessel expand and recoil (sometimes visibly) with the pres-

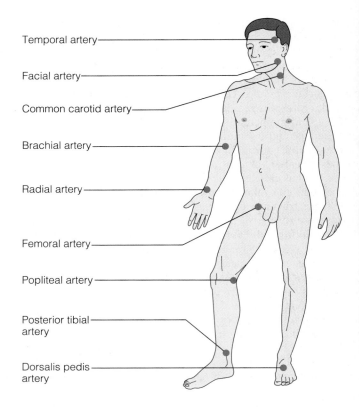

Figure 34.3 Body sites where the pulse is most easily palpated.

sure waves? Can you feel it strongly, or is it difficult to detect? Is it regular like the ticking of a clock, or does it seem to skip beats?

A c t i v i t y :
Palpating Superficial Pulse Points

The pulse may be felt easily on any superficial artery when the artery is compressed over a bone or firm tissue. Palpate the following pulse or pressure points on your partner by placing the fingertips of the first two or three fingers of one hand over the artery. It helps to compress the artery firmly as you begin your palpation and then immediately ease up on the pressure slightly. In each case, notice the regularity of the pulse, and assess the degree of tension or amplitude. Figure 34.3 illustrates the superficial pulse points to be palpated.

Common carotid artery: At the side of the neck

Temporal artery: Anterior to the ear, in the temple region

Facial artery: Clench the teeth, and palpate the pulse just anterior to the masseter muscle on the mandible (in line with the corner of the mouth)

Brachial artery: In the antecubital fossa, at the point where it bifurcates into the radial and ulnar arteries

Radial artery: At the lateral aspect of the wrist, above the thumb

Femoral artery: In the groin

Popliteal artery: At the back of the knee

Posterior tibial artery: Just above the medial malleolus

Dorsalis pedis artery: On the dorsum of the foot

Which pulse point had the greatest amplitude?

_____

Which the least? _____

Can you offer any explanation for this? _____

_____

_____

_____

Because of its easy accessibility, the pulse is most often taken on the radial artery. With your partner sitting quietly, practice counting the radial pulse for 1 minute. Make three counts and average the results.

count 1 _____   count 2 _____

count 3 _____   average _____ ∎

Apical-Radial Pulse

The correlation between the apical and radial pulse rates can be determined by simultaneously counting them. The **apical pulse** (actually the counting of heartbeats) may be slightly faster than the radial because of a slight lag in time as the blood rushes from the heart into the large arteries where it can be palpated. However, any *large* difference between the values observed, referred to as a **pulse deficit,** may indicate cardiac impairment (a weakened heart that is unable to pump blood into the arterial tree to a normal extent—low cardiac output) or abnormal heart rhythms. In the case of atrial fibrillation or ectopic heartbeats, for instance, the second beat may follow the first so quickly that no second pulse is felt even though the apical pulse can still be heard (auscultated). Apical pulse counts are routinely ordered for those with cardiac decompensation.

Activity:
Taking an Apical Pulse

With the subject sitting quietly, one student, using a stethoscope, should determine the apical pulse rate while another simultaneously counts the radial pulse rate. The stethoscope should be positioned over the fifth left intercostal space. The person taking the radial pulse should determine the starting point for the count and give the stop-count signal exactly 1 minute later. Record your values below.

apical count _____ beats/min

radial count _____ pulses/min

pulse deficit _____ /min ∎

Blood Pressure Determinations

Blood pressure is defined as the pressure the blood exerts against any unit area of the blood vessel walls, and it is generally measured in the arteries. Because the heart alternately contracts and relaxes, the resulting rhythmic flow of blood into the arteries causes the blood pressure to rise and fall during each beat. Thus you must take two blood pressure readings: the **systolic pressure,** which is the pressure in the arteries at the peak of ventricular ejection, and the **diastolic pressure,** which reflects the pressure during ventricular relaxation. Blood pressures are reported in millimeters of mercury (mm Hg), with the systolic pressure appearing first; 120/80 translates to 120 over 80, or a systolic pressure of 120 mm Hg and a diastolic pressure of 80 mm Hg. Normal blood pressure varies considerably from one person to another.

In this procedure, you will measure arterial and venous pressures by indirect means and under various conditions. You will investigate and demonstrate factors affecting blood pressure, the rapidity of blood pressure changes, and the large differences between arterial and venous pressures.

Activity:
Using a Sphygmomanometer to Measure Arterial Blood Pressure Indirectly

The **sphygmomanometer,** commonly called a *blood pressure cuff,* is an instrument used to obtain blood pressure readings by the auscultatory method (Figure 34.4). It consists of an inflatable cuff with an attached pressure gauge. The cuff is placed around the arm and inflated to a pressure higher than systolic pressure to occlude circulation to the forearm. As cuff pressure is gradually released, the examiner listens with a stethoscope for characteristic sounds called the **sounds of Korotkoff,** which indicate the resumption of blood flow into the forearm. The pressure at which the first soft tapping sounds can be detected is recorded as the systolic pressure. As the pressure is reduced further, blood flow becomes more turbulent, and the sounds become louder. As the pressure is reduced still further, below the diastolic pressure, the artery is no longer compressed; and blood flows freely and without turbulence. At this point, the sounds of Korotkoff can no longer be detected. The pressure at which the sounds disappear is recorded as the diastolic pressure.

1. Work in pairs to obtain radial artery blood pressure readings. Obtain a stethoscope, alcohol swabs, and a sphygmomanometer. Clean the earpieces of the stethoscope with the alcohol swabs, and check the cuff for the presence of trapped air by compressing it against the laboratory table. (A partially inflated cuff will produce erroneous measurements.)

2. The subject should sit in a comfortable position with one arm resting on the laboratory table (approximately at heart level if possible). Wrap the cuff around the subject's arm, just above the elbow, with the inflatable area on the medial arm surface. The cuff may be marked with an arrow; if so, the arrow should be positioned over the brachial artery (Figure

(a) (b) (c) (d)

Figure 34.4 Procedure for measurement of blood pressure.
(a) The course of the brachial artery of the arm. Assume a blood pressure of 120/70. **(b)** The blood pressure cuff is wrapped snugly around the arm just above the elbow and inflated until blood flow into the forearm is stopped and a brachial pulse cannot be felt or heard. **(c)** The pressure in the cuff is gradually reduced while the examiner listens (auscultates) carefully for sounds (of Korotkoff) in the brachial artery with a stethoscope. The pressure read as the first soft tapping sounds are heard (the first point at which a small amount of blood is spurting through the constricted artery) is recorded as the systolic pressure. **(d)** As the pressure is reduced still further, the sounds become louder and more distinct, but when the artery is no longer restricted and blood flows freely, the sounds can no longer be heard. The pressure at which the sounds disappear is routinely recorded as the diastolic pressure.

34.4). Secure the cuff by tucking the distal end under the wrapped portion or by bringing the Velcro areas together.

3. Palpate the brachial pulse, and lightly mark its position with a felt pen. Don the stethoscope, and place its diaphragm over the pulse point.

⚠️ *The cuff should not be kept inflated for more than 1 minute.* If you have any trouble obtaining a reading within this time, deflate the cuff, wait 1 or 2 minutes, and try again. (A prolonged interference with BP homeostasis can lead to fainting.)

4. Inflate the cuff to approximately 160 mm Hg pressure, and slowly release the pressure valve. Watch the pressure gauge as you listen carefully for the first soft thudding sounds of the blood spurting through the partially occluded artery. Mentally note this pressure (systolic pressure), and continue to release the cuff pressure. You will notice first an increase, then a muffling, of the sound. Note, as the diastolic pressure,

the pressure at which the sound becomes muffled or disappears. Controversy exists over which of the two points should be recorded as the diastolic pressure; so in some cases you may see readings such as 120/80/78, which indicates the systolic pressure followed by the *first* and *second diastolic end points*. The first diastolic end point is the pressure at which the sound muffles; the second is the pressure at which the sound disappears. It makes little difference here which of the two diastolic pressures is recorded, but be consistent. Make two blood pressure determinations, and record your results below.

First trial:

systolic pressure _____

diastolic pressure _____

Second trial:

systolic pressure _____

diastolic pressure _____

5. Compute the **pulse pressure** for each trial. The pulse pressure is the difference between the systolic and diastolic pressures, and indicates the amount of blood forced from the heart during systole, or the actual "working" pressure.

Pulse pressure:

first trial _____   second trial _____

6. Compute the **mean arterial pressure (MAP)** for each trial using the following equation:

$$MAP = \text{diastolic pressure} + \frac{\text{pulse pressure}}{3}$$

first trial _____   second trial _____ ■

Activity:
Estimating Venous Pressure

It is not possible to measure venous pressure with the sphygmomanometer. The methods available for measuring it produce estimates at best, because venous pressures are so much lower than arterial pressures. The difference in pressure becomes obvious when these vessels are cut. If a vein is cut, the blood flows evenly from the cut. A lacerated artery produces rapid spurts of blood.

1. Ask your lab partner to stand with his or her right side toward the blackboard, arms hanging freely at the sides. On the board, mark the approximate level of the right atrium. (This will be just slightly higher than the point at which you auscultated the apical pulse.)

2. Observe the superficial veins on the dorsum of the right hand as the subject alternately raises and lowers it. Notice the collapsing and filling of the veins as internal pressures change. Have the subject repeat this action until you can determine the point at which the veins have just collapsed. Then measure, in millimeters, the distance in the vertical plane from this point to the level of the right atrium (previously marked). Record this value. Distance of right arm from right atrium at point of venous collapse:

_____ mm

3. Compute the venous pressure (P_V), in millimeters of mercury, with the following formula:

$$P_V = \frac{1.056 \ (\text{specific gravity of blood}) \times \text{mm (measured)}}{13.6 \ (\text{specific gravity of Hg})}$$

Venous pressure computed: _____ mm Hg

Normal venous pressure varies from approximately 30 to 90 mm Hg. That of the hand ranges between 30 and 40 mm Hg. How does your computed value compare?

_____

4. Because venous walls are so thin, pressure within them is readily affected by external factors such as muscle activity, deep pressure, and pressure changes occurring in the thorax during breathing. The Valsalva maneuver, which increases intrathoracic pressure, is used to demonstrate the effect of thoracic pressure changes on venous pressure.

To perform this maneuver take a deep breath, and then mimic the motions of exhaling forcibly, but without actually exhaling. In reaction to this, the glottis will close; and intrathoracic pressure will increase. (Most of us have performed this maneuver unknowingly in acts of defecation in which there is "straining at stool.") Measure and record below the distance of the right arm from right atrium at the point of venous collapse while the subject is performing the Valsalva maneuver. Compute the venous pressure and record it below.

_____ mm   Venous pressure: _____ mm Hg

How does this value compare with the venous pressure measurement computed for the relaxed state?

_____

Explain: _____

_____

_____ ■

Activity:
Observing the Effect of Various Factors on Blood Pressure and Heart Rate

Arterial blood pressure is directly proportional to cardiac output (amount of blood pumped out of the left ventricle per unit time) and peripheral resistance to blood flow, that is

$$BP = CO \times PR$$

Peripheral resistance is increased by constriction of blood vessels (most importantly the arterioles), by an increase in blood viscosity or volume, and by a loss of elasticity of the arteries (seen in arteriosclerosis). Any factor that increases either the cardiac output or the peripheral resistance causes an almost immediate reflex rise in blood pressure. A close examination of these relationships reveals that many factors—age, weight, time of day, exercise, body position, emotional state, and various drugs, for example—alter blood pressure. The influence of a few of these factors is investigated here.

The following tests are done most efficiently if one student acts as the subject; two are examiners (one taking the radial pulse and the other auscultating the brachial blood pressure); and a fourth student collects and records data. The sphygmomanometer cuff should be left on the subject's arm throughout the experiments (in a deflated state, of course) so that, at the proper times, the blood pressure can be taken quickly. In each case, take the measurements at least twice.

Chart 1   Posture

	Trial 1		Trial 2	
	BP	Pulse	BP	Pulse
Sitting quietly	_____	_____	_____	_____
Reclining (after 2 to 3 min)	_____	_____	_____	_____
Immediately on standing from the reclining position ("at attention" stance)	_____	_____	_____	_____
After standing for 3 min	_____	_____	_____	_____

Posture

To monitor circulatory adjustments to changes in position, take blood pressure and pulse measurements under the conditions noted in Chart 1. Also record your results on that chart.

Exercise

Blood pressure and pulse changes during and after exercise provide a good yardstick for measuring one's overall cardiovascular fitness. Although there are more sophisticated and more accurate tests that evaluate fitness according to a specific point system, the *Harvard step test* described here is a quick way to compare the relative fitness level of a group of people.

You will be working in groups of four, duties assigned as indicated above, except that student 4, in addition to recording the data, will act as the timer and call the cadence.

⚠️ Any student with a known heart problem should refuse to participate as the subject.

All four students may participate as the subject in turn, if desired, but the bench stepping is to be performed *at least twice* in each group—once with a well-conditioned person acting as the subject, and once with a poorly conditioned subject.

Bench stepping is the following series of movements repeated sequentially:

1. Place one foot on the step.

2. Step up with the other foot so that both feet are on the platform. Straighten the legs and the back.

3. Step down with one foot.

4. Bring the other foot down.

The pace for the stepping will be set by the "timer" (student 4), who will repeat "Up-2-3-4, up-2-3-4" at such a pace that each "up-2-3-4" sequence takes 2 sec (i.e., 30 cycles/min).

1. Student 4 should obtain the step (20-in. height for male subject, or 16 in. for a female subject) while baseline measurements are being obtained on the subject.

2. Once the baseline pulse and blood pressure measurements have been recorded on Chart 2, the subject is to stand quietly at attention for 2 min to allow his or her blood pressure to stabilize before beginning to step.

3. The subject is to perform the bench stepping for as long as possible, up to a maximum of 5 min, according to the cadence called by the timer. The subject is to be watched for and warned against crouching (posture must remain erect). If he or she is unable to keep the pace for a span of 15 sec, the test is to be terminated.

4. When the subject is stopped by the pacer, stops voluntarily because he or she is unable to continue, or has completed 5 min of bench stepping, he or she is to sit down. The duration of exercise (in seconds) is to be recorded, and the blood pressure and pulse are to be measured immediately and thereafter at 1-min intervals for 3 min post-exercise.

5. The subject's *index of physical fitness* is to be calculated using the formula given below:

$$\text{Index} = \frac{\text{duration of exercise in seconds} \times 100}{2 \times \text{sum of the 3 pulse counts in recovery}}$$

Scores are interpreted according to the following scale:

below 55	poor physical condition
55 to 62	low average
63 to 71	average
72 to 79	high average
80 to 89	good
90 and over	excellent

6. Record the test values on Chart 2, and repeat the testing and recording procedure with the second subject.

When did you notice a greater elevation of blood pressure and pulse?

Chart 2	Exercise									
Harvard step test for 5 min at 30/min	**Baseline**		**Interval Following Test**							
			Immediately		**1 min**		**2 min**		**3 min**	
	BP	P	BP	P	BP	P	BP	P	BP	P
Well-conditioned individual	___	___	___	___	___	___	___	___	___	___
Poorly conditioned individual	___	___	___	___	___	___	___	___	___	___

Chart 3	Nicotine						
Baseline		**After smoking for 2 min**		**After smoking for 3 min**		**After smoking for 4 min**	
BP	P	BP	P	BP	P	BP	P
___	___	___	___	___	___	___	___

Explain: _____

_____

Was there a sizable difference between the after-exercise values for well-conditioned and poorly conditioned individuals?

_____ Explain: _____

_____

Did the diastolic pressure also increase? _____

Explain: _____

_____

Nicotine (Optional)

To investigate the effects of nicotine on blood pressure and pulse, take a baseline blood pressure and pulse of a smoker who is in a standing position. Ask the subject to light up and smoke in the usual manner. After 2 min, take blood pressure

and pulse readings. Repeat the measurements at 1-min intervals for the next 2 min, and record your values on Chart 3.

How do the effects of nicotine bring about the changes noted? (Hint: nicotine has vasoconstrictor effects initially.)

_____

_____

_____

A Noxious Sensory Stimulus (Cold)

There is little question that blood pressure is affected by emotions and pain. This lability of blood pressure will be investigated through use of the **cold pressor test,** in which one hand will be immersed in unpleasantly (even painfully) cold water.

Measure the blood pressure and pulse of the subject as he or she sits quietly. Obtain a basin and thermometer, fill the basin with ice cubes, and add water. When the temperature of the ice bath has reached 5°C, immerse the subject's other hand (the noncuffed limb) in the ice water. With the hand still

Chart 4		A Noxious Sensory Stimulus (Cold)					
Baseline		**1 min**		**2 min**		**3 min**	
BP	**P**	**BP**	**P**	**BP**	**P**	**BP**	**P**
_____	_____	_____	_____	_____	_____	_____	_____

immersed, take blood pressure and pulse readings at 1-min intervals for a period of 3 min, and record the values on Chart 4.

How did the blood pressure change during cold exposure?

_____

_____

Was there any change in pulse? _____

Subtract the respective baseline readings of systolic and diastolic blood pressure from the highest single reading of systolic and diastolic pressure obtained during cold immersion. (For example, if the highest experimental reading is 140/88 and the baseline reading is 120/70, then the differences in blood pressure would be systolic pressure, 20 mm Hg, and diastolic pressure, 18 mm Hg.) These differences are called the index of response. According to their index of response, subjects can be classified as follows:

Hyporeactors (stable blood pressure): Exhibit a rise of diastolic and/or systolic pressure ranging from 0 to 22 mm Hg; or a drop in pressures

Hyperreactors (labile blood pressure): Exhibit a rise of 23 mm Hg or more in the diastolic and/or systolic blood pressure

Is the subject tested a hypo- or hyperreactor?

_____ ■

Skin Color as an Indicator of Local Circulatory Dynamics

Skin color reveals with surprising accuracy the state of the local circulation, and allows inferences concerning the larger blood vessels and the circulation as a whole. The experiments on local circulation outlined below illustrate a number of factors that affect blood flow to the tissues.

Clinical expertise often depends upon good observation skills, accurate recording of data, and logical interpretation of the findings. A single example will be given to demonstrate this statement: A massive hemorrhage may be internal and hidden (thus, not obvious), but will still threaten the blood delivery to the brain and other vital organs. One of the earliest compensatory responses of the body to such a threat is constriction of cutaneous blood vessels, which reduces blood flow to the skin and diverts it into the circulatory mainstream to serve other, more vital tissues. As a result, the skin of the face and particularly of the extremities becomes pale, cold, and eventually moist with perspiration. Therefore, pale, cold, clammy skin should immediately lead the careful diagnostician to suspect that the circulation is dangerously inefficient. Other conditions, such as local arterial obstruction and venous congestion, as well as certain pathologies of the heart and lungs, also alter skin texture, color, and circulation in characteristic ways.

Activity:
Examining the Effect of Local Chemical and Physical Factors on Skin Color

The local blood supply to the skin (indeed, to any tissue) is influenced by (1) local metabolites, (2) oxygen supply, (3) local temperature, (4) autonomic nervous system impulses, (5) local vascular reflexes, (6) certain hormones, and (7) substances released by injured tissues. A number of these factors are examined in the simple experiments that follow. Each experiment should be conducted by students in groups of three or four. One student will act as the subject; the others will conduct the tests and make and record observations.

Vasodilation and Flushing of the Skin Due to Local Metabolites

1. Obtain a blood pressure cuff (sphygmomanometer) and stethoscope. You will also need a watch with a second hand.

2. The subject should bare both arms by rolling up the sleeves as high as possible and then lay the forearms side by side on the bench top.

3. Observe the general color of the subject's forearm skin, and the normal contour and size of the veins. Notice whether skin color is bilaterally similar. Record your observations:

_____

_____

_____

4.   Apply the blood pressure cuff to one arm, and inflate it to 250 mm Hg. Keep it inflated for 1 min. During this period, repeat the observations made above and record the results:

_____

_____

5.   Release the pressure in the cuff (leaving the deflated cuff in position), and again record the forearm skin color and the condition of the forearm veins. Make this observation immediately after deflation and then again 30 sec later.

Immediately after deflation _____

_____

30 sec after deflation _____

_____

The above observations constitute your baseline information. Now conduct the following tests.

6.   Instruct the subject to raise the cuffed arm above his or her head and to clench the fist as tightly as possible. While the hand and forearm muscles are tightly contracted, rapidly inflate the cuff to 240 mm Hg or more. This maneuver partially empties the hand and forearm of blood, and stops most blood flow to the hand and forearm. Once the cuff has been inflated, the subject is to relax the fist and return the forearm to the bench top so that it can be compared to the other forearm.

7.   Leave the cuff inflated for exactly 1 min. During this interval, compare the skin color in the "ischemic" (blood-deprived) hand to that of the "normal" (non-cuffed-limb) hand. Quickly release the pressure immediately after 1 min.

What are the subjective effects* of stopping blood flow to the arm and hand for 1 min?

_____

_____

What are the objective effects (color of skin and condition of veins)?

_____

_____

How long does it take for the subject's ischemic hand to regain its normal color?

_____

*Subjective effects are sensations—such as pain, coldness, warmth, tingling, and weakness—experienced by the subject. They are "symptoms" of a change in function.

Effects of Venous Congestion

1.   Again, but with a different subject, observe and record the appearance of the skin and veins on the forearms resting on the bench top. This time, pay particular attention to the color of the fingers, particularly the distal phalanges, and the nail beds. Record this information:

_____

_____

2.   Wrap the blood pressure cuff around one of the subject's arms, and inflate it to 40 mm Hg. Maintain this pressure for 5 min. Make a record of the subjective and objective findings just before the 5 min are up, and then again immediately after release of the pressure at the end of 5 min.

Subjective (arm cuffed) _____

_____

Objective (arm cuffed) _____

_____

Subjective (pressure released) _____

_____

Objective (pressure released) _____

_____

3.   With still another subject, conduct the following simple experiment: Raise one arm above the head, and let the other hang by the side for 1 min. After 1 min, quickly lay both arms on the bench top, and compare their color.

Color of raised arm _____

Color of dependent arm _____

From this and the two preceding observations, analyze the factors that determine tint of color (pink or blue) and intensity of skin color (deep pink or blue as opposed to light pink or blue). Record your conclusions.

_____

_____

_____

_____

Collateral Blood Flow

In some diseases, blood flow to an organ through one or more arteries may be completely and irreversibly obstructed. Fortunately, in most cases a given body area is supplied both by one main artery and by anastomosing channels connecting the main artery with one or more neighboring blood vessels. Consequently, an organ may remain viable even though its main arterial supply is occluded, as long as the **collateral vessels** are still functional.

The effectiveness of collateral blood flow in preventing ischemia can be easily demonstrated.

1. Check the subject's hands to be sure they are *warm* to the touch. If not, choose another, "warm-handed" subject, or warm the subject's hands in 35°C water for 10 min before beginning.

2. Palpate the subject's radial and ulnar arteries approximately 1 in. above the wrist flexure, and mark their locations with a felt marker.

3. Instruct the subject to supinate one forearm and to hold it in a partially flexed (about a 30° angle) position, with the elbow resting on the bench top.

4. Face the subject and grasp his or her forearm with both of your hands, the thumb and fingers of one hand compressing the marked radial artery and the thumb and fingers of the other hand compressing the ulnar artery. Maintain the pressure for 5 min, noticing the progression of the subject's hand to total ischemia.

5. At the end of 5 min, release the pressure abruptly. Record the subject's sensations, as well as the intensity and duration of the flush in the previously occluded hand. (Use the other hand as a baseline for comparison.)

_____

_____

_____

6. Allow the subject to relax for 5 min; then repeat the maneuver, but this time *compress only the radial artery.* Record your observations.

_____

_____

How do the results of the first test differ from those of the second test with respect to color changes during compres-

sion, and the intensity and duration of reactive hyperemia (redness of the skin)?

_____

_____

_____

_____

7. Once again allow the subject to relax for 5 min. Then repeat the maneuver, *with only the ulnar artery compressed.* Record your observations:

_____

_____

What can you conclude about the relative sizes of, and hand areas served by, the radial and ulnar arteries?

_____

_____

Effect of Mechanical Stimulation of Blood Vessels of the Skin

With moderate pressure, draw the blunt end of your pen across the skin of a subject's forearm. Wait 3 min to observe the effects, and then repeat with firmer pressure.

What changes in skin color do you observe with light-to-moderate pressure?

_____

With heavy pressure? _____

The redness, or *flare,* observed after mechanical stimulation of the skin results from a local inflammatory response promoted by chemical mediators released by injured tissues. These mediators stimulate increased blood flow into the area and leaking of fluid (from the capillaries) into the local tissues. (Note: People differ considerably in skin sensitivity. Those most sensitive will show **dermographism,** a condition in which the direct line of stimulation will swell quite obviously. This excessively swollen area is called a *wheal.*) ■

Frog Cardiovascular Physiology: Wet Lab

Objectives

1. To list the properties of cardiac muscle as automaticity and rhythmicity and define each.

2. To explain the statement "Cardiac muscle has an intrinsic ability to beat."

3. To compare the intrinsic rate of contraction of the "pacemaker" of the frog heart (sinus venosus) to that of the atria and ventricle.

4. To compare the relative length of the refractory period of cardiac muscle with that of skeletal muscle, and explain why it is not possible to tetanize cardiac muscle.

5. To define *extrasystole* and to explain at what point in the cardiac cycle (and on an ECG tracing) an extrasystole can be induced.

6. To describe the effect of the following on heart rate: cold, heat, vagal stimulation, pilocarpine, digitalis, atropine sulfate, epinephrine, and potassium, sodium, and calcium ions.

7. To define *vagal escape* and discuss its value.

8. To define *ectopic pacemaker.*

9. To define *partial* and *total heart block.*

10. To describe how heart block was induced in the laboratory and explain the results, and to explain how heart block might occur in the body.

11. To list the components of the microcirculatory system.

12. To identify an arteriole, venule, and capillaries in a frog's web, and to cite differences between relative size, rate of blood flow, and regulation of blood flow in these vessels.

13. To describe the effect of heat, cold, local irritation, and histamine on the rate of blood flow in the microcirculation, and to explain how these responses help maintain homeostasis.

Materials

- ❑ Frogs*
- ❑ Dissecting pan and instruments
- ❑ Disposable plastic gloves
- ❑ Petri dishes
- ❑ Medicine dropper
- ❑ Millimeter ruler
- ❑ Frog Ringer's solutions (at room temperature, 5°C, and 32°C)
- ❑ Thread, large rubber bands, fine common pins
- ❑ Frog board
- ❑ Cotton balls
- ❑ Freshly prepared solutions (using frog Ringer's solution as the solvent) of the following in dropper bottles:
  - 2.5% pilocarpine
  - 2% digitalis
  - 5% atropine sulfate
  - 1% epinephrine
  - 5% potassium chloride (KCl)
  - 2% calcium chloride ($CaCl_2$)
  - 0.7% sodium chloride (NaCl)
  - 0.01% histamine
  - 0.01 *N* HCl
- ❑ Apparatus A or B:
  - A: Kymograph, kymograph paper (smoking stand, burner and glazing fluid if using a smoke-writing apparatus), ring stand, right-angle clamps, heart lever, signal magnet, electronic stimulator, platinum electrodes
  - B: Physiograph (polygraph), physiograph paper and ink, myograph transducer, transducer cable, transducer stand, stimulator output extension cable, electrodes

*Instructor will double-pith frogs as required for student experimentation.

nvestigations of human cardiovascular physiology are very interesting, but many areas obviously do not lend themselves to experimentation. It would be tantamount to murder to inject a human subject with various drugs to observe their effects on heart activity, or to expose the human heart in order to study the length of its refractory period. However, this type of investigation can be done on frogs or small laboratory animals and provides valuable data, because the physiologic mechanisms in these animals are similar if not identical to those in humans.

In this exercise, you will conduct the cardiac investigations just mentioned and others. In addition, you will observe the microcirculation in a frog's web and subject it to various chemical and thermal agents to demonstrate their influence on local blood flow.

Special Electrical Properties of Cardiac Muscle: Automaticity and Rhythmicity

Cardiac muscle differs from skeletal muscle both functionally and in its fine structure. Skeletal muscle must be electrically stimulated to contract. In contrast, heart muscle can and does depolarize spontaneously in the absence of external stimulation. This property, called **automaticity,** is due to plasma membranes that have reduced permeability to potassium ions, but still allow sodium ions to slowly leak into the cells. This leakage causes the muscle cells to gradually depolarize until the action potential threshold is reached and *fast calcium channels* open, allowing Ca^{2+} entry from the extracellular fluid. Shortly thereafter, contraction occurs.

Also, the spontaneous depolarization-repolarization events occur in a regular and continuous manner in cardiac muscle, a property referred to as **rhythmicity.**

In the following experiment, you will observe these properties of cardiac muscle in an *in vitro* (outside, or removed from the body) situation. Work together in groups of three or four. (The instructor may choose to demonstrate this procedure if time or frogs are at a premium.)

Activity:
Investigating the Automaticity and Rhythmicity of Heart Muscle

1. Obtain a dissecting pan and instruments, disposable plastic gloves, two petri dishes, frog Ringer's solution, and a medicine dropper, and bring them to your laboratory bench.

⚠ 2. Don the gloves, and then request and obtain a doubly pithed frog from your instructor. Quickly open the thoracic cavity and observe the heart rate *in situ* (at the site or within the body).

Record the heart rate: _____ beats/min

3. Dissect out the heart and gastrocnemius muscle of the calf and place the removed organs in separate petri dishes containing frog Ringer's solution. (Note: Just in case you don't remember, the procedure for removing the gastrocnemius muscle is provided on pp. 156–159 in Exercise 16A. The extreme care used in that procedure for the removal of the gastrocnemius muscle need not be exercised here.)

4. Observe the activity of the two organs for a few seconds.

Which is contracting? _____

At what rate? _____ beats/min

Is the contraction rhythmic? _____

5. Sever the sinus venosus from the heart (see Figure 35A.1). The sinus venosus of the frog's heart corresponds to the SA node of the human heart. Does the sinus venosus

continue to beat? _____

If not, lightly touch it with a probe to stimulate it. Record its rate of contraction.

Rate: _____ beats/min

6. Sever the right atrium from the heart; then remove the left atrium. Does each atrium continue to beat?

_____ Rate: _____ beats/min

Does the ventricle continue to beat? _____

Rate: _____ beats/min

7. Continue to fragment the ventricle to determine how small the ventricular fragments must be before the automaticity of ventricular muscle is abolished. Measure these fragments and record their approximate size.

_____ mm × _____ mm × _____ mm

Which portion of the heart exhibited the most marked

automaticity? _____

Which the least? _____

8. Properly dispose of the frog and heart fragments before continuing with the next procedure. ■

Baseline Frog Heart Activity

The heart's effectiveness as a pump is dependent both on intrinsic (within the heart) and extrinsic (external to the heart) controls. In this experiment, you will investigate some of these factors.

The nodal system, in which the "pacemaker" imposes its depolarization rate on the rest of the heart, is one intrinsic factor that influences the heart's pumping action. If its impulses fail to reach the ventricles (as in heart block), the ventricles continue to beat but at their own inherent rate, which is much slower than that usually imposed on them. Although heart contraction does not depend on nerve impulses, its rate can be modified by extrinsic impulses reaching it through the autonomic nerves. Additionally, cardiac activity is modified by various chemicals, hormones, ions, and metabolites. The effects of several of these chemical factors are examined in the next experimental series.

The frog heart has two atria and a single, incompletely divided ventricle (see Figure 35A.1). The pacemaker is located in the sinus venosus, an enlarged region between the venae cavae and the right atrium. The SA node of mammals may have evolved from the sinus venosus.

Activity:
Recording Baseline Frog Heart Activity

To record baseline frog heart activity, work in groups of four—two students handling the equipment setup and two preparing the frog for experimentation. Two sets of instructions are provided for apparatus setup—one for the kymograph (Figure 35A.2), the other for the physiograph (Figure 35A.3). Follow the procedure outlined for the apparatus you will be using.

Kymograph Apparatus Setup

1. Obtain a kymograph, recording paper, ring stand, frog board, two right-angle clamps, signal magnet, stimulator, platinum electrodes, and a heart lever.

2. If you are not using an ink-writing kymograph apparatus, smoke the drum lightly; it is important that the recording be made with minimum pressure against the drum.

3. As in the setup depicted in Figure 35A.2, connect the electrodes to the two output terminals on the stimulator, and attach the signal magnet to the two binding posts on the back of the stimulator. Connect the signal magnet to the ring stand with a clamp so it will record close to the bottom of the drum. Attach the heart lever to the stand above and in line with the signal magnet.

Physiograph Apparatus Setup

1. Obtain a myograph transducer, transducer cable, and stand, and bring them to the recording site.

2. Attach the myograph transducer to the transducer stand as shown in Figure 35A.3.

(a)

(b)

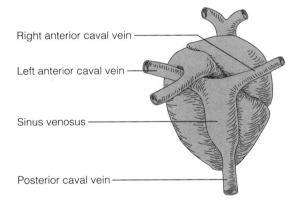

(c)

Figure 35A.1 Anatomy of the frog heart.
(a) Ventral view showing the single truncus arteriosus leaving the undivided ventricle. **(b)** Longitudinal section showing the two atrial and single ventricular chambers. **(c)** Dorsal view showing the sinus venosus (pacemaker).

Figure 35A.2 Kymograph apparatus setup for recording the activity of the frog heart.

3. Then attach the transducer cable to the transducer coupler (input) on the amplifier of the physiograph and to the myograph transducer.

4. Attach the stimulator output extension cable to output on the stimulator panel (red to red, black to black).

Preparation of the Frog

1. Obtain room-temperature frog Ringer's solution, a medicine dropper, dissecting instruments and pan, disposable gloves, fine common pins, and some thread, and bring them to your bench.

2. Don the gloves and obtain a doubly pithed frog from your instructor.

3. Make a longitudinal incision through the abdominal and thoracic walls with scissors, and then cut through the sternum to expose the heart.

4. Grasp the pericardial sac with forceps, and cut it open so that the beating heart can be observed.

Is the sequence an atrial-ventricular one? _____

5. Locate the vagus nerve, which runs down the lateral aspect of the neck and parallels the trachea and carotid artery.

(In good light, it appears to be striated.) Slip an 18-inch length of thread under the vagus nerve so that it can later be lifted away from the surrounding tissues by the thread. Then place a saline-soaked cotton ball over the nerve to keep it moistened until you are ready to stimulate it later in the procedure.

6. Using a medicine dropper, flush the heart with the saline (Ringer's) solution. *From this point on the heart must be kept continually moistened with room-temperature Ringer's solution unless other solutions are being used for the experimentation.*

7. Attach the frog to the frog board.

8. Bend a common pin to a 90° angle, and tie an 18- to 20-inch length of thread to its head. Force the pin through the apex of the heart (but do not penetrate the ventricular chamber) until the apex is well secured in the angle of the pin.

9. Tie the thread to the end of the muscle lever directly over the heart if using the kymograph, or to the hook on the myograph transducer if using the physiograph apparatus. Do not pull the thread too tightly. It should be taut enough to lift the heart apex upward, away from the thorax, but should *not* stretch the heart. Adjust the muscle lever or myograph transducer as necessary.

Figure 35A.3 Physiograph setup for recording the activity of the frog heart.

10. If using a kymograph, adjust the writing tip of the heart lever so that it is directly in line with the stylus of the signal magnet. Leave enough room between the two writing tips to give each adequate working distance. When properly set up, the heart lever should be in a horizontal position.

Making the Baseline Recording

1. Kymograph users should set the drum speed at 7, and briefly turn on the kymograph to ensure that the styluses are recording properly. If using the physiograph, turn the amplifier on and balance the apparatus according to instructions provided by your instructor. Set the paper speed at 0.5 cm/sec. Press the record and paper advance buttons.

2. Set the signal magnet or time marker at 1/sec.

3. Record several normal heartbeats (12 to 15). Be sure you can distinguish atrial and ventricular contractions (see Figure 35A.4). Then adjust the paper speed so that the peaks of ventricular contractions are approximately 2 cm apart. (Peaks indicate systole; troughs indicate diastole.) Pay attention to the relative force of heart contractions while recording.

4. Count the number of ventricular contractions per minute and record:

(a)   One-second time line

(b)   One-second time line

Figure 35A.4 Recording of contractile activity of a frog heart. (a) Normal heartbeat. **(b)** Induction of an extrasystole.

5.   Compute the A-V interval (period from the beginning of atrial contraction to the beginning of ventricular contraction).

_____ sec

How do the two tracings compare in time?

_____

_____

_____

6.   Mark the atrial and ventricular systoles on the record. _Remember to keep the heart moistened with Ringer's solution._ ■

A c t i v i t y :

Investigating the Refractory Period of Cardiac Muscle

In conducting Exercise 16A, you saw that repeated rapid stimuli could cause skeletal muscle to remain in a contracted state. In other words, the muscle could be tetanized. This was possible because of the relatively short refractory period of skeletal muscle. In this experiment, you will investigate the refractory period of cardiac muscle and its response to stimulation. During the procedure, one student should keep the stimulating electrodes in constant contact with the frog heart ventricle.

1.   Set the stimulator to deliver 20-V shocks of 2-ms duration, and begin recording.

2.   Deliver single shocks at the beginning of ventricular contraction, the peak of ventricular contraction, and then later and later in the cardiac cycle.

3.   Observe the recording for **extrasystoles,** which are extra beats that show up riding on the ventricular contraction peak. Also note the **compensatory pause,** which allows the heart to get back on schedule after an extrasystole. (See Figure 35A.4b.)

During which portion of the cardiac cycle was it possible to induce an extrasystole?

_____

_____

4.   Attempt to tetanize the heart by stimulating it at the rate of 20 to 30 impulses per second. What is the result?

_____

_____

Considering the function of the heart, why is it important that heart muscle cannot be tetanized?

_____

_____ ■

A c t i v i t y :

Assessing Physical and Chemical Modifiers of Heart Rate

Now that you have observed normal frog heart activity, you will have an opportunity to investigate the effects of various factors that modify heart activity. In each case, record a few normal heartbeats before introducing the modifying factor. After removing the agent, allow the heart to return to its normal rate before continuing with the testing. On each record, indicate the point of introduction and removal of the modifying agent.

Temperature

1.   Obtain 5°C and 32°C frog Ringer's solutions and medicine droppers.

2.   Increase the drum or paper speed so that heartbeats appear as spikes 4 to 5 mm apart.

3.   Bathe the heart with 5°C saline, and continue to record until the recording indicates a change in cardiac activity.

4.   Stop recording, pipette off the cold Ringer's solution (remove the fluid by sucking it into the barrel of a medicine dropper), and flood the heart with room-temperature saline.

5.   Start recording again to determine the resumption of the normal heart rate. When this has been achieved, flood the heart with 32°C Ringer's solution and again record until a change is noted.

6.   Stop the recording, pipette off the warm saline, and bathe the heart with room-temperature Ringer's solution once again.

What change occurred with the cold (5°C) Ringer's solution?

_____

What change occurred with the warm (32°C) Ringer's solution?

_____

7.   Count the heart rate at the two temperatures and record below.

_____ beats/min at 5°C; _____ beats/min at 32°C

Vagus Nerve Stimulation

The vagus nerve carries parasympathetic impulses to the heart, which modify heart activity.

1.   Remove the cotton placed over the vagus nerve. Using the previously tied thread, lift the nerve away from the tissues and place the nerve on the stimulating electrodes.

2. Using a duration of 0.5 msec at a voltage of 1 mV, stimulate the nerve at a rate of 50/sec. Continue stimulation until the heart stops momentarily and then begins to beat again (vagal escape). If no effect is observed, increase stimulus intensity and try again. If no effect is observed after a substantial increase in stimulus voltage, reexamine your "vagus nerve" to make sure that it is not simply strands of connective tissue.

3. Discontinue stimulation after you observe vagal escape, and flush the heart with saline until the normal heart rate resumes. What is the effect of vagal stimulation on heart rate?

_____

The phenomenon of vagal escape demonstrates that many factors are involved in heart regulation and that any deleterious factor (in this case, excessive vagal stimulation) will be overcome, if possible, by other physiologic mechanisms such as activation of the sympathetic division of the autonomic nervous system (ANS).

Pilocarpine

Flood the heart with a 2.5% solution of pilocarpine. Record until a change in the pattern of the ECG is noticed. Pipette off the excess pilocarpine solution, and proceed immediately to the next test, which uses atropine as the testing solution. What happened when the heart was bathed in the pilocarpine solution?

_____

Pilocarpine simulates the effect of parasympathetic nerve (hence, vagal) stimulation by enhancing acetylcholine release; such drugs are called parasympathomimetic drugs.

Atropine Sulfate

Apply a few drops of atropine sulfate to the frog's heart and observe the recording. If no changes are observed within 2 minutes, apply a few more drops. When you observe a response, pipette off the excess atropine sulfate and flood the heart with Ringer's solution. What happens when the atropine sulfate is added?

_____

Atropine is a drug that blocks the effect of the neurotransmitter acetylcholine, which is liberated by the parasympathetic nerve endings. Do your results accurately reflect this effect of atropine?

_____

Are pilocarpine and atropine agonists or antagonists in their effects on heart activity?

Epinephrine

Flood the frog heart with epinephrine solution, and continue to record until a change in heart activity is noted.

What are the results? _____

_____

Which division of the autonomic nervous system does its effect imitate?

_____

Digitalis

Pipette off the excess epinephrine solution, and rinse the heart with room-temperature Ringer's solution. Continue recording, and when the heart rate returns to baseline values, bathe it in digitalis solution. What is the effect of digitalis on the heart?

_____

_____

Digitalis is a drug commonly prescribed for heart patients with congestive heart failure. It slows heart rate, providing more time for venous return and decreasing the load on the weakened heart. These effects are thought to be due to inhibition of the Na^+-K^+ pump and enhancement of Ca^{2+} entry into the myocardial fibers.

Various Ions

To test the effect of various ions on the heart, apply the designated solution until you observe a change in heart rate or in strength of contraction. Pipette off the solution, flush with Ringer's solution, and allow the heart to resume its normal rate before continuing. *Do not allow the heart to stop.* If the rate should decrease dramatically, flood the heart with room-temperature Ringer's solution.

Effect of Ca^{2+} (use 2% $CaCl_2$) _____

_____

Effect of Na^+ (use 0.7% NaCl) _____

_____

Effect of K^+ (use 5% KCl) _____

_____

Potassium ion concentration is normally higher within cells than in the extracellular fluid. *Hyperkalemia* decreases the resting potential of plasma membranes, thus decreasing the force of heart contraction. In some cases, the conduction rate of the heart is so depressed that **ectopic pacemakers** (pacemakers appearing erratically and at abnormal sites in the heart muscle) appear in the ventricles, and fibrillation may occur. Was there any evidence of premature beats in the recording of potassium ion effects?

_____

Was arrhythmia produced with any of the ions tested?

_____ If so, which? _____

Intrinsic Conduction System Disturbance (Heart Block)

1. Moisten a 10-inch length of thread and make a Stannius ligature (loop the thread around the heart at the junction of the atria and ventricle).

2. Decrease the drum or paper speed to achieve intervals of approximately 2 cm between the ventricular contractions, and record a few normal heartbeats.

3. Tighten the ligature in a stepwise manner while observing the atrial and ventricular contraction curves. As heart block occurs, the atria and ventricle will no longer show a 1:1 contraction ratio. Record a few beats each time you observe a different degree of heart block—a 2:1 atria to ventricle, 3:1, 4:1, and so on. As long as you can continue to count a whole number ratio between the two chamber types, the heart is in **partial heart block.** When you can no longer count a whole number ratio, the heart is in **total, or complete, heart block.**

4. When total heart block occurs, release the ligature to see if the normal A-V rhythm is reestablished. What is the result?

_____

_____

5. Attach properly labeled recordings (or copies of the recordings) made during this procedure to the last page of this exercise for future reference.

6. Dispose of the frog remains and gloves in appropriate containers, and dismantle the experimental apparatus before continuing. ■

The Microcirculation and Local Blood Flow

The thin web of a frog's foot provides an excellent opportunity to observe the flow of blood to, from, and within the capillary beds, where the real business of the circulatory system occurs. The collection of vessels involved in the exchange mechanism is referred to as the **microcirculation;** it consists of arterioles, venules, capillaries, and vascular shunts called *metarteriole-thoroughfare channels* (Figure 35A.5).

The total cross-sectional area of the capillaries in the body is much greater than that of the veins and arteries combined. Thus, the flow through the capillary beds is quite slow.

(a) Sphincters open

(b) Sphincters closed

Figure 35A.5 Anatomy of a capillary bed. The composite metarteriole—thoroughfare channels act as shunts to bypass the true capillaries when precapillary sphincters controlling blood entry into the true capillaries are constricted.

Capillary flow is also intermittent, because if all capillary beds were filled with blood at the same time, there would be no blood at all in the large vessels. The flow of blood into the capillary beds is regulated by the activity of muscular *terminal arterioles,* which feed the beds, and by *precapillary sphincters* at entrances to the true capillaries. The amount of blood flowing into the true capillaries of the bed is regulated most importantly by local chemical controls (local concentrations of carbon dioxide, histamine, pH). Thus a capillary bed may be flooded with blood or almost entirely bypassed depending on what is happening within the body or in a particular body region at any one time. You will investigate some of the local controls in the next group of experiments.

Activity:
Investigating the Effect of Various Factors on the Microcirculation

1. Obtain a frog board (with a hole at one end), dissecting pins, disposable gloves, frog Ringer's solution, 0.01 *N* HCl,

0.01% histamine solution, 1% epinephrine solution, a large rubber band, and some paper towels.

2.   Put on the gloves and obtain a frog (alive and hopping, *not* pithed). Moisten several paper towels with room-temperature Ringer's solution, and wrap the frog's body securely with them. One hind leg should be left unsecured and extending beyond the paper cocoon.

3.   Attach the frog to the frog board (or other supporting structure) with a large rubber band and then carefully spread (but do not stretch) the web of the exposed hindfoot over the hole in the support. Have your partners hold the edges of the web firmly for viewing. Alternatively, secure the toes to the board with dissecting pins.

4.   Obtain a compound microscope, and observe the web under low power to find a capillary bed. Focus on the vessels in high power. Keep the web moistened with Ringer's solution as you work. If the circulation seems to stop during your observations, massage the hind leg of the frog gently to restore blood flow.

5.   Observe the red blood cells of the frog. Notice that, unlike human RBCs, they are nucleated. Watch their movement through the smallest vessels—the capillaries. Do they move in single file or do they flow through two or three cells abreast?

_____

Are they flexible? _____ Explain. _____

Can you see any white blood cells in the capillaries?

_____ If so, which types? _____

6.   Notice the relative speed of blood flow through the blood vessels. Differentiate between the arterioles, which feed the capillary bed, and the venules, which drain it. This may be tricky, because images are reversed in the microscope. Thus, the vessel that appears to feed into the capillary bed will actually be draining it. You can distinguish between the vessels, however, if you consider that the flow is more pulsating and turbulent in the arterioles and smoother and steadier in the venules. How does the *rate* of flow in the arterioles compare with that in the venules?

_____

In the capillaries? _____

What is the relative difference in the diameter of the arterioles and capillaries?

_____

Temperature

To investigate the effect of temperature on blood flow, flood the web with cold saline two or three times to chill the entire area. Is a change in vessel diameter noticeable?

_____ Which vessels are affected? _____

How? _____

Blot the web gently with a paper towel, and then bathe the web with warm saline. Record your observations.

_____

Inflammation

Pipette 0.01 *N* HCl onto the frog's web. Hydrochloric acid will act as an irritant and cause a localized inflammatory response. Is there an increase or decrease in the blood flow into the capillary bed following the application of HCl?

_____

What purpose do these local changes serve during a localized inflammatory response?

_____

_____

Flush the web with room-temperature Ringer's solution and blot.

Histamine

Histamine, which is released in large amounts during allergic responses, causes extensive vasodilation. Investigate this effect by adding a few drops of histamine solution to the frog web. What happens?

_____

_____

How does this response compare to that produced by HCl?

_____

Blot the web and flood with warm saline as before. Now add a few drops of 1% epinephrine solution, and observe the web. What are epinephrine's effects on the blood vessels?

_____

Epinephrine is used clinically to reverse the vasodilation seen in severe allergic attacks (such as asthma) which are mediated by histamine and other vasoactive molecules.

   Return the dropper bottles to the supply area and the frog to the terrarium. Properly clean your work area before leaving the lab. ∎

Frog Cardiovascular Physiology: Computer Simulation

Objectives

1. To list the properties of cardiac muscle as automaticity and rhythmicity and define each.

2. To explain the statement "Cardiac muscle has an intrinsic ability to beat."

3. To compare the relative length of the refractory period of cardiac muscle with that of skeletal muscle, and explain why it is not possible to tetanize cardiac muscle.

4. To define *extrasystole* and to explain at what point in the cardiac cycle (and on an ECG tracing) an extrasystole can be induced.

5. To describe the effect of the following on heart rate: cold, heat, vagal stimulation, pilocarpine, digitalis, atropine, epinephrine, and potassium, sodium, and calcium ions.

6. To define *vagal escape* and discuss its value.

7. To define *ectopic pacemaker.*

Materials

Minimum equipment required:

Windows:

❑ 486/66MHz or better strongly recommended

❑ Windows 95/98 or higher recommended

❑ SVGA display (256 colors at 640 x 480)

❑ Minimum 16 MB available RAM

❑ Double-speed CD-ROM drive (quad-speed or higher recommended)

❑ Sound card

❑ Speakers or headphones

❑ Printer

Macintosh:

❑ 68040 or Power Macintosh strongly recommended

❑ System 7.1 or higher

❑ Minimum 16 MB available RAM

❑ 13-inch or larger color monitor (640 x 480 resolution)

❑ Double-speed CD-ROM drive (quad-speed or higher recommended)

❑ Printer

Software:

❑ Benjamin/Cummings PhysioEx CD-ROM—Frog Cardiovascular Physiology module nvestigation of human cardiovascular physiology is very interesting, but many areas obviously do not lend themselves to experimentation. It would be tantamount to murder to inject a human subject with various drugs to observe their effects on heart activity, or to expose the human heart in order to study the length of its refractory period. However, this type of investigation can be done on frogs or computer simulations, and provides valuable data, because the physiologic mechanisms in these animals, or programmed into the computer simulation, are similar if not identical to those in humans.

In this exercise, you will conduct the cardiac investigations just mentioned.

Special Electrical Properties of Cardiac Muscle: Automaticity and Rhythmicity

Cardiac muscle differs from skeletal muscle both functionally and in its fine structure. Skeletal muscle must be electrically stimulated to contract. In contrast, heart muscle can and does depolarize spontaneously in the absence of external stimulation. This property, called **automaticity,** is due to plasma membranes that have reduced permeability to potassium ions, but still allow sodium ions to slowly leak into the cells. This leakage causes the muscle cells to slowly depolarize until the action potential threshold is reached and *fast calcium channels* open, allowing Ca^{2+} entry from the extracellular fluid. Shortly thereafter, contraction occurs.

Additionally, the spontaneous depolarization-repolarization events occur in a regular and continuous manner in cardiac muscle, a property referred to as **rhythmicity.**

In the following experiment, you will observe these properties of cardiac muscle in a computer simulation. Additionally, your instructor may demonstrate this procedure using a real frog.

Getting Started

Begin by making sure you have the computer equipment and software listed in the Materials section on the opposite page.

1.   Quit all applications currently running on your computer.

2.   Insert the PhysioEx CD-ROM into the CD-ROM drive and keep PhysioEx in the drive during the entire time you use the program.

3.   Follow the instructions below for your computer type.

Windows 95/Windows 98
The program will launch automatically after you load the CD-ROM into the drive.
If autorun is disabled on your machine follow these instructions:
1.   Double-click the My Computer icon on the Windows desktop.
2.   Double-click the PhysioEx CD icon.
3.   Look for the icon that looks like a tiny movie projector and double-click it.

Macintosh
Double-click the PhysioEx CD icon that appears in the window on your desktop to launch the program.

Baseline Frog Heart Activity

The heart's effectiveness as a pump is dependent both on intrinsic (within the heart) and extrinsic (external to the heart) controls. In this experiment, you will investigate some of these factors.

The nodal system, in which the "pacemaker" imposes its depolarization rate on the rest of the heart, is one intrinsic factor that influences the heart's pumping action. If its impulses fail to reach the ventricles (as in heart block), the ventricles continue to beat but at their own inherent rate, which is much slower than that usually imposed on them. Although heart contraction does not depend on nerve impulses, its rate can be modified by extrinsic impulses reaching it through the autonomic nerves. Additionally, cardiac activity is modified by various chemicals, hormones, ions, and metabolites. The effects of several of these chemical factors are examined in the next experimental series.

The frog heart has two atria and a single, incompletely divided ventricle. (see Figure 35A.1, p. 351.) The pacemaker is located in the sinus venosus, an enlarged region between the venae cavae and the right atrium. The SA node of mammals may have evolved from the sinus venosus.

Choose Frog Cardiovascular Physiology from the main menu. The opening screen will appear in a few seconds (Figure 35B.1). When the program starts, you will see a tracing of the frog's heart beat on the *oscilloscope display* in the upper right part of the screen. Because the simulation automatically adjusts itself to your computer's speed, you may not see the heart tracing appear in real-time. If you want to increase the speed of the tracing (at the expense of tracing quality), click the **Tools** menu, choose **Modify Display,** and then select **Increase Speed.**

Figure 35B.1 Opening screen of the Electrical Stimulation experiment.

The oscilloscope display shows the ventricular contraction rate in the Heart rate window. The *heart activity window* to the right of the Heart Rate display provides the following messages:

- Heart Rate Normal—displayed when the heart is beating under resting conditions.

- Heart Rate Changing—displayed when the heart rate is increasing or decreasing.

- Heart Rate Stable—displayed when the heart rate is steady, but higher or lower than normal. For example, if you applied a chemical that increased heart rate to a stable but higher-than-normal rate, you would see this message.

The *electrical stimulator* is below the oscilloscope display. In the experiment, clicking **Single Stimulus** delivers a single electrical shock to the frog heart. Clicking **Multiple Stimulus** delivers repeated electrical shocks at the rate indicated in the Stimuli/sec window just below the Multiple Stimulus button. When the **Multiple Stimulus** button is clicked, it changes to a **Stop Stimulus** button that allows you to stop electrical stimulation as desired. Clicking the (+) or (−) buttons next to the Stimuli/sec window adjusts the stimulus rate. The voltage delivered when Single Stimulus or Multiple Stimulus is clicked is displayed in the Voltage window. The simulation automatically adjusts the voltage for the experiment. The post-like apparatus extending upward from the electrical stimulator is the *electrode holder* into which you will drag-and-drop electrodes from the supplies cabinet in the bottom left corner of the screen.

The left side of the screen contains the apparatus that sustains the frog heart. The heart has been lifted away from the body of the frog by a hook passed through the apex of the heart. Although the frog can not be seen because it is in the dissection tray, its heart has not been removed from its circulatory system. A thin string connects the hook in the heart to the force transducer at the top of the support bracket. As the heart contracts, the string exerts tension on the force transducer that converts the contraction into the oscilloscope tracing. The slender white strand extending from the heart toward the right side of the dissection tray is the vagus nerve. In the simulation, room temperature frog Ringer's solution continuously drips onto the heart to keep it moist and responsive so that a regular heart beat is maintained.

The two electrodes you will use during the experiment are located in the supplies cabinet beneath the dissection tray. The Direct Heart Stimulation electrode is used to stimulate the ventricular muscle directly. The Vagus Nerve Stimulation electrode is used to stimulate the vagus nerve. To position either electrode, click and drag the electrode to the two-pronged plug in the electrode holder and then release the mouse button.

Activity:
Recording Baseline Frog Heart Activity

1. Before beginning to stimulate the frog heart experimentally, watch several heart beats. Be sure you can distinguish atrial and ventricular contraction (Figure 35B.2a).

2. Record the number of ventricular contractions per minute displayed in the Heart Rate window under the oscilloscope.

_____ bpm (beats per minute) ∎

(a) One-second time line

(b) One-second time line

Figure 35B.2 Recording of contractile activity of a frog heart. (a) Normal heartbeat. **(b)** Induction of an extrasystole.

Activity:
Investigating the Refractory Period of Cardiac Muscle

In Exercise 16B you saw that repeated rapid stimuli could cause skeletal muscle to remain in a contracted state. In other words, the muscle could be tetanized. This was possible because of the relatively short refractory period of skeletal muscle. In this experiment you will investigate the refractory period of cardiac muscle and its response to stimulation.

1. Click and hold the mouse button on the Direct Heart Stimulation electrode and drag it to the electrode holder.

2. Release the mouse button to lock the electrode in place. The electrode will touch the ventricular muscle tissue.

3. Deliver single shocks by clicking **Single Stimulus** at each of the following times. You may need to practice to acquire the correct technique.

- Near the beginning of ventricular contraction
- At the peak of ventricular contraction
- During the relaxation part of the cycle

Watch for *extrasystoles,* which are extra beats that show up riding on the ventricular contraction peak. Also note the compensatory pause, which allows the heart to get back on schedule after an extrasystole. (Figure 35B.2b).

During which portion of the cardiac cycle was it possible to induce an extrasystole?

_____

4. Attempt to tetanize the heart by clicking **Multiple Stimulus.** Electrical shocks will be delivered to the muscle at a rate of 20 stimuli/sec. What is the result?

_____

Considering the function of the heart, why is it important that the heart muscle cannot be tetanized?

_____

5. Click **Stop Stimulus** to stop the electrical stimulation. ■

Activity:
Examining the Effect of Vagus Nerve Stimulation

The vagus nerve carries parasympathetic impulses to the heart, which modify heart activity.

1. Click the Direct Heart Stimulation electrode to return it to the supplies cabinet.

2. Click and drag the Vagus Nerve Stimulation electrode to the electrode holder.

3. Release the mouse button to lock the electrode in place. The vagus nerve will automatically be draped over the electrode contacts.

4. Adjust the stimulator to 50 stimuli/sec by clicking the (+) or (−) buttons.

5. Click **Multiple Stimulus.** Allow the vagal stimulation to continue until the heart stops momentarily and then begins to beat again (vagal escape) and then click **Stop Stimulus.**

What is the effect of vagal stimulation on heart rate?

_____

_____

_____

The phenomenon of vagal escape demonstrates that many factors are involved in heart regulation and that any deleterious factor (in this case, excessive vagal stimulation) will be overcome, if possible, by other physiological mechanisms such as activation of the sympathetic division of the autonomic nervous system (ANS). ■

Assessing Physical and Chemical Modifiers of Heart Rate

Now that you have observed normal frog heart activity, you will have an opportunity to investigate the effects of various modifying factors on heart activity. After removing the agent in each activity, allow the heart to return to its normal rate before continuing with the testing.

Choose **Modifiers of Heart Rate** from the Experiment menu. The opening screen will appear in a few seconds (Figure 35B.3). The appearance and functionality of the *oscilloscope display* is the same as it was in the Electrical Stimulation experiment. The *solutions shelf* above the oscilloscope display contains the chemicals you'll use to modify heart rate in the experiment. You can choose the temperature of the Ringer's solution dispensed by clicking the appropriate button in the Ringer's dispenser at the left part of the screen. The

doors to the supplies cabinet are closed during this experiment because the electrical stimulator is not used.

When you click **Record Data** in the *data control unit* below the oscilloscope, your data is stored in the computer's memory and is displayed in the data grid at the bottom of the screen; data displayed include the solution used and the resulting heart rate. If you are not satisfied with a trial, you can click **Delete Line.** Click **Clear Table** if you wish to repeat the entire experiment.

Figure 35B.3 Opening screen of the Modifiers of Heart Rate experiment.

Activity:
Assessing the Effect of Temperature

1. Click the **5°C Ringer's** button to bathe the frog heart in cold Ringer's solution. Watch the recording for a change in cardiac activity.

2. When the heart activity window displays the message: Heart Rate Stable, click **Record Data** to retain your data in the data grid.

What change occurred with the cold (5°C) Ringer's solution?

_____

3. Now click the **23°C Ringer's** button to flood the heart with fresh room temperature Ringer's solution.

4. After you see the message: Heart Rate Normal in the heart activity window, click the **32°C Ringer's** button.

5. When the heart activity window displays the message: Heart Rate Stable, click **Record Data** to retain your data.

What change occurred with the warm (32°C) Ringer's solution?

_____

Record the heart rate at the two temperatures below.

_____ bpm at 5°C; _____ bpm at 32°C

What can you say about temperature and heart rate?

_____

_____

6. Click the **23°C Ringer's** button to flush the heart with fresh Ringer's solution. Watch the heart activity window for the message: Heart Rate Normal before beginning the next test. ∎

Activity:
Assessing the Effect of Pilocarpine

1. Click and hold the mouse on the Pilocarpine dropper cap.

2. Drag the dropper cap to a point about an inch above the heart and release the mouse.

3. Pilocarpine solution will be dispensed onto the heart and the dropper cap will automatically return to the Pilocarpine bottle.

4. Watch the heart activity window for the message: Heart Rate Stable, indicating that the heart rate has stabilized under the effects of pilocarpine.

5. After the heart rate stabilizes, record the heart rate in the space provided below, and click **Record Data** to retain your data in the grid.

_____ bpm

What happened when the heart was bathed in the pilocarpine solution?

_____

6. Click the **23°C Ringer's** button to flush the heart with fresh Ringer's solution. Watch the heart activity window for the message: Heart Rate Normal, an indication that the heart is ready for the next test. ∎

Pilocarpine simulates the effect of parasympathetic nerve (hence, vagal) stimulation by enhancing acetylcholine release; such drugs are called parasympathomimetic drugs.

Activity:
Assessing the Effect of Atropine

1. Drag-and-drop the Atropine dropper cap to a point about an inch above the heart.

2. Atropine solution will automatically drip onto the heart and the dropper cap will return to its position in the Atropine bottle.

3. Watch the heart activity window for the message: Heart Rate Stable.

4. After the heart rate stabilizes, record the heart rate in the space below, and click **Record Data** to retain your data in the grid.

_____ bpm

What is the effect of atropine on the heart?

_____

Atropine is a drug that blocks the effect of the neurotransmitter acetylcholine, liberated by the parasympathetic nerve endings. Do your results accurately reflect this effect of atropine?

_____

Are pilocarpine and atropine agonists or antagonists in their effects on heart activity?

_____

5. Click the **23°C Ringer's** button to flush the heart with fresh Ringer's solution. Watch the heart activity window for the message: Heart Rate Normal, before beginning the next test. ■

Activity:
Assessing the Effect of Epinephrine

1. Drag-and-drop the Epinephrine dropper cap to a point about an inch above the heart.

2. Epinephrine solution will be dispensed onto the heart and the dropper cap will return to the Epinephrine bottle.

3. Watch the heart activity window for the message: Heart Rate Stable.

4. After the heart rate stabilizes, record the heart rate in the space provided below, and click **Record Data** to retain your data in the grid.

_____ bpm

What happened when the heart was bathed in the epinephrine solution?

_____

Which division of the autonomic nervous system does its effect imitate?

_____

5. Click the **23°C Ringer's** button to flush the heart with fresh Ringer's solution. Watch the heart activity window for the message: Heart Rate Normal, meaning that the heart is ready for the next test. ■

A c t i v i t y :
Assessing the Effect of Digitalis

1. Drag-and-drop the Digitalis dropper cap to a point about an inch above the heart.

2. Digitalis solution will automatically drip onto the heart and then the dropper will return to the Digitalis bottle.

3. Watch the heart activity window to the right of the Heart Rate window for the message: Heart Rate Stable.

4. After the heart rate stabilizes, record the heart rate in the space provided below, and click **Record Data** to retain your data in the grid.

_____ bpm

What is the effect of digitalis on the heart?

_____

5. Click the **23°C Ringer's** button to flush the heart with fresh Ringer's solution. Watch the heart activity window for the message: Heart Rate Normal, then proceed to the next test. ■

Digitalis is a drug commonly prescribed for heart patients with congestive heart failure. It slows heart rate, providing more time for venous return and decreasing the workload on the weakened heart. These effects are thought to be due to inhibition of the $Na^+ - K^+$ pump and enhancement of Ca^{2+} entry into myocardial fibers.

A c t i v i t y :
Assessing the Effect of Various Ions

To test the effect of various ions on the heart, apply the desired solution using the following method.

1. Drag-and-drop the Calcium Ions dropper cap to a point about an inch above the heart.

2. Calcium Ions will automatically be dripped onto the heart and the dropper cap will return to the Calcium Ions bottle.

3. Watch the heart activity window for the message: Heart Rate Stable.

4. After the heart rate stabilizes, record the heart rate in the space provided below, and click **Record Data** to retain your data in the grid.

5. Click the **23°C Ringer's** button to flush the heart with fresh Ringer's solution. Watch the heart activity window for the message: Heart Rate Normal, which means that the heart is ready for the next test.

6. Repeat steps 1 through 5 for Sodium Ions and then Potassium Ions.

Effect of Ca^{2+}:

Heart rate _____ bpm

Describe the results

_____

_____

Effect of Na^+:

Heart rate _____ bpm

Describe the results

_____

_____

Effect of K^+:

Heart rate _____ bpm

Describe the results

_____

_____ ■

Potassium ion concentration is normally higher within cells than in the extracellular fluid. *Hyperkalemia* decreases the resting potential of plasma membranes, thus decreasing the force of heart contraction. In some cases, the conduction rate of the heart is so depressed that **ectopic pacemakers** (pacemakers appearing erratically and at abnormal sites in the heart muscle) appear in the ventricle, and fibrillation may occur.

Was there any evidence of premature beats in the recording of potassium ion effects?

_____

Was arrhythmia produced with any of the ions tested?

_____ If so, which?

_____ ■

Activity:
Using Your Data

Save Your Data

When you have finished with all experiments, you can save your data set for later review and additional analysis.

1. Select **Save Data to File** from the **File** menu.

2. When the dialog box opens, click a drive and/or directory location.

3. Type a file name into the File name box (your instructor may provide you with a name).

4. Click **Save** to complete the save. If you want to review previously-saved data, select **Open Data File** from the **File** menu, and choose your file.

Print Your Data

If your instructor requires a hard copy of your experiment results, choose **Print Data** from the **File** menu.

1. Type your name into the provided space.

2. When you are ready to print, click **OK**. ■

The Lymphatic System and Immune Response

Objectives

1. To name the components of the lymphatic system.
2. To relate the function of the lymphatic system to that of the blood vascular system.
3. To describe the formation and composition of lymph, and to describe how it is transported through the lymphatic vessels.
4. To relate immunologic memory, specificity, and differentiation of self from nonself to immune function.
5. To differentiate between the roles of B cells and T cells in the immune response.
6. To describe the structure and function of lymph nodes, and to indicate the localization of T cells, B cells, and macrophages in a typical lymph node.
7. To describe (or draw) the structure of the immunoglobulin monomer, and to name the five immunoglobulin subclasses.
8. To discuss the immunologic "abnormality" evidenced by many rheumatoid arthritis sufferers.

Materials

❑ Large anatomical chart of the human lymphatic system
❑ Prepared microscope slides of lymph nodes
❑ Compound microscope
❑ Microscope slide
❑ Wax marking pencil
❑ Applicator sticks
❑ Arthritis screening test (Wampole)
❑ Plastic (disposable) gloves
❑ Disposable autoclave bag

 For instructions on animal dissections, see the dissection exercises starting on p. 733 in the cat and fetal pig editions of this manual.

 See Appendix C, Exercise 36 for links to A.D.A.M. Interactive Anatomy.

The Lymphatic System

General Description

The **lymphatic system** consists of a network of lymphatic vessels (lymphatics), lymph nodes, and a number of other lymphoid organs, such as the tonsils, thymus, and spleen (Figure 36.1). We will focus on the lymphatic vessels and lymph nodes in this section. The overall function of the lymphatic system is twofold. It transports tissue fluid (lymph) to the blood vessels. In addition, it protects the body by removing foreign material such as bacteria from the lymphatic stream and by serving as a site for lymphocyte "policing" of body fluids and lymphocyte multiplication. The function of these white blood cells in body immunity is described later in this exercise.

Distribution and Function of Lymphatic Vessels and Lymph Nodes

As blood circulates through the body, the hydrostatic and osmotic pressures operating at the capillary beds result in fluid outflow at the arterial end of the bed and in its return at the venous end. However, not all of the lost fluid is returned to the bloodstream by this mechanism, and the fluid that lags behind in the tissue spaces must eventually return to the blood if the vascular system is to operate properly. (If it does not, fluid accumulates in the tissues, producing a condition called edema.) It is the microscopic, blind-ended **lymphatic capillaries,** which ramify through nearly all the tissues of the body, that pick up this leaked fluid (primarily water and a small amount of dissolved proteins) and carry it through successively larger vessels—**lymphatic collecting vessels** to **lymphatic trunks**—until the lymph finally returns to the blood vascular system through one of the two large ducts in the thoracic region. The **right lymphatic duct** drains lymph from the right upper extremity, head, and thorax; the large **thoracic duct** receives lymph from the rest of the body (Figure 36.1a). In humans, both ducts empty the lymph into the venous circulation at the junction of the internal jugular vein and the subclavian vein, on their respective sides of the body. Notice that the lymphatic system, lacking both a contractile "heart" and arteries, is a one-way system; it carries lymph only toward the heart.

Like veins of the blood vascular system, the lymphatic collecting vessels have three tunics and are equipped with valves (see Plate 28 in the Histology Atlas). However, the lymphatics tend to be thinner-walled, to have *more* valves, and to anastomose more than veins. Since the lymphatic system is a pumpless system, lymph transport depends largely on the milking action of the skeletal muscles and on pressure changes within the thorax that occur during breathing.

As lymph is transported, it filters through bean-shaped **lymph nodes,** which cluster along the lymphatic vessels of

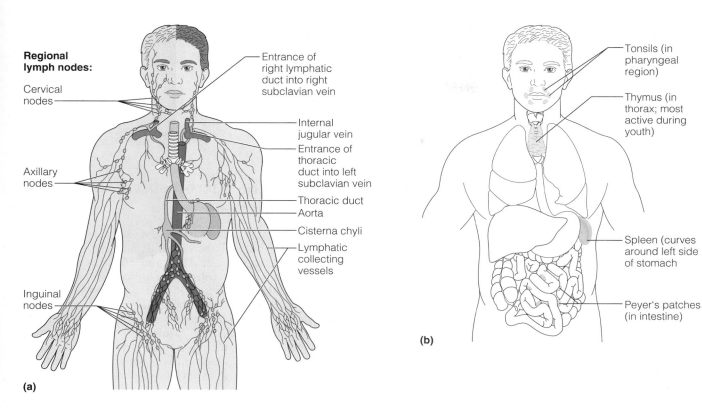

Regional lymph nodes:

Cervical nodes

Axillary nodes

Inguinal nodes

Entrance of right lymphatic duct into right subclavian vein

Internal jugular vein

Entrance of thoracic duct into left subclavian vein

Thoracic duct

Aorta

Cisterna chyli

Lymphatic collecting vessels

(a)

Tonsils (in pharyngeal region)

Thymus (in thorax; most active during youth)

Spleen (curves around left side of stomach

Peyer's patches (in intestine)

(b)

Figure 36.1   Lymphatic system. (a) Distribution of lymphatic vessels and lymph nodes. The green shaded area represents body area drained by the right lymphatic duct. **(b)** Body location of the tonsils, thymus, spleen, and Peyer's patches.

the body. There are thousands of lymph nodes; but because they are usually embedded in connective tissue, they are not ordinarily seen. Within the lymph nodes are **macrophages,** phagocytes that destroy bacteria, cancer cells, and other foreign matter in the lymphatic stream, thus rendering many harmful substances or cells harmless before the lymph enters the bloodstream. Particularly large collections of lymph nodes are found in the inguinal, axillary, and cervical regions of the body. Although we are not usually aware of the filtering and protective nature of the lymph nodes, most of us have experienced "swollen glands" during an active infection. This swelling is a manifestation of the trapping function of the nodes.

Other lymphoid organs—the tonsils, thymus, and spleen (Figure 36.1b)—resemble the lymph nodes histologically, and house similar cell populations (lymphocytes and macrophages).

Activity:
Identifying the Organs of the Lymphatic System

Study the large anatomical chart to observe the general plan of the lymphatic system. Notice the distribution of lymph nodes, various lymphatics, the lymphatic trunks, and the location of the right lymphatic duct and the thoracic duct. Also identify the **cisterna chyli,** the enlarged terminus of the thoracic duct that receives lymph from the digestive viscera. ■

 For instructions on animal dissections, see the dissection exercises starting on p. 733 in the cat and fetal pig editions of this manual.

The Immune Response

The **immune system** is a functional system that recognizes something as foreign and acts to destroy or neutralize it. This is known as the **immune response.** When operating effectively, the immune response protects us from bacterial and viral infections, bacterial toxins, and cancer. When it fails or malfunctions, the body is quickly devastated by pathogens or its own assaults.

Major Characteristics of the Immune Response

The most important characteristics of the immune response are its (1) **memory,** (2) **specificity,** and (3) **ability to differentiate self from nonself.** Not only does the immune system have a "memory" for previously encountered antigens (molecules foreign to the body—the chicken pox virus for example), but this memory is also remarkably accurate and highly specific.

An almost limitless variety of macromolecules is *antigenic*—that is, capable of provoking an immune response and reacting with its products. Nearly all foreign proteins, many polysaccharides, and many small molecules (haptens), when linked to our own body proteins, exhibit this capability. The cells that recognize antigens and initiate the immune response are lymphocytes. (As described in Exercise 29, lymphocytes are the second most numerous members of the leukocyte or white blood cell [WBC] population.) Each immunocompetent lymphocyte is virtually monospecific; that is, it has receptors on its surface allowing it to bind with only one or a few very similar antigens.

As a rule, our own proteins are tolerated, a fact that reflects the ability of the immune system to distinguish our own tissues (self) from foreign antigens (nonself). Nevertheless, an inability to recognize self can and does occasionally happen and our own tissues are attacked by the immune system. This phenomenon is called *autoimmunity*.

Organs, Cells, and Cell Interactions of the Immune Response

The immune system utilizes as part of its arsenal the **lymphoid organs,** including the thymus, lymph nodes, spleen, tonsils, and bone marrow. Of these, the thymus and bone marrow are considered to be the *primary lymphoid organs.* The others are *secondary lymphoid areas.*

The stem cells that give rise to the immune system arise in the bone marrow. Their subsequent differentiation into one of the two populations of immunocompetent lymphocytes occurs in the primary lymphoid organs. The **B cells** differentiate in bone marrow, and the **T cells** differentiate in the thymus. While in their "programming organs" the lymphocytes become *immunocompetent,* an event indicated by the appearance of specific cell-surface proteins that enable the lymphocytes to respond (by binding) to a particular antigen.

After differentiation, the B and T cells leave the bone marrow and thymus, respectively; enter the bloodstream; and travel to peripheral (secondary) lymphoid organs, where clonal selection occurs. **Clonal selection** is triggered when an antigen binds to the specific cell-surface receptors of a T or B cell. This event causes the lymphocyte to proliferate rapidly, forming a clone of like cells, all bearing the same antigen-specific receptors. Then, in the presence of certain regulatory signals, the members of the clone specialize, or differentiate—some forming memory cells and others becoming effector or regulatory cells. Upon subsequent meetings with the same antigen, the immune response proceeds considerably faster because the troops are already mobilized and awaiting further orders, so to speak.

In the case of B cell clones, some become memory B cells; the others form antibody-producing **plasma cells.** Because the B cells act indirectly through the antibodies that their progeny release into the bloodstream (or other body fluids), they are said to provide **humoral immunity.** T cell clones are more diverse. Although all T cell clones also contain memory cells, some clones contain *cytotoxic,* or *killer,* T cells (effector cells that directly attack virus-infected tissue cells). Others contain regulatory cells such as the *helper cells* (that help activate the B cells and cytotoxic T cells) and still others contain *suppressor cells* that can inhibit the immune response. Because the T cells act directly to destroy virus-infected cells, many bacteria, and cancer cells, and to reject foreign grafts, T cells are said to mediate **cellular immunity.**

Absence or failure of thymic differentiation of T lymphocytes results in a marked depression of both anti-

body and cell-mediated immune functions. Additionally, the observation that the thymus naturally involutes with age has been correlated with the relatively immune-deficient status of elderly individuals. ■

All lymphoid tissues except the thymus and bone marrow contain both T and B cell–dependent regions. The lymph node is used as the example in the following microscopic study.

Activity:
Studying the Microscopic Anatomy of a Lymph Node

Obtain a prepared slide of a lymph node and a compound microscope. As you examine the slide, notice the following anatomical features, depicted in Figure 36.2. The node is enclosed within a fibrous **capsule,** from which connective tissue septa (**trabeculae**) extend inward to divide the node into several compartments. Very fine strands of reticular connective tissue issue from the trabeculae, forming the stroma of the gland within which cells are found.

In the outer region of the node, the **cortex,** some of the cells are arranged in globular masses, referred to as germinal centers. The **germinal centers** contain rapidly dividing B lymphocytes. The rest of the cortical cells are primarily T lymphocytes that circulate continuously, moving from the blood into the node and then exiting from the node in the lymphatic stream.

In the internal portion of the gland, the **medulla,** the cells are arranged in cordlike fashion. Most of the medullary cells are macrophages. Macrophages are important not only for their phagocytic function but also because they play an essential role in "presenting" the antigens to the T cells.

Lymph enters the node through a number of afferent vessels, circulates through sinuses within the node, and leaves the node through efferent vessels at the **hilus.** Since each node has fewer efferent than afferent vessels, the lymph flow stagnates somewhat within the node. This allows time for the generation of an immune response and for the macrophages to remove debris from the lymph before it reenters the blood vascular system.

In the space provided here, draw a pie-shaped section of a lymph node, showing the detail of cells in a germinal center, sinusoids, and afferent and efferent vessels. Label all elements.

Before leaving the topic of the lymphoid organs, compare and contrast the structure of the lymph node examined here with the microscopic anatomy of the spleen and tonsils illustrated in Plates 29 and 30 in the Histology Atlas. ■

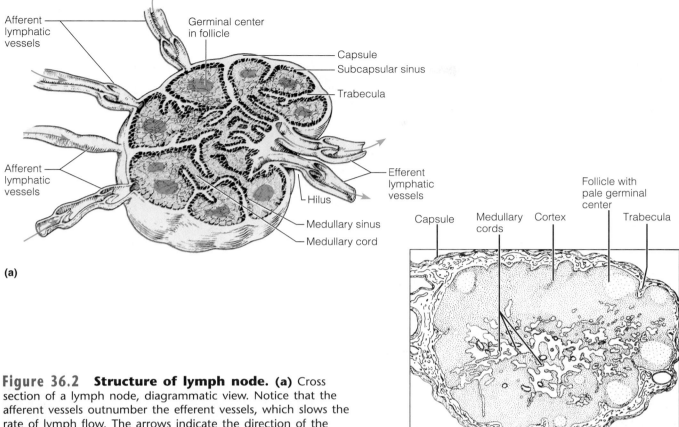

(a)

(b)

Figure 36.2 **Structure of lymph node. (a)** Cross section of a lymph node, diagrammatic view. Notice that the afferent vessels outnumber the efferent vessels, which slows the rate of lymph flow. The arrows indicate the direction of the lymph flow. **(b)** Line drawing of a photomicrograph showing part of a lymph node. (See also Plate 27 in the Histology Atlas.)

Antibodies and Tests for Their Presence

Antibodies, or **immunoglobulins**, produced by sensitized B cells and their plasma cell offspring in response to an antigen, are a heterogenous group of proteins that comprise the general class of plasma proteins called **gamma globulins.** Antibodies are found not only in plasma but also (to greater or lesser extents) in all body secretions. Five major classes of immunoglobulins (Igs) have been identified: IgM, IgG, IgD, IgA, and IgE. The immunoglobulin classes share a common basic structure, but differ functionally and in their localization in the body.

All Igs are composed of one or more monomers (structural units). A monomer consists of four protein chains bound together by disulfide bridges (Figure 36.3). Two of the chains are quite large and have a high molecular weight; these are the **heavy chains.** The other two chains are only half as long and have a low molecular weight. These are called **light chains.** The two heavy chains have a *constant (C) region,* in which the amino acid sequence is identical in both chains, and a *variable (V) region,* which differs in the Igs formed in response to different antigens. The same is true of the two light chains; each has a constant and a variable region.

The intact Ig molecule has a three-dimensional shape that is generally Y-shaped. Together, the variable regions of the light and heavy chains in each "arm" construct one **anti-**gen-binding site uniquely shaped to "fit" a specific *antigenic determinant* (portion) of an antigen. Thus, each Ig monomer bears two identical sites that bind to a specific (and the same) antigen. Binding of the immunoglobulins to their complementary antigen(s) effectively immobilizes the antigens until they can be phagocytized or lysed by complement fixation.

Although the role of the immune system is to protect the body, symptoms of certain diseases involve excessively high antibody synthesis (as in multiple myeloma, a cancer of the bone marrow and adjacent bony structures) and/or the production of abnormal antibodies (such as **rheumatoid factor** present in the blood of many rheumatoid arthritis sufferers). The simple experiment described below tests for the presence of the rheumatoid factor in plasma.

Test for Rheumatoid Factor

Although the precise cause of rheumatoid arthritis is not yet known, the disease process itself involves an astonishing number of the body's immune elements—T cells, rheumatoid factor, and antirheumatoid antibodies, to name a few. Rheumatoid arthritis is an autoimmune reaction, and rheumatoid factor (while not fully diagnostic for the disease) is found in the serum of many rheumatoid arthritis victims. Perhaps the initial disease trigger is a virus (or bacterium) bearing molecules similar to some naturally present in the joints, which in turn prompts the clustering, at certain body joints, of T cells, protective antibodies, and other inflammatory cells.

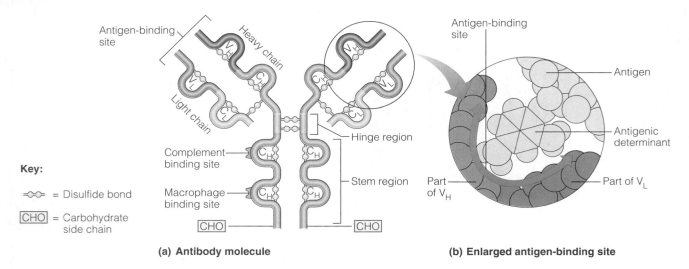

(a) Antibody molecule

(b) Enlarged antigen-binding site

Figure 36.3 Structure of an immunoglobulin monomer. (a) Each monomer is composed of four protein chains (two heavy chains and two light chains) connected by disulfide bonds. Both the heavy and light chains have regions of constant amino acid sequence (C regions) and regions of variable amino acid sequence (V regions). The variable regions differ in each type of antibody and construct the antigen-binding sites. Each immunoglobulin monomer has two such antigen-specific sites. **(b)** Enlargement of an antigen-binding site of an immunoglobulin.

Although rheumatoid arthritis is a systemic disease that may involve many body organs and tissues, it is characterized by the manner in which the joints are affected. The synovial membranes become inflamed and then thicken as inflammatory cells accumulate. This is followed by the formation of **pannus,** an abnormal tissue that clings to the articular cartilages and eventually destroys them. The final stage involves the elaboration of fibrous tissue that connects the bone ends. Because this fibrous tissue eventually ossifies, the bone ends become firmly fused and often deformed (Figure 36.4). Not all cases progress to joint immobilization, the severely crippling stage, but all cases do involve some restriction of joint movement and extreme pain.

The test for the rheumatoid factor involves mixing a drop of plasma with serum containing antibodies against the abnormal antibody (that is, antirheumatoid-factor antibodies). If rheumatoid factor is present, antigen-antibody complexes will form, causing agglutination.

Activity:
Testing for Rheumatoid Factor

1. Obtain the arthritis screening test materials, a microscope slide, a wax marker, two applicator sticks, and plastic gloves.

2. With the wax marker, label one end of the slide "control" and the other end "test." Don the gloves.

3. Place a drop of test plasma on the end of the slide marked "test" and a drop of the positive control plasma on the end marked "control."

4. Add one drop of reagent 1 to each sample, and mix with an applicator stick. (Use opposite ends of the applicator stick to mix each sample, and then discard the applicator stick in the disposable autoclave bag provided at the supply area.)

Figure 36.4 Photograph of a hand deformed by rheumatoid arthritis.

5. Add 2 drops of reagent 2 to each sample. Using opposite ends of a fresh applicator stick, mix each sample thoroughly; and then spread the samples over a 1-inch-square area on the slide.

6. Gently rock the slide back and forth for 3 minutes, and then check each sample for the presence of agglutination.

Which sample gave a positive test for the presence of rheumatoid factor?

_____

7. Dispose of the second applicator stick and the gloves in the autoclave bag. ■

Anatomy of the Respiratory System

Body cells require an abundant and continuous supply of oxygen. As the cells use oxygen, they release carbon dioxide, a waste product that the body must get rid of. These oxygen-using cellular processes, collectively referred to as *cellular respiration,* are more appropriately described in conjunction with the topic of cellular metabolism. The major role of the **respiratory system,** our focus in this exercise, is to supply the body with oxygen and dispose of carbon dioxide. To fulfill this role, at least four distinct processes, collectively referred to as **respiration,** must occur:

Pulmonary ventilation: The tidelike movement of air into and out of the lungs so that the gases in the alveoli are continuously changed and refreshed. Also more simply called *ventilation,* or *breathing.*

External respiration: The gas exchanges to and from the pulmonary circuit blood that occur in the lungs (oxygen loading/carbon dioxide unloading).

Transport of respiratory gases: The transport of respiratory gases between the lungs and tissue cells of the body accomplished by the cardiovascular system, using blood as the transport vehicle.

Internal respiration: Exchange of gases to and from the blood capillaries of the systemic circulation (oxygen unloading and carbon dioxide loading).

Only the first two processes are the exclusive province of the respiratory system, but all four must occur for the respiratory system to "do its job." Hence, the respiratory and circulatory systems are irreversibly linked. If either system fails, cells begin to die from oxygen starvation and accumulation of carbon dioxide. Uncorrected, this situation soon causes death of the entire organism.

Upper Respiratory System Structures

The upper respiratory system structures—the nose, pharynx, and larynx—are shown in Figure 37.1 and described below. As you read through the descriptions, identify each structure in the figure.

Air generally passes into the respiratory tract through the **external nares** (nostrils), and enters the **nasal cavity** (divided by the **nasal septum**). It then flows posteriorly over three pairs of lobelike structures, the **inferior, superior,** and

Objectives

1. To define the following terms: *respiratory system, pulmonary ventilation, external respiration,* and *internal respiration.*
2. To label the major respiratory system structures on a diagram (or identify them on a model) and to describe the function of each.
3. To recognize the histologic structure of the trachea (cross section) and lung tissue on prepared slides, and describe the functions the observed structural modifications serve.

Materials

- ❑ Resin cast of the respiratory tree (if available)
- ❑ Human torso model
- ❑ Respiratory organ system model and/or chart of the respiratory system
- ❑ Larynx model (if available)
- ❑ Preserved inflatable lung preparation (obtained from a biological supply house) or sheep pluck fresh from the slaughterhouse
- ❑ Source of compressed air*
- ❑ 2-foot length of laboratory rubber tubing
- ❑ Dissection tray
- ❑ Protective skin cream or disposable gloves
- ❑ Disposable autoclave bag
- ❑ Histologic slides of the following (if available): trachea (cross section), lung tissue, both normal and pathologic specimens (e.g., sections taken from lung tissues exhibiting bronchitis, pneumonia, emphysema, or lung cancer)
- ❑ Compound and dissecting microscopes

 For instructions on animal dissections, see the dissection exercises starting on p. 733 in the cat and fetal pig editions of this manual.

 See Appendix C, Exercise 37 for links to A.D.A.M. Interactive Anatomy.

---

* If a compressed air source is not available, cardboard mouthpieces that fit the cut end of the rubber tubing should be available for student use. Disposable autoclave bags should also be provided for discarding the mouthpiece.

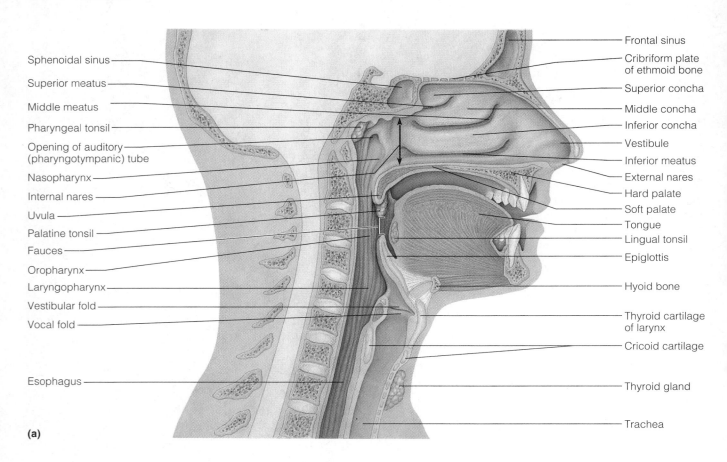

Sphenoidal sinus
Superior meatus
Middle meatus
Pharyngeal tonsil
Opening of auditory (pharyngotympanic) tube
Nasopharynx
Internal nares
Uvula
Palatine tonsil
Fauces
Oropharynx
Laryngopharynx
Vestibular fold
Vocal fold
Esophagus

Frontal sinus
Cribriform plate of ethmoid bone
Superior concha
Middle concha
Inferior concha
Vestibule
Inferior meatus
External nares
Hard palate
Soft palate
Tongue
Lingual tonsil
Epiglottis
Hyoid bone
Thyroid cartilage of larynx
Cricoid cartilage
Thyroid gland
Trachea

(a)

Olfactory epithelium
Mucosa of pharynx
Tubal tonsil
Pharyngo-tympanic (auditory) tube
Nasopharynx

Olfactory bulb
Superior concha and superior meatus
Middle concha and middle meatus
Inferior concha and inferior meatus
Hard palate
Soft palate
Uvula

(b)

Figure 37.1 Structures of the upper respiratory tract (sagittal section).
(a) Diagrammatic view. **(b)** Photograph.

middle nasal conchae, which increase the air turbulence. As the air passes through the nasal cavity, it is also warmed, moistened, and filtered by the nasal mucosa. The air that flows directly beneath the superior part of the nasal cavity may chemically stimulate the olfactory receptors located in the mucosa of that region. The nasal cavity is surrounded by the **paranasal sinuses** in the frontal, sphenoid, ethmoid, and maxillary bones. These sinuses act as resonance chambers in speech and their mucosae, like that of the nasal cavity, warm and moisten the incoming air.

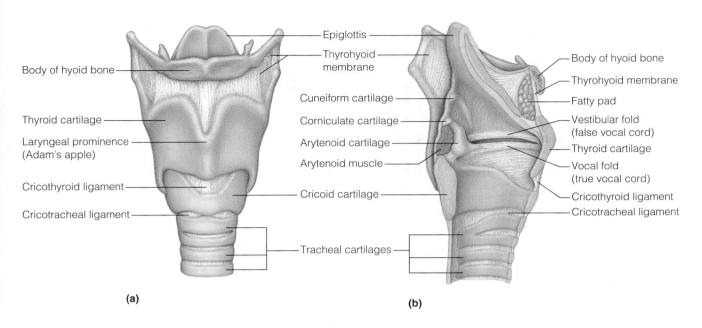

Figure 37.2 Structure of the larynx. (a) Anterior view. **(b)** Sagittal section.

The nasal passages are separated from the oral cavity below by a partition composed anteriorly of the **hard palate** and posteriorly by the **soft palate.**

The genetic defect called **cleft palate** (failure of the palatine bones and/or the palatine processes of the maxillary bones to fuse medially) causes difficulty in breathing and oral cavity functions such as sucking and, later, mastication and speech. ■

Of course, air may also enter the body via the mouth. From there it passes through the oral cavity to move into the pharynx posteriorly, where the oral and nasal cavities are joined temporarily.

Commonly called the *throat,* the funnel-shaped **pharynx** connects the nasal and oral cavities to the larynx and esophagus inferiorly. It has three named parts (Figure 37.1):

1. The **nasopharynx** lies posterior to the nasal cavity and is continuous with it via the **internal nares.** It lies above the soft palate; hence, it serves only as an air passage. High on its posterior wall are the *pharyngeal tonsils,* paired masses of lymphoid tissue that help to protect the respiratory passages from invading pathogens. The *pharyngotympanic (auditory) tubes,* which allow middle ear pressure to become equalized to atmospheric pressure, drain into the lateral aspects of the nasopharynx.

Because of the continuity of the middle ear and nasopharyngeal mucosae, nasal infections may invade the middle ear cavity causing *otitis media,* which is difficult to treat. ■

2. The **oropharynx** is continuous posteriorly with the oral cavity. Since it extends from the soft palate to the epiglottis of the larynx inferiorly, it serves as a common conduit for food and air. In its lateral walls are the *palatine tonsils.* The *lingual tonsils* cover the base of the tongue.

3. The **laryngopharynx,** like the oropharynx, accommodates both ingested food and air. It lies directly posterior to the upright epiglottis and extends to the larynx, where the common pathway divides into the respiratory and digestive channels. From the laryngopharynx, air enters the lower respiratory passageways by passing through the larynx (voice box) and into the trachea below.

The **larynx** (Figure 37.2) consists of nine cartilages. The two most prominent are the large shield-shaped **thyroid cartilage,** whose anterior medial prominence is commonly referred to as *Adam's apple,* and the inferiorly located, ring-shaped **cricoid cartilage,** whose widest dimension faces posteriorly. All the laryngeal cartilages are composed of hyaline cartilage except the flaplike **epiglottis,** a flexible elastic cartilage located superior to the opening of the larynx. The epiglottis, sometimes referred to as the "guardian of the airways," forms a lid over the larynx when we swallow. This closes off the respiratory passageways to incoming food or drink, which is routed into the posterior esophagus, or food chute.

● Palpate your larynx by placing your hand on the anterior neck surface approximately halfway down its length. Swallow. Can you feel the cartilaginous larynx rising?

If anything other than air enters the larynx, a cough reflex attempts to expel the substance. Note that this reflex operates only when a person is conscious. Therefore, you should never try to feed or pour liquids down the throat of an unconscious person.

The mucous membrane of the larynx is thrown into two pairs of folds—the upper **vestibular folds,** also called the **false vocal cords,** and the lower **vocal folds,** or **true vocal cords,** which vibrate with expelled air for speech. The vocal cords are attached posterolaterally to the small triangular **arytenoid cartilages** by the *vocal ligaments.* The slitlike passageway between the folds is called the **glottis.**

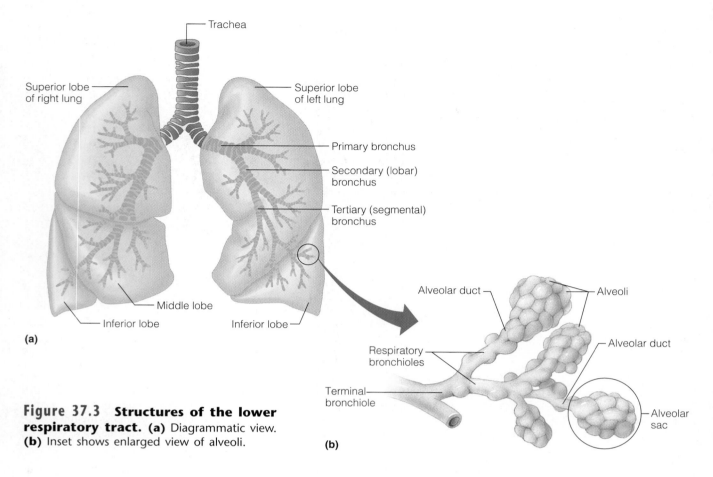

Figure 37.3 Structures of the lower respiratory tract. (a) Diagrammatic view. (b) Inset shows enlarged view of alveoli.

Lower Respiratory System Structures

Air entering the **trachea,** or windpipe, from the larynx travels down its length (about 11.0 cm) to the level of the *sternal angle* (or the disc between the fourth and fifth thoracic vertebrae). There the passageway divides into the right and left **primary bronchi** (Figure 37.3), which plunge into their respective lungs at an indented area called the **hilus** (see Figure 37.5b). The right primary bronchus is wider, shorter, and more vertical than the left. As a result, foreign objects that enter the respiratory passageways are more likely to become lodged in it.

The trachea is lined with a ciliated mucus-secreting, pseudostratified columnar epithelium as are many of the other respiratory system passageways. The cilia propel mucus (produced by goblet cells) laden with dust particles, bacteria, and other debris away from the lungs and toward the throat, where it can be expectorated or swallowed. The walls of the trachea are reinforced with C-shaped cartilage rings, the incomplete portion located posteriorly. These C-shaped cartilages serve a double function: The incomplete parts allow the esophagus to expand anteriorly when a large food bolus is swallowed. The solid portions reinforce the trachea walls to maintain its open passageway regardless of the pressure changes that occur during breathing.

The primary bronchi further divide into smaller and smaller branches (the secondary, tertiary, on down), finally becoming the **bronchioles,** which have terminal branches called **respiratory bronchioles** (Figure 37.3b). All but the most minute branches have cartilaginous reinforcements in their walls, usually in the form of small plates of hyaline cartilage rather than cartilaginous rings. As the respiratory tubes get smaller and smaller, the relative amount of smooth muscle in their walls increases as the amount of cartilage declines and finally disappears. The complete layer of smooth muscle present in the bronchioles enables them to provide considerable resistance to air flow under certain conditions (asthma, hay fever, etc.). The continuous branching of the respiratory passageways in the lungs is often referred to as the **respiratory tree.** The comparison becomes much more meaningful if you observe a resin cast of the respiratory passages. Do so, if one is available for observation in the laboratory.

The respiratory bronchioles in turn subdivide into several **alveolar ducts,** which terminate in alveolar sacs that rather resemble clusters of grapes. **Alveoli,** tiny balloonlike expansions along the alveolar sacs, and occasionally found protruding from the alveolar ducts and respiratory bronchioles, are composed of a single thin layer of squamous epithelium overlying a wispy basal lamina. The external surfaces of the alveoli are densely spiderwebbed with a network of pulmonary capillaries (Figure 37.4). Together, the alveolar and capillary walls and their fused basal laminas form the **respi-**

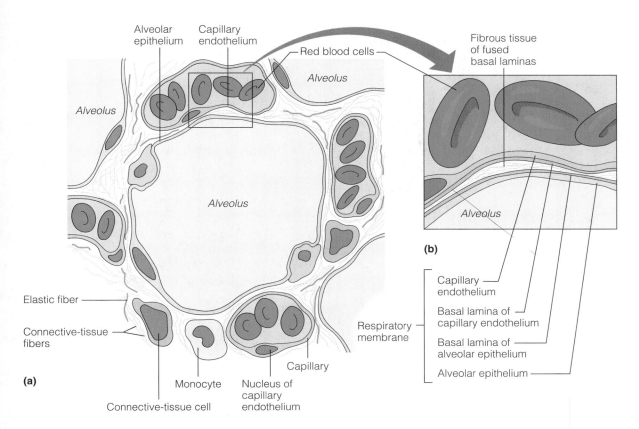

Figure 37.4   Diagrammatic view of the relationship between the alveoli and pulmonary capillaries involved in gas exchange. **(a)** One alveolus surrounded by capillaries. **(b)** Enlargement of the respiratory membrane.

ratory membrane, also called the **air-blood barrier.** Because gas exchanges occur by simple diffusion across the respiratory membrane—oxygen passing from the alveolar air to the capillary blood and carbon dioxide leaving the capillary blood to enter the alveolar air—the alveolar sacs, alveolar ducts, and respiratory bronchioles are referred to collectively as **respiratory zone structures.** All other respiratory passageways (from the nasal cavity to the bronchioles), simply serve as access or exit routes to and from these gas exchange chambers, and are called **conducting zone structures.** Because the conducting zone structures have no exchange function, they are also referred to as *anatomical dead space.*

The Lungs and Their Pleural Coverings

The paired lungs are soft, spongy organs that occupy the entire thoracic cavity except for the *mediastinum,* which houses the heart, bronchi, esophagus, and other organs (Figure 37.5). Each lung is connected to the mediastinum by a *root* containing its vascular and bronchial attachments. The structures of the root enter (or leave) the lung via a medial indentation called the *hilus.* All structures distal to the primary bronchi are found within the lung substance. A lung's *apex,* the narrower superior aspect, lies just deep to the clavicle, and its *base,* the inferior concave surface, rests on the diaphragm. Anterior, lateral, and posterior lung surfaces are in close contact with the ribs and, hence, are collectively called the *costal*

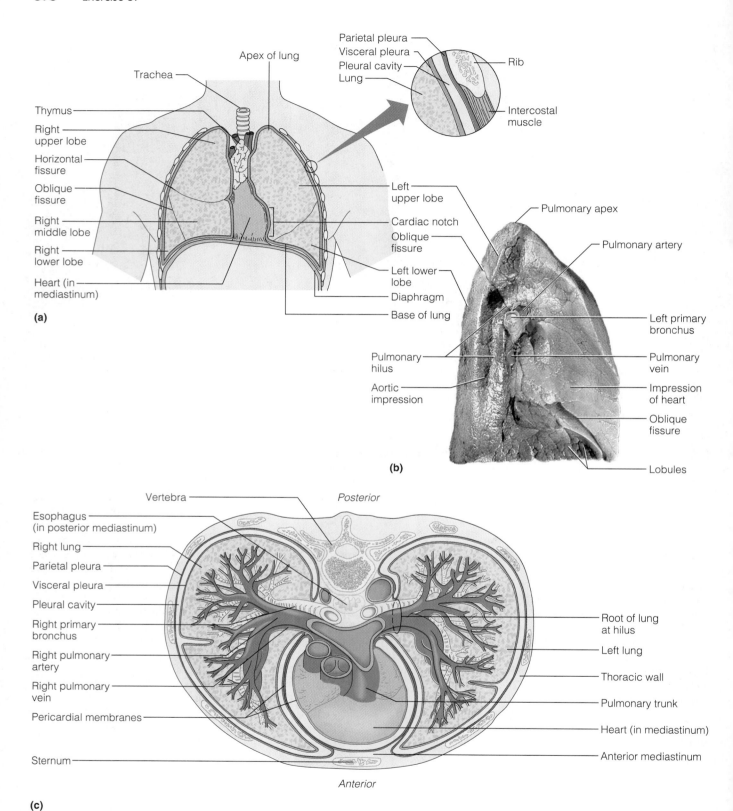

(a) Anterior view of the thoracic organs. The lungs flank the central mediastinum. The inset at upper right depicts the pleura and the pleural cavity. **(b)** Photograph of medial aspect of left lung. **(c)** Transverse section through the superior part of the thorax, showing the lungs and the main organs in the mediastinum.

Figure 37.5 Anatomical relationships of organs in the thoracic cavity.

surface. The medial surface of the left lung exhibits a concavity called the *cardiac notch,* which accommodates the heart where it extends left from the body midline. Fissures divide the lungs into a number of *lobes*—two in the left lung and three in the right. Other than the respiratory passageways and air spaces that make up the bulk of their volume, the lungs are mostly elastic connective tissue, which allows them to recoil passively during expiration.

Each lung is enclosed in a double-layered sac of serous membrane called the **pleura.** The outer layer, the **parietal pleura,** is attached to the thoracic walls and the **diaphragm;** the inner layer, covering the lung tissue, is the **visceral pleura.** The two pleural layers are separated by the *pleural space,* which is more of a potential space than an actual one. The pleural layers produce lubricating serous fluid that causes them to adhere closely to one another, holding the lungs to the thoracic wall and allowing them to move easily against one another during the movements of breathing.

Activity:
Identifying Respiratory System Organs

Before proceeding, be sure to locate on the torso model, thoracic cavity structures model, or an anatomical chart all the respiratory structures described—both upper and lower respiratory system organs. ■

 For instructions on animal dissections, see the dissection exercises starting on p. 733 in the cat and fetal pig editions of this manual.

Activity:
Demonstrating Lung Inflation in a Sheep Pluck

A *sheep pluck* includes the larynx, trachea with attached lungs, the heart and pericardium, and portions of the major blood vessels found in the mediastinum (aorta, pulmonary artery and vein, venae cavae).

⚠ Don plastic gloves, obtain a fresh sheep pluck (or a preserved pluck of another animal), and identify the lower respiratory system organs. Once you have completed your observations, insert a hose from an air compressor (vacuum pump) into the trachea and alternately allow air to flow in and out of the lungs. Notice how the lungs inflate. This observation is educational in a preserved pluck but it is a spectacular sight in a fresh one. Another advantage of using a fresh pluck is that the lung pluck changes color (becomes redder) as hemoglobin in trapped RBCs becomes loaded with oxygen.

If air compressors are not available, the same effect may be obtained by using a length of laboratory rubber tubing to blow into the trachea. Obtain a cardboard mouthpiece and fit it into the cut end of the laboratory tubing before attempting to inflate the lungs.

⚠ Dispose of the mouthpiece and gloves in the autoclave bag immediately after use. ■

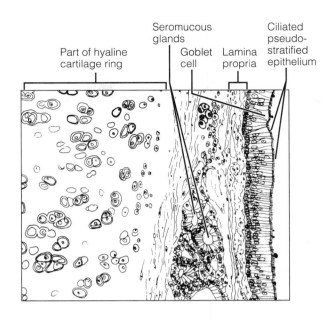

Figure 37.6 **Microscopic structure of the trachea, cross-sectional view.** (See also Plate 32 in the Histology Atlas.)

Activity:
Examining Prepared Slides of Trachea and Lung Tissue

1. Obtain a compound microscope and a slide of a cross section of the trachea wall. Identify the smooth muscle layer, the hyaline cartilage supporting rings, and the pseudostratified ciliated epithelium. Using Figure 37.6 as a guide, also try to identify a few goblet cells in the epithelium. In the space below, draw a section of the trachea wall and label all tissue layers.

2.   Obtain a slide of lung tissue for examination. The alveolus is the main structural and functional unit of the lung and is the actual site of gas exchange. Identify a bronchiole (Figure 37.7a) and the thin squamous epithelium of the alveolar walls (Figure 37.7b). Draw your observations of a small section of the alveolar tissue in the space below and label the alveoli.

(a)

3.   Examine slides of pathologic lung tissues, and compare them to the normal lung specimens. Record your observations in the Exercise 37 Review Sheets. ■

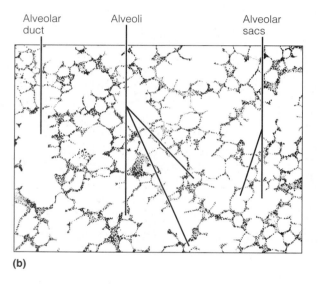

(b)

Figure 37.7 Microscopic structure of a bronchiole and alveoli. (a) Cross-sectional view of a bronchiole. (See also Plate 34 in the Histology Atlas.) **(b)** Alveoli. (See also Plate 31 in the Histology Atlas.)

Respiratory System Physiology

Objectives

1. To define the following (and be prepared to provide volume figures if applicable):

   inspiration   *expiratory reserve volume*
   expiration   *inspiratory reserve volume*
   tidal volume   *minute respiratory volume*
   vital capacity

2. To explain the role of muscles and volume changes in the mechanical process of breathing.

3. To demonstrate proper usage of the spirometer.

4. To explain the relative importance of various mechanical and chemical factors in producing respiratory variations.

5. To describe bronchial and vesicular breathing sounds.

6. To explain the importance of the carbonic acid–bicarbonate buffer system in maintaining blood pH.

Materials

- ❏ Model lung (bell jar demonstrator)
- ❏ Tape measure
- ❏ Spirometer
- ❏ Disposable mouthpieces
- ❏ Nose clips
- ❏ Alcohol swabs
- ❏ 70% ethanol solution in a battery jar
- ❏ Disposable autoclave bag
- ❏ Clear adhesive tape
- ❏ Paper bag
- ❏ Pneumograph and recording attachments
- ❏ Recording apparatus (kymograph or physiograph)
- ❏ Stethoscope
- ❏ Table (on chalkboard) for recording class data
- ❏ pH meter (standardized with buffer of pH 7)
- ❏ Buffer solution (pH 7)
- ❏ Concentrated HCl and NaOH
- ❏ 0.01 M HCl
- ❏ 250- and 50-ml beakers
- ❏ Graduated cylinder (100 ml)

- ❏ Glass stirring rod
- ❏ Plastic wash bottles containing distilled water
- ❏ Animal plasma
- ❏ *Human Respiratory System* Videotape*

*Available to qualified adopters from Benjamin/Cummings.

Mechanics of Respiration

Pulmonary ventilation, or **breathing,** consists of two phases: **inspiration,** during which air is taken into the lungs, and **expiration,** during which air passes out of the lungs. As the inspiratory muscles (external intercostals and diaphragm) contract during inspiration, the size of the thoracic cavity increases. The diaphragm moves from its relaxed dome shape to a flattened position, increasing the superoinferior volume. The external intercostals lift the rib cage, increasing the anteroposterior and lateral dimensions (Figure 38.1). Since the lungs adhere to the thoracic walls like flypaper because of the presence of serous fluid in the intrapleural space, the intrapulmonary volume (volume within the lungs) also increases, lowering the air (gas) pressure inside the lungs. The gases then expand to fill the available space, creating a partial vacuum that causes air to flow into the lungs—constituting the act of inspiration. During expiration, the inspiratory muscles relax, and the natural tendency of the elastic lung tissue to recoil acts to decrease the intrathoracic and intrapulmonary volumes. As the gas molecules within the lungs are forced closer together, the intrapulmonary pressure rises to a point higher than atmospheric pressure. This causes gases to flow from the lungs to equalize the pressure inside and outside the lungs—the act of expiration.

Activity:
Operating the Model Lung

Observe the model lung, which demonstrates the principles involved in gas flows into and out of the lungs. It is a simple apparatus with a bottle "thorax," a rubber membrane "diaphragm," and "balloon lungs."

1. Go to the demonstration area and work the model lung by moving the rubber diaphragm up and down. Notice the changes in balloon (lung) size as the volume of the thoracic cavity is alternately increased and decreased.

2. Check the appropriate columns in the chart concerning these observations in the Exercise 38 Review Sheet on p. 687 at the back of this book.

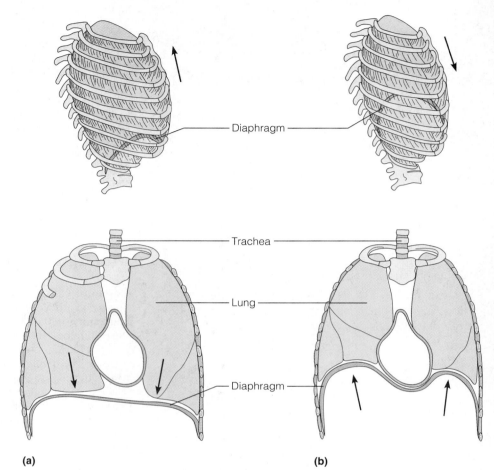

Inspiration

Expiration

Diaphragm

Trachea

Lung

Diaphragm

Figure 38.1 Rib cage and diaphragm positions during breathing. (a) At the end of a normal inspiration; chest expanded, diaphragm depressed. **(b)** At the end of a normal expiration; chest depressed, diaphragm elevated.

(a)

(b)

3. After observing the operation of the model lung, conduct the following tests on your lab partner. Use the tape measure to determine his or her chest circumference by placing the tape around the chest as high up under the armpits as possible. Record the measurements in inches in the appropriate space for each of the conditions below.

Quiet breathing:

Inspiration _____ Expiration _____

Forced breathing:

Inspiration _____ Expiration _____

Do the results coincide with what you expected on the basis

of what you have learned thus far? _____ ■

Respiratory Volumes and Capacities—Spirometry

A person's size, sex, age, and physical condition produce variations in respiratory volumes. Normal quiet breathing moves about 500 ml of air in and out of the lungs with each breath. As you have seen in the previous experiment, a person can usually forcibly inhale or exhale much more air than is exchanged in normal quiet breathing. The terms given to the measurable respiratory volumes are defined just below. These terms and their normal values for an adult male should be memorized.

Tidal volume (TV): Amount of air inhaled or exhaled with each breath under resting conditions (500 ml)

Inspiratory reserve volume (IRV): Amount of air that can be forcefully inhaled after a normal tidal volume inhalation (3100 ml)

Expiratory reserve volume (ERV): Amount of air that can be forcefully exhaled after a normal tidal volume exhalation (1200 ml)

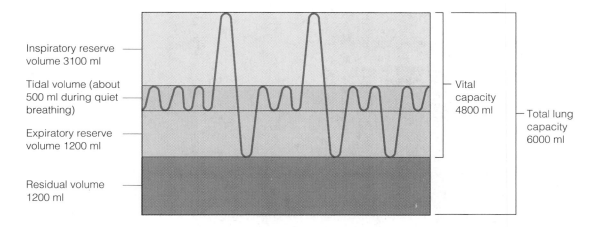

Figure 38.2   Idealized tracing of the various respiratory volumes.

Figure 38.3   The Wright handheld dry spirometer. Reset to zero prior to each test.

Vital capacity (VC): Maximum amount of air that can be exhaled after a maximal inspiration (4800 ml)

$$VC = TV + IRV + ERV$$

An idealized tracing of the various respiratory volumes and their relationships to each other are shown in Figure 38.2.

Respiratory volumes will be measured with an apparatus called a **spirometer.** There are two major types of spirometers, which give comparable results—the handheld dry, or wheel, spirometers (such as the Wright spirometer illustrated in Figure 38.3) and "wet" spirometers, such as the Phipps and Bird spirometer and the Collins spirometer (which is available in both recording and nonrecording varieties). The somewhat more sophisticated wet spirometer consists of a

plastic or metal *bell* within a rectangular or cylindrical tank that air can be added to or removed from (Figure 38.4). The outer tank contains water and has a tube running through it to carry air above the water level. The floating bottomless bell is inverted over the water-containing tank and connected to a volume indicator.

In nonrecording spirometers, an indicator moves as air is *exhaled,* and only expired air volumes can be measured directly. By contrast, recording spirometers allow both inspired and expired gas volumes to be measured. Directions for both types of apparatus are provided below.

Activity:
Measuring Respiratory Volumes

Follow the steps in Procedure A if using a nonrecording spirometer, and those in Procedure B if using a wet recording spirometer.

Procedure A

1.   Without using the spirometer, count and record the subject's normal respiratory rate.

Respirations per minute _____

2.   Identify the parts of the spirometer you will be using by comparing it to the illustration in Figure 38.3 or 38.4a. Examine the spirometer volume indicator *before beginning* to make sure you know how to read the scale. Work in pairs, with one person acting as the subject while the other records the data of the volume determinations. The subject should stand erect during testing. Reset the indicator to zero before beginning each trial. If you are using the handheld spirometer, make sure its dial faces upward so that the volumes can be easily read during the tests.

Obtain a disposable cardboard mouthpiece. Insert it in the open end of the valve assembly (attached to the flexible tube) of the wet spirometer or over the fixed stem of the handheld dry spirometer. Before beginning, the subject

Volume indicator scale

Floating bell

Drum retainer

Water tank

(a)

Drum retainer

Reset knob

Mouthpiece

Free–breathing valve

Floating bell

Flexible arm

Kymograph drum with chart paper

Water tank

Ventilometer pen

Thermometer

Power switch

Oxygen inlet

(b)

Speed switch (fast, medium, slow)

Figure 38.4 Wet spirometers. (a) The Phipps and Bird "wet" spirometer.
(b) The Collins-9L "wet" recording spirometer.

should practice exhaling through the mouthpiece without exhaling through the nose, or prepare to use the nose clips (clean them first with an alcohol swab).

3. Conduct the test three times for each required measurement. Record the data here, and then find the average volume figure for that respiratory measurement. After you have completed the trials and computed the averages, enter the average values on the table prepared on the chalkboard for tabulation of class data,* and copy all averaged data onto the Exercise 38 Review Sheet.

4. **Tidal volume (TV).** The volume of air inhaled and exhaled with each normal respiration is approximately 500 ml. To conduct the test, inhale a normal breath, and then exhale a normal breath of air into the spirometer mouthpiece. (Do not force the expiration!) Record the volume and repeat the test twice.

trial 1 _____ ml   trial 2 _____ ml

trial 3 _____ ml   average TV _____ ml

5. Compute the subject's **minute respiratory volume (MRV)** using the following formula:

$$MRV = TV \times \text{respirations/min}$$

MRV _____ ml/min

6. **Expiratory reserve volume (ERV).** The volume of air that can be forcibly exhaled after a normal expiration ranges between 1000 and 1200 ml.

Inhale and exhale normally two or three times, then insert the spirometer mouthpiece and exhale forcibly as much of the additional air as you can. Record your results, and repeat the test twice again.

trial 1 _____ ml   trial 2 _____ ml

trial 3 _____ ml   average ERV _____ ml

The ERV is dramatically reduced in conditions in which the elasticity of the lungs is decreased by a chronic obstructive pulmonary disease (COPD) such as **emphysema.** Since energy must be used to *deflate* the lungs in such conditions, expiration is physically exhausting to individuals suffering from COPD. ■

7. **Vital capacity (VC).** The total exchangeable air of the lungs (the sum of TV + IRV + ERV) is normally around 4500 ml with a range of 3600 ml to 4800 ml. Breathe in and out normally two or three times, and then bend over and ex-

* Note to the Instructor: The format of class data tabulation can be similar to that shown here. However, it would be interesting to divide the class into smokers and nonsmokers and then compare the mean average VC and ERV for each group. Such a comparison might help to determine if smokers are handicapped in any way. It also might be a good opportunity for an informal discussion of the early warning signs of bronchitis and emphysema, which are primarily smokers' diseases.

hale all the air possible. Then, as you raise yourself to the upright position, inhale as fully as possible. (It is very important to *strain* to inhale the maximum amount of air that you can.) Quickly insert the mouthpiece, and exhale as forcibly as you can. Record your results and repeat the test twice again.

trial 1 _____ ml   trial 2 _____ ml

trial 3 _____ ml   average VC _____ ml

8. **Inspiratory reserve volume (IRV).** The IRV, or volume of air that can be forcibly inhaled following a normal inspiration, can now be computed using the average values obtained for TV, ERV, and VC and plugging them into the equation:

$$IRV = VC - (TV + ERV)$$

Record your average IRV: _____ ml

The normal IRV is substantial, ranging from 2100 to 3100 ml. How does your computed value compare?

_____

Steps 9 and 10, which provide common directions for both nonrecording and recording spirometers, continue after the Procedure B directions.

Procedure B

1. In preparation for recording, familiarize yourself with the spirometer by comparing it to the equipment illustrated in Figure 38.4b.

2. Examine the chart paper, noting that its horizontal lines represent milliliter units. To apply the chart paper to the recording drum, lift the drum retainer and then remove the kymograph drum. Wrap a sheet of chart paper around the drum, making sure that the right edge overlaps the left. Fasten it with tape, and then replace the kymograph drum and lower the drum retainer into its original position in the hole in the top of the drum.

3. Raise and lower the floating bell several times, noting as you do so that the *ventilometer pen* moves up and down on the drum. This pen, which writes in black ink, will be used for recording and should be positioned or adjusted so that it records in the approximate middle of the chart paper. This adjustment is made by repositioning the floating bell using the *reset knob* on the metal pulley at the top of the spirometer apparatus. The other pen, the respirometer pen, which records in red ink, will not be used for these tests and should be moved away from the drum's recording surface.

4. Clean the nose clips with an alcohol swab. While you wait for the alcohol to air dry, count and record your normal respiratory rate.

Respirations per minute _____

After the alcohol has air dried, apply the nose clips to your nose. This will enforce mouth breathing.

Figure 38.5 A typical spirometry recording of tidal volume, inspiratory capacity, expiratory reserve volume, and vital capacity. At a drum speed of 32 mm/min, each vertical column of the chart represents a time interval of 1 minute.

5. Open the *free-breathing valve.* Insert a disposable cardboard mouthpiece into the end (valve assembly) of the breathing tube, and then insert the mouthpiece into your mouth. Practice breathing for several breaths to get used to the apparatus. At this time, you are still breathing room air.

6. Set the spirometer switch to SLOW (32 mm/min). Close the free-breathing valve, and breathe in a normal manner for 2 minutes to record your tidal volume—the amount of air inspired or expired with each normal respiratory cycle. This recording should show a regular pattern of inspiration-expiration spikes and should gradually move upward on the chart paper. (A downward slope indicates that there is an air leak somewhere in the system—most likely at the mouthpiece.) Notice that on an apparatus using a counterweighted

pen, such as the Collins Vitalometer shown in Fig. 38.4b, inspirations are recorded by upstrokes and expirations are recorded by downstrokes.*

7. To record your vital capacity, take the deepest possible inspiration you can and then exhale to the greatest extent possible (really *push* the air out). The recording obtained should resemble that shown in Figure 38.5. Repeat the vital capacity maneuver twice again. Then turn off the spirometer and remove the chart paper from the kymograph drum.

* If a Collins survey spirometer is used, the situation is exactly opposite: Upstrokes are expirations and downstrokes are inspirations.

8. Determine and record your measured, averaged, and corrected respiratory volumes here. Because the pressure and temperature inside the spirometer are influenced by room temperature and differ from those in the body, all measured values are to be multiplied by a **BTPS** (body temperature, atmospheric pressure, and water saturation) **factor.** At room temperature, the BTPS factor is typically 1.1 or very close to that value. Hence, you will multiply your measured values by 1.1 to obtain your corrected respiratory volume values. Copy the averaged and corrected values onto the Exercise 38 Review Sheet.

• Tidal volume (TV). Select a typical resting tidal breath recording. Subtract the millimeter value of the trough (exhalation) from the millimeter value of the peak (inspiration). Record this value below as *measured TV 1*. Select two other TV tracings to determine the TV values for the TV 2 and TV 3 measurements. Then, determine your average TV and multiply it by 1.1 to obtain the BTPS-corrected average TV value.

measured TV 1 _____ ml   average TV _____ ml

measured TV 2_____ ml   corrected average TV .

measured TV 3_____ ml   _____ ml

Also compute your **minute respiratory volume (MRV)** using the following formula:

$$MRV = TV \times \text{respirations/min}$$

MRV _____ ml/min

• Inspiratory capacity (IC). In the first vital capacity recording, find the expiratory trough immediately preceding the maximal inspiratory peak achieved during vital capacity determination. Subtract the milliliter value of that expiration from the value corresponding to the peak of the maximal inspiration that immediately follows. For example, according to Figure 38.5, these values would be

$$6600 - 3650 = 2950 \text{ ml}$$

Record your computed value and the results of the two subsequent tests on the appropriate lines below. Then calculate the measured and corrected inspiratory capacity averages and record.

measured IC 1 _____ ml   average IC _____ ml

measured IC 2_____ ml   corrected average IC

measured IC 3_____ ml   _____ ml

• Inspiratory reserve volume (IRV). Subtract the corrected average tidal volume from the corrected average for the inspiratory capacity and record below.

$$IRV = \text{corrected average IC} - \text{corrected average TV}$$

corrected average IRV _____ ml

• Expiratory reserve volume (ERV). Subtract the number of milliliters corresponding to the trough of the maximal expiration obtained during the vital capacity maneuver from milliliters corresponding to the last *normal* expiration before the VC maneuver is performed. For example, according to Figure 38.5, these values would be

$$3650 \text{ ml} - 2050 \text{ ml} = 1650 \text{ ml}$$

Record your measured and averaged values (three trials) below.

measured ERV 1_____ ml   average ERV _____ ml

measured ERV 2_____ ml   corrected average ERV

measured ERV 3 _____ ml   _____ ml

• Vital capacity (VC). Add your corrected values for ERV and IC to obtain the corrected average VC. Record below and on the Exercise 38 Review Sheet.

corrected average VC _____ ml

[*Now continue with step 9, whether you are following Procedure A or Procedure B.*]

9. Figure out how closely your measured average vital capacity volume compares with the *predicted values* for someone your age, sex, and height. Obtain the predicted figure from Appendix B either from Table 1 (male values) or Table 2 (female values). Notice that you will have to convert your height in inches to centimeters (cm) to find the corresponding value. This is easily done by multiplying your height in inches by 2.54.

Computed height: _____ cm

Predicted VC value (obtained from the appropriate table):

_____ ml

Use the following equation to compute your VC as a percentage of the predicted VC value:

$$\% \text{ of predicted VC} = \frac{\text{averaged measured VC}}{\text{predicted value}} \times 100$$

% predicted VC value: _____ %

Figure 38.2 is an idealized tracing of the respiratory volumes described and tested in this exercise. Examine it carefully. How closely do your test results compare to the values in the tracing?

---

10. A respiratory volume that cannot be experimentally demonstrated here is the residual volume (RV), which is the amount of air remaining in the lungs after a maximal expiratory effort. The presence of residual air (usually about 1200 ml) that cannot be voluntarily flushed from the lungs is important because it allows gas exchange to go on continuously—even between breaths.

Although the residual volume cannot be measured directly, it can be approximated by using one of the following factors:

For ages 16–34   Factor = 0.250

For ages 35–49   Factor = 1.305

For ages 50–69   Factor = 1.445

Compute your predicted RV using the following equation:

$$RV = VC \times factor$$

⚠ 11. Recording is finished for this subject. Before continuing with the next member of your group:

● Dispose of used cardboard mouthpieces in the autoclave bag.

● Swish the valve assembly (if removable) in the 70% ethanol solution, then rinse with tap water.

● Put a fresh mouthpiece into the valve assembly (or on the stem of the handheld spirometer). Using the procedures outlined above measure and record the respiratory volumes for all members of your group. ■

Forced Expiratory Volume (FEV$_T$) Measurement

While they are not really diagnostic, pulmonary function tests can help the clinician to distinguish between obstructive and restrictive pulmonary diseases. (In obstructive disorders, like chronic bronchitis and asthma, airway resistance is increased, whereas in restrictive diseases, such as polio and tuberculosis, total lung capacity declines.) Two highly useful pulmonary function tests used for this purpose are the FVC and the FEV$_T$.

The **FVC** (forced vital capacity) measures the amount of gas expelled when the subject takes the deepest possible breath and then exhales forcefully and rapidly. This volume is reduced in those with restrictive pulmonary disease. The **FEV$_T$** (forced expiratory volume) involves the same basic testing procedure, but it specifically looks at the percentage of the vital capacity that is exhaled during specific time intervals of the FVC test. FEV$_1$, for instance, is the amount exhaled during the first second. Healthy individuals can expire 75% to 85% of their FVC in the first second. The FEV$_1$ is low in those with obstructive disease.

Activity:
Measuring the FVC and FEV$_1$

Directions provided here for the FEV$_T$ determination apply only to the recording spirometer.

1. Prepare to make your recording as described in Procedure B, steps 1–6 just above.

2. At a signal agreed upon by you and your lab partner, take the deepest inspiration possible and hold it for 1 to 2 seconds. As the inspiratory peak levels off, your partner is to change the drum speed to FAST (1920 mm/ min) so that the distance between the vertical lines on the chart represents a time of 1 second.

3. Once the drum speed has been changed, exhale as much air as rapidly and forcibly as possible.

4. When the tracing plateaus (bottoms out), stop recording and determine your FVC. Subtract the milliliter reading in the expiration trough (the bottom plateau) from the preceding inhalation peak (the top plateau). Record this value.

FVC = _____ ml

5. Prepare to calculate the FEV$_1$. Draw a vertical line intersecting with the spirogram tracing at the precise point that exhalation began. Identify this line as *line 1*. From line 1, measure 32 mm horizontally to the left, and draw a second vertical line. Label this as *line 2*. The distance between the two lines represents 1 second, and the volume exhaled in the first second is read where line 2 intersects the spirogram tracing. Subtract that milliliter value from the milliliter value of the inhalation peak (at the intersection of line 1), to determine the volume of gas expired in the first second. According to the values given in Figure 38.6, that figure would be 3400 ml (6800 ml − 3400 ml). Record your measured value below.

Milliliters of gas expired in second 1: _____ ml

6. To compute the FEV$_1$ use the following equation:

$$FEV_1 = \frac{\text{volume expired in second 1}}{\text{FVC volume}} \times 100\%$$

Record your calculated value below and on the Exercise 38 Review Sheet.

FEV$_1$ = _____ % of FVC ■

Figure 38.6 **A recording of the forced vital capacity (FVC) and forced expiratory volume (FEV) or timed vital capacity test.**

Use of the Pneumograph to Determine Factors Influencing Rate and Depth of Respiration*

The neural centers that control respiratory rhythm and maintain a rate of 12 to 18 respirations/min are located in the medulla and pons. On occasion, input from the stretch receptors in the lungs (via the vagus nerve to the medulla) modifies the respiratory rate, as in cases of extreme overinflation of the lungs (Hering-Breuer reflex).

Death occurs when medullary centers are completely suppressed, as from an overdose of sleeping pills or gross overindulgence in alcohol, and respiration ceases completely.

Although the nervous system centers initiate the basic rhythm of breathing, there is no question that physical phenomena such as talking, yawning, coughing, and exercise can modify the rate and depth of respiration. So too can chemical factors such as changes in oxygen or carbon dioxide concentrations in the blood or fluctuations in blood pH. Changes in carbon dioxide blood levels seem to act directly on the medulla control centers, whereas changes in pH and oxygen concentrations are monitored by chemoreceptor regions in the aortic and carotid bodies, which in turn send input to the medulla. The experimental sequence in this section is designed to test the relative importance of various physical and chemical factors in the process of respiration.

The **pneumograph,** an apparatus that records variations in breathing patterns, is the best means of observing respiratory variations resulting from physical and chemical factors. The chest pneumograph is a coiled rubber hose that is attached around the thorax. As the subject breathes, chest movements produce pressure changes within the pneumograph that are transmitted to a recorder.

The instructor will demonstrate the method of setting up the pneumograph and discuss the interpretation of the results. Work in pairs so that one person can mark the record to identify the test for later interpretation. Ideally, the student being tested should face away from the recording apparatus to prevent voluntary modification of the record.

Activity:
Using the Pneumograph to Visualize Respiratory Variations

1. Attach the pneumograph tubing firmly, but not restrictively, around the thoracic cage at the level of the sixth rib, leaving room for chest expansion during testing. If the subject is female, position the tubing above the breasts to prevent slippage during testing. Set the pneumograph speed at 1 or 2, and the time signal at 10-second intervals. Record quiet breathing for 1 minute with the subject in a sitting position.

* Note to the Instructor: This exercise may also be done without using the recording apparatus by simply having the students count the respiratory rate visually.

Record breaths per minute. _____

2. Make a vital capacity tracing: Record a maximal inhalation followed by a maximal exhalation. This should correlate to the vital capacity measurement obtained earlier and will provide a baseline for comparison during the rest of the pneumograph testing. Stop the recording apparatus and indicate the following on the graph by marking the graph appropriately: tidal volume, expiratory reserve volume, inspiratory reserve volume, and vital capacity (the total of the three measurements). Also mark, with arrows, the direction the recording stylus moves during inspiration and during expiration.

Measure in mm the height of the vital capacity recording. Divide the vital capacity measurement average recorded on the Exercise 38 Review Sheet by the millimeter figure to obtain the volume (in milliliters of air) represented by 1 mm on the recording. For example, if your vital capacity reading is 4000 ml and the vital capacity tracing occupies a vertical distance of 40 mm on the pneumograph recording, then a vertical distance of 1 mm equals 100 ml of air.

Record your computed value. _____ ml air/mm

3. Record the subject's breathing as he or she performs activities from the list below. Make sure the record is marked accurately to identify each test conducted. Record your results on the Exercise 38 Review Sheet.

talking	swallowing water
yawning	coughing
laughing	lying down
standing	doing a math problem (concentrating)
running in place	

4. Without recording, have the subject breathe normally for 2 minutes, then inhale deeply and hold his or her breath for as long as he or she can.

Time the breath-holding interval. _____ sec

As the subject exhales, turn on the recording apparatus and record the recovery period (time to return to normal breathing—usually slightly over 1 minute):

Time of recovery period. _____ sec

Did the subject have the urge to inspire *or* expire during breath holding?

_____

Without recording, repeat the above experiment, but this time exhale completely and forcefully *after* taking the deep breath. What was observed this time?

_____

_____

Explain the results. (Hint: the vagus nerve is the sensory nerve of the lungs and plays a role here.)

_____

_____

_____

5. Have the subject hyperventilate (breathe deeply and forcefully at the rate of 1 breath/4 sec) for about 30 seconds.* Record both during and after hyperventilation. How does the pattern obtained during hyperventilation compare with that recorded during the vital capacity tracing?

_____

_____

Is the respiratory rate after hyperventilation faster _or_ slower than during normal quiet breathing?

_____

_____

6. Repeat the above test, but do not record until after hyperventilating. After hyperventilation, the subject is to hold his or her breath as long as he or she can. Can the breath be held for a longer or shorter time after hyperventilating?

_____

_____

7. Without recording, have the subject breathe into a paper bag for 3 minutes, then record his or her breathing movements.

⚠ _Caution_: During the bag-breathing exercise the subject's partner should watch the subject carefully for any untoward reactions.

Is the breathing rate faster _or_ slower than that recorded during normal quiet breathing?

_____

_____

After hyperventilating? _____

8. Run in place for 2 minutes, and then have your partner determine the length of time that you can hold your breath.

Length of breath-holding. _____ sec

9. To prove that respiration has a marked effect on circulation, conduct the following test. Have your lab partner record the rate and relative force of your radial pulse before you begin.

Rate _____ beats/min   Relative force _____

* A sensation of dizziness may develop. (As the carbon dioxide is washed out of the blood by overventilation, the blood pH increases, leading to a decrease in blood pressure and reduced cerebral circulation.) The subject may experience a lack of desire to breathe after forced breathing is stopped. If the period of breathing cessation—apnea—is extended, cyanosis of the lips may occur.

Inspire forcibly. Immediately close your mouth and nose to retain the inhaled air, and then make a forceful and prolonged expiration. Your lab partner should observe and record the condition of the blood vessels of your neck and face, and again immediately palpate the radial pulse.

Observations _____

_____

Radial pulse _____ beats/min   Relative force _____

Explain the changes observed. _____

_____

_____

⚠ Dispose of the paper bag in the autoclave bag. Keep the pneumograph records to interpret results and hand them in if requested by the instructor. Observation of the test results should enable you to determine which chemical factor, carbon dioxide or oxygen, has the greatest effect on modifying the respiratory rate and depth. ■

Respiratory Sounds

As air flows in and out of the respiratory tree, it produces two characteristic sounds that can be picked up with a stethoscope (auscultated). The **bronchial sounds** are produced by air rushing through the large respiratory passageways (the trachea and the bronchi). The second sound type, **vesicular breathing sounds,** apparently results from air filling the alveolar sacs and resembles the sound of a rustling or muffled breeze.

Activity:
Auscultating Respiratory Sounds

1. Obtain a stethoscope and clean the earpieces with an alcohol swab. Allow the alcohol to dry before donning the stethoscope.

2. Place the diaphragm of the stethoscope on the throat of the test subject just below the larynx. Listen for bronchial sounds on inspiration and expiration. Move the stethoscope down toward the bronchi until you can no longer hear sounds.

3. Place the stethoscope over the following chest areas and listen for vesicular sounds during respiration (heard primarily during inspiration).

• At various intercostal spaces

• At the _triangle of auscultation_ (a small depressed area of the back where the muscles fail to cover the rib cage; located just medial to the inferior part of the scapula)

• Under the clavicle ■

Diseased respiratory tissue, mucus, or pus can produce abnormal chest sounds such as rales (a rasping sound) and wheezing (a whistling sound). ■

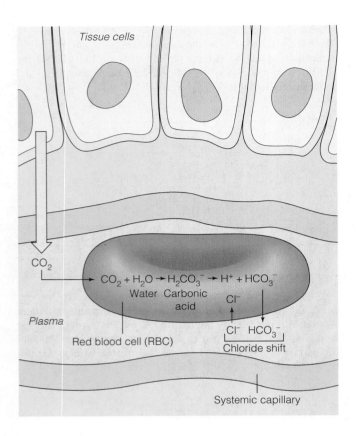

Figure 38.7 Carbon dioxide loading in the tissues. As carbon dioxide is loaded into the blood, it enters red blood cells where it combines with water to yield carbonic acid, which immediately splits to HCO_3^- (bicarbonate ions) and H^+. HCO_3^- diffuses out of the RBC into the plasma, and chloride ions enter the RBC in exchange (the chloride shift).

Role of the Respiratory System in Acid-Base Balance of Blood

As you have already learned, pulmonary ventilation is necessary for continuous oxygenation of the blood and removal of carbon dioxide (a waste product of cellular respiration) from the blood. Blood pH must be relatively constant for the cells of the body to function optimally. The carbonic acid–bicarbonate buffer system of the blood is extremely important because it helps stabilize arterial blood pH at 7.4 ± 0.02.

When carbon dioxide diffuses into the blood from the tissue cells, much of it enters the red blood cells, where it combines with water to form carbonic acid (Figure 38.7):

$$H_2O + CO_2 \xrightarrow[\text{enzyme present in RBC}]{\text{Carbonic anhydrase}} H_2CO_3$$

Some carbonic acid is also formed in the plasma, but its formation there is very slow because of the lack of the carbonic anhydrase enzyme. Shortly after it forms, carbonic acid dissociates to release bicarbonate (HCO_3^-) and hydrogen (H^+) ions. The hydrogen ions that remain in the cells are neutralized, or buffered, when they combine with hemoglobin molecules. If they were not neutralized, the intracellular pH would become very acidic as H^+ ions accumulated. The bicarbonate ions diffuse out of the red blood cells into the plasma, where they become part of the carbonic acid–bicarbonate buffer system. As HCO_3^- follows its concentration gradient into the plasma, an electrical imbalance develops in the RBCs that draws Cl^- into them from the plasma. This exchange phenomenon is called the *chloride shift*.

Acids (more precisely, H^+) released into the blood by the body cells tend to lower the pH of the blood and to cause it to become acidic. On the other hand, basic substances that enter the blood tend to cause the blood to become more alkaline and the pH to rise. Both of these tendencies are resisted in large part by the carbonic acid–bicarbonate buffer system. If H^+ concentration in the blood begins to increase, the H^+ ions combine with bicarbonate ions to form carbonic acid (a weak acid that does not tend to dissociate at physiologic or acid pH) and are thus removed.

$$H^+ + HCO_3^- \rightarrow H_2CO_3$$

Likewise, as blood H^+ concentration drops below what is desirable and blood pH rises, H_2CO_3 dissociates to release bicarbonate ions and H^+ ions to the blood. The released H^+ lowers the pH again. The bicarbonate ions, being *weak* bases, are poorly functional under alkaline conditions and have little effect on blood pH unless and until blood pH drops toward acid levels.

$$H_2CO_3 \rightarrow H^+ + HCO_3^-$$

In the case of excessively slow or shallow breathing (hypoventilation) or fast deep breathing (hyperventilation), the amount of carbonic acid in the blood can be greatly modified—increasing dramatically during hypoventilation and decreasing substantially during hyperventilation. In either situation, if the buffering ability of the blood is inadequate, respiratory acidosis or alkalosis can result. Therefore, maintaining the normal rate and depth of breathing is important for proper control of blood pH.

Activity:
Observing the Operation of Standard Buffers

1. To observe the ability of a buffer system to stabilize the pH of a solution, obtain five 250-ml beakers, and a wash bottle containing distilled water. Set up the following experimental samples:

Beaker 1
(150 ml distilled water)   pH _____

Beaker 2
(150 ml distilled water and
1 drop concentrated HCl)   pH _____

Beaker 3
(150 ml distilled water and
1 drop concentrated NaOH)   pH _____

Beaker 4
(150 ml standard buffer solution
[pH 7] and 1 drop concentrated HCl)   pH _____

Beaker 5
(150 ml standard buffer solution
[pH 7] and 1 drop concentrated NaOH)   pH _____

2. Using a pH meter standardized with a buffer solution of pH 7, determine the pH of the contents of each beaker and record above. After *each and every* pH recording, the pH meter switch should be turned to *standby,* and the electrodes rinsed thoroughly with a stream of distilled water from the wash bottle.

3. Add 3 more drops of concentrated HCl to beaker 4,

stir, and record the pH: _____

4. Add 3 more drops of concentrated NaOH to beaker 5,

stir, and record the pH: _____

How successful was the buffer solution in resisting pH changes when a strong acid (HCl) or a strong base (NaOH) was added?

_____ ■

Activity:
Exploring the Operation of the Carbonic Acid–Bicarbonate Buffer System

To observe the ability of the carbonic acid–bicarbonate buffer system of blood to resist pH changes, perform the following simple experiment.

1. Obtain two small beakers (50 ml), animal plasma, graduated cylinder, glass stirring rod, and a dropper bottle of 0.01 *M* HCl. Using the pH meter standardized with the buffer solution of pH 7.0, measure the pH of the animal plasma. Use only enough plasma to allow immersion of the electrodes and measure the volume used carefully.

pH of the animal plasma: _____

2. Add 2 drops of the 0.01 *M* HCl solution to the plasma; stir and measure the pH again.

pH of plasma plus 2 drops of HCl: _____

3. Turn the pH meter switch to *standby,* rinse the electrodes, and then immerse them in a quantity of distilled water (pH 7) exactly equal to the amount of animal plasma used. Measure the pH of the distilled water.

pH of distilled water: _____

4. Add 2 drops of 0.01 *M* HCl, swirl, and measure the pH again.

pH of distilled water plus the two drops of HCl: _____

Is the plasma a good buffer? _____

What component of the plasma carbonic acid–bicarbonate buffer system was acting to counteract a change in pH when HCl was added?

_____ ■

exercise
39
Anatomy of the Digestive System

Objectives

1. To state the overall function of the digestive system.

2. To identify on an appropriate diagram or torso model the organs comprising the alimentary canal; and to name their subdivisions if any.

3. To name and/or identify the accessory digestive organs.

4. To describe the general functions of the digestive system organs or structures.

5. To describe the general histologic structure of the wall of the alimentary canal and/or label a cross-sectional diagram of the wall with the following terms: mucosa, submucosa, muscularis externa, and serosa or adventitia.

6. To list and explain the specializations of anatomical structure of the stomach and small intestine that contribute to their functional roles.

7. To list the major enzymes or enzyme groups produced by each of the following organs: salivary glands, stomach, small intestine, pancreas.

8. To name human deciduous and permanent teeth and describe the anatomy of the generalized tooth.

9. To recognize by microscopic inspection, or by viewing an appropriate diagram or photomicrograph, the histologic structure of the following organs:

small intestine	pancreas	stomach
salivary glands	tooth	liver

Materials

❑ Dissectible torso model

❑ Anatomical chart of the human digestive system

❑ Model of a villus and liver lobules (if available)

❑ Jaw model or human skull

❑ Prepared microscope slides of the liver, pancreas, and mixed salivary glands; of longitudinal sections of the gastroesophageal junction and a tooth; and of cross sections of the stomach, duodenum, and ileum

❑ Compound microscope

❑ *Human Digestive System* Videotape*

 For instructions on animal dissections, see the dissection exercises starting on p. 733 in the cat and fetal pig editions of this manual.

 See Appendix C, Exercise 39 for links to A.D.A.M. Interactive Anatomy.

*Available to qualified adopters from Benjamin/Cummings.

The **digestive system** provides the body with the nutrients, water, and electrolytes essential for health. The organs of this system ingest, digest, and absorb food and eliminate the undigested remains as feces.

The digestive system consists of a hollow tube extending from the mouth to the anus, into which various accessory organs or glands empty their secretions (Figure 39.1). Food material within this tube, the *alimentary canal,* is technically outside the body because it has contact only with the cells lining the tract. For ingested food to become available to the body cells, it must first be broken down *physically* (chewing, churning) and *chemically* (enzymatic hydrolysis) into its smaller diffusible molecules—a process called **digestion.** The digested end products can then pass through the epithelial cells lining the tract into the blood for distribution to the body cells—a process termed **absorption.** In one sense, the digestive tract can be viewed as a disassembly line, in which food is carried from one stage of its digestive processing to the next by muscular activity, and its nutrients are made available to the cells of the body en route.

The organs of the digestive system are traditionally separated into two major groups: the **alimentary canal,** or **gastrointestinal (GI) tract,** and the **accessory digestive organs.** The alimentary canal is approximately 30 feet long in a cadaver but is considerably shorter in a living person. It consists of the mouth, pharynx, esophagus, stomach, small and large intestines, rectum, and anus. The accessory structures include the salivary glands, gallbladder, liver, and pancreas, which secrete their products into the alimentary canal. These individual organs are described shortly.

General Histological Plan of the Alimentary Canal

Because the alimentary canal has a shared basic structural plan (particularly from the esophagus to the anus), it makes sense to review that structure as we begin studying this group

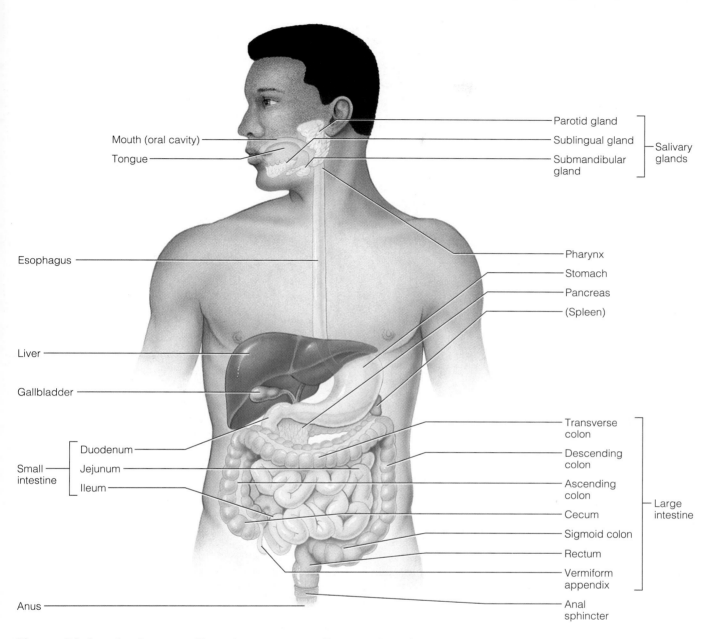

Figure 39.1   The human digestive system: alimentary tube and accessory organs. (Liver and gallbladder are reflected superiorly and to the right.)

of organs. Once done, all that need be emphasized as the individual organs are described is their specializations for unique functions in the digestive process.

Essentially the alimentary canal walls have four basic **tunics** (layers). From the lumen outward, these are the *mucosa,* the *submucosa,* the *muscularis externa,* and the *serosa* or *adventitia* (Figure 39.2). Each of these tunics has a predominant tissue type and a specific function in the digestive process.

Mucosa (mucous membrane): The mucosa is the wet epithelial membrane abutting the alimentary canal lumen. It consists of a surface *epithelium* (in most cases, a simple columnar), a *lamina propria* (areolar connective tissue on which the epithelial layer rests), and a *muscularis mucosae* (a scant layer of smooth muscle fibers that enable local move-

ments of the mucosa). The major functions of the mucosa are secretion (of enzymes, mucus, hormones, etc.), absorption of digested foodstuffs, and protection (against bacterial invasion). A particular mucosal region may be involved in one or all three functions.

Submucosa: Superficial to the mucosa, the submucosa is moderately dense connective tissue containing blood and lymphatic vessels, scattered lymph nodules, and nerve fibers. Its intrinsic nerve supply is called the *submucosal plexus.* Its major functions are nutrition and protection.

Muscularis externa: The muscularis externa, also simply called the *muscularis,* typically is a bilayer of smooth muscle, with the deeper layer running circularly and the superficial layer running longitudinally. Another important intrinsic nerve plexus, the *myenteric plexus,* is associated with this

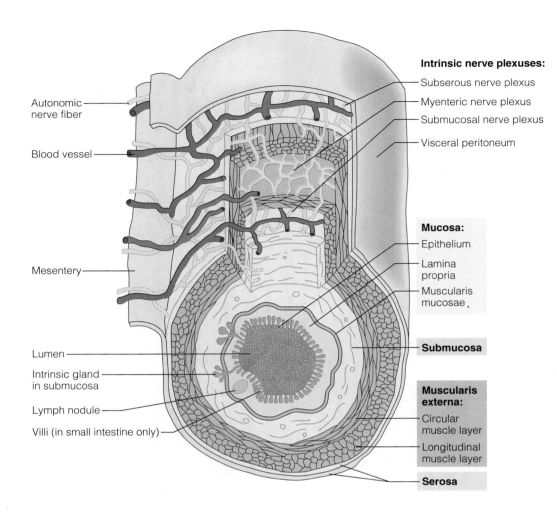

Autonomic nerve fiber

Blood vessel

Mesentery

Lumen

Intrinsic gland in submucosa

Lymph nodule

Villi (in small intestine only)

Intrinsic nerve plexuses:

Subserous nerve plexus

Myenteric nerve plexus

Submucosal nerve plexus

Visceral peritoneum

Mucosa:

Epithelium

Lamina propria

Muscularis mucosae.

Submucosa

Muscularis externa:

Circular muscle layer

Longitudinal muscle layer

Serosa

Figure 39.2 Basic structural pattern of the alimentary canal wall. (See also Plate 38 in the Histology Atlas)

tunic. By controlling the smooth muscle of the muscularis, this plexus is the major regulator of GI motility.

Serosa: The outermost serosa is equal to the *visceral peritoneum.* It consists of mesothelium associated with a thin layer of areolar connective tissue and houses a third intrinsic nerve plexus, the small *subserous plexus.* In areas *outside* the abdominopelvic cavity, the serosa is replaced by an **adventitia,** a layer of coarse fibrous connective tissue that binds the organ to surrounding tissues. (This is the case with the esophagus.) The serosa reduces friction as the mobile GI tract organs work and slide across one another and the cavity walls. The adventitia anchors and protects the surrounded GI tract organ.

Organs of the Alimentary Canal

A c t i v i t y :
Identifying Alimentary Canal Organs

The sequential pathway and fate of food as it passes through the alimentary canal organs are described in the next sections. Identify each structure in Figure 39.1 and on the torso model as you work. ∎

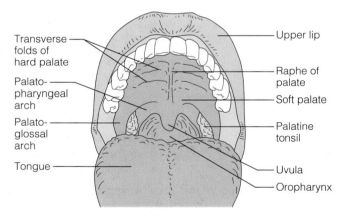

Transverse folds of hard palate

Palato-pharyngeal arch

Palato-glossal arch

Tongue

Upper lip

Raphe of palate

Soft palate

Palatine tonsil

Uvula

Oropharynx

Figure 39.3   Anterior view of the oral cavity.

Oral Cavity or Mouth

Food enters the digestive tract through the **oral cavity,** or **mouth** (Figure 39.3). Within this mucous membrane–lined cavity are the gums, teeth, tongue, and openings of the ducts of the salivary glands. The **lips** (**labia**) protect the opening of the chamber anteriorly, the **cheeks** form its lateral walls, and the **palate,** its roof. The anterior portion of the palate is re-

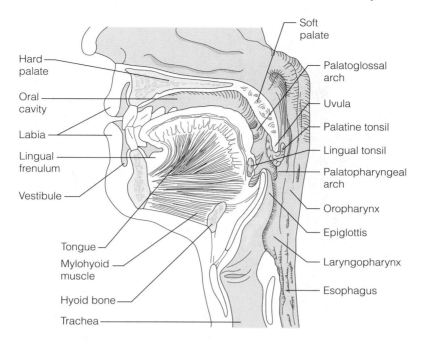

Figure 39.4 Sagittal view of the head showing oral, nasal, and pharyngeal cavities.

ferred to as the **hard palate** because bone (the palatine processes of the maxillae and the palatine bones) underlies it. The posterior **soft palate** is a fibromuscular structure that is unsupported by bone. The **uvula,** a fingerlike projection of the soft palate, extends inferiorly from its posterior margin. The soft palate rises to close off the oral cavity from the nasal and pharyngeal passages during swallowing. The floor of the oral cavity is occupied by the muscular **tongue,** which is largely supported by the **mylohyoid muscle** (Figure 39.4) and attaches to the hyoid bone, mandible, styloid processes, and pharynx. A membrane called the **lingual frenulum** secures the inferior midline of the tongue to the floor of the mouth. The space between the lips and cheeks and the teeth is the **vestibule;** the area that lies within the teeth and gums is the **oral cavity** proper.

On each side of the mouth at its posterior end are masses of lymphoid tissue, the **palatine tonsils** (Figure 39.4). Each lies in a concave area bounded anteriorly and posteriorly by membranes, the **palatoglossal arch** (anterior membrane) and the **palatopharyngeal arch** (posterior membrane). Another mass of lymphoid tissue, the **lingual tonsil,** covers the base of the tongue, posterior to the oral cavity proper. The tonsils, in common with other lymphoid tissues, are part of the body's defense system.

Very often in young children, the palatine tonsils become inflamed and enlarge, partially blocking the entrance to the pharynx posteriorly and making swallowing difficult and painful. This condition is called **tonsilitis.** ■

Three pairs of salivary glands duct their secretion, saliva, into the oral cavity. One component of saliva, salivary amylase, begins the digestion of starchy foods within the oral cavity. (The salivary glands are discussed in more detail on p. 405.)

As food enters the mouth, it is mixed with saliva and masticated (chewed). The cheeks and lips help hold the food between the teeth during mastication, and the highly mobile tongue manipulates the food during chewing and initiates swallowing. Thus the mechanical and chemical breakdown of food begins before the food has left the oral cavity. As noted in Exercise 26, the surface of the tongue is covered with papillae, many of which contain taste buds, receptors for taste sensation. So, in addition to its manipulative function, the tongue provides for the enjoyment and appreciation of the food ingested.

Pharynx

When the tongue initiates swallowing, the food passes posteriorly into the pharynx, a common passageway for food, fluid, and air (Figure 39.4). The pharynx is subdivided anatomically into three parts—the **nasopharynx** (behind the nasal cavity), the **oropharynx** (behind the oral cavity extending from the soft palate to the epiglottis overlying the larynx), and the **laryngopharynx** (extending from the epiglottis to the base of the larynx), which is continuous with the esophagus.

The walls of the pharynx consist largely of two layers of skeletal muscles: an inner layer of longitudinal muscle (the levator muscles) and an outer layer of circular constrictor muscles, which initiate wavelike contractions that propel the food inferiorly into the esophagus. Its mucosa, like that of the oral cavity, contains a friction-resistant stratified squamous epithelium.

Esophagus

The **esophagus,** or gullet, extends from the pharynx through the diaphragm to the gastroesophageal sphincter in the superior aspect of the stomach. It is approximately 10 inches long in humans and is essentially a food passageway that conducts food to the stomach in a wavelike peristaltic motion. The esophagus has no digestive or absorptive function. The walls

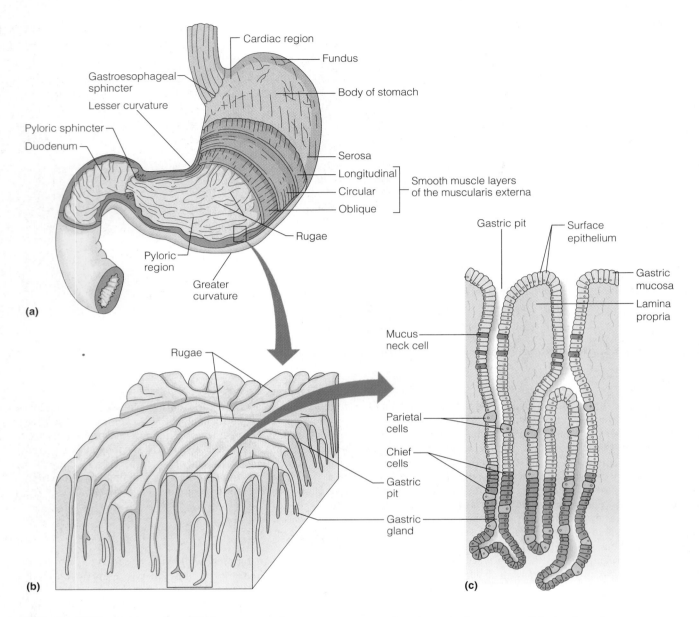

Figure 39.5 **Anatomy of the stomach.** **(a)** Gross internal and external anatomy. **(b)** Section of the stomach wall showing rugae and gastric pits. **(c)** Enlarged view of gastric pits (longitudinal section).

at its superior end contain skeletal muscle, which is replaced by smooth muscle in the area nearing the stomach. The **gastroesophageal sphincter,** a thickening of the smooth muscle layer at the esophagus-stomach junction, controls food passage into the stomach (see Figure 39.5). Since the esophagus is located in the thoracic rather than the abdominal cavity, its outermost layer is an *adventitia,* rather than the serosa.

Stomach

The **stomach** (Figures 39.1 and 39.5) is on the left side of the abdominal cavity and is hidden by the liver and diaphragm. Different regions of the saclike stomach are the **cardiac region** (the area surrounding the cardiac orifice through which food enters the stomach from the esophagus), the **fundus** (the expanded portion of the stomach, superolateral to the cardiac

region), the **body** (midportion of the stomach, inferior to the fundus), and the **pyloric region** (the terminal part of the stomach, which is continuous with the small intestine through the **pyloric sphincter**).

The concave medial surface of the stomach is called the **lesser curvature;** its convex lateral surface is the **greater curvature.** Extending from these curvatures are two mesenteries, called *omenta.* The **lesser omentum** extends from the liver to attach to the lesser curvature of the stomach. The **greater omentum,** a saclike mesentery, extends from the greater curvature of the stomach, reflects downward over the abdominal contents to cover them in an apronlike fashion, and then blends with the **mesocolon** attaching the transverse colon to the posterior body wall. Figure 39.7 (p. 400) illustrates the omenta as well as the other peritoneal attachments of the abdominal organs.

The stomach is a temporary storage region for food as well as a site for mechanical and chemical breakdown of food. It contains a third *obliquely* oriented layer of smooth muscle in its muscularis externa that allows it to churn, mix, and pummel the food, physically reducing it to smaller fragments. Gastric glands of the mucosa secrete hydrochloric acid (HCl) and hydrolytic enzymes (primarily pepsinogen, the inactive form of *pepsin,* a protein-digesting enzyme), which begin the enzymatic, or chemical, breakdown of protein foods. The mucosal glands also secrete a viscous mucus that helps prevent the stomach itself from being digested by the proteolytic enzymes. Most digestive activity occurs in the pyloric region of the stomach. After the food is processed in the stomach, it resembles a creamy mass (chyme), which enters the small intestine through the pyloric sphincter.

Activity:
Studying the Histological Structure of Selected GI Tract Organs

To prepare for the histological study you will be conducting now and later in the lab, obtain a microscope and the following slides: salivary glands (submandibular or sublingual); pancreas; liver; cross sections of the duodenum, ileum, and stomach, and longitudinal sections of a tooth and the gastroesophageal junction.

1. **Stomach:** The stomach slide will be viewed first. Refer to Figure 39.6a as you scan the tissue under low power to locate the muscularis externa; then move to high power to more closely examine this layer. Try to pick out the three smooth muscle layers. How does the extra (oblique) layer of smooth muscle found in the stomach correlate with the stomach's churning movements?

_____

_____

_____

Identify the gastric glands and the gastric pits (see Figures 39.5 and 39.6b). If the section is taken from the stomach fundus and is appropriately stained, you can identify, in the gastric glands, the blue-staining **chief,** or **zymogenic, cells,** which produce pepsinogen, and the red-staining **parietal cells,** which secrete HCl. Draw a small section of the stomach wall and label it appropriately.

(a)

(b)

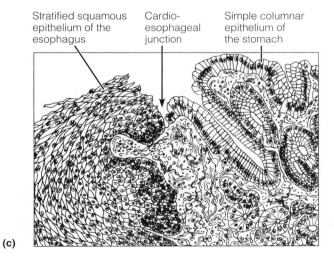

(c)

Figure 39.6   Histology of selected regions of the stomach and gastroesophageal junction. **(a)** Stomach wall. **(b)** Gastric pits and glands. **(c)** Gastroesophageal junction, longitudinal section. (See also Plates 36, 37, and 35 in the Histology Atlas.)

(a)

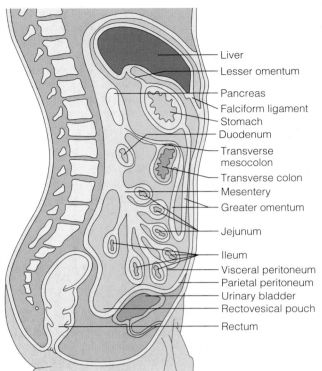

(b)

Figure 39.7 Peritoneal attachments of the abdominal organs. (a) Superficial anterior view of abdominal cavity with the greater omentum in place. (b) Sagittal view of a male torso. (c) Mesentery of the small intestine.

2. **Gastroesophageal junction:** Scan the slide under low power to locate the mucosal junction between the end of the esophagus and the beginning of the stomach, the gastroesophageal junction. Compare your observations to Figure 39.6c. What is the functional importance of the epithelial differences seen in the two organs?

_____

_____

_____ ∎

Small Intestine

The **small intestine** is a convoluted tube, 6 to 7 meters (about 20 feet) long in a cadaver but only about 2 m long during life because of its muscle tone. It extends from the pyloric sphincter to the ileocecal valve. The small intestine is suspended by a double layer of peritoneum, the fan-shaped **mesentery,** from the posterior abdominal wall (Figure 39.7), and it lies, framed laterally and superiorly by the large intestine, in the abdominal cavity. The small intestine has three subdivisions (see Figure 39.1): (1) the **duodenum** extends from the pyloric sphincter for about 25 cm (10 inches) and curves around the head of the pancreas; most of the duodenum lies in a retroperitoneal position. (2) The **jejunum,** continuous with the duodenum, extends for 2.5 m (about 8 feet). Most of the jejunum occupies the umbilical region of the ab-

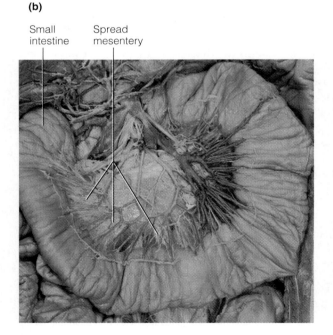

(c)

dominal cavity. (3) The **ileum,** the terminal portion of the small intestine, is about 3.6 m (12 feet) long and joins the large intestine at the **ileocecal valve.** It is located inferiorly and somewhat to the right in the abdominal cavity, but its major portion lies in the hypogastric region.

Brush border enzymes, hydrolytic enzymes bound to the microvilli of the columnar epithelial cells, and, more importantly, enzymes produced by the pancreas and ducted into the duodenum via the **pancreatic duct** complete the enzymatic digestion process in the small intestine. Bile (formed in

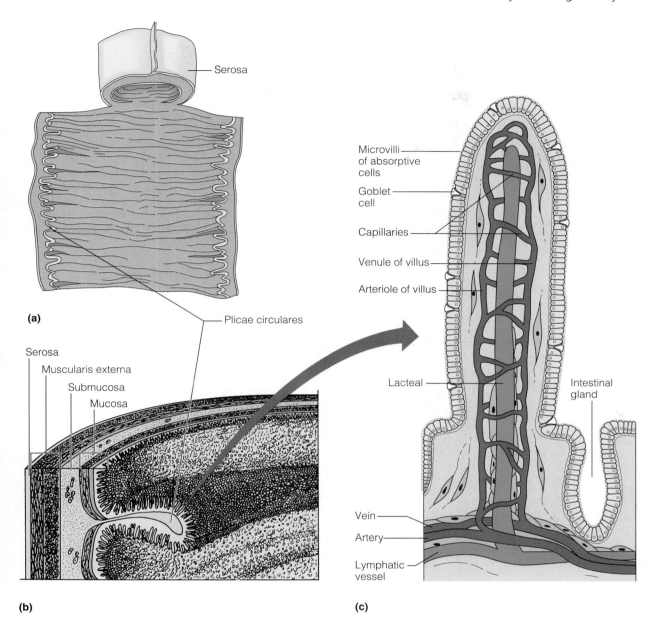

(a)

Serosa

Plicae circulares

Serosa
Muscularis externa
Submucosa
Mucosa

(b)

Microvilli of absorptive cells
Goblet cell
Capillaries
Venule of villus
Arteriole of villus
Lacteal
Intestinal gland
Vein
Artery
Lymphatic vessel

(c)

Figure 39.8 Structural modifications of the small intestine.
(a) Plicae circulares (circular folds) seen on the internal surface of the small intestine.
(b) Enlargement of one plica circulare to show villi. **(c)** Detailed anatomy of a villus.

the liver) also enters the duodenum via the **bile duct** in the same area (see Figure 39.1). At the duodenum, the ducts join to form the bulblike **hepatopancreatic ampulla** and empty their products into the duodenal lumen through the **major duodenal papilla,** an orifice controlled by a muscular valve called the **hepatopancreatic sphincter (sphincter of Oddi).**

Nearly all nutrient absorption occurs in the small intestine, where three structural modifications that increase the mucosa absorptive area appear—the microvilli, villi, and plicae circulares (Figure 39.8). **Microvilli** are minute projections of the surface plasma membrane of the columnar epithelial lining cells of the mucosa. **Villi** are the fingerlike projections of the mucosa tunic that give it a velvety appearance and texture. The **plicae circulares** are deep folds of the

mucosa and submucosa layers that force chyme to spiral through the intestine, mixing it and slowing its progress. These structural modifications, which increase the surface area, decrease in frequency and elaboration toward the end of the small intestine. Any residue remaining undigested and unabsorbed at the terminus of the small intestine enters the large intestine through the ileocecal valve. In contrast, the amount of lymphoid tissue in the submucosa of the small intestine (especially the aggregated lymphoid nodules called **Peyer's patches**) increases along the length of the small intestine and is very apparent in the ileum. This reflects the fact that the remaining undigested food residue contains large numbers of bacteria that must be prevented from entering the bloodstream.

Lamina propria — Duodenal glands (of Brunner) — Goblet cells — Villi

(a)

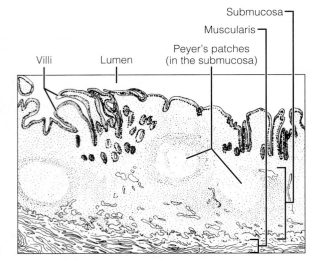

Submucosa
Muscularis
Villi — Lumen — Peyer's patches (in the submucosa)

(b)

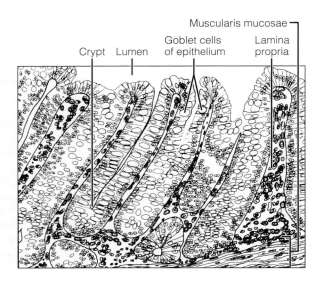

Muscularis mucosae
Goblet cells of epithelium — Lamina propria
Crypt — Lumen

(c)

Figure 39.9 Histology of selected regions of the small and large intestines (cross-sectional views). **(a)** Duodenum of the small intestine. **(b)** Ileum of the small intestine. **(c)** Large intestine. (See also Plates 39, 40, and 41 in the Histology Atlas.)

Activity:
Observing the Histological Structure of the Small Intestine

1. **Duodenum:** Secure the slide of the duodenum (cross section) to the microscope stage. Observe the tissue under low power to identify the four basic tunics of the intestinal wall—that is, the **mucosa** lining (and its three sublayers), the **submucosa** (areolar connective tissue layer deep to the mucosa), the **muscularis externa** (composed of circular and longitudinal smooth muscle layers), and the **serosa** (the outermost layer, also called the *visceral peritoneum*). Consult Figure 39.9a to help you identify the scattered mucus-producing **duodenal glands** (Brunner's glands) in the submucosa.

What type of epithelium do you see here? _____

_____

Examine the large leaflike *villi,* which increase the surface area for absorption. Note also the *intestinal crypts* (crypts of Lieberkühn), invaginated areas of the mucosa between the villi containing the cells that produce intestinal juice, a watery mucus-containing mixture that serves as a carrier fluid for absorption of nutrients from the chyme. Sketch and label a small section of the duodenal wall, showing all layers and villi.

2. **Ileum:** The structure of the ileum resembles that of the duodenum, except that the villi are less elaborate (most of the absorption has occurred by the time the ileum is reached). Obtain and secure a slide of the ileum for viewing. Observe the villi, and identify the four layers of the wall and the large generally spherical Peyer's patches (Figure 39.9b). What tissue comprises Peyer's patches?

_____

3. If a *villus model* is available, identify the following cells or regions before continuing: absorptive epithelium, goblet cells, lamina propria, slips of the muscularis mucosae, capillary bed, and lacteal. If possible, also identify the intestinal crypts that lie between the villi. ∎

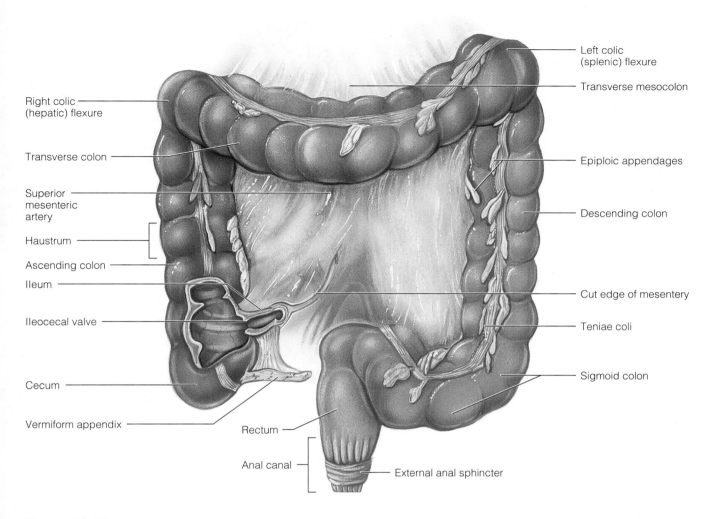

Right colic (hepatic) flexure

Transverse colon

Superior mesenteric artery

Haustrum

Ascending colon

Ileum

Ileocecal valve

Cecum

Vermiform appendix

Rectum

Anal canal

External anal sphincter

Left colic (splenic) flexure

Transverse mesocolon

Epiploic appendages

Descending colon

Cut edge of mesentery

Teniae coli

Sigmoid colon

Figure 39.10   The large intestine. (Section of the cecum removed to show the ileocecal valve.)

Large Intestine

The **large intestine** (see Figure 39.10) is about 1.5 m (5 feet) long and extends from the ileocecal valve to the anus. It encircles the small intestine on three sides and consists of the following subdivisions: the **cecum, appendix, colon, rectum,** and **anal canal.**

The blind tubelike appendix, which hangs from the cecum, is a trouble spot in the large intestine. Since it is generally twisted, it provides an ideal location for bacteria to accumulate and multiply. Inflammation of the appendix, or appendicitis, is the result. ■

The colon is divided into several distinct regions. The **ascending colon** travels up the right side of the abdominal cavity and makes a right-angle turn at the **right colic (hepatic) flexure** to cross the abdominal cavity as the **transverse colon.** It then turns at the **left colic (splenic) flexure** and continues down the left side of the abdominal cavity as the **descending colon,** where it takes an S-shaped course as the **sigmoid colon.** The sigmoid colon, rectum, and the anal canal lie in the pelvis anterior to the sacrum and thus are not considered abdominal cavity structures. Except for the transverse and sigmoid colons, which are secured to the dorsal body wall by mesocolons (see Figure 39.7), the colon is retroperitoneal.

The anal canal terminates in the **anus,** the opening to the exterior of the body. The anus, which has an external sphincter of skeletal muscle (the voluntary sphincter) and an internal sphincter of smooth muscle (the involuntary sphincter), is normally closed except during defecation when the undigested remains of the food and bacteria are eliminated from the body as feces.

In the large intestine, the longitudinal muscle layer of the muscularis externa is reduced to three longitudinal muscle bands called the **teniae coli.** Since these bands are shorter than the rest of the wall of the large intestine, they cause the wall to pucker into small pocketlike sacs called **haustra.** Fat-filled pouches of visceral peritoneum, called *epiploic appendages* hang from the colon's surface.

The major function of the large intestine is to consolidate and propel the unusable fecal matter toward the anus and eliminate it from the body. While it does that chore, it (1) provides a site for the manufacture, by intestinal bacteria, of some vitamins (B and K), which it then absorbs into the bloodstream; and (2) reclaims most of the remaining water (and some of the electrolytes) from undigested food, thus conserving body water.

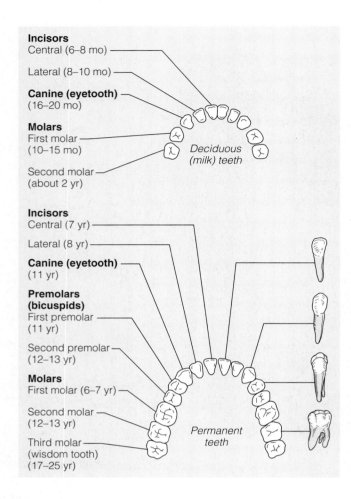

Figure 39.11 Human deciduous teeth and permanent teeth. (Approximate time of teeth eruption shown in parentheses.)

Watery stools, or diarrhea, result from any condition that rushes undigested food residue through the large intestine before it has had sufficient time to absorb the water (as in irritation of the colon by bacteria). Conversely, when food residue remains in the large intestine for extended periods (as with atonic colon or failure of the defecation reflex), excessive water is absorbed and the stool becomes hard and difficult to pass (constipation). ■

Activity:

Examining the Histological Structure of the Large Intestine

Examine Figure 39.9c to compare the histology of the large intestine to that of the small intestine just studied. ■

Accessory Digestive Organs

Teeth

By the age of 21, two sets of teeth have developed (Figure 39.11). The initial set, called the **deciduous,** or **milk teeth,** normally appears between the ages of 6 months and 2½ years. The first of these to erupt are the lower central incisors. The child begins to shed the deciduous teeth around the age of 6, and a second set of teeth, the **permanent teeth,** gradually replaces them. As the deeper permanent teeth progressively enlarge and develop, the roots of the deciduous teeth are resorbed, leading to their final shedding. During years 6 to 12, the child has mixed dentition—both permanent and deciduous teeth. Generally, by the age of 12, all of the deciduous teeth have been shed, or exfoliated.

Teeth are classified as **incisors, canines** (eye teeth), **premolars** (bicuspids), and **molars.** Teeth names reflect differences in relative structure and function. The incisors are chisel-shaped and exert a shearing action used in biting. Canines are cone-shaped or fanglike, the latter description being much more applicable to the canines of animals whose teeth are used for the tearing of food. Incisors, canines, and premolars typically have single roots, though the first upper premolars may have two. The lower molars have two roots but the upper molars usually have three. The premolars have two cusps (grinding surfaces); the molars have broad crowns with rounded cusps specialized for the fine grinding of food.

Dentition is described by means of a **dental formula,** which designates the numbers, types, and position of the teeth in one side of the jaw. (Because tooth arrangement is bilaterally symmetrical, it is only necessary to designate one side of the jaw.) The complete dental formula for the deciduous teeth from the medial aspect of each jaw and proceeding posteriorly is as follows:

$$\frac{\text{Upper teeth: 2 incisors, 1 canine, 0 premolars, 2 molars}}{\text{Lower teeth: 2 incisors, 1 canine, 0 premolars, 2 molars}} \times 2$$

This formula is generally abbreviated to read as follows:

$$\frac{2,1,0,2}{2,1,0,2} \times 2 = 20 \text{ (number of deciduous teeth)}$$

The 32 permanent teeth are then described by the following dental formula:

$$\frac{2,1,2,3}{2,1,2,3} \times 2 = 32 \text{ (number of deciduous teeth)}$$

Although 32 is designated as the normal number of permanent teeth, not everyone develops a full complement. In many people, the No. 3 molars, commonly called *wisdom teeth,* never erupt.

Activity:

Identifying Types of Teeth

Identify the four types of teeth (incisors, canines, premolars, and molars) on the jaw model or human skull. ■

A tooth consists of two major regions, the **crown** and the **root.** A longitudinal section made through a tooth shows the following basic anatomical plan (Figure 39.12). The crown is the superior portion of the tooth. The portion of the crown visible above the **gum,** or **gingiva,** is referred to as the **clinical crown.** The entire area covered by **enamel** is called the **anatomical crown.** Enamel is the hardest substance in the body and is fairly brittle. It consists of 95% to 97% inorganic calcium salts (chiefly CaPO$_4$) and thus is heavily mineralized. The crevice between the end of the anatomical crown and the upper margin of the gingiva is referred to as the **gingival sulcus** and its apical border is the **gingival margin.**

That portion of the tooth embedded in the alveolar portion of the jaw is the root, and the root and crown are connected by a slight constriction, the **neck.** The outermost surface of the root is covered by **cementum,** which is similar to bone in composition and less brittle than enamel. The cementum attaches the tooth to the **periodontal ligament** which holds the tooth in the alveolar socket and exerts a cushioning effect. **Dentin,** which comprises the bulk of the tooth, is the bonelike material medial to the enamel and cementum.

The **pulp cavity** occupies the central portion of the tooth. **Pulp,** connective tissue liberally supplied with blood vessels, nerves, and lymphatics, occupies this cavity and provides for tooth sensation and supplies nutrients to the tooth tissues. **Odontoblasts,** specialized cells which reside in the outer margins of the pulp cavity, produce the dentin. The pulp cavity extends into distal portions of the root and becomes the **root canal.** An opening at the root apex, the **apical foramen,** provides a route of entry into the tooth for blood vessels, nerves, and other structures from the tissues beneath.

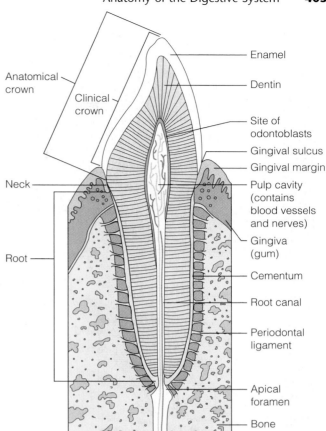

Figure 39.12 **Longitudinal section of human canine tooth.**

A c t i v i t y :
Studying Microscopic Tooth Anatomy

Observe a slide of a longitudinal section of a tooth, and compare your observations with the structures detailed in Figure 39.12. Identify as many of these structures as possible. ■

Salivary Glands

Three pairs of major **salivary glands** (see Figure 39.1) empty their secretions into the oral cavity.

Parotid glands: Large glands located anterior to the ear and ducting into the mouth over the second upper molar through the parotid duct

Submandibular glands: Located along the medial aspect of the mandibular body in the floor of the mouth, and ducting under the tongue to the base of the lingual frenulum

Sublingual glands: Small glands located most anteriorly in the floor of the mouth and emptying under the tongue via several small ducts

Food in the mouth and mechanical pressure (even chewing rubber bands or wax) stimulate the salivary glands to secrete saliva. Saliva consists primarily of mucin (a viscous glycoprotein), which moistens the food and helps to bind it together into a mass called a **bolus,** and a clear serous fluid containing the enzyme *salivary amylase.* Salivary amylase begins the digestion of starch (a large polysaccharide), breaking it down into disaccharides, or double sugars, and glucose. Parotid gland secretion is mainly serous, whereas the submandibular and sublingual glands are mixed glands that produce both mucin and serous components.

A c t i v i t y :
Examining Salivary Gland Tissue

Examine salivary gland tissue under low power and then high power to become familiar with the appearance of a glandular tissue. Notice the clustered arrangement of the cells around their ducts. The cells are basically triangular, with their pointed ends facing the duct orifice. If possible, differentiate between mucus-producing cells, which look hollow or have a clear cytoplasm, and serous cells, which produce the clear, enzyme-containing fluid and have granules in their cytoplasm. The serous cells often form *demilunes* ("caps") around the more central mucous cells. The photo in Plate 42 in the Histology Atlas may be helpful in this task. In the space below, draw your version of a small portion of the salivary gland tissue and label it appropriately. ■

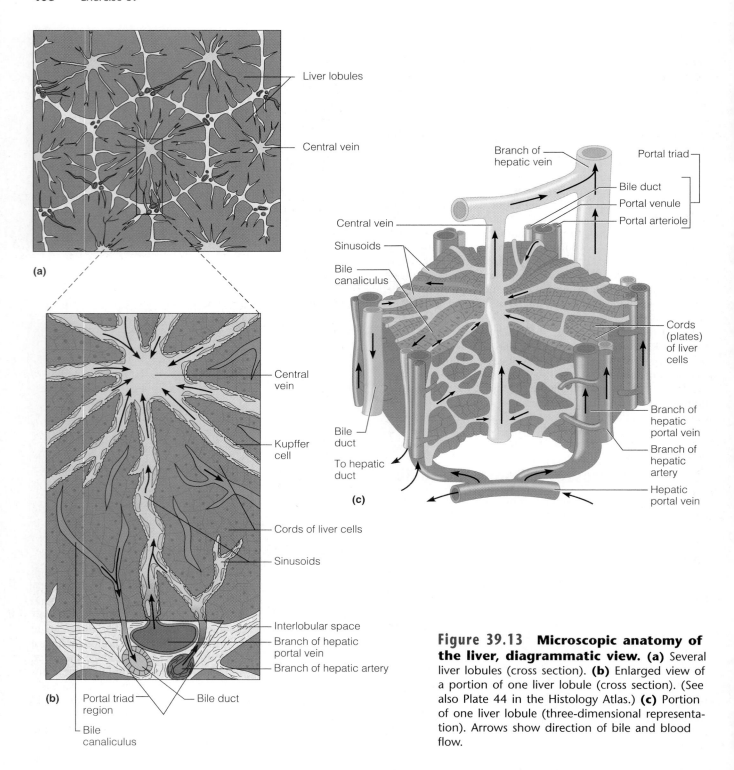

(a)

(b)

(c)

Liver lobules

Central vein

Central vein

Kupffer cell

Cords of liver cells

Sinusoids

Interlobular space

Branch of hepatic portal vein

Branch of hepatic artery

Portal triad region

Bile duct

Bile canaliculus

Branch of hepatic vein

Central vein

Sinusoids

Bile canaliculus

Bile duct

To hepatic duct

Portal triad

Bile duct

Portal venule

Portal arteriole

Cords (plates) of liver cells

Branch of hepatic portal vein

Branch of hepatic artery

Hepatic portal vein

Figure 39.13 Microscopic anatomy of the liver, diagrammatic view. (a) Several liver lobules (cross section). **(b)** Enlarged view of a portion of one liver lobule (cross section). (See also Plate 44 in the Histology Atlas.) **(c)** Portion of one liver lobule (three-dimensional representation). Arrows show direction of bile and blood flow.

Liver and Gallbladder

The **liver** (see Figure 39.1), the largest gland in the body, is located inferior to the diaphragm, more to the right than the left side of the body. As noted earlier, it hides the stomach from view in a superficial observation of abdominal contents. The human liver has four lobes and is suspended from the diaphragm and anterior abdominal wall by the **falciform ligament** (see Figure 39.7a).

The liver is one of the body's most important organs, and it performs many metabolic roles. However, its digestive function is to produce bile, which leaves the liver through the **common hepatic duct** and then enters the duodenum through the **bile duct.** Bile has no enzymatic action but emulsifies fats (spreads thin or breaks up large fat particles into smaller ones), thus creating a larger surface area for more efficient lipase activity. Without bile, very little fat digestion or absorption occurs.

When digestive activity is not occurring in the digestive tract, bile backs up the **cystic duct** and enters the **gallbladder,** a small, green sac on the inferior surface of the liver. It is stored there until needed for the digestive process. While in the gallbladder, bile is concentrated by the removal of water and some ions. When fat-rich food enters the duodenum, a hormonal stimulus causes the gallbladder to contract, releasing the stored bile and making it available to the duodenum.

If the common hepatic or bile duct is blocked (for example, by wedged gallstones), bile is prevented from entering the small intestine, accumulates, and eventually backs up into the liver. This exerts pressure on the liver cells, and bile begins to enter the bloodstream. As the bile circulates through the body, the tissues become yellow or jaundiced. ■

Blockage of the ducts is just one cause of jaundice. More often it results from actual liver problems such as **hepatitis** (an inflammation of the liver) or **cirrhosis,** a condition in which the liver is severely damaged and becomes hard and fibrous. Cirrhosis is almost guaranteed in those who drink excessive alcohol for many years. ■

As demonstrated by its highly organized anatomy, the liver (Figure 39.13) is very important in the initial processing of the nutrient-rich blood draining the digestive organs. Its structural and functional units are called **lobules.** Each lobule is a basically cylindrical structure consisting of cordlike arrays of parenchyma cells, which radiate outward from a central vein running upward in the longitudinal axis of the lobule. At each of the six corners of the lobule is a **portal triad,** so named because three basic structures are always present there: a branch of the *hepatic artery* (the functional blood supply of the liver), a branch of the *hepatic portal vein* (carrying nutrient-rich blood from the digestive viscera), and a *bile duct.* Between the liver parenchyma cells are blood-filled spaces, or **sinusoids,** through which blood from the hepatic portal vein and hepatic artery percolates past the parenchyma cells. Special phagocytic cells, **Kupffer cells,** line the sinusoids and remove debris such as bacteria from the blood as it flows past, while the parenchyma cells pick up oxygen and nutrients. Much of the glucose transported to the liver from the digestive system is stored as glycogen in the liver for later use, and amino acids are taken from the blood by the liver cells and utilized to make plasma proteins. The sinusoids empty into the central vein, and the blood ultimately drains from the liver via the *hepatic vein.*

Bile is continuously being made by the parenchyma cells. It flows through tiny canals, the **bile canaliculi,** which run between adjacent parenchyma cells toward the bile duct branches in the triad regions, where the bile eventually leaves the liver. Notice that the directions of blood and bile flow in the liver lobule are exactly opposite.

Activity:
Examining the Histology of the Liver

Examine a slide of liver tissue and identify as many of the structural features illustrated in Figure 39.13 and Histology Atlas Plates 44 and 45 as possible. Also examine a three-dimensional model of liver lobules if this is available.

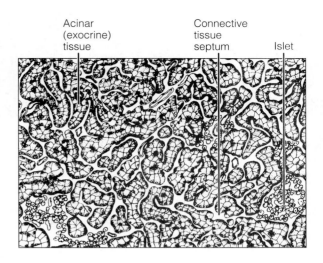

Acinar (exocrine) tissue   Connective tissue septum   Islet

Figure 39.14   Histology of the pancreas. The pancreatic islet cells produce insulin and glucagon (hormones). The acinar cells synthesize digestive enzymes for "export" to the duodenum. (See also Plate 43 in the Histology Atlas.)

Pancreas

The **pancreas** is a soft, triangular gland that extends horizontally across the posterior abdominal wall from the spleen to the duodenum (see Figure 39.1). Like the duodenum, it is a retroperitoneal organ (see Figure 39.7). As noted in Exercise 27, the pancreas has both an endocrine function (it produces the hormones insulin and glucagon) and an exocrine (enzyme-producing) function. It produces a whole spectrum of hydrolytic enzymes, which it secretes in an alkaline fluid into the duodenum through the pancreatic duct. Pancreatic juice is very alkaline. Its high concentration of bicarbonate ion (HCO_3^-) neutralizes the acidic chyme entering the duodenum from the stomach, enabling the pancreatic and intestinal enzymes to operate at their optimal pH. (Optimal pH for digestive activity to occur in the stomach is very acidic and results from the presence of HCl; that for the small intestine is slightly alkaline.)

Activity:
Examining the Histology of the Pancreas

Observe pancreas tissue under low power and then high power to distinguish between the lighter-staining, endocrine-producing clusters of cells (pancreatic islets) and the deeper-staining acinar cells, which produce the hydrolytic enzymes and form the major portion of the pancreatic tissue (see Figure 39.14). Notice the arrangement of the exocrine cells around their central ducts. ■

 For instructions on animal dissections, see the dissection exercises starting on p. 733 in the cat and fetal pig editions of this manual.

Chemical and Physical Processes of Digestion: Wet Lab

Objectives

1. To list the digestive system enzymes involved in the digestion of proteins, fats, and carbohydrates; to state their site of origin; and to summarize the environmental conditions promoting their optimal functioning.

2. To recognize the variation between different types of enzyme assays.

3. To name the end products of protein, fat, and carbohydrate digestion.

4. To perform the appropriate chemical tests to determine if digestion of a particular foodstuff has occurred.

5. To cite the function(s) of bile in the digestive process.

6. To discuss the possible role of temperature and pH in the regulation of enzyme activity.

7. To define *enzyme, catalyst, control, substrate,* and *hydrolase.*

8. To explain why swallowing is both a voluntary and a reflex activity.

9. To discuss the role of the tongue, larynx, and gastroesophageal sphincter in swallowing.

10. To compare and contrast segmentation and peristalsis as mechanisms of propulsion.

Materials

Part I: Enzyme Action

General supply area:

❑ Test tubes and test tube rack
❑ Wax markers
❑ Hot plates
❑ 250-ml beakers
❑ Boiling chips
❑ Water bath set at 37°C (if not available, incubate at room temperature and double the time)
❑ Ice water bath
❑ Chart on chalkboard for recording class results

Supply area 1:

❑ Spot plates
❑ Dropper bottles of distilled water
❑ Dropper bottles of
   1% alpha-amylase solution*
   1% boiled starch solution, freshly prepared†
   1% maltose solution
   Lugol's IKI (Lugol's iodine)
   Benedict's solution

Supply area 2:

❑ Dropper bottles of
   1% trypsin
   1% BAPNA solution

Supply area 3:

❑ Dropper bottles of
   1% pancreatin solution
   Litmus cream (fresh cream to which powdered litmus is added to achieve a blue color)
   0.1 N HCl
   Vegetable oil
❑ Bile salts (sodium taurocholate)
❑ Parafilm (small squares to cover the test tubes)

Part II: Physical Processes

Supply area 4:

❑ Water pitcher
❑ Paper cups
❑ Stethoscope
❑ Alcohol swabs
❑ Disposable autoclave bag

Videotape viewing station:

❑ VHS cassette illustrating deglutition and peristalsis (*Passage of Food Through the Digestive Tract,* available from Ward's Biology)
❑ Television and VCR for independent viewing of videocassette by student.

*The alpha-amylase must be a low-maltose preparation for good results.

†Prepare by adding 1 g starch to 100 ml distilled water; boil and cool; add a pinch of salt (NaCl). Prepare fresh daily.

Path of absorption	Foodstuff	Enzyme(s) and source	Site of action

Carbohydrate digestion

Absorption: Monosaccharides—glucose, galactose, and fructose—are absorbed into the capillary blood in the villi and transported to the liver via the hepatic portal vein.

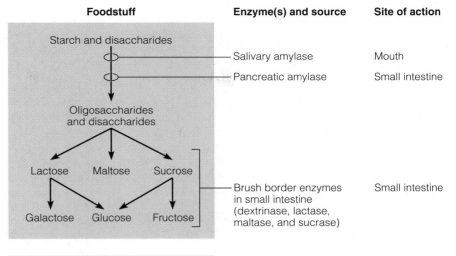

	Salivary amylase	Mouth
	Pancreatic amylase	Small intestine
	Brush border enzymes in small intestine (dextrinase, lactase, maltase, and sucrase)	Small intestine

Protein digestion

Absorption: Amino acids are absorbed into the capillary blood in the villi and transported to the liver via the hepatic portal vein.

	Pepsin (stomach glands) in the presence of HCl	Stomach
	Pancreatic enzymes (trypsin, chymotrypsin, carboxypeptidase)	Small intestine
	Brush border enzymes (aminopeptidases, carboxypeptidase, and dipeptidases)	Small intestine

Fat digestion

Absorption: Absorbed primarily into the lacteals of the villi and transported to the systemic circulation via the lymph in the thoracic duct. (Glycerol and short–chain fatty acids are absorbed into the capillary blood in the villi and transported to the liver via the hepatic portal vein.)

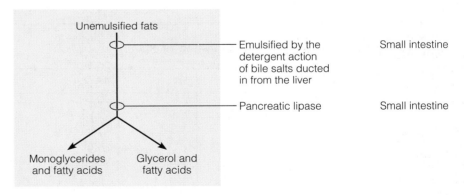

	Emulsified by the detergent action of bile salts ducted in from the liver	Small intestine
	Pancreatic lipase	Small intestine

Figure 40A.1   Flowchart of digestion and absorption of foodstuffs.

Chemical Digestion of Foodstuffs: Enzymatic Action

Because nutrients can only be absorbed when broken down to their monomers, food digestion is a prerequisite to food absorption. You have already studied mechanisms of passive and active absorption in Exercises 5A and 5B. Before proceeding, review that material on pages 33–50.

Enzymes are large protein molecules produced by body cells. They are biologic *catalysts* that increase the rate of a chemical reaction without themselves becoming part of the product. The digestive enzymes are hydrolytic enzymes, or *hydrolases.* Their **substrates,** or the molecules on which they act, are organic food molecules which they break down by adding water to the molecular bonds, thus cleaving the bonds between the subunits or monomers.

The various hydrolytic enzymes are highly specific in their action. Each enzyme hydrolyzes only one or a small group of substrate molecules, and specific environmental conditions are necessary for it to function optimally. Since digestive enzymes actually function outside the body cells in the digestive tract, their hydrolytic activity can also be studied in a test tube. Such an *in vitro* study provides a convenient laboratory environment for investigating the effect of such variations on enzymatic activity.

Figure 40A.1 is a flowchart of the progressive digestion of proteins, fats, and carbohydrates. It indicates specific enzymes involved, their site of formation, and their site of action. Acquaint yourself with the flowchart before beginning this experiment, and refer to it as necessary during the laboratory session.

Work in groups of 3 or 4, with each group taking responsibility for setting up and conducting one of the following experiments. In each of the digestive procedures being studied (starch, protein, and fat digestion) and in the amylase assay (p. 412), you are directed to boil the contents of one or more test tubes. To do this, obtain a 250-ml beaker, boiling chips, and a hot plate from the general supply area. Place a few boiling chips into the beaker and add about 125 ml of water, and bring to a boil. Place the specimen in the boiling water for the time (min.) specified in the directions.

Upon completion of the experiments, each group should communicate its results to the rest of the class by recording them in a chart on the chalkboard. All members of the class should observe the controls* as well as the positive and negative examples of all experimental results. Additionally, all members of the class should be able to explain the tests used and the results observed and anticipated for each experiment. Note that water baths and hot plates are at the general supply area.

*The specimens or standards against which experimental samples are compared.

Activity:
Assessing Starch Digestion by Salivary Amylase

1. From the general supply area, obtain a test tube rack, 10 test tubes, and a wax marking pencil. From supply area 1, obtain a dropper bottle of distilled water and dropper bottles of maltose, amylase, and starch solutions.

2. Since in this experiment you will investigate the hydrolysis of starch to maltose by **salivary amylase** (the enzyme produced by the salivary glands and secreted into the mouth), it is important to be able to identify the presence of these substances to determine to what extent the enzymatic activity has occurred. Thus controls must be prepared to provide a known standard against which comparisons can be made. Starch decreases and sugar increases as digestion occurs, according to the following equation:

$$\text{Starch} + \text{water} \xrightarrow{\text{Amylase}} \text{Maltose}$$

Two students should prepare the controls (tubes 1A to 3A) while the other two prepare the experimental samples (tubes 4A to 6A).

• Mark each tube with a wax pencil and load the tubes as indicated in Chart 1, using 3 drops (gtt) of each indicated substance.

Chart 1	Salivary Amylase Digestion of Starch					
Tube no.	**1A**	**2A**	**3A**	**4A**	**5A**	**6A**
Additives (3 gtt ea)						
Incubation condition	37°C	37°C	37°C	Boil 4 min first, then incubate at 37°C	37°C	0°C
IKI test						
Benedict's test						

Additive Key:

■ = Amylase   ■ = Starch   □ = Maltose   □ = Water

- Place all tubes in a rack in the appropriate water bath for approximately 1 hour. Shake the rack gently from time to time to keep the contents evenly mixed.

- While these tubes are incubating, proceed to the next section. ■

Protein Digestion by Trypsin

Trypsin, an enzyme produced by the pancreas, hydrolyzes proteins to small fragments (proteoses, peptones, and peptides). BAPNA is a synthetic (man-made) protein substrate consisting of a dye covalently bound to an amino acid. Trypsin hydrolysis of BAPNA cleaves the dye molecule from the amino acid, causing the solution to change from colorless to bright yellow. Since the covalent bond between the dye molecule and the amino acid is the same as the peptide bonds that link amino acids together, the appearance of a yellow color indicates the presence and activity of an enzyme that is capable of peptide bond hydrolysis. Because the color change from clear to yellow is direct evidence of hydrolysis, additional tests are not required when determining trypsin activity using BAPNA.

Activity:
Assessing Protein Digestion by Trypsin

1. From the general supply area, obtain 5 test tubes and a test tube rack, and from supply area 2 get a dropper bottle of trypsin and one of BAPNA and bring them to your bench.

2. Two students should prepare the controls (tubes 1T and 2T) while the other two prepare the experimental samples (tubes 3T to 5T).

- Mark each tube with a wax pencil and load the tubes as indicated in Chart 2, using 3 drops (gtt) of each indicated substance.

- Place all tubes in a rack in the appropriate water bath for approximately 1 hour. Shake the rack occasionally to keep the contents well mixed.

- While these tubes are incubating, proceed to the next section. ■

Pancreatic Lipase Digestion of Fats and the Action of Bile

The treatment that fats and oils go through during digestion in the small intestine is a bit more complicated than that of carbohydrates or proteins—pretreatment with bile to physically emulsify the fats is required. Hence, two sets of reactions occur.

$$\text{First: Fats/oils} \xrightarrow[\text{(emulsification)}]{\text{Bile}} \text{minute fat/oil droplets}$$

$$\text{Then: Fat/oil droplets} \xrightarrow{\text{Lipase}} \text{monoglycerides and fatty acids}$$

The term **pancreatin** describes the enzymatic product of the pancreas, which includes protein, carbohydrate, nucleic acid, and fat-digesting enzymes. It is used here to investigate the properties of **pancreatic lipase,** which hydrolyzes fats and oils to their component monoglycerides and two fatty acids (and occasionally to glycerol and three fatty acids).

The fact that some of the end products of fat digestion (fatty acids) are organic acids that decrease the pH provides an easy way to recognize that digestion is ongoing or completed. You will be using a pH indicator called *litmus blue* to follow these changes; it changes from *blue* to *pink* as the test tube contents become acid.

Chart 2	Trypsin Digestion of Protein				
Tube no.	**1T**	**2T**	**3T**	**4T**	**5T**
Additives (3 gtt ea)					
Incubation condition	37°C	37°C	Boil 4 min first, then incubate at 37°C	37°C	0°C
Color change					

Additive Key:

■ = Trypsin   ▨ = BAPNA   ▢ = Water

Activity:
Assessing Fat Digestion by Lipase

1. From the general supply area, obtain 9 test tubes and a test tube rack, plus one dropper bottle of each of the solutions in supply area 3.

2. Although *bile,* a secretory product of the liver, is not an enzyme, it is important to fat digestion because of its emulsifying action (the physical breakdown of larger particles into smaller ones) on fats. Emulsified fats provide a larger surface area for enzymatic activity. To demonstrate the action of bile on fats, prepare two test tubes and mark them 1E and 2E.

• To tube 1E add 10 drops of water and 2 drops of vegetable oil.

• To tube 2E add 10 drops of water, 2 drops of vegetable oil, and a pinch of bile salts.

• Cover each tube with a small square of Parafilm, shake vigorously, and allow the tubes to stand at room temperature.

After 10–15 minutes, observe both tubes. If emulsification has not occurred, the oil will be floating on the surface of the water. If emulsification has occurred, the fat droplets will be suspended throughout the water, forming an emulsion.

In which tube has emulsification occurred? _____

3. Two students should prepare the controls (1L and 2L), while the other two students in the group set up the experimental samples (3L to 5L, 4B, and 5B) as illustrated in Chart 3.

• Mark each tube with a wax pencil and load the tubes using 5 drops (gtt) of each indicated solution.

• Place a pinch of bile salts in tubes 4B and 5B.

• Cover each tube with a small square of Parafilm and shake to mix the contents of the tube.

• Remove the Parafilm and place all tubes in a rack in the appropriate water bath for approximately 1 hour. Shake the test tube rack from time to time to keep the contents well mixed. ■

Activity:
Conducting Enzyme Assays to Determine Results

Amylase Assay

1. From supply area 1 obtain a spot plate and dropper bottles of IKI and Benedict's solution. Set up your boiling water bath using a hot plate, boiling chips, and a 250-ml beaker obtained from the general supply area.

2. While the water is heating, mark the spot plate *A* (for amylase) and number six of its depressions 1–6 for sample identification.

3. Pour about a drop of the sample from each of the tubes 1A–6A into the appropriately numbered spot. Into each sample droplet, place a drop of IKI solution. A blue-black color indicates the presence of starch and is referred to as a **positive starch test.** If starch is not present, the mixture will not turn blue, which is referred to as a **negative starch test.** Record your results (+ for positive, − for negative) in Chart 1 and on the chalkboard.

4. Into the remaining mixture in each tube, place 3 drops of Benedict's solution. Put each tube into the beaker of boiling water for about 5 minutes. If a green-to-orange precipitate

Chart 3	Pancreatic Lipase Digestion of Fats						
Tube no.	**1L**	**2L**	**3L**	**4L**	**5L**	**4B**	**5B**
Additives (5 gtt ea)							
Incubation condition	37°C	37°C	Boil 4 min first, then incubate at 37°C	37°C	0°C	37°C	0°C
Color change							

Additive Key:

 = Lipase   = Litmus cream   = Water   = Pinch bile salts

forms, maltose is present; this is a **positive sugar test**. A **negative sugar test** is indicated by no color change. Record your results in Chart 1 and on the chalkboard.

Trypsin Assay

Since BAPNA is a synthetic colorigenic (color-producing) substrate, the presence of yellow color indicates a **positive hydrolysis test;** the dye molecule has been cleaved from the amino acid. If the sample mixture remains clear, no detectable hydrolysis has occurred.

Record the color of the experimental tubes in Chart 2 and on the chalkboard.

Lipase Assay

The basis of this assay is a pH change that is detected by a litmus powder indicator. Alkaline or neutral solutions containing litmus are blue but will turn reddish in the presence of acid. Since fats are digested to fatty acids (organic acids) during hydrolysis, they lower the pH of the sample they are in. Litmus cream (fresh cream providing the fat substrate to which litmus powder was added) will turn from blue to pink if the solution is acid. Because the effect of hydrolysis is directly seen, additional assay reagents are not necessary.

1.   To prepare a color control, add 0.1 N HCl drop by drop to tubes 1L and 2L (covering the tubes with a square of Parafilm after each addition and shaking to mix) until the cream turns pink.

2.   Record the color of the tubes in Chart 3 and on the chalkboard. ■

Physical Processes: Mechanisms of Food Propulsion and Mixing

Although enzyme activity is a very important part of the overall digestion process, foods must also be processed physically (churning and chewing), and moved by mechanical means along the tract if digestion and absorption are to be completed. Just about any time organs exhibit mobility, muscles are involved, and movements of and in the gastrointestinal tract are no exception. Although we tend to think only of smooth muscles when visceral activities are involved, both skeletal and smooth muscles are involved in digestion. This fact is amply demonstrated by the simple activities that follow.

Deglutition (Swallowing)

Swallowing, or **deglutition,** which is largely the result of skeletal muscle activity, occurs in two phases: *buccal* (mouth) and *pharyngeal-esophageal.* The initial phase—the buccal—is voluntarily controlled and initiated by the tongue. Once begun, the process continues involuntarily in the pharynx and esophagus, through peristalsis, resulting in the delivery of the swallowed contents to the stomach.

Observing Movements and Sounds of Digestion

1.   Obtain a pitcher of water, a stethoscope, a paper cup, an alcohol swab, and an autoclave bag in preparation for making the following observations.

2.   While swallowing a mouthful of water, consciously note the movement of your tongue during the process. Record your observations.

_____

_____

_____

3.   Repeat the swallowing process while your laboratory partner watches the externally visible movements of your larynx. (This movement is more obvious in a male, who has a larger Adam's apple.) Record your observations.

_____

_____

What do these movements accomplish? _____

_____

4.   Before donning the stethoscope your lab partner should clean the earpieces with an alcohol swab. Then, he or she should place the diaphragm of the stethoscope over your abdominal wall, approximately 1 inch below the xiphoid process and slightly to the left, to listen for sounds as you again take two or three swallows of water. There should be two audible sounds—one when the water splashes against the gastroesophageal sphincter and the second when the peristaltic wave of the esophagus arrives at the sphincter and the sphincter opens, allowing water to gurgle into the stomach. Determine, as accurately as possible, the time interval between these two sounds and record it below.

Interval between arrival of water at the sphincter and the

opening of the sphincter: _____ sec

This interval gives a fair indication of the time it takes for the peristaltic wave to travel down the 10-inch-long esophagus. (Actually the time interval is slightly less than it seems, because pressure causes the sphincter to relax before the peristaltic wave reaches it.)

  Dispose of the used paper cup in the autoclave bag. ■

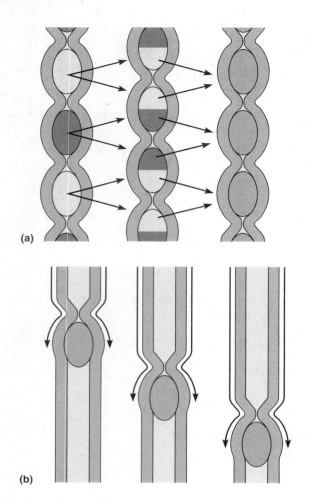

(a)

(b)

Figure 40A.2 Segmental and peristaltic movements of the digestive tract. (a) Segmentation: single segments of intestine alternately contract and relax. Because inactive segments exist between active segments, food mixing occurs to a greater degree than food movement. **(b)** Peristalsis: superimposed on segmentation. Neighboring segments of the intestine alternately contract and relax, moving food along the tract.

Segmentation and Peristalsis

Although several types of movements occur in the digestive tract organs, segmentation and peristalsis are most important as mixing and propulsive mechanisms (Figure 40A.2).

Segmental movements are local constrictions of the organ wall that occur rhythmically. They serve mainly to mix the foodstuffs with digestive juices and to increase the rate of absorption by continually moving different portions of the chyme over adjacent regions of the intestinal wall. However, segmentation is also an important means of food propulsion in the small intestine, and slow segmenting movements called haustral contractions are frequently seen in the large intestine.

Peristaltic movements are the major means of propelling food through most of the digestive viscera. Essentially they are waves of contraction followed by waves of relaxation that squeeze foodstuffs through the alimentary canal, and they are superimposed on segmental movements.

Activity:
Viewing Segmental and Peristaltic Movements

If a videotape showing some of the propulsive movements is available, go to a viewing station to view it before leaving the laboratory. ■

Chemical and Physical Processes of Digestion: Computer Simulation

Objectives

1. To list the digestive system enzymes involved in the digestion of proteins, fats, and carbohydrates; to state their site of origin; and to summarize the environmental conditions promoting their optimal functioning.

2. To recognize the variation between different types of enzyme assays.

3. To name the end products of digestion of proteins, fats, and carbohydrates.

4. To perform the appropriate chemical tests to determine if digestion of a particular food has occurred.

5. To cite the function(s) of bile in the digestive process.

6. To discuss the possible role of temperature and pH in the regulation of enzyme activity.

7. To define *enzyme, catalyst, control, substrate,* and *hydrolase.*

8. To explain why swallowing is both a voluntary and a reflex activity.

9. To discuss the role of the tongue, larynx, and gastroesophageal sphincter in swallowing.

10. To compare and contrast segmentation and peristalsis as mechanisms of propulsion.

Materials

Part I—Enzyme Action

Minimum equipment required:

Windows:

- ❑ 486/66 MHz or better recommended
- ❑ Windows 95/98 recommended
- ❑ SVGA display (256 colors at 640 × 480)
- ❑ Minimum 16 MB available RAM
- ❑ Double-speed CD-ROM drive (quad-speed or higher recommended)
- ❑ Sound card
- ❑ Speakers or headphones
- ❑ Printer

Macintosh:

- ❑ 68040 processor or Power Macintosh recommended
- ❑ System 7.1 or higher
- ❑ Minimum 16 MB available RAM
- ❑ 13-inch or larger color monitor (640 × 480 resolution)
- ❑ Double-speed CD-ROM drive (quad-speed or higher recommended)
- ❑ Printer

Software:

- ❑ Benjamin/Cummings PhysioEx CD-ROM—Chemical Processes of Digestion module

Part II—Physical Processes

- ❑ Water pitcher
- ❑ Paper cups
- ❑ Stethoscope
- ❑ Alcohol swabs
- ❑ Disposable autoclave bag
- ❑ VHS tape viewing station:

  VHS tape illustrating deglutition and peristalsis (*Passage of Food through the Digestive Tract,* available from Ward's Biology)

  VCR and television or monitor set up for independent viewing of film loop by students

The digestive system is a physiological marvel, composed of finely orchestrated chemical and physical activities. The food we ingest must be broken down to its molecular form for us to get the nutrients we need, and digestion involves a complex sequence of mechanical and chemical processes designed to achieve this goal as efficiently as possible. As food passes through the gastrointestinal tract, it is progressively broken down by the mechanical action of smooth muscle and the chemical action of enzymes until most nutrients have been extracted and absorbed into the blood.

Chemical Digestion of Foodstuffs: Enzymatic Action

Nutrients can only be absorbed when broken down into their monomer form, so food digestion is a prerequisite to food absorption. You have already studied mechanisms of passive and active absorption in Exercise 5. Before proceeding, review that material on pp. 35–50.

Enzymes are large protein molecules produced by body cells. They are biological **catalysts** that increase the rate of a chemical reaction without becoming part of the product. The digestive enzymes are hydrolytic enzymes, or **hydrolases**, which break down organic food molecules or **substrates** by adding water to the molecular bonds, thus cleaving the bonds between the subunits or monomers.

A hydrolytic enzyme is highly specific in its action. Each enzyme hydrolyzes one or, at most, a small group of substrate molecules, and specific environmental conditions are necessary for an enzyme to function optimally. For example, temperature and pH have a large effect on the degree of enzymatic hydrolysis, and each enzyme has its preferred environment.

Because digestive enzymes actually function outside the body cells in the digestive tract lumen, their hydrolytic activity can also be studied in a test tube. Such *in vitro* studies provide a convenient laboratory environment for investigating the effect of various factors on enzymatic activity.

Figure 40A.1, p. 409, is a flowchart of the progressive digestion of proteins, fats, and carbohydrates. It indicates the specific enzymes involved, their site of formation, and their site of action. Acquaint yourself with this flowchart before beginning this experiment, and refer to it as necessary during the laboratory session.

Getting Started

Begin by making sure you have the computer equipment and software listed in the Materials section on p. 415.

1. Quit all applications currently running on your computer.

2. Insert the PhysioEx CD-ROM into the CD-ROM drive and keep PhysioEx in the drive during the entire time you use the program.

3. Follow the instructions below for your computer type.

Windows 95/Windows 98
The program will launch automatically when you load the CD-ROM into the drive.
If autorun is disabled on your machine follow these instructions:
1. Double-click the My Computer icon on the Windows desktop.
2. Double-click the PhysioEx CD icon.
3. Look for the icon that looks like a tiny movie projector and double-click it.

Macintosh
Double-click the PhysioEx CD icon that appears in the window on your desktop to launch the program.

Starch Digestion by Salivary Amylase

In this experiment you will investigate the hydrolysis of starch to maltose by salivary amylase, the enzyme produced by the salivary glands and secreted into the mouth. For you to be able to detect whether or not enzymatic action has occurred, you need to be able to identify the presence of these substances to determine to what extent hydrolysis has occurred. Thus, **controls** must be prepared to provide a known standard against which comparisons can be made. The controls will vary for each experiment, and will be discussed in each enzyme section in this exercise.

Starch decreases and sugar increases as digestion proceeds according to the following equation:

$$\text{Starch} + \text{water} \xrightarrow{\text{amylase}} \text{X maltose}$$

Because the chemical changes that occur as starch is digested to maltose cannot be seen by the naked eye, you need to conduct an *enzyme assay,* the chemical method of detecting the presence of digested substances. You will perform two enzyme assays on each sample. The IKI assay detects the presence of starch and the Benedict's assay tests for the presence of maltose, which is the digestion product of starch. Normally a caramel-colored solution, IKI turns blue-black in the presence of starch. Benedict's reagent is a bright blue solution that changes to green to orange to reddish-brown with increasing amounts of maltose. It is important to understand that enzyme assays only indicate the presence or absence of substances. It is up to you to analyze the results of the experiments to decide if enzymatic hydrolysis has occurred.

Choose **Chemical and Physical Processes of Digestion** from the main menu. The opening screen will appear in a few seconds (Figure 40B.1). The *solutions shelf* in the upper right part of the screen contains the substances to be used in the ex-

periment. The *incubation unit* beneath the solutions shelf contains a rack of test tube holders and apparatus needed to run the experiments. Test tubes from the *test tube washer* on the left part of the screen are loaded into the rack in the incubation unit by clicking and holding the mouse button on the first tube, and then releasing (dragging-and-dropping) it into any position in the rack. The substances in the dropper bottles on the solutions shelf are dispensed by dragging-and-dropping the dropper cap to a position over any test tube in the rack and then releasing it. During each dispensing event, five drops of solution drip into the test tube; then the dropper cap automatically returns to its position in the bottle.

Each test tube holder in the incubation unit not only supports but also allows you to boil the contents of a single test

Figure 40B.1 Opening screen of the Amylase experiment.

tube. Clicking the numbered button at the base of a test tube holder causes that single tube to descend into the incubation unit. To boil the contents of all tubes inside the incubation unit, click **Boil.** After they have been boiled, the tubes automatically rise. You can adjust the incubation temperature for the experiment by clicking the (**+**) or (**−**) buttons next to the Temperature window. Set the incubation time by clicking the (**+**) or (**−**) buttons next to the Timer window. Clicking the **Incubate** button starts the timer and causes the entire rack of tube holders to descend into the incubation unit where the tubes will be incubated at the temperature and the time indicated. While incubating, the tubes are automatically agitated to ensure that their contents are well mixed. During the experiment elapsed time is displayed in the Elapsed Time window.

The cabinet doors in the *assay cabinet* above the test tube washer are closed at the beginning of the experiment, but they automatically open when the set time for incubation has elapsed. The assay cabinet contains the reagents and glassware needed to assay your experimental samples.

When you click the **Record Data** button in the *data control unit* at the bottom of the screen, your data is recorded in the computer's memory and displayed in the data grid at the bottom of the screen. Data displayed in the grid include the tube number, the three substances dispensed into each tube, the time and incubation temperature, and (**+**) or (**−**) marks indicating enzyme assay results and whether or not a sample was boiled. If you are not satisfied with a single run, you can click **Delete Run** to erase an experiment.

Once an experimental run is completed and you have recorded your data, discard the test tubes to prepare for a new run by dragging the used tubes to the large opening in the test tube washer. The test tubes will automatically be prepared for the next experiment.

A c t i v i t y :
Assessing Starch Digestion by Salivary Amylase

Amylase Incubation

1. Individually drag seven test tubes to the test tube holders in the incubation unit.

2. Prepare tubes 1 through 7 with the substances indicated in Chart 1 using the following approach.

- Click and hold the mouse button on the dropper cap of the desired substance on the solutions shelf.

- While still holding the mouse button down, drag the dropper cap to the top of the desired test tube.

- Release the mouse button to dispense the substance. The dropper cap automatically returns to its bottle.

3. When all tubes are prepared, click the number (**1**) under the first test tube. The tube will descend into the incubation unit. All other tubes should remain in the raised position.

4. Click **Boil** to boil the number 1 tube. After boiling for a few moments, the tube will automatically rise.

5. Now adjust the incubation temperature to 37°C and the timer to 60 min (compressed time) by clicking the (**+**) or (**−**) buttons.

6. Click **Incubate** to start the run. The incubation unit will gently agitate the test tube rack, evenly mixing the contents of all test tubes throughout the incubation. Notice that the computer compresses the 60-minute time period into 60 seconds of real time, so what would be a 60-minute incubation in

Chart 1	Salivary Amylase Digestion of Starch						
Tube no.	**1**	**2**	**3**	**4**	**5**	**6**	**7**
Additives	Amylase Starch pH 7.0 buffer	Amylase Starch pH 7.0 buffer	Amylase Deionized water pH 7.0 buffer	Deionized water Starch pH 7.0 buffer	Deionized water Maltose pH 7.0 buffer	Amylase Starch pH 2.0 buffer	Amylase Starch pH 9.0 buffer
Incubation condition	Boil first, then incubate at 37°C	37°C	37°C	37°C	37°C	37°C	37°C
Benedict's test							
IKI test							

real life will take only 60 seconds in the simulation. When the incubation time elapses, the test tube rack will automatically rise, and the doors to the assay cabinet will open.

Amylase Assays

After the assay cabinet doors open, notice the two reagents in the assay cabinet. IKI tests for the presence of starch and Benedict's detects the presence of maltose, the digestion product of starch. Below the reagents are seven small assay tubes into which you will dispense a small amount of test solution from the incubated samples in the incubation unit, plus a drop of IKI.

1. Click and hold the mouse on the first tube in the incubation unit. Notice that the mouse pointer is now a miniature test tube tilted to the left.

2. While still holding the mouse button down, move the mouse pointer to the first small assay tube on the left side of the assay cabinet. Release the mouse button. Watch the first test tube automatically decant approximately half of its contents into the first assay tube on the left.

3. Repeat steps 1 and 2 for the remaining tubes in the incubation unit, moving to a fresh assay tube each time.

4. Next, click and hold the mouse on the IKI dropper cap and drag it to the first assay tube. Release the mouse button to dispense a drop of IKI into the first assay tube on the left. You will see IKI drip into the tube, which may cause a color change in the solution. A blue-black color indicates a **positive starch test.** If starch is not present, the mixture will look like diluted IKI, a **negative starch test.** Intermediate starch amounts result in a pale gray color.

5. Now dispense IKI into the remaining assay tubes. Record your results (+ for positive, − for negative) in Chart 1.

6. Dispense Benedict's reagent into the remaining mixture in each tube in the incubation unit by dragging-and-dropping the Benedict's dropper cap to the top of each test tube.

7. After Benedict's reagent has been delivered to each tube in the incubation unit, click **Boil.** The entire tube rack will descend into the incubation unit and automatically boil the tube contents for a few moments.

8. When the rack of tubes rises, inspect the tubes for color change. A green-to-reddish color indicates that maltose is present; this is a **positive sugar test.** An orange-colored sample contains more maltose than a green sample. A reddish-brown color indicates even more maltose. A negative sugar test is indicated by no color change from the original bright blue. Record your results in Chart 1.

9. Click **Record Data** to display your results in the grid and retain your data in the computer's memory for later analysis. To repeat the experiment, drag all test tubes to the test tube washer and start again.

10. Answer the following questions, referring to Chart 1 (or the data grid in the simulation) as necessary. Hint: closely examine the IKI and Benedict's results for each tube.

What do tubes 2, 6, and 7 reveal about pH and amylase activity?

_____

_____

Which pH buffer allowed the highest amylase activity?

_____

Which tube indicates that the amylase did not contain mal-

tose? _____

Which tubes indicate that the deionized water did not contain

starch or maltose? _____

If we left out control tubes 3, 4, or 5, what objections could be raised to the statement: "Amylase digests starch to maltose"? (Hint: think about the purity of the chemical solutions.)

_____

_____

Would the amylase present in saliva be active in the stomach? Explain your answer.

_____

_____

What effect does boiling have on enzyme activity?

_____

_____ ■

Figure 40B.2 **Opening screen of the Pepsin experiment.**

Protein Digestion by Pepsin

The chief cells of the stomach glands produce pepsin, a protein-digesting enzyme. Pepsin hydrolyzes proteins to small fragments (proteoses, peptones, and peptides). In this experiment, you will use BAPNA, a synthetic "protein" that is transparent and colorless when in solution. However, if an active, protein-digesting enzyme such as pepsin is present, the solution will become yellow. You can use this characteristic to detect pepsin activity: the solution turns yellow if the enzyme digests the BAPNA substrate; it remains colorless if pepsin is not active or not present. One advantage of using a

synthetic substrate is that you do not need any additional indicator reagents to see enzyme activity.

Choose **Pepsin** from the **Experiment** menu. The opening screen will appear in a few seconds (Figure 40B.2). The solutions shelf, test tube washer, and incubation equipment are the same as in the amylase experiment; only the solutions have changed.

Data displayed in the grid include the tube number, the three substances dispensed into each tube, a (+) or (−) mark indicating whether or not a sample was boiled, the time and temperature of the incubation, and the optical density measurement indicating enzyme assay results.

Activity:
Assessing Protein Digestion by Pepsin

Pepsin Incubation

1. Individually drag six test tubes to the test tube holders in the incubation unit.

2. Prepare the tubes with the substances indicated in Chart 2 using the following method.

- Click and hold the mouse button on the dropper cap of the desired substance and drag the dropper cap to the top of the desired test tube.

- Release the mouse button to dispense the substance.

3. Once all tubes are prepared, click the number (1) under the first test tube. The tube will descend into the incubation unit. All other tubes should remain in the raised position.

4. Click **Boil** to boil the number 1 tube. After boiling for a few moments, the tube will automatically rise.

5. Adjust the incubation temperature to 37°C and the timer to 60 min (compressed time) by clicking the (+) or (−) buttons.

6. Click **Incubate** to start the run. The incubation unit will gently agitate the test tube rack, evenly mixing the contents of all test tubes throughout the incubation. The computer is compressing the 60-minute time period into 60 seconds of real time. When the incubation time elapses, the test tube rack will automatically rise, and the doors to the assay cabinet will open.

Pepsin Assay

After the assay cabinet doors open, you will see an instrument called a spectrophotometer, which you will use to measure how much yellow dye was liberated by pepsin digestion of BAPNA. When a test tube is dragged to the holder in the spectrophotometer and the **Analyze** button is clicked, the instrument will shine a light through a specimen to measure the amount of light absorbed by the sample within the tube. The measure of the amount of light absorbed by the solution is known as its *optical density*. A colorless solution does not absorb light, whereas a colored solution has a relatively high light absorbance. For example, a colorless solution has an optical density of 0.0. A colored solution, however, absorbs some of the light emitted by the spectrophotometer, resulting in an optical density reading greater than zero.

In this experiment a yellow-colored solution is a direct indication of the amount of BAPNA digested by pepsin. Although you can visually estimate the yellow color produced by pepsin digestion of BAPNA, the spectrophotometer precisely measures how much BAPNA digestion occurred in the experiment.

1. Click and hold the mouse on the first tube in the incubation unit and drag it to the holder in the spectrophotometer.

2. Release the mouse button to drop the tube into the holder.

3. Click **Analyze.** You will see light shining through the solution in the test tube as the spectrophotometer measures its optical density. The optical density of the sample will be displayed in the optical density window below the Analyze button.

Chart 2	Pepsin Digestion of Protein					
Tube no.	**1**	**2**	**3**	**4**	**5**	**6**
Additives	Pepsin BAPNA pH 2.0 buffer	Pepsin BAPNA pH 2.0 buffer	Pepsin Deionized water pH 2.0 buffer	Deionized water BAPNA pH 2.0 buffer	Pepsin BAPNA pH 7.0 buffer	Pepsin BAPNA pH 9.0 buffer
Incubation condition	Boil first, then incubate at 37°C	37°C	37°C	37°C	37°C	37°C
Optical density						

4. Record the optical density in Chart 2.

5. Drag the tube to its original position in the incubation unit and release the mouse button.

6. Repeat steps 1 through 5 for the remaining test tubes in the incubation unit.

7. Click **Record Data** to display your results in the grid and retain your data in the computer's memory for later analysis. To repeat the experiment, you must drag all test tubes to the test tube washer and start again.

8. Answer the following questions, referring to Chart 2 (or the data grid in the simulation) as necessary.

Which pH provided the highest pepsin activity? _____

Would pepsin be active in the mouth? Explain your answer.

_____

_____

How did the results of tube 1 compare with those of tube 2?

_____

Tubes 1 and 2 contained the same substances. Explain why their optical density measurements were different.

_____

_____

If you had not run the tube 2 and 3 samples, what argument could be made against the statement "Pepsin digests BAPNA"?

_____

_____

What do you think would happen if you reduced the incubation time to 30 minutes? Use the simulation to help you answer this question if you are not sure.

_____

_____

What do you think would happen if you decreased the temperature of incubation to 10°C? Use the simulation to help you answer this question if you are not sure.

_____

_____

Fat Digestion by Pancreatic Lipase and the Action of Bile

The treatment that fats and oils undergo during digestion in the small intestine is a bit more complicated than that of carbohydrates or proteins. Fats and oils require pretreatment with bile to physically emulsify the fats. As a result, two sets of reactions must occur.

First:

$$\text{Fats/oils} \xrightarrow[\text{(emulsification)}]{\text{bile}} \text{minute fat/oil droplets}$$

Then:

$$\text{Fat/oil droplets} \xrightarrow{\text{lipase}} \text{monoglycerides and fatty acids}$$

Lipase hydrolyzes fats and oils to their component monoglycerides and two fatty acids. Occasionally lipase hydrolyzes fats and oils to glycerol and three fatty acids.

The fact that some of the end products of fat digestion (fatty acids) are organic acids that decrease the pH provides an easy way to recognize that digestion is ongoing or completed. You will be using a pH meter in the assay cabinet to record the drop in pH as the test tube contents become acid.

Choose **Lipase** from the **Experiment** menu. The opening screen will appear in a few seconds (Figure 40B.3). The solutions shelf, test tube washer, and incubation equipment are the same as in the previous two experiments; only the solutions have changed.

Data displayed in the grid include the tube number, the four reagents dispensed into each tube, a (+) or (−) mark indicating whether or not a sample was boiled, the time and temperature of the incubation, and the pH measurement indicating enzyme assay results.

Activity:

Assessing Fat Digestion by Pancreatic Lipase and the Action of Bile

Lipase Incubation

1. Individually drag 6 test tubes to the test tube holders in the incubation unit.

2. Prepare the tubes with the solutions indicated in Chart 3 on page 424 by using the following method.

• Click and hold the mouse button on the dropper cap of the desired substance.

• While holding the mouse button down, drag the dropper cap to the top of the desired test tube.

• Release the mouse button to dispense the substance.

Figure 40B.3 Opening screen of the Lipase experiment.

3. Adjust the incubation temperature to 37°C and the timer to 60 min (compressed time) by clicking the (+) or (−) buttons.

4. Click **Incubate** to start the run. The incubation unit will gently agitate the test tube rack, evenly mixing the contents of all test tubes throughout the incubation. The computer is compressing the 60-minute time period into 60 seconds of real time. When the incubation time elapses, the test tube rack automatically rises, and the doors to the assay cabinet open.

Lipase Assay

After the assay cabinet doors open, you will see a pH meter that you will use to measure the relative acidity of your test solutions. When a test tube is dragged to the holder in the pH meter and the Measure pH button is clicked, a probe will descend into the sample, take a pH reading, and then retract. The pH of the sample will be displayed in the pH window below the Measure pH button. A solution containing fatty acids liberated from fat by the action of lipase will exhibit a lower pH than one without fatty acids.

Chart 3		Pancreatic Lipase Digestion of Fats and the Action of Bile					
Tube no.	**1**	**2**	**3**	**4**	**5**	**6**	
Additives	Lipase Vegetable oil Bile salts pH 7.0 buffer	Lipase Vegetable oil Deionized water pH 7.0 buffer	Lipase Deionized water Bile salts pH 9.0 buffer	Deionized water Vegetable oil Bile salts pH 7.0 buffer	Lipase Vegetable oil Bile salts pH 2.0 buffer	Lipase Vegetable oil Bile salts pH 9.0 buffer	
Incubation condition	37°C	37°C	37°C	37°C	37°C	37°C	
pH							

1. Click and hold the mouse on the first tube in the incubation unit and drag it to the holder in the pH meter. Release the mouse button to drop the tube into the holder.

2. Click **Measure pH.**

3. In Chart 3, record the pH displayed in the pH window.

4. Drag the test tube in the pH meter to its original position in the incubation unit and release the mouse button.

5. Repeat steps 1 through 4 for the remaining test tubes in the incubation unit.

6. Click **Record Data** to display your results in the grid and retain your data in the computer's memory for later analysis. To repeat the experiment, you must drag all test tubes to the test tube washer and begin again.

7. Answer the following questions, referring to Chart 3 (or the data grid in the simulation) as necessary.

Explain the difference in activity between tubes 1 and 2.

_____

_____

Can we determine if fat hydrolysis has occurred in tube 6?

_____ Explain your answer. _____

_____

Which pH resulted in maximum lipase activity? _____

Is this method of assay sufficient to determine if the optimum

activity of lipase is at pH 2.0? _____

In theory, would lipase be active in the mouth? _____

Would it be active in the stomach? _____

Explain your answers. _____

_____

Based on the enzyme pH optima you determined, where in the body would we expect to find the enzymes in these experiments?

Amylase _____

Pepsin _____

Lipase _____ ∎

Activity:
Using Your Data

Save Your Data

When you have finished with all experiments, you can save your data set for later review and additional analysis.

1. Select **Save Data to File** from the **File** menu.
2. When the dialog box opens, click a drive and/or directory location.
3. Type a file name into the **File name** box (your instructor may provide you with a name).
4. Click **Save** to complete the save. If you want to review previously saved data, select **Open Data File** from the **File** menu, and choose your file.

Print Your Data

If your instructor requires a hard copy of your experiment results, choose **Print Data** from the **File** menu.

1. Type your name into the provided space.
2. When you are ready to print, click **OK.** ■

Physical Processes: Mechanisms of Food Propulsion and Mixing

Although enzyme activity is an essential part of the overall digestion process, food must also be processed physically by churning and chewing, and moved by mechanical means along the tract if digestion and absorption are to be completed. Just about any time organs exhibit mobility, muscles are involved, and movements of and in the gastrointestinal tract are no exception. Although we tend to think only of smooth muscles for visceral activities, both skeletal and smooth muscles are necessary in digestion. This fact is demonstrated by the simple activities that follow.

Activity:
Studying Mechanisms of Food Propulsion and Mixing

Deglutition (Swallowing)

Swallowing, or *deglutition,* which is largely the result of skeletal muscle activity, occurs in two phases: *buccal* (mouth) and *pharyngeal-esophageal.* The initial phase—the buccal—is voluntarily controlled and initiated by the tongue. Once begun, the process continues involuntarily in the pharynx and esophagus, through peristalsis, resulting in the delivery of the swallowed contents to the stomach.

Obtain a pitcher of water, a stethoscope, a paper cup, alcohol swabs, and an autoclave bag in preparation for making the following observations.

1. While swallowing a mouthful of water, consciously note the movement of your tongue during the process. Record your observations.

_____

_____

_____

2. Repeat the swallowing process while your laboratory partner watches the externally visible movements of your larynx. This movement is more obvious in a male, who has a larger Adam's apple. Record your observations.

_____

_____

What do these movements accomplish? _____

_____

_____

3. Your lab partner should clean the ear pieces of a stethoscope with an alcohol swab and don the stethoscope. Then your lab partner should place the diaphragm of the stethoscope on your abdominal wall, approximately 1 inch below the xiphoid process and slightly to the left, to listen for sounds as you again take two or three swallows of water. There should be two audible sounds. The first sound occurs when the water splashes against the gastroesophageal sphincter. The second occurs when the peristaltic wave of the esophagus arrives at the sphincter and the sphincter opens, allowing water to gurgle into the stomach. Determine, as accurately as possible, the time interval between these two sounds and record it below.

Interval between arrival of water at the sphincter and the opening of the sphincter:

_____ sec.

This interval gives a fair indication of the time it takes for the peristaltic wave to travel down the 10-inch-long esophagus. Actually the time interval is slightly less than it seems because pressure causes the sphincter to relax before the peristaltic wave reaches it.

Dispose of the used paper cup in the autoclave bag. ■

Segmentation and Peristalsis

Although several types of movement occur in the digestive tract organs, segmentation and peristalsis are most important as mixing and propulsive mechanisms (Figure 40A.6, p. 414).

Segmental movements are local constrictions of the organ wall that occur rhythmically. They serve mainly to mix the foodstuffs with digestive juices and to increase the rate of absorption by continually moving different portions of the chyme over adjacent regions of the intestinal wall. However, segmentation is an important means of food propulsion in the small intestine, and slow segmenting movements called haustral contractions are common in the large intestine.

Peristaltic movements are the major means of propelling food through most of the digestive viscera. Essentially they are waves of contraction followed by waves of relaxation that squeeze foodstuffs through the alimentary canal, and they are superimposed on segmental movements.

Activity:
Viewing Propulsive Movements

If a videotape showing some of the propulsive movements is available, go to a viewing station to view it before leaving the laboratory. ■

Anatomy of the Urinary System

M etabolism of nutrients by the body produces wastes (carbon dioxide, nitrogenous wastes, ammonia, and so on) that must be eliminated from the body if normal function is to continue. Although excretory processes involve several organ systems (the lungs excrete carbon dioxide and skin glands excrete salts and water), it is the **urinary system** that is primarily concerned with the removal of nitrogenous wastes from the body. In addition to this purely excretory function, the kidney maintains the electrolyte, acid-base, and fluid balances of the blood and is thus a major, if not *the* major, homeostatic organ of the body.

To perform its functions, the kidney acts first as a blood "filter," and then as a blood "processor." It allows toxins, metabolic wastes, and excess ions to leave the body in the urine, while simultaneously retaining needed substances and returning them to the blood. Malfunction of the urinary system, particularly of the kidneys, leads to a failure in homeostasis which, unless corrected, is fatal.

Gross Anatomy of the Human Urinary System

The urinary system (Figure 41.1) consists of the paired kidneys and ureters and the single urinary bladder and urethra. The kidneys perform the functions described above and manufacture urine in the process. The remaining organs of the system provide temporary storage reservoirs or transportation channels for urine.

Activity:
Identifying Urinary System Organs

Examine the human torso model, a large anatomical chart, or a three-dimensional model of the urinary system to locate and study the anatomy and relationships of the urinary organs.

1. Locate the paired **kidneys** on the dorsal body wall in the superior lumbar region. Note that they are not positioned at exactly the same level. Because it is "crowded" by the liver, the right kidney is slightly lower than the left kidney. In a living person, fat deposits (the *adipose capsules*) hold the kidneys in place in a retroperitoneal position.

When the fatty material surrounding the kidneys is reduced or too meager in amount (in cases of rapid weight loss or in very thin individuals), the kidneys are less securely anchored to the body wall and may drop to a lower or more inferior position in the abdominal cavity. This phenomenon is called **ptosis**. ■

2. Observe the **renal arteries** as they diverge from the descending aorta and plunge into the indented medial region

Objectives

1. To describe the function of the urinary system.

2. To identify, on an appropriate diagram or torso model, the urinary system organs and to describe the general function of each.

3. To define *micturition,* and to explain pertinent differences in the control of the two bladder sphincters (internal and external).

4. To compare the course and length of the urethra in males and females.

5. To identify these regions of the dissected kidney (longitudinal section): hilus, cortex, medulla, medullary pyramids, major and minor calyces, pelvis, renal columns, and renal and adipose capsules.

6. To trace the blood supply of the kidney from the renal artery to the renal vein.

7. To define the nephron as the physiological unit of the kidney and to describe its anatomy.

8. To define *glomerular filtration, tubular reabsorption,* and *tubular secretion,* and to indicate the nephron areas involved in these processes.

9. To recognize microscopic or diagrammatic views of the histologic structure of the kidney and bladder.

Materials

- Human dissectible torso model and/or anatomical chart of the human urinary system
- 3-dimensional model of the cut kidney and of a nephron (if available)
- Dissection tray and instruments
- Pig or sheep kidney, doubly or triply injected
- Protective skin cream or disposable gloves
- Prepared histologic slides of a longitudinal section of kidney and cross sections of the bladder
- Compound microscope

 For instructions on animal dissections, see the dissection exercises starting on p. 733 in the cat and fetal pig editions of this manual.

 See Appendix C, Exercise 41 for links to A.D.A.M. Interactive Anatomy.

(a)

Figure 41.1 Organs of the urinary system.
(a) Anterior view of the female urinary organs. Most unrelated abdominal organs have been removed. **(b)** Posterior in situ view showing the position of the kidneys relative to the twelfth rib pair.

(b)

(**hilus**) of each kidney. Note also the **renal veins,** which drain the kidneys (circulatory drainage) and the two **ureters,** which drain urine from the kidneys and conduct it by peristalsis to the bladder for temporary storage.

3.   Locate the **urinary bladder,** and observe the point of entry of the two ureters into this organ. Also locate the single **urethra,** which drains the bladder. The triangular region of the bladder, which is delineated by these three openings (two ureteral and one urethral orifice), is referred to as the **trigone** (Figure 41.2).

4.   Follow the course of the urethra to the body exterior. In the male, it is approximately 20 cm (8 inches) long, travels the length of the **penis,** and opens at its tip. Its three named regions—the *prostatic, membranous,* and *spongy (penile) urethrae*—are described in more detail in Exercise 43 and illustrated in Figure 43.1 (p. 441). The urethra of males has a dual function: it is a urine conduit to the body exterior, and it provides a passageway for the ejaculation of semen. Thus, in the male, the urethra is part of both the urinary and reproductive systems. In females, the urethra is very short, approximately 4 cm (1½ inches) long. There are no common urinary-reproductive pathways in the female, and the female's urethra serves only to transport urine to the body exterior. Its external opening, the **external urethral orifice,** lies anterior to the vaginal opening. ■

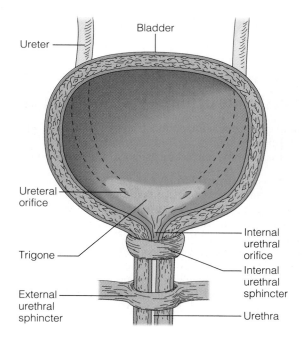

Figure 41.2 Detailed structure of the urinary bladder and urethral sphincters.

Dissection:
Gross Internal Anatomy of the Pig or Sheep Kidney

1.   Obtain a preserved sheep or pig kidney, dissecting pan, and instruments. Observe the kidney to identify the **renal capsule,** a smooth transparent membrane that adheres tightly to the external aspect of the kidney.

2.   Find the ureter, renal vein, and renal artery at the hilus (indented) region. The renal vein has the thinnest wall and will be collapsed. The ureter is the largest of these structures and has the thickest wall.

3.   In preparation for dissection, don gloves or apply protective skin cream. Make a cut through the longitudinal axis (frontal section) of the kidney and locate the anatomical areas described below and depicted in Figure 41.3.

Kidney cortex: The superficial kidney region, which is lighter in color. If the kidney is doubly injected with latex, you will see a predominance of red and blue latex specks in this region indicating its rich vascular supply.

Medullary region: Deep to the cortex; a darker, reddish-brown color. The medulla is segregated into triangular regions that have a striped, or striated, appearance—the **medullary (renal) pyramids.** The base of each pyramid faces toward the cortex. Its more pointed **apex,** or **papilla,** points to the innermost kidney region.

Renal columns: Areas of tissue, more like the cortex in appearance, which segregate and dip inward between the pyramids.

Renal pelvis: Medial to the hilus; a relatively flat, basinlike cavity that is continuous with the **ureter,** which exits from the hilus region. Fingerlike extensions of the pelvis should be visible. The larger, or primary, extensions are called the **major calyces;** subdivisions of the major calyces are the **minor calyces.** Notice that the minor calyces terminate in cuplike areas that enclose the apexes of the medullary pyramids and collect urine draining from the pyramidal tips into the pelvis.

4.   If the preserved kidney is doubly or triply injected, follow the renal blood supply from the renal artery to the **glomeruli.** The glomeruli appear as little red and blue specks in the cortex region. (See Figures 41.3 and 41.4.)

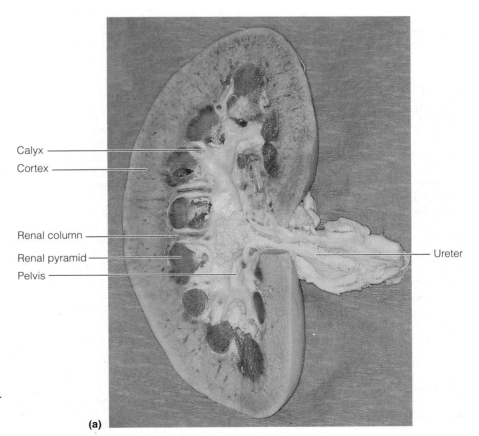

Figure 41.3   Frontal section of a kidney. (a) Photograph of a triple-injected pig kidney. Arteries and veins injected with red and blue latex respectively; pelvis and ureter injected with yellow latex.

Calyx

Cortex

Renal column

Renal pyramid

Pelvis

Ureter

(a)

Interlobular vein
Interlobular artery
Arcuate vein
Arcuate artery
Interlobar vein
Interlobar artery
Renal hilus
Renal artery
Renal vein
Segmental artery
Lobar artery
Ureter

Medullary (renal) pyramid

Major calyx
Minor calyx
Renal pelvis
Papilla of pyramid
Renal column
Cortex
Renal capsule

Figure 41.3 Frontal section of a kidney (continued). **(b)** Diagrammatic view showing the larger arteries supplying the kidney tissue.

(b)

Approximately a fourth of the total blood flow of the body is delivered to the kidneys each minute by the large **renal arteries.** As a renal artery approaches the kidney, it breaks up into five branches called **segmental arteries,** which enter the hilus. Each segmental artery, in turn, divides into several **lobar arteries.** The lobar arteries branch to form **interlobar arteries,** which ascend toward the cortex in the renal column areas. At the top of the medullary region, these arteries give off arching branches, the **arcuate arteries,** which curve over the bases of the medullary pyramids. Small **interlobular arteries** branch off the arcuate arteries and ascend into the cortex, giving off the individual **afferent arterioles,** which provide the capillary networks (**glomeruli** and **peritubular capillary beds**) that supply the nephrons, or functional units, of the kidney. Blood draining from the nephron capillary networks in the cortex enters the **interlobular veins** and then drains through the **arcuate veins** and the **interlobar veins** to finally enter the **renal vein** in the pelvis region. (There are no lobar or segmental veins.) ▪

Functional Microscopic Anatomy of the Kidney and Bladder

Kidney

Each kidney contains over a million nephrons, which are the anatomical units responsible for forming urine. Figure 41.4 depicts the detailed structure and the relative positioning of the nephrons in the kidney.

Each nephron consists of two major structures: a **glomerulus** (a capillary knot) and a **renal tubule.** During embryologic development, each renal tubule begins as a blind-ended tubule that gradually encloses an adjacent capillary cluster, or glomerulus. The enlarged end of the tubule encasing the glomerulus is the **glomerular (Bowman's) capsule,** and its inner, or visceral, wall consists of highly specialized cells called **podocytes.** Podocytes have long, branching processes (*foot processes*) that interdigitate with those of other podocytes and cling to the endothelial wall of the glomerular capillaries, thus forming a very porous epithelial membrane surrounding the glomerulus. The glomerulus-capsule complex is sometimes called the **renal corpuscle.**

The rest of the tubule is approximately 3 cm (1.25 inches) long. As it emerges from the glomerular capsule, it becomes highly coiled and convoluted, drops down into a long hairpin loop, and then again coils and twists before entering a collecting duct. In order from the glomerular capsule, the anatomical areas of the renal tubule are: the **proximal convoluted tubule, loop of Henle** (descending and ascending limbs), and the **distal convoluted tubule.** The wall of the renal tubule is composed almost entirely of cuboidal epithelial cells, with the exception of part of the descending limb (and sometimes part of the ascending limb) of the loop of Henle, which is simple squamous epithelium. The lumen surfaces of the cuboidal cells in the proximal convoluted tubule have dense microvilli (a cellular modification that greatly increases the surface area exposed to the lumen contents, or filtrate). Microvilli also occur on cells of the distal convoluted tubule but in greatly reduced numbers, revealing its less significant role in reclaiming filtrate contents.

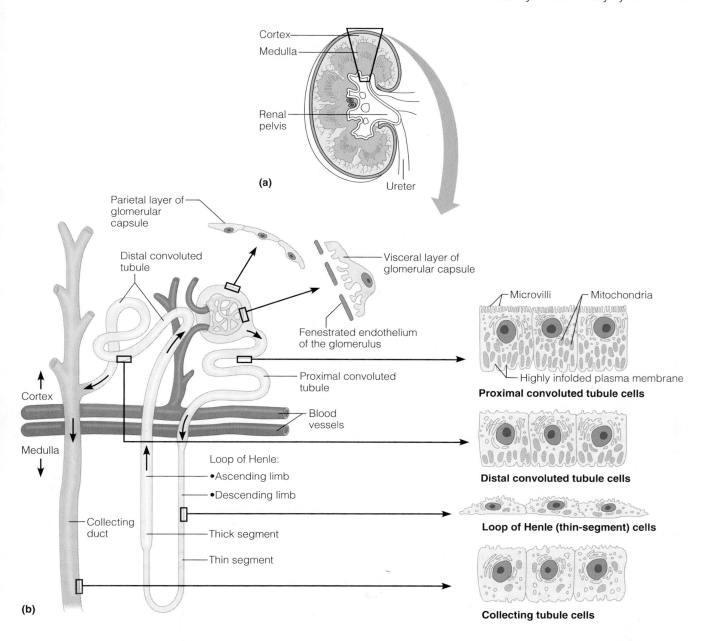

Figure 41.4 **Structure of a nephron. (a)** Wedge-shaped section of kidney tissue, indicating the position of the nephrons in the kidney. **(b)** Detailed nephron anatomy and associated blood supply.

Most nephrons, called **cortical nephrons,** are located entirely within the cortex. However, parts of the loops of Henle of the **juxtamedullary nephrons** (located close to the cortex-medulla junction) penetrate well into the medulla. The **collecting ducts,** each of which receives urine from many nephrons, run downward through the medullary pyramids, giving them their striped appearance. As the collecting ducts approach the renal pelvis, they fuse to form larger *papillary ducts,* which empty the final urinary product into the calyces and pelvis of the kidney.

The function of the nephron depends on several unique features of the renal circulation. The capillary vascular supply consists of two distinct capillary beds, the *glomerulus* and the *peritubular capillary bed.* Vessels leading to and from the

glomerulus, the first capillary bed, are both arterioles: the **afferent arteriole** feeds the bed while the **efferent arteriole** drains it. The glomerular capillary bed has no parallel elsewhere in the body. It is a high-pressure bed along its entire length. Its high pressure is a result of two major factors: (1) the bed is *fed and drained* by arterioles (arterioles are high-resistance vessels as opposed to venules, which are low-resistance vessels), and (2) the afferent feeder arteriole is larger in diameter than the efferent arteriole draining the bed. The high hydrostatic pressure created by these two anatomical features forces out fluid and blood components smaller than proteins from the glomerulus into the glomerular capsule. That is, it forms the filtrate which is processed by the nephron tubule.

Key:

a ![arrow] = Filtration

b ![arrow] = Reabsorption

c ![arrow] = Secretion

Urine

Figure 41.5 The kidney depicted as a single, large nephron. A kidney actually has millions of nephrons acting in parallel. The three major mechanisms by which the kidneys adjust the composition of plasma are **(a)** glomerular filtration, **(b)** tubular reabsorption, and **(c)** tubular secretion. Red arrows show the path of blood flow through the renal microcirculation.

The **peritubular capillary bed** arises from the efferent arteriole draining the glomerulus. This set of capillaries clings intimately to the renal tubule and empties into the interlobular veins that leave the cortex. The peritubular capillaries are *low-pressure* very porous capillaries adapted for absorption rather than filtration and readily take up the solutes and water reabsorbed from the filtrate by the tubule cells. The juxtamedullary nephrons have additional looping vessels, called the **vasa recta** ("straight vessels"), that parallel the long loops of Henle in the medulla (see Figure 41.4). Hence, the two capillary beds of the nephron have very different, but complementary, roles: The glomerulus produces the filtrate and the peritubular capillaries reclaim most of that filtrate.

Urine formation is a result of three processes: *filtration, reabsorption,* and *secretion* (Figure 41.5). **Filtration,** the role of the glomerulus, is largely a passive process in which a portion of the blood passes from the glomerular bed into the glomerular capsule. This filtrate then enters the proximal convoluted tubule where tubular reabsorption and secretion begin. During **tubular reabsorption,** many of the filtrate components move through the tubule cells and return to the blood in the peritubular capillaries. Some of this reabsorption is passive, such as that of water which passes by osmosis, but the reabsorption of most substances depends on active transport processes and is highly selective. Which substances are reabsorbed at a particular time depends on the composition of

the blood and needs of the body at that time. Substances that are almost entirely reabsorbed from the filtrate include water, glucose, and amino acids. Various ions are selectively reabsorbed or allowed to go out in the urine according to what is required to maintain appropriate blood pH and electrolyte composition. Waste products (urea, creatinine, uric acid, and drug metabolites) are reabsorbed to a much lesser degree or not at all. Most (75% to 80%) of tubular reabsorption occurs in the proximal convoluted tubule. The balance occurs in other areas, especially the distal convoluted tubules and collecting ducts.

Tubular secretion is essentially the reverse process of tubular reabsorption. Substances such as hydrogen and potassium ions and creatinine move either from the blood of the peritubular capillaries through the tubular cells or from the tubular cells into the filtrate to be disposed of in the urine. This process is particularly important for the disposal of substances not already in the filtrate (such as drug metabolites), and as a device for controlling blood pH.

A c t i v i t y :
Studying Nephron Structure

1. Begin your study of nephron structure by identifying the glomerular capsule, proximal and distal convoluted tubule regions, and the loop of Henle on a model of the nephron. Then, obtain a compound microscope and a prepared slide of kidney tissue to continue with the microscope study of the kidney.

2. Hold the longitudinal section of the kidney up to the light to identify cortical and medullary areas. Then secure the slide on the microscope stage and scan the slide under low power.

3. Move the slide so that you can see the cortical area. Identify a glomerulus, which appears as a ball of tightly packed material containing many small nuclei (Figure 41.6). It is usually delineated by a vacant-appearing region (corresponding to the space between the visceral and parietal layers of the glomerular capsule) that surrounds it.

4. Notice that the renal tubules are cut at various angles. Also try to differentiate between the thin-walled loop of Henle portion of the tubules and the cuboidal epithelium of the proximal convoluted tubule, which has dense microvilli. ■

Bladder

Although urine production by the kidney is a continuous process, urine is usually removed from the body when voiding is convenient. In the meantime the bladder, which receives urine via the ureters and discharges it via the urethra, stores it temporarily.

Voiding, or **micturition,** is the process in which urine empties from the bladder. Two sphincter muscles or valves (Figure 41.2), the **internal urethral sphincter** (more superiorly located) and the **external urethral sphincter** (more inferiorly located) control the outflow of urine from the bladder. Ordinarily, the bladder continues to collect urine until about 200 ml have accumulated, at which time the stretching of the bladder wall activates stretch receptors. Impulses transmitted to the central nervous system subsequently produce reflex contractions of the bladder wall through parasympathetic nervous system pathways (i.e., via the

(a)

Renal tubules   Glomeruli   Lumen of the glomerular capsule

(b)

Parietal layer of the glomerular capsule
Lumen of the glomerular capsule
Cuboidal epithelium of the renal tubule
Juxtaglomerular cells
Macula densa
Glomerulus

Figure 41.6 Microscopic structure of kidney tissue. (a) Low-power view of the renal cortex. **(b)** Detailed structure of the glomerulus. (See also Plates 46 and 47 in the Histology Atlas.)

pelvic splanchnic nerves). As the contractions increase in force and frequency, the stored urine is forced past the internal sphincter, which is a smooth muscle involuntary sphincter, into the superior part of the urethra. It is then that a person feels the urge to void. The inferior external sphincter consists of skeletal muscle and is voluntarily controlled. If it is not convenient to void, the opening of this sphincter can be inhibited. Conversely, if the time is convenient, the sphincter may be relaxed and the stored urine flushed from the body. If voiding is inhibited, the reflex contractions of the bladder cease temporarily and urine continues to accumulate in the bladder. After another 200 to 300 ml of urine have been collected, the *micturition reflex* will again be initiated.

Lack of voluntary control over the external sphincter is referred to as **incontinence.** Incontinence is normal in children 2 years old or younger, as they have not yet gained control over the voluntary sphincter. In adults and older children, incontinence is generally a result of spinal cord injury, emotional problems, bladder irritability, or some other pathology of the urinary tract. ∎

Activity:
Studying Bladder Structure

1.   Return the kidney slide to the supply area and obtain a slide of bladder tissue. Scan the bladder tissue. Identify its three layers: mucosa, muscular layer, and fibrous adventitia.

2.   Study the mucosa with its highly specialized transitional epithelium. The plump, transitional epithelial cells have the ability to slide over one another, thus decreasing the thickness of the mucosa layer as the bladder fills and stretches to accommodate the increased urine volume. Depending on the degree of stretching of the bladder, the mucosa may be three to eight cell layers thick. Compare the transitional epithelium of the mucosa to that shown in Figure 6.3h (p. 57).

3.   Examine the heavy muscular wall (detrusor muscle), which consists of three irregularly arranged muscular layers. The innermost and outermost muscle layers are arranged longitudinally; the middle layer is arranged circularly. Attempt to differentiate the three muscle layers.

4.   Draw a small section of the bladder wall, and label all regions or tissue areas.

5.   Compare your sketch of the bladder wall to the structure of the ureter wall shown in Plate 48 in the Histology Atlas. How are the two organs similar histologically?

_____

_____

_____

What is/are the most obvious differences?

_____

_____ ∎

 For instructions on animal dissections, see the dissection exercises starting on p. 733 in the cat and fetal pig editions of this manual.

exercise

42

Urinalysis

Objectives

1. To list the physical characteristics of urine, and to indicate the normal pH and specific gravity ranges.

2. To list substances that are normal urinary constituents.

3. To conduct various urinalysis tests and procedures and use them to determine the substances present in a urine specimen.

4. To define the following urinary conditions:

   albuminuria   *hematuria*

   calculi   *hemoglobinuria*

   casts   *ketonuria*

   glycosuria   *pyuria*

5. To explain the implications and possible causes of conditions listed in Objective 4.

Materials

❑ "Normal" artificial urine provided by the instructor*

❑ Numbered "pathologic" urine specimens provided by the instructor*

❑ Urinometer

❑ Hot plate

❑ 500-ml beakers

❑ Wide-range pH paper

❑ Dip sticks: individual (Clinistix, Ketostix, Albustix, Hemastix, Bilistix) or combination (Chemstrip or Multistix)

❑ Ictotest reagent

❑ Test reagents for sulfates: 10% barium chloride solution, dilute HCl (hydrochloric acid)

❑ Test reagent for glucose (Clinitest tablets)

❑ Test reagent for phosphates (dilute nitric acid, dilute ammonium molybdate)

❑ Test reagent for urea (concentrated nitric acid in dropper bottle)

❑ Test reagent for Cl: 3.0% silver nitrate solution ($AgNO_3$), freshly prepared

*Directions for making artificial urine are provided in the Instructor's Guide for this manual.

434

❑ Test tubes, test tube rack, and test tube holders

❑ 10-cc graduated cylinders

❑ Glass stirring rods

❑ Medicine droppers

❑ Wax marking pencils

❑ Microscope slides

❑ Coverslips

❑ Compound microscope

❑ Centrifuge and centrifuge tubes

❑ Sedi-stain

❑ Flasks and laboratory buckets containing 10% bleach solution

❑ Disposable plastic gloves

❑ Disposable autoclave bags

❑ *Demonstration:* Instructor-prepared specimen of urine sediment set up for microscopic analysis

B lood composition depends on three major factors: diet, cellular metabolism, and urinary output. In 24 hours, the kidneys' 2 million nephrons filter approximately 150 to 180 liters of blood plasma through their glomeruli into the tubules, where it is selectively processed by tubular reabsorption and secretion. In the same period, urinary output, which contains by-products of metabolism and excess ions, is 1.0 to 1.8 liters. In healthy individuals, the kidneys can maintain blood constancy despite wide variations in diet and metabolic activity. With certain pathologic conditions, urine composition often changes dramatically.

Characteristics of Urine

Freshly voided urine is generally clear and pale yellow to amber in color. This normal yellow color is due to *urochrome,* a pigment metabolite arising from the body's destruction of hemoglobin (via bilirubin or bile pigments). As a rule, color variations from pale yellow to deeper amber indicate the relative concentration of solutes to water in the urine. The greater the solute concentration, the deeper the color. Abnormal urinary color may be due to certain foods, such as beets, various drugs, bile, or blood.

The odor of freshly voided urine is characteristic and slightly aromatic, but bacterial action gives it an ammonia-like odor when left standing. Some drugs, vegetables (such as asparagus), and various disease processes (such as diabetes mellitus) alter the characteristic odor of urine. For example, the urine of a person with uncontrolled diabetes mellitus (and elevated levels of ketones) smells fruity or acetonelike.

The pH of urine ranges from 4.5 to 8.0, but its average value, 6.0, is slightly acidic. Diet may markedly influence the pH of the urine. For example, a diet high in protein (meat, eggs, cheese) and whole wheat products increases the acidity of urine. Such foods are called acid *ash foods.* On the other hand, a vegetarian diet (*alkaline ash diet*) increases the alkalinity of the urine. A bacterial infection of the urinary tract may also result in urine with a high pH.

Specific gravity is the relative weight of a specific volume of liquid compared with an equal volume of distilled water. The specific gravity of distilled water is 1.000, because 1 ml weighs 1 g. Since urine contains dissolved solutes, it weighs more than water, and its customary specific gravity ranges from 1.001 to 1.030. Urine with a specific gravity of 1.001 contains few solutes and is considered very dilute. Dilute urine commonly results when a person drinks excessive amounts of water, uses diuretics, or suffers from diabetes insipidus or chronic renal failure. Conditions that produce urine with a high specific gravity include limited fluid intake, fever, and kidney inflammation, called *pyelonephritis.* If urine becomes excessively concentrated, some of the substances normally held in solution begin to precipitate or crystallize, forming **kidney stones,** or **renal calculi.**

Normal constituents of urine (in order of decreasing concentration) include water; urea;* sodium,† potassium, phosphate, and sulfate ions; creatinine,* and uric acid.* Much smaller but highly variable amounts of calcium, magnesium, and bicarbonate ions are also found in the urine. Abnormally high concentrations of any of these urinary constituents may indicate a pathological condition.

Abnormal Urinary Constituents

Abnormal urinary constituents are substances not normally present in the urine when the body is operating properly.

Glucose

The presence of glucose in the urine, a condition called **glycosuria,** indicates abnormally high blood sugar levels. Normally blood sugar levels are maintained between 80 and 100 mg/100 ml of blood. At this level all glucose in the filtrate is reabsorbed by the tubular cells and returned to the blood. Glycosuria may result from carbohydrate intake so excessive that normal physiologic and hormonal mechanisms cannot

---

* Urea, uric acid, and creatinine are the most important nitrogenous wastes found in urine. Urea is an end product of protein breakdown; uric acid is a metabolite of purine breakdown; and creatinine is associated with muscle metabolism of creatine phosphate.

† Sodium ions appear in relatively high concentration in the urine because of reduced urine volume, not because large amounts are being secreted. Sodium is the major positive ion in the plasma; under normal circumstances, most of it is actively reabsorbed.

clear it from the blood quickly enough. In such cases, the active transport reabsorption mechanisms of the renal tubules for glucose are exceeded, but only temporarily.

Pathologic glycosuria occurs in conditions such as uncontrolled diabetes mellitus, in which the body cells are unable to absorb glucose from the blood because the pancreatic islet cells produce inadequate amounts of the hormone insulin, or there is some abnormality of the insulin receptors. Under such circumstances, the body cells increase their metabolism of fats, and the excess and unusable glucose spills out in the urine.

Albumin

Albuminuria, or the presence of albumin in urine, is an abnormal finding. Albumin is the single most abundant blood protein and is very important in maintaining the osmotic pressure of the blood. Albumin, like other blood proteins, normally is too large to pass through the glomerular filtration membrane. Thus, albuminuria is generally indicative of an abnormally increased permeability of the glomerular membrane. Certain nonpathologic conditions, such as excessive exertion, pregnancy, or overabundant protein intake, can temporarily increase the membrane permeability, leading to **physiologic albuminuria.** Pathologic conditions resulting in albuminuria include events that damage the glomerular membrane, such as kidney trauma due to blows, the ingestion of heavy metals, bacterial toxins, glomerulonephritis, and hypertension.

Ketone Bodies

Ketone bodies (acetoacetic acid, beta-hydroxybutyric acid, and acetone) normally appear in the urine in very small amounts. **Ketonuria,** the presence of these intermediate products of fat metabolism in excessive amounts, usually indicates that abnormal metabolic processes are occurring. The result may be *acidosis* and its complications. Ketonuria is an expected finding during starvation, when inadequate food intake forces the body to use its fat stores. Ketonuria coupled with a finding of glycosuria is generally diagnostic for diabetes mellitus.

Red Blood Cells

Hematuria, the appearance of red blood cells, or erythrocytes, in the urine, almost always indicates pathology of the urinary tract, because erythrocytes are too large to pass through the glomerular pores. Possible causes include irritation of the urinary tract organs by calculi (kidney stones), which produces frank bleeding; infections; or physical trauma to the urinary organs. In healthy menstruating females, it may reflect accidental contamination of the urine sample with the menstrual flow.

Hemoglobin

Hemoglobinuria, the presence of hemoglobin in the urine, is a result of the fragmentation, or hemolysis, of red blood cells. As a result, hemoglobin is liberated into the plasma and subsequently appears in the kidney filtrate. Hemoglobinuria indicates various pathologic conditions including hemolytic anemias, transfusion reactions, burns, or renal disease.

TABLE 42.1	Urinalysis Results*		
Observation or test	**Normal values**	**Standard urine specimen**	**Unknown specimen (#)**
Physical characteristics			
Color	Pale yellow	Yellow: pale medium dark other _____	Yellow: pale medium dark other _____
Transparency	Transparent	Clear slightly cloudy cloudy	Clear slightly cloudy cloudy
Odor	Characteristic	Describe _____	Describe _____
pH	4.5–8.0	_____	_____
Specific gravity	1.001–1.030	_____	_____
Inorganic components			
Sulfates	Present	Present Absent	Present Absent
Phosphates	Present	Present Absent	Present Absent
Chlorides	Present	Present Absent	Present Absent
Organic components			
Urea	Present	Present Absent	Present Absent
Glucose Dip stick:	Negative	Record results: _____	Record results: _____
_____			
Clinitest	Negative	Record results: _____	Record results: _____
_____			
Albumin Dip stick:	Negative	Record results: _____	Record results: _____
_____			
Ketone bodies Dip stick:	Negative	Record results: _____	Record results: _____
_____			
RBCs/hemoglobin Dip stick:	Negative	Record results: _____	Record results: _____
_____			
Bilirubin Dip stick:	Negative	Record results: _____	Record results: _____
_____			
Ictotest	Negative (no color change)	Negative Positive (purple)	Negative Positive (purple)

* In recording urinalysis data, circle the appropriate description if provided; otherwise record the results you observed. Identify dip sticks used.

Bile Pigments

Bilirubinuria, the appearance of bilirubin (bile pigments) in urine, is an abnormal finding and usually indicates liver pathology, such as hepatitis or cirrhosis. Bilirubinuria is signaled by a yellow foam that forms when the urine sample is shaken.

White Blood Cells

Pyuria is the presence of white blood cells or other pus constituents in the urine. It indicates inflammation of the urinary tract.

Casts

Any complete discussion of the varieties and implications of casts is beyond the scope of this exercise. However, because they always represent a pathologic condition of the kidney or urinary tract, they should at least be mentioned. **Casts** are hardened cell fragments, usually cylindrical, which are flushed out of the urinary tract. *White blood cell casts* are a typical finding with pyelonephritis, *red blood cell casts* are commonly seen with glomerulonephritis, and *fatty casts* indicate severe renal damage.

A c t i v i t y :
Analyzing Urine Samples

In this part of the exercise, you will use prepared dip sticks and perform chemical tests to determine the characteristics of normal urine as well as to identify abnormal urinary components. You will investigate both "normal" urine (designated as the *standard urine specimen* in Table 42.1) *and* an unknown specimen of urine provided by your instructor. Make the following determinations on both samples and record your results by circling the appropriate item or description or by adding data to complete Table 42.1. If you have more than one unknown sample, accurately identify each sample by number.

⚠ *Obtain and wear plastic disposable gloves throughout this laboratory session.* Although the instructor-provided urine samples are actually artificial urine (concocted in the laboratory to resemble real urine), the techniques of safe handling of body fluids should still be observed as part of your learning process. When you have completed the laboratory procedures: (1) dispose of the gloves and used pH paper strips in the autoclave bag; (2) put used glassware in the bleach-containing laboratory bucket; (3) wash the lab bench down with 10% bleach solution.

Determination of the Physical Characteristics of Urine

1. Determine the color, transparency, and odor of your "normal" sample and one of the numbered pathologic samples, and circle the appropriate descriptions in Table 42.1.

2. Obtain a roll of wide-range pH paper to determine the pH of each sample. Use a fresh piece of paper for each test, and dip the strip into the urine to be tested two or three times before comparing the color obtained with the chart on the dispenser. Record your results in Table 42.1. (If you will be using one of the combination dip sticks—Chemstrip or Multistix—this pH determination can be done later.)

3. To determine specific gravity, obtain a urinometer cylinder and float. Mix the urine well, and fill the urinometer cylinder about two-thirds full with urine.

4. Examine the urinometer float to determine how to read its markings. In most cases, the scale has numbered lines separated by a series of unnumbered lines. The numbered lines give the reading for the first two decimal places. You must determine the third decimal place by reading the lower edge of the meniscus—the curved surface representing the urine-air junction—on the stem of the float.

5. Carefully lower the urinometer float into the urine. Make sure it is floating freely before attempting to take the reading. Record the specific gravity of both samples in the table. *Do not dispose of this urine if the samples that you have are less than 200 ml in volume* because you will need to make several more determinations.

Determination of Inorganic Constituents in Urine

Sulfates Add 5 ml of urine to a test tube, and then add a few drops of dilute hydrochloric acid and 2 ml of 10% barium chloride solution. The appearance of a white precipitate (barium sulfate) indicates the presence of sulfates in the sample. Clean the test tubes well after use. Record your results. Are sulfates a *normal* constituent of urine?

Phosphates Obtain a hot plate and a 500 ml beaker. To prepare the hot water bath, half fill the beaker with tap water and heat it on the hot plate. Add 5 ml of urine to a test tube, and then add three or four drops of dilute nitric acid and 3 ml of ammonium molybdate. Mix well with a glass stirring rod, and then heat gently in a hot water bath. Formation of a yellow precipitate indicates the presence of phosphates in the sample.

Chlorides Place 5 ml of urine in a test tube, and add several drops of silver nitrate ($AgNO_3$). The appearance of a white precipitate (silver chloride) is a positive test for chlorides. Record your results.

Determination of Organic Constituents in Urine

Individual dip sticks or combination dip sticks (Chemstrip or Multistix) may be used for many of the tests in this section. If combination dip sticks are used, be prepared to take the readings on several factors (pH, protein [albumin], glucose, ketones, and blood/hemoglobin) at the same time. Generally speaking, results for all of these tests may be read *during* the second minute after immersion, but readings taken after 2 minutes have passed should be considered invalid. Pay careful attention to the directions for method and time of immersion and disposal of excess urine from the strip, regardless of the dip stick used.

Urea Put two drops of urine on a clean microscope slide and *carefully* add one drop of concentrated nitric acid to the urine. Slowly warm the mixture on a hot plate until it begins to dry at the edges, but do not allow it to boil or to evaporate to dryness. When the slide has cooled, examine the edges of the preparation under low power to identify the rhombic or hexagonal crystals of urea nitrate, which form when urea and nitric acid react chemically. Keep the light low for best contrast. Record your results.

Glucose Use a combination dip stick or obtain a vial of Clinistix, and conduct the dip stick test according to the instructions on the vial. Record the results in Table 42.1.

Because the Clinitest reagent is routinely used in clinical agencies for glucose determinations in pediatric patients (children), it is worthwhile to conduct this test as well. Obtain the Clinitest tablets and the associated color chart. Using a medicine dropper, put 5 drops of urine into a test tube; then rinse the dropper and add 10 drops of water to the tube. Add a Clinitest tablet. Wait 15 seconds and then compare the color obtained to the color chart. Record the results.

Albumin Use a combination dip stick or obtain the Albustix dip sticks, and conduct the determinations as indicated on the vial. Record your results.

Ketones Use a combination dip stick or obtain the Ketostix dip sticks. Conduct the determinations as indicated on the vial. Record your results.

Blood/Hemoglobin Test your urine samples for the presence of hemoglobin by using a Hemastix dip stick or a combination dip stick according to the directions on the vial. Usually a short drying period is required before making the reading, so read the directions carefully. Record your results.

Bilirubin Using a Bilistix dip stick, determine if there is any bilirubin in your urine samples. Record your results.

Also conduct the Ictotest for the presence of bilirubin. Using a medicine dropper, place one drop of urine in the center of one of the special test mats provided with the Ictotest reagent tablets. Place one of the reagent tablets over the drop of urine, and then add two drops of water directly to the tablet. If the mixture turns purple when you add water, bilirubin is present. Record your results. ■

A c t i v i t y :
Analyzing Urine Sediment Microscopically (Optional)

If desired by your instructor, a microscopic analysis of urine sediment (of "real" urine) can be done. The urine sample to be analyzed microscopically has been centrifuged to spin the more dense urine components to the bottom of a tube, and some of the sediment (mounted on a slide) has been stained with Sedi-stain to make the components more visible.

Go to the demonstration microscope to conduct this study. Using the lowest light source possible, examine the slide under low power to determine if any of the sediments illustrated in Figures 42.1 and 42.2 (and described here) can be seen.

Unorganized sediments (Figure 42.1): Chemical substances that form crystals or precipitate from solution; for example, calcium oxalates, carbonates, and phosphates; uric acid; ammonium ureates; and cholesterol. Also, if one has been taking antibiotics or certain drugs such as sulfa drugs, these may be detectable in the urine in crystalline form. Normal urine contains very small amounts of crystals, but conditions such as urinary retention or urinary tract infection may cause the appearance of much larger amounts (and their possible consolidation into calculi). The high-power lens may be needed to view the various crystals, which tend to be much more minute than the organized (cellular) sediments.

Organized sediments (Figure 42.2): Include epithelial cells (rarely of any pathologic significance), pus cells (white blood cells), red blood cells, and casts. Urine is normally negative for organized sediments, and the presence of the last three categories mentioned, other than trace amounts, always indicates kidney pathology.

Draw a few of your observations below and attempt to identify them by comparing them with Figures 42.1 and 42.2. ■

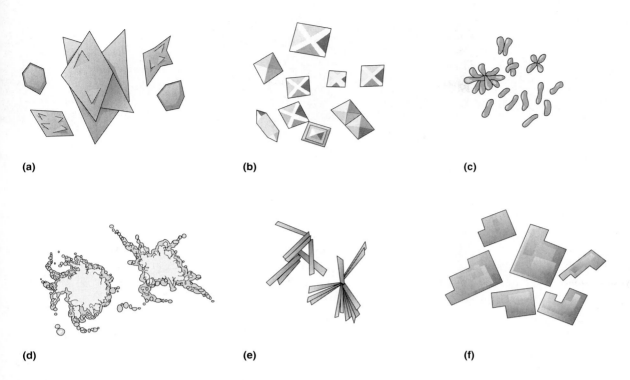

Figure 42.1   Examples of unorganized sediments. (a) Uric acid crystals, **(b)** calcium oxalate crystals, **(c)** calcium carbonate crystals, **(d)** ammonium ureate crystals, **(e)** calcium phosphate crystals, **(f)** cholesterol crystals.

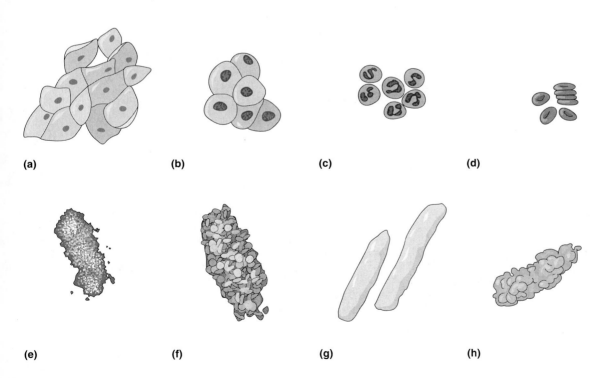

Figure 42.2   Examples of organized sediments. (a) Squamous epithelial cells, **(b)** transitional epithelial cells, **(c)** white blood cells, largely neutrophils (pus), **(d)** red blood cells, **(e)** granular casts, **(f)** red blood cell casts, **(g)** hyaline casts, **(h)** fatty casts.

Anatomy of the Reproductive System

Objectives

1. To discuss the general function of the reproductive system.

2. To identify and name the structures of the male and female reproductive systems when provided with an appropriate model or diagram, and to discuss the general function of each.

3. To define *semen,* discuss its composition, and name the organs involved in its production.

4. To trace the pathway followed by a sperm from its site of formation to the external environment.

5. To name the exocrine and endocrine products of the testes and ovaries, indicating the cell types or structures responsible for the production of each.

6. To identify homologous structures of the male and female systems.

7. To discuss the microscopic structure of the penis, epididymis, uterine tube, and uterus, and to relate structure to function.

8. To define *ejaculation, erection,* and *gonad.*

9. To discuss the function of the fimbriae and ciliated epithelium of the uterine (fallopian) tubes.

10. To identify the fundus, body, and cervical regions of the uterus.

11. To define *endometrium, myometrium,* and *ovulation.*

Materials

❑ Models or large laboratory charts of the male and female reproductive tracts

❑ Prepared microscope slides of cross sections of the penis, epididymis, uterine tube, and uterus showing endometrium (proliferative phase)

❑ Compound microscope

 For instructions on animal dissections, see the dissection exercises starting on p. 733 in the cat and fetal pig editions of this manual.

 See Appendix C, Exercise 43 for links to A.D.A.M. Interactive Anatomy.

Other organ systems of the body function primarily to sustain the existing individual, but the reproductive system is unique. Most simply stated, the biologic function of the **reproductive system** is to perpetuate the species.

The essential organs of reproduction are those that produce the germ cells—the testes and the ovaries. The reproductive role of the male is to manufacture sperm and to deliver them to the female reproductive tract. The female, in turn, produces eggs. If the time is suitable, the combination of sperm and egg produces a fertilized egg, which is the first cell of a new individual. Once fertilization has occurred, the female uterus provides a nurturing, protective environment in which the embryo, later called the fetus, develops until birth.

Although the drive to reproduce is strong in all animals, in humans this drive is also intricately related to nonbiologic factors. Emotions and social considerations often enhance or thwart its expression.

Gross Anatomy of the Human Male Reproductive System

The primary reproductive organs of the male are the **testes,** the male *gonads,* which have both an exocrine (sperm production) and an endocrine (testosterone production) function. All other reproductive structures are conduits or sources of secretions, which aid in the safe delivery of the sperm to the body exterior or female reproductive tract.

Activity:
Identifying Male Reproductive Organs

As the following organs and structures are described, locate them on Figure 43.1, and then identify them on a three-dimensional model of the male reproductive system or on a large laboratory chart. ■

The paired oval testes lie in the **scrotal sac** outside the abdominopelvic cavity. The temperature there (approximately 94°F, or 34°C) is slightly lower than body temperature, a requirement for producing viable sperm.

The accessory structures forming the *duct system* are the epididymis, the ductus deferens, the ejaculatory duct, and the urethra. The **epididymis** is an elongated structure running up the posterolateral aspect of the testis and capping its superior aspect. The epididymis forms the first portion of the duct system and provides a site for immature sperm entering it from the testis to complete their maturation process. The **ductus deferens** (sperm duct) arches superiorly from the epididymis,

Ejaculatory duct
Prostatic urethra
Urinary bladder

Symphysis pubis

Ductus deferens

Corpus cavernosum
Corpus spongiosum

Penile urethra

Glans penis
Prepuce

Seminal vesicle
Ampulla

Prostate gland
Membranous urethra
Rectum
Bulbourethral gland

Epididymis
Testis

(a)

Crura of penis

Bulb of penis

Glans

Corpus
spongiosum

Corpus
cavernosum

Bulbourethral
glands

(b)

Figure 43.1   Reproductive system of the human male. (a) Midsagittal section. **(b)** Inferior view of the penis with corpus spongiosum partially reflected.

passes through the inguinal canal into the pelvic cavity, and courses over the superior aspect of the urinary bladder. In life, the ductus deferens (also called the *vas deferens*) is enclosed along with blood vessels and nerves in a connective tissue sheath called the **spermatic cord.** The terminus of the ductus deferens enlarges to form the region called the **ampulla,** which empties into the **ejaculatory duct.** Contraction of the ejaculatory duct propels the sperm through the prostate gland to the **prostatic urethra,** which in turn empties into the **membranous urethra** and then into the **penile urethra,** which runs through the length of the penis to the body exterior.

The spermatic cord is easily palpated through the skin of the scrotum. When a *vasectomy* is performed, a small inci-

sion is made in each side of the scrotum, and each ductus deferens is cut through or cauterized. Although sperm are still produced, they can no longer reach the body exterior; thus a man is sterile after this procedure (and 12 to 15 ejaculations to clear the conducting tubules).

The *accessory glands* include the prostate gland, the paired seminal vesicles, and the bulbourethral glands. These glands produce **seminal fluid,** the liquid medium in which sperm leave the body. The **seminal vesicles,** which produce about 60% of seminal fluid, lie at the posterior wall of the urinary bladder close to the terminus of the ductus deferens. They produce a viscous alkaline secretion containing fructose (a simple sugar) and other substances that nourish the sperm passing through the tract or promote the fertilizing capability of sperm in some way. The duct of each seminal vesicle merges with a ductus deferens to form the ejaculatory duct (mentioned above); thus sperm and seminal fluid enter the urethra together.

The **prostate gland** encircles the urethra just inferior to the bladder. It secretes a milky fluid into the urethra, which plays a role in activating the sperm.

Hypertrophy of the prostate gland, a troublesome condition commonly seen in elderly men, constricts the urethra so that urination is difficult. ■

The **bulbourethral glands** are tiny, pea-shaped glands inferior to the prostate. They produce a thick, clear, alkaline mucus that drains into the membranous urethra. This secretion acts to wash residual urine out of the urethra when ejaculation of **semen** (sperm plus seminal fluid) occurs. The relative alkalinity of seminal fluid also buffers the sperm against the acidity of the female reproductive tract.

The **penis,** part of the external genitalia of the male along with the scrotal sac, is the copulatory organ of the male. Designed to deliver sperm into the female reproductive tract, it consists of a shaft, which terminates in an enlarged tip, the **glans** (see Figure 43.1b). The skin covering the penis is loosely applied, and it reflects downward to form a circular fold of skin, the **prepuce,** or **foreskin,** around the proximal end of the glans. (The foreskin is removed in the surgical procedure called *circumcision.*) Internally, the penis consists primarily of three elongated cylinders of erectile tissue, which engorge with blood during sexual excitement. This causes the penis to become rigid and enlarged so that it may more adequately serve as a penetrating device. This event is called **erection.** The paired dorsal cylinders are the **corpora cavernosa.** The single ventral **corpus spongiosum** surrounds the penile urethra.

Microscopic Anatomy of Selected Male Reproductive Organs

Testis

Each testis is covered by a dense connective tissue capsule called the **tunica albuginea** (literally, "white tunic"). Extensions of this sheath enter the testis, dividing it into a number of lobes, each of which houses one to four highly coiled **seminiferous tubules,** the sperm-forming factories (Figure 43.2). The seminiferous tubules of each lobe converge to empty the sperm into another set of tubules, the **rete testis,** at the mediastinum of the testis. Sperm traveling through the rete testis then enter the epididymis, located on the exterior aspect of the testis, as previously described. Lying between the seminiferous tubules and softly padded with connective tissue are the **interstitial cells,** which produce testosterone, the hormonal product of the testis. You will conduct a microscopic study of testes tissue in Exercise 44.

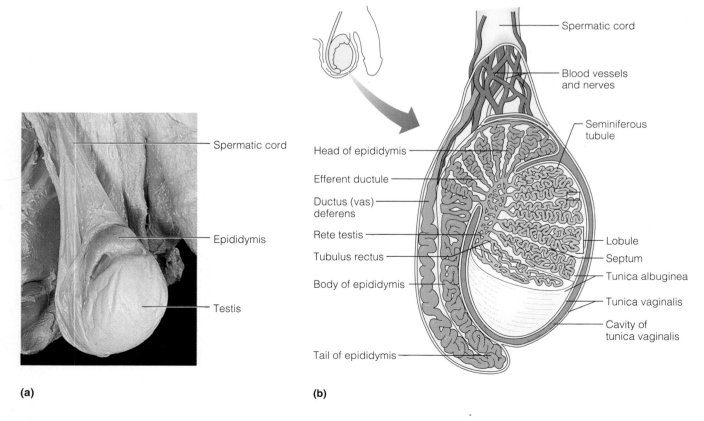

(a)

(b)

Figure 43.2 Structure of the testis. (a) External view of a testis from a cadaver. **(b)** Longitudinal section of the testis showing seminiferous tubules. Same orientation as (a). (Epididymis and part of the ductus deferens also shown.)

Epididymis

Activity:

Examining the Epididymis Through the Microscope

Obtain a microscope and a cross section of the epididymis. Notice the abundant tubule cross sections resulting from the fact that the coiling epididymis tubule has been cut through many times in the specimen. Look for sperm in the lumen of the tubule. Using Figure 43.3a as a guide, examine the composition of the tubule wall carefully. Identify the *stereocilia* of the pseudostratified columnar epithelial lining. These non-motile microvilli absorb excess fluid and pass nutrients to the sperm in the lumen. Now identify the smooth muscle layer. What do you think the function of the smooth muscle is?

_____

_____

_____ ■

Penis

Activity:

Examining the Penis Microscopically

Obtain a cross section of the penis. Scan the tissue under low power to identify the urethra and the cavernous bodies. Compare your observations to Figure 43.3b. Observe the lumen of the urethra carefully. What type of epithelium do you see?

_____

Explain the function of this type of epithelium.

_____

_____

_____

_____ ■

Gross Anatomy of the Human Female Reproductive System

The **ovaries** (female gonads) are the primary reproductive organs of the female. Like the testes of the male, the ovaries produce both an exocrine product (eggs, or ova) and endocrine products (estrogens and progesterone). The other accessory structures of the female reproductive system transport, house, nurture, or otherwise serve the needs of the reproductive cells and/or the developing fetus.

The reproductive structures of the female are generally considered in terms of internal organs and external organs, or external genitalia.

Activity:

Identifying Female Reproductive Organs

As you read the descriptions of these structures, locate them on Figures 43.4 and 43.5 and then on the female reproductive system model or large laboratory chart. ■

(a)

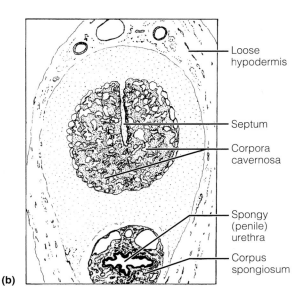

(b)

Figure 43.3 Microscopic anatomy of selected organs of the male reproductive duct system. (a) Epididymis **(b)** Cross-sectional view of the penis. (See also Plates 49 and 50 in the Histology Atlas.)

The **external genitalia (vulva)** consist of the mons pubis, the labia majora and minora, the clitoris, the urethral and vaginal orifices, the hymen, and the greater vestibular glands. The **mons pubis** is a rounded fatty eminence overlying the pubic symphysis. Running inferiorly and posteriorly from the mons pubis are two elongated, pigmented, hair-covered skin folds, the **labia majora,** which are homologous to the scrotum of the male. These enclose two smaller hair-free folds, the **labia minora.** (Terms indicating only one of the two folds in each case are *labium majus* and *minus,* respectively.) The labia minora, in turn, enclose a region called the **vestibule,** which contains many structures—the clitoris, most anteriorly, followed by the urethral orifice and the vaginal orifice. The diamond-shaped region between the anterior end of the labial folds, the ischial tuberosities laterally, and the anus posteriorly is called the **perineum.**

The **clitoris** is a small protruding structure, homologous to the male penis. Like its counterpart, it is composed of highly sensitive, erectile tissue. It is hooded by skin folds of the anterior labia minora, referred to as the **prepuce of the clitoris.** The urethral orifice, which lies posterior to the clitoris, is the outlet for the urinary system and has no reproductive function in the female. The vaginal opening is partially closed by a thin fold of mucous membrane called the **hymen** and is flanked by the pea-sized, mucus-secreting **greater vestibular glands.** These glands (not illustrated) lubricate the distal end of the vagina during coitus.

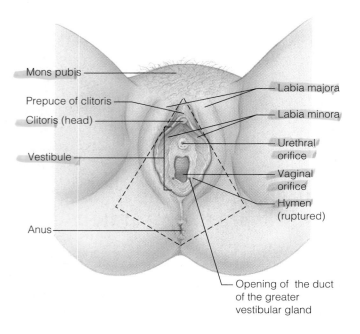

Figure 43.4 **External genitalia of the human female.** (The region enclosed by dashed lines is the perineum.)

Figure 43.5 **Midsagittal section of the human female reproductive system.**

The internal female organs include the vagina, uterus, uterine tubes, ovaries, and the ligaments and supporting structures that suspend these organs in the pelvic cavity. The **vagina** extends for approximately 10 cm (4 inches) from the vestibule to the uterus superiorly. It serves as a copulatory organ and birth canal, and permits passage of the menstrual flow. The pear-shaped **uterus,** situated between the bladder and the rectum, is a muscular organ with its narrow end, the **cervix,** directed inferiorly. The major portion of the uterus is referred to as the **body;** its superior rounded region above the entrance of the uterine tubes is called the **fundus.** A fertilized egg is implanted in the uterus, which houses the embryo or fetus during its development.

In some cases, the fertilized egg may implant in a uterine tube or even on the abdominal viscera, creating an **ectopic pregnancy.** Such implantations are usually unsuccessful and may even endanger the mother's life because the uterine tubes cannot accommodate the increasing size of the fetus. ■

The apical region of the **endometrium,** the thick mucosal lining of the uterus, sloughs off periodically (about every 28 days) in response to cyclic changes in the levels of ovarian hormones in the woman's blood. This sloughing-off process, which is accompanied by bleeding, is referred to as **menstruation,** or **menses.**

The **uterine,** or **fallopian, tubes** enter the superior region of the uterus and extend laterally for about 10 cm (4 inches) toward the **ovaries** in the peritoneal cavity. The distal ends of the tubes are funnel-shaped and have fingerlike projections called **fimbriae.** Unlike the male duct system, there is no actual contact between the female gonad and the initial part of the female duct system—the uterine tube.

Because of this open passageway between the female reproductive organs and the peritoneal cavity, reproductive system infections, such as gonorrhea, can cause widespread inflammations of the pelvic viscera, a condition called **pelvic inflammatory disease** or **PID.** ■

The internal female organs are all retroperitoneal, except the ovaries. They are supported and suspended somewhat freely by ligamentous folds of peritoneum. The peritoneum takes an undulating course. From the pelvic cavity floor it moves superiorly over the top of the bladder, reflects over the anterior and posterior surfaces of the uterus, and then over the rectum, and up the posterior body wall. The fold that encloses the uterine tubes and uterus and secures them to the lateral body walls is the **broad ligament.** The part of the broad ligament specifically anchoring the uterus is called the **mesometrium** and that anchoring the uterine tubes, the **mesosalpinx.** The **round ligaments,** fibrous cords that run from the uterus to the labia majora, and the **uterosacral ligaments,** which course posteriorly to the sacrum, also help attach the uterus to the body wall. The ovaries are supported medially by the **ovarian ligament** (extending from the uterus to the ovary), laterally by the **suspensory ligaments,** and posteriorly by a fold of the broad ligament, the **mesovarium.**

Within the ovaries, the female gametes (eggs) begin their development in saclike structures called *follicles*. The growing follicles also produce *estrogens*. When a developing egg has reached the appropriate stage of maturity, it is ejected from the ovary in an event called **ovulation.** The ruptured follicle is then converted to a second type of endocrine gland, called a *corpus luteum*, which secretes progesterone (and some estrogens).

Figure 43.6   Cross-sectional view of a uterine tube. Notice its highly folded mucosa. (See also Plate 53 in the Histology Atlas.)

The flattened almond-shaped ovaries lie adjacent to the uterine tubes but are not connected to them; consequently, an ovulated egg* enters the pelvic cavity. The waving fimbriae of the uterine tubes create fluid currents that, if successful, draw the egg into the lumen of the uterine tube, where it begins its passage to the uterus, propelled by the cilia of the tubule walls. The usual and most desirable site of fertilization is the uterine tube, because the journey to the uterus takes about 3 to 4 days and an egg is viable for up to 24 hours after it is expelled from the ovary. Thus, sperm must swim upward through the vagina and uterus and into the uterine tubes to reach the egg. This must be an arduous journey, because they must swim against the downward current created by ciliary action—rather like swimming against the tide!

Activity:
Conducting a Microscopic Study of Selected Female Reproductive Organs[†]

Uterine Tube

Obtain a prepared slide of a cross-sectional view of a uterine tube for examination. Notice the highly folded mucosa (the folds nearly fill the tubule lumen) as illustrated in Figure 43.6. Then switch to high power to examine the ciliated secretory epithelium.

---

* To simplify this discussion, the ovulated cell is called an egg. What is actually expelled from the ovary is an earlier stage of development called a secondary oocyte. These matters are explained in more detail in Exercise 44.

[†] A microscopic study of the ovary is described in Exercise 44, p. 452. If that exercise is not to be conducted in its entirety, the instructor might want to include the ovary study in the scope of this exercise.

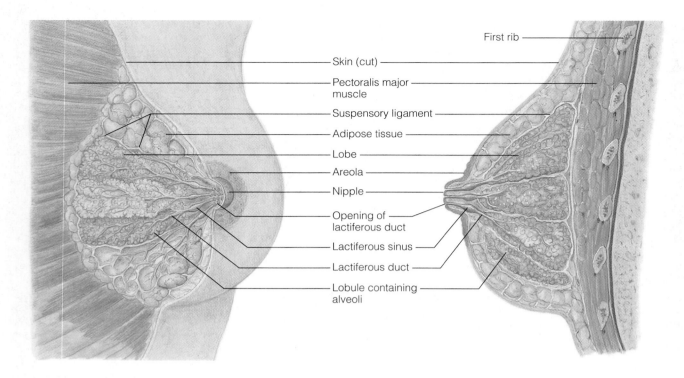

Figure 43.7 **Female mammary gland.** **(a)** Anterior view. **(b)** Sagittal section.

Wall of the Uterus

Obtain a cross-sectional view of the uterine wall. Identify the three layers of the uterine wall—the endometrium, myometrium, and serosa. Also identify the two strata of the endometrium: the **functional layer** (stratum functionalis) and the **basal layer** (stratum basalis), which forms a new functional layer each month. Figure 44.6a (p. 453) of the proliferative endometrium may be of some help in this study.

As you study the slide, notice that the bundles of smooth muscle are oriented in several different directions. What is the function of the myometrium (smooth muscle layer) during the birth process?

_____

_____ ■

Sketch a small portion of the uterine wall in the space below. Label the functional and basal layers.

The Mammary Glands

The **mammary glands** or breasts exist, of course, in both sexes, but they have a reproduction-related function only in females. Since the function of the mammary glands is to produce milk to nourish the newborn infant, their importance is more closely associated with events that occur when reproduction has already been accomplished. Periodic stimulation by the female sex hormones, especially estrogens, increases the size of the female mammary glands at puberty. During this period, the duct system becomes more elaborate, and fat is deposited—fat deposition being the more important contributor to increased breast size.

The rounded, skin-covered mammary glands lie anterior to the pectoral muscles of the thorax, attached to them by connective tissue. Slightly below the center of each breast is a pigmented area, the **areola,** which surrounds a centrally protruding **nipple** (Figure 43.7).

Internally each mammary gland consists of 15 to 20 **lobes** which radiate around the nipple and are separated by fibrous connective tissue and adipose, or fatty, tissue. Within each lobe are smaller chambers called **lobules,** containing the glandular **alveoli** that produce milk during lactation. The alveoli of each lobule pass the milk into a number of **lactiferous ducts,** which join to form an expanded storage chamber, the **lactiferous sinus,** as they approach the nipple. The sinuses open to the outside at the nipple.

 For instructions on animal dissections, see the dissection exercises starting on p. 733 in the cat and fetal pig editions of this manual.

Physiology of Reproduction: Gametogenesis and the Female Cycles

Objectives

1. To define *meiosis, gametogenesis, oogenesis, spermatogenesis, spermiogenesis, synapsis, haploid, diploid,* and *menses.*

2. To relate the stages of spermatogenesis to the cross-sectional structure of the seminiferous tubule.

3. To discuss the microscopic structure of the ovary and to be prepared to identify primary, secondary, and vesicular follicles and the corpus luteum, and to state the hormonal products of the last two structures.

4. To relate the stages of oogenesis to follicle development in the ovary.

5. To cite similarities and differences between mitosis and meiosis, and between spermatogenesis and oogenesis.

6. To describe sperm anatomy and relate it to function.

7. To discuss the phases and control of the menstrual cycle.

8. To discuss the effect of FSH and LH on the ovary and to describe the feedback relationship between anterior pituitary gonadotropins and ovarian hormones.

9. To describe the effect of FSH and LH (ICSH) on testicular function.

Materials

❏ Prepared microscope slides of testis, ovary, human sperm, and uterine endometrium (showing menstrual, proliferative, and secretory stages)

❏ Three-dimensional models illustrating meiosis, spermatogenesis, and oogenesis

❏ Compound microscope

❏ *Demonstration Area:* microscopes set up to demonstrate the following stages of oogenesis in *Ascaris megalocephala:*

1. Primary oocyte with fertilization membrane, sperm nucleus, and aligned tetrads apparent

2. Formation of the first polar body

3. Secondary oocyte with dyads aligned

4. Formation of the ovum and second polar body

5. Fusion of the male and female pronuclei to form the fertilized egg

Meiosis

Every human being, so far, has developed from the union of gametes. Gametes, produced only in the testis or ovary, are unique cells. They have only half the normal chromosome number (designated as *n,* or the **haploid complement**) seen in all other body cells. In humans, gametes have 23 chromosomes instead of the 46 of other tissue cells. Theoretically, every gamete has a full set of genetic instructions, a conclusion borne out by the observation that some animals can develop from an egg that is artificially stimulated, as by a pinprick, rather than by sperm entry. (This is less true of the sperm, because it has an incomplete sex chromosome.)

Gametogenesis, the process of gamete formation, involves reduction of the chromosome number by half. This event is important to maintain the characteristic chromosomal number of the species generation after generation. Otherwise there would be a doubling of chromosome number with each succeeding generation and the cells would become so chock-full of genetic material there would be little room for anything else.

Egg and sperm chromosomes that carry genes for the same traits are called **homologous chromosomes.** When the sperm and egg fuse to form the **zygote,** or fertilized egg, it is said to contain 23 pairs of homologous chromosomes, or the **diploid (2*n*)** chromosome number of 46. The zygote, once formed, then divides to produce the cells needed to construct the multicellular human body. All cells of the developing human body have a chromosome content exactly identical in quality and quantity to that of the fertilized egg. This is assured by the nuclear division process called **mitosis.** (Mitosis was considered in depth in Exercise 4. You may want to review it at this time.)

To produce gametes with the reduced (haploid) chromosomal number, **meiosis,** a specialized type of nuclear division, occurs in the ovaries and testes during gametogenesis. Before meiosis begins, the chromosomes are replicated in the *mother cell* or stem cell just as they are before mitosis. As a result, the mother cell briefly has double the normal diploid genetic complement. The stem cell then undergoes two consecutive nuclear divisions, termed *meiosis I and II,* or the first and second maturation divisions, *without replicating the chromosomes before the second division.* The result is that four haploid daughter cells are produced, rather than the two diploid daughter cells produced by mitotic division.

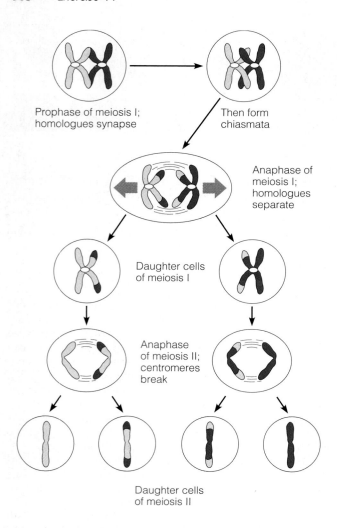

Prophase of meiosis I; homologues synapse

Then form chiasmata

Anaphase of meiosis I; homologues separate

Daughter cells of meiosis I

Anaphase of meiosis II; centromeres break

Daughter cells of meiosis II

Figure 44.1 Events of meiosis involving one pair of homologous chromosomes. (Male homologue is purple; female homologue is pink.)

The entire process of meiosis is complex and is dealt with here only to the extent necessary to reveal important differences between this type of nuclear division and mitosis. Essentially, each meiotic division involves the same phases and events seen in mitosis (prophase, metaphase, anaphase, and telophase), but during the first maturation division (meiosis I) an event not seen in mitosis occurs during prophase. The homologous chromosomes, each now a duplicated structure, begin to pair so that they become closely aligned along their entire length. This pairing is called **synapsis.** As a result, 23 **tetrads** (groupings of four chromatids) form, become attached to the spindle fibers, and begin to align themselves on the spindle equator. While in synapsis, two of the four strands in each tetrad (one from each homologue) coil around each other, forming many points of **crossover,** or **chiasmata.** (Perhaps this could be called the conjugal bed of the cell!) When anaphase of meiosis I begins, the homologues separate from one another, breaking and exchanging parts at points of crossover, and move apart toward opposite poles of the cell. The centromeres holding the "sister" chromatids or **dyads** together do *not* break at this point (Figure 44.1).

During the second maturation division, events parallel those in mitosis, except that the daughter cells do *not* replicate their chromosomes before this division, and each daughter cell has only half of the homologous chromosomes rather than a complete set. The crossover events and the way in which the homologues align on the spindle equator during the first maturation division introduce an immense variability in the resulting gametes, which explains why we are all unique.

Activity:
Identifying Meiotic Phases and Structures

Obtain a model depicting the events of meiosis, and follow the sequence of events during the first and second maturation divisions. Identify prophase, metaphase, anaphase, and telophase in each. Also identify tetrads and chiasmata during the first maturation division and dyads (groupings of two chromatids connected by centromeres) in the second maturation division. Note ways in which the daughter cells resulting from meiosis I differ from the mother cell and how the gametes differ from both cell populations. (Use the key on the model, your textbook, or an appropriate reference as necessary to aid you in these observations.) ■

Spermatogenesis

Human sperm production or **spermatogenesis** begins at puberty and continues without interruption throughout life. The average male ejaculation contains about a quarter-billion sperm. Because only one sperm fertilizes an ovum, it seems that nature has tried to ensure that the perpetuation of the species will not be endangered for lack of sperm.

As explained in Exercise 43, spermatogenesis (the process of gametogenesis in males) occurs in the seminiferous tubules of the testes. The primitive stem cells or **spermatogonia,** found at the tubule periphery, divide extensively to build up the stem cell line. Before puberty, all divisions are mitotic divisions that produce more spermatogonia. At puberty, however, under the influence of FSH (follicle-stimulating hormone) secreted by the anterior pituitary gland, each mitotic division of a spermatogonium produces one spermatogonium and one **primary spermatocyte,** which is destined to undergo meiosis. As meiosis occurs, the dividing cells approach the lumen of the tubule. Thus the progression of meiotic events can be followed from the tubule periphery to the lumen. It is important to recognize that **spermatids,** haploid cells that are the actual product of meiosis, are not functional gametes. They are nonmotile cells and have too much excess baggage to function well in a reproductive capacity. Another process, called **spermiogenesis,** which follows meiosis, strips away the extraneous cytoplasm from the spermatid, converting it to a motile, streamlined sperm.

Activity:
Examining Events of Spermatogenesis

1. Obtain a slide of the testis and a microscope. Examine the slide under low power to identify the cross-sectional views of the cut seminiferous tubules. Then rotate the high power lens into position and observe the wall of one of the cut tubules. As you work, refer to Figure 44.2 and to Plate 51 in the Histology Atlas to make the following identifications.

2. Scrutinize the cells at the periphery of the tubule. The cells in this area are the spermatogonia. About half of these will form primary spermatocytes, which begin meiosis. These are recognizable by their pale-staining nuclei with centrally located nucleoli. The remaining daughter cells resulting from mitotic divisions of spermatogonia remain at the tubule periphery to maintain the germ cell line.

3. Observe the cells in the middle of the tubule wall. There you should see a large number of cells (spermatocytes) that are obviously undergoing a nuclear division process. Look for coarse clumps of chromatin or threadlike chromosomes (visible only during nuclear division) that have the appearance of coiled springs. Attempt to differentiate between the larger primary spermatocytes and the somewhat smaller secondary spermatocytes. Once formed, the secondary spermatocytes quickly undergo mitosis and so are more difficult to find.

Can you see tetrads? _____

Evidence of crossover? _____

Where would you expect to see the tetrads, closer to the spermatogonia or closer to the lumen?

_____

In the primary or secondary spermatocytes? _____

_____

4. Examine the cells at the tubule lumen. Identify the small round-nucleated spermatids, many of which may appear lopsided and look as though they are starting to lose their cytoplasm. See if you can find a spermatid embedded in an elongated cell type, a **sustentacular (Sertoli) cell,** which extends inward from the periphery of the tubule. The sustentacular cells nourish the spermatids as they begin their transformation into sperm. Also in the adluminal area, locate sperm, which can be identified by their tails. The sperm develop directly from the spermatids by the loss of extraneous cytoplasm and the development of a propulsive tail.

5. Identify the **interstitial cells** lying external to and between the seminiferous tubules. LH (luteinizing hormone), also called *interstitial cell-stimulating hormone (ICSH)* in males, prompts these cells to produce testosterone which acts synergistically with FSH to stimulate sperm production.

In the next stage of sperm development, spermiogenesis, all the superficial cytoplasm is sloughed off, and the remaining cell organelles are compacted into the three regions of the

(a)

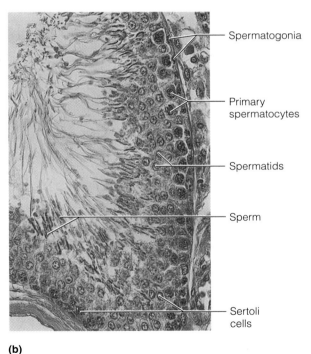

(b)

Figure 44.2 Spermatogenesis. (a) Flowchart of meiotic and spermiogenesis events. **(b)** Micrograph of an active seminiferous tubule.

mature sperm. At the risk of oversimplifying, these anatomical regions are the *head,* the *midpiece,* and the *tail,* which correspond roughly to the activating and genetic region, the metabolic region, and the locomotor region respectively. The mature sperm is a streamlined cell equipped with an organ of locomotion and a high rate of metabolism that enable it to move long distances in jig time to get to the egg. It is a prime example of the correlation of form and function.

The pointed sperm head contains the DNA, or genetic material, of the chromosomes. Essentially it is the nucleus of the spermatid. Anterior to the nucleus is the **acrosome,** which contains enzymes involved in sperm penetration of the egg.

In the midpiece of the sperm is a centriole which gives rise to the filaments that structure the sperm tail. Wrapped tightly around the centriole are mitochondria that provide the ATP needed for contractile activity of the tail filaments.

The tail filaments are constructed of contractile proteins, much like those in muscle. When powered by ATP, these filaments propel the sperm.

6.   Obtain a prepared slide of human sperm and view it with the oil immersion lens. Compare what you see to the photograph of sperm in Plate 52 in the Histology Atlas. Identify the head, acrosome, and tail regions. Draw and appropriately label two or three sperm in the space below.

Deformed sperm, for example sperm with multiple heads or tails, are sometimes present in such preparations. Did you observe any?

_____ If so, describe them. _____

_____

_____

7.   Examine the model of spermatogenesis to identify the spermatogonia, the primary and secondary spermatocytes, the spermatids, and the functional sperm.

If your instructor wishes you to observe meiosis in *Ascaris* to provide cellular material for comparison, continue with the microscopic study described next. Otherwise, skip to the study of human oogenesis. ■

Demonstration of Oogenesis in *Ascaris* (Optional)

Generally speaking, oogenesis (the process of gametogenesis resulting in egg production) in mammals is difficult to demonstrate in a laboratory situation. However, the process

of oogenesis and mechanics of meiosis* may be studied rather easily in the transparent eggs of *Ascaris megalocephala,* an invertebrate roundworm parasite found in the intestine of mammals. Since its diploid chromosome number is 4, the chromosomes are easily counted.

A c t i v i t y :
Examining Meiotic Events Microscopically

Go to the demonstration area where the slides are set up and make the following observations:

1.   Scan the first demonstration slide to identify a *primary oocyte,* the cell type that begins the meiotic process. It will have what appears to be a relatively thick cell membrane; this is the *fertilization membrane* that the oocyte produces after sperm penetration. Find and study a primary oocyte that is undergoing the first maturation division. Look for a barrel-shaped spindle with two tetrads (two groups of four beadlike chromosomes) in it. Most often the spindle is located at the periphery of the cell. (The sperm nucleus may or may not be seen, depending on how the cell was cut.)

2.   Observe slide 2. Locate a cell in which half of each tetrad (a dyad) is being extruded from the cell surface into a smaller cell called the *first polar body.*

3.   On slide 3, attempt to locate a *secondary oocyte* (a daughter cell produced during meiosis I) undergoing the second maturation division. In this view, two dyads (two groups of two beadlike chromosomes) will be seen on the spindle.

4.   On slide 4, locate a cell in which the *second polar body* is being formed. In this case, both it and the ovum will now contain two chromosomes, the haploid number for *Ascaris.*

5.   On the fifth slide, identify a *fertilized egg* or a cell in which the sperm and ovum nuclei (actually *pronuclei*) are fusing to form a single nucleus containing four chromosomes. ■

Human Oogenesis and the Ovarian Cycle

Gonadotropic hormones produced by the anterior pituitary influence the development of ova in the ovaries and their cyclic production of female sex hormones. Within an ovary, each immature ovum develops within a saclike structure called a *follicle,* where it is encased by one or more layers of smaller cells called **follicle cells** (when one layer is present) or **granulosa cells** (when there is more than one layer).

---

* In *Ascaris,* meiosis does not begin until the sperm has penetrated the primary oocyte, whereas in humans, meiosis I occurs before sperm penetration.

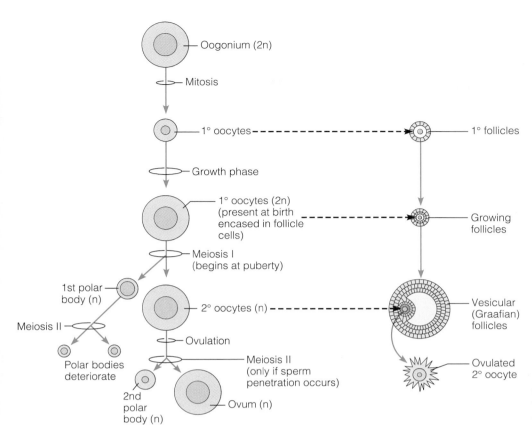

Figure 44.3 Oogenesis. Left, flowchart of meiotic events. Right, correlation with follicular development and ovulation in the ovary.

The process of **oogenesis,** or female gamete formation, which occurs in the ovary, is similar to spermatogenesis occurring in the testis, but there are some important differences. The process, schematically outlined in Figure 44.3, begins with primitive stem cells called **oogonia,** located in the ovarian cortices of the developing female fetus. During fetal development, the oogonia undergo mitosis thousands of times until their number reaches 700,000 or more. They then become encapsulated by a single layer of squamouslike follicle cells and form the **primordial follicles** of the ovary. By the time the female child is born, most of her oogonia have increased in size and have become **primary oocytes,** which are in the prophase stage of meiosis I. Thus at birth, the total potential for producing germ cells in the female is already determined; the primitive stem-cell line no longer exists or will exist for only a brief period after birth.

From birth until puberty, the primary oocytes are quiescent. Then, under the influence of FSH, one or sometimes more of the follicles begin to undergo maturation approximately every 28 days.

As a follicle grows, its epithelium changes from squamous to cuboidal cells and it comes to be called a **primary follicle** (see also Figure 44.4). The primary follicle begins to produce estrogens, and the primary oocyte completes its first maturation division, producing two haploid daughter cells that are very disproportionate in size. One of these is the **secondary oocyte,** which contains nearly all of the cytoplasm in the primary oocyte. The other is the tiny **first polar body.** The first polar body then completes the second maturation division, producing two more polar bodies. These eventually disintegrate for lack of sustaining cytoplasm.

As the follicle containing the secondary oocyte continues to enlarge, blood levels of estrogens rise. Initially, estrogen exerts a negative feedback influence on the release of gonadotropins by the anterior pituitary. However, approximately in the middle of the 28-day cycle, as the follicle reaches the mature **vesicular,** or **Graafian, follicle** stage, rising estrogen levels become highly stimulatory and a sudden burstlike release of LH (and, to a lesser extent, FSH) by the anterior pituitary triggers ovulation. The secondary oocyte is extruded and begins its journey down the uterine tube to the uterus. If penetrated en route by a sperm, the secondary oocyte will undergo meiosis II, producing one large **ovum** and a tiny **second polar body.** When the second maturation division is complete, the chromosomes of the egg and sperm combine to form the diploid nucleus of the fertilized egg. If sperm penetration does not occur, the secondary oocyte simply disintegrates without ever producing the female gamete in human females.

Thus in the female, meiosis produces only one functional gamete, in contrast to the four produced in the male. Another major difference is in the relative size and structure of the functional gametes. Sperm are tiny and equipped with tails for locomotion. They have few organelles and virtually no nutrient-containing cytoplasm; hence the nutrients contained in semen are essential to their survival. In contrast, the egg is a relatively large nonmotile cell, well stocked with cytoplasmic reserves that nourish the developing embryo until implantation can be accomplished. Essentially all the zygote's organelles are "delivered" by the egg.

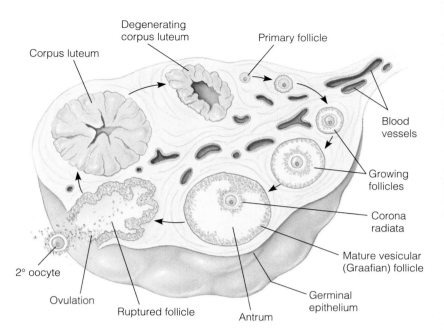

Figure 44.4 Anatomy of the human ovary.

Labels: Corpus luteum; Degenerating corpus luteum; Primary follicle; Blood vessels; Growing follicles; Corona radiata; Mature vesicular (Graafian) follicle; Germinal epithelium; Antrum; Ruptured follicle; Ovulation; 2° oocyte

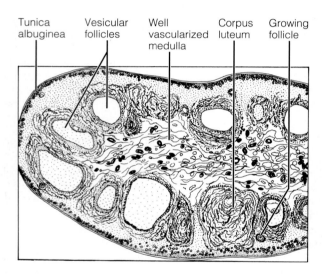

Labels: Tunica albuginea; Vesicular follicles; Well vascularized medulla; Corpus luteum; Growing follicle

Figure 44.5 Line drawing of a photomicrograph of the human ovary. (See also Plate 54 in the Histology Atlas.)

Once the secondary oocyte has been expelled from the ovary, LH transforms the ruptured follicle into the **corpus luteum,** which begins producing progesterone and estrogen. Rising blood levels of the two ovarian hormones inhibit FSH release by the anterior pituitary. As FSH declines, its stimulatory effect on follicular production of estrogens ends, and estrogen blood levels begin to decline. Since rising estrogen levels triggered LH release by the anterior pituitary, falling estrogen levels result in declining levels of LH in the blood. Corpus luteum secretory function is maintained by high blood levels of LH. Thus as LH blood levels begin to drop toward the end of the 28-day cycle, progesterone production ends and the corpus luteum begins to degenerate and is replaced by scar tissue **(corpus albicans).** The graphs in Figure 44.6 depict the hormone relationships described here.

Activity:
Examining Oogenesis in the Ovary

Because many different stages of ovarian development exist within the ovary at any one time, a single microscopic preparation will contain follicles at many different stages of development. Obtain a cross section of ovary tissue, and identify the following structures. Refer to Figures 44.4 and 44.5 as you work.

Germinal epithelium: Outermost layer of the ovary.

Primary follicle: One or a few layers of cuboidal follicle cells surrounding the larger central developing ovum.

Secondary (growing) follicles: Follicles consisting of several layers of follicle (granulosa) cells surrounding the central developing ovum, and beginning to show evidence of fluid accumulation and **antrum** (central cavity) formation.

Vesicular (Graafian) follicle: At this stage of development, the follicle has a large antrum containing fluid produced by the granulosa cells. The developing ovum is pushed to one side of the follicle and is surrounded by a capsule of several layers of granulosa cells called the **corona radiata** (radiating crown). When the immature ovum (secondary oocyte) is released, it enters the uterine tubes with its corona radiata intact. The connective tissue stroma (background tissue) adjacent to the mature follicle forms a capsule, called the **theca,** that encloses the follicle. (See also Plate 24 in the Histology Atlas.)

Corpus luteum: A solid glandular structure or a structure containing a scalloped lumen that develops from the ovulated follicle. (See Plate 25 in the Histology Atlas.)

Examine the model of oogenesis and compare it with the spermatogenesis model. Note differences in the size and structure of the functional gametes. ■

(a)

(b)

(c)

Figure 44.6   Endometrial changes during the menstrual cycle. (a) Early proliferative phase. **(b)** Early secretory phase. **(c)** Onset of menstruation.

The Menstrual Cycle

The **menstrual cycle,** sometimes referred to as the **uterine cycle,** is hormonally controlled by estrogens and progesterone secreted by the ovary. It is normally divided into three stages: menstrual, proliferative, and secretory. The endometrial changes are described next and shown in Figure 44.6. Figure 44.7 shows how they correlate with hormonal and ovarian changes.

Menstrual stage (menses): Approximately days 1 to 5. Sloughing off of the thick functional layer of the endometrial lining of the uterus, accompanied by bleeding.

Proliferative stage: Approximately days 6 to 14. Under the influence of estrogens produced by the growing follicle of the ovary, the endometrium is repaired, glands and blood vessels proliferate, and the endometrium thickens. Ovulation occurs at the end of this stage.

Secretory stage: Approximately days 15 to 28. Under the influence of progesterone produced by the corpus luteum, the vascular supply to the endometrium increases further. The glands increase in size and begin to secrete nutrient substances to sustain a developing embryo, if present, until implantation can occur. If fertilization has occurred, the embryo will produce a hormone much like LH, which will maintain the function of the corpus luteum. Otherwise, as the corpus luteum begins to deteriorate, lack of ovarian hormones in the blood causes blood vessels supplying the endometrium to kink and become spastic, setting the stage for menses to begin by the 28th day.

Although the foregoing explanation assumes a classic 28-day cycle, the length of the menstrual cycle is highly variable, sometimes as short as 21 days or as long as 38. Only one interval is relatively constant in all females: the time from ovulation to the onset of menstruation is almost always 14 days.

A c t i v i t y :

Observing Histological Changes in the Endometrium During the Menstrual Cycle

Obtain slides showing the menstrual, secretory, and proliferative phases of the uterine endometrium. Observe each carefully, comparing their relative thicknesses and vascularity. As you work, refer to the corresponding photomicrographs in Figure 44.6. ■

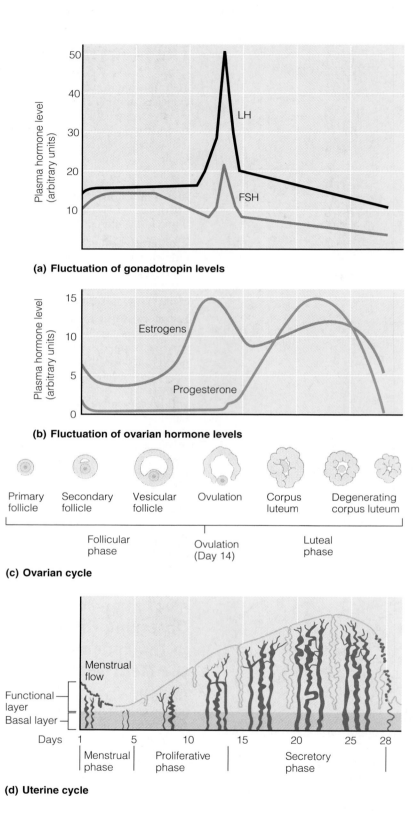

(a) Fluctuation of gonadotropin levels

(b) Fluctuation of ovarian hormone levels

(c) Ovarian cycle

(d) Uterine cycle

Figure 44.7 Hormonal interactions of the female cycles.
Relative level of anterior pituitary hormones correlated with follicular and hormonal changes in the ovary. The menstrual cycle is also depicted.

Survey of Embryonic Development

Because reproduction is such a familiar event, we tend to lose sight of the wonder of the process. One part of that process, the development of the embryo, is the concern of embryologists who study the changes in structure that occur from the time of fertilization until the time of birth.

Early development in all animals involves three basic types of activities, which are integrated to ensure the formation of a viable offspring: (1) an increase in cell number and subsequent cell growth; (2) cellular specialization; and (3) morphogenesis, the formation of functioning organ systems. This exercise provides a rather broad overview of the changes in structure that take place during embryonic development in humans.

Developmental Stages of the Human

Activity:
Examining the Stages of Human Development

Go to the demonstration area where the models of human development are on display. If these are not available, use Figure 45.1 for this study. Observe the models to identify the various stages of human development as they are described, and respond to the questions posed below.

1. Observe the fertilized egg, or zygote, which appears as a single cell immediately surrounded by a jellylike *zona pellucida* and then a crown of granulosa cells (the *corona radiata*). After a secondary oocyte is penetrated by a sperm, and meiosis is completed to yield the ovum (egg) nucleus, the egg and the sperm nuclei fuse to form a single nucleus. This process is called **fertilization.** Shortly after sperm penetration, a granular barrier forms beneath the zona pellucida to prevent the entry of additional sperm.

2. Next, observe the cleavage stages. Once fertilization has occurred, the zygote begins to divide, forming a mass of successively smaller and smaller cells, called **blastomeres.** This series of mitotic divisions without intervening growth periods is referred to as **cleavage,** and it provides a large number of building blocks (cells) with which to build the forming body. If this is a little difficult to understand, consider trying to erect a building with one huge block of granite rather than

Objectives

1. To define *fertilization* and *zygote.*
2. To define and discuss the function of *cleavage* and *gastrulation.*
3. To name the three primary germ layers and discuss the importance of each.
4. To identify the following structures of a human chorionic vesicle when provided with an appropriate diagram, and to state the function of each.

   inner cell mass   *trophoblast*

   amnion   *allantois*

   yolk sac   *chorionic villi*
5. To describe the process and timing of implantation in the human.
6. To define *decidua basalis* and *decidua capsularis.*
7. To state the germ-layer origin of several body organs and organ systems of the human.
8. To describe developmental direction.
9. To describe the gross anatomy and general function of the human placenta.

Materials

- ❏ Human development models or plaques (if available)
- ❏ Pregnant cat, rat, or pig uterus (one per laboratory session) with uterine wall dissected to allow student examination
- ❏ Dissecting instruments
- ❏ Model of pregnant human torso
- ❏ Fresh or formalin-preserved placenta (obtained from a clinical agency)
- ❏ Microscope slide of placenta tissue
- ❏ Compound microscope
- ❏ Demonstration: Phases of human development in *A Colour Atlas of Life Before Birth*: *Normal Fetal Development*

(a) Zygote (fertilized egg)

(b) Early cleavage 4-cell stage

(c) Morula

(d) Early blastocyst

(e) Late blastocyst (implanting)

Blastocyst cavity

Inner cell mass

Trophoblast

Fertilization

Uterine tube

Secondary oocyte

Ovulation

Ovary

Uterus

Endometrium

Uterine cavity

Amnion

Chorion

Body stalk (umbilical cord)

Chorionic villi

Ectoderm

Forming mesoderm

Endoderm

Embryo

Figure 45.1 Early embryonic development of the human.
Top **(a–e)**, from fertilization to blastocyst implantation in the uterus. Below, embryo of approximately 22 days. Embryonic membranes and germ layers present.

with small bricks. The product of early cleavage is a solid ball of cells. At the 32-cell stage, it is called the **morula,** and the embryo resembles a raspberry in form. Then the cell mass hollows out to become the embryonic form called the **blastula,** which is a ball of cells surrounding a central cavity. The blastula, more commonly called the **blastocyst** in humans, is the final product of cleavage.

Only a portion of the human blastocyst cells contribute to the formation of the body—those seen on the top of the blastocyst forming the so-called **inner cell mass (ICM).** The rest of the blastocyst—that portion enclosing the central cavity and overriding the ICM—is referred to as the **trophoblast.** The trophoblast becomes an extraembryonic membrane called the **chorion,** which forms the fetal portion of the *placenta.*

3. Observe the *implanting* blastocyst shown on the model or in the figure. By approximately the seventh day after ovulation, a developing human embryo (blastocyst) is floating free in the uterine cavity. About that time, it adheres to the uterine wall over the ICM area, and implantation begins. The trophoblast cells secrete enzymes that erode the uterine mucosa at the point of attachment to reach the vascular supply in the submucosa. By the fourteenth day after ovulation, implantation is completed and the uterine mucosa has grown over the burrowed-in embryo. The portion of the uterine wall beneath the ICM, destined to take part in placenta formation, is called the **decidua basalis** and that surrounding the rest of the blastocyst is called the **decidua capsularis.** Identify these regions.

By the time implantation is complete, the blastocyst has undergone **gastrulation.** As a result of gastrulation, a three-layered embryo called a **gastrula** forms. Each of the gastrula's three layers corresponds to a **primary germ layer** from which specific body tissues develop. Within the next 6 weeks, virtually all of the body organ systems will have been laid down at least in rudimentary form by the germ layers. The outermost layer, **ectoderm,** gives rise to the epidermis of the skin and the nervous system. The deepest layer, the **endoderm,** forms the mucosa of the digestive and respiratory tracts and associated glands. **Mesoderm,** the middle layer, forms virtually everything lying between the two (skeleton, walls of the digestive organs, urinary system, skeletal muscles, circulatory system, and others).

All the groundwork has been completed by the eighth week. By the ninth week of development, the embryo is referred to as a **fetus,** and from this point on, the major activities are growth and tissue and organ specialization.

4. Again observe the blastocyst to follow the formation of the embryonic membranes and the placenta (see Figure 45.1). Notice the villus extensions of the trophoblast. By the time implantation is complete, the trophoblast has differentiated into the **chorion,** and its large elaborate villi are lying in the blood-filled sinusoids in the uterine tissue. This composite of uterine tissue and **chorionic villi** is called the **placenta,** and all exchanges to and from the embryo occur through the chorionic membranes.

Three embryonic membranes (originating in the ICM) have also formed by this time—the amnion, the allantois, and the yolk sac. Attempt to identify each.

- The **amnion** encases the young embryonic body in a fluid-filled chamber that protects the embryo against mechanical trauma and temperature extremes.

- The **yolk sac** in humans has lost its original function, which was to pass nutrients to the embryo after digesting the yolk mass. The placenta has taken over that task; also, the human egg has very little yolk. However, the yolk sac is not totally useless. The embryo's first blood cells originate here, and the primordial germ cells migrate from it into the embryo's body to seed the gonadal tissue.

- The **allantois,** which protrudes from the posterior end of the yolk sac, is also largely redundant in humans because of the placenta. In birds and reptiles, it is a repository for embryonic wastes. In humans, it is the structural basis on which the mesoderm migrates to form the body stalk, or **umbilical cord,** which attaches the embryo to the placenta. (Refer ahead to p. 459 to refresh your memory of the structure of the umbilical cord if necessary.)

5. Go to the demonstration area to view the photographic series, *A Colour Atlas of Life Before Birth: Normal Fetal Development.* These photographs illustrate human development in a way you will long remember. After viewing them, respond to the following questions.

In your own words, what do the chorionic villi look like?

_____

_____

_____

What organs or organ systems appear *very* early in embryonic development?

_____

_____

_____

Does development occur in a rostral to caudal (head to toe) direction, or vice versa?

_____

Does development occur in a distal to proximal direction, or vice versa?

_____

Does spontaneous movement occur *in utero*? _____

How does the mother recognize this? _____

_____

The very young embryo has been described as resembling "an astronaut suspended and floating in space." Do you think this definition is appropriate?

_____ Why or why not? _____

_____

_____

What is vernix caseosa? _____

_____

_____

What is lanugo? _____

_____

_____ ■

In Utero Development

A c t i v i t y :
Identifying Fetal Structures

1. Go to the appropriate demonstration area and observe the fetuses in the Y-shaped animal uterus. Identify the following fetal or fetal-related structures:

● **Placenta** (a composite structure formed from the uterine mucosa and the fetal chorion).

Describe its appearance. _____

_____

_____

_____

● **Umbilical cord.** Describe its relationship to the placenta and fetus.

_____

_____

_____

● **Amniotic sac.** Identify the transparent amnion surrounding a fetus. Open one amniotic sac and note the amount, color, and consistency of the fluid.

_____

_____

_____

Remove a fetus and observe the degree of development of the head, body, and extremities. Is the skin thick or thin?

_____

_____

What is the basis for your response? _____

_____

_____

2. Observe the model of a pregnant human torso. Identify the placenta. How does it differ in shape from the animal placenta observed?

_____

_____

_____

Identify the umbilical cord. In what region of the uterus does implantation usually occur, as indicated by the position of the placenta?

_____

_____

What might be the consequence if it occurred lower?

_____

_____

_____

Why would a feet-first position (breech presentation) be less desirable than the positioning of the model?

_____

_____

_____ ■

Gross and Microscopic Anatomy of the Placenta

The placenta is a remarkable temporary organ. Composed of maternal and fetal tissues, it is responsible for providing nutrients and oxygen to the embryo and fetus while removing carbon dioxide and metabolic wastes.

Maternal blood pools within intervillus space

Maternal venule

Myometrium

Maternal arteriole

Fetal arteriole

Fetal venule

Umbilical arteries

Umbilical vein

Umbilical cord

Fetal portion of placenta (chorion)

Maternal portion of placenta (decidua basalis)

Figure 45.2 Diagrammatic representation of the structure of the placenta.

Activity:
Studying Placental Structure

1.   Notice that the placenta on display has two very different-appearing surfaces—one smooth and the other spongy, roughened, and torn-looking.

Which is the fetal side? _____

What is the basis for your conclusion? _____

_____

_____

Identify the umbilical cord. Within the cord, identify the umbilical vein and two umbilical arteries. What is the function of the umbilical vein?

_____

_____

The umbilical arteries? _____

_____

Are any of the fetal membranes still attached?

_____

If so, which? _____

2.   Obtain a microscope slide of placental tissue. Observe the tissue carefully, comparing it to Figure 45.2. Identify the *intervillous spaces* (maternal sinusoids), which are blood-filled in life. Identify the villi, and notice their rich vascular supply. Draw a small representative diagram of your observations below and label it appropriately. ■

46

Principles of Heredity

Objectives

1. To define *allele, dominance, genotype, heterozygous, homozygous, incomplete dominance, phenotype,* and *recessiveness.*
2. To gain practice working out simple genetics problems, using a Punnett square.
3. To become familiar with basic laws of probability.
4. To observe selected human phenotypes and determine their genotype basis.

Materials

❑ Pennies (for coin tossing)
❑ PTC (phenylthiocarbamide) taste strips
❑ Sodium benzoate taste strips
❑ Chart drawn on chalkboard for tabulation of class results of human phenotype/genotype determinations

Blood typing supplies:
❑ Anti-A and Anti-B sera, slides, toothpicks, wax pencils, sterile lancets, alcohol swabs
❑ Beaker containing 10% bleach solution
❑ Disposable autoclave bag

T he field of genetics is bristling with excitement. Complex gene-splicing techniques have allowed researchers to precisely isolate genes coding for specific proteins and then to use those genes to harvest large amounts of particular proteins and even to cure some dreaded human diseases. At present, growth hormone, insulin, erythropoietin, and interferon produced by these genetic engineering techniques are available for clinical use, and the list is growing daily.

Comprehending genetics relative to such studies requires arduous training. However, a basic understanding and appreciation of how genes regulate our various traits (dimples and hair color, for example) can be gained by anyone. The thrust of this exercise is to provide a "genetics sampler" or relatively simple introduction to the principles of heredity.

Introduction to the Language of Genetics

In humans all cells, except eggs and sperm, contain 46 chromosomes, that is, the diploid number. This number is established when fertilization occurs and the egg and sperm fuse, combining the 23 chromosomes (or haploid complement) each is carrying. The diploid chromosomal number is maintained throughout life in nearly all cells of the body by the precise process of mitosis. As explained in Exercise 44, the diploid chromosomal number actually represents two complete (or nearly complete) sets of genetic instructions—one from the egg and the other from the sperm—or 23 pairs of *homologous chromosomes.*

Genes coding for the same traits on each pair of homologous chromosomes are called **alleles.** The alleles may be identical or different in their influence. For example, the members of the gene pair, or alleles, coding for hairline shape on your forehead may specify either straight across or widow's peak. When both alleles in a homologous chromosome pair have the same expression, the individual is **homozygous** for that trait. When the alleles differ in their expression, the individual is **heterozygous** for the given trait; and typically only one of the alleles, called the **dominant gene,** will exert its effects. The allele with less potency, the **recessive gene,** will be present but masked. Whereas dominant genes, or alleles, exert their effects in both homozygous and heterozygous conditions, as a rule recessive alleles *must* be present in double dose (homozygosity) to exert their influence. An individual's actual genetic makeup, that is, whether he is homozygous or heterozygous for the various alleles, is called **genotype.** The manner in which genotype is expressed (for example, the presence of a widow's peak or not, blue vs. brown eyes) is referred to as **phenotype.**

The complete story of heredity is much more complex than just outlined, and in actuality the expression of many traits (for example, eye color) is determined by the interaction of many allele pairs. However, our emphasis here will be to investigate only the less complex aspects of genetics.

Dominant-Recessive Inheritance

One of the best ways to master the terminology and learn the principles of heredity is to work out the solutions to some genetic crosses in much the same manner Gregor Mendel did in his classic experiments on pea plants. (Mendel, an Austrian monk of the mid-1800s, found evidence in these experiments that each gamete contributes just one allele to each pair in the zygote.)

To work out the various simple monohybrid (one pair of alleles) crosses in this exercise, you will be given the genotype of the parents. You will then determine the possible genotypes of their offspring by using a grid called the *Punnett square,* and you will record both genotype and phenotype percentages. To illustrate the procedure, an example of one of Mendel's pea plant crosses is outlined next.

Alleles: *T*   (determines   *tallness;*   dominant)
   t (determines *dwarfness; recessive*)

Genotypes of parents: *TT* (♂) × *tt* (♀)

Phenotypes of parents: Tall × dwarf

To use the Punnett, or checkerboard, square, write the alleles (actually gametes) of one parent across the top and the gametes of the other parent down the left side. Then combine the gametes across and down to achieve all possible combinations as shown below:

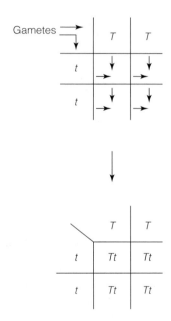

Results: Genotypes 100% *Tt* (all heterozygous)
   Phenotypes 100% tall (because *T,* which determines tallness, is dominant and all contain the *T* allele)

Activity:
Working Out Crosses Involving Dominant and Recessive Genes

1.   Using the technique outlined above, determine the genotypes and phenotypes of the offspring of the following crosses:

   a.   Genotypes of parents: *Tt* (♂) × *tt* (♀)

   % of each genotype: _____

   % of each phenotype: _____ % tall

   _____ % dwarf

   b.   Genotypes of parents: *Tt* (♂) × *Tt* (♀)

   % of each genotype: _____

   % of each phenotype: _____ % tall

   _____ % dwarf

2.   In guinea pigs, rough coat (*R*) is dominant over smooth coat (*r*). What will be the genotypes and phenotypes of the following monohybrid crosses?

   a.   Genotypes of the parents: *RR* × *rR*

   % of each genotype: _____

   % of each phenotype: _____ % rough

   _____ % smooth

   b.   Genotypes of the parents: *Rr* × *rr*

   % of each genotype: _____

   % of each phenotype: _____ % rough

   _____ % smooth

c.   Genotypes of the parents: *RR* × *rr*

% of each genotype: _____

% of each phenotype: _____ % rough

_____ % smooth

Incomplete Dominance

The concepts of dominance and recessiveness are somewhat arbitrary and artificial in some instances because so-called dominant genes may be expressed differently in homozygous and heterozygous individuals. This produces a condition called *incomplete dominance* or *intermediate inheritance*. In such cases, both alleles express themselves in the offspring. The crosses are worked out in the same manner as indicated previously, but heterozygous offspring exhibit a phenotype intermediate between that of the homozygous individuals. Some examples follow.

Activity:
Working Out Crosses Involving Incomplete Dominance

1.   The inheritance of flower color in snapdragons illustrates the principle of incomplete dominance. The genotype *RR* is expressed as a red flower, *Rr* yields pink flowers, and *rr* produces white flowers. Work out the following crosses to determine phenotypes seen and both genotype and phenotype percentages.

a.   Genotypes of parents: *RR* × *rr*

Genotypes and %: _____

Phenotypes and %: _____

b.   Genotypes of parents: *Rr* × *rr*

Genotypes and %: _____

Phenotypes and %: _____

_____

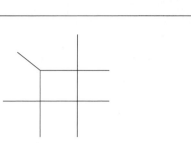

c.   Genotypes of parents *Rr* × *Rr*

Genotypes and %: _____

Phenotypes and %: _____

_____

2.   In humans, the inheritance of sickle-cell anemia/trait is determined by a single pair of alleles that exhibit incomplete dominance. Individuals homozygous for the sickling gene (*s*) have *sickle-cell anemia* (SCA). In double dose (*ss*) the sickling gene causes production of a very abnormal hemoglobin, which crystallizes and becomes sharp and spiky under conditions of oxygen deficit. This, in turn, leads to clumping and hemolysis of red blood cells in the circulation, which causes a great deal of pain and can be fatal. Heterozygous individuals (*Ss*) have the *sickle-cell trait* (SCT), which is much less severe. However, they are carriers for the abnormal gene and may pass it on to their offspring. Individuals with the genotype *SS* form normal hemoglobin. Work out the following crosses:

a.   Parental genotypes: *SS* × *ss*

Genotypes and %: _____

Phenotypes and %: _____

(a)   (b)

Figure 46.1 **Karyotype (chromosomal complement) of human male.** Each pair of homologous chromosomes is numbered (1–22) except the sex chromosomes, which are identified by their letters, X and Y.

b.   Parental genotypes: *Ss × Ss*

Genotypes and %: _____

Phenotypes and %: _____

_____

c.   Parental genotypes: *ss × Ss*

Genotypes and %: _____

Phenotypes and %: _____

_____

Sex-Linked Inheritance

Of the 23 pairs of homologous chromosomes, 22 pairs are referred to as **autosomes.** Autosomes contain genes that determine most body (somatic) characteristics. The 23rd pair, the **sex chromosomes,** determine the sex of an individual, that is, whether an individual will be male or female. Normal females possess two sex chromosomes that look alike, the X chromosomes. Males possess two dissimilar sex chromosomes, referred to as X and Y. Possession of the Y chromosome determines maleness. A photomicrograph of a male's chromosome complement (male karyotype) is shown in Figure 46.1. The Y sex chromosome is only about a third the size of the X sex chromosome, and it lacks many of the genes (directing characteristics other than sex) that are found on the X.

Genes present *only* on the X sex chromosome are called *sex-linked* (or X-linked) genes. Some examples of X-linked genes include those that determine normal color vision (or, conversely, color blindness), and normal clotting ability (as opposed to hemophilia, or bleeder's disease). The alleles that determine color blindness and hemophilia are recessive alleles. In females, *both* X chromosomes must carry the recessive alleles for a woman to express either of these conditions, and thus they tend to be infrequently seen. However, should a male receive even one sex-linked recessive allele for these conditions, he will exhibit the recessive phenotype because his Y chromosome lacks any genes that might dominate or mask the recessive allele.

The critical understanding of X-linked inheritance is the *absence* of male to male (that is, father to son) transmission of X-linked genes. The X of the father *will* pass to each of his daughters but to none of his sons. Males always inherit sex-linked conditions from their mothers (via the X chromosome).

Activity:
Working Out Crosses Involving Sex-Linked Inheritance

1. A heterozygous woman carrying the recessive gene for color blindness marries a man who is color-blind. Assume the dominant gene is X^C (allele for normal color vision) and the recessive gene is X^c (determines color blindness). The mother's genotype is $X^C X^c$ and the father's $X^c Y$. Do a Punnett square to determine the answers to the following questions.

According to the laws of probability, what percent of their children will be color-blind?

_____ %

What is the proportion of color-blind individuals by sex?

_____ male; _____ female

What percentage will be carriers? _____ %

What is the sex of the carriers? _____

2. A heterozygous woman carrying the recessive gene for hemophilia marries a man who is not a hemophiliac. Assume the dominant gene is X^H and the recessive gene is X^h. The woman's genotype is $X^H X^h$ and her husband's genotype is $X^H Y$. What is the potential percentage and sex of their offspring that will be hemophiliacs?

_____ % males; _____ % females

What percentage can be expected to neither exhibit nor carry the allele for hemophilia?

_____ %

What is the anticipated sex and percentage of individuals that will be carriers for hemophilia?

_____ %; _____ sex ■

Probability

Segregation (or parceling out) of chromosomes to daughter cells (gametes) during meiosis and the combination of egg and sperm are random or chance events. Hence, the possibility that certain genomes will arise and be expressed is based on the laws of probability. The randomness of gene recombination from each parent determines individual uniqueness and explains why siblings, however similar, never have totally corresponding traits (unless, of course, they are identical twins). The Punnett square method that you have been using to work out the genetics problems actually provides information on the *probability* of the appearance of certain genotypes considering all possible events. Probability (*P*) is defined as:

$$P = \frac{\text{number of specific events or cases}}{\text{total number of events or cases}}$$

If an event is certain to happen, its probability is 1. If it happens one out of every two times, its probability is $\frac{1}{2}$; if one out of four times, its probability is $\frac{1}{4}$, and so on.

When figuring the probability of separate events occurring together (or consecutively), the probability of each event must be multiplied together to get the final probability figure. For example, the probability of a penny coming up "heads" in each toss is $\frac{1}{2}$ (because it has two sides—heads and tails). But the probability of a tossed penny coming up heads four times in a row is $\frac{1}{2} \times \frac{1}{2} \times \frac{1}{2} \times \frac{1}{2} = \frac{1}{16}$.

Activity:
Exploring Probability

1. Obtain two pennies and perform the following simple experiment to explore the laws of probability.

a. Toss one penny into the air 10 times, and record the number of heads/tails observed.

_____ heads _____ tails

Probability: _____ /10 tails; _____ /10 heads

b. Now simultaneously toss two pennies into the air for 24 tosses, and record the results of each toss below. In each case, report the probability in the lowest fractional terms.

#HH _____ Probability _____

#HT _____ Probability _____

#TT _____ Probability _____

Does the first toss have any influence on the second?

_____

Does the third toss have any influence on the fourth?

_____

c.   Do a Punnett square using HT for one coin and HT for the "alleles" of the other.

Probability of HH: _____

Probability of HT: _____

Probability of TT: _____

How closely do your coin-tossing results correlate with the percentages obtained from the Punnett square results?

_____

_____

2.   Determine the probability of having a boy or girl offspring for each conception.

Parental genotypes: XY × XX

Probability of males: _____ %

Probability of females: _____ %

3.   Dad wants a baseball team! What are the chances of his having nine sons in a row?

_____ (Sorry, Dad!) ∎

Genetic Determination of Selected Human Characteristics

Most human traits are determined by multiple alleles or the interaction of several gene pairs. However, a few visible human traits or phenotypes can be traced to a single gene pair. It is some of these that will be investigated here.

Using Phenotype to Determine Genotype

For each of the characteristics described here, determine (as best you can) both your own phenotype and genotype, and record this information on Table 46.1. Since it is impossible to know if you are homozygous or heterozygous for a non-detrimental trait when you exhibit its dominant expression, you are to record your genotype as *A*— (or *B*—, and so on, depending on the letter used to indicate the alleles) in such cases. As you will see, all the traits examined here are non-harmful characteristics. Generally speaking, in humans dominant gene disorders are presumed to be heterozygous (one dominant and one recessive allele), because having two such defective mutated alleles is usually not compatible with life. On the other hand, if you exhibit the recessive trait, you are homozygous for the recessive allele and should record it accordingly as *aa* (*bb, cc,* and so on). When you have completed your observations, also record your data on the chart for tabulation of class results, on the chalkboard.

Interlocking fingers: Clasp your hands together by interlocking your fingers. Observe your clasped hands. Which thumb is uppermost? If the left thumb is uppermost, you possess a dominant allele (*I*) for this trait. If you clasped your right over your left thumb, you are illustrating the homozygous recessive (*ii*) phenotype.

PTC taste: Obtain a PTC taste strip. PTC or phenylthiocarbamide is a harmless chemical that some people can taste and others find tasteless. Chew the strip. If it tastes slightly bitter, you are a "taster" and possess the dominant gene (*P*) for this trait. If you cannot taste anything, you are a nontaster and are homozygous recessive (*pp*) for the trait. Approximately 70% of the people in the United States are tasters.

Sodium benzoate taste: Obtain a sodium benzoate taste strip and chew it. A different pair of alleles (from that determining PTC taste) determines the ability to taste sodium benzoate. If you can taste it, you have at least one of the dominant alleles (*S*). If not, you are homozygous recessive (*ss*) for the trait. Also record whether sodium benzoate tastes salty, bitter, or sweet to you (if a taster). Even though PTC and sodium benzoate taste are inherited independently, they interact to determine a person's taste sensations. Individuals who find PTC bitter and sodium benzoate salty tend to be devotees of sauerkraut, buttermilk, spinach, and other slightly bitter or salty foods.

Sex: The genotype XX determines the female phenotype, whereas XY determines the male phenotype.

Dimpled cheeks: The presence of dimples in one or both cheeks is due to a dominant gene (*D*). Absence of dimples indicates the homozygous recessive condition (*dd*).

Tongue rolling: Extend your tongue and attempt to roll it into a U shape longitudinally. People with this ability have the dominant allele for this trait. Use *T* for the dominant allele, and *t* for the recessive allele.

Table 46.1   Record of Human Genotypes/Phenotypes

Characteristic	Phenotype	Genotype
Interlocking fingers (*I,i*)		
PTC taste (*P,p*)		
Sodium benzoate taste (*S,s*)		
Sex (X,Y)		
Dimples (*D,d*)		
Tongue rolling (*T,t*)		
Attached earlobes (*E,e*)		
Widow's peak (*W,w*)		
Double-jointed thumb (*J,j*)		*I*
Bent little finger (*L,l*)		
Middigital hair (*H,h*)		
Freckles (*F,f*)		
Blaze (*B,b*)		
ABO blood type (*I^A,I^B,i*)		

Attached earlobes: Have your lab partner examine your earlobes. If no portion of the lobe hangs free inferior to its point of attachment to the head, you are homozygous recessive (*ee*) for attached earlobes. If part of the lobe hangs free below the point of attachment, you possess at least one dominant gene (*E*) (see Figure 46.2).

Widow's peak: A distinct downward V-shaped hairline at the middle of the forehead is referred to as a widow's peak. It is determined by a dominant allele (*W*), whereas the straight or continuous forehead hairline is determined by the homozygous recessive condition (*ww*) (see Figure 46.2).

Double-jointed thumb: A dominant gene determines a condition of loose ligaments that allows one to throw the thumb out of joint. The homozygous recessive condition determines tight joints. Use *J* for the dominant allele and *j* for the recessive allele (see Figure 46.2).

Bent little finger: Examine your little finger on each hand. If its terminal phalanx angles toward the ring finger, you are dominant for this trait. If one or both terminal digits are essentially straight, you are homozygous recessive for the trait. Use *L* for the dominant allele and *l* for the recessive allele.

Middigital finger hair: Critically examine the dorsum of the middle segment (phalanx) of fingers 3 and 4. If no hair is obvious, you are recessive (*hh*) for this condition. If hair is seen, you have the dominant gene (*H*) for this trait (which, however, is determined by multigene inheritance) (see Figure 46.2).

Freckles: Freckles are the result of a dominant gene. Use *F* as the dominant allele and *f* as the recessive allele (see Figure 46.2).

Blaze: A lock of hair different in color from the rest of scalp hair is called a blaze; it is determined by a dominant gene. Use *B* for the dominant gene and *b* for the recessive gene.

Blood type: Inheritance of the ABO blood type is based on the existence of three alleles designated as I^A, I^B, and *i*. Both I^A and I^B are dominant over *i*, but neither is dominant over the other. Thus the possession of I^A and I^B will yield type AB blood, whereas the possession of the I^A and *i* alleles will yield type A blood, and so on as explained in Exercise 29. There are four ABO blood groups or phenotypes, A, B, AB, and O, and their correlation to genotype is indicated as follows:

ABO blood group	Genotype
A	$I^A I^A$ or $I^A i$
B	$I^B I^B$ or $I^B i$
AB	$I^A I^B$
O	ii

Assuming you have previously typed your blood, record your phenotype and genotype in Table 46.1. If not, type your blood following the instructions on pp. 292–293, and then enter your results in the table.

⚠ Dispose of any blood-soiled supplies by placing the glassware in the bleach-containing beaker and all other items in the autoclave bag. ■

Once class data have been tabulated, scrutinize the results. Is there a single trait that is expressed in an identical manner by all members of the class?

_____

Because all human beings have 23 pairs of homologues and each pair segregates independently at meiosis, the number of possible combinations at segregation is more than 8 million! On the basis of this information, what would you guess are the chances of any two individuals in the class having identical phenotypes for all 14 traits investigated?

_____   ■

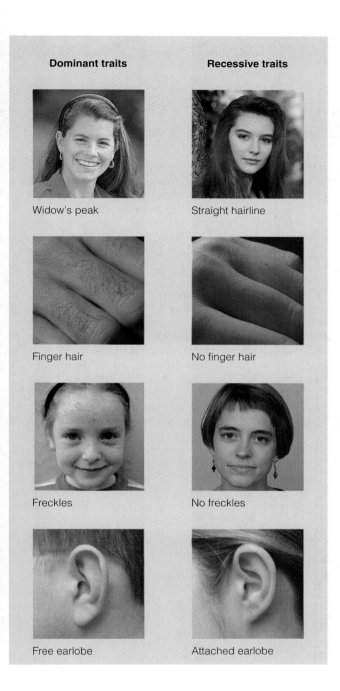

Dominant traits

Widow's peak

Finger hair

Freckles

Free earlobe

Recessive traits

Straight hairline

No finger hair

No freckles

Attached earlobe

Figure 46.2   Selected examples of human phenotypes.

47

Surface Anatomy Roundup

Objectives

1. Define *surface anatomy,* and explain why it is an important field of study. Define *palpation.*

2. Describe and palpate the major surface features of the cranium, face, and neck.

3. Describe the easily palpated bony and muscular landmarks of the back. Locate the vertebral spines on the living body.

4. List the bony surface landmarks of the thoracic cage, and explain how they relate to the major soft organs of the thorax. Explain how to find the second to eleventh ribs.

5. Name and palpate the important surface features on the anterior abdominal wall, and explain how to palpate a full bladder.

6. Define and explain the following: *linea alba, umbilical hernia,* examination for an inguinal hernia, *linea semilunaris,* and *McBurney's point.*

7. Locate and palpate the main surface features of the upper limb.

8. Explain the significance of the antecubital fossa, pulse points in the distal forearm, and the anatomical snuff box.

9. Describe and palpate the surface landmarks of the lower limb.

10. Explain exactly where to administer an injection in the gluteal region and in the other major sites of intramuscular injection.

Materials

- ❏ Articulated skeletons
- ❏ Charts or models of the skeletal muscles of the body
- ❏ Washable markers
- ❏ Hand mirror
- ❏ Stethoscope
- ❏ Alcohol swabs

Surface anatomy is a valuable branch of anatomical and medical science. True to its name, surface anatomy does indeed study the *external surface* of the body, but more importantly, it also studies *internal* organs as they relate to external surface landmarks and as they are seen and felt through the skin. Feeling internal structures through the skin with the fingers is called **palpation** (literally, "touching").

Surface anatomy is living anatomy, better studied in live people than in cadavers. It can provide a great deal of information about the living skeleton (almost all bones can be palpated) and about the muscles and blood vessels that lie near the body surface. Furthermore, a skilled examiner can learn a good deal about the heart, lungs, and other deep organs by performing a surface assessment. Thus, surface anatomy serves as the basis of the standard physical examination. For those planning a career in the health sciences or physical education, a study of surface anatomy will show you where to take pulses, where to insert tubes and needles, where to locate broken bones and inflamed muscles, and where to listen for the sounds of the lungs, heart, and intestines.

We will take a regional approach to surface anatomy, exploring the head first and proceeding to the trunk and the limbs. You will be observing and palpating your own body as you work through the exercise, because your body is the best learning tool of all. To aid your exploration of living anatomy, skeletons and muscle models or charts are provided around the lab so that you can review the bones and muscles you will encounter. Where you are asked to mark with a washable marker areas of skin that are personally unreachable, it probably would be best to choose a male student as a subject.

Activity:
Palpating Landmarks of the Head

The head (Figures 47.1 and 47.2) is divided into the cranium and the face.

Cranium

1. Run your fingers over the superior surface of your head. Notice that the underlying cranial bones lie very near the surface. Proceed to your forehead and palpate the *superciliary arches* ("brow ridges") directly superior to your orbits (Figure 47.1).

2. Move your hand to the posterior surface of your skull, where you can feel the knoblike *external occipital protuberance.* Run your finger directly laterally from this projection to feel the ridgelike *superior nuchal line* on the occipital bone. This line, which marks the superior extent of the muscles of the posterior neck, serves as the boundary between the head and the neck. Now feel the prominent *mastoid process* on each side of the cranium just posterior to your ear.

3. Place a hand on your temple, and clench your teeth together in a biting action. You should be able to feel the *temporalis muscle* bulge as it contracts. Next, raise your eyebrows, and feel your forehead wrinkle, an action produced by

Figure 47.1 Surface anatomy of the lateral aspect of the head.

Temporalis muscle

External occipital protuberance

Superficial temporal artery (pulse point)

Auricle

Mastoid process

Angle of mandible

Superciliary arch

Zygomatic arch

Temporomandibular joint

Ramus of mandible

Body of mandible

Facial artery (pulse point)

Figure 47.2 Surface structures of the face.

Frontalis muscle

Lacrimal fossa

Ala of nose

Root and bridge of nose

Dorsum nasi

Apex of nose

Philtrum

the *frontalis muscle.* The frontalis inserts superiorly onto the broad aponeurosis called the *galea aponeurotica* (Table 15.1, p. 130) that covers the superior surface of the cranium. This aponeurosis binds tightly to the overlying subcutaneous tissue and skin to form the true **scalp.** Push on your scalp, and confirm that it slides freely over the underlying cranial bones. Because the scalp is only loosely bound to the skull, people can easily be "scalped" (in industrial accidents, for example). The scalp is richly vascularized by a large number of arteries running through its subcutaneous tissue. Most arteries of the body constrict and close after they are cut or torn, but those in

the scalp are unable to do so because they are held open by the dense connective tissue surrounding them.

What do these facts suggest about the amount of bleeding that accompanies scalp wounds?

_____

_____

_____

Face

The surface of the face is divided into many different regions, including the *orbital, nasal, oral* (mouth), and *auricular* (ear) areas (Figure 47.2).

1. Trace a finger around the entire margin of the bony orbit. The *lacrimal fossa,* which contains the tear-gathering lacrimal sac, may be felt on the medial side of the eye socket.

2. Touch the most superior part of your nose, its **root,** which lies between the eyebrows (Figure 47.2). Just inferior to this, between your eyes, is the **bridge** of the nose formed by the nasal bones. Continue your finger's progress inferiorly along the nose's anterior margin, the **dorsum nasi,** to its tip, the **apex.** Place one finger in a nostril and another finger on the flared winglike **ala** that defines the nostril's lateral border. Then feel the **philtrum,** the shallow vertical groove on the upper lip below the nose.

3. Grasp your *auricle,* the shell-like part of the external ear that surrounds the opening of the *external auditory canal* (Figure 47.1). Now trace the ear's outer rim, or *helix,* to the *lobule* (earlobe) inferiorly. The lobule is easily pierced, and since it is not highly sensitive to pain, it provides a convenient place to hang an earring or obtain a drop of blood for clinical blood analysis. Next, place a finger on your temple just anterior to the auricle (Figure 47.1). There, you will be able to feel the pulsations of the *superficial temporal artery,* which ascends to supply the scalp.

4. Run your hand anteriorly from your ear toward the orbit, and feel the *zygomatic arch* just deep to the skin. This bony arch is easily broken by blows to the face. Next, place your fingers on the skin of your face, and feel it bunch and stretch as you contort your face into smiles, frowns, and grimaces. You are now monitoring the action of several of the subcutaneous *muscles of facial expression* (Table 15.1, p. 130).

5. On your lower jaw, palpate the parts of the bony *mandible:* its anterior body and its posterior ascending *ramus.* Press on the skin over the mandibular ramus, and feel the *masseter muscle* bulge when you clench your teeth. Palpate the anterior border of the masseter, and trace it to the mandible's inferior margin. At this point, you will be able to detect the pulse of your *facial artery* (Figure 47.1). Finally, to feel the *temporomandibular joint,* place a finger directly anterior to the external auditory canal of your ear, and open and close your mouth several times. The bony structure you feel moving is the *head of the mandible.* ■

Activity:
Palpating Landmarks of the Neck

Bony Landmarks

1. Run your fingers inferiorly along the back of your neck, in the posterior midline, to feel the *spinous processes* of the cervical vertebrae. The spine of G$_7$, the *vertebra prominens,* is especially prominent.

2. Now, beginning at your chin, run a finger inferiorly along the anterior midline of your neck (Figure 47.3). The first hard structure you encounter will be the U-shaped *hyoid bone,* which lies in the angle between the floor of the mouth and the vertical part of the neck (Figure 47.4). Directly inferior to this, you will feel the *laryngeal prominence* (Adam's apple) of the thyroid cartilage. Just inferior to the laryngeal prominence, your finger will sink into a soft depression (formed by the *cricothyroid ligament*) before proceeding onto the rounded surface of the *cricoid cartilage.* Now swallow several times, and feel the whole larynx move up and down.

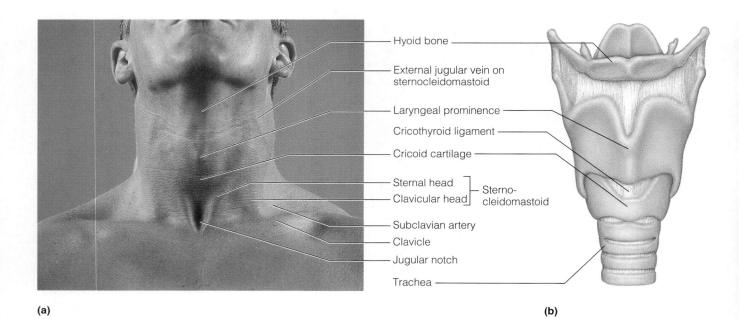

(a)   (b)

Figure 47.3 Anterior surface of the neck. (a) Photograph.
(b) Diagram of the underlying skeleton of the larynx.

External carotid artery (pulse point)

Trapezius

Subclavian artery

Hyoid bone

Laryngeal prominence

Sternal head

Clavicular head

Sternocleido-mastoid

Jugular notch

Figure 47.4   Lateral surface of the neck.

3. Continue inferiorly to the trachea. Attempt to palpate the *isthmus of the thyroid gland,* which feels like a spongy cushion over the second to fourth tracheal rings (see Figure 47.3b). Then, try to palpate the two soft lateral *lobes* of your thyroid gland along the sides of the trachea.

4. Move your finger all the way inferiorly to the root of the neck, and rest it in the *jugular notch,* the depression in the superior part of the sternum between the two clavicles. By pushing deeply at this point, you can feel the cartilage rings of the trachea.

Muscles

The *sternocleidomastoid* is the most prominent muscle in the neck and the neck's most important surface landmark. You can best see and feel it when you turn your head to the side.

1. Obtain a hand mirror, hold it in front of your face, and turn your head sharply from right to left several times. You will be able to see both heads of this muscle, the *sternal head* medially and the *clavicular head* laterally (Figures 47.3 and 47.4). Several important structures lie beside or beneath the sternocleidomastoid:

● The *cervical lymph nodes* lie both superficial and deep to this muscle. (Swollen cervical nodes provide evidence of infections or cancer of the head and neck.)

● The *common carotid artery* and *internal jugular vein* lie just deep to the sternocleidomastoid, a relatively superficial location that exposes these vessels to danger in slashing wounds to the neck.

● Just lateral to the inferior part of the sternocleidomastoid is the large *subclavian artery* on its way to supply the upper limb. By pushing on the subclavian artery at this point, one can stop the bleeding from a wound anywhere in the associated limb.

● Just anterior to the sternocleidomastoid, superior to the level of your larynx, you can feel a carotid pulse—the pulsations of the *external carotid artery* (Figure 47.4).

● The *external jugular vein* descends vertically, just superficial to the sternocleidomastoid and deep to the skin (Figure 47.3). To make this vein "appear" on your neck, stand before the mirror, and gently compress the skin superior to your clavicle with your fingers.

2. Another large muscle in the neck, on the posterior aspect, is the *trapezius* (Figure 47.4). You can feel this muscle contracting just deep to the skin as you shrug your shoulders.

Triangles of the Neck

The sternocleidomastoid muscles divide each side of the neck into the posterior and anterior triangles (Figure 47.5a).

1. The **posterior triangle** is defined by the sternocleidomastoid anteriorly, the trapezius posteriorly, and the clavicle inferiorly. Palpate the borders of the posterior triangle.

The **anterior triangle** is defined by the inferior margin of the mandible superiorly, the midline of the neck anteriorly, and the sternocleidomastoid posteriorly.

2. The contents of these two triangles are shown in Figure 47.5b. The posterior triangle contains many important nerves and blood vessels, including the *accessory nerve* (cranial nerve XI), most of the *cervical plexus,* and the *phrenic nerve.* In the inferior part of the triangle are the external jugular vein, the trunks of the *brachial plexus,* and the subclavian artery. These structures are relatively superficial and are easily cut or injured by wounds to the neck. In the neck's anterior triangle, important structures include the *submandibular gland,* the *suprahyoid* and *infrahyoid muscles,* and parts of the carotid arteries and jugular veins that lie superior to the sternocleidomastoid.

● Palpate your carotid pulse. ■

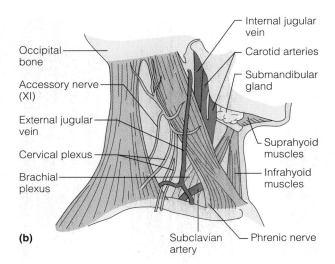

Figure 47.5 Anterior and posterior triangles of the neck.
(a) Boundaries of the triangles. **(b)** Some contents of the triangles.

A wound to the posterior triangle of the neck can lead to long-term loss of sensation in the skin of the neck and shoulder, as well as partial paralysis of the sternocleidomastoid and trapezius muscles. Can you explain these effects?

_____

_____

_____ ■

Activity:
Palpating Landmarks of the Trunk

The trunk of the body consists of the thorax, abdomen, pelvis, and perineum. The *back* includes parts of all of these regions, but for convenience it is treated separately.

The Back

Bones 1.   The vertical groove in the center of the back is called the **posterior median furrow** (Figure 47.6a). The *spinous processes* of the vertebrae are visible in the furrow when the spinal column is flexed. Palpate a few of these processes on your partner's back (C_7 and T_1 are the most prominent and the easiest to find). Also palpate the posterior parts of some ribs, as well as the prominent *spine of the scapula* and the scapula's long *medial border.* The scapula lies superficial to ribs 2 to 7; its *inferior angle* is at the level of the spinous process of vertebra T_7. The medial end of the scapular spine lies opposite the T_3 spinous process.

2.   Now feel the *iliac crests* (superior margins of the iliac bones) in your own lower back. You can find these crests effortlessly by resting your hands on your hips. Locate the most superior point of each crest, a point that lies roughly halfway between the posterior median furrow and the lateral side of

the body (Figure 47.6a). A horizontal line through these two superior points, the **supracristal line,** intersects L_4, providing a simple way to locate that vertebra. The ability to locate L_4 is essential for performing a *lumbar puncture,* a procedure in which the clinician inserts a needle into the vertebral canal of the spinal column directly superior or inferior to L_4 and withdraws cerebrospinal fluid.

3.   The *sacrum* is easy to palpate just superior to the cleft in the buttocks. You can feel the *coccyx* in the extreme inferior part of that cleft, just posterior to the anus.

Muscles   The largest superficial muscles of the back are the *trapezius* superiorly and *latissimus dorsi* inferiorly (Figure 47.6b). Furthermore, the deeper *erector spinae* muscles are very evident in the lower back, flanking the vertebral column like thick vertical cords.

1.   Feel your partner's erector spinae muscles contract and bulge as he straightens his spine from a slightly bent-over position.

The superficial muscles of the back fail to cover a small area of the rib cage called the **triangle of auscultation** (Figure 47.6a). This triangle lies just medial to the inferior part of the scapula. Its three boundaries are formed by the trapezius medially, the latissimus dorsi inferiorly, and the scapula laterally. The physician places a stethoscope over the skin of this triangle to listen for lung sounds (*auscultation* = listening). To hear the lungs clearly, the doctor first asks the patient to fold the arms together in front of the chest and then flex the trunk.

What do you think is the precise reason for having the patient take this action?

_____

_____

_____

(a)

(b)

Figure 47.6 Surface anatomy of the back. Two different poses, **(a)** and **(b)**.

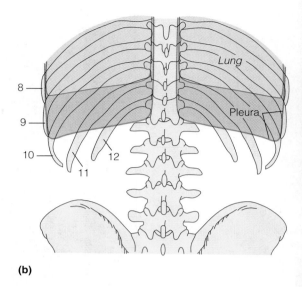

Figure 47.7 The bony rib cage as it relates to the underlying lungs and pleural cavities. Both the pleural cavities (blue) and the lungs (red) are outlined. **(a)** Anterior view. **(b)** Posterior view.

2. Have your partner assume the position just described. After cleaning the earpieces with an alcohol swab, use the stethoscope to auscultate the lung sounds. Compare the clarity of the lung sounds heard over the triangle of auscultation to that over other areas of the back.

The Thorax

Bones 1. Start exploring the anterior surface of your partner's bony *thoracic cage* (Figures 47.7 and 47.8) by defining the extent of the *sternum*. Use a finger to trace the sternum's triangular *manubrium* inferior to the jugular notch, its flat *body,* and the tongue-shaped *xiphoid process.* Now palpate the ridgelike *sternal angle,* where the manubrium meets the body of the sternum. Locating the sternal angle is important because it directs you to the second ribs (which attach to it). Once you find the second rib, you can count down to identify every other rib in the thorax (except the first and sometimes the twelfth rib, which lie too deep to be palpated). The sternal angle is a highly reliable landmark—it is easy to locate, even in overweight people.

2. By locating the individual ribs, you can mentally "draw" a series of horizontal lines of "latitude" that you can use to map and locate the underlying visceral organs of the thoracic cavity. Such mapping also requires lines of "longitude," so let us construct some vertical lines on the wall of your partner's trunk. As he lifts an arm straight up in the air, extend a line inferiorly from the center of the axilla onto his lateral thoracic wall. This is the **midaxillary line** (Figure 47.7a). Now estimate the midpoint of his *clavicle,* and run a vertical line inferiorly from that point toward the groin. This is the **midclavicular line,** and it will pass about 1 cm medial to the nipple.

3. Next, feel along the V-shaped inferior edge of the rib cage, the *costal margin*. At the **infrasternal angle,** the superior angle of the costal margin, lies the *xiphisternal joint.* Deep to the xiphisternal joint, the heart lies on the diaphragm.

4. The thoracic cage provides many valuable landmarks for locating the vital organs of the thoracic and abdominal cavities. On the anterior thoracic wall, ribs 2–6 define the superior-to-inferior extent of the female breast, and the fourth intercostal space indicates the location of the *nipple* in men, children, and small-breasted women. The right costal margin runs across the anterior surface of the liver and gallbladder. Surgeons must be aware of the inferior margin of the *pleural cavities* because if they accidentally cut into one of these cavities, a lung collapses. The inferior pleural margin lies adjacent to vertebra T_{12} near the posterior midline (Figure 47.7b) and runs horizontally across the back to reach rib 10 at the midaxillary line. From there, the pleural margin ascends to rib 8 in the midclavicular line (Figure 47.7a) and to the level of the xiphisternal joint near the anterior midline. The *lungs* do not fill the inferior region of the pleural cavity. Instead, their inferior borders run at a level that is two ribs superior to the pleural margin, until they meet that margin near the xiphisternal joint.

5. The relationship of the *heart* to the thoracic cage is considered in Exercise 30 (p. 295). We will review that information here. In essence, the superior right corner of the heart lies at the junction of the third rib and the sternum; the superior left corner lies at the second rib, near the sternum; the inferior left corner lies in the fifth intercostal space in the midclavicular line; and the inferior right corner lies at the sternal border of the sixth rib. You may wish to outline the heart on your chest or that of your lab partner by connecting the four corner points with a washable marker.

Muscles The main superficial muscles of the anterior thoracic wall are the *pectoralis major* and the anterior slips of the *serratus anterior.*

Using Figure 47.8 as a guide, try to palpate these two muscles on your chest. They both contract during push-ups, and you can confirm this by pushing yourself up from your

Jugular notch

Clavicle

Sternal angle

Xiphisternal joint

Fifth intercostal space

Costal margin

Linea semilunaris

McBurney's point

Pubic tubercle

Pubic symphysis

Anterior axillary fold

Base of axilla

Posterior axillary fold

Pectoralis major

Xiphoid process

Serratus anterior

Linea alba

Tendinous insertions
in the rectus abdominis

External oblique

Iliac crest

Anterior superior
iliac spine

Groove for inguinal
ligament

Figure 47.8   The anterior thorax and abdomen.

desk with one arm while palpating the muscles with your opposite hand. ■

Activity:
Palpating Landmarks of the Abdomen

Bony Landmarks

The anterior abdominal wall (Figure 47.8) extends inferiorly from the costal margin to an inferior boundary that is defined by several landmarks. Palpate these landmarks as they are described below.

1. **Iliac crest.** Recall that the iliac crests are the superior margins of the iliac bones, and you can locate them by resting your hands on your hips.

2. **Anterior superior iliac spine.** Representing the most anterior point of the iliac crest, this spine is a prominent landmark. It can be palpated in everyone, even those who are overweight. Run your fingers anteriorly along the iliac crest to its end.

3. **Inguinal ligament.** The inguinal ligament, indicated by a groove on the skin of the groin, runs medially from the an-

terior superior iliac spine to the pubic tubercle of the pubic bone.

4. **Pubic crest.** You will have to press deeply to feel this crest on the pubic bone near the median *pubic symphysis*. The *pubic tubercle,* the most lateral point of the pubic crest, is easier to palpate, but you will still have to push deeply.

   Inguinal hernias occur immediately superior to the inguinal ligament and may exit from a medial opening called the *superficial inguinal ring*. To locate this ring, one would palpate the pubic tubercle (Figure 47.9). The procedure used by the physician to test whether a male has an inguinal hernia is depicted in Figure 47.9. ■

Muscles and Other Surface Features

The central landmark of the anterior abdominal wall is the *umbilicus* (navel). Running superiorly and inferiorly from the umbilicus is the *linea alba* ("white line"), represented in the skin of lean people by a vertical groove (Figure 47.8). The linea alba is a tendinous seam that extends from the xiphoid process to the pubic symphysis, just medial to the rectus abdominis muscles (Table 15.3, p. 137). The linea alba is a favored site for surgical entry into the abdominal cavity

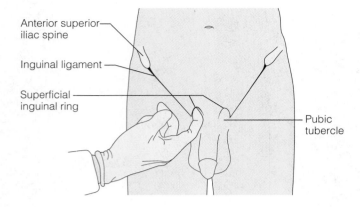

Figure 47.9 **Clinical examination for an inguinal hernia in a male.** The examiner palpates the patient's pubic tubercle, pushes superiorly to invaginate the scrotal skin into the superficial inguinal ring, and asks the patient to cough. If an inguinal hernia exists, it will push inferiorly and touch the examiner's fingertip.

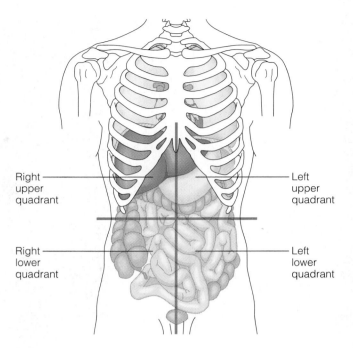

Figure 47.10 **The four abdominopelvic quadrants.**

because the surgeon can make a long cut through this line with no muscle damage and minimal bleeding.

Several kinds of hernias involve the umbilicus and the linea alba. In an **acquired umbilical hernia,** the linea alba weakens until intestinal coils push through it just superior to the navel. The herniated coils form a bulge just deep to the skin.

Another type of umbilical hernia is a **congenital umbilical hernia,** present in some infants: The umbilical hernia is seen as a cherry-sized bulge deep to the skin of the navel that enlarges whenever the baby cries. Congenital umbilical hernias are usually harmless, and most correct themselves automatically before the child's second birthday. ■

1.   **McBurney's point** is the spot on the anterior abdominal skin that lies directly superficial to the base of the appendix (Figure 47.8). It is located one-third of the way along a line between the right anterior superior iliac spine and the umbilicus. Try to find it on your body.

McBurney's point is the most common site of incision in appendectomies, and it is often the place where the pain of appendicitis is experienced most acutely. Pain at McBurney's point after the pressure is removed (rebound tenderness) can indicate appendicitis. This is not a *precise* method of diagnosis, however.

2.   Flanking the linea alba are the vertical straplike *rectus abdominis* muscles (Figure 47.8). Feel these muscles contract just deep to your skin as you do a bent-knee sit-up (or as you bend forward after leaning back in your chair). In the skin of lean people, the lateral margin of each rectus muscle makes a groove known as the **linea semilunaris** ("half-moon line"). On your right side, estimate where your linea semilunaris crosses the costal margin of the rib cage. The *gallbladder* lies just deep to this spot, so this is the standard point of incision for gallbladder surgery. In muscular people, three horizontal grooves can be seen in the skin covering the rectus abdominis. These grooves represent the *tendinous insertions* (or *intersections* or *inscriptions*), fibrous bands that subdivide the rectus muscle. Because of these subdivisions, each rectus ab-

dominis muscle presents four distinct bulges. Try to identify these insertions on yourself or your partner.

3.   The only other major muscles that can be seen or felt through the anterior abdominal wall are the lateral *external obliques.* Feel these muscles contract as you cough, strain, or raise your intra-abdominal pressure in some other way.

4.   Recall that the anterior abdominal wall can be divided into four quadrants (Figure 47.10). A clinician listening to a patient's **bowel sounds** places the stethoscope over each of the four abdominal quadrants, one after another. Normal bowel sounds, which result as peristalsis moves air and fluid through the intestine, are high-pitched gurgles that occur every 5 to 15 seconds.

•   Use the stethoscope to listen to your own or your partner's bowel sounds.

Abnormal bowel sounds can indicate intestinal disorders. Absence of bowel sounds indicates a halt in intestinal activity, which follows long-term obstruction of the intestine, surgical handling of the intestine, peritonitis, or other conditions. Loud tinkling or splashing sounds, by contrast, indicate an increase in intestinal activity. Such loud sounds may accompany gastroenteritis (inflammation and upset of the GI tract) or a partly obstructed intestine. ■

The Pelvis and Perineum

The bony surface features of the *pelvis* are considered with the bony landmarks of the abdomen (p. 475) and the gluteal region (p. 481). Most *internal* pelvic organs are not palpable through the skin of the body surface. A full *bladder,* however, becomes firm and can be felt through the abdominal wall just superior to the pubic symphysis. A bladder that can be palpated more than a few centimeters above this symphysis is

Acromion

Spine of scapula

Deltoid

Lateral head of
the triceps brachii

Lateral epicondyle
of the humerus

Olecranon process

Acrominoclavicular joint

Clavicle

Greater tubercle
of the humerus

Biceps brachii

Head of radius

(a)

Figure 47.11   Shoulder and arm. (a) Lateral view.

retaining urine and dangerously full, and it should be drained by catheterization. ∎

Activity:
Palpating Landmarks of the Upper Limb

Axilla

The **base of the axilla** is the groove in which the underarm hair grows (Figure 47.8). Deep to this base lie the axillary *lymph nodes* (which swell and can be palpated in breast cancer), the large *axillary vessels* serving the upper limb, and much of the brachial plexus. The base of the axilla forms a "valley" between two thick, rounded ridges, the **axillary folds.** Just anterior to the base, clutch your **anterior axillary fold,** formed by the pectoralis major muscle. Then grasp your **posterior axillary fold.** This fold is formed by the latissimus dorsi and teres major muscles of the back as they course toward their insertions on the humerus.

Shoulder

1.   Relocate the prominent spine of the scapula posteriorly (Figure 47.11). Follow the spine to its lateral end, the flattened *acromion* on the shoulder's summit. Then, palpate the *clavicle* anteriorly, tracing this bone from the sternum to the shoulder. Notice the clavicle's curved shape.

2.   Now locate the junction between the clavicle and the acromion on the superolateral surface of your shoulder, at the *acromioclavicular joint.* To find this joint, thrust your arm anteriorly repeatedly until you can palpate the precise point of pivoting action.

3.   Next, place your fingers on the *greater tubercle* of the humerus. This is the most lateral bony landmark on the superior surface of the shoulder. It is covered by the thick *deltoid muscle,* which forms the rounded superior part of the shoulder. Intramuscular injections are often given into the deltoid, about 5 cm (2 inches) inferior to the greater tubercle (refer to Figure 47.19a, p. 482).

Arm

Remember, according to anatomists, the arm runs only from the shoulder to the elbow, and not beyond.

1.   In the arm, palpate the humerus along its entire length, especially along its medial and lateral sides.

2.   Feel the *biceps brachii* muscle contract on your anterior arm when you flex your forearm against resistance. The medial boundary of the biceps is represented by the **medial bicipital furrow** (Figure 47.11b). This groove contains the large *brachial artery,* and by pressing on it with your fingertips you can feel your *brachial pulse.* Recall that the brachial artery is the artery routinely used in measuring blood pressure with a sphygmomanometer.

Acrominoclavicular joint

Deltoid

Biceps brachii

Medial bicipital furrow

Medial epicondyle of the humerus

Clavicle

Greater tubercle of the humerus

Cephalic vein

(b)

Figure 47.11 Shoulder and arm (continued).
(b) Anterior and medial view.

Triceps brachii:
Lateral head
Long head
Tendon
Medial head

Dimple in which head of radius is felt

Head of ulna

Figure 47.12 Surface anatomy of the upper limb, posterior view.

3. Extend your forearm against resistance, and feel your *triceps brachii* muscle bulge in the posterior arm. All three heads of the triceps (lateral, long, and medial) are visible through the skin of a muscular person (Figure 47.12).

Elbow Region

1. In the distal part of your arm, near the elbow, palpate the two projections of the humerus, the *lateral* and *medial epicondyles* (Figure 47.11). Midway between the epicondyles, on the posterior side, feel the *olecranon process* of the ulna, which forms the point of the elbow.

2. Confirm that the two epicondyles and the olecranon all lie in the same horizontal line when the elbow is extended. If these three bony processes do not line up, the elbow is dislocated.

3. Now feel along the posterior surface of the medial epicondyle. You are palpating your ulnar nerve.

4. On the anterior surface of the elbow is a triangular depression called the **antecubital fossa** or **cubital fossa** (Figure 47.13). The triangle's superior *base* is formed by a horizontal line between the humeral epicondyles; its two inferior sides are defined by the *brachioradialis* and *pronator teres* muscles (Figure 47.13). Try to define these boundaries on your

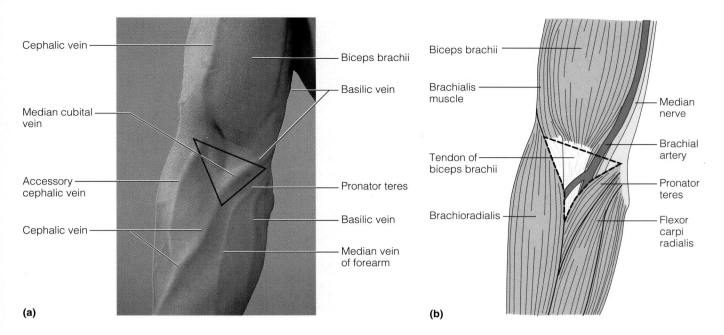

Cephalic vein

Biceps brachii

Basilic vein

Median cubital vein

Accessory cephalic vein

Pronator teres

Cephalic vein

Basilic vein

Median vein of forearm

(a)

Biceps brachii

Brachialis muscle

Median nerve

Brachial artery

Tendon of biceps brachii

Pronator teres

Brachioradialis

Flexor carpi radialis

(b)

Figure 47.13   The antecubital (cubital) fossa on the anterior surface of the right elbow (outlined by the triangle).
(a) Photograph. **(b)** Diagram of deeper structures in the fossa.

own limb. To find the brachioradialis muscle, flex your forearm against resistance, and watch this muscle bulge through the skin of your lateral forearm. To feel your pronator teres contract, palpate the antecubital fossa as you pronate your forearm against resistance. (Have your partner provide the resistance.)

Superficially, the antecubital fossa contains the *median cubital* vein (Figure 47.13a). Clinicians often draw blood from this superficial vein and insert intravenous (IV) catheters into it to administer medications, transfused blood, and nutrient fluids. The large *brachial artery* lies just deep to the median cubital vein (Figure 47.13b), so a needle must be inserted into the vein from a shallow angle (almost parallel to the skin) to avoid puncturing the artery. Other structures that lie deep in the fossa are also shown in Figure 47.13b.

5.   The median cubital vein interconnects the larger cephalic and basilic veins of the upper limb. Recall from Exercise 32 that the *cephalic vein* ascends along the lateral side of the forearm and arm, whereas the *basilic vein* ascends through the limb's medial side. These veins are visible through the skin of lean people (Figure 47.13a). Examine your arm to see if your cephalic and basilic veins are visible.

Forearm and Hand

The two parallel bones of the forearm are the medial *ulna* and the lateral *radius.*

1.   Feel the ulna along its entire length as a sharp ridge on the posterior forearm (confirm that this ridge runs inferiorly from the olecranon process). As for the radius, you can feel its distal half, but most of its proximal half is covered by muscle. You can, however, feel the rotating *head* of the radius. To do this, extend your forearm, and note that a dimple forms on the posterior lateral surface of the elbow region

(Figure 47.12). Press three fingers into this dimple, and rotate your free hand as if you were turning a doorknob. You will feel the head of the radius rotate as you perform this action.

2.   Both the radius and ulna have a knoblike *styloid process* at their distal ends. Figure 47.14 shows a way to locate these processes. Do not confuse the ulna's styloid process with the conspicuous *head of the ulna,* from which the styloid process stems. Confirm that the styloid process of the radius lies about 1 cm (0.4 inch) distal to that of the ulna.

Styloid process of radius

Head of ulna

Styloid process of ulna

Figure 47.14   A way to locate the styloid processes of the ulna and radius. The right hand is palpating the left hand in this picture. Note that the head of the ulna is not the same as its styloid process. The styloid process of the radius lies about 1 cm distal to the styloid process of the ulna.

(a)

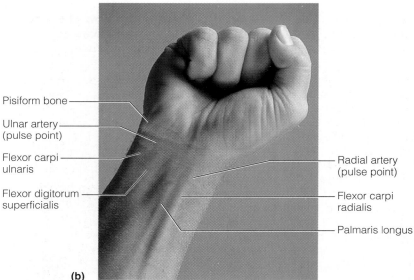

(b)

Figure 47.15 The anterior surface of the forearm and fist.
(a) The entire forearm. **(b)** Enlarged view of the distal forearm and hand. The tendons of the flexor muscles guide the clinician to several sites for pulse taking.

Colles' fracture of the wrist is an impacted fracture in which the distal end of the radius is pushed proximally into the shaft of the radius. This sometimes occurs when someone falls on outstretched hands, and it most often happens to elderly women with osteoporosis. Colles' fracture bends the wrist into curves that resemble those on a fork. ■ Can you deduce how physicians use palpation to diagnose a Colles' fracture?

_____

_____

3. Next, feel the major groups of muscles within your forearm. Flex your hand and fingers against resistance, and feel the anterior *flexor muscles* contract. Then extend your hand at the wrist, and feel the tightening of the posterior *extensor muscles.*

4. Near the wrist, the anterior surface of the forearm reveals many significant features (Figure 47.15). Flex your fist against resistance; the tendons of the main wrist flexors will bulge the skin of the distal forearm. The tendons of the *flexor carpi radialis* and *palmaris longus* muscles are most obvious. (The palmaris longus, however, is absent from at least one arm in 30% of all people, so your forearm may exhibit just one prominent tendon instead of two.) The *radial artery* lies just lateral to (on the thumb side of) the flexor carpi radialis tendon, where the pulse is easily detected (Figure 47.15b). Feel your radial pulse here. The *median nerve* (which innervates the thumb) lies deep to the palmaris longus tendon. Finally, the *ulnar artery* lies on the medial side of the forearm, just lateral to the tendon of the *flexor carpi ulnaris.* Using Figure 47.15b as a guide, locate and feel your ulnar arterial pulse.

Figure 47.16 **The dorsum of the hand.** Note especially the anatomical snuff box and dorsal venous network.

5.   Extend your thumb and point it posteriorly to form a triangular depression in the base of the thumb on the back of your hand. This is the **anatomical snuff box** (Figure 47.16). Its two elevated borders are defined by the tendons of the thumb extensor muscles, *extensor pollicis brevis* and *extensor pollicis longus.* The radial artery runs within the snuff box, so this is another site for taking a radial pulse. The main bone on the floor of the snuff box is the scaphoid bone of the wrist, but the styloid process of the radius is also present here. (If displaced by a bone fracture, the radial styloid process will be felt outside of the snuff box rather than within it.) The "snuff box" took its name from the fact that people once put snuff (tobacco for sniffing) in this hollow before lifting it up to the nose.

6.   On the dorsum of your hand, observe the superficial veins just deep to the skin. This is the *dorsal venous network,* which drains superiorly into the cephalic vein. This venous network provides a site for drawing blood and inserting intravenous catheters and is preferred over the median cubital vein for these purposes. Next, extend your hand and fingers, and observe the tendons of the *extensor digitorum* muscle.

7.   The anterior surface of the hand also contains some features of interest (Figure 47.17). These features include the *epidermal ridges* ("fingerprints") and many *flexion creases* in the skin. Grasp your *thenar eminence* (the bulge on the palm that contains the thumb muscles) and your *hypothenar eminence* (the bulge on the medial palm that contains muscles that move the little finger). ■

Activity:
Palpating Landmarks of the Lower Limb

Gluteal Region

Dominating the gluteal region are the two *prominences* ("cheeks") of the buttocks. These are formed by subcutaneous fat and by the thick *gluteus maximus* muscles. The midline groove between the two prominences is called the

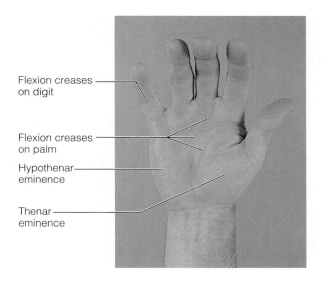

Figure 47.17 **The palmar surface of the hand.**

natal cleft (*natal* = rump) or *gluteal cleft.* The inferior margin of each prominence is the horizontal **gluteal fold,** which roughly corresponds to the inferior margin of the gluteus maximus.

1.   Try to palpate your *ischial tuberosity* just above the medial side of each gluteal fold (it will be easier to feel if you sit down or flex your thigh first). The ischial tuberosities are the robust inferior parts of the ischial bones, and they support the body's weight during sitting.

2.   Next, palpate the *greater trochanter* of the femur on the lateral side of your hip (Figures 47.18 and 47.20). This trochanter lies just anterior to a hollow and about 10 cm (one hand's breadth) inferior to the iliac crest. To confirm that you have found the greater trochanter, alternately flex and extend your thigh. Because this trochanter is the most superior point on the lateral femur, it moves with the femur as you perform this movement.

Figure 47.18 **The gluteal region.** The region extends from the iliac crests superiorly to the gluteal folds inferiorly. Therefore, it includes more than just the prominences of the buttock.

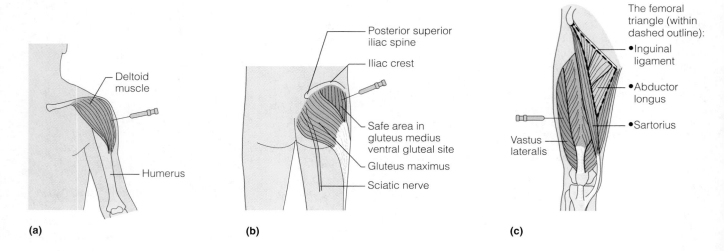

Figure 47.19 **Three major sites of intramuscular injections.**
(a) Deltoid muscle of the arm. **(b)** Ventral gluteal site (gluteus medius).
(c) Vastus lateralis in the lateral thigh. The femoral triangle is also shown.

3. To palpate the sharp *posterior superior iliac spine* (Figure 47.18), locate your iliac crests again, and trace each to its most posterior point. You may have difficulty feeling this spine, but it is indicated by a distinct dimple in the skin that is easy to find. This dimple lies two to three fingers' breadths lateral to the midline of the back. The dimple also indicates the position of the *sacroiliac joint,* where the hip bone attaches to the sacrum of the spinal column. (You can check *your* "dimples" out in the privacy of your home.)

The gluteal region is a major site for administering intramuscular injections. When giving such injections, extreme care must be taken to avoid piercing a major nerve that lies just deep to the gluteus maximus muscle. Can you guess what nerve this is?

It is the thick *sciatic nerve,* which innervates much of the lower limb. Furthermore, the needle must avoid the gluteal nerves and gluteal blood vessels, which also lie deep to the gluteus maximus.

To avoid harming these structures, the injections are most often applied to the gluteus *medius* (not maximus) muscle superior to the cheeks of the buttocks, in a safe area called the **ventral gluteal site** (Figure 47.19b). To locate this site, mentally draw a line laterally from the posterior superior iliac spine (dimple) to the greater trochanter; the injection would be given 5 cm (2 inches) superior to the midpoint of that line. Another safe way to locate the ventral gluteal site is to approach the lateral side of the patient's left hip with your extended right hand (or the right hip with your left hand). Then,

Greater trochanter of femur

Hollow posterior to the greater trochanter

Hamstring muscles

Patella

Head of fibula

Peroneal muscles

Medial malleolus

Lateral malleolus

Figure 47.20   Lateral surface of the lower limb.

place your thumb on the anterior superior iliac spine and your index finger as far posteriorly on the iliac crest as it can reach. The heel of your hand comes to lie on the greater trochanter, and the needle is inserted in the angle of the V formed between your thumb and index finger about 4 cm (1.5 inches) inferior to the iliac crest.

Gluteal injections are not given to small children because their "safe area" is too small to locate with certainty and because the gluteal muscles are thin at this age. Instead, infants and toddlers receive intramuscular shots in the prominent vastus lateralis muscle of the thigh.

Thigh

The thigh is pictured in Figures 47.20, 47.21, and 47.22. Much of the femur is clothed by thick muscles, so the thigh has few palpable bony landmarks.

1. Distally, feel the *medial* and *lateral condyles of the femur* and the *patella* anterior to the condyles (see Figure 47.21c and a).

2. Next, palpate your three groups of thigh muscles (Figures 47.20, 47.21a, and 47.21b)—the *quadriceps femoris muscles* anteriorly, the *adductor muscles* medially, and the *hamstrings* posteriorly. The *vastus lateralis,* the lateral muscle of the quadriceps group, is a site for intramuscular injections. Such injections are administered about halfway down the length of this muscle (see Figure 47.19c).

3. The anterosuperior surface of the thigh exhibits a three-sided depression called the **femoral triangle** (Figure 47.21a). As shown in Figure 47.19c, the superior border of this triangle is formed by the inguinal ligament, and its two inferior borders are defined by the *sartorius* and *adductor longus* muscles. The large *femoral artery* and *vein* descend vertically through the center of the femoral triangle. To feel the pulse of your femoral artery, press inward just inferior to your midinguinal point (halfway between the anterior superior iliac spine and the pubic tubercle). Be sure to push hard, because the artery lies somewhat deep. By pressing very hard on this point, one can stop the bleeding from a hemorrhage in the lower limb. The femoral triangle also contains most of the *inguinal lymph nodes* (which are easily palpated if swollen).

Figure 47.21 **Anterior surface of the lower limb.** **(a)** Both limbs, with the right limb revealing its medial aspect. The femoral triangle is outlined on the right limb. **(b)** Enlarged view of the left thigh. **(c)** The left knee region. **(d)** The dorsum of the left foot.

Leg and Foot

1. Locate your patella again, then follow the thick *patellar ligament* inferiorly from the patella to its insertion on the superior tibia (Figure 47.21c). Here you can feel a rough projection, the *tibial tuberosity.* Continue running your fingers inferiorly along the tibia's sharp *anterior crest* and its flat *medial surface*—bony landmarks that lie very near the surface throughout their length.

2. Now, return to the superior part of your leg, and palpate the expanded *lateral* and *medial condyles of the tibia* just in-

ferior to the knee. (You can distinguish the tibial condyles from the femoral condyles because you can feel the tibial condyles move with the tibia during knee flexion.) Feel the bulbous *head of the fibula* in the superolateral region of the leg (Figures 47.20 and 47.21c). Try to feel the *common peroneal nerve* (nerve to the anterior leg and foot) where it wraps around the fibula's *neck* just inferior to its head. This nerve is often bumped against the bone here and damaged.

3. In the most distal part of the leg, feel the *lateral malleolus* of the fibula as the lateral prominence of the ankle (Figure 47.21d). Notice that this lies slightly inferior to the *medial*

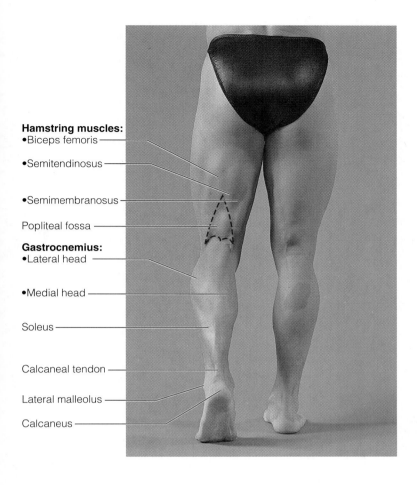

Hamstring muscles:
•Biceps femoris

•Semitendinosus

•Semimembranosus

Popliteal fossa

Gastrocnemius:
•Lateral head

•Medial head

Soleus

Calcaneal tendon

Lateral malleolus

Calcaneus

Figure 47.22 Posterior surface of the lower limb. Notice the diamond-shaped popliteal fossa posterior to the knee.

malleolus of the tibia, which forms the ankle's medial prominence. Place your finger just posterior to the medial malleolus to feel the pulse of your *posterior tibial artery.*

4. On the posterior aspect of the knee is a diamond-shaped hollow called the **popliteal fossa** (Figure 47.22). Palpate the large muscles that define the four borders of this fossa: The *biceps femoris* forming the superolateral border, the *semitendinosus* and *semimembranosus* defining the superomedial border, and the two heads of the *gastrocnemius* forming the inferior border. The *popliteal artery* and *vein* (main vessels to the leg) lie deep within this fossa. To feel a popliteal pulse, flex your leg at the knee and push your fingers firmly into the popliteal fossa. If a physician is unable to feel a patient's popliteal pulse, the femoral artery may be narrowed by atherosclerosis.

5. Next, palpate the main muscle groups of your leg, starting with the calf muscles posteriorly (Figure 47.22). Standing on tiptoes will help you feel the *lateral* and *medial heads of the gastrocnemius* and, inferior to these, the broad *soleus* muscle. Also feel the tension in your *calcaneal (Achilles) tendon* and at the point of insertion of this tendon onto the calcaneus bone of the foot.

6. Return to the anterior surface of the leg, and palpate the *anterior muscle compartment* (Figure 47.21c and d) while alternately dorsiflexing then plantar flexing your foot. You will feel the tibialis anterior and extensor digitorum muscles con-

tracting then relaxing. Then, palpate the *peroneal muscles* that cover most of the fibula laterally (Figure 47.20). The tendons of these muscles pass posterior to the lateral malleolus and can be felt at a point posterior and slightly superior to that malleolus.

7. Observe the dorsum (superior surface) of your foot. You may see the superficial *dorsal venous arch* overlying the proximal part of the metatarsal bones (Figure 47.21d). This arch gives rise to both saphenous veins (the main superficial veins of the lower limb). Visible in lean people, the *great saphenous vein* ascends along the medial side of the entire limb (see Figure 32.8, p. 317). The *small saphenous vein* ascends through the center of the calf.

  As you extend your toes, observe the tendons of the *extensor digitorum longus* and *extensor hallucis longus* muscles on the dorsum of the foot. Finally, place a finger on the extreme proximal part of the space between the first and second metatarsal bones. Here you should be able to feel the pulse of the *dorsalis pedis artery.* ■

Mucus secretion

Microvilli (brush border)

Goblet cells

Underlying connective tissue

Plate 1 Simple columnar epithelium containing unicellular glands—the goblet cells. Secretion of mucus by the goblet cells is obvious in this view (1092X).

Astrocytes

Dendrites

Nucleus

Axon hillock

Axon

Plate 4 Light micrograph of a multipolar neuron (1005X).

Muscle fibers, longitudinal view

Nuclei of muscle fibers

Muscle fibers, cross–sectional view

Plate 2 Skeletal muscle, transverse and longitudinal views shown (576X).

Dendrites

Cell body

Plate 5 Silver-stained Purkinje cells of the cerebellum (2730X).

Branches of axon to motor unit

Axonal terminals at neuromuscular junctions

Muscle fibers

Plate 3 Part of a motor unit (273X).

Cell bodies of unipolar neurons

Satellite cells

Nerve fibers

Plate 6 Dorsal root ganglion displaying unipolar neuron cell bodies surrounded by satellite cells (1092X).

— Fibroblast nucleus

— Myelin sheath

— Schwann cell nucleus

— Nodes of Ranvier

Plate 7 Longitudinal view of myelinated axons (1000X). Myelin sheaths appear "bubbly" because some of the fatty myelin is dissolved during slide preparation.

— Meissner's corpuscle

— Epidermal cells

— Dermal papilla

Plate 10 Meissner's corpuscle in a dermal papilla (2730X).

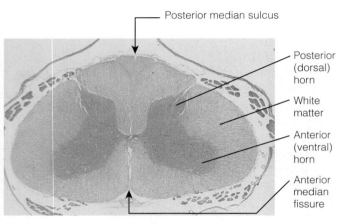

Posterior median sulcus

— Posterior (dorsal) horn

— White matter

— Anterior (ventral) horn

— Anterior median fissure

Plate 8 Adult spinal cord, cross-sectional view (453X).

— Epidermal cells

— Free dendritic endings

— Dermis

Plate 11 Free dendritic endings at the dermal–epidermal junction (251X).

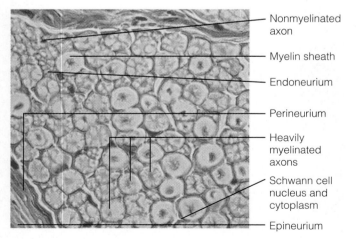

— Nonmyelinated axon

— Myelin sheath

— Endoneurium

— Perineurium

— Heavily myelinated axons

— Schwann cell nucleus and cytoplasm

— Epineurium

Plate 9 Cross section of a portion of a peripheral nerve (560X). Heavily myelinated fibers are identified by a centrally located axon, surrounded by an unstained ring of myelin.

— Pacinian corpuscle

— Dense irregular connective tissue

Plate 12 Pacinian corpuscle in the hypodermis (450X).

Extrafusal muscle fibers

Capsule

Intrafusal fibers of the muscle spindle (receptor)

Plate 13 Longitudinal section of a muscle spindle (382X).

Fibers of the optic nerve

Ganglion cell layer

Nuclei of bipolar neurons

Nuclei of rods and cones

Outer segments of rods and cones

Choroid

Sclera

Plate 14 Structure of the retina of the eye (1092X).

Vestibular membrane

Scala vestibuli

Hair (receptor) cells

Tectorial membrane

Afferent fibers of the cochlear nerve

Scala tympani

Basilar membrane

Plate 15 The organ of Corti (109X).

Trough between adjacent papillae

Taste buds

Foliate papillae

Plate 16 Location of the taste buds on the lateral aspects of foliate papillae of the tongue (130X).

Lamina propria containing mucus secreting glands

Basal cell nucleus

Supporting cell nucleus

Olfactory cell nucleus

Cilia of olfactory receptor cells

Lumen of nasal cavity

Plate 17 Olfactory epithelium. From the lamina propria to the nasal cavity, the general arrangement of cells in this pseudostratified epithelium is basal cells, olfactory receptor cells, and supporting cells (560X).

Plate 18 The thyroid gland (355X).

Follicle cells

Colloid–filled follicles

Blood vessel

Plate 19 Parathyroid gland tissue (560X).

Oxyphil cells

Chief cells

Plate 20 Pancreatic islet stained differentially to allow identification of the glucagon-secreting alpha cells and the insulin-secreting beta cells (206X).

Exocrine (acinar) tissue of the pancreas

Alpha cells

Beta cells

Acidophils

Chromophobes

Basophils

Plate 21 Anterior pituitary gland. Differential staining allows the acidophils, basophils, and chromophobes to be distinguished (560X).

Nuclei of pituicytes

Fenestrated capillaries

Plate 22 Posterior pituitary. The axons of neurosecretory cells are indistinguishable from the cytoplasm of the pituicytes (449X).

Capsule (torn)

Zona glomerulosa

Zona fasciculata

Zona reticularis

Medulla cells

Medullary vein

Plate 23 Histologically distinct regions of the adrenal gland (214X).

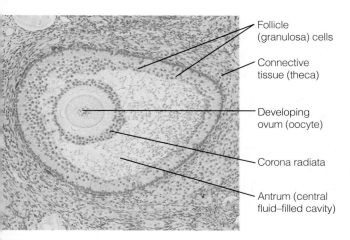

Follicle (granulosa) cells

Connective tissue (theca)

Developing ovum (oocyte)

Corona radiata

Antrum (central fluid–filled cavity)

Plate 24 A vesicular follicle of the ovary (108X).

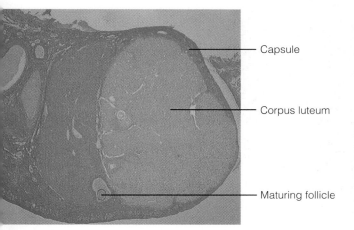

Capsule

Corpus luteum

Maturing follicle

Plate 25 The glandular corpus luteum of an ovary (55X).

Cross sections of nerves

Medium vein

Tunica media

Adventitia

External elastic lamina

Small artery

Plate 26 Cross-sectional view of a small artery, a medium vein, and nerves (273X).

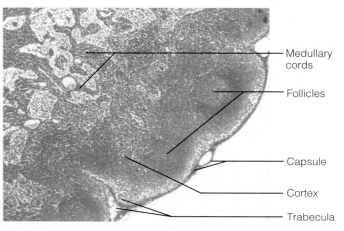

Medullary cords

Follicles

Capsule

Cortex

Trabecula

Plate 27 Main structural features of a lymph node (68X).

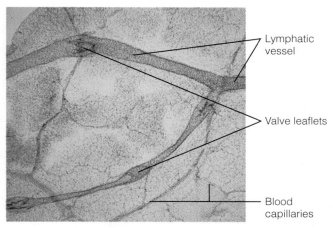

Lymphatic vessel

Valve leaflets

Blood capillaries

Plate 28 Lymphatic vessels and blood capillaries. The valves in the lymphatic vessels are clearly visible (109X).

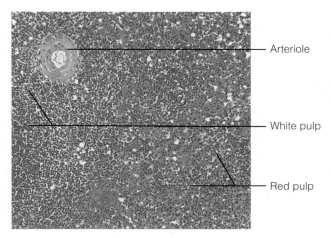

Arteriole

White pulp

Red pulp

Plate 29 Microscopic anatomy of a portion of the spleen showing the red and white pulp regions (273X).

Plate 30 Histology of a palatine tonsil. The luminal surface is covered with epithelium that invaginates deeply to form crypts (50X).

Tonsil

Crypt

Alveolar sacs

Alveolar duct

Alveoli

Plate 31 Photomicrograph of part of the lung showing alveoli and alveolar ducts and sacs (36X).

Hyaline cartilage ring

Goblet cells

Ciliated pseudostratified epithelium

Seromucous glands

Lamina propria

Plate 32 Cross section through the trachea showing the pseudostratified ciliated epithelium, glands, and part of the supporting ring of hyaline cartilage (155X).

Hyaline cartilage ring

Lumen of trachea

Tracheal glands

Trachealis muscle

Follicles of thyroid gland

Lumen of esophagus

Stratified squamous epithelium

Submucosa

Muscularis

Plate 33 Cross section through the trachea and esophagus (124X).

Alveolar sacs

Smooth muscle layer

Ciliated columnar epithelium

Lumen

Lamina propria

Plate 34 Bronchiole, cross-sectional view (109X).

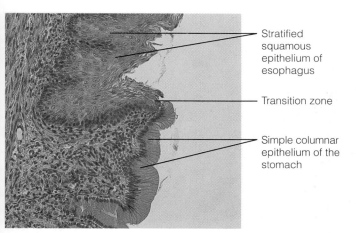

Stratified squamous epithelium of esophagus

Transition zone

Simple columnar epithelium of the stomach

Plate 35   Gastroesophageal junction showing the meeting of the simple columnar epithelium of the stomach and the stratified squamous epithelium of the esophagus (633X).

Simple columnar epithelium

Lamina propria

Gastric pit

Gastric glands

Plate 37   Detailed structure of the gastric glands and pits (218X).

Muscularis externa
- Longitudinal layer
- Circular layer
- Oblique layer

Gastric glands

Muscularis mucosae

Submucosa

Mucosa

Plate 36   Stomach. Low-power longitudinal view through its wall showing four tunics (124X).

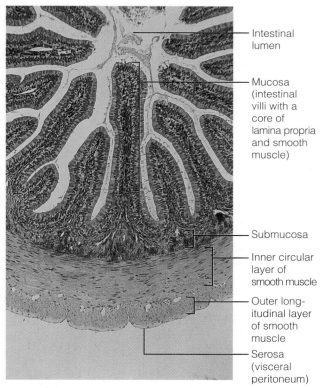

Intestinal lumen

Mucosa (intestinal villi with a core of lamina propria and smooth muscle)

Submucosa

Inner circular layer of smooth muscle

Outer longitudinal layer of smooth muscle

Serosa (visceral peritoneum)

Plate 38   Cross section through the wall of the small intestine showing the arrangement of its layers or tunics (129X). The villi of its mucosa are large and obvious.

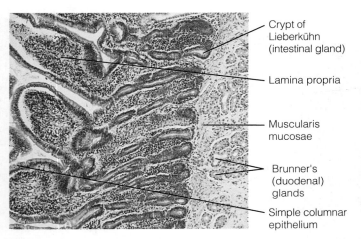

Crypt of Lieberkühn (intestinal gland)

Lamina propria

Muscularis mucosae

Brunner's (duodenal) glands

Simple columnar epithelium

Plate 39 Cross-sectional view of the duodenum showing villi and duodenal glands (273X).

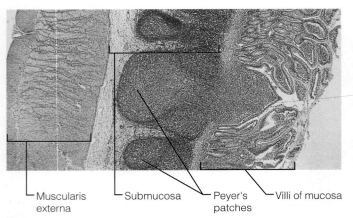

Muscularis externa

Submucosa

Peyer's patches

Villi of mucosa

Plate 40 Cross section through the ileum, showing Peyer's patches (109X).

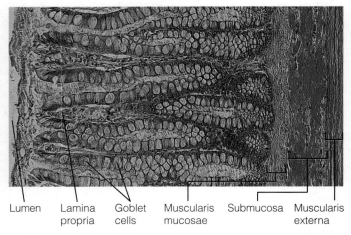

Lumen

Lamina propria

Goblet cells

Muscularis mucosae

Submucosa

Muscularis externa

Plate 41 Large intestine. Cross-sectional view showing the abundant goblet cells of the mucosa (546X).

Mucous cells

Serous demilunes

Duct

Plate 42 Sublingual salivary glands (361X).

Acinar (exocrine) tissue

Islets (endocrine tissue)

Connective tissue septa

Plate 43 Pancreas tissue. Exocrine and endocrine (islets) areas clearly visible (110X).

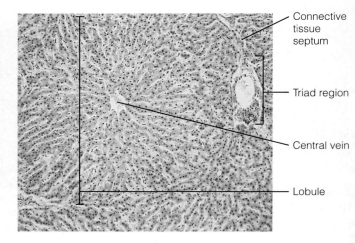

Connective tissue septum

Triad region

Central vein

Lobule

Plate 44 Pig liver. Structure of the liver lobules (64X).

Plate 45 Liver stained to show the location of the phagocytic cells (Kupffer cells) lining the sinusoids (255X).

Sinusoids

Kupffer cells containing dark deposits

Liver parenchyma (hepatocytes)

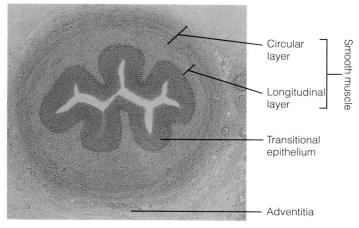

Plate 48 Cross section of the ureter (43X).

Circular layer

Longitudinal layer

Smooth muscle

Transitional epithelium

Adventitia

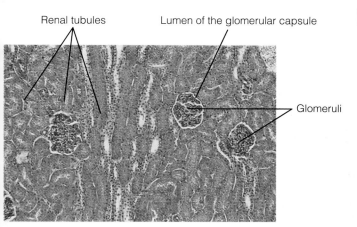

Plate 46 Renal cortex of the kidney (86X).

Renal tubules

Lumen of the glomerular capsule

Glomeruli

Plate 49 Cross section of the epididymis (1092X). Stereocilia of the epithelial lining are obvious.

Pseudostratified columnar epithelium

Spermatozoa

Stereocilia

Connective tissue

Plate 47 Detailed structure of a glomerulus (337X).

Cuboidal epithelium of the renal tubule

Lumen of the glomerular capsule

Glomerulus

Juxtaglomerular cells

Parietal layer of the glomerular capsule

Macula densa

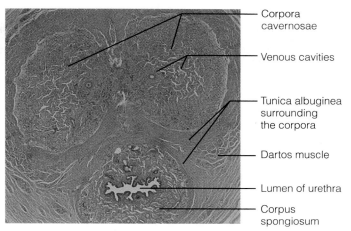

Plate 50 Penis, transverse section (51X).

Corpora cavernosae

Venous cavities

Tunica albuginea surrounding the corpora

Dartos muscle

Lumen of urethra

Corpus spongiosum

Immature sperm in lumen

Spermatogenic cells in tubule wall

Cytoplasm of Sertoli cell

Areolar connective tissue containing interstitial cells

Plate 51 Cross section of a portion of a seminiferous tubule (36X).

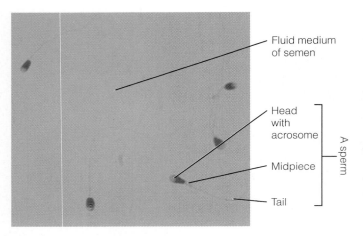

Fluid medium of semen

Head with acrosome

Midpiece

Tail

A sperm

Plate 52 Semen, the product of ejaculation, consisting of sperm and fluids secreted by the accessory glands (particularly the prostate and seminal vesicles) (332X).

Serosa

Smooth muscle

Highly folded mucosa

Lumen

Plate 53 Cross-sectional view of the uterine tube (523X).

Medulla

Primary follicle

Oocyte

Granulosa cells

Cortex of ovary

Antrum of vesicular (Graafian) follicle

Germinal epithelium

Plate 54 The ovary showing its follicles in various stages of development (104X).

Inactive alveoli

Plate 55 Breast alveoli of a non-pregnant woman (273X).

Active (secreting) alveoli

Plate 56 Breast alveoli (lactating) of a pregnant woman (273X).

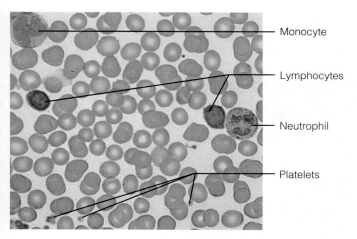

Plate 57   Human blood smear (656X).

Plate 60   A monocyte surrounded by erythrocytes (842X).

Plate 58   Two neutrophils surrounded by erythrocytes (842X).

Plate 61   An eosinophil surrounded by erythrocytes (842X).

Plate 59   A lymphocyte surrounded by erythrocytes (842X).

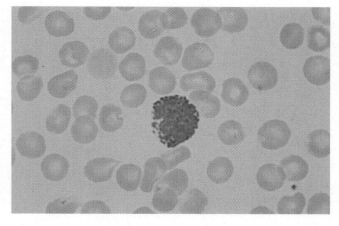

Plate 62   A basophil surrounded by erythrocytes (842X).

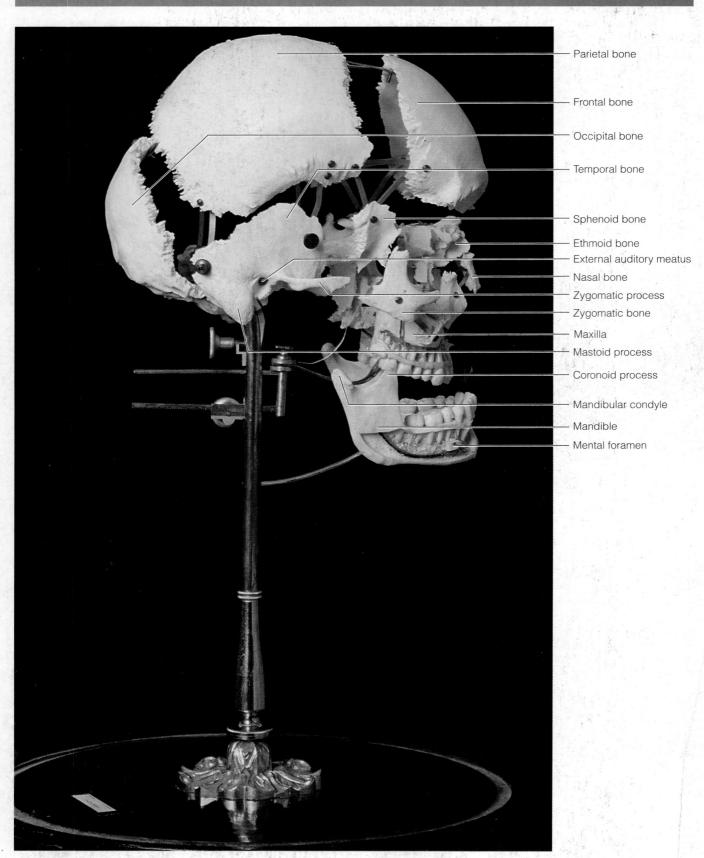

- Parietal bone
- Frontal bone
- Occipital bone
- Temporal bone
- Sphenoid bone
- Ethmoid bone
- External auditory meatus
- Nasal bone
- Zygomatic process
- Zygomatic bone
- Maxilla
- Mastoid process
- Coronoid process
- Mandibular condyle
- Mandible
- Mental foramen

Plate A Beauchene skull, lateral view.

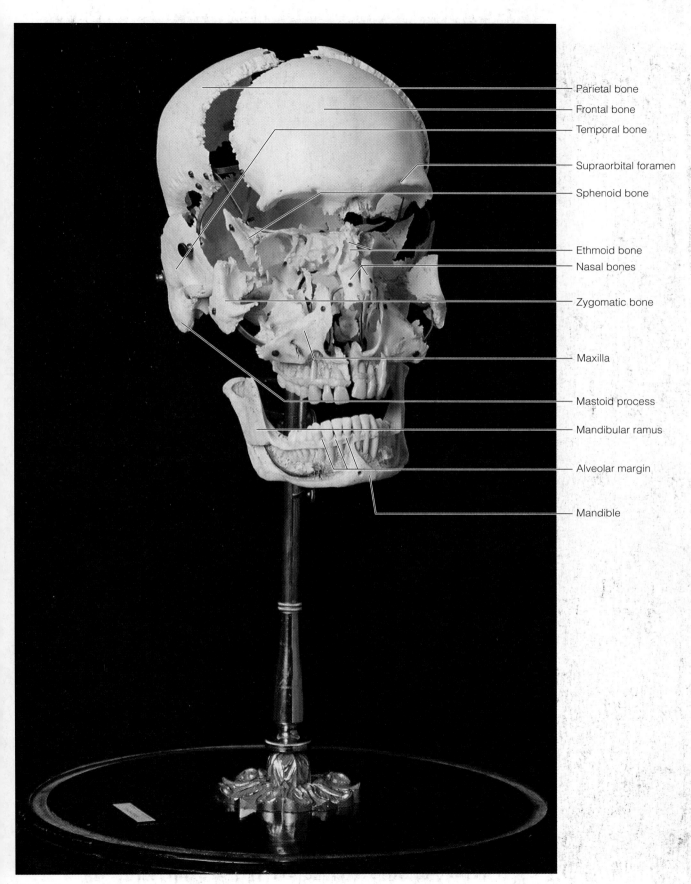

Parietal bone

Frontal bone

Temporal bone

Supraorbital foramen

Sphenoid bone

Ethmoid bone

Nasal bones

Zygomatic bone

Maxilla

Mastoid process

Mandibular ramus

Alveolar margin

Mandible

Plate B   Beauchene skull, frontal view.

Longitudinal fissure

Olfactory tract

Optic chiasma

Pons

Medulla

Spinal column

Cerebellum

Occipital pole

Frontal pole

Lateral fissure

Inferior temporal gyrus

Uncus

Glossopharyngeal (IX), vagus (X), and accessory (XI) nerves

Arachnoid mater

Transverse cerebral fissure

Plate C   Ventral view of the brain.

Anterior cerebral
artery

Middle cerebral
artery

Internal carotid
artery

Posterior
communicating
artery

Posterior cerebral
artery

Basilar artery

Vertebral
artery

Olfactory tract

Anterior
communicating artery

Optic chiasma

Optic tract

Infundibulum

Mammillary body

Cerebral peduncle

Pons

Medulla
oblongata

Plate D Circle of Willis.

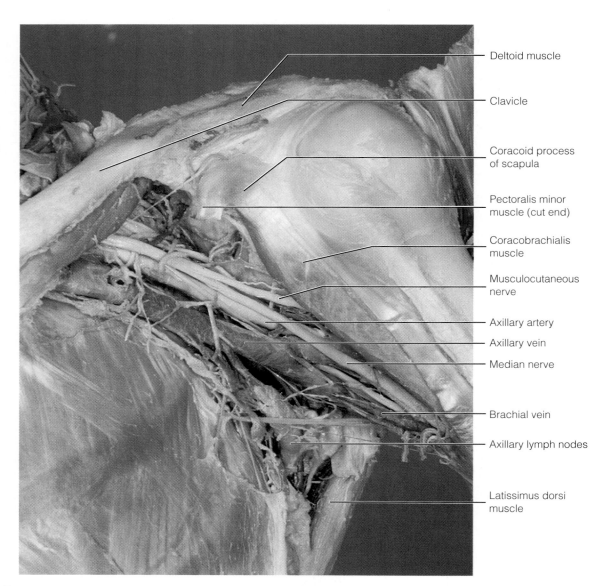

Deltoid muscle

Clavicle

Coracoid process
of scapula

Pectoralis minor
muscle (cut end)

Coracobrachialis
muscle

Musculocutaneous
nerve

Axillary artery

Axillary vein

Median nerve

Brachial vein

Axillary lymph nodes

Latissimus dorsi
muscle

Plate E   Brachial plexus and axilla.

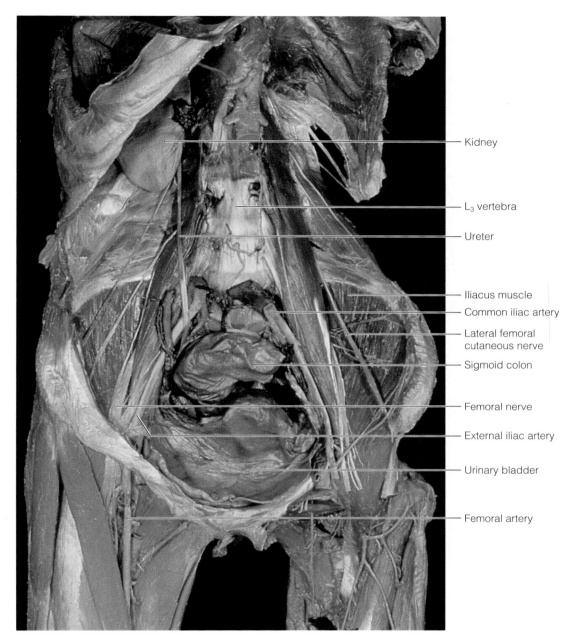

Kidney

L₃ vertebra

Ureter

Iliacus muscle

Common iliac artery

Lateral femoral
cutaneous nerve

Sigmoid colon

Femoral nerve

External iliac artery

Urinary bladder

Femoral artery

Plate F Lumbar plexus.

Sciatic nerve

Ischial tuberosity

Gracilis muscle

Muscular branch of
common peroneal
nerve (to short head of
biceps femoris)

Biceps femoris muscle

Tibial nerve

Popliteal artery

Popliteal fossa

Gastrocnemius muscle

Plate G Course of sciatic nerve along posterior thigh
and knee.

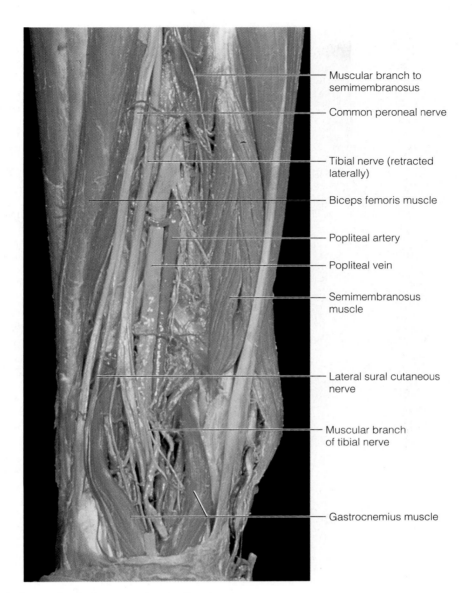

Muscular branch to
semimembranosus

Common peroneal nerve

Tibial nerve (retracted
laterally)

Biceps femoris muscle

Popliteal artery

Popliteal vein

Semimembranosus
muscle

Lateral sural cutaneous
nerve

Muscular branch
of tibial nerve

Gastrocnemius muscle

Plate H   Deep relations of sciatic nerve in popliteal
fossa.

Left subclavian
artery and vein

First rib

Internal thoracic
artery and vein

Pulmonary trunk

Left pulmonary
veins

Left phrenic
nerve

Left ventricle

Pericardial
cavity

Pericardium

Left subclavian
artery

Left vagus
nerve

Left pulmonary
artery

Left primary
bronchus

Vertebral
body

Accessory
hemiazygos
vein

Thoracic
sympathetic
ganglion

Sympathetic trunk

Hemiazygos vein

Thoracic aorta

Diaphragm

Plate I Lateral view of mediastinum.

The Language of Anatomy

Surface Anatomy

1. Match each of the following descriptions with a key equivalent, and record the key letter or term in front of the description.

   Key: a. buccal   c. deltoid   e. patellar
         b. calcaneal   d. digital   f. scapular

   _____ 1. cheek   _____ 4. anterior aspect of knee

   _____ 2. pertaining to the fingers _____ 5. heel of foot

   _____ 3. shoulder blade region   _____ 6. curve of shoulder

2. Indicate the following body areas on the accompanying diagram by placing the correct key letter at the end of each line.

   Key:

   a. abdominal
   b. antecubital
   c. axillary
   d. brachial
   e. cervical
   f. femoral
   g. gluteal
   h. inguinal
   i. lumbar
   j. occipital
   k. oral
   l. popliteal
   m. pubic
   n. sural
   o. thoracic
   p. umbilical

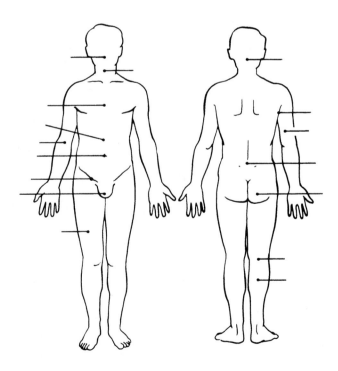

3. Classify each of the surface anatomy terms in the key of question 2 above, into one of the body regions indicated below. Insert the appropriate key letters on the answer blanks.

   _____ 1. Appendicular   _____ 2. Axial

Body Orientation, Direction, Planes, and Sections

1. Describe completely the standard human anatomical position. _____

   _____

2. Define *section:* _____

3. Several incomplete statements are listed below. Correctly complete each statement by choosing the appropriate anatomical term from the key. Record the key letters and/or terms on the correspondingly numbered blanks below.

   Key: a. anterior   e. lateral   i. sagittal
         b. distal   f. medial   j. superior
         c. frontal   g. posterior   k. transverse
         d. inferior   h. proximal

   In the anatomical position, the face and palms are on the __1__ body surface; the buttocks and shoulder blades are on the __2__ body surface; and the top of the head is the most __3__ part of the body. The ears are __4__ and __4__ to the shoulders and __5__ to the nose. The heart is __6__ to the vertebral column (spine) and __7__ to the lungs. The elbow is __8__ to the fingers but __9__ to the shoulder. The abdominopelvic cavity is __10__ to the thoracic cavity and __11__ to the spinal cavity. In humans, the dorsal surface can also be called the __12__ surface; however, in quadruped animals, the dorsal surface is the __13__ surface.

   If an incision cuts the heart into right and left parts, the section is a __14__ section; but if the heart is cut so that superior and inferior portions result, the section is a __15__ section. You are told to cut a dissection animal along two planes so that the kidneys are observable in both sections. The two sections that meet this requirement are the __16__ and __17__ sections.

   1. _____   7. _____   13. _____

   2. _____   8. _____   14. _____

   3. _____   9. _____   15. _____

   4. _____   10. _____   16. _____

   5. _____   11. _____   17. _____

   6. _____   12. _____

4. Correctly identify each of the body planes by inserting the appropriate term for each on the answer line below the drawing.

   1. _____   2. _____   3. _____

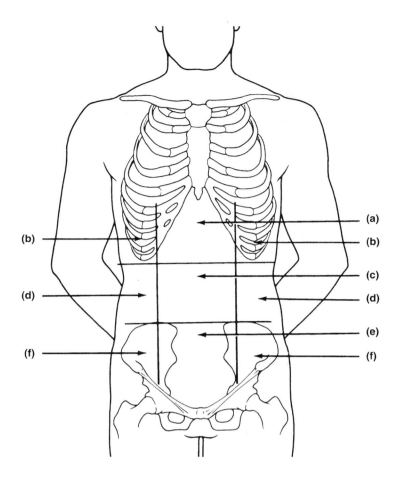

5. Correctly identify each of the nine areas of the abdominal surface by inserting the appropriate term for each of the letters indicated in the drawing just above.

a. _____   d. _____

b. _____   e. _____

c. _____   f. _____

Body Cavities

1. Which body cavity would have to be opened for the following types of surgery? (Insert letter of key choice in same-numbered blank. More than one choice may apply.)

Key: a. abdominopelvic   c. dorsal   e. thoracic
      b. cranial   d. spinal   f. ventral

1. surgery to remove a cancerous lung lobe
2. removal of the uterus or womb
3. removal of a brain tumor
4. appendectomy
5. stomach ulcer operation

     The abdominopelvic and thoracic cavities are subdivisions of the __6__ body cavity, while the cranial and spinal cavities are subdivisions of the __7__ body cavity. The __8__ body cavity is totally surrounded by bone, and thus affords its contained structures very good protection.

1. _____

2. _____

3. _____

4. _____

5. _____

6. _____

7. _____

8. _____

2. Name the serous membranes covering the lungs (#1), the heart (#2), and the organs of the abdominopelvic cavity (#3), and insert your responses in the blanks on the right.

1. _____

2. _____

3. _____

3. Name the muscle that subdivides the ventral body cavity. _____

4. Which of the following organ systems are represented in all three subdivisions of the ventral body cavity? (Circle all appropriate responses.)

respiratory	circulatory	reproductive	lymphatic
nervous	excretory (urinary)	muscular	integumentary

5. Which organ system would not be represented in any of the body cavities? _____

6. What are the bony landmarks of the abdominopelvic cavity? _____

_____

7. Which body cavity affords the least protection to its internal structures? _____

8. What is the function of the serous membranes of the body? _____

_____

_____

9. A nurse informs you that she is about to take blood from the antecubital region. What portion of your body should you present to her? _____

10. What do peritonitis, pleurisy, and pericarditis (pathologic conditions) have in common? _____

_____

11. Why are these conditions accompanied by a great deal of pain? _____

_____

12. The mouth, or buccal cavity, and its extension, which stretches through the body inside the digestive system, is not listed as an internal body cavity. Why is this so? _____

_____

_____

Organ Systems Overview

1. Use the key below to indicate the body systems that perform the following functions for the body.

Key: a. cardiovascular   d. integumentary   g. nervous   j. skeletal
    b. digestive   e. lymphatic   h. reproductive   k. urinary
    c. endocrine   f. muscular   i. respiratory

_____ 1. rids the body of nitrogen-containing wastes

_____ 2. is affected by removal of the thyroid gland

_____ 3. provides support and levers on which the muscular system acts

_____ 4. includes the heart

_____ 5. causes the onset of the menstrual cycle

_____ 6. protects underlying organs from drying out and from mechanical damage

_____ 7. protects the body; destroys bacteria and tumor cells

_____ 8. breaks down ingested food into its building blocks

_____ 9. removes carbon dioxide from the blood

_____ 10. delivers oxygen and nutrients to the tissues

_____ 11. moves the limbs; facilitates facial expression

_____ 12. conserves body water or eliminates excesses

_____ and _____ 13. facilitate conception and childbearing

_____ 14. controls the body by means of chemical molecules called hormones

_____ 15. is damaged when you cut your finger or get a severe sunburn

2. Using the above key, choose the *organ system* to which each of the following sets of organs or body structures belong:

_____ 1. thymus, spleen,
lymphatic vessels

_____ 2. bones, cartilages,
tendons

_____ 3. pancreas, pituitary,
adrenals

_____ 4. trachea, bronchi, alveoli

_____ 5. kidneys, bladder,
ureters

_____ 6. testis, vas deferens,
urethra

_____ 7. esophagus, large
intestine, rectum

_____ 8. arteries, veins, heart

3. Using the key below, place the following organs in their proper body cavity.

Key: a. abdominopelvic   b. cranial   c. spinal   d. thoracic

_____ 1. stomach   _____ 6. urinary bladder

_____ 2. esophagus   _____ 7. heart

_____ 3. large intestine   _____ 8. trachea

_____ 4. liver   _____ 9. brain

_____ 5. spinal cord   _____ 10. rectum

4. Using the organs listed in item 3 above, record, by number, which would be found in the abdominal regions listed below:

_____ 1. hypogastric region   _____ 4. epigastric region

_____ 2. right lumbar region   _____ 5. left iliac region

_____ 3. umbilical region   _____ 6. left hypochondriac region

5. The five levels of organization of a living body are cell, _____,

_____, _____, and organism.

6. Define *organ*. _____

_____

_____

7. Using the terms provided, correctly identify all of the body organs provided with leader lines in the drawings shown below. Then name the organ systems by entering the name of each on the answer blank below each drawing.

Blood vessels   Heart   Nerves   Spinal cord   Urethra
Brain   Kidney   Sensory receptor   Ureter   Urinary bladder

1. _____   2. _____   3. _____

The Microscope

Care and Structure of the Compound Microscope

1. Label all indicated parts of the microscope.

2. The following statements are true or false. If true, write *T* on the answer blank. If false, correct the statement by writing on the blank the proper word or phrase to replace the one that is underlined.

_____ 1. The microscope lens may be cleaned with any soft tissue.

_____ 2. The coarse adjustment knob may be used in focusing with all three objectives.

_____ 3. The microscope should be stored with the oil immersion lens in position over the stage.

_____ 4. When beginning to focus, the low-power lens should be used.

_____ 5. In low power, always focus toward the specimen.

_____ 6. A coverslip should always be used with the high-power and oil lenses.

_____ 7. The greater the amount of light delivered to the objective lens, the less the resolution.

3. Match the microscope structures given in column B with the statements in column A that identify or describe them.

Column A

_____ 1. platform on which the slide rests for viewing

_____ 2. lens located at the superior end of the body tube

_____ 3. secure(s) the slide to the stage

_____ 4. delivers a concentrated beam of light to the specimen

_____ 5. used for precise focusing once initial focusing has been done

_____ 6. carries the objective lenses; rotates so that the different objective lenses can be brought into position over the specimen

_____ 7. used to increase the amount of light passing through the specimen

Column B

a. coarse adjustment knob

b. condenser

c. fine adjustment knob

d. iris diaphragm

e. mechanical stage or spring clips

f. movable nosepiece

g. objective lenses

h. ocular

i. stage

4. Explain the proper technique for transporting the microscope.

_____

_____

5. Define the following terms.

real image: _____

_____

virtual image: _____

_____

6. Define _total magnification:_ _____

_____

7. Define _resolution:_ _____

Viewing Objects Through the Microscope

1. Complete, or respond to, the following statements:

_____ The distance from the bottom of the objective lens in use to the specimen is called the _____.

_____ The resolution of the human eye is approximately _____ μm.

_____ The area of the specimen seen when looking through the microscope is the _____.

_____ If a microscope has a 10× ocular and the total magnification at a particular time is 950×, the objective lens in use at that time is _____ ×.

_____ Why should the light be dimmed when looking at living (nearly transparent) cells?

_____ If, after focusing in low power, only the fine adjustment need be used to focus the specimen at the higher powers, the microscope is said to be _____.

_____ If, when using a 10× ocular and a 15× objective, the field size is 1.5 mm, the approximate field size with a 30× objective is _____ mm.

_____ If the size of the high-power field is 1.2 mm, an object that occupies approximately a third of that field has an estimated diameter of _____ mm.

_____ Assume there is an object on the left side of the field that you want to bring to the center (that is, toward the apparent right). In what direction would you move your slide?

_____ If the object is in the top of the field and you want to move it downward to the center, you would move the slide _____.

2. You have been asked to prepare a slide with the letter *k* on it (as below). In the circle below, draw the *k* as seen in the low-power field.

k

3. The numbers for the field sizes below are too large to represent the typical compound microscope lens system, but the relationships depicted are accurate. Figure out the magnification of fields 1 and 3, and the field size of 2. (*Hint:* Use your ruler.)

5 mm
1. →○←
_____ ×

_____ mm
2. →○←
100×

0.5 mm
3. →∘←
_____ ×

4. Say you are observing an object in the low-power field. When you switch to high-power, it is no longer in your field of view.

Why might this occur? _____

_____

What should be done initially to prevent this from happening? _____

_____

5. Do the following factors increase or decrease as one moves to higher magnifications with the microscope?

   resolution _____   amount of light needed _____

   working distance _____   depth of field _____

6. A student has the high-dry lens in position and appears to be intently observing the specimen. The instructor, noting a working distance of about 1 cm, knows the student isn't actually seeing the specimen.

   How so? _____

7. Why is it important to be able to use your microscope to perceive depth when studying slides?

   _____

   _____

8. If you are observing a slide of tissue two cell layers thick, how can you determine which layer is superior?

   _____

   _____

   _____

9. Describe the proper procedure for preparing a wet mount.

   _____

   _____

   _____

10. Give two reasons why the light should be dimmed when viewing living or unstained material.

    _____

    _____

11. Indicate the probable cause of the following situations arising during use of a microscope.

    a. Only half of the field is illuminated: _____

    _____

    b. Field does not change as mechanical stage is moved: _____

    _____

The Cell—Anatomy and Division

Anatomy of the Composite Cell

1. Define the following:

organelle: _____

_____

cell: _____

2. Although cells have differences that reflect their specific functions in the body, what functional capabilities do all cells

exhibit? _____

3. Identify the following cell parts:

_____ 1. external boundary of cell; regulates flow of materials into and out of the cell; site of cell signaling

_____ 2. contains digestive enzymes of many varieties; "suicide sac" of the cell

_____ 3. scattered throughout the cell; major site of ATP synthesis

_____ 4. slender extensions of the plasma membrane that increase its surface area

_____ 5. stored glycogen granules, crystals, pigments, and so on

_____ 6. membranous system consisting of flattened sacs and vesicles; packages proteins for export

_____ 7. control center of the cell; necessary for cell division and cell life

_____ 8. two rod-shaped bodies near the nucleus; direct formation of the mitotic spindle

_____ 9. dense, darkly staining nuclear body; packaging site for ribosomes

_____ 10. contractile elements of the cytoskeleton

_____ 11. membranous system; involved in intracellular transport of proteins and synthesis of membrane lipids

_____ 12. attached to membrane systems or scattered in the cytoplasm; synthesize proteins

_____ 13. threadlike structures in the nucleus; contain genetic material (DNA)

_____ 14. site of free radical detoxification

4. In the following diagram, label all parts provided with a leader line.

Differences and Similarities in Cell Structure

1. For each of the following cell types, list (a) *one* important structural characteristic observed in the laboratory, and (b) the function that the structure complements or ensures.

squamous epithelium   a. _____

                           b. _____

sperm   a. _____

                           b. _____

smooth muscle   a. _____

                           b. _____

red blood cells   a. _____

                      b. _____

2. What is the significance of the red blood cell being anucleate (without a nucleus)? _____

_____

Did it ever have a nucleus? _____ When? _____

Cell Division: Mitosis and Cytokinesis

1. Identify the three phases of mitosis in the following photomicrographs.

1. _____

2. _____

3. _____

2. What is the importance of mitotic cell division? _____

_____

3. Complete or respond to the following statements:

Division of the __1__ is referred to as mitosis. Cytokinesis is division of the __2__. The major structural difference between chromatin and chromosomes is that the latter is __3__. Chromosomes attach to the spindle fibers by undivided structures called __4__. If a cell undergoes mitosis but not cytokinesis, the product is __5__. The structure that acts as a scaffolding for chromosomal attachment and movement is called the __6__. __7__ is the period of cell life when the cell is not involved in division. Two cell populations in the body that do not undergo cell division are __8__ and __9__. The implication of an inability of a cell population to divide is that when some of its members die, they are replaced by __10__.

1. _____

2. _____

3. _____

4. _____

5. _____

6. _____

7. _____

8. _____

9. _____

10. _____

4. Using the key, categorize each of the events described below according to the phase in which it occurs.

Key: a. prophase   b. anaphase   c. telophase   d. metaphase   e. none of these

_____ 1. Chromatin coils and condenses, forming chromosomes.

_____ 2. The chromosomes (chromatids) are V-shaped.

_____ 3. The nuclear membrane re-forms.

_____ 4. Chromosomes stop moving toward the poles.

_____ 5. Chromosomes line up in the center of the cell.

_____ 6. The nuclear membrane fragments.

_____ 7. The mitotic spindle forms.

_____ 8. DNA synthesis occurs.

_____ 9. Centrioles replicate.

_____ 10. Chromosomes first appear to be duplex structures.

_____ 11. Chromosomal centromeres are attached to the kinetochore fibers.

_____ 12. Cleavage furrow forms.

_____ 13. The nuclear membrane(s) is absent.

5. What is the physical advantage of the chromatin coiling and condensing to form short chromosomes at the onset of mitosis?

_____

_____

Cell Transport Mechanisms and Permeability: Wet Lab

Choose all answers that apply to items 1 and 2, and place their letters on the response blanks to the right.

1. Molecular motion _____

   a. reflects the kinetic energy of molecules
   b. reflects the potential energy of molecules
   c. is ordered and predictable
   d. is random and erratic

2. Kinetic energy _____

   a. is higher in larger molecules
   b. is lower in larger molecules
   c. increases with increasing temperature
   d. decreases with increasing temperature
   e. is reflected in the speed of molecular movement

3. The following refer to the laboratory experiment using dialysis sacs 1 through 4 to study diffusion through nonliving membranes:

   Sac 1: 40% glucose suspended in distilled water

   Did glucose pass out of the sac? _____

   Test used to determine presence of glucose: _____

   Did the sac weight change? _____

   If so, explain the reason for its weight change: _____

   _____

   _____

   Sac 2: 40% glucose suspended in 40% glucose

   Was there net movement of glucose in either direction? _____

   Explanation: _____

   Did the sac weight change? _____   Explanation: _____

   _____

   Sac 3: 10% NaCl in distilled water

   Was there net movement of NaCl out of the sac? _____

   Test used to determine the presence of NaCl: _____

   _____

   Direction of net osmosis: _____

Sac 4: Boiled starch in distilled water

Was there net movement of starch out of the sac? _____

Test used to determine presence of starch: _____

Direction of net osmosis: _____

4. What single characteristic of the semipermeable membranes used in the laboratory determines the substances that can pass

through them? _____

In addition to this characteristic, what other factors influence the passage of substances through living membranes?

_____

_____

5. A semipermeable sac containing 4% NaCl, 9% glucose, and 10% albumin is suspended in a solution with the following composition: 10% NaCl, 10% glucose, and 40% albumin. Assume that the sac is permeable to all substances except albumin.
State whether each of the following will (a) move into the sac, (b) move out of the sac, or (c) not move.

glucose _____   albumin _____

water _____   NaCl _____

6. The diagrams below represent three microscope fields containing red blood cells. Arrows show the direction of net osmosis.

Which field contains a hypertonic solution? _____ The cells in this field are said to be _____.

Which field contains an isotonic bathing solution? _____ Which field contains a hypotonic solution? _____ What is

happening to the cells in this field? _____

_____

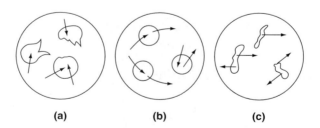

(a)   (b)   (c)

7. Assume you are conducting the experiment illustrated below. Both hydrochloric acid (HCl) with a molecular weight of about
36.5 and ammonium hydroxide (NH_4OH) with a molecular weight of 35 are volatile and easily enter the gaseous state. When
they meet, the following reaction will occur:

$$HCl + NH_4OH \rightarrow H_2O + NH_4Cl$$

Ammonium chloride (NH$_4$Cl) will be deposited on the glass tubing as a smoky precipitate where the two gases meet. Predict which gas will diffuse more quickly and indicate to which end of the tube the smoky precipitate will be closer.

a. The faster diffusing gas is _____.

b. The precipitate forms closer to the _____ end.

Rubber stopper   Cotton wad with HCl   Cotton wad with NH$_4$OH

Support

8. When food is pickled for human consumption, as much water as possible is removed from the food. What method is used to

achieve this dehydrating effect? _____

_____

9. What determines whether a transport process is active or passive? _____

_____

10. Characterize passive and active transport as fully as possible by choosing all the phrases that apply and inserting their letters on the answer blanks.

Passive transport: _____   Active transport: _____

a. accounts for the movement of fats and respiratory gases through the plasma membrane
b. explains solute pumping, phagocytosis, and pinocytosis
c. includes osmosis, simple diffusion, and filtration
d. may occur against concentration and/or electrical gradients
e. uses hydrostatic pressure or molecular energy as the driving force
f. moves ions, amino acids, and some sugars across the plasma membrane

11. For the osmometer demonstration (#2), explain why the level of the water column rose during the laboratory session.

_____

_____

_____

_____

_____

12. Define the following terms:

diffusion: _____

_____

_____

osmosis: _____

_____

_____

simple diffusion: _____

_____

_____

filtration: _____

_____

_____

active transport: _____

_____

_____

phagocytosis: _____

_____

_____

pinocytosis: _____

_____

_____

Cell Transport Mechanisms and Permeability: Computer Simulation

Choose all answers that apply to items 1 and 2, and place their letters on the response blanks to the right.

1. Differential permeability: _____

   a. is also called selective permeability
   b. refers to the ability of the plasma membrane to select what passes through it
   c. implies that all substances pass through membranes without hindrance
   d. keeps wastes inside the cell and nutrients outside the cell

2. Passive transport includes: _____

   a. osmosis   b. simple diffusion   c. phagocytosis   d. pinocytosis   e. facilitated diffusion

3. The following refer to the dialysis simulation.

   Did the 20 MWCO membrane exclude any solute(s)? _____

   Which solute(s) passed through the 100 MWCO membrane? _____

   Which solute exhibited the highest diffusion rate through the 100 MWCO membrane? _____

   Explain why this is so: _____

4. The following refer to the facilitated diffusion simulation.

   Are substances able to travel against their concentration gradient? _____

   Name two ways to increase the rate of glucose transport. _____

   Did NaCl affect glucose transport? _____

   Does NaCl require a transport protein for diffusion? _____

5. The following refer to the osmosis simulation.

   Does osmosis require energy? _____

   Is water excluded by any of the dialysis membranes? _____

   Is osmotic pressure generated if solutes freely diffuse? _____

   Explain how solute concentration affects osmotic pressure. _____

   _____

6. The following refer to the filtration simulation.

What does the simulated filtration membrane represent in a living organism? _____

_____

What characteristic of a solute determines whether or not it passes through a filtration membrane?

_____

Would filtration occur if we equalized the pressure on both sides of a filtration membrane? _____

7. The following questions refer to the active transport simulation.

Does the presence of glucose carrier proteins affect Na^+ transport? _____

Can Na^+ be transported against its concentration gradient? _____

Are Na^+ and K^+ transported in the same direction? _____

The ratio of Na^+ to K^+ transport is _____ Na^+ transported out of the cell for every _____ K^+ transported into the cell.

8. What single characteristic of the semipermeable membranes used in the simple diffusion and filtration experiments deter-

mines which substances pass through them? _____

In addition to this characteristic, what other factors influence the passage of substances through living membranes?

_____

9. Assume the left beaker contains 4 mM NaCl, 9 mM glucose, and 10 mM albumin. The right beaker contains 10 mM NaCl, 10 mM glucose, and 40 mM albumin. Furthermore, the dialysis membrane is permeable to all substances except albumin. State whether the substance will move (a) to the right beaker, (b) to the left beaker, or (c) not move.

Glucose _____   Albumin _____

Water _____   NaCl _____

10. Assume you are conducting the experiment illustrated below. Both hydrochloric acid (HCl) with a molecular weight of about 36.5 and ammonium hydroxide (NH_4OH) with a molecular weight of 35 are volatile and easily enter the gaseous state. When they meet, the following reaction will occur:

$$HCl + NH_4OH \rightarrow H_2O + NH_4Cl$$

Ammonium chloride (NH_4Cl) will be deposited on the glass tubing as a smoky precipitate where the two gases meet. Predict which gas will diffuse more quickly and indicate to which end of the tube the smoky precipitate will be closer.

a.   The faster diffusing gas is _____.

b.   The precipitate forms closer to the _____ end.

Rubber stopper   Cotton wad with HCl   Cotton wad with NH_4OH

Support

11. When food is pickled for human consumption, as much water as possible is removed from the food. What method is used to

achieve this dehydrating effect? _____

_____

12. What determines whether a transport process is active or passive? _____

_____

13. Characterize passive and active transport as fully as possible by choosing all the phrases that apply and inserting their letters on the answer blanks.

Passive transport _____   Active transport _____

    a.   accounts for the movement of fats and respiratory gases through the plasma membrane

    b.   explains solute pumping, phagocytosis, and pinocytosis

    c.   includes osmosis, simple diffusion, and filtration

    d.   may occur against concentration and/or electrical gradients

    e.   uses hydrostatic pressure or molecular energy as the driving force

    f.   moves ions, amino acids, and some sugars across the plasma membrane

14. For the osmometer demonstration, explain why the level of the water column rose during the laboratory session.

_____

_____

_____

15. Define the following:

diffusion: _____

_____

osmosis: _____

_____

simple diffusion: _____

_____

filtration: _____

_____

active transport: _____

_____

phagocytosis: _____

_____

pinocytosis: _____

_____

facilitated diffusion: _____

_____

Classification of Tissues

Tissue Structure and Function—General Review

1. Define *tissue:* _____

_____

2. Use the key choices to identify the major tissue types described below.

Key: a. connective tissue   b. epithelium   c. muscle   d. nervous tissue

_____ 1. lines body cavities and covers the body's external surface

_____ 2. pumps blood, flushes urine out of the body, allows one to swing a bat

_____ 3. transmits electrochemical impulses

_____ 4. anchors, packages, and supports body organs

_____ 5. cells may absorb, secrete, and protect

_____ 6. most involved in regulating and controlling body functions

_____ 7. major function is to contract

_____ 8. synthesizes hormones

_____ 9. the most durable tissue type

_____ 10. abundant nonliving extracellular matrix

_____ 11. most widespread tissue in the body

_____ 12. forms nerves and the brain

Epithelial Tissue

1. Describe the general characteristics of epithelial tissue. _____

_____

_____

_____

2. On what bases are epithelial tissues classified? _____

3. What are the major functions of epithelium in the body? (Give examples.) _____

_____

_____

_____

4. How is the function of epithelium reflected in its arrangement? _____

_____

_____

5. Where is ciliated epithelium found? _____

_____

What role does it play? _____

_____

_____

6. Transitional epithelium is actually stratified squamous epithelium, but there is something special about it.

How does it differ structurally from other stratified squamous epithelia? _____

_____

How does this reflect its function in the body? _____

_____

_____

7. How do the endocrine and exocrine glands differ in structure and function? _____

_____

_____

_____

8. Respond to the following with the key choices.

Key: a. pseudostratified ciliated columnar   c. simple cuboidal   e. stratified squamous
       b. simple columnar   d. simple squamous   f. transitional

_____ 1. lining of the esophagus

_____ 2. lining of the stomach

_____ 3. lung tissue, alveolar sacs

_____ 4. tubules of the kidney

_____ 5. epidermis of the skin

_____ 6. lining of bladder; peculiar cells that have the ability to slide over each other

_____ 7. forms the thin serous membranes; a single layer of flattened cells

Connective Tissue

1. What are the general characteristics of connective tissues? _____

_____

_____

_____

_____

2. What functions are performed by connective tissue? _____

_____

3. How are the functions of connective tissue reflected in its structure? _____

_____

_____

4. Using the key, choose the best response to identify the connective tissues described below.

Key: a. adipose connective tissue   e. fibrocartilage
      b. areolar connective tissue   f. hematopoietic tissue
      c. dense connective tissue   g. hyaline cartilage
      d. elastic cartilage   h. osseous tissue

_____ 1. attaches bones to bones and muscles to bones

_____ 2. acts as a storage depot for fat

_____ 3. the dermis of the skin

_____ 4. makes up the intervertebral discs

_____ 5. forms your hip bone

_____ 6. composes basement membranes; a soft packaging tissue with a jellylike matrix

_____ 7. forms the larynx, the costal cartilages of the ribs, and the embryonic skeleton

_____ 8. provides a flexible framework for the external ear

_____ 9. firm, structurally amorphous matrix heavily invaded with fibers; appears glassy and smooth

_____ 10. matrix hard owing to calcium salts; provides levers for muscles to act on

_____ 11. insulates against heat loss

5. Why are adipose cells called "signet ring" cells? _____

_____

Muscle Tissue

The three types of muscle tissue exhibit similarities as well as differences. Check the appropriate space in the chart to indicate which muscle types exhibit each characteristic.

Characteristic	Skeletal	Cardiac	Smooth
Voluntarily controlled			
Involuntarily controlled			
Striated			
Has a single nucleus in each cell			
Has several nuclei per cell			
Found attached to bones			
Allows you to direct your eyeballs			
Found in the walls of the stomach, uterus, and arteries			
Contains spindle-shaped cells			
Contains branching cylindrical cells			
Contains long, nonbranching cylindrical cells			
Has intercalated discs			
Concerned with locomotion of the body as a whole			
Changes the internal volume of an organ as it contracts			
Tissue of the heart			

Nervous Tissue

1. What two physiological characteristics are highly developed in nervous tissue? _____

_____

2. In what ways are nerve cells similar to other cells? _____

_____

How are they different? _____

3. Sketch a neuron, recalling in your diagram the most important aspects of its structure. Below the diagram, describe how its particular structure relates to its function in the body.

_____

_____

For Review

Label the following tissue types here and on the next pages, and identify all major structures.

(a) _____   **(b)** _____

(c) _____

(e) _____

(d) _____

(f) _____

(g) _____

(i) _____

(h) _____

(j) _____

(k) _____

(l) _____

The Integumentary System

Basic Structure of the Skin

1. Complete the following statements by writing the appropriate word or phrase on the correspondingly numbered blank:

    The two basic tissues of which the skin is composed are dense irregular connective tissue, which makes up the dermis, and __1__, which forms the epidermis. The tough water-repellent protein found in the epidermal cells is called __2__. The pigments, melanin and __3__, contribute to skin color. A localized concentration of melanin is referred to as a __4__.

1. _____

2. _____

3. _____

4. _____

2. Four protective functions of the skin are _____

_____.

3. Using the key choices, choose all responses that apply to the following descriptions.

Key: a. stratum basale   d. stratum lucidum   g. reticular layer
      b. stratum corneum   e. stratum spinosum   h. epidermis as a whole
      c. stratum granulosum   f. papillary layer   i. dermis as a whole

_____ 1. translucent cells containing keratin fibrils

_____ 2. dead cells

_____ 3. dermis layer responsible for fingerprints

_____ 4. vascular region

_____ 5. major skin area where derivatives (nails and hair) arise

_____ 6. epidermal region exhibiting rapid cell division

_____ 7. scalelike dead cells, full of keratin, that constantly slough off

_____ 8. also called the stratum germinativum

_____ 9. has abundant elastic and collagenic fibers

_____ 10. site of melanin formation

_____ 11. area where weblike prekeratin filaments first appear

4. Label the skin structures and areas indicated in the accompanying diagram of skin. Then, complete the statements that follow.

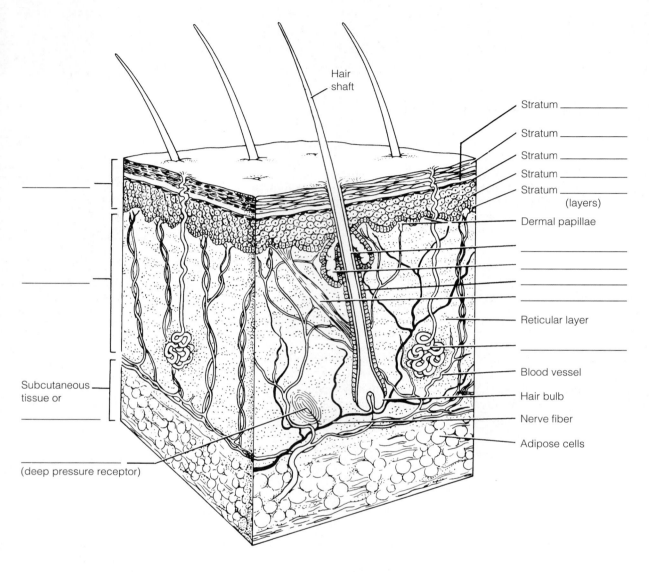

Hair shaft

Stratum _____

Stratum _____

Stratum _____

Stratum _____

Stratum _____
(layers)

Dermal papillae

_____

_____

_____

Reticular layer

_____

Blood vessel

Hair bulb

Nerve fiber

Adipose cells

Subcutaneous tissue or

_____

_____

(deep pressure receptor)

1. _____ granules extruded from the keratinocytes prevent water loss by diffusion through

   the epidermis.

2. The origin of the fibers in the dermis is _____.

3. Glands that respond to rising androgen levels are the _____ glands.

4. Phagocytic cells that occupy the epidermis are called _____.

5. A unique touch receptor formed from a stratum basale cell and a nerve fiber is a _____.

5. What substance is manufactured in the skin (but is not a secretion) to play a role elsewhere in the body?

_____

6. What sensory receptors are found in the skin? _____

_____

_____

7. A nurse tells a doctor that a patient is cyanotic. What is cyanosis? _____

What does its presence imply? _____

8. What is the mechanism of a suntan? _____

_____

9. What is a decubitus ulcer? _____

Why does it occur? _____

_____

10. Some injections hurt more than others. On the basis of what you have learned about skin structure, can you determine why

this is so? _____

_____

Appendages of the Skin

1. Using key choices, respond to the following descriptions.

Key: a. arrector pili   d. hair follicle   g. sweat gland—apocrine
       b. cutaneous receptors   e. nail   h. sweat gland—eccrine
       c. hair   f. sebaceous glands

_____ 1. A blackhead is an accumulation of oily material that is produced by _____.

_____ 2. Tiny muscles, attached to hair follicles, that pull the hair upright during fright or cold.

_____ 3. More numerous variety of perspiration gland.

_____ 4. Sheath formed of both epithelial and connective tissues.

_____ 5. Less numerous type of perspiration-producing gland; found mainly in the pubic and axillary regions.

_____ 6. Found everywhere on body except palms of hands and soles of feet.

_____ 7. Primarily dead/keratinized cells.

_____ 8. Specialized nerve endings that respond to temperature, touch, etc.

_____ 9. Its secretion contains cell fragments.

_____ 10. "Sports" a lunula and a cuticle.

2. How does the skin help in regulating body temperature? (Describe two different mechanisms.) _____

_____

_____

_____

_____

3. Several structures or skin regions are lettered in the photomicrograph below. Identify each by matching its letter with the appropriate description that follows.

| _____ | Adipose cells | _____ | Hair follicle |

_____ Adipose cells   _____ Hair follicle

_____ Dermis   _____ Sebaceous gland

_____ Epidermis   _____ Sloughing stratum corneum cells

Plotting the Distribution of Sweat Glands

1. With what substance in the bond paper does the iodine painted on the skin react? _____

2. Which skin area—the forearm or palm of hand—has more sweat glands? _____

Which other body areas would, if tested, prove to have a high density of sweat glands? _____

_____

3. What organ system controls the activity of the eccrine sweat glands? _____

Classification of Body Membranes

Classification of Body Membranes

1. Complete the following chart:

Membrane	Tissue type (epithelial/connective)	Common locations	Functions
cutaneous			
mucous			
serous			
synovial			

2. Respond to the following statements by choosing an answer from the key.

Key: a. cutaneous   b. mucous   c. serous   d. synovial

_____ 1. membrane type associated with skeletal system structures

_____ 2. always formed of simple squamous epithelium

_____, _____ 3. membrane types *not* found in the ventral body cavity

_____ 4. the only membrane type in which goblet cells are found

_____ 5. the only *external* membrane

_____ 6. "wet" membranes

_____ 7. adapted for absorption and secretion

_____ 8. has parietal and visceral layers

3. Using terms from the list above the figure, identify specifically the different types of body membranes (cutaneous, mucous, serous, and synovial) by writing in the terms at the end of the appropriate leader lines.

Cutaneous membrane (skin)
Esophageal mucosa
Nasal mucosa
Oral mucosa

Mucosa of lung bronchi
Parietal pleura
Synovial membrane of joint
Visceral pleura

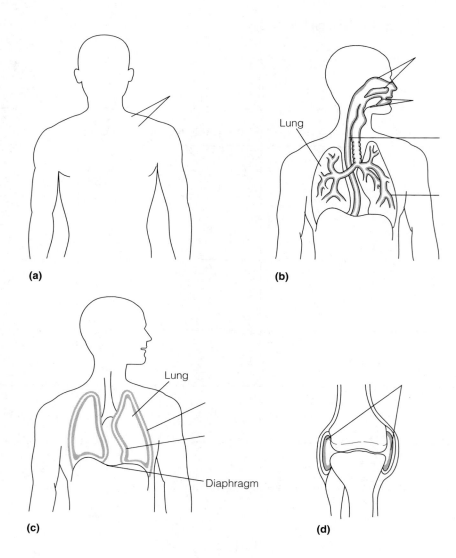

(a)

(b)

Lung

(c)

Lung

Diaphragm

(d)

Overview of the Skeleton

Bone Markings

Match the terms in column B with the appropriate description in column A:

Column A   **Column B**

_____   1. sharp, slender process*   a. condyle

_____   2. small rounded projection*   b. crest

_____   3. narrow ridge of bone*   c. epicondyle

_____   4. large rounded projection*   d. fissure

_____   5. structure supported on neck†   e. foramen

_____   6. armlike projection†   f. fossa

_____   7. rounded, convex projection†   g. head

_____   8. narrow depression or opening‡   h. meatus

_____   9. canal-like structure‡   i. ramus

_____   10. opening through a bone‡   j. sinus

_____   11. shallow depression†   k. spine

_____   12. air-filled cavity   l. trochanter

_____   13. large, irregularly shaped projection*   m. tubercle

_____   14. raised area of a condyle*   n. tuberosity

* A site of muscle attachment.
† Takes part in joint formation.
‡ A passageway for nerves or blood vessels.

Classification of Bones

1. The four major anatomical classifications of bones are long, short, flat, and irregular. Which category has the least amount

    of spongy bone relative to its total volume? _____

2. Classify each of the bones in the next chart into one of the four major categories by checking the appropriate column. Use
    appropriate references as necessary.

	Long	Short	Flat	Irregular
humerus				
phalanx				
parietal				
calcaneus				
rib				
vertebra				
ulna				

Gross Anatomy of the Typical Long Bone

1. Use the terms below to identify the structures marked by leader lines and braces in the diagrams (some terms are used more than once). After labeling the diagrams, use the listed terms to characterize the statements following the diagrams.

Key: a.   articular cartilage   f.   epiphysis   i.   red marrow cavity
          b.   compact bone   g.   medullary cavity   j.   trabeculae of spongy bone
          c.   diaphysis   h.   periosteum   k.   yellow marrow
          d.   endosteum
          e.   epiphyseal line

Nutrient artery

_____ 1. contains spongy bone in adults

_____ 2. made of compact bone

_____ 3. site of blood cell formation

_____ 4. major submembranous site of osteoclasts

_____ 5. scientific term for bone shaft

_____ 6. contains fat in adult bones

_____ 7. growth plate remnant

_____ 8. major submembranous site of osteoblasts

2. What differences between compact and spongy bone can be seen with the naked eye? _____

_____

3. What is the function of the periosteum? _____

_____

Microscopic Structure of Compact Bone

1. Trace the route taken by nutrients through a bone, starting with the periosteum and ending with an osteocyte in a lacuna.

Periosteum ___→_____ →_____

_____ →_____ →___ osteocyte

2. Several descriptions of bone structure are given in column B. Identify the structure involved by choosing the appropriate term from column A and placing its letter in the blank.

Column A

a. canaliculi

b. central canal

c. concentric lamellae

d. lacunae

Column B

_____ 1. layers of bony matrix around a central canal

_____ 2. site of osteocytes

_____ 3. longitudinal canal carrying blood vessels, lymphatics, and nerves

_____ 4. minute canals connecting osteocytes of an osteon

3. On the photomicrograph of bone below (208×), identify all structures named in column A in question 2 above.

Chemical Composition of Bone

1. What is the function of the organic matrix in bone? _____

2. Name the important organic bone components. _____

3. Calcium salts form the bulk of the inorganic material in bone. What is the function of the calcium salts?

    _____

4. Which is responsible for bone structure? (circle the appropriate response)

    inorganic portion   organic portion   both contribute

Cartilages of the Skeleton

1. Using key choices, identify each type of cartilage described (in terms of its body location or function) below.

    Key: a. elastic   b. fibrocartilage   c. hyaline

    _____ 1. supports the external ear   _____ 6. meniscus in a knee joint

    _____ 2. between the vertebrae   _____ 7. connects the ribs to the sternum

    _____ 3. forms the walls of the   _____ 8. most effective at resisting
                       voice box (larynx)   compression

    _____ 4. the epiglottis   _____ 9. most springy and flexible

    _____ 5. articular cartilages   _____ 10. most abundant

2. Identify the two types of cartilage diagrammed below. On each, label the *chondrocytes in the lacunae* and the *matrix*.

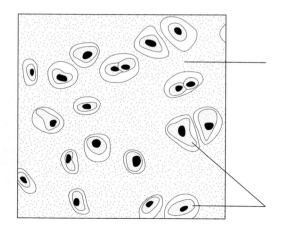

(a) _____   (b) _____

The Axial Skeleton

The Skull

1. The skull is one of the major components of the axial skeleton. Name the other two:

   _____ and _____

   What structures do each of these areas protect? _____

   _____

   _____

2. Define *suture:* _____

3. With one exception, the skull bones are joined by sutures. Name the exception. _____

   _____

4. What are the four major sutures of the skull, and what bones do they connect? _____

   _____

   _____

   _____

   _____

5. Name the eight bones composing the cranium.

   _____ _____ _____ _____

   _____ _____ _____ _____

6. Give two possible functions of the sinuses. _____

   _____

7. What is the orbit? _____

   What bones contribute to the formation of the orbit? _____

   _____

8. Why can the sphenoid bone be called the keystone of the cranial floor? _____

   _____

   _____

9. What is a cleft palate? _____

_____

10. Match the bone names in column B with the descriptions in column A.

Column A		Column B
_____ 1. forehead bone		
_____ 2. cheekbone		a. ethmoid
_____ 3. lower jaw		b. frontal
_____ 4. bridge of nose		c. hyoid
_____ 5. posterior bones of the hard palate		d. lacrimal
_____ 6. much of the lateral and superior cranium		e. mandible
_____ 7. most posterior part of cranium		f. maxilla
_____ 8. single, irregular, bat-shaped bone forming part of the cranial floor		g. nasal
_____ 9. tiny bones bearing tear ducts		h. occipital
_____ 10. anterior part of hard palate		i. palatine
_____ 11. superior and medial nasal conchae formed from its projections		j. parietal
_____ 12. site of mastoid process		k. sphenoid
_____ 13. site of sella turcica		l. temporal
_____ 14. site of cribriform plate		m. vomer
_____ 15. site of mental foramen		n. zygomatic
_____ 16. site of styloid processes		

_____, _____, _____, and

_____ 17. four bones containing paranasal sinuses

_____ 18. condyles here articulate with the atlas

_____ 19. foramen magnum contained here

_____ 20. small U-shaped bone in neck, where many tongue muscles attach

_____ 21. middle ear found here

_____ 22. nasal septum

_____ 23. bears an upward protrusion, the "cock's comb," or crista galli

_____, _____ 24. contain alveoli bearing teeth

11. Using choices from column B in question 10 and from the numbered key to the right, identify all bones and bone markings provided with leader lines in the two diagrams below.

1. carotid canal

2. coronal suture

3. external occipital protuberance

4. foramen lacerum

5. foramen magnum

6. foramen ovale

7. glabella

8. incisive foramen

9. inferior nasal concha

10. inferior orbital fissure

11. infraorbital foramen

12. jugular foramen

13. mandibular fossa

14. mandibular symphysis

15. mastoid process

16. mental foramen

17. middle nasal concha of ethmoid

18. occipital condyle

19. palatine process of maxilla

20. sagittal suture

21. styloid process

22. stylomastoid foramen

23. superior orbital fissure

24. supraorbital foramen

25. zygomatic process of temporal bone

The Vertebral Column

1. Using the key, correctly identify the vertebral parts/areas described below. (More than one choice may apply in some cases.) Also use the key letters to correctly identify the vertebral areas in the diagram.

Key: a. body
      b. intervertebral foramina
      c. lamina

d. pedicle
e. spinous process
f. superior articular process

g. transverse process
h. vertebral arch
i. vertebral foramen

_____ 1. cavity enclosing the nerve cord

_____ 2. weight-bearing portion of the vertebra

_____ 3. provide levers against which muscles pull

_____ 4. provide an articulation point for the ribs

_____ 5. openings providing for exit of spinal nerves

_____ 6. structures that form an enclosure for the spinal cord

2. The distinguishing characteristics of the vertebrae composing the vertebral column are noted below. Correctly identify each described structure/region by choosing a response from the key.

Key: a. atlas
      b. axis
      c. cervical vertebra—typical

d. coccyx
e. lumbar vertebra

f. sacrum
g. thoracic vertebra

_____ 1. vertebral type containing foramina in the transverse processes, through which the vertebral arteries ascend to reach the brain

_____ 2. dens here provides a pivot for rotation of the first cervical vertebra (C_1)

_____ 3. transverse processes faceted for articulation with ribs; spinous process pointing sharply downward

_____ 4. composite bone; articulates with the hip bone laterally

_____ 5. massive vertebrae; weight-sustaining

_____ 6. "tail bone"; vestigial fused vertebrae

_____ 7. supports the head; allows a rocking motion in conjunction with the occipital condyles

_____ 8. seven components; unfused

_____ 9. twelve components; unfused

3. Identify specifically each of the vertebra types shown in the diagrams below. Also identify and label the following markings on each: transverse processes, spinous process, body, superior articular processes.

_____   _____

4. Describe how a spinal nerve exits from the vertebral column. _____

_____

5. Name two factors/structures that allow for flexibility of the vertebral column.

_____ and _____

6. What kind of tissue comprises the intervertebral discs? _____

7. What is a herniated disc? _____

   What problems might it cause? _____

_____

8. Which two spinal curvatures are obvious at birth? _____ and

_____

   Under what conditions do the secondary curvatures develop? _____

_____

_____

9. On this illustration of an articulated vertebral column, identify each curvature indicated as a primary or a secondary curvature. Also identify by letter the terms in the column below and to the left.

a. atlas
b. axis
c. a disc
d. two thoracic vertebrae
e. two lumbar vertebrae
f. sacrum
g. vertebra prominens

(curvature)

(curvature)

(curvature)

(curvature)

10. Diagram the abnormal spinal curvatures named below. (Use posterior or lateral views as necessary.)

Lordosis   Scoliosis   Kyphosis

The Bony Thorax

1. The major components of the thorax (excluding the vertebral column) are the _____

and the _____.

2. Differentiate between a true rib and a false rib. _____

_____

_____

Is a floating rib a true or a false rib? _____

3. What is the general shape of the thoracic cage? _____

_____

_____

4. Provide the more scientific name for the following rib types.

a. True ribs _____

b. False ribs (not including c) _____

c. Floating ribs _____

5. Using the terms at the right, identify the regions and landmarks of the bony thorax.

L₁ vertebra

a. body

b. clavicular notch

c. costal cartilage

d. false ribs

e. floating ribs

f. jugular notch

g. manubrium

h. sternal angle

i. sternum

j. true ribs

k. xiphisternal joint

l. xiphoid process

The Appendicular Skeleton

Bones of the Pectoral Girdle and Upper Extremity

1. Match the bone names or markings in column B with the descriptions in column A.

Column A

_____ 1. raised area on lateral surface of humerus to which deltoid muscle attaches

_____ 2. arm bone

_____, _____ 3. bones of the shoulder girdle

_____, _____ 4. forearm bones

_____ 5. scapular region to which the clavicle connects

_____ 6. shoulder girdle bone that is unattached to the axial skeleton

_____ 7. shoulder girdle bone that transmits forces from the upper limb to the bony thorax

_____ 8. depression in the scapula that articulates with the humerus

_____ 9. process above the glenoid cavity that permits muscle attachment

_____ 10. the "collarbone"

_____ 11. distal condyle of the humerus that articulates with the ulna

_____ 12. medial bone of forearm in anatomical position

_____ 13. rounded knob on the humerus; adjoins the radius

_____ 14. anterior depression, superior to the trochlea, which receives part of the ulna when the forearm is flexed

_____ 15. forearm bone involved in formation of the elbow joint

_____ 16. wrist bones

_____ 17. finger bones

_____ 18. heads of these bones form the knuckles

_____, _____ 19. bones that articulate with the clavicle

Column B

a. acromion

b. capitulum

c. carpals

d. clavicle

e. coracoid process

f. coronoid fossa

g. deltoid tuberosity

h. glenoid cavity

i. humerus

j. metacarpals

k. olecranon fossa

l. olecranon process

m. phalanges

n. radial tuberosity

o. radius

p. scapula

q. sternum

r. styloid process

s. trochlea

t. ulna

2. Why is the clavicle at risk to fracture when a person falls on his or her shoulder? _____

_____

3. Why is there generally no problem in the arm clearing the widest dimension of the thoracic cage?

_____

_____

4. What is the total number of phalanges in the hand? _____

5. What is the total number of carpals in the wrist? _____

In the proximal row, the carpals are (medial to lateral) _____

In the distal row, they are (medial to lateral) _____

6. Using items from the list at the right, identify the anatomical landmarks and regions of the scapula.

a. acromion

b. coracoid process

c. glenoid cavity

d. inferior angle

e. infraspinous fossa

f. lateral border

g. medial border

h. spine

i. superior angle

j. superior border

k. suprascapular notch

l. supraspinous fossa

7. Match the terms in the key with the appropriate leader lines on the photos of the humerus and the radius and ulna. Also decide whether these bones are right or left bones.

Key:

a. anatomical neck

b. coronoid process

c. distal radioulnar joint

d. greater tubercle

e. head of humerus

f. head of radius

g. head of ulna

h. lateral epicondyle

i. medial epicondyle

j. olecranon fossa

k. olecranon process

l. proximal radioulnar joint

m. radial groove

n. radial notch

o. radial tuberosity

p. styloid process of radius

q. styloid process of ulna

r. surgical neck

s. trochlea

t. trochlear notch

The humerus is a _____ bone; the radius and ulna are _____ bones.

Bones of the Pelvic Girdle and Lower Limb

1. Compare the pectoral and pelvic girdles by choosing appropriate descriptive terms from the key.

Key: a. flexibility most important   d. insecure axial and limb attachments
       b. massive   e. secure axial and limb attachments
       c. lightweight   f. weight-bearing most important

Pectoral: _____, _____, _____   Pelvic: _____, _____, _____

2. What organs are protected, at least in part, by the pelvic girdle? _____

_____

3. Distinguish between the true pelvis and the false pelvis. _____

_____

_____

4. Use letters from the key to identify the bone markings on this illustration of an articulated pelvis. Make an educated guess as to whether the illustration shows a male or female pelvis and provide two reasons for your decision.

Key:

a. acetabulum

b. ala

c. anterior superior iliac spine

d. iliac crest

e. iliac fossa

f. ischial spine

g. pelvic brim

h. pubic crest

i. pubic symphysis

j. sacroiliac joint

k. sacrum

This is a _____ (female/male) pelvis because:

_____

_____

5. Deduce why the pelvic bones of a four-legged animal such as the cat or pig are much less massive than those of the human.

_____

_____

6. A person instinctively curls over his abdominal area in times of danger. Why? _____

_____

7. For what anatomical reason do many women appear to be slightly knock-kneed? _____

_____

_____

8. What does *fallen arches* mean? _____

_____

9. Match the bone names and markings in column B with the descriptions in column A.

Column A

_____, _____, and

_____ 1. fuse to form the coxal bone

_____ 2. inferoposterior "bone" of the coxal bone

_____ 3. point where the coxal bones join anteriorly

_____ 4. superiormost margin of the coxal bone

_____ 5. deep socket in the coxal bone that receives the head of the thigh bone

_____ 6. joint between axial skeleton and pelvic girdle

_____ 7. longest, strongest bone in body

_____ 8. thin lateral leg bone

_____ 9. heavy medial leg bone

_____, _____ 10. bones forming knee joint

_____ 11. point where the patellar ligament attaches

_____ 12. kneecap

_____ 13. shin bone

_____ 14. medial ankle projection

_____ 15. lateral ankle projection

_____ 16. largest tarsal bone

_____ 17. ankle bones

_____ 18. bones forming the instep of the foot

_____ 19. opening in hip bone formed by the pubic and ischial rami

_____ and _____ 20. sites of muscle attachment on the proximal femur

_____ 21. tarsal bone that "sits" on the calcaneus

Column B

a. acetabulum

b. calcaneus

c. femur

d. fibula

e. gluteal tuberosity

f. greater sciatic notch

g. greater and lesser trochanters

h. iliac crest

i. ilium

j. ischial tuberosity

k. ischium

l. lateral malleolus

m. lesser sciatic notch

n. linea aspera

o. medial malleolus

p. obturator foramen

q. metatarsals

r. patella

s. pubic symphysis

t. pubis

u. sacroiliac joint

v. talus

w. tarsals

x. tibia

y. tibial tuberosity

10. Match the terms in the key with the appropriate leader lines on the photos of the femur and the tibia and fibula. Also decide if these bones are right or left bones.

Key:

a. distal tibiofibular joint

b. fovea capitus

c. gluteal tuberosity

d. greater trochanter

e. head of femur

f. head of fibula

g. intercondylar eminence

h. intertrochanteric crest

i. lateral condyle

j. lateral epicondyle

k. lateral malleolus

l. lesser trochanter

m. medial condyle

n. medial epicondyle

o. medial malleolus

p. neck of femur

q. proximal tibiofibular joint

r. tibial anterior crest

s. tibial tuberosity

The femur (the diagram on the _____ side) is the _____ member of the two femurs.

The tibia and fibula (the diagram on the _____ side) are _____ bones.

Summary of Skeleton

1. Identify all indicated bones (or groups of bones) in the diagram of the articulated skeleton on page 561.

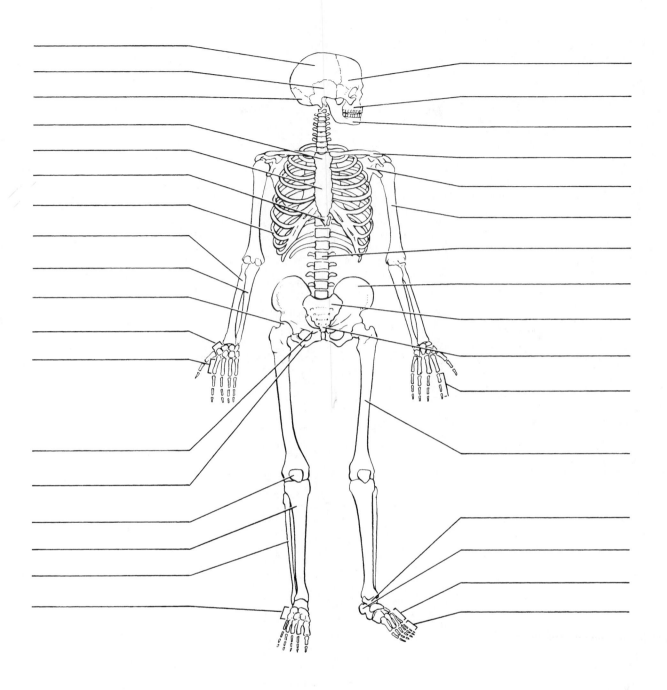

The Fetal Skeleton

1. Are the same skull bones seen in the adult found in the fetal skull? _____

2. How does the size of the fetal face compare to its cranium? _____

    _____

    How does this compare to the adult skull? _____

    _____

    _____

3. What are the outward conical projections in some of the fetal cranial bones? _____

    _____

4. What is a fontanel? _____

    What is its fate? _____

    What is the function of the fontanels in the fetal skull? _____

    _____

5. Describe how the fetal skeleton compares with the adult skeleton in the following areas:

    vertebrae _____

    _____

    os coxae _____

    _____

    carpals and tarsals _____

    sternum _____

    frontal bone _____

    patella _____

    rib cage _____

    _____

6. How does the size of the fetus's head compare to the size of its body? _____

_____

7. Using the terms listed, identify each of the fontanels shown on the fetal skull below.

    a. anterior fontanel   b. mastoid fontanel   c. posterior fontanel   d. sphenoidal fontanel

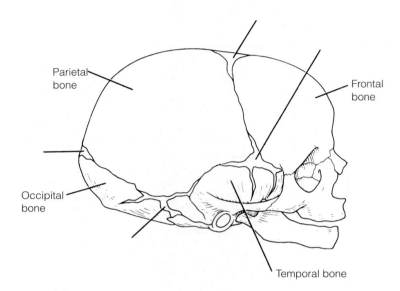

Articulations and Body Movements

Types of Joints

1. Use key responses to identify the joint types described below.

Key: a. cartilaginous   b. fibrous   c. synovial

_____ 1. typically allows a slight degree of movement

_____ 2. includes joints between the vertebral bodies and the pubic symphysis

_____ 3. essentially immovable joints

_____ 4. sutures are the most remembered examples

_____ 5. characterized by cartilage connecting the bony portions

_____ 6. all characterized by a fibrous articular capsule lined with a synovial membrane surrounding a joint cavity

_____ 7. all are freely movable or diarthrotic

_____ 8. bone regions are united by fibrous connective tissue

_____ 9. include the hip, knee, and elbow joints

2. Match the joint subcategories in column B with their descriptions in column A, and place an asterisk (*) beside all choices that are examples of synovial joints.

Column A	Column B
_____ 1. joint between skull bones	a. ball and socket
_____ 2. joint between the axis and atlas	b. condyloid
_____ 3. hip joint	c. gliding
_____ 4. intervertebral joints (between articular processes)	d. hinge
_____ 5. joint between forearm bones and wrist	e. pivot
_____ 6. elbow	f. saddle
_____ 7. interphalangeal joints	g. suture
_____ 8. intercarpal joints	h. symphysis
_____ 9. joint between tarsus and tibia/fibula	i. synchondrosis
_____ 10. joint between skull and vertebral column	j. syndesmosis

_____ 11. joint between jaw and skull

_____ 12. joints between proximal phalanges and metacarpal bones

_____ 13. epiphyseal plate of a child's long bone

_____ 14. a multiaxial joint

_____, _____ 15. biaxial joints

_____, _____ 16. uniaxial joints

3. What characteristics do all joints have in common? _____

_____

4. Describe the structure and function of the following structures or tissues in relation to a synovial joint and label the structures indicated by leader lines in the diagram.

ligament _____

_____

tendon _____

_____

articular cartilage _____

_____

synovial membrane _____

_____

_____

bursa _____

5. Which joint, the hip or the knee, is more stable? _____

Name two important factors that contribute to the stability of the hip joint.

_____ and _____

Name two important factors that contribute to the stability of the knee.

_____ and _____

6. The diagram shows a frontal section of the hip joint. Identify its major structural elements by using the key letters.

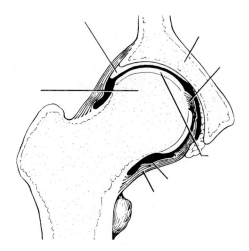

Key:

a. acetabular labrum

b. articular capsule

c. articular cartilage

d. coxal bone

e. head of femur

f. ligamentum teres

g. synovial cavity

Movements Allowed by Synovial Joints

1. Label the *origin* and *insertion* points on the diagram below and complete the following statement:

During muscle contraction, the _____

moves toward the _____.

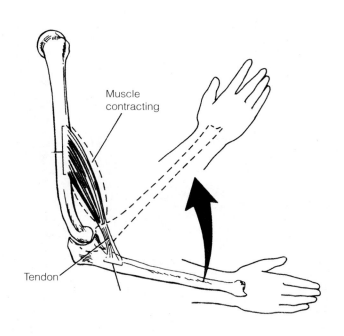

2. Complete the statements below the stick diagrams by inserting the missing words in the answer blanks.

1. _____

2. _____

(continues on page 4)

1. _____ of hand   2. _____ of head

3. _____

4. _____

5. _____

6. _____

7. _____

8. _____

9. _____

10. _____

11. _____

3. _____ of the arm

4. _____ of hip
5. _____ of knee

6. _____ of thigh
7. _____ of arm

8. _____ of hip

9. _____ of foot

10. _____ of elbow

11. _____ of foot

Joint Disorders

1. What structural joint changes are common to the elderly? _____

_____

2. Define:

sprain _____

dislocation _____

Microscopic Anatomy, Organization, and Classification of Skeletal Muscle

Skeletal Muscle Cells and Their Packaging into Muscles

1. What capability is most highly expressed in muscle tissue? _____

2. Use the items on the right to correctly identify the structures described on the left.

_____ 1. connective tissue ensheathing a bundle of muscle cells

_____ 2. bundle of muscle cells

_____ 3. contractile unit of muscle

_____ 4. a muscle cell

_____ 5. thin reticular connective tissue investing each muscle cell

_____ 6. plasma membrane of the muscle fiber

_____ 7. a long filamentous organelle with a banded appearance found within muscle cells

_____ 8. actin- or myosin-containing structure

_____ 9. cord of collagen fibers that attaches a muscle to a bone

a. endomysium

b. epimysium

c. fascicle

d. fiber

e. myofilament

f. myofibril

g. perimysium

h. sarcolemma

i. sarcomere

j. sarcoplasm

k. tendon

3. Why are the connective tissue wrappings of skeletal muscle important? (Give at least three reasons.)

_____

_____

_____

_____

4. Why are indirect—that is, tendinous—muscle attachments to bone seen more often than direct attachments?

_____

_____

_____

5. How does an aponeurosis differ from a tendon? _____

_____

_____

6. The diagram illustrates a small portion of a muscle myofibril. Using letters from the key, correctly identify each structure indicated by a leader line or a bracket. Below the diagram make a sketch of how this segment of the myofibril would look if contracted.

Key: a. actin filament   d. myosin filament
      b. A band   e. sarcomere
      c. I band   f. Z disc

The Neuromuscular Junction

Complete the following statements:

The junction between a motor neuron's axon and the muscle cell membrane is called a neuromuscular junction or a __1__ junction. A motor neuron and all of the skeletal muscle cells it stimulates is called a __2__. The actual gap between the axonal terminal and the muscle cell is called a __3__. Within the axonal terminal are many small vesicles containing a neurotransmitter substance called __4__. When the __5__ reaches the ends of the axon, the neurotransmitter is released and diffuses to the muscle cell membrane to combine with receptors there. The combining of the neurotransmitter with the muscle membrane receptors causes the membrane to become permeable to sodium, which results in the influx of sodium ions and __6__ of the membrane. Then contraction of the muscle cell occurs. Before a muscle cell can be stimulated to contract again, __7__ must occur.

1. _____

2. _____

3. _____

4. _____

5. _____

6. _____

7. _____

Classification of Skeletal Muscles

1. Several criteria were given relative to the naming of muscles. Match the criteria (column B) to the muscle names (column A). Note that more than one criterion may apply in some cases.

   Column A

   _____ 1. gluteus maximus

   _____ 2. adductor magnus

   _____ 3. biceps femoris

   _____ 4. abdominis transversus

   _____ 5. extensor carpi ulnaris

   _____ 6. trapezius

   _____ 7. rectus femoris

   _____ 8. external oblique

   Column B

   a. action of the muscle

   b. shape of the muscle

   c. location of the origin and/or insertion of the muscle

   d. number of origins

   e. location of the muscle relative to a bone or body region

   f. direction in which the muscle fibers run relative to some imaginary line

   g. relative size of the muscle

2. When muscles are discussed relative to the manner in which they interact with other muscles, the terms shown in the key are often used. Match the key terms with the appropriate definitions.

   Key: a. antagonist   b. fixator   c. prime mover   d. synergist

   _____ 1. agonist

   _____ 2. postural muscles, for the most part

   _____ 3. reverses and/or opposes the action of a prime mover

   _____ 4. stabilizes a joint so that the prime mover may act at more distal joints

   _____ 5. performs the same movement as the prime mover

   _____ 6. immobilizes the origin of a prime mover

Gross Anatomy of the Muscular System

Muscles of the Head and Neck

1. Using choices from the list at the right, correctly identify muscles provided with leader lines on the diagram.

Cranial aponeurosis
(galea aponeurotica)

a. buccinator

b. corrugator supercilii

c. depressor anguli oris

d. depressor labii inferioris

e. epicranius frontalis

f. epicranius occipitalis

g. levator labii superioris

h. masseter

i. mentalis

j. platysma

k. orbicularis oculi

l. orbicularis oris

m. zygomaticus

2. Using the terms provided above, identify the muscles described next.

_____ 1. used in smiling

_____ 2. used to suck in your cheeks

_____ 3. used in blinking and squinting

_____ 4. used to pout (pulls the corners of the mouth downward)

_____ 5. raises your eyebrows for a questioning expression

_____ 6. used to form the vertical frown crease on the forehead

_____ 7. your "kisser"

_____ 8. prime mover to raise the lower jawbone

_____ 9. tenses skin of the neck during shaving

Muscles of the Trunk

1. Correctly identify both intact and transected (cut) muscles depicted in the diagram, using the terms given at the right. (Not all terms will be used in this identification.)

Coracoid process

Bursa

Clavicle

Sternum

6th rib

Radius

a. biceps brachii

b. coracobrachialis

c. deltoid (cut)

d. external intercostals

e. external oblique

f. internal intercostals

g. internal oblique

h. latissimus dorsi

i. pectoralis major (cut)

j. pectoralis minor

k. rectus abdominis

l. rhomboids

m. serratus anterior

n. subscapularis

o. teres major

p. teres minor

q. transversus abdominis

r. trapezius

2. Using choices from the terms provided in question 1 above, identify the major muscles described next:

_____ 1. a major spine flexor

_____ 2. prime mover for pulling the arm posteriorly

_____ 3. prime mover for shoulder flexion

_____ 4. assume major responsibility for forming the abdominal girdle (three pairs of muscles)

_____ 5. pulls the shoulder backward and downward

_____ 6. prime mover of shoulder abduction

_____ 7. important in shoulder adduction; antagonists of the shoulder abductor (two muscles)

_____ 8. moves the scapula forward and downward

_____ 9. small, inspiratory muscles between the ribs; elevate the ribs

_____ 10. extends the head

_____ 11. pull the scapulae medially

Muscles of the Upper Limb

1. Using terms from the list on the right, correctly identify all muscles provided with leader lines in the diagram. Note that not all the listed terms will be used in this exercise.

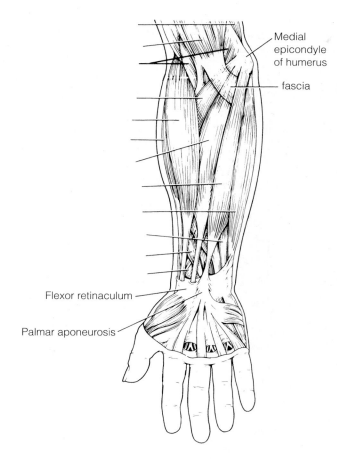

Medial
epicondyle
of humerus

— fascia

Flexor retinaculum

Palmar aponeurosis

a. biceps brachii

b. brachialis

c. brachioradialis

d. extensor carpi radialis longus

e. extensor digitorum

f. flexor carpi radialis

g. flexor carpi ulnaris

h. flexor digitorum superficialis

i. flexor pollicis longus

j. palmaris longus

k. pronator quadratus

l. pronator teres

m. supinator

n. triceps brachii

2. Use the terms provided in question 1 to identify the muscles described next.

_____ 1. places the palm upward (two muscles)

_____ 2. flexes the forearm and supinates the hand

_____ 3. forearm flexors; no role in supination (two muscles)

_____ 4. elbow extensor

_____ 5. power wrist flexor and abductor

_____ 6. flexes wrist and distal phalanges

_____ 7. pronate the hand (two muscles)

_____ 8. flexes the thumb

_____ 9. extends and abducts the wrist

_____ 10. extends the wrist and digits

_____ 11. flat muscle that is a weak wrist flexor

Muscles of the Lower Limb

1. Using the terms listed to the right, correctly identify all muscles provided with leader lines in the diagram below. Not all listed terms will be used in this exercise.

Head of fibula

Superior extensor retinaculum

Interior extensor retinaculum

Lateral malleolus

5th metatarsal

a.   adductor group

b.   biceps femoris

c.   extensor digitorum longus

d.   flexor hallucis longus

e.   gastrocnemius

f.   gluteus maximus

g.   gluteus medius

h.   peroneus brevis

i.   peroneus longus

j.   rectus femoris

k.   semimembranosus

l.   semitendinosus

m.   soleus

n.   tensor fasciae latae

o.   tibialis anterior

p.   tibialis posterior

q.   vastus muscles

2.   Use the key terms in exercise 1 to respond to the descriptions below.

_____ 1. flexes the great toe and inverts the ankle

_____ 2. lateral compartment muscles that plantar flex and evert the ankle (two muscles)

_____ 3. move the thigh laterally to take the "at ease" stance (two muscles)

_____ 4. used to extend the hip when climbing stairs

_____ 5. prime movers of ankle plantar flexion (two muscles)

_____ 6. major foot inverter

_____ 7. prime mover of ankle dorsiflexion

_____ 8. allow you to draw your legs to the midline of your body, as when standing at attention

_____ 9. extends the toes

_____ 10. extend thigh and flex knee (three muscles)

_____ 11. extends knee and flexes thigh

General Review: Muscle Recognition

1. Identify the lettered muscles in the diagram of the human anterior superficial musculature by matching the letter with one of the following muscle names:

_____ 1. orbicularis oris

_____ 2. pectoralis major

_____ 3. external oblique

_____ 4. sternocleidomastoid

_____ 5. biceps brachii

_____ 6. deltoid

_____ 7. vastus lateralis

_____ 8. brachioradialis

_____ 9. frontalis

_____ 10. rectus femoris

_____ 11. pronator teres

_____ 12. rectus abdominis

_____ 13. sartorius

_____ 14. gracilis

_____ 15. flexor carpi ulnaris

_____ 16. adductor longus

_____ 17. palmaris longus

_____ 18. flexor carpi radialis

_____ 19. latissimus dorsi

_____ 20. orbicularis oculi

_____ 21. gastrocnemius

_____ 22. masseter

_____ 23. trapezius

_____ 24. tibialis anterior

_____ 25. extensor digitorum longus

_____ 26. tensor fasciae latae

_____ 27. pectineus

_____ 28. sternohyoid

_____ 29. serratus anterior

_____ 30. adductor magnus

_____ 31. vastus medialis

_____ 32. transversus abdominis

_____ 33. peroneus longus

_____ 34. iliopsoas

_____ 35. temporalis

_____ 36. zygomaticus

_____ 37. coracobrachialis

_____ 38. triceps brachii

_____ 39. internal oblique

2. Identify each of the lettered muscles in this diagram of the human posterior superficial musculature by matching its letter to one of the following muscle names:

_____ 1. gluteus maximus

_____ 2. semimembranosus

_____ 3. gastrocnemius

_____ 4. latissimus dorsi

_____ 5. deltoid

_____ 6. iliotibial tract (tendon)

_____ 7. teres major

_____ 8. semitendinosus

_____ 9. trapezius

_____ 10. biceps femoris

_____ 11. triceps brachii

_____ 12. external oblique

_____ 13. gluteus medius

_____ 14. gracilis

_____ 15. flexor carpi ulnaris

_____ 16. extensor carpi ulnaris

_____ 17. extensor digitorum communis

_____ 18. extensor carpi radialis longus

_____ 19. occipitalis

_____ 20. extensor carpi radialis brevis

_____ 21. sternocleidomastoid

_____ 22. adductor magnus

_____ 23. anconeus

General Review: Muscle Descriptions

1. Identify the muscles described below by completing the statements:

_____, _____, and _____

are commonly used for intramuscular injections (three muscles).

The insertion tendon of the _____ group contains a large sesamoid bone, the patella.

The triceps surae insert in common into the _____ tendon.

The bulk of the tissue of a muscle tends to lie _____ to the part of the body it causes to move.

The extrinsic muscles of the hand originate on the _____.

Most flexor muscles are located on the _____ aspect of the body; most extensors are

located _____. An exception to this generalization is the extensor-flexor musculature

of the _____.

Skeletal Muscle Physiology—
Frog Experimentation: Wet Lab

Muscle Activity

1. The following group of incomplete statements refers to a muscle cell in the resting or polarized state just before stimulation. Complete each statement by choosing the correct response from the key items below.

Key: a. Na^+ diffuses out of the cell
    b. K^+ diffuses out of the cell
    c. Na^+ diffuses into the cell
    d. K^+ diffuses into the cell
    e. inside the cell
    f. outside the cell
    g. relative ionic concentrations on the two sides of the membrane

    h. electrical conditions
    i. activation of the sodium-potassium pump, which moves K^+ into the cell and Na^+ out of the cell
    j. activation of the sodium-potassium pump, which moves Na^+ into the cell and K^+ out of the cell

There is a greater concentration of Na^+ _____; there is a greater concentration of K^+ _____. When the stimulus

is delivered, the permeability of the membrane at that point is changed; and _____ initiating the depolarization of the

membrane. Almost as soon as the depolarization wave has begun, a repolarization wave follows it across the membrane.

This occurs as _____. Repolarization restores the _____ of the resting cell membrane. The _____ is (are) reestab-

lished by _____.

2. Number the following statements in the proper sequence to describe the contraction mechanism in a skeletal muscle cell. Number 1 has already been designated.

____1____ Acetylcholine is released into the neuromuscular junction by the axonal terminal.

_____ The action potential, carried deep into the cell by the T system, triggers the release of calcium ions from the sarcoplasmic reticulum.

_____ The muscle cell relaxes and lengthens.

_____ Acetylcholine diffuses across the neuromuscular junction and binds to receptors on the sarcolemma.

_____ The calcium ion concentrations at the myofilaments increase; the myofilaments slide past one another, and the cell shortens.

_____ Depolarization occurs, and the action potential is generated.

_____ The concentration of the calcium ions at the myofilaments decreases as they are actively transported into the sarcoplasmic reticulum.

3. Muscle contraction is commonly explained by the sliding filament hypothesis. What are the essential points of this hypothesis? _____

_____

_____

4. Relative to your observations of muscle fiber contraction (pp. 155–156):

    a. What percentage of contraction was observed with the solution containing ATP, K^+, and Mg^{2+}? _____%

    With *just* ATP? _____%   With *just* Mg^{2+} and K^+? _____%

    b. *Explain* your observations fully. _____

    _____

    c. What zones or bands disappear when the muscle cell contracts? _____

    d. Draw a relaxed and a contracted sarcomere below.

                    Relaxed   Contracted

Induction of Contraction in the Frog Gastrocnemius Muscle

1. Why is it important to destroy the brain and spinal cord of a frog before conducting physiological experiments on muscle contraction? _____

_____

2. What sources of stimuli, other than electrical shocks, cause a muscle to contract? _____

_____

3. What is the most common stimulus for muscle contraction in the body? _____

4. Name the three phases of the muscle twitch, and describe what is happening during each phase:

    (1) _____, _____

    _____

    (2) _____, _____

    (3) _____, _____

5. Use the terms given on the right to identify the conditions described on the left:

_____ 1. sustained contraction without any evidence of relaxation

_____ 2. stimulus that results in no perceptible contraction

_____ 3. stimulus at which the muscle first contracts perceptibly

_____ 4. increasingly stronger contractions in the absence of increased stimulus intensity

_____ 5. increasingly stronger contractions owing to stimulation at a rapid rate

_____ 6. increasingly stronger contractions owing to increased stimulus strength

_____ 7. weakest stimulus at which all muscle cells in the muscle are contracting

a. maximal stimulus

b. multiple motor unit summation

c. subthreshold stimulus

d. tetanus

e. threshold stimulus

f. treppe

g. wave summation

6. With brackets and labels, identify the portions of the tracing below that best correspond to three of the phenomena listed in the preceding key. Assume that only the timing of the stimulus has changed.

7. Complete the following statements by writing the appropriate words on the correspondingly numbered blanks at the right.

The "all or none" law applies to skeletal muscle functions at the __1__ level. When a weak but smooth muscle contraction is desired, a few motor units are stimulated at a __2__ rate. Treppe is referred to as the "warming up" process. It is believed that muscles contract more strongly after the first few contractions because the __3__ become more efficient. If blue litmus paper is pressed to the cut surface of a fatigued muscle, the paper color changes to red, indicating low pH. This situation is caused by the accumulation of __4__ in the muscle. Within limits, as the load on a muscle is increased, the muscle contracts __5__ strongly. The relative refractory period is the time when the muscle cell will not respond to a stimulus because __6__ is occurring.

1. _____

2. _____

3. _____

4. _____

5. _____

6. _____

8. During the experiment on muscle fatigue, how did the muscle contraction pattern change as the muscle began to fatigue?

_____

_____

How long was stimulation continued before fatigue was apparent? _____

If the sciatic nerve that stimulates the living frog's gastrocnemius muscle had been left attached to the muscle and the stimulus had been applied to the nerve rather than the muscle, would fatigue have become apparent sooner or later?

_____

Explain your answer. _____

_____

9. Explain how the weak but sustained (smooth) muscle contractions of precision movements are produced.

_____

10. What do you think happens to a muscle in the body when its nerve supply is destroyed or badly damaged?

_____

_____

11. Explain the relationship between the load on a muscle and its strength of contraction. _____

_____

_____

12. The skeletal muscles are maintained in a slightly stretched condition for optimal contraction. How is this accomplished?

_____

Why does overstretching a muscle drastically reduce its ability to contract? (Include an explanation of the events at the level

of the myofilaments.) _____

_____

_____

13. If the length but not the tension of a muscle is changed, the contraction is called an isotonic contraction. In an isometric contraction the tension is increased but the muscle does not shorten. Which type of contraction did you observe most often

during the laboratory experiments? _____

What is the role of isometric contractions in normal body functioning? _____

_____

_____

Skeletal Muscle Physiology: Computer Simulation

Electrical Stimulation

1. Complete the following statements by filling in your answer on the lines provided below.

   A motor unit consists of a __a__ and all the __b__ it innervates. If a single motor unit is stimulated, it will respond in a(n) __c__ fashion, whereas whole muscle contraction is a(n) __d__ response. In order for muscles to work in a practical sense, __e__ is the method used to produce a slow, steady increase in muscle force.
   When we see the slightest evidence of force production on a tracing, the stimulus applied must have reached __f__.
   The weakest stimulation that will elicit the strongest contraction that a muscle is capable of is called the __g__. That level of contraction is called the __h__.
   When the __i__ of stimulation is so high that the muscle tracing shows fused peaks, __j__ has been achieved.

   a. _____   f. _____

   b. _____   g. _____

   c. _____   h. _____

   d. _____   i. _____

   e. _____   j. _____

2. Name each phase of a typical muscle twitch and describe what is happening in each phase.

   a. _____

   _____

   b. _____

   _____

   c. _____

   _____

3. Explain how the PhysioEx experimental muscle stimulation differs from the *in vivo* stimulation via the nervous system. (Note that the graded muscle response following both stimulation methods is similar.)

   _____

   _____

   _____

4. What are the two *experimental* ways in which mode of stimulation can affect the muscle force?

_____ and _____

Explain your answer.

_____

_____

_____

Isometric Contraction

1. Identify the following conditions by choosing one of the key terms listed on the right.

_____ is generated by muscle tissue when it is being stretched

_____ requires the input of energy

_____ is measured by recording instrumentation during contraction

Key:

a. Total force

b. Resting force

c. Active force

2. Circle the correct response in the parentheses for each statement.

An increase in resting length results in an (increase/decrease) in passive force.

The active force initially (increased/decreased) and then (increased/decreased) as the resting length was increased from minimum to maximum.

As the total force increased, the active force (increased/decreased).

3. Explain what happens to muscle force production at extremes of length (too short or too long). Hint: think about sarcomere structure.

Muscle too short: _____

_____

Muscle too long: _____

_____

Isotonic Contraction

1. Assuming a fixed starting length, describe the effect resistance has on the initial velocity of shortening, and explain why it has this effect.

_____

_____

_____

2. A muscle has just been stimulated under conditions that will allow both isometric and isotonic contractions. Describe what is happening in terms of length and force.

Isometric: _____

_____

Isotonic: _____

_____

Terms

Select the condition from column B that most correctly identifies the term in column A.

Column A

_____ 1. muscle twitch

_____ 2. wave summation

_____ 3. multiple motor unit summation

_____ 4. resting length

_____ 5. resistance

_____ 6. initial velocity of shortening

_____ 7. isotonic shortening

_____ 8. isotonic lengthening

_____ 9. motor unit

_____ 10. whole muscle

_____ 11. tetanus

_____ 12. maximal response

Column B

a. response is all-or-none

b. affects the force a muscle can generate

c. a single contraction of intact muscle

d. recruitment

e. increasing force produced by increasing stimulus frequency

f. muscle length changing due to relaxation

g. caused by application of maximal stimulus

h. weight

i. exhibits graded response

j. high values with low resistance values

k. changing muscle length due to active forces

l. recording shows no evidence of muscle relaxation

Histology of Nervous Tissue

1. The cellular unit of the nervous system is the neuron. What is the major function of this cell type?

_____

2. Name four types of neuroglia and list at least four functions of these cells. (You will need to consult your textbook for this.)

Types   **Functions**

_____   _____

                                 _____

_____   _____

                                 _____

_____   _____

                                 _____

_____   _____

3. Match each statement with a response chosen from the key.

Key: a. afferent neuron   e. ganglion   i. nuclei
      b. association neuron   f. neuroglia   j. peripheral nervous system
      c. central nervous system   g. neurotransmitters   k. synapse
      d. efferent neuron   h. nerve   l. tract

_____ 1. the brain and spinal cord collectively

_____ 2. specialized supporting cells in the CNS

_____ 3. junction or point of close contact between neurons

_____ 4. a bundle of nerve processes inside the central nervous system

_____ 5. neuron serving as part of the conduction pathway between sensory and motor neurons

_____ 6. spinal and cranial nerves and ganglia

_____ 7. collection of nerve cell bodies found outside the CNS

_____ 8. neuron that conducts impulses away from the CNS to muscles and glands

_____ 9. neuron that conducts impulses toward the CNS from the body periphery

_____ 10. chemicals released by neurons that stimulate or inhibit other neurons or effectors

Neuron Anatomy

1. Match the following anatomical terms (column B) with the appropriate description or function (column A).

   Column A

   _____ 1. region of the cell body from which the axon originates

   _____ 2. secretes neurotransmitters

   _____ 3. receptive region of a neuron

   _____ 4. insulates the nerve fibers

   _____ 5. is site of the nucleus and the most important metabolic area

   _____ 6. may be involved in the transport of substances within the neuron

   _____ 7. essentially rough endoplasmic reticulum, important metabolically

   _____ 8. impulse generator and transmitter

   Column B

   a. axon

   b. axonal terminal

   c. axon hillock

   d. dendrite

   e. myelin sheath

   f. neuronal cell body

   g. neurofibril

   h. Nissl bodies

2. Draw a "typical" neuron in the space below. Include and label the following structures on your diagram: cell body, nucleus, Nissl bodies, dendrites, axon, axon collaterals, myelin sheath, and nodes of Ranvier.

3. How is one-way conduction at synapses assured? _____

   _____

4. What anatomical characteristic determines whether a particular neuron is classified as unipolar, bipolar, or multipolar?

   _____

   Make a simple line drawing of each type here.

   Unipolar neuron   Bipolar neuron   Multipolar neuron

5. Correctly identify the sensory (afferent) neuron, association neuron (interneuron), and motor (efferent) neuron in the figure below.

Which of these neuron types is/are unipolar? _____

Which is/are most likely multipolar? _____

Receptors (thermal
and pain in the skin)

Effector (biceps
brachii muscle)

6. Describe how the Schwann cells form the myelin sheath and the neurilemma encasing the nerve processes. (You may want to diagram the process.)

_____

_____

_____

Structure of a Nerve

1. What is a nerve? _____

_____

2. State the location of each of the following connective tissue coverings:

endoneurium _____

perineurium _____

epineurium _____

3. What is the value of the connective tissue wrappings found in a nerve? _____

_____

4. Define *mixed nerve:* _____

_____

5. Identify all indicated parts of the nerve section.

Neurophysiology of Nerve Impulses

The Nerve Impulse

1. Match each of the terms in column B to the appropriate definition in column A.

Column A

_____ 1. period of depolarization of the neuron membrane during which it cannot respond to a second stimulus

_____ 2. reversal of the resting potential owing to an influx of sodium ions

_____ 3. period during which potassium ions diffuse out of the neuron owing to a change in membrane permeability

_____ 4. self-propagated transmission of the depolarization wave along the neuronal membrane

_____ 5. mechanism in which ATP is used to move sodium out of the cell and potassium into the cell; restores the resting membrane voltage and intracellular ionic concentrations

Column B

a. action potential

b. absolute refractory period

c. depolarization

d. relative refractory period

e. repolarization

f. sodium-potassium pump

2. Respond appropriately to each statement below either by completing the statement or by answering the question raised. Insert your responses in the corresponding numbered blanks on the right.

1. The cellular unit of the nervous system is the neuron. What is the major function of this cell type?

2 and 3. What characteristics are highly developed to allow the neuron to perform this function?

4. Would a substance that decreases membrane permeability to sodium increase or decrease the probability of generating a nerve impulse?

1. _____

2. _____

3. _____

4. _____

3. Why don't the terms *depolarization* and *action potential* mean the same thing? (*Hint:* under which conditions will a local depolarization *not* lead to the action potential?) _____

_____

_____

4. A nerve generally contains many thickly myelinated fibers that typically exhibit nodes of Ranvier. An action potential is generated along these fibers by "saltatory conduction." Use an appropriate reference to explain how saltatory conduction differs from conduction along unmyelinated fibers.

_____

_____

_____

Physiology of Nerve Fibers: Eliciting and Inhibiting the Nerve Impulse

1. Respond appropriately to each question posed below. Insert your responses in the corresponding numbered blanks to the right.

   1–3. Name three types of stimuli that resulted in action potential generation in the sciatic nerve of the frog during the laboratory experiments.

   4. Which of the stimuli resulted in the most effective nerve stimulation?

   5. Which of the stimuli employed in that experiment might represent types of stimuli to which nerves in the human body are subjected?

   6. What is the usual mode of stimulus transfer in neuron-to-neuron interactions?

   7. Since the action potentials themselves were not visualized with an oscilloscope during this initial set of experiments, how did you recognize that impulses were being transmitted?

   1. _____

   2. _____

   3. _____

   4. _____

   5. _____

   6. _____

   7. _____

2. Describe the observed effects of ether on nerve-muscle interaction. _____

   _____

   Does ether exert its blocking effects on nerve *or* muscle? _____ What observations made during the experiment support your conclusion? _____

   _____

   _____

3. At what site did the tubocurarine block the impulse transmission? _____ Provide evidence from the experiment to substantiate your conclusion. _____

   _____

   _____

   Why was one of the frog's legs ligated in this experiment? _____

   _____

   _____

Visualizing the Action Potential with an Oscilloscope

1. What is a stimulus artifact? _____

_____

2. Explain why the amplitude of the action potential recorded from the frog sciatic nerve increased when the voltage of the stimulus was increased above the threshold value. _____

_____

3. What was the effect of cold temperature (flooding the nerve with iced Ringer's solution) on the functioning of the sciatic nerve tested? _____

_____

4. When the nerve was reversed in position, was the impulse conducted in the opposite direction? _____

   How can this result be reconciled with the concept of one-way conduction in neurons? _____

_____

_____

_____

_____

_____

Gross Anatomy of the Brain and Cranial Nerves

NAME _____ LAB TIME/DATE _____

The Human Brain

1. Match the letters on the diagram of the human brain (right lateral view) to the appropriate terms listed at the left:

_____ 1. frontal lobe

_____ 2. parietal lobe

_____ 3. temporal lobe

_____ 4. precentral gyrus

_____ 5. parieto-occipital sulcus

_____ 6. postcentral gyrus

_____ 7. lateral sulcus _____ 10. medulla

_____ 8. central sulcus _____ 11. occipital lobe

_____ 9. cerebellum _____ 12. pons

2. In which of the cerebral lobes would the following functional areas be found?

auditory area _____   olfactory area _____

primary motor area _____   visual area _____

primary sensory area _____   Broca's area _____

3. Which of the following structures are not part of the brain stem? (Circle the appropriate response or responses.)

cerebral hemispheres   pons   midbrain   cerebellum   medulla   diencephalon

4. Complete the following statements by writing the proper word or phrase on the corresponding blanks at the right.

A(n) __1__ is an elevated ridge of cerebral tissue. The convolutions seen in the cerebrum are important because they increase the __2__. Gray matter is composed of __3__. White matter is composed of __4__. A fiber tract that provides for communication between different parts of the same cerebral hemisphere is called a(n) __5__, whereas one that carries impulses to and from the cerebrum from and to lower CNS areas is called a(n) __6__ tract. The lentiform nucleus along with the amygdaloid and caudate nuclei are collectively called the __7__.

1. _____

2. _____

3. _____

4. _____

5. _____

6. _____

7. _____

5. Identify the structures on the following sagittal view of the human brain by matching the lettered areas to the proper terms at the left:

_____ 1. cerebellum

_____ 2. cerebral aqueduct

_____ 3. cerebral hemisphere

_____ 4. cerebral peduncle

_____ 5. choroid plexus

_____ 6. corpora quadrigemina

_____ 7. corpus callosum

_____ 8. fornix

_____ 9. fourth ventricle

_____ 10. hypothalamus

_____ 11. mammillary bodies   _____ 14. optic chiasma   _____ 17. pons

_____ 12. massa intermedia   _____ 15. pineal body   _____ 18. septum pellucidum

_____ 13. medulla oblongata   _____ 16. pituitary gland   _____ 19. thalamus

6. Using the letters from the diagram in item 5, match the appropriate structures with the descriptions given below:

_____ 1. site of regulation of body temperature and water balance; most important autonomic center

_____ 2. consciousness depends on the function of this part of the brain

_____ 3. located in the midbrain; contains reflex centers for vision and audition

_____ 4. responsible for regulation of posture and coordination of complex muscular movements

_____ 5. important synapse site for afferent fibers traveling to the sensory cortex

_____ 6. contains autonomic centers regulating blood pressure, heart rate, and respiratory rhythm, as well as coughing, sneezing, and swallowing centers

_____ 7. large commissure connecting the cerebral hemispheres

_____ 8. fiber tract involved with olfaction

_____ 9. connects the third and fourth ventricles

_____ 10. encloses the third ventricle

7. Embryologically, the brain arises from the rostral end of a tubelike structure that quickly becomes divided into three major regions. Groups of structures that develop from the embryonic brain are listed below. Designate the embryonic origin of each group as the hindbrain, midbrain, or forebrain.

_____ 1. the diencephalon, including the thalamus, optic chiasma, and hypothalamus

_____ 2. the medulla, pons, and cerebellum

_____ 3. the cerebral hemispheres

8. What is the function of the basal nuclei? _____

_____

9. What is the corpus striatum, and how is it related to the fibers of the internal capsule? _____

_____

_____

10. A brain hemorrhage within the region of the right internal capsule results in paralysis of the left side of the body.

Explain why the left side (rather than the right side) is affected. _____

_____

11. Explain why trauma to the base of the brain is often much more dangerous than trauma to the frontal lobes. (*Hint:* Think about the relative functioning of the cerebral hemispheres and the brain stem structures. Which contain centers more vital to life?)

_____

_____

_____

12. In "split brain" experiments, the main commissure connecting the cerebral hemispheres is cut. First, name this commissure:

_____

Then, describe what results (in terms of behavior) can be anticipated in such experiments. (Use an appropriate reference if you need help with this one!)

_____

_____

_____

_____

_____

Meninges of the Brain

Identify the meningeal (or associated) structures described below:

_____ 1. outermost meninx covering the brain; composed of tough fibrous connective tissue

_____ 2. innermost meninx covering the brain; delicate and highly vascular

_____ 3. structures instrumental in returning cerebrospinal fluid to the venous blood in the dural sinuses

_____ 4. structure that forms the cerebrospinal fluid

_____ 5. middle meninx; like a cobweb in structure

_____ 6. its outer layer forms the periosteum of the skull

_____ 7. a dural fold that attaches the cerebrum to the crista galli of the skull

_____ 8. a dural fold separating the cerebrum from the cerebellum

Cerebrospinal Fluid

Fill in the following flowchart by delineating the circulation of cerebrospinal fluid from its formation site (assume that this is one of the lateral ventricles) to the site of its reabsorption into the venous blood:

Lateral ventricle ------------- > _____ ------------- > Third ventricle ---- >

------------------------- > _____ ------------- > _____

_____ ------------- > _____

_____ > _____ --- >

_____ surrounding the brain and cord ------------- > Arachnoid villi -------- >
(and central canal of the cord)

------------------------- > _____ containing venous blood

Now label appropriately the structures involved with circulation of CS fluid on the accompanying diagram. (These structures are identified by leader lines.)

Cranial Nerves

1. Using the terms below, correctly identify all structures indicated by leader lines on the diagram below.

   a. abducens nerve (VI)

   b. accessory nerve (XI)

   c. cerebellum

   d. cerebral peduncle

   e. decussation of the pyramids

   f. facial nerve (VII)

   g. frontal lobe of cerebral hemisphere

   h. glossopharyngeal nerve (IX)

   i. hypoglossal nerve (XII)

   j. longitudinal fissure

   k. mammillary body

   l. medulla oblongata

   m. oculomotor nerve (III)

   n. olfactory bulb

   o. olfactory tract

   p. optic chiasma

   q. optic nerve (II)

   r. optic tract

   s. pituitary gland

   t. pons

   u. spinal cord

   v. temporal lobe of cerebral hemisphere

   w. trigeminal nerve (V)

   x. trochlear nerve (IV)

   y. vagus nerve (X)

   z. vestibulocochlear nerve (VIII)

2. Provide the name and number of the cranial nerves involved in each of the following activities, sensations, or disorders:

_____ 1. shrugging the shoulders

_____ 2. smelling a flower

_____ 3. raising the eyelids; focusing the lens of the eye for accommodation; and pupillary constriction

_____ 4. slows the heart; increases the mobility of the digestive tract

_____ 5. involved in Bell's palsy (facial paralysis)

_____ 6. chewing food

_____ 7. listening to music; seasickness

_____ 8. secretion of saliva; tasting well-seasoned food

_____ 9. involved in "rolling" the eyes (three nerves—provide numbers only)

_____ 10. feeling a toothache

_____ 11. reading *Mad* magazine

_____ 12. purely sensory in function (three nerves—provide numbers only)

Dissection of the Sheep Brain

1. In your own words, describe the relative hardness of the sheep brain tissue as observed when cutting into it.

_____

Because formalin hardens all tissue, what conclusions might you draw about the relative hardness and texture of living brain

tissue? _____

2. How does the relative size of the cerebral hemispheres compare in sheep and human brains? _____

_____

What is the significance? _____

_____

3. What is the significance of the fact that the olfactory bulbs are much larger in the sheep brain than in the human

brain? _____

_____

Electroencephalography

Brain Wave Patterns and the Electroencephalogram

1. Define *EEG*. _____

2. What are the four major types of brain wave patterns? _____

   Match each statement below to a type of brain wave pattern:

   _____ below 4 Hz; slow, large waves; normally seen during deep sleep

   _____ rhythm generally apparent when an individual is in a relaxed, nonattentive state with the eyes closed

   _____ correlated to the alert state; usually about 15 to 25 Hz

3. What is meant by the term *alpha block*? _____

   _____

   _____

4. List at least four types of brain lesions that may be determined by EEG studies.

   _____

5. What is the common result of hypoactivity or hyperactivity of the brain neurons? _____

Observing Brain Wave Patterns

1. How was alpha block demonstrated in the laboratory experiment? _____

   _____

2. What was the effect of mental concentration on the brain wave pattern? _____

   _____

   _____

3. What effect on the brain wave pattern did hyperventilation have? _____

   Why? _____

   _____

Spinal Cord, Spinal Nerves, and the Autonomic Nervous System

Anatomy of the Spinal Cord

1. Match the descriptions given below to the proper anatomical term:

a. cauda equina   b. conus medullaris   c. filum terminale   d. foramen magnum

_____ 1.   most superior boundary of the spinal cord

_____ 2.   meningeal extension beyond the spinal cord terminus

_____ 3.   spinal cord terminus

_____ 4.   collection of spinal nerves traveling in the vertebral canal below the terminus of the spinal cord

2. Using the terms below, correctly identify on the diagram all structures provided with leader lines.

a. anterior (ventral) horn   f. dorsal root of spinal nerve   k. posterior (dorsal) horn

b. arachnoid mater   g. dura mater   l. spinal nerve

c. central canal   h. gray commissure   m. ventral ramus of spinal nerve

d. dorsal ramus of spinal nerve   i. lateral horn   n. ventral root of spinal nerve

e. dorsal root ganglion   j. pia mater   o. white matter

3. Choose the proper answer from the following key to respond to the descriptions relating to spinal cord anatomy.

Key: a.   afferent   b.   efferent   c.   both afferent and efferent   d.   association

_____ 1.   neuron type found in dorsal horn   _____ 4.   fiber type in ventral root

_____ 2.   neuron type found in ventral horn   _____ 5.   fiber type in dorsal root

_____ 3.   neuron type in dorsal root ganglion   _____ 6.   fiber type in spinal nerve

4. Where in the vertebral column is a lumbar puncture generally done? _____

Why is this the site of choice? _____

_____

5. The spinal cord is enlarged in two regions, the _____ and the _____ regions.

What is the significance of these enlargements? _____

_____

6. How does the position of the gray and white matter differ in the spinal cord and the cerebral hemispheres?

_____

_____

7. Choose the name of the tract, from the following key, that might be damaged when the following conditions are observed. (More than one choice may apply.)

_____ 1.   uncoordinated movement   Key: a.   fasciculus gracilis
                                                                              b.   fasciculus cuneatus
_____ 2.   lack of voluntary movement   c.   lateral corticospinal tract
                                                                              d.   anterior corticospinal tract
_____ 3.   tremors, jerky movements   e.   tectospinal tract
                                                                              f.   rubrospinal tract
_____ 4.   diminished pain perception   g.   lateral spinothalamic tract
                                                                              h.   anterior spinothalamic tract
_____ 5.   diminished sense of touch   i.   posterior spinocerebellar tract
                                                                              j.   vestibulospinal tract
                                                                              k.   olivospinal tract
                                                                              l.   anterior spinocerebellar tract

8. Use an appropriate reference to describe the functional significance of an upper motor neuron and a lower motor neuron:

upper motor neuron _____

_____

lower motor neuron _____

_____

Will contraction of a muscle occur if the lower motor neurons serving it have been destroyed? _____ If the upper motor

neurons serving it have been destroyed? _____ Using an appropriate reference, differentiate between flaccid and spastic

paralysis and note the possible causes of each. _____

_____

_____

_____

_____

Spinal Nerves and Nerve Plexuses

1. In the human, there are 31 pairs of spinal nerves named according to the region of the vertebral column from which they issue. The spinal nerves are named below; note, by number, the vertebral level at which they emerge:

   cervical nerves _____   sacral nerves _____

   lumbar nerves _____   thoracic nerves _____

2. The ventral rami of spinal nerves C_1 through T_1 and T_{12} through S_4 take part in forming _____,

   which serve the _____ of the body. The ventral rami of T_2 through T_{12} run

   between the ribs to serve the _____. The dorsal rami of the spinal nerves

   serve _____.

3. What would happen if the following structures were damaged or transected? (Use key choices for responses.)

   Key: a. loss of motor function   b. loss of sensory function   c. loss of both motor and sensory function

   _____ 1. dorsal root of a spinal nerve   _____ 3. anterior ramus of a spinal nerve

   _____ 2. ventral root of a spinal nerve

4. Define _plexus:_ _____

   _____

5. Name the major nerves that serve the following body areas:

   _____ 1. head, neck, shoulders (name plexus only)

   _____ 2. diaphragm

   _____ 3. posterior thigh

   _____ 4. leg and foot (name two)

   _____ 5. most anterior forearm muscles

   _____ 6. arm muscles (name two)

   _____ 7. abdominal wall (name plexus only)

   _____ 8. anterior thigh

   _____ 9. medial side of the hand

The Autonomic Nervous System

1. For the most part, sympathetic and parasympathetic fibers serve the same organs and structures. How can they exert antagonistic effects? (After all, nerve impulses are nerve impulses—aren't they?)

   _____

2. Name three structures that receive sympathetic but not parasympathetic innervation.

   _____

   _____

3. A pelvic splanchnic nerve contains (circle one):

   a. preganglionic sympathetic fibers   c. preganglionic parasympathetic fibers

   b. postganglionic sympathetic fibers   d. postganglionic parasympathetic fibers

4. The following chart states a number of conditions. Use a check mark to show which division of the autonomic nervous system is involved in each.

Sympathetic division	Condition	Parasympathetic division
	Secretes norepinephrine; adrenergic fibers	
	Secretes acetylcholine; cholinergic fibers	
	Long preganglionic axon; short postganglionic axon	
	Short preganglionic axon; long postganglionic axon	
	Arises from cranial and sacral nerves	
	Arises from spinal nerves T$_1$ through L$_3$	
	Normally in control	
	"Fight or flight" system	
	Has more specific control (Look it up!)	

5. You are alone in your home late in the evening, and you hear an unfamiliar sound in your backyard. List four physiologic events promoted by the sympathetic nervous system that would aid you in coping with this rather frightening situation:

   _____

   _____

   _____

   _____

   Often after surgery, people are temporarily unable to urinate, and bowel sounds are absent. What division of the ANS is

   affected by the anesthesia? _____

Human Reflex Physiology

The Reflex Arc

1. Define *reflex:* _____

    _____

2. Name five essential components of a reflex arc: _____, _____,

    _____, _____, and _____

3. In general, what is the importance of reflex testing in a routine physical examination? _____

    _____

    _____

Somatic and Autonomic Reflexes

1. Use the key terms to complete the statements given below.

    Key: a. abdominal reflex   d. corneal reflex   g. patellar reflex
          b. Achilles jerk   e. crossed extensor reflex   h. plantar reflex
          c. ciliospinal reflex   f. gag reflex   i. pupillary light reflex

    Reflexes classified as somatic reflexes include a _____, _____, _____, _____, _____, _____, and _____.

    Of these, the simple stretch reflexes are _____ and _____, and the superficial cord reflexes are _____ and _____.

    Reflexes classified as autonomic reflexes include _____ and _____.

2. In what way do cord-mediated reflexes differ from those involving higher brain centers? _____

    _____

    Name two cord-mediated reflexes: _____ and _____

    _____

    Name two somatic reflexes in which the higher brain centers participate: _____

    and _____

3. Can the stretch reflex be elicited in a pithed animal? _____

    Explain your answer. _____

    _____

4. Trace the reflex arc, naming efferent and afferent nerves, receptors, effectors, and integration centers, for the following reflexes:

patellar reflex _____

_____

_____

Achilles reflex _____

_____

_____

5. Three factors that influence the rapidity and effectiveness of reflex arcs were investigated in conjunction with patellar reflex testing—mental distraction, effect of simultaneous muscle activity in another body area, and fatigue.

Which of these factors increases the excitatory level of the spinal cord? _____

Which factor decreases the excitatory level of the muscles? _____

When the subject was concentrating on an arithmetic problem, did the change noted in the patellar reflex indicate that brain

activity is necessary for the patellar reflex or only that it may modify it? _____

_____

6. Name the division of the autonomic nervous system responsible for each of the following reflexes:

ciliospinal reflex _____ salivary reflex _____

pupillary light reflex _____

7. The pupillary light reflex, the crossed extensor reflex, and the corneal reflex illustrate the purposeful nature of reflex activity. Describe the protective aspect of each:

pupillary light reflex _____

corneal reflex _____

crossed extensor reflex _____

_____

_____

_____

_____

8. Was the pupillary consensual response contralateral or ipsilateral? _____

Why would such a response be of significant value in this particular reflex? _____

_____

9. Differentiate between the types of activities accomplished by somatic and autonomic reflexes. _____

_____

_____

_____

10. Several types of reflex activity were not investigated in this exercise. The most important of these are autonomic reflexes, which are difficult to illustrate in a laboratory situation. To rectify this omission, complete the following chart, using references as necessary.

Reflex	Organ involved	Receptors stimulated	Action
Micturition (urination)			
Hering-Breuer			
Defecation			
Carotid sinus			

Reaction Time of Unlearned Responses

1. Name at least three factors that may modify reaction time to a stimulus. _____

    _____

2. In general, how did the response time for the unlearned activity performed in the laboratory compare to that for the simple

    patellar reflex? _____

3. Did the response time without verbal stimuli decrease with practice? _____ Explain the reason for this.

    _____

4. Explain, in detail, why response time increased when the subject had to react to a word stimulus.

    _____

    _____

General Sensation

Structure of General Sensory Receptors

1. Differentiate between interoceptors and exteroceptors relative to location and stimulus source:

   Interoceptor: _____

   Exteroceptor: _____

2. A number of activities and sensations are listed in the chart below. For each, check whether the receptors would be exteroceptors or interoceptors; and then name the specific receptor types. (Because visceral receptors were not described in detail in this exercise, you need only indicate that the receptor is a visceral receptor if it falls into that category.)

Activity or sensation	Exteroceptor	Interoceptor	Specific receptor type
Backing into a sun-heated iron railing			
Someone steps on your foot			
Reading a book			
Leaning on your elbows			
Doing sit-ups			
The "too full" sensation			
Seasickness			

Receptor Physiology

1. Explain how the sensory receptors act as transducers: _____

   _____

2. Define *stimulus:* _____

   _____

3. What was demonstrated by the two-point discrimination test? _____

   _____

   How did the accuracy of the subject's tactile localization correlate with the results of the two-point discrimination test?

   _____

4. Define *punctate distribution:* _____

_____

5. Several questions regarding general sensation are posed below. Answer each by placing your response in the appropriately numbered blanks to the right.

    1.   Which cutaneous receptors are the most numerous?

    2–3.   Which two body areas tested were most sensitive to touch?

    4–5.   Which two body areas tested were least sensitive to touch?

    6.   Which appear to be more numerous—receptors that respond to cold or to heat?

    7–9.   Where would referred pain appear if the following organs were receiving painful stimuli—(7) gallbladder, (8) kidneys, and (9) appendix? (Use your textbook if necessary.)

    10.   Where was referred pain felt when the elbow was immersed in ice water during the laboratory experiment?

    11.   What region of the cerebrum interprets the kind and intensity of stimuli that cause cutaneous sensations?

1. _____

2. _____

3. _____

4. _____

5. _____

6. _____

7. _____

8. _____

9. _____

10. _____

11. _____

6. Define *adaptation:* _____

_____

7. Why is it advantageous to have pain receptors that are sensitive to all vigorous stimuli, whether heat, cold, or pressure?

_____

Why is the nonadaptability of pain receptors important? _____

_____

8. Imagine yourself without any cutaneous sense organs. Why might this be very dangerous? _____

_____

_____

9. Define *referred pain:* _____

_____

What is the probable explanation for referred pain? (Consult your textbook or an appropriate reference if necessary.)

_____

_____

Special Senses: Vision

Anatomy of the Eye

1. Several accessory eye structures contribute to the formation of tears and/or aid in lubrication of the eyeball. Name three of these and then name the major secretory product of each. Indicate which has antibacterial properties by circling the correct secretory product. Students can choose any three of the following five answers.

Accessory structures	Product

2. The eyeball is wrapped in adipose tissue within the orbit. What is the function of the adipose tissue?

_____

What seven bones form the bony orbit? (Think! If you can't remember, check a skull or your text.)

_____   _____   _____

_____   _____

_____   _____

3. Why does one often have to blow one's nose after having a good cry? _____

_____

4. Identify the extrinsic eye muscle predominantly responsible for the actions described below.

_____ 1.   turns the eye laterally

_____ 2.   turns the eye medially

_____ 3.   turns the eye up and laterally

_____ 4.   turns the eye inferiorly

_____ 5.   turns the eye superiorly

_____ 6.   turns the eye down and laterally

5. What is a sty? _____

_____

Conjunctivitis? _____

6. Using the terms listed on the right, correctly identify all structures provided with leader lines in the diagram.

Blowup of
photosensitive
retina

Pigmented
epithelium

a. anterior chamber

b. anterior segment
    containing aqueous
    humor

c. bipolar neurons

d. ciliary body and
    processes

e. ciliary muscle

f. choroid

g. cornea

h. dura mater

i. fovea centralis

j. ganglion cells

k. iris

l. lens

m. optic disc

n. optic nerve

o. photoreceptors

p. posterior chamber

q. retina

r. sclera

s. scleral venous sinus

t. suspensory
    ligaments

u. vitreous body in
    posterior segment

Notice the arrows drawn close to the left side of the iris in the diagram opposite. What do they indicate?

_____

_____

_____

7. Match the key responses with the descriptive statements that follow.

Key: a. aqueous humor   e. cornea   j. retina
           b. choroid   f. fovea centralis   k. sclera
           c. ciliary body   g. iris   l. scleral venous sinus
           d. ciliary processes of   h. lens   m. suspensory ligament
               the ciliary body   i. optic disc   n. vitreous humor

_____ 1. attaches the lens to the ciliary body

_____ 2. fluid filling the anterior segment of the eye

_____ 3. the "white" of the eye

_____ 4. part of the retina that lacks photoreceptors

_____ 5. modification of the choroid that controls the shape of the crystalline lens

_____ 6. contains the ciliary muscle

_____ 7. drains the aqueous humor from the eye

_____ 8. tunic containing the rods and cones

_____ 9. substance occupying the posterior segment of the eyeball

_____ 10. forms the bulk of the heavily pigmented vascular tunic

_____, _____ 11. smooth muscle structures

_____ 12. area of critical focusing and discriminatory vision

_____ 13. form (by filtration) the aqueous humor

_____, _____, _____,

_____ 14. light-bending media of the eye

_____ 15. anterior continuation of the sclera—your "window on the world"

_____ 16. composed of tough, white, opaque, fibrous connective tissue

8. The iris is composed primarily of two smooth muscle layers, one arranged radially and the other circularly.

Which of these dilates the pupil? _____

9. You would expect the pupil to be dilated in which of the following circumstances? Circle the correct response(s).

a. in brightly lit surroundings   c. during focusing for near vision

b. in dimly lit surroundings   d. in observing distant objects

10. The intrinsic eye muscles are under the control of which of the following? (Circle the correct response.)

                autonomic nervous system   somatic nervous system

Microscopic Anatomy of the Retina

1. The two major layers of the retina are the epithelial and nervous layers. In the nervous layer, the neuron populations are arranged as follows from the epithelial layer to the vitreous humor. (Circle all proper responses.)

   bipolar cells, ganglion cells, photoreceptors   photoreceptors, ganglion cells, bipolar cells

   ganglion cells, bipolar cells, photoreceptors   photoreceptors, bipolar cells, ganglion cells

2. The axons of the _____ cells form the optic nerve, which exits from the eyeball.

3. Complete the following statements by writing either *rods* or *cones* on each blank:

   The dim light receptors are the _____. Only _____ are

   found in the fovea centralis, whereas mostly _____ are found in the periphery of the retina.

   _____ are the photoreceptors that operate best in bright light and allow for color vision.

Visual Pathways to the Brain

1. The visual pathway to the occipital lobe of the brain consists most simply of a chain of five neurons. Beginning with the photoreceptor cell of the retina, name them and note their location in the pathway.

   (1) _____   (4) _____

   (2) _____   (5) _____

   (3) _____   _____

2. Visual field tests are done to reveal destruction along the visual pathway from the retina to the optic region of the brain. Note where the lesion is likely to be in the following cases:

   Normal vision in left eye visual field; absence of vision in right eye visual field: _____

   Normal vision in both eyes for right half of the visual field; absence of vision in both eyes for left half of the visual

   field: _____

3. How is the right optic *tract* anatomically different from the right optic *nerve*? _____

   _____

   _____

   _____

Dissection of the Cow (Sheep) Eye

1. What modification of the choroid that is not present in humans is found in the cow eye? _____

   _____

   What is its function? _____

   _____

2. What is the anatomical appearance of the retina? _____

_____

At what point is it attached to the posterior aspect of the eyeball? _____

Visual Tests and Experiments

1. Match the terms in column B with the descriptions in column A:

Column A	Column B
_____ 1. light bending	a. accommodation
_____ 2. ability to focus for close (under 20 ft) vision	b. astigmatism
_____ 3. normal vision	c. convergence
_____ 4. inability to focus well on close objects (farsightedness)	d. emmetropia
_____ 5. nearsightedness	e. hyperopia
_____ 6. blurred vision due to unequal curvatures of the lens or cornea	f. myopia
_____ 7. medial movement of the eyes during focusing on close objects	g. refraction

2. Complete the following statements:

In farsightedness, the light is focused __1__ the retina. The lens required to treat myopia is a __2__ lens. The "near point" increases with age because the __3__ of the lens decreases as we get older. A convex lens, like that of the eye, produces an image that is upside down and reversed from left to right. Such an image is called a __4__ image.

1. _____

2. _____

3. _____

4. _____

3. Use terms from the key to complete the statements concerning near and distance vision.

Key:　a. contracted　　b. decreased　　c. increased　　d. relaxed　　e. taut

During distance vision: The ciliary muscle is _____, the suspensory ligament is _____, the convexity of the lens

is _____, and light refraction is _____. During close vision: The ciliary muscle is _____, the suspensory ligament is

_____, lens convexity is _____, and light refraction is _____.

4. Explain why vision is lost when light hits the blind spot. _____

5. What is meant by the term *negative afterimage* and what does this phenomenon indicate? _____

_____

_____

_____

6. Record your Snellen eye test results below:

Left eye (without glasses) _____ (with glasses) _____

Right eye (without glasses) _____ (with glasses) _____

Is your visual acuity normal, less than normal, or better than normal? _____

Explain. _____

_____

Explain why each eye is tested separately when using the Snellen eye chart. _____

_____

Explain 20/40 vision. _____

_____

Explain 20/10 vision. _____

_____

7. Define *astigmatism:* _____

_____

How can it be corrected? _____

8. Record the distance of your near point of accommodation as tested in the laboratory:

right eye _____ left eye _____

Is your near point within the normal range for your age? _____

9. Define *presbyopia:* _____

_____

What causes it? _____

10. To which wavelengths of light do the three cone types of the retina respond maximally?

_____, _____, and _____

11. How can you explain the fact that we see a great range of colors even though only three cone types exist?

_____

_____

12. From what condition does color blindness result? _____

_____

13. Record the results of the demonstration of the relative positioning of rods and cones in the circle below (use appropriately colored pencils).

14. Explain the difference between binocular and panoramic vision. _____

_____

_____

_____

_____

What is the advantage of binocular vision? _____

What factor(s) are responsible for binocular vision? _____

_____

_____

15. In the experiment on the convergence reflex, what happened to the position of the eyeballs as the object was moved closer

to the subject's eyes? _____

What extrinsic eye muscles control the movement of the eyes during this reflex? _____

What is the value of this reflex? _____

_____

What would be the visual result of an inability of these muscles to function? _____

_____

16. In the experiment on the photopupillary reflex, what happened to the pupil of the eye exposed to light?

_____ What happened to the pupil of the nonilluminated eye?_____

Explanation? _____

_____

_____

17. Why is the ophthalmoscopic examination an important diagnostic tool? _____

_____

18. Many college students struggling through mountainous reading assignments are told that they need glasses for "eyestrain." Why is it more of a strain on the extrinsic and intrinsic eye muscles to look at close objects than at far objects?

_____

_____

Special Senses: Hearing and Equilibrium

Anatomy of the Ear

1. Select the terms from column B that apply to the column A descriptions. Some terms are used more than once.

Column A

_____, _____,

_____ 1. structures comprising the outer or external ear

_____, _____,

_____ 2. structures composing the inner ear

_____, _____,

_____ 3. collectively called the ossicles

_____, _____ 4. ear structures not involved with audition

_____ 5. involved in equalizing the pressure in the middle ear with atmospheric pressure

_____ 6. vibrates at the same frequency as sound waves hitting it; transmits the vibrations to the ossicles

_____, _____ 7. contain receptors for the sense of balance

_____ 8. transmits the vibratory motion of the stirrup to the fluid in the scala vestibuli of the inner ear

_____ 9. acts as a pressure relief valve for the increased fluid pressure in the scala tympani; bulges into the tympanic cavity

_____ 10. passage between the throat and the tympanic cavity

_____ 11. fluid contained within the membranous labyrinth

_____ 12. fluid contained within the osseous labyrinth and bathing the membranous labyrinth

Column B

a. auditory (pharyngotympanic) tube

b. anvil (incus);

c. cochlea

d. endolymph

e. external auditory canal

f. hammer (malleus)

g. oval window

h. perilymph

i. pinna

j. round window

k. semicircular canals

l. stirrup (stapes)

m. tympanic membrane

n. vestibule

2. Sound waves hitting the eardrum initiate its vibratory motion. Trace the pathway through which vibrations and fluid currents are transmitted to finally stimulate the hair cells in the organ of Corti. (Name the appropriate ear structures in their correct

sequence.) Eardrum → _____

_____

_____

3. Identify all indicated structures and ear regions in the following diagram.

4. Match the membranous labyrinth structures listed in column B with the descriptive statements in column A:

Column A

_____, _____ 1. sacs found within the vestibule

_____ 2. contains the organ of Corti

_____, _____ 3. sites of the maculae

_____ 4. positioned in all spatial planes

_____ 5. hair cells of organ of Corti rest on this membrane

_____ 6. gelatinous membrane overlying the hair cells of the organ of Corti

_____ 7. contains the crista ampullaris

_____, _____, _____, _____ 8. function in static equilibrium

_____, _____, _____, _____ 9. function in dynamic equilibrium

Column B

a. ampulla

b. basilar membrane

c. cochlear duct

d. cochlear nerve

e. cupula

f. otoliths

g. saccule

h. semicircular ducts

i. tectorial membrane

_____ 10. carries auditory information to the brain

_____ 11. gelatinous cap overlying hair cells of the crista ampullaris

_____ 12. grains of calcium carbonate in the maculae

j.   utricle

k.   vestibular nerve

5. Describe how sounds of different frequency (pitch) are differentiated in the cochlea. _____

_____

_____

_____

6. Explain the role of the endolymph of the semicircular canals in activating the receptors during angular motion.

_____

_____

_____

7. Explain the role of the otoliths in perception of static equilibrium (head position). _____

_____

_____

_____

Laboratory Tests

1. Was the auditory acuity measurement made during the experiment on page 256 the same or different for both ears?

_____ What factors might account for a difference in the acuity of the two ears?

_____

_____

2. During the sound localization experiment on page 256, in which position(s) was the sound least easily located?

_____

How can this phenomenon be explained? _____

_____

3. In the experiment on page 256, which tuning fork was the most difficult to hear? _____ Hz

What conclusion can you draw? _____

_____

4. When the tuning fork handle was pressed to your forehead during the Weber test, where did the sound seem to originate?

_____

Where did it seem to originate when one ear was plugged with cotton? _____

How do sound waves reach the cochlea when conduction deafness is present? _____

_____

5. Indicate whether the following conditions relate to conduction deafness (C) or sensorineural deafness (S):

_____ 1.   can result from the fusion of the ossicles

_____ 2.   can result from a lesion on the cochlear nerve

_____ 3.   sound heard in one ear but not in the other during bone and air conduction

_____ 4.   can result from otitis media

_____ 5.   can result from impacted cerumen or a perforated eardrum

_____ 6.   can result from a blood clot in the auditory cortex

6. The Rinne test evaluates an individual's ability to hear sounds conducted by air or bone. Which is more indicative of normal

hearing? _____

7. Define *nystagmus*: _____

_____

Define *vertigo*: _____

_____

8. The Barany test investigated the effect that rotatory acceleration had on the semicircular canals. Explain *why* the subject still

had the sensation of rotation immediately after being stopped. _____

_____

9. What is the usual reason for conducting the Romberg test? _____

_____

Was the degree of sway greater with the eyes open or closed? _____

Why? _____

10. Normal balance, or equilibrium, depends on input from a number of sensory receptors. Name them.

_____

_____

Special Senses: Olfaction and Taste

Localization and Anatomy of Taste Buds

1. Name three sites where receptors for taste are found, and circle the predominant site:

   _____ , _____ , and

   _____

2. Describe the cellular makeup and arrangement of a taste bud. (Use a diagram, if helpful.) _____

   _____

Localization and Anatomy of the Olfactory Receptors

1. Describe the cellular composition and the location of the olfactory epithelium. _____

   _____

   _____

2. How and why does sniffing improve your sense of smell? _____

   _____

Laboratory Experiments

1. Taste and smell receptors are both classified as _____ , because they both

   respond to _____

2. Why is it impossible to taste substances with a dry tongue? _____

   _____

3. State the most important sites of the taste-specific receptors, as determined during the plotting exercise in the laboratory:

   salt _____   sour _____

   bitter _____   sweet _____

4. The basic taste sensations are elicited by specific chemical substances or groups. Name them:

   salt _____   sour _____

   bitter _____   sweet _____

5. Name three factors that influence our appreciation of foods. Substantiate each choice with an example from the laboratory experience.

_____   Substantiation _____

                                                    _____

                                                    _____

_____   Substantiation _____

                                                    _____

                                                    _____

_____   Substantiation _____

                                                    _____

                                                    _____

Which of the factors chosen is most important? _____

Substantiate your choice with an example from everyday life. _____

_____

Expand on your explanation and choices by explaining why a cold, greasy hamburger is unappetizing to most people.

_____

_____

6. Babies tend to favor bland foods, whereas adults tend to like highly seasoned foods. What is the basis for this phenomenon?

_____

_____

7. How palatable is food when you have a cold? _____

Explain. _____

_____

8. What is the mechanism of olfactory adaptation? _____

_____

_____

In your opinion, is olfactory adaptation desirable? _____ Explain your answer.

_____

_____

Anatomy and Basic Function of the Endocrine Glands

Gross Anatomy and Basic Function of the Endocrine Glands

1. Both the endocrine and nervous systems are major regulating systems of the body; however, the nervous system has been compared to an airmail delivery system and the endocrine system to the pony express. Briefly explain this comparison.

_____

_____

_____

2. Define hormone: _____

_____

3. Chemically, hormones belong chiefly to two molecular groups, the _____

and the _____ .

4. What do all hormones have in common? _____

_____

5. Define *target organ:* _____

_____

6. Why don't all tissues respond to all hormones? _____

_____

7. Identify the endocrine organ described by the following statements:

_____ 1. located in the throat; bilobed gland connected by an isthmus

_____ 2. found close to the kidney

_____ 3. a mixed gland, located close to the stomach and small intestine

_____ 4. paired glands suspended in the scrotum

_____ 5. ride "horseback" on the thyroid gland

_____ 6. found in the pelvic cavity of the female, concerned with ova and female hormone production

_____ 7. found in the upper thorax overlying the heart; large during youth

_____ 8. found in the roof of the third ventricle

8. For each statement describing hormonal effects, identify the hormone(s) involved by choosing a number from key A, and note the hormone's site of production with a letter from key B. More than one hormone may be involved in some cases.

Key A:

1. ACTH
2. ADH
3. aldosterone
4. cortisone
5. epinephrine
6. estrogens
7. FSH
8. glucagon
9. GH
10. insulin
11. LH
12. melatonin

13. MSH
14. oxytocin
15. progesterone
16. prolactin
17. PTH
18. serotonin
19. testosterone
20. thymosin
21. thyrocalcitonin/calcitonin
22. T_4 / T_3
23. TSH

Key B:

a. adrenal cortex
b. adrenal medulla
c. anterior pituitary
d. hypothalamus
e. ovaries
f. pancreas
g. parathyroid glands
h. pineal gland
i. posterior pituitary
j. testes
k. thymus gland
l. thyroid gland

_____, _____ 1. basal metabolism hormone

_____, _____ 2. programming of T lymphocytes

_____, _____ and _____, _____ 3. regulate blood calcium levels

_____, _____ and _____, _____ 4. released in response to stressors

_____, _____ and _____, _____ 5. drives development of secondary sexual characteristics

_____, _____; _____, _____; _____, _____; and _____, _____ 6. regulate the function of another endocrine gland

_____, _____ 7. mimics the sympathetic nervous system

_____, _____ and _____, _____ 8. regulate blood glucose levels; produced by the same "mixed" gland

_____, _____ and _____, _____ 9. directly responsible for regulation of the menstrual cycle

_____, _____ and _____, _____ 10. regulate the ovarian cycle

_____, _____ and _____, _____ 11. maintenance of salt and water balance in the ECF

_____, _____ and _____, _____ 12. directly involved in milk production and ejection

_____, _____ and _____, _____ 13. questionable function; may stimulate the melanocytes of the skin

9. Although the pituitary gland is often referred to as the master gland of the body, the hypothalamus exerts some control over the pituitary gland. How does the hypothalamus control both anterior and posterior pituitary functioning?

_____

_____

_____

_____

_____

10. Indicate whether the release of the hormones listed below is stimulated by (A) another hormone; (B) the nervous system (neurotransmitters, or releasing factors); or (C) humoral factors (the concentration of specific nonhormonal substances in the blood or extracellular fluid):

_____ 1. T_4 / T_3 _____ 4. parathyroid hormone   _____ 7. ADH

_____ 2. insulin   _____ 5. testosterone   _____ 8. TSH, FSH

_____ 3. estrogens   _____ 6. norepinephrine   _____ 9. aldosterone

11. Name the hormone that would be produced in _inadequate_ amounts under the following conditions. (Use your textbook as necessary.)

_____ 1. sexual immaturity

_____ 2. tetany

_____ 3. excessive diuresis without high blood glucose levels

_____ 4. polyurea, polyphagia, and polydipsia

_____ 5. abnormally small stature, normal proportions

_____ 6. miscarriage

_____ 7. lethargy, hair loss, low BMR, obesity

12. Name the hormone that is produced in _excessive_ amounts in the following conditions. (Use your textbook as necessary.)

_____ 1. lantern jaw and large hands and feet in the adult

_____ 2. bulging eyeballs, nervousness, increased pulse rate

_____ 3. demineralization of bones, spontaneous fractures

Microscopic Anatomy of Selected Endocrine Glands (Optional)

1. Choose a response from the key below to name the hormone(s) produced by the cell types listed:

Key: a. insulin   d. calcitonin   g. glucagon
       b. GH, prolactin   e. TSH, ACTH, FSH, LH   h. PTH
       c. T_4 / T_3 f. mineralocorticoids   i. glucocorticoids

_____ 1. parafollicular cells of the thyroid   _____ 6. zona fasciculata cells

_____ 2. follicular epithelial cells of the thyroid   _____ 7. zona glomerulosa cells

_____ 3. beta cells of the islets of Langerhans   _____ 8. chief cells

_____ 4. alpha cells of the islets of Langerhans   _____ 9. acidophil cells of the anterior pituitary

_____ 5. basophil cells of the anterior pituitary

2. Five diagrams of the microscopic structures of the endocrine glands are presented here. Identify each and name all indicated structures.

Experiments on Hormonal Action

Determining the Effect of Thyroid Hormone on Metabolic Rate

1. Relative to the measurement of oxygen consumption in rats, which group had the highest metabolic rate?

   _____ Which group had the lowest metabolic rate? _____

   Correlate these observations with the pretreatment these animals received. _____

   _____

   _____

   Which group of rats was hyperthyroid? _____

   Which euthyroid? _____ Which hypothyroid? _____

2. Since oxygen used = carbon dioxide evolved, how were you able to measure the oxygen consumption in the experiments?

   _____

   _____

3. What did changes in the fluid levels in the manometer arms indicate? _____

   _____

4. The techniques used in this set of laboratory experiments probably allowed for several inaccuracies. One was the inability to control the activity of the rats. How would changes in their activity levels affect the results observed?

   _____

   _____

   Another possible source of error was the lack of control over the amount of food consumed by the rats in the 14-day period preceding the laboratory session. If each of the rats had been force-fed equivalent amounts of food in that 14-day period, which group (do you think) would have gained the most weight?

   _____ Which the least? _____ Explain your answers.

   _____

   _____

   _____

5. TSH, produced by the anterior pituitary, prods the thyroid gland to release thyroid hormone to the blood. Which group of rats can be assumed to have the *highest* blood levels of TSH? _____

Which the lowest? _____ Explain your reasoning. _____

_____

_____

6. Use an appropriate reference to determine how each of the following factors modifies metabolic rate. Indicate increase by ↑ and decrease by ↓.

increased exercise _____   aging _____   infection/fever _____

small/slight stature _____   obesity _____   sex (♂ or ♀) _____

Determining the Effect of Pituitary Hormones on the Ovary

1. In the experiment on the effects of pituitary hormones, two anterior pituitary hormones caused ovulation to occur in the experimental animal. Which of these actually triggered ovulation or egg expulsion?

_____ The normal function of the second hormone involved, _____,

is to _____

2. Why was a second frog injected with saline? _____

Observing the Effects of Hyperinsulinism

1. Briefly explain what was happening within the fish's system when the fish was immersed in the insulin solution.

_____

_____

2. What is the mechanism of the recovery process observed? _____

_____

3. What would you do to help a friend who had inadvertently taken an overdose of insulin? _____

_____ Why? _____

_____

4. What is a glucose tolerance test? (Use an appropriate reference, as necessary, to answer this question.)

_____

_____

_____

5. How does diabetic coma differ from insulin shock? _____

_____

_____

_____

_____

Testing the Effect of Epinephrine on the Heart

1. Based on your observations, what is the effect of epinephrine on the force and rate of the heartbeat?

_____

2. What is the role of this effect in the "fight or flight" response?

_____

Blood

Composition of Blood

1. What is the blood volume of an average-size adult? _____ liters

2. What determines whether blood is bright red or a dull brick-red? _____

_____

3. Use the key to identify the cell type(s) or blood elements that fit the following descriptive statements.

Key: a. red blood cell   d. basophil   g. lymphocyte
       b. megakaryocyte   e. monocyte   h. formed elements
       c. eosinophil   f. neutrophil   i. plasma

_____ 1. most numerous leukocyte

_____, _____, and _____

_____ 2. granulocytes

_____ 3. also called an erythrocyte; anucleate formed element

_____, _____ 4. actively phagocytic leukocytes

_____, _____ 5. agranulocytes

_____ 6. ancestral cell of platelets

_____ 7. (a) through (g) are all examples of these

_____ 8. number rises during parasite infections

_____ 9. releases histamine; promotes inflammation

_____ 10. many formed in lymphoid tissue

_____ 11. transports oxygen

_____ 12. primarily water, noncellular; the fluid matrix of blood

_____ 13. increases in number during prolonged infections

_____, _____, _____,

_____, _____ 14. also called white blood cells

4. List three classes of nutrients normally found in plasma: _____ ,

_____ , and _____

Name two gases. _____ and _____

Name three ions. _____ , _____ , and _____

5. Describe the consistency and color of the plasma you observed in the laboratory. _____

_____

6. What is the average life span of a red blood cell? How does its anucleate condition affect this life span?

_____

_____

_____

7. From memory, describe the structural characteristics of each of the following blood cell types as accurately as possible, and note the percentage of each in the total white blood cell population.

eosinophils _____

_____

neutrophils _____

_____

lymphocytes _____

_____

basophils _____

_____

monocytes _____

_____

8. Correctly identify the blood pathologies described in column A by matching them with selections from column B:

Column A	Column B
_____ 1. abnormal increase in the number of WBCs	a. anemia
_____ 2. abnormal increase in the number of RBCs	b. leukocytosis
_____ 3. condition of too few RBCs or of RBCs with hemoglobin deficiencies	c. leukopenia
_____ 4. abnormal decrease in the number of WBCs	d. polycythemia

Hematologic Tests

1. Broadly speaking, why are hematologic studies of blood so important in the diagnosis of disease?

_____

_____

2. In the chart below, record information from the blood tests you conducted. Complete the chart by recording values for healthy male adults and indicating the significance of high or low values for each test.

Test	Student test results	Normal values (healthy male adults)	Significance High values	Low values
Total WBC count				
Total RBC count				
Hematocrit				
Hemoglobin determination				
Sedimentation rate				
Coagulation time				

3. Why is a differential WBC count more valuable than a total WBC count when trying to pin down the specific

source of pathology? _____

_____

_____

4. Explain the reasons for using different diluents for the RBC and WBC counts. _____

_____

_____

5. What name is given to the process of RBC production? _____

What acts as a stimulus for this process? _____

What organ provides this stimulus and under what conditions? _____

_____

_____

6. Discuss the effect of each of the following factors on RBC count. Consult an appropriate reference as necessary, and explain your reasoning.

athletic training (for example, running 4 to 5 miles per day over a period of 6 to 9 months) _____

_____

_____

a permanent move from sea level to a high-altitude area _____

_____

_____

7. Define *hematocrit:* _____

8. If you had a high hematocrit, would you expect your hemoglobin determination to be high or low?

_____ Why? _____

_____

9. What is an anticoagulant? _____

Name two anticoagulants used in conducting the hematologic tests. _____

and _____

What is the body's natural anticoagulant? _____

10. If your blood clumped with both anti-A and anti-B sera, your ABO blood type would be _____

To what ABO blood groups could you give blood? _____

From which ABO donor types could you receive blood? _____

Which ABO blood type is most common? _____ Least common? _____

11. Explain why an Rh-negative person does not have a transfusion reaction on the first exposure to Rh-positive blood but *does*

have a reaction on the second exposure. _____

_____

What happens when an ABO blood type is mismatched for the first time? _____

_____

12. Record your observations of the five demonstration slides viewed.

a.   Macrocytic hypochromic anemia: _____

b.   Microcytic hypochromic anemia: _____

_____

c.   Sickle-cell anemia: _____

d.   Lymphocytic leukemia (chronic): _____

_____

e.   Eosinophilia: _____

Which of slides a through e above corresponds with the following conditions?

_____ 1.   iron-deficient diet   _____ 4.   lack of vitamin B$_{12}$

_____ 2.   a type of bone marrow cancer   _____ 5.   a tapeworm infestation in the body

_____ 3.   genetic defect that causes hemoglobin _____ 6.   a bleeding ulcer
              to become sharp/spiky

13. Provide the normal, or at least "desirable," range for plasma cholesterol concentration:

_____ mg/100 ml

14. Describe the relationship between high blood cholesterol levels and cardiovascular diseases such as hypertension, heart attacks, and strokes.

_____

_____

_____

_____

_____

Anatomy of the Heart

Gross Anatomy of the Human Heart

1. An anterior view of the heart is shown here. Identify each numbered structure by writing its name on the correspondingly numbered line:

1. _____

2. _____

3. _____

4. _____

5. _____

6. _____

7. _____

8. _____

9. _____

10. _____

11. _____

12. _____

13. _____

14. _____

15. _____

16. _____

17. _____

18. _____

19. _____

20. _____

21. _____

2. What is the function of the fluid that fills the pericardial sac? _____

_____

3. Match the terms in the key to the descriptions provided below.

_____ 1. location of the heart in the thorax

_____ 2. superior heart chambers

_____ 3. inferior heart chambers

_____ 4. visceral pericardium

_____ 5. "anterooms" of the heart

_____ 6. equals cardiac muscle

_____ 7. provide nutrient blood to the heart muscle

_____ 8. lining of the heart chambers

_____ 9. actual "pumps" of the heart

_____ 10. drains blood into the right atrium

Key:

a. atria

b. coronary arteries

c. coronary sinus

d. endocardium

e. epicardium

f. mediastinum

g. myocardium

h. ventricles

4. What is the function of the valves found in the heart? _____

_____

5. Can the heart function with leaky valves? (Think! Can a water pump function with leaky valves?) _____

6. What is the role of the chordae tendineae? _____

_____

7. Define:

angina pectoris _____

_____

pericarditis _____

Pulmonary, Systemic, and Cardiac Circulations

1. A simple schematic of a so-called general circulation is shown on the opposite page. What part of the circulation is missing

from this diagram? _____

Add to the diagram as best you can to make it depict a complete systemic/pulmonary circulation and reidentify "general circulation" as the correct subcirculation.

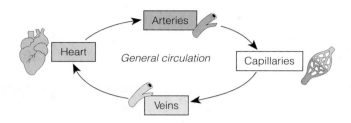

2. Differentiate clearly between the roles of the pulmonary and systemic circulations. _____

_____

_____

3. Complete the following scheme of circulation in the human body:

Right atrium through the tricuspid valve to the _____ through the _____

_____ valve to the pulmonary trunk to the _____ to the capillary

beds of the lungs to the _____ to the _____ of the heart through

the _____ valve to the _____ through the _____

valve to the _____ to the systemic arteries to the _____ of the

tissues to the systemic veins to the _____ and _____ entering

the right atrium of the heart.

4. If the mitral valve does not close properly, which circulation is affected? _____

5. Why might a thrombus (blood clot) in the anterior descending branch of the left coronary artery cause sudden death?

_____

_____

Microscopic Anatomy of Cardiac Muscle

1. How would you distinguish cardiac muscle from skeletal muscle? _____

    _____

    _____

    _____

2. Add the following terms to the photo
    of cardiac muscle at the right:

    a.   intercalated disc

    b.   nucleus of cardiac fiber

    c.   striations

    d.   cardiac muscle fiber

3. What role does the unique structure of cardiac muscle play in its function? (Note: before attempting a response, *describe* the

    unique anatomy.) _____

    _____

    _____

    _____

Dissection of the Sheep Heart

1. During the sheep heart dissection, you were asked initially to identify the right and left ventricles without cutting into the
    heart. During this procedure, what differences did you observe between the two chambers?

    _____

    _____

Knowing that structure and function are related, how would you say this structural difference reflects the relative functions of these two heart chambers? _____

_____

_____

_____

2. Semilunar valves prevent backflow into the _____; AV valves prevent backflow into the

_____. Using your own observations, explain how the operation of the semilunar valves

differs from that of the AV valves. _____

_____

_____

_____

_____

3. Differentiate clearly between the location and appearance of pectinate muscle and trabeculae carneae. _____

_____

_____

4. Two remnants of fetal structures are observable in the heart—the ligamentum arteriosum and the fossa ovalis. What were they called in the fetal heart, where was each located, and what common purpose did they serve as functioning fetal structures?

_____

_____

_____

_____

Conduction System of the Heart and Electrocardiography: Wet Lab

The Intrinsic Conduction System

1. List the elements of the intrinsic conduction system in order starting from the SA node.

   SA node $\longrightarrow$ _____ $\longrightarrow$ _____ $\longrightarrow$

   _____ $\longrightarrow$ _____

   Which of those structures is replaced when an artificial pacemaker is installed? _____

   At what structure in the transmission sequence is the impulse temporarily delayed? _____

   Why? _____

   _____

2. Even though cardiac muscle has an inherent ability to beat, the nodal system plays a critical role in heart physiology. What

   is that role? _____

   _____

   _____

3. How does the "all-or-none" law apply to normal heart operation? _____

   _____

   _____

Electrocardiography

1. Define *ECG*. _____

   _____

2. Draw an ECG wave form representing one heartbeat. Label the P, QRS, and T waves; the P-R interval; the S-T segment, and the Q-T interval.

3. What changes from *baseline* were noted in the ECG recorded during running? _____

_____

_____

Explain. _____

_____

_____

In that recorded during breath holding? _____

_____

Explain. _____

_____

_____

4. Describe what happens in the cardiac cycle in the following situations:

1. during the P wave _____

2. immediately before the P wave _____

3. immediately after the P wave _____

4. during the QRS wave _____

5. immediately after the QRS wave (S-T interval) _____

6. during the T wave _____

5. Define the following terms:

1. *tachycardia* _____

2. *bradycardia* _____

3. *flutter* _____

4. *fibrillation* _____

5. *myocardial infarction* _____

6. Which would be more serious, atrial or ventricular fibrillation? _____

Why? _____

_____

7. Abnormalities of heart valves can be detected more accurately by auscultation than by electrocardiography. Why is this so?

_____

_____

_____

Anatomy of Blood Vessels

Microscopic Structure of the Blood Vessels

1. Use key choices to identify the blood vessel tunic described.

Key: a.   tunica intima   b.   tunica media   c.   tunica externa

_____ 1.   most internal tunic

_____ 2.   bulky middle tunic contains smooth muscle and elastin

_____ 3.   its smooth surface decreases resistance to blood flow

_____ 4.   tunic(s) of capillaries

_____, _____, _____ 5.   tunic(s) of arteries and veins

_____ 6.   is especially thick in elastic arteries

_____ 7.   most superficial tunic

2. Servicing the capillaries is the essential function of the organs of the circulatory system. Explain this statement.

_____

_____

_____

3. Cross-sectional views of an artery and of a vein are shown here. Identify each; and on the lines beneath, note the structural details that enabled you to make these identifications:

_____   _____

_____   _____

4. Why are valves present in veins but not in arteries? _____

_____

_____

5. Name two events *occurring within the body* that aid in venous return:

_____

and _____

6. Why are the walls of arteries proportionately thicker than those of the corresponding veins? _____

_____

_____

Major Systemic Arteries and Veins of the Body

1. Use the key on the right to identify the arteries or veins described on the left.   Key:

_____ 1. the arterial system has one of these; the venous system has two

_____ 2. these arteries supply the myocardium

_____, _____ 3. two paired arteries serving the brain

_____ 4. longest vein in the lower limb

_____ 5. artery on the dorsum of the foot checked after leg surgery

_____ 6. serves the posterior thigh

_____ 7. supplies the diaphragm

_____ 8. formed by the union of the radial and ulnar veins

_____, _____ 9. two superficial veins of the arm

_____ 10. artery serving the kidney

_____ 11. veins draining the liver

_____ 12. artery that supplies the distal half of the large intestine

_____ 13. drains the pelvic organs

_____ 14. what the external iliac artery becomes on entry into the thigh

_____ 15. major artery serving the arm

_____ 16. supplies most of the small intestine

_____ 17. join to form the inferior vena cava

_____ 18. an arterial trunk that has three major branches, which run to the liver, spleen, and stomach

_____ 19. major artery serving the tissues external to the skull

_____, _____, _____ 20. three veins serving the leg

_____ 21. artery generally used to take the pulse at the wrist

a. anterior tibial

b. basilic

c. brachial

d. brachiocephalic

e. celiac trunk

f. cephalic

g. common carotid

h. common iliac

i. coronary

j. deep femoral

k. dorsalis pedis

l. external carotid

m. femoral

n. greater saphenous

o. hepatic

p. inferior mesenteric

q. internal carotid

r. internal iliac

s. peroneal

t. phrenic

u. posterior tibial

v. radial

w. renal

x. subclavian

y. superior mesenteric

z. vertebral

2. The human arterial and venous systems are diagrammed on the next two pages. Identify all indicated blood vessels.

Arteries

Veins

3. Trace the blood flow for the following situations:

a. From the capillary beds of the left thumb to the capillary beds of the right thumb _____

_____

_____

_____

_____

_____

b. From the bicuspid valve to the tricuspid valve by way of the great toe _____

_____

_____

_____

_____

c. From the pulmonary vein to the pulmonary artery by way of the right side of the brain _____

_____

_____

_____

_____

_____

_____

Special Circulations

Pulmonary circulation:

1. Trace the pathway of a carbon dioxide gas molecule in the blood from the inferior vena cava until it leaves the bloodstream. Name all structures (vessels, heart chambers, and others) passed through en route.

_____

_____

_____

2. Trace the pathway of an oxygen gas molecule from an alveolus of the lung to the right atrium of the heart. Name all structures through which it passes. _____

_____

_____

3. Most arteries of the adult body carry oxygen-rich blood, and the veins carry oxygen-depleted, carbon dioxide–rich blood. What is different about the pulmonary arteries and veins? _____

_____

_____

4. How do the arteries of the pulmonary circulation differ structurally from the systemic arteries? What condition is indicated by this anatomical difference? _____

_____

_____

Hepatic portal circulation:

1. What is the source of blood in the hepatic portal system? _____

_____

2. Why is this blood carried to the liver before it enters the systemic circulation? _____

_____

_____

3. The hepatic portal vein is formed by the union of the _____, which drains the

_____, _____, _____,

and the _____, which drains the _____ and _____

_____. The _____ vein, which drains the lesser curvature of the

stomach, empties directly into the hepatic portal vein.

4. Trace the flow of a drop of blood from the small intestine to the right atrium of the heart, noting all structures encountered or passed through on the way. _____

_____

_____

Arterial supply of the brain and the circle of Willis:

1. What two paired arteries enter the skull to supply the brain?

   _____ and _____

2. Branches of the paired arteries just named cooperate to form a ring of blood vessels encircling the pituitary gland, at the base

   of the brain. What name is given to this communication network? _____

   What is its function? _____

   _____

3. What portion of the brain is served by the anterior and middle cerebral arteries? _____

   _____

   Both the anterior and middle cerebral arteries arise from the _____ arteries.

4. Trace the pathway of a drop of blood from the aorta to the left occipital lobe of the brain, noting all structures through which

   it flows. _____

   _____

Fetal circulation:

1. The failure of two of the fetal bypass structures to become obliterated after birth can cause congenital heart disease, in which
   the youngster would have improperly oxygenated blood. Which two structures are these?

   _____ and _____

2. For each of the following structures, first indicate its function in the fetus; and then note what happens to it or what it is converted to after birth. Circle the blood vessel that carries the most oxygen-rich blood.

Structure	Function in fetus	Fate
Umbilical artery		
Umbilical vein		
Ductus venosus		
Ductus arteriosus		
Foramen ovale		

3. What organ serves as a respiratory/digestive/excretory organ for the fetus? _____

Cardiovascular Dynamics: Computer Simulation

For numbers 1 and 2 below, choose all answers that apply and place their letters on the response blanks to the right of the statement.

1. The circulation of blood through the vascular system is influenced by _____.

   a. blood viscosity
   b. the length of blood vessels
   c. the driving pressure behind the blood
   d. the radius or diameter of blood vessels

2. Peripheral resistance depends on _____.

   a. blood viscosity
   b. blood pressure
   c. vessel length
   d. vessel radius

3. Complete the following statements.

   The volume of blood remaining in the heart after ventricular contraction is called the _____.

   Cardiac output is defined as _____.

   The amount of blood pumped by the heart in a single beat is called the _____ volume.

   The human heart is actually two individual pumps working in _____.

   Stroke volume is calculated by _____.

   If stroke volume decreased, heart rate would _____ in order to maintain blood flow.

4. How could the heart compensate to maintain proper blood flow for the following conditions?

   High peripheral resistance: _____

   A leaky atrioventricular valve: _____

   A constricted semilunar valve: _____

5. How does the size of the heart change under conditions of chronic high peripheral resistance?

   _____

6. The following questions refer to the Vessel Resistance experiment.

   How was the flow rate affected when the radius of the flow tube was increased? _____

   Which of the adjustable parameters had the greatest effect on fluid flow? _____

   How does vessel length affect fluid flow? _____

If you increased fluid viscosity, what parameter(s) could you adjust to keep fluid flow constant?

_____

Explain your answer. _____

_____

If the driving pressure in the left beaker was 100 mm Hg, how could you adjust the conditions of the experiment to com-

pletely stop fluid flow? _____

7. The following questions refer to the Pump Mechanics experiment.

What would happen if the right side of the heart pumped faster than the left side of the heart?

_____

_____

When you change the radius of the right flow tube in Pump Mechanics the resulting plot looks different than the radius plot in the Vessel Resistance experiment. How would the plot look if you changed the radius of both flow tubes in Pump Mechanics instead of just the right flow tube?

_____

_____

Why are valves needed in the Pump Mechanics equipment? _____

What happens to blood flow if peripheral resistance equals pump pressure? _____

Theoretically, what would happen to the pumping ability of the heart if the end systolic volume was equal to the end dia-

stolic volume? _____

8. Match the part in the simulation equipment to the analogous cardiac structure or physiological term listed in the key below.

Simulation equipment: Key:

_____ 1. valve leading to the right beaker   a. pulmonary veins

_____ 2. valve leading to the pump   b. bicuspid valve

_____ 3. left flow tube   c. ventricular filling pressure

_____ 4. right flow tube   d. peripheral resistance

_____ 5. pump end volume   e. aortic valve

_____ 6. pump starting volume   f. end diastolic volume

_____ 7. pressure in the right beaker   g. aorta

_____ 8. pressure in the left beaker   h. end systolic volume

_____ 9. pump pressure   i. systolic pressure

Define the following terms:

blood flow: _____

_____

peripheral resistance: _____

_____

viscosity: _____

_____

radius: _____

_____

end diastolic volume: _____

_____

systole: _____

_____

diastole: _____

_____

Human Cardiovascular Physiology—Blood Pressure and Pulse Determinations

Cardiac Cycle

1. Correctly identify valve closings and openings, chamber pressures, and volume lines, and the ECG and heart sound scan lines on the diagram below by using the lettered terms from the list to the right of the diagram.

Key:
a. aortic pressure

b. aortic valve closes

c. aortic valve opens

d. atrial pressure

e. AV valve closes

f. AV valve opens

g. ECG

h. heart sounds

i. ventricular pressure

j. ventricular volume

2. Define the following terms:

systole _____

diastole _____

cardiac cycle _____

3. Answer the following questions, which concern events of the cardiac cycle:

When are the AV valves closed? _____

Open? _____

What event within the heart causes the AV valves to open? _____

What causes them to close? _____

When are the semilunar valves closed? _____

_____

Open? _____

What event causes the semilunar valves to open? _____

_____

To close? _____

Are both sets of valves closed during any part of the cycle? _____

If so, when? _____

Are both sets of valves open during any part of the cycle? _____

At what point in the cardiac cycle is the pressure in the heart highest? ___

Lowest? _____

What event results in the pressure deflection called the dicrotic notch? ____

_____

4. Using the key below, indicate the time interval occupied by the following events of the cardiac cycle.

Key: a. 0.4 sec   b. 0.3 sec   c. 0.1 sec   d. 0.8 sec

_____ 1.   the length of the normal cardiac cycle   _____ 3.   the quiescent period, or pause

_____ 2.   the time interval of atrial systole   _____ 4.   the ventricular contraction period

5. If an individual's heart rate is 80 beats/min, what is the length of the cardiac cycle? _____ What portion of

the cardiac cycle is shortened by this more rapid heart rate? _____

_____

6. What two factors promote the movement of blood through the heart? _____

_____ and _____

_____

Heart Sounds

1. Complete the following statements:

      The monosyllables describing the heart sounds are __1__ . The first heart sound is a result of closure of the __2__ valves, whereas the second is a result of closure of the __3__ valves. The heart chambers that have just been filled when you hear the first heart sound are the__4__, and the chambers that have just emptied are the__5__. Immediately after the second heart sound, the__6__are filling with blood, and the__7__are empty.

1. _____

2. _____

3. _____

4. _____

5. _____

6. _____

7. _____

2. As you listened to the heart sounds during the laboratory session, what differences in pitch, length, and amplitude (loudness)

of the two sounds did you observe? _____

_____

_____

3. Indicate where you would place your stethoscope to auscultate most accurately the following:

closure of the tricuspid valve _____

closure of the aortic semilunar valve _____

apical heartbeat _____

Which valve is heard most clearly when the apical heartbeat is auscultated? _____

4. No one expects you to be a full-fledged physician on such short notice; but on the basis of what you have learned about heart sounds, how might abnormal sounds be used to diagnose heart problems?

_____

_____

_____

The Pulse

1. Define *pulse*. _____

_____

2. Describe the procedure used to take the pulse. _____

_____

_____

3. Identify the artery palpated at each of the following pressure points:

at the wrist_____ on the dorsum of the foot _____

in front of the ear_____ at the side of the neck _____

4. When you were palpating the various pulse or pressure points, which appeared to have the greatest amplitude or tension?

_____ Why do you think this was so? _____

_____

5. Assume someone has been injured in an auto accident and is hemorrhaging badly. What pressure point would you compress to help stop bleeding from each of the following areas?

the thigh_____ the calf _____

the forearm_____ the thumb _____

6. How could you tell by simple observation whether bleeding is arterial or venous? _____

_____

7. You may sometimes observe a slight difference between the value obtained from an apical pulse (beats/min) and that from an arterial pulse taken elsewhere on the body. What is this difference called?

_____

Blood Pressure Determinations

1. Define *blood pressure*. _____

2. Identify the phase of the cardiac cycle to which each of the following apply:

systolic pressure _____ diastolic pressure _____

3. What is the name of the instrument used to compress the artery and record pressures in the auscultatory method of deter-

mining blood pressure? _____

4. What are the sounds of Korotkoff? _____

_____

What causes the systolic sound? _____

_____

The disappearance of sound? _____

5. Interpret 145/85/82. _____

_____

_____

6. Assume the following BP measurement was recorded for an elderly patient with severe arteriosclerosis:170/110/–. Explain the inability to obtain the third reading.

_____

_____

7. Define *pulse pressure:* _____

Why is this measurement important? _____

_____

8. How do venous pressures compare to arterial pressures? _____

Why? _____

9. What maneuver to increase the thoracic pressure illustrates the effect of external factors on venous pressure?

_____ How was it performed? _____

_____

_____

10. What might an abnormal increase in venous pressure indicate? (Think!) _____

_____

Observing the Effect of Various Factors on Blood Pressure and Heart Rate

1. What effect do the following have on blood pressure? (Indicate increase by I and decrease by D.)

_____ 1. increased diameter of the arterioles _____ 4. hemorrhage

_____ 2. increased blood viscosity _____ 5. arteriosclerosis

_____ 3. increased cardiac output _____ 6. increased pulse rate

2. In which position (sitting, reclining, or standing) is the blood pressure normally the highest?

_____ The lowest? _____

What immediate changes in blood pressure did you observe when the subject stood up after being in the sitting or reclining

position? _____

_____

What changes in the blood vessels might account for the change? _____

_____

_____

After the subject stood for 3 minutes, what changes in blood pressure were observed? _____

_____

How do you account for this change? _____

3. What was the effect of exercise on blood pressure? _____

   On pulse? _____ Do you think these effects reflect changes in cardiac output *or* in

   peripheral resistance? _____

   Why are there normally no significant increases in diastolic pressure after exercise? _____

   _____

4. What effects of the following did you observe on blood pressure in the laboratory?

   nicotine _____   cold temperature _____

   What do you think the effect of heat would be? _____

   Why? _____

5. Differentiate between a hypo- and a hyperreactor relative to the cold pressor test. _____

   _____

Skin Color as an Indicator of Local Circulatory Dynamics

1. Describe normal skin color and the appearance of the veins in the subject's forearm before any testing was conducted.

   _____

2. What changes occurred when the subject emptied his forearm of blood (by raising his arm and making a fist) and the flow

   was occluded with the cuff? _____

   What changes occurred during venous congestion? _____

   _____

3. What is the importance of collateral blood supplies? _____

   _____

4. Explain the mechanism by which mechanical stimulation of the skin produced a flare. _____

   _____

Frog Cardiovascular Physiology: Wet Lab

Special Electrical Properties of Cardiac Muscle: Automaticity and Rhythmicity

1. Define the following terms:

automaticity _____

rhythmicity _____

2. Discuss the anatomical differences you observed between frog and human hearts. _____

_____

_____

3. Which region of the dissected frog heart had the highest intrinsic rate of contraction? _____

The greatest automaticity? _____

The greatest regularity or rhythmicity? _____ How do these properties correlate with the

duties of a pacemaker? _____

_____

Is this region the pacemaker of the frog heart? _____

Which region had the lowest intrinsic rate of contraction? _____

Investigating the Refractory Period of Cardiac Muscle

1. Define *extrasystole:* _____

2. In responding to the following questions, refer to the recordings you made during this exercise:

What was the effect of stimulation of the heart during ventricular contraction? _____

During ventricular relaxation (first portion)? _____

During the pause interval? _____

What does this indicate about the refractory period of cardiac muscle? _____

_____

Can cardiac muscle be tetanized? _____ Why or why not? _____

Assessing Physical and Chemical Modifiers of Heart Rate

1. Describe the effect of thermal factors on the frog heart:

   cold _____ heat _____

2. What was the effect of vagal stimulation on heart rate? _____

   Which of the following factors cause the same (or very similar) heart rate–reducing effects? Epinephrine, acetylcholine, atropine sulfate, pilocarpine, sympathetic nervous system activity, digitalis, potassium ions?

   _____

   Which of the factors listed above would reverse or antagonize vagal effects? _____

   _____

3. What is vagal escape? _____

   _____

   Why is vagal escape valuable in maintaining homeostasis? _____

   _____

4. Once again refer to your recordings. Did the administration of the following produce any changes in force of contraction (shown by peaks of increasing or decreasing height)? If so, explain the mechanism.

   epinephrine _____

   _____

   acetylcholine _____

   _____

   _____

   calcium ions _____

   _____

   _____

5. Excessive amounts of each of the following ions would most likely interfere with normal heart activity. Note the type of changes caused in each case.

   K^+ _____

   Ca^{2+} _____

   Na^+ _____

6. How does the Stannius ligature used in the laboratory produce heart block? _____

_____

_____

7. Define *partial heart block,* and describe how it was recognized in the laboratory. _____

_____

_____

8. Define *total heart block,* and describe how it was recognized in the laboratory. _____

_____

_____

_____

9. What do your heart block experiment results indicate about the spread of impulses from the atria to the ventricles?

_____

_____

Observing the Microcirculation Under Various Conditions

1. In what way are the red blood cells of the frog different from those of the human? _____

_____

On the basis of this one factor, would you expect their life spans to be longer or shorter? _____

2. The following statements refer to your observation of one or more of the vessel types observed in the microcirculation in the frog's web. Characterize each statement by choosing one or more responses from the key.

Key: a. arteriole   b. venule   c. capillary

_____ 1. smallest vessels observed

_____ 2. vessel within which the blood flow is rapid, pulsating

_____ 3. vessel in which blood flow is least rapid

_____ 4. red blood cells pass through these vessels in single file

_____ 5. blood flow smooth and steady

_____ 6. most numerous vessels

_____ 7. vessels that deliver blood to the capillary bed

_____ 8. vessels that serve the needs of the tissues via exchanges

_____ 9. vessels that drain the capillary beds

3. Which of the vessel diameters changed most? _____

What division of the nervous system controls the vessels? _____

4. Discuss the effects of the following on blood vessel diameter (state specifically the blood vessels involved) and rate of blood flow. Then explain the importance of the reaction observed to the general well-being of the body.

local application of cold _____

_____

_____

local application of heat _____

_____

_____

inflammation (or application of HCl) _____

_____

_____

histamine _____

_____

_____

Frog Cardiovascular Physiology: Computer Simulation

Special Electrical Properties of Cardiac Muscle: Automaticity and Rhythmicity

1. Define the following terms:

   Automaticity _____

   Rhythmicity _____

2. Explain the anatomical differences between frog and human hearts.

   _____

   _____

   _____

Baseline Frog Heart Activity

1. Define the following terms:

   Intrinsic heart control _____

   Extrinsic heart control _____

2. Why is it necessary to keep the frog heart moistened with Ringer's solution? _____

   _____

Refractory Period of Cardiac Muscle

1. Define *extrasystole* _____

   _____

2. Refer to the exercise to answer the following questions.

   What was the effect of stimulating the heart during ventricular contraction? _____

   During ventricular relaxation _____

   During the pause interval _____

   What does this information indicate about the refractory period of cardiac muscle?

   _____

   Can cardiac muscle be tetanized? _____ Why or why not? _____

The Effect of Vagus Nerve Stimulation

1. What was the effect of vagal stimulation on heart rate? _____ .

2. What is vagal escape? _____

3. Why is the vagal escape valuable in maintaining homeostasis? _____

    _____

Physical and Chemical Modifiers of Heart Rate

1. Describe the effect of thermal factors on the frog heart.

    Cold _____ Heat _____

2. Which of the following factors caused the same, or very similar, heart rate-reducing effects: epinephrine, atropine, pilocarpine, digitalis, potassium ions.

    _____

    _____

    Which of the factors listed above would reverse or antagonize vagal effects? _____

    _____

3. Did administering any the following produce any changes in force of contraction (shown by peaks of increasing or decreasing height)? If so, explain the mechanism.

    Epinephrine _____

    _____

    Calcium ions _____

    _____

4. Excessive amounts of each of the following ions would most likely interfere with normal heart activity. Explain the type of changes caused in each case.

    K^+ _____

    Ca^{2+} _____

    Na^+ _____

5. Define the following:

    Parasympathomimetic _____

    Ectopic pacemaker _____

6. Explain how digitalis works. _____

    _____

The Lymphatic System and Immune Response

The Lymphatic System

1. Identify all structures that have leader lines on the diagram below. Use the key for your selection.

Key: a. bone marrow   d. lymphatic vessels   g. thoracic duct
           b. cisterna chyli   e. right lymphatic duct   h. thymus
           c. lymph nodes   f. spleen   i. tonsils

1. _____

2. _____

3. _____

4. _____

5. _____

6. _____

7. _____

8. _____

9. _____

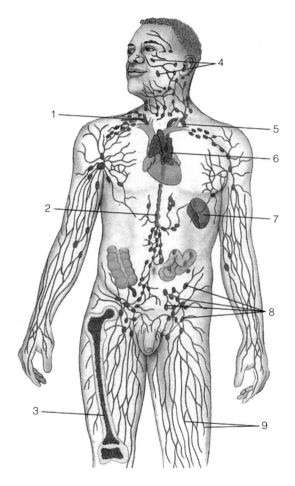

2. Explain why the lymphatic system is a one-way system, whereas the blood vascular system is a two-way system.

_____

_____

3. How do lymphatic vessels resemble veins? _____

How do lymphatic capillaries differ from blood capillaries? _____

_____

4. What is the function of the lymphatic vessels? _____

_____

5. What is lymph? _____

6. What factors are involved in the flow of lymphatic fluid? _____

_____

7. What name is given to the terminal duct draining most of the body? _____

8. What is the cisterna chyli? _____

_____

How does the composition of lymph in the cisterna chyli differ from that in the general lymphatic stream?

_____

9. Which portion of the body is drained by the right lymphatic duct? _____

10. Note three areas where lymph nodes are densely clustered: _____,

_____, and _____

11. What are the two major functions of the lymph nodes? _____

_____

12. The radical mastectomy is an operation in which a cancerous breast, surrounding tissues, and the underlying muscles of the anterior thoracic wall, plus the axillary lymph nodes, are removed. After such an operation, the arm usually swells, or becomes edematous, and is very uncomfortable—sometimes for months. Why?

_____

The Immune Response

1. Define the following terms related to the operation of the immune system:

Immunologic memory _____

_____

Specificity _____

_____

Recognition of self from nonself _____

_____

2. What is the function of B cells in the immune response? _____

_____

3. What is the role of T cells? _____

_____

_____

Studying the Microscopic Anatomy of a Lymph Node

1. In the space below, make a rough drawing of the structure of a lymph node. Identify the cortex area, germinal centers, and medulla. For each identified area, note the cell type (T cell, B cell, or macrophage) most likely to be found there.

2. What structural characteristic ensures a *slow* flow of lymph through a lymph node? _____

_____

Why is this desirable? _____

_____

Antibodies and Tests for Their Presence

1. Describe the structure of the immunoglobulin monomer. _____

_____

2. Are the genes coding for one antibody entirely different from those coding for a different antibody? _____

Explain. _____

_____

_____

3. Rheumatoid arthritis is said to involve an autoimmune response. What is an autoimmune response?

_____

4. In the test to identify the presence of rheumatoid factor in plasma, a clumping (agglutination), indicated a positive response (that is, the presence of the abnormal antibody). What event did the clumping reflect?

_____

Anatomy of the Respiratory System

Upper and Lower Respiratory System Structures

1. Complete the labeling of the diagram of the upper respiratory structures (sagittal section).

Frontal sinus

Cribriform plate of ethmoid bone

Superior meatus

Middle meatus

Sphenoidal sinus

Opening of auditory tube

Inferior meatus

Nasopharynx

Hard palate

Tongue

Hyoid bone

Thyroid cartilage of larynx

Cricoid cartilage

Thyroid gland

2. Two pairs of vocal folds are found in the larynx. Which pair are the true vocal cords (superior or inferior)?

_____

3. What is the significance of the fact that the human trachea is reinforced with cartilage rings?

_____

Of the fact that the rings are incomplete posteriorly? _____

_____

4. Name the specific cartilages in the larynx that correspond to the following descriptions:

1. forms the Adam's apple_____

3. shaped like a signet ring _____

2. a "lid" for the larynx_____

4. vocal cord attachment _____

5. Trace a molecule of oxygen from the external nares to the pulmonary capillaries of the lungs: External nares →

_____

_____

_____

6. What is the function of the pleural membranes? _____

_____

_____

7. Name two functions of the nasal cavity mucosa: _____

8. The following questions refer to the primary bronchi:

Which is longer?_____   Larger in diameter?_____   More horizontal? _____

The more common site for lodging of a foreign object that has entered the respiratory passageways? _____

9. Appropriately label all structures provided with leader lines on the diagrams below.

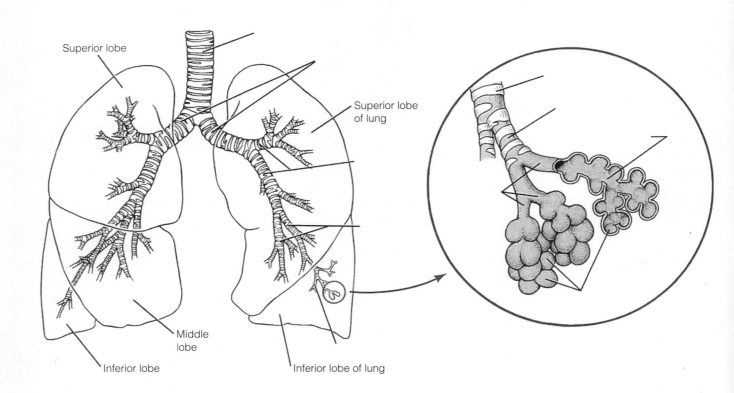

Superior lobe

Superior lobe
of lung

Middle
lobe

Inferior lobe

Inferior lobe of lung

10. Match the terms in column B to those in column A.

Column A

_____ 1. nerve that activates the diaphragm during inspiration

_____ 2. "floor" of the nasal cavity

_____ 3. food passageway posterior to the trachea

_____ 4. flaps over the glottis during swallowing of food

_____ 5. contains the vocal cords

_____ 6. part of the conducting pathway between the larynx and the primary bronchi

_____ 7. pleural layer lining the walls of the thorax

_____ 8. site from which oxygen enters the pulmonary blood

_____ 9. autonomic nervous system nerve serving the thoracic region

_____ 10. opening between the vocal folds

_____ 11. increases air turbulence in the nasal cavity

Column B

a. alveolus

b. bronchiole

c. concha

d. epiglottis

e. esophagus

f. glottis

g. larynx

h. palate

i. parietal pleura

j. phrenic nerve

k. primary bronchi

l. trachea

m. vagus nerve

n. visceral pleura

11. What portions of the respiratory system are referred to as anatomical dead space? _____

_____

_____

Why? _____

12. Define _external respiration:_ _____

_____

internal respiration: _____

_____

13. On the diagram below identify alveolar epithelium, capillary endothelium, alveoli, and red blood cells, and bracket the respiratory membrane.

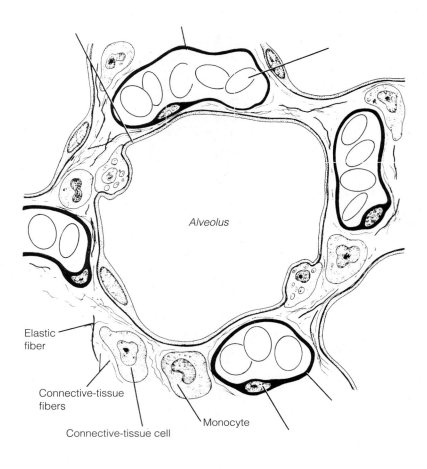

Alveolus

Elastic fiber

Connective-tissue fibers

Connective-tissue cell

Monocyte

Demonstrating Lung Inflation in a Sheep Pluck

Does the lung inflate part by part or as a whole, like a balloon? _____

What happened when the pressure was released? _____

What type of tissue ensures this phenomenon? _____

Examining Prepared Slides of Lung and Trachea Tissue

1. The tracheal epithelium is ciliated and has goblet cells. What is the function of each of these modifications?

Cilia? _____

Goblet cells? _____

2. The tracheal epithelium is said to be pseudostratified. Why? _____

_____

_____

3. What structural characteristics of the alveoli make them an ideal site for the diffusion of gases?

_____

Why does oxygen move from the alveoli into the pulmonary capillary blood? _____

_____

_____

4. If you observed pathologic lung sections, what were the responsible conditions and how did the tissue differ from normal lung tissue?

Slide Type	Observations

Respiratory System Physiology

Mechanics of Respiration

1. For each of the following cases, check the column appropriate to your observations on the operation of the model lung.

Change	Diaphragm pushed up		Diaphragm pulled down	
	Increased	Decreased	Increased	Decreased
In internal volume of the bell jar (thoracic cage)				
In internal pressure				
In the size of the balloons (lungs)				
In direction of air flow	Into lungs	Out of lungs	Into lungs	Out of lungs

2. Base your answers to the following on your observations in question 1.

   Under what internal conditions does air tend to flow into the lungs? _____

   _____

   Under what internal conditions does air tend to flow out of the lungs? Explain. _____

   _____

3. Activation of the diaphragm and the external intercostal muscles begins the inspiratory process. What results from the contraction of these muscles, and how is this accomplished? _____

   _____

   _____

   _____

4. What was the approximate increase in diameter of chest circumference during a quiet inspiration?

   _____ inches   During forced inspiration? _____ inches

What temporary physiologic advantage does the substantial increase in chest circumference during forced inspiration

create? _____

_____

_____

5. The presence of a partial vacuum between the pleural membranes is integral to normal breathing movements. What would happen if an opening were made into the chest cavity, as with a puncture wound?

_____

How is this condition treated medically? _____

Respiratory Volumes and Capacities—Spirometry

1. Write the respiratory volume term and the normal value that is described by the following statements:

Volume of air present in the lungs after a forceful expiration _____

Volume of air that can be expired forcibly after a normal expiration _____

Volume of air that is breathed in and out during a normal respiration _____

Volume of air that can be inspired forcibly after a normal inspiration _____

Volume of air corresponding to TV + IRV + ERV _____

2. Record experimental respiratory volumes as determined in the laboratory. (Corrected values are for Procedure B only.)

Average tidal volume _____ ml   Average VC _____ ml

Corrected value for TV _____ ml   Corrected value for VC _____ ml

Average IRV _____ ml   % predicted VC _____ %

Corrected value for IRV _____ ml FEV_1 _____ % FVC

Minute respiratory volume _____ ml/min

Average ERV _____ ml

Corrected value for ERV _____ ml

3. Would your vital-capacity measurement differ if you performed the test while standing? _____ While lying down?

_____ Explain. _____

_____

4. Which respiratory ailments can respiratory volume tests be used to detect?

_____

_____

5. Using an appropriate reference, complete the chart below:

		O_2	CO_2	N_2
% of composition of air	Inspired			
	Expired			

Use of the Pneumograph to Determine Factors Influencing Rate and Depth of Respiration

1. Where are the neural control centers of respiratory rhythm? _____ and _____

2. Based on pneumograph reading of respiratory variation, what was the rate of quiet breathing?

Initial testing _____ breaths/min

Record observations of how the initial pneumograph recording was modified during the various testing procedures described below. Indicate the respiratory rate, and include comments on the relative depth of the respiratory peaks observed.

Test performed	Observations
Talking	
Yawning	
Laughing	
Standing	
Concentrating	
Swallowing water	
Coughing	
Lying down	
Running in place	

3. Student data:

Breath-holding interval after a deep inhalation _____ sec   length of recovery period _____ sec

Breath-holding interval after a forceful expiration _____ sec   length of recovery period _____ sec

After breathing quietly and taking a deep breath (which you held), was your urge to inspire or expire? _____

After exhaling and then holding one's breath, was the desire for inspiration or expiration? _____

Explain these results. (*Hint:* what reflex is involved here?) _____

_____

_____

4. Observations after hyperventilation: _____

_____

5. Length of breath holding after hyperventilation: _____ sec

Why does hyperventilation produce apnea or a reduced respiratory rate? _____

_____

_____

6. Observations for rebreathing breathed air: _____

_____

Why does rebreathing breathed air produce an increased respiratory rate? _____

_____

7. What was the effect of running in place (exercise) on the duration of breath holding? _____

Explain: _____

_____

8. Relative to the test illustrating the effect of respiration on circulation: *(student data)*

Radial pulse before beginning test _____ /min   Radial pulse after testing _____ /min

Relative pulse force before beginning test _____   Relative force of radial pulse after testing _____

Condition of neck and facial veins after testing _____

Explain: _____

_____

9. Do the following factors generally increase (indicate with I) or decrease (indicate with D) the respiratory rate and depth?

1. increase in blood CO_2 _____   3. increase in blood pH _____

2. decrease in blood O_2 _____   4. decrease in blood pH _____

Did it appear that CO_2 or O_2 had a more marked effect on modifying the respiratory rate? _____

10. Sensory receptors sensitive to changes in blood pressure are located where? _____

_____

11. Where are sensory receptors sensitive to changes in O_2 levels in the blood located? _____

_____

12. What is the primary factor that initiates breathing in a newborn infant? _____

_____

13. Blood CO_2 levels and blood pH are related. When blood CO_2 levels increase, does the pH increase or decrease?

_____ Explain why. _____

_____

_____

Respiratory Sounds

1. Which of the respiratory sounds is heard during both inspiration and expiration? _____

Which is heard primarily during inspiration? _____

2. Where did you best hear the vesicular respiratory sounds? _____

Role of the Respiratory System in Acid-Base Balance of Blood

1. Define *buffer*. _____

2. How successful was the laboratory buffer (pH 7) in resisting changes in pH when the acid was added? _____

_____

When the base was added? _____

How successful was the buffer in resisting changes in pH when the additional aliquots (3 more drops) of the acid and base

were added to the original samples? _____

3. What buffer system operates in blood plasma? _____

Which of its species resists a *drop* in pH? _____

Which resists a *rise* in pH? _____

4. Explain how the carbonic acid–bicarbonate buffer system of the blood operates. _____

_____

_____

_____

_____

Anatomy of the Digestive System

General Histological Plan of the Alimentary Canal

1. The general anatomical features of the digestive tube have been presented. Fill in the table below to complete the information listed.

Wall layer	Subdivisions of the layer	Major functions
mucosa		
submucosa		
muscularis externa		
serosa or adventitia		

Organs of the Alimentary Canal

1. The tubelike digestive system canal that extends from the mouth to the anus is the _____ canal.

2. How is the muscularis externa of the stomach modified? _____

    _____

    How does this modification relate to the function of the stomach? _____

    _____

3. What transition in epithelium type exists at the gastroesophageal junction? _____

    _____

    How do the epithelia of these two organs relate to their specific functions? _____

    _____

    _____

4. Match the items in column B with the descriptive statements in column A.

Column A

_____ 1. structure that suspends the small intestine from the posterior body wall

_____ 2. fingerlike extensions of the intestinal mucosa that increase the surface area for absorption

_____ 3. large collections of lymphoid tissue found in the submucosa of the small intestine

_____ 4. deep folds of the mucosa and submucosa that extend completely or partially around the circumference of the small intestine

_____, _____ 5. regions that break down foodstuffs mechanically

_____ 6. mobile organ that manipulates food in the mouth and initiates swallowing

_____ 7. conduit for both air and food

_____, _____, _____ 8. three structures continuous with and representing modifications of the peritoneum

_____ 9. the "gullet"; no digestive/absorptive function

_____ 10. folds of the gastric mucosa

_____ 11. sacculations of the large intestine

_____ 12. projections of the plasma membrane of a mucosal epithelial cell

_____ 13. valve at the junction of the small and large intestines

_____ 14. primary region of food and water absorption

_____ 15. membrane securing the tongue to the floor of the mouth

_____ 16. absorbs water and forms feces

_____ 17. area between the teeth and lips/cheeks

_____ 18. wormlike sac that outpockets from the cecum

_____ 19. initiates protein digestion

_____ 20. structure attached to the lesser curvature of the stomach

_____ 21. organ distal to the stomach

_____ 22. valve controlling food movement from the stomach into the duodenum

_____ 23. posterosuperior boundary of the oral cavity

_____ 24. location of the hepatopancreatic sphincter through which pancreatic secretions and bile pass

_____ 25. serous lining of the abdominal cavity wall

_____ 26. principal site for the synthesis of vitamin K by microorganisms

_____ 27. region containing two sphincters through which feces are expelled from the body

_____ 28. bone-supported anterosuperior boundary of the oral cavity

Column B

a. anus

b. appendix

c. esophagus

d. frenulum

e. greater omentum

f. hard palate

g. haustra

h. ileocecal valve

i. large intestine

j. lesser omentum

k. mesentery

l. microvilli

m. oral cavity

n. parietal peritoneum

o. Peyer's patches

p. pharynx

q. plicae circulares

r. pyloric valve

s. rugae

t. small intestine

u. soft palate

v. stomach

w. tongue

x. vestibule

y. villi

z. visceral peritoneum

5. Correctly identify all organs depicted in the diagram below.

6. You have studied the histologic structure of a number of organs in this laboratory. Three of these are diagrammed below. Identify each.

_____ _____ _____

Accessory Digestive Organs

1. Correctly label all structures provided with leader lines in the diagram of a molar below. (Note: some of the terms in the key for item 2 may be helpful in this task.)

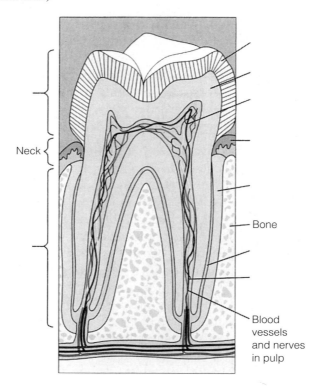

Neck

Bone

Blood vessels and nerves in pulp

2. Use the key to identify each tooth area described below.

_____ 1. visible portion of the tooth *in situ*

_____ 2. material covering the tooth root

_____ 3. hardest substance in the body

_____ 4. attaches the tooth to bone and surrounding alveolar structures

_____ 5. portion of the tooth embedded in bone

_____ 6. forms the major portion of tooth structure; similar to bone

_____ 7. produces the dentin

_____ 8. site of blood vessels, nerves, and lymphatics

_____ 9. entire portion of the tooth covered with enamel

Key: a. anatomical crown

b. cementum

c. clinical crown

d. dentin

e. enamel

f. gingiva

g. odontoblast

h. periodontal ligament

i. pulp

j. root

3. In the human, the number of deciduous teeth is _____; the number of permanent teeth is _____.

4. The dental formula for permanent teeth is $\dfrac{2,\,1,\,2,\,3}{2,\,1,\,2,\,3}$

Explain what this means: _____

_____

What is the dental formula for the deciduous teeth? _____

5. What teeth are the "wisdom teeth"? _____

_____

_____

6. Various types of glands form a part of the alimentary tube wall or duct their secretions into it. Match the glands listed in column B with the function/locations described in column A.

Column A

_____ 1. produce(s) mucus; found in the submucosa of the small intestine

_____ 2. produce(s) a product containing amylase that begins starch breakdown in the mouth

_____ 3. produce(s) a whole spectrum of enzymes and an alkaline fluid that is secreted into the duodenum

_____ 4. produce(s) bile that it secretes into the duodenum via the bile duct

_____ 5. produce(s) HCl and pepsinogen

_____ 6. found in the mucosa of the small intestine; produce(s) intestinal juice

Column B

a. duodenal glands

b. gastric glands

c. intestinal crypts

d. liver

e. pancreas

f. salivary glands

7. Which of the salivary glands produces a secretion that is mainly serous? _____

8. What is the role of the gallbladder? _____

9. Name three structures always found in the portal triad regions of the liver. _____,

_____ and _____.

10. Where would you expect to find the Kupffer cells of the liver? _____

What is their function? _____

11. Why is the liver so dark red in the living animal? _____

_____

12. The pancreas has two major populations of secretory cells—those in the islets and the acinar cells. Which population serves

the digestive process? _____

Chemical and Physical Processes of Digestion: Wet Lab

Chemical Digestion of Foodstuffs: Enzymatic Action

1. Match the following definitions with the proper choices from the key.

Key: a. catalyst b. control c. enzyme d. substrate

_____ 1. increases the rate of a chemical reaction without becoming part of the product

_____ 2. provides a standard of comparison for test results

_____ 3. biologic catalyst: protein in nature

_____ 4. substance on which a catalyst works

2. List the three characteristics of enzymes. _____

_____

_____

3. The enzymes of the digestive system are classified as hydrolases. What does this mean?

_____

_____

4. Fill in the following chart about the various digestive system enzymes encountered in this exercise.

Enzyme	Organ producing it	Site of action	Substrate(s)	Optimal pH
Salivary amylase				
Trypsin				
Lipase (pancreatic)				

5. Name the end products of digestion for the following types of foods:

proteins: _____ carbohydrates: _____

fats: _____ and _____

6. You used several indicators or tests in the laboratory to determine the presence or absence of certain substances. Choose the correct test or indicator from the key to correspond to the condition described below.

Key: a. IKI (Lugol's iodine)   b. Benedict's solution   c. litmus   d. BAPNA

_____ 1. used to test for protein hydrolysis, which was indicated by a yellow color

_____ 2. used to test for the presence of starch, which was indicated by blue-black color

_____ 3. used to test for the presence of fatty acids, which was evidenced by a color change from blue to pink

_____ 4. used to test for the presence of reducing sugars (maltose, sucrose, glucose) as indicated by a blue to green color change

7. What conclusions can you draw when an experimental sample gives both a positive starch test and a positive maltose test

after incubation? _____

Why was 37°C the optimal incubation temperature? _____

Why did very little, if any, starch digestion occur in test tube 4A? _____

_____

Why did very little, if any, starch digestion occur in test tube 6A? _____

_____

Assume you have made the statement to a group of your peers that amylase is capable of starch hydrolysis to maltose. If you

had not done control tube 1A, what objection to your statement could be raised? _____

_____

What if you had not done tube 2A? _____

8. In the exercise concerning trypsin function, why was an enzyme assay like Benedict's or Lugol's IKI (which test for the pres-

ence of a reaction product) not necessary? _____

Why was tube 1T necessary? _____

Why was tube 2T necessary? _____

Trypsin is a protease similar to pepsin, the protein-digesting enzyme in the stomach. Would trypsin work well in the

stomach? _____ Why? _____

9. In the procedure concerning pancreatic lipase digestion of fats and the action of bile salts, how did the appearance of tubes

1E and 2E differ? _____

_____

Can you explain the difference? _____

_____

Why did the litmus indicator change from blue to pink during fat hydrolysis? _____

_____

Why is bile not considered an enzyme? _____

How did the tubes containing bile compare with those not containing bile? _____

What role does bile play in fat digestion? _____

_____

10. The three-dimensional structure of a functional protein is altered by intense heat or nonphysiological pH even though peptide bonds may not break. Such inactivation is called denaturation, and denatured enzymes are nonfunctional. Explain why.

_____

_____

What specific experimental conditions resulted in denatured enzymes? _____

_____

11. Pancreatic and intestinal enzymes operate optimally at a pH that is slightly alkaline, yet the chyme entering the duodenum from the stomach is very acid. How is the proper pH for the functioning of the pancreatic-intestinal enzymes ensured?

_____

_____

12. Assume you have been chewing a piece of bread for 5 or 6 minutes. How would you expect its taste to change during this

interval? _____

Why? _____

13. Note the mechanism of absorption (passive or active transport) of the following food breakdown products, and indicate by a check mark (✓) whether the absorption would result in their movement into the blood capillaries or the lymph capillaries (lacteals).

Substance	Mechanism of absorption	Blood	Lymph
Monosaccharides			
Fatty acids and glycerol			
Amino acids			
Water			
Na^+, Cl^-, Ca^{2+}			

14. People on a strict diet to lose weight begin to metabolize stored fats at an accelerated rate. How does this condition affect

blood pH? _____

15. Trace the pathway of a ham sandwich (ham = protein and fat; bread = starch) from the mouth to the site of absorption of its breakdown products, noting where digestion occurs and what specific enzymes are involved.

16. Some of the digestive organs have groups of secretory cells that liberate hormones (parahormones) into the blood. These exert an effect on the digestive process by acting on other cells or structures and causing them to release digestive enzymes, expel bile, or increase the mobility of the digestive tract. For each hormone below, note the organ producing the hormone and its effects on the digestive process. Include the target organs affected.

Hormone	Produced by	Target organ(s) and effects
Secretin		
Gastrin		
Cholecystokinin		

Physical Processes: Mechanisms of Food Propulsion and Mixing

Complete the following statements.

Swallowing, or__1__, occurs in two phases—the__2__and __3__. One of these phases, the__4__phase, is voluntary. During the voluntary phase, the__5__is used to push the food into the back of the throat. During swallowing, the__6__rises to ensure that its passageway is covered by the epiglottis so that the ingested substances don't enter the respiratory passageways. It is possible to swallow water while standing on your head because the water is carried along the esophagus involuntarily by the process of__7__. The pressure exerted by the foodstuffs on the__8__sphincter causes it to open, allowing the foodstuffs to enter the stomach.

The two major types of propulsive movements that occur in the small intestine are__9__and__10__. One of these movements, the __11__, acts to continually mix the foods and to increase the absorption rate by moving different parts of the chyme mass over the intestinal mucosa, but has less of a role in moving foods along the digestive tract.

1. _____
2. _____
3. _____
4. _____
5. _____
6. _____
7. _____
8. _____
9. _____
10. _____
11. _____

Chemical and Physical Processes of Digestion

Chemical Digestion of Foodstuffs: Enzymatic Action

1. Match the following definitions with the proper choices from the key.

Key: a.   catalyst   b.   control   c.   enzyme   d.   substrate

_____ 1.   increases the rate of a chemical reaction without becoming part of the product

_____ 2.   provides a standard of comparison for test results

_____ 3.   biological catalyst: protein in nature

_____ 4.   substance on which an enzyme works

2. Name three characteristics of enzymes. _____

_____

3. Explain the following statement: The enzymes of the digestive system are classified as hydrolases.

_____

_____

4. Fill in the chart below with what you have learned about the various digestive system enzymes encountered in this exercise.

Enzyme	Organ producing it	Site of action	Substrate(s)	Optimal pH
Salivary amylase				
Pepsin				
Lipase (pancreatic)				

5. Name the end products of digestion for the following types of foods.

Proteins: _____   Carbohydrates: _____

Fats: _____ and _____

6. You used several different indicators or tests in the laboratory to determine the presence or absence of certain substances. Choose the correct test or indicator from the key to correspond to the condition described below.

Key: a. IKI   b. Benedict's solution   c. pH meter   d. BAPNA

_____ 1. used to test for protein hydrolysis, which was indicated by a yellow color

_____ 2. used to test for the presence of starch, which was indicated by a blue-black color

_____ 3. used to test for the presence of fatty acids

_____ 4. used to test for the presence of maltose, which was indicated by a blue to green (or to rust) color change

7. The three-dimensional structure of a functional protein is altered by intense heat or nonphysiological pH even though peptide bonds may not break. Such a change in protein structure is called denaturation, and denatured enzymes are not functional. Explain why.

_____

_____

8. What experimental conditions in the simulation resulted in denatured enzymes? _____

_____

9. Complete the mechanism of absorption section in the chart below for each of the substances listed. Use a check mark to indicate whether the absorption would result in the movement of a substance into the blood capillaries or the lymph capillaries (lacteals).

Substance	Mechanism of absorption	Blood	Lymph
Monosaccharides			
Fatty acids and glycerol			
Amino acids			

10. Imagine that you have been chewing a piece of bread for 5 to 6 minutes. How would you expect its taste to change during

this time? _____

11. People on a strict diet to lose weight begin to metabolize stored fats at an accelerated rate. How could this condition affect

blood pH? _____

Starch Digestion by Salivary Amylase

1. What conclusions can you draw when an experimental sample gives both a positive starch test and a positive maltose test?

2. Why was 37°C the optimal incubation temperature? _____

3. Why did very little, if any, starch digestion occur in tube 1? _____

4. Why did very little starch digestion occur in tubes 6 and 7? _____

5. Imagine that you have told a group of your peers that amylase is capable of digesting starch to maltose. If you had not run the experiment in control tubes 3, 4, and 5, what objections to your statement could be raised?

Protein Digestion by Pepsin

1. Why is an indicator reagent such as IKI or Benedict's solution not necessary when using a substrate like BAPNA?

2. Trypsin is a pancreatic hydrolase present in the small intestine during digestion. Would trypsin work well in the stomach? Explain your answer.

3. How does the optical density of a solution containing BAPNA relate to enzyme activity? _____

4. What happens to pepsin activity as it reaches the small intestine? _____

Fat Digestion by Pancreatic Lipase and the Action of Bile

1. Why does the pH of a fatty solution decrease as enzymatic hydrolysis increases? _____

2. How does bile affect fat digestion? _____

3. Why is it not possible to determine the activity of lipase in the pH 2.0 buffer using the pH meter assay method?

4. Why is bile not considered an enzyme? _____

Physical Processes: Mechanisms of Food Propulsion and Mixing

Complete the following statements. Write your answers in the numbered spaces below.

Swallowing, or __1__, occurs in two phases—the __2__ and __3__. One of these phases, the __4__ phase, is voluntary. During the voluntary phase, the __5__ is used to push the food into the back of the throat. During swallowing, the __6__ rises to ensure that its passageway is covered by the epiglottis so that the ingested substances do not enter the respiratory passageways. It is possible to swallow water while standing on your head because the water is carried along the esophagus involuntarily by the process of __7__. The pressure exerted by the foodstuffs on the __8__ sphincter causes it to open, allowing the food to enter the stomach.

The two major types of propulsive movements that occur in the small intestine are __9__ and __10__. One of these movements, __11__, acts to continually mix the foods and to increase the absorption rate by moving different parts of the chyme mass over the intestinal mucosa, but it has less of a role in moving foods along the digestive tract.

1. _____

2. _____

3. _____

4. _____

5. _____

6. _____

7. _____

8. _____

9. _____

10. _____

11. _____

Anatomy of the Urinary System

Gross Anatomy of the Human Urinary System

1. Complete the following statements:

    The kidney is referred to as an excretory organ because it excretes __1__ wastes. It is also a major homeostatic organ because it maintains the electrolyte, __2__ , and __3__ balance of the blood.

    Urine is continuously formed by the __4__ and is routed down the __5__ by the mechanism of __6__ to a storage organ called the __7__ . Eventually, the urine is conducted to the body __8__ by the urethra. In the male, the urethra is __9__ inches long and transports both urine and __10__ . The female urethra is __11__ inches long and transports only urine.

    Voiding or emptying the bladder is called __12__ . Voiding has both voluntary and involuntary components. The voluntary sphincter is the __13__ sphincter. An inability to control this sphincter is referred to as __14__ .

1. _____

2. _____

3. _____

4. _____

5. _____

6. _____

7. _____

8. _____

9. _____

10. _____

11. _____

12. _____

13. _____

14. _____

2. What is the function of the fat cushion that surrounds the kidneys in life? _____

_____

3. Define *ptosis.* _____

_____

4. Why is incontinence a normal phenomenon in the child under 1½ to 2 years old? _____

_____

What events may lead to its occurrence in the adult? _____

_____

_____

5. Complete the labeling of the diagram to correctly identify the urinary system organs.

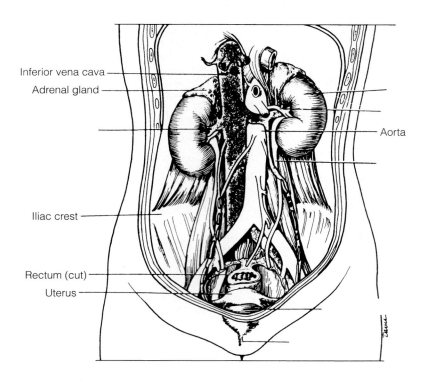

Inferior vena cava

Adrenal gland

Aorta

Iliac crest

Rectum (cut)

Uterus

Gross Internal Anatomy of the Pig or Sheep Kidney

Match the appropriate structure in column B to its description in column A.

Column A	Column B
_____ 1. smooth membrane, tightly adherent to the kidney surface	a. cortex
_____ 2. portion of the kidney containing mostly collecting ducts	b. medulla
_____ 3. portion of the kidney containing the bulk of the nephron structures	c. minor calyx
_____ 4. superficial region of kidney tissue	d. renal capsule
_____ 5. basinlike area of the kidney, continuous with the ureter	e. renal column
_____ 6. a cup-shaped extension of the pelvis that encircles the apex of a pyramid	f. renal pelvis
_____ 7. area of cortical tissue running between the medullary pyramids	

Functional Microscopic Anatomy of the Kidney and Bladder

1. Match each of the lettered structures on the diagram of the nephron (and associated renal blood supply) on the left with the terms on the right:

_____ 1. collecting duct

_____ 2. glomerulus

_____ 3. peritubular capillaries

_____ 4. distal convoluted tubule

_____ 5. proximal convoluted tubule

_____ 6. interlobar artery

_____ 7. interlobular artery

_____ 8. arcuate artery

_____ 9. interlobular vein

_____ 10. efferent arteriole

_____ 11. arcuate vein

_____ 12. loop of Henle

_____ 13. afferent arteriole

_____ 14. interlobar vein

_____ 15. glomerular capsule

2. Using the terms provided in item 1, identify the following:

_____ 1. site of filtrate formation

_____ 2. primary site of tubular reabsorption

_____ 3. secondarily important site of tubular reabsorption

_____ 4. structure that conveys the processed filtrate (urine) to the renal pelvis

_____ 5. blood supply that directly receives substances from the tubular cells

_____ 6. its inner (visceral) membrane forms part of the filtration membrane

3. Explain _why_ the glomerulus is such a high-pressure capillary bed. _____

_____

How does its high-pressure condition aid its function of filtrate formation? _____

_____

4. What structural modification of certain tubule cells enhances their ability to reabsorb substances from the filtrate?

_____

5. Explain the mechanism of tubular secretion and explain its importance in the urine formation process. _____

_____

_____

_____

6. Compare and contrast the composition of blood plasma and glomerular filtrate. _____

_____

7. Trace a drop of blood from the time it enters the kidney in the renal artery until it leaves the kidney through the renal vein.

Renal artery → _____

_____

_____→ renal vein

8. Trace the anatomical pathway of a molecule of creatinine (metabolic waste) from the glomerular capsule to the urethra. Note each microscopic and/or gross structure it passes through in its travels. Name the subdivisions of the renal tubule.

Glomerular capsule → _____

_____

_____→ urethra

9. What is important functionally about the specialized epithelium (transitional epithelium) in the bladder?

_____

_____

Urinalysis

Characteristics of Urine

1. What is the normal volume of urine excreted in a 24-hour period? _____

    a. 0.1–0.5 liters   b. 0.5–1.2 liters   c. 1.0–1.8 liters

2. Assuming normal conditions, note whether each of the following substances would be (a) in greater relative concentration in the urine than in the glomerular filtrate, (b) in lesser concentration in the urine than in the glomerular filtrate, or (c) absent in both the urine and the glomerular filtrate.

   _____ 1. water   _____ 6. amino acids   _____ 10. urea

   _____ 2. phosphate ions   _____ 7. glucose   _____ 11. uric acid

   _____ 3. sulfate ions   _____ 8. albumin   _____ 12. creatinine

   _____ 4. potassium ions   _____ 9. red blood cells   _____ 13. pus (WBC)

   _____ 5. sodium ions

3. Explain why urinalysis is a routine part of any good physical examination. _____

_____

4. What substance is responsible for the normal yellow color of urine? _____

5. Which has a greater specific gravity: 1 ml of urine or 1 ml of distilled water? _____

    Explain. _____

_____

6. Explain the relationship between the color, specific gravity, and volume of urine. _____

_____

Abnormal Urinary Constituents

1. A microscopic examination of urine may reveal the presence of certain abnormal urinary constituents.

    Name three constituents that might be present if a urinary tract infection exists. _____,

_____, and _____

2. How does a urinary tract infection influence urine pH? _____

    How does starvation influence urine pH? _____

3. Several specific terms have been used to indicate the presence of abnormal urine constituents. Identify each of the abnormalities described below by inserting a term from the list at the right that names the condition.

_____ 1. presence of erythrocytes in the urine

_____ 2. presence of hemoglobin in the urine

_____ 3. presence of glucose in the urine

_____ 4. presence of albumin in the urine

_____ 5. presence of ketone bodies (acetone and others) in the urine

_____ 6. presence of pus (white blood cells) in the urine

a. albuminuria

b. glycosuria

c. hematuria

d. hemoglobinuria

e. ketonuria

f. pyuria

4. What are renal calculi and what conditions favor their formation? _____

_____

5. All urine specimens become alkaline and cloudy on standing at room temperature. Explain. _____

_____

6. Glucose and albumin are both normally absent in the urine, but the reason for their exclusion differs. Explain the reason for

the absence of glucose. _____

Explain the reason for the absence of albumin. _____

7. Several conditions (both pathologic and nonpathologic) are named below. Using the key provided, characterize the probable abnormal constituents or conditions of the urinary product of each. More than one choice may be necessary to fully characterize the condition in most cases.

_____ 1. glomerulonephritis

_____ 2. diabetes mellitus

_____ 3. pregnancy, exertion

_____ 4. hepatitis, cirrhosis of the liver

_____ 5. pyelonephritis

_____ 6. gonorrhea

_____ 7. starvation

_____ 8. diabetes insipidus

_____ 9. kidney stones

_____ 10. eating a 5-lb box of candy at one sitting

_____ 11. hemolytic anemias

_____ 12. cystitis (inflammation of the bladder)

Key:

a. albumin
b. hemoglobin
c. blood cells
d. glucose
e. ketone bodies
f. bilirubin
g. pus
h. high specific gravity
i. low specific gravity
j. casts

8. Name the three major nitrogenous wastes found in the urine. _____,

_____, and _____

9. Explain the difference between organized and unorganized sediments. _____

_____

_____

Anatomy of the Reproductive System

Gross Anatomy of the Human Male Reproductive System

1. List the two principal functions of the testis: _____

2. Identify all indicated structures or portions of structures on the diagrammatic view of the male reproductive system below.

3. A common part of any physical examination of the male is palpation of the prostate gland. How is this accomplished?

(Think!) _____

4. How might enlargement of the prostate gland interfere with urination or the reproductive ability of the male?

_____

_____

5. Match the terms in column B to the descriptive statements in column A.

Column A

_____ 1. copulatory organ/penetrating device

_____ 2. site of sperm/androgen production

_____ 3. muscular passageway conveying sperm to the ejaculatory duct; in the spermatic cord

_____ 4. transports both sperm and urine

_____ 5. sperm maturation site

_____ 6. location of the testis in adult males

_____ 7. loose fold of skin encircling the glans penis

_____ 8. portion of the urethra between the prostate gland and the penis

_____ 9. empties a secretion into the prostatic urethra

_____ 10. empties a secretion into the membranous urethra

Column B

a. bulbourethral glands

b. epididymis

c. glans penis

d. membranous urethra

e. penile urethra

f. penis

g. prepuce

h. prostate gland

i. prostatic urethra

j. seminal vesicles

k. scrotum

l. testes

m. vas (ductus) deferens

6. Why are the testes located in the scrotum? _____

_____

7. Describe the composition of semen and name all structures contributing to its formation. _____

_____

_____

8. Of what importance is the fact that seminal fluid is alkaline? _____

_____

9. What structures comprise the spermatic cord? _____

_____

Where is it located? _____

_____

10. Using the following terms, trace the pathway of sperm from the testes to the urethra: rete testis, epididymis, seminiferous tubule, ductus deferens.

_____ → _____ → _____ → _____

11. Using an appropriate reference, define *cryptorchidism* and discuss its significance.

_____

_____

Gross Anatomy of the Human Female Reproductive System

1. On the diagram of a frontal section of a portion of the female reproductive system seen below, identify all indicated structures.

2. Identify the female reproductive system structures described below:

_____ 1. site of fetal development

_____ 2. copulatory canal

_____ 3. "fertilized egg" typically formed here

_____ 4. becomes erectile during sexual excitement

_____ 5. duct extending superolaterally from the uterus

_____ 6. partially closes the vaginal canal; a membrane

_____ 7. produces eggs, estrogens, and progesterone

_____ 8. fingerlike ends of the fallopian tube

3. Do any sperm enter the pelvic cavity of the female? Why or why not? _____

_____

4. What is an ectopic pregnancy, and how can it happen? _____

_____

_____

_____

5. Name the structures composing the external genitalia, or vulva, of the female. _____

_____

_____

6. Put the following vestibular-perineal structures in their proper order from the anterior to the posterior aspect: vaginal orifice, anus, urethral opening, and clitoris.

Anterior limit: _____ → _____ → _____ → _____

7. Name the male structure that is homologous to the female structures named below.

labia majora _____   clitoris _____

8. Assume a couple has just consummated the sex act and the male's sperm have been deposited in the woman's vagina. Trace the pathway of the sperm through the female reproductive tract.

_____

9. Define *ovulation:* _____

Microscopic Anatomy of Selected Male and Female Reproductive Organs

1. The testis is divided into a number of lobes by connective tissue. Each of these lobes contains one to four _____

_____, which converge on a tubular region at the testis hilus called the _____

_____.

2. What is the function of the cavernous bodies seen in the male penis? _____

_____

_____

3. Name the three layers of the uterine wall from the inside out.

_____, _____, _____

Which of these is sloughed during menses? _____

Which contracts during childbirth? _____

4. What is the function of the stereocilia exhibited by the epithelial cells of the mucosa of the epididymis? _____

_____

5. On the diagram showing the sagittal section of the human testis, correctly identify all structures provided with leader lines.

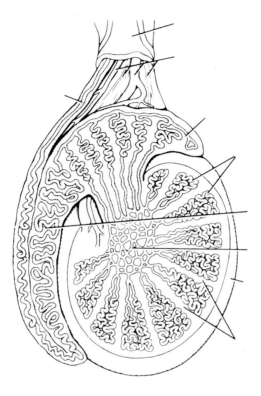

The Mammary Glands

1. To describe breast function, complete the following sentences:

Milk is formed by _____ within the _____ of the breast.

Milk is then excreted into enlarged storage regions called _____ and then finally through the

_____.

2. Using the key terms, correctly identify breast structures.

Key: a. adipose tissue
       b. alveoli
       c. areola
       d. lactiferous duct
       e. lactiferous sinus
       f. nipple

Clavicle

1st rib

Pectoralis muscles

Intercostal muscles

Fibrous connective
tissue stroma

Skin

6th rib

3. Describe the procedure for self-examination of the breasts. (Men are not exempt from breast cancer, you know!)

_____

_____

_____

_____

_____

_____

Physiology of Reproduction: Gametogenesis and the Female Cycles

Meiosis

1. The following statements refer to events occurring during mitosis and/or meiosis. For each statement, decide if the event occurs in (a) mitosis only, (b) meiosis only, or (c) both mitosis and meiosis.

_____ 1. dyads are visible

_____ 2. tetrads are visible

_____ 3. product is two diploid daughter cells

_____ 4. product is four haploid daughter cells

_____ 5. involves the phases prophase, metaphase, anaphase, and telophase

_____ 6. occurs throughout the body

_____ 7. occurs only in the ovaries and testes

_____ 8. provides cells for growth and repair

_____ 9. homologues synapse and chiasmata are seen

_____ 10. daughter cells are quantitatively and qualitatively different from the mother cell

_____ 11. daughter cells are genetically identical to the mother cell

_____ 12. chromosomes are replicated before the division process begins

_____ 13. provides cells for replication of the species

_____ 14. consists of two consecutive nuclear divisions, without chromosomal replication occurring before the second division

2. Describe the process of synapsis. _____

_____

3. How does crossover introduce variability in the daughter cells? _____

_____

_____

4. Define *homologous chromosomes:* _____

_____

Spermatogenesis

1. The cell types seen in the seminiferous tubules are listed in the key. Match the correct cell type(s) with the descriptions given below.

   Key: a. primary spermatocyte
         b. secondary spermatocyte
         c. spermatogonium
         d. sustentacular cell
         e. spermatid
         f. sperm

   _____ 1. primitive stem cell

   _____ 2. haploid

   _____ 3. provides nutrients to
                               developing sperm

   _____ 4. products of meiosis II

   _____ 5. product of spermiogenesis

   _____ 6. product of meiosis I

2. Why are spermatids not considered functional gametes? _____

   _____

3. Define *spermatogenesis:* _____

   Define *spermiogenesis:* _____

   _____

4. Draw a sperm below and identify the *acrosome, head, midpiece,* and *tail.* Then beside each label, note the composition and function of each of these sperm structures.

5. The life span of a sperm is very short. What anatomical characteristics might lead you to suspect this even if you didn't know

   its life span? _____

Oogenesis, the Ovarian Cycle, and the Menstrual Cycle

1. The sequence of events leading to germ cell formation in the female begins during fetal development. By the time the child

   is born, all the oogonia have been converted to _____.

   In view of this fact, how does the total germ cell potential of the female compare to that of the male?

   _____

2. The female gametes develop in structures called *follicles.* What is a follicle? _____

   _____

How are primary and vesicular follicles anatomically different? _____

_____

_____

_____

What is a corpus luteum? _____

_____

3. What hormone is produced by the vesicular follicle? _____

By the corpus luteum? _____

4. Use the key to identify the cell type you would expect to find in the following structures.

Key: a.   oogonium   b.   primary oocyte   c.   secondary oocyte   d.   ovum

_____ 1.   forming part of the primary follicle in the   _____ 3.   in the mature vesicular follicle of the ovary
ovary

_____ 2.   in the uterine tube before fertilization   _____ 4.   in the uterine tube shortly after sperm penetration

5. The cellular product of spermatogenesis is four _____; the final product of oogenesis is one

_____ and three _____. What is the function of this unequal cytoplasmic

division seen during oogenesis in the female? _____

_____

_____

What is the fate of the three tiny cells produced during oogenesis? _____

Why? _____

6. The following statements deal with anterior pituitary and ovarian hormonal interrelationships. Name the hormone(s)
described in each statement.

_____ 1.   stimulates ovarian follicles to grow and to produce estrogens

_____ 2.   ovulation occurs after its burstlike release

_____   and _____ 3.   exert negative feedback on the anterior pituitary
relative to FSH secretion

_____ 4.   stimulates LH release by the anterior pituitary

_____ 5.   stimulates the corpus luteum to produce progesterone and estrogen

_____ 6.   maintains the hormonal production of the corpus luteum in a nonpregnant woman

7. Why does the corpus luteum deteriorate toward the end of the ovarian cycle? _____

_____

8. For each statement below dealing with hormonal blood levels during the female ovarian and menstrual cycles, decide whether the condition in column A is usually (a) greater than, (b) less than, or (c) essentially equal to the condition in column B.

Column A

	Column B
_____ 1. amount of estrogen in the blood during menses	amount of estrogen in the blood at ovulation
_____ 2. amount of progesterone in the blood on the fourteenth day	amount of progesterone in the blood on the twenty-third day
_____ 3. amount of LH in the blood during menses	amount of LH in the blood at ovulation
_____ 4. amount of FSH in the blood on day 6 of the cycle	amount of FSH in the blood on day 20 of the cycle
_____ 5. amount of estrogen in the blood on the tenth day	amount of progesterone in the blood on the tenth day

9. Ovulation and menstruation usually cease by the age of _____.

10. What uterine tissue undergoes dramatic changes during the menstrual cycle? _____

11. When during the female menstrual cycle would fertilization be unlikely? _____

_____

12. Assume that a woman could be an "on demand" ovulator like the rabbit, in which copulation stimulates the hypothalamic-anterior pituitary axis and causes LH release, and an oocyte was ovulated and fertilized on day 26 of her 28-day cycle. Why

would a successful pregnancy be unlikely at this time? _____

_____

_____

13. The menstrual cycle depends on events within the female ovary. The stages of the menstrual cycle are listed below. For each, note its approximate time span and the related events in the uterus; and then to the right, record the ovarian events occurring simultaneously. Pay particular attention to hormonal events.

Menstrual cycle stage	Uterine events	Ovarian events
Menstruation		
Proliferative		
Secretory		

Survey of Embryonic Development

Developmental Stages of the Human

1. Use the key choices to identify the embryonic stage or process described below.

Key: a. cleavage   c. zygote   e. blastula
     b. morula   d. fertilization f. gastrulation

_____ 1. the event most immediately following sperm penetration

_____ 2. solid ball of embryonic cells

_____ 3. process of rapid mitotic cell division without intervening growth periods

_____ 4. combination of egg and sperm

_____ 5. process involving cell rearrangements to form the three primary germ layers

_____ 6. embryonic stage in which the embryo consists of a hollow ball of cells

2. What is the importance of cleavage in embryonic development? _____

_____

How is cleavage different from mitotic cell division, which occurs late in life? _____

_____

_____

3. The cells of the human blastula (more commonly called the blastocyst or chorionic vesicle) have various fates. Which blastocyst structures have the following fates?

_____ 1. produces the embryonic body

_____ 2. becomes the chorion and cooperates with uterine tissues to form the placenta

_____ 3. produces the amnion, yolk sac, and allantois

_____ 4. produces the primordial germ cells (an embryonic membrane)

_____ 5. an embryonic membrane that provides the structural basis for the body stalk or umbilical cord

4. Using the letters on the diagram, correctly identify each of the following maternal or embryonic structures.

_____ amnion _____ chorion _____ decidua basalis _____ endoderm

_____ body stalk _____ chorionic villi _____ decidua capsularis _____ mesoderm

_____ ectoderm _____ uterine cavity

5. Explain the importance of gastrulation. _____

_____

_____

6. What is the function of the amnion and the amniotic fluid? _____

_____

_____

7. Describe the process of implantation, noting the role of the trophoblast cells. _____

_____

_____

8. How many days after fertilization is implantation generally completed? _____ What event in the female menstrual cycle

ordinarily occurs just about this time if implantation does not occur? _____

_____

9. What name is given to the part of the uterine wall directly under the implanting embryo? _____

That surrounding the rest of the embryonic structure? _____

10. Using an appropriate reference, find out what *decidua* means and state the definition. _____

_____

How is this terminology applicable to the deciduas of pregnancy? _____

_____

11. Referring to the illustrations and text of *A Colour Atlas of Life Before Birth: Normal Fetal Development,* answer the following:

Which two organ systems are extensively developed in the *very young* embryo?

_____ and _____

Describe the direction of development by circling the correct descriptions below:

proximal-distal   distal-proximal   caudal-rostral   rostral-caudal

Does bodily control during infancy develop in the same directions? Think! Can an infant pick up a common pin (pincer grasp) or wave his arms earlier? Is arm-hand or leg-foot control achieved earlier?

_____

12. Note whether each of the following organs or organ systems develops from the (a) ectoderm, (b) endoderm, or (c) meso-derm. Use an appropriate reference as necessary.

_____ 1.   skeletal muscle   _____ 4.   respiratory mucosa   _____ 7.   nervous system

_____ 2.   skeleton   _____ 5.   circulatory system   _____ 8.   serosa membrane

_____ 3.   lining of gut   _____ 6.   epidermis of skin   _____ 9.   liver, pancreas

In Utero Development

1.   Make the following comparisons between a human and the pregnant dissected animal structures.

Comparison object	Human	Dissected animal
Shape of the placenta		
Shape of the uterus		

2.   Where in the human uterus do implantation and placentation ordinarily occur? _____

_____

3.   Describe the function(s) of the placenta. _____

_____

What embryonic membranes has the placenta more or less "put out of business"? _____

_____

4. When does the human embryo come to be called a fetus? _____

5. What is the usual and most desirable fetal position in utero? _____

   Why is this the most desirable position? _____

   _____

   _____

   _____

Gross and Microscopic Anatomy of the Placenta

1. Describe fully the gross structure of the human placenta as observed in the laboratory. _____

   _____

   _____

2. What is the tissue origin of the placenta: fetal, maternal, or both? _____

3. What are the placental barriers that must be crossed to exchange materials? _____

   _____

Principles of Heredity

Introduction to the Language of Genetics

1. Match the key choices with the definitions given below.

Key: a. alleles   d. genotype   g. phenotype
      b. autosomes   e. heterozygous   h. recessive
      c. dominant   f. homozygous   i. sex chromosomes

_____ 1. actual genetic makeup

_____ 2. chromosomes determining maleness/femaleness

_____ 3. situation in which an individual has identical alleles for a particular trait

_____ 4. genes not expressed unless they are present in homozygous condition

_____ 5. expression of a genetic trait

_____ 6. situation in which an individual has different alleles making up his genotype for a particular trait

_____ 7. genes for the same trait that may have different expressions

_____ 8. chromosomes regulating most body characteristics

_____ 9. the more-potent gene allele; masks the expression of the less-potent allele

Dominant-Recessive Inheritance

1. In humans, farsightedness is inherited by possession of a dominant gene. If a man who is homozygous for normal vision (*aa*) marries a woman who is heterozygous for farsightedness, what proportion of their children would be expected to be

farsighted? _____%

2. A metabolic disorder called PKU is due to an abnormal recessive gene (*p*). Only homozygous recessive individuals exhibit

this disorder. What percentage of the offspring will be anticipated to have PKU if the parents are *Pp* and *pp*? _____%

3. A man obtained 32 spotted and 10 solid-color rabbits from a mating of two spotted rabbits.

Which trait is dominant? _____ Recessive? _____

What is the probable genotype of the rabbit parents? _____ × _____

4. Assume that the allele controlling brown eyes (B) is dominant over that controlling blue eyes (b) in human beings. (In actuality, eye color in humans is an example of multigene inheritance, which is much more complex than this.) A blue-eyed man marries a brown-eyed woman; and they have six children, all brown-eyed. What is the most likely genotype of the father?

_____ Of the mother? _____ If the seventh child had *blue* eyes, what could you conclude about the parents' genotypes?

_____

_____

Incomplete Dominance

1. Tail length on a bobcat is controlled by incomplete dominance. The alleles are T for normal tail length and t for tail-less.

What name could/would you give to the tails of heterozygous (Tt) cats? _____

How would their tail length compare with that of TT or tt bobcats? _____

2. If curly-haired individuals are genotypically CC, straight-haired individuals are cc, and wavy-haired individuals are heterozygotes (Cc), what percentage of the various phenotypes would be anticipated from a cross between a CC woman and a cc man?

_____% curly   _____% wavy   _____% straight

Sex-Linked Inheritance

1. What does it mean when someone says a particular characteristic is sex-linked? _____

_____

2. You are a male, and you have been told that hemophilia "runs in your genes." Whose ancestors, your mother's or your

father's, should you investigate? _____ Why? _____

_____

_____

3. An $X^C X^c$ female marries an $X^C Y$ man. Do a Punnett square for this match.

What is the probability of producing a color-blind son? _____

A color-blind daughter? _____

A daughter that is a carrier for the color-blind gene? _____

4. Why are consanguineous marriages (marriages between blood relatives) prohibited in most cultures?

_____

_____

Probability

1. What is the probability of having three daughters in a row? _____

2. A man and a woman, each of seemingly normal intellect, marry. Although neither is aware of the fact, each is a heterozygote

   for the allele for feeblemindedness. Is the allele for feeblemindedness dominant or recessive? _____

   What are the chances of their having one feebleminded child? _____

   What are the chances that all of their children (they plan a family of four) will be feebleminded? _____

Genetic Determination of Selected Human Characteristics

1. Look back at your data to complete this section. For each of the situations described here, determine if an offspring with the characteristics noted is possible with the parental genotypes listed. Check (✓) the appropriate column.

		Possibility	
Parental genotypes	**Phenotype of child**	**Yes**	**No**
Jj × jj	Double-jointed thumbs		
FF × Ff	Straight little finger		
EE × ee	Detached ear lobes		
HH × Hh	Middigital hair		
$I^A i \times I^B i$	Type O blood		
$I^A I^B \times ii$	Type B blood		

2. You have dimples, and you would like to know if you are homozygous or heterozygous for this trait. You have six brothers and sisters. By observing your siblings, how could you tell, with some degree of certainty, that you are a heterozygote?

   _____

   _____

   _____

   _____

   _____

Surface Anatomy Roundup

_____ 1. A blow to the cheek is most likely to break what superficial bone or bone part? (a) superciliary arches, (b) the philtrum, (c) zygomatic arch, (d) the tragus.

_____ 2. Rebound tenderness (a) occurs in appendicitis, (b) is whiplash of the neck, (c) is a sore foot from playing basketball, (d) occurs when the larynx falls back into place after swallowing.

_____ 3. The anatomical snuff box (a) is in the nose, (b) contains the styloid process of the radius, (c) is defined by tendons of the flexor carpi radialis and palmaris longus, (d) cannot really hold snuff.

_____ 4. Some landmarks on the body surface can be seen or felt, but others are abstractions that you must construct by drawing imaginary lines. Which of the following pairs of structures is abstract and invisible? (a) umbilicus and costal margin, (b) anterior superior iliac spine and natal cleft, (c) linea alba and linea semilunaris, (d) McBurney's point and midaxillary line, (e) philtrum and sternocleidomastoid.

_____ 5. Many pelvic organs can be palpated by placing a finger in the rectum or the vagina, but only one pelvic organ is readily palpated through the skin. This is the (a) nonpregnant uterus, (b) prostate gland, (c) full bladder, (d) ovaries, (e) rectum.

_____ 6. A muscle that contributes to the posterior axillary fold is the (a) pectoralis major, (b) latissimus dorsi, (c) trapezius, (d) infraspinatus, (e) pectoralis minor, (f) a and e.

_____ 7. Which of the following is not a pulse point? (a) anatomical snuff box, (b) inferior margin of mandible anterior to masseter muscle, (c) center of distal forearm at palmaris longus tendon, (d) medial bicipital furrow on arm, (e) dorsum of foot between the first two metatarsals.

_____ 8. Which pair of ribs inserts on the sternum at the sternal angle? (a) first, (b) second, (c) third, (d) fourth, (e) fifth.

_____ 9. The inferior angle of the scapula is at the same level as the spinous process of which vertebra? (a) C_5, (b) C_7, (c) T_3, (d) T_7, (e) L_4.

_____ 10. An important bony landmark that can be recognized by a distinct dimple in the skin is the (a) posterior superior iliac spine, (b) styloid process of the ulna, (c) shaft of the radius, (d) acromion.

_____ 11. A nurse missed a patient's median cubital vein while trying to withdraw blood and then inserted the needle far too deeply into the cubital fossa. This error could cause any of the following problems, except this one: (a) paralysis of the ulnar nerve, (b) paralysis of the median nerve, (c) bruising the insertion tendon of the biceps brachii muscle, (d) blood spurting from the brachial artery.

_____ 12. Which of these organs is almost impossible to study with surface anatomy techniques? (a) heart, (b) lungs, (c) brain, (d) nose.

_____ 13. A preferred site for inserting an intravenous medication line into a blood vessel is the (a) medial bicipital furrow on arm, (b) external carotid artery, (c) dorsal venous arch of hand, (d) popliteal fossa.

_____ 14. One listens for bowel sounds with a stethoscope placed (a) on the four quadrants of the abdominal wall; (b) in the triangle of auscultation; (c) in the right and left midaxillary line, just superior to the iliac crests; (d) inside the patient's bowels (intestines), on the tip of an endoscope.

Dissection and Identification of Cat Muscles

The skeletal muscles of all mammals are named in a similar fashion. However, some muscles that are separate in lower animals are fused in humans, and some muscles present in lower animals are lacking in humans. This exercise involves dissection of the cat musculature to enhance your knowledge of the human muscular system. Since the aim is to become familiar with the muscles of the human body, you should pay particular attention to the similarities between cat and human muscles. However, pertinent differences will be pointed out as they are encountered.

Activity:
Preparing the Cat for Dissection

The preserved laboratory animals purchased for dissection have been embalmed with a solution that prevents deterioration of the tissues. The animals are generally delivered in plastic bags that contain a small amount of the embalming fluid. Do not dispose of this fluid when you remove the cat; the fluid prevents the cat from drying out. It is very important to keep the cat's tissues moist because you will probably use the same cat from now until the end of the course. The embalming fluid may cause your eyes to smart and may dry your skin, but these small irritants are preferable to working with a cat that has become hard and odoriferous due to bacterial action. You can alleviate the possibility of skin irritation by using disposable gloves or applying skin cream before beginning the dissection.

1. Obtain a cat, dissection tray, dissection instruments, and a name tag. Mark the name tag with the names of the members of your group and set it aside. The name tag will be attached to the plastic bag at the end of the dissection so that you may identify your animal in subsequent laboratories.

Objectives

1. To name and locate muscles on a dissected cat.
2. To recognize similarities and differences between human and cat musculature.

Materials

- ❑ Disposable (plastic) gloves or protective skin cream
- ❑ Preserved and injected cat (one for every two to four students)
- ❑ Dissecting tray and instruments
- ❑ Name tag and large plastic bag
- ❑ Paper towels
- ❑ Embalming fluid

2. To begin removing the skin, place the cat ventral side down on the dissecting tray. With a scalpel, make a short, shallow incision in the midline of the neck, just to penetrate the skin. From this point on, use scissors. Continue the cut the length of the back to the sacrolumbar region, stopping at the tail (Figure D1.1).

3. From the dorsal surface of the tail region, continue the incision around the tail, encircling the anus and genital organs. The skin will not be removed from this region.

See Exercise 15 of this manual for a discussion of the anatomy of the human muscular system.

Figure D1.1 Incisions to be made in skinning a cat. Numbers indicate sequence.

4. Beginning again at the dorsal tail region, make an incision through the skin down each hind leg nearly to the ankle. Continue the cuts completely around the ankles.

5. Return to the neck. Cut the skin around the neck.

6. Cut down each foreleg to the wrist. Completely cut through the skin around the wrists.

7. Now free the skin from the loose connective tissue (superficial fascia) that binds it to the underlying structures. With one hand, grasp the skin on one side of the midline dorsal incision. Then, using your fingers or a blunt probe, break through the "cottony" connective tissue fibers to release the skin from the muscle beneath. Work toward the ventral surface and then toward the neck. As you pull the skin from the body, you should see small, white, cordlike structures extending from the skin to the muscles at fairly regular intervals. These are the cutaneous nerves that serve the skin. You will also see (particularly as you approach the ventral surface) that a thin layer of muscle fibers remains adhered to the skin. This is the **cutaneous maximus** muscle, which enables the cat to move its skin rather like our facial muscles allow us to express emotion. Where the cutaneous maximus fibers cling to those of the deeper muscles, they should be carefully cut free.

8. You will notice as you start to free the skin in the neck that it is more difficult to remove. Take extra care and time in this area. The large flat **platysma** muscle in the ventral neck region (a skin muscle like the cutaneous maximus) will remain attached to the skin. The skin will not be removed from the head since the cat's muscles are not sufficiently similar to human head muscles to merit study.

9. Complete the skinning process by freeing the skin from the forelimbs, the lower torso, and the hindlimbs in the same manner.

10. Inspect your skinned cat. Notice that it is difficult to see any cleavage lines between the muscles because of the overlying connective tissue, which is white or yellow. If time allows, carefully remove as much of the fat and fascia from the surface of the muscles as possible, using forceps or your fingers. The muscles, when exposed, look grainy or threadlike and are light brown. If this clearing process is done carefully and thoroughly, you will be ready to begin your identification of the superficial muscles.

11. If the muscle dissection exercises are to be done at a later laboratory session, carefully rewrap the cat's skin around its body. Then dampen several paper towels in embalming fluid and wrap these around the animal. Return the cat to the plastic bag (add more embalming fluid to the bag if necessary), seal the bag, and attach your name tag. Prepare your cat for storage in this way every time the cat is used. Place the cat in the storage container designated by your instructor.

12. Before leaving the laboratory, dispose of any tissue remnants in the organic debris container, wash the dissecting tray and instruments with soapy water, rinse and dry, and wash down the laboratory bench. ■

Activity:
Dissecting Trunk and Neck Muscles

The proper dissection of muscles involves careful separation of one muscle from another and transection of superficial muscles in order to study those lying deeper. In general, when directions are given to transect a muscle, it should be completely freed from all adhering connective tissue and then cut through about halfway between its origin and insertion points.

As a rule, all the fibers of one muscle are held together by a connective tissue sheath (epimysium) and run in the same general direction. Before you begin dissection, observe your skinned cat. If you look carefully, you can see changes

Mandible

Digastric muscles

Mylohyoid

Sternohyoid

Sternothyroid

Sternomastoid

External jugular vein

Clavotrapezius

Figure D1.2 **Superficial muscles of the anterior neck of the cat.** The sternomastoids are pulled slightly laterally to reveal the deeper sternothyroid muscles.

in the direction of the muscle fibers, which will help you to locate the muscle borders. Pulling in slightly different directions on two adjacent muscles will usually allow you to expose the normal cleavage line between them. Once cleavage lines have been identified, use a blunt probe to break the connective tissue between muscles and to separate them. If the muscles separate as clean, distinct bundles, your procedure is probably correct. If they appear ragged or chewed up, you are probably tearing a muscle apart rather than separating it from adjacent muscles. Only the muscles that are most easily identified and separated out will be identified in this exercise because of time considerations.

Anterior Neck Muscles

1. Using Figure D1.2 as a guide, examine the anterior neck surface of the cat and identify the following superficial neck muscles. (The platysma belongs in this group but was probably removed during the skinning process.) The **sternomastoid** muscle and the more lateral and deeper **cleidomastoid** muscle (not visible in Figure D1.2) are joined in humans to form the sternocleidomastoid. The large external jugular vein, which drains the head, should be obvious crossing the anterior aspect of these muscles. The **mylohyoid** muscle parallels the bottom aspect of the chin, and the **digastric** muscles form a V over the mylohyoid muscle. Although it is not one of the neck muscles, you can now identify the fleshy **masseter** muscle, which flanks the digastric muscle laterally. Finally, the **sternohyoid** is a narrow muscle between the mylohyoid (superiorly) and the inferior sternomastoid.

2. The deeper muscles of the anterior neck of the cat are small and straplike and hardly worth the effort of dissection. However, one of these deeper muscles can be seen with a minimum of extra effort. Transect the sternomastoid and sternohyoid muscles approximately at midbelly. Reflect the cut ends to reveal the **sternothyroid** muscle, which runs along the anterior surface of the throat just deep and lateral to the sternohyoid muscle. The cleidomastoid muscle, which lies deep to the sternomastoid, is also more easily identified now.

Superficial Chest Muscles

In the cat, the chest or pectoral muscles adduct the arm, just as they do in humans. However, humans have only two pectoral muscles, and cats have four—the pectoralis major, pectoralis minor, xiphihumeralis, and pectoantebrachialis (see Figure D1.3). However, because of their relatively great degree of fusion, the cat's pectoral muscles appear to be a single muscle. The pectoral muscles are rather difficult to dissect and identify, as they do not separate from one another easily.

The **pectoralis major** is 2 to 3 inches wide and can be seen arising on the manubrium, just inferior to the sternomastoid muscle of the neck, and running to the humerus. Its fibers run at right angles to the longitudinal axis of the cat's body.

The **pectoralis minor** lies beneath the pectoralis major and extends posterior to it on the abdominal surface. It originates on the sternum and inserts on the humerus. Its fibers run obliquely to the long axis of the body, which helps to distinguish it from the pectoralis major. Contrary to what its name

Figure D1.3 **Superficial thorax muscles, ventral view.** Latissimus dorsi is reflected away from the thorax. (Compare to human thorax muscles, anterior view, Figure 15.5.)

implies, the pectoralis minor is a larger and thicker muscle than the pectoralis major.

The **xiphihumeralis** can be distinguished from the posterior edge of the pectoralis minor only by virtue of the fact that its origin is lower—on the xiphoid process of the sternum. Its fibers run parallel to and are fused with those of the pectoralis minor.

The **pectoantebrachialis** is a thin, straplike muscle, about ½ inch wide, lying over the pectoralis major. It originates from the manubrium, passes laterally over the pectoralis major, and merges with the muscles of the forelimb approximately halfway down the humerus. It has no homologue in humans.

Identify, free, and trace out the origin and insertion of the cat's chest muscles. Refer to Figure D1.3 as you work.

Muscles of the Abdominal Wall

The superficial trunk muscles include those of the abdominal wall (see Figure D1.4). Cat musculature in this area is quite similar in function to that of humans.

1.   Complete the dissection of the more superficial anterior trunk muscles of the cat by identifying the origins and insertions of the muscles of the abdominal wall. Work carefully here. These muscles are very thin, and it is easy to miss their boundaries. Begin with the **rectus abdominis,** a long band of muscle approximately 1 inch wide running immediately lateral to the midline of the body on the abdominal surface. Humans have four transverse *tendinous intersections* in the rectus abdominis (see Figure 15.7a), but they are absent or difficult to identify in the cat. Identify the **linea alba,** which separates the rectus abdominis muscles. Note the relationship of the rectus abdominis to the other abdominal muscles and their fascia.

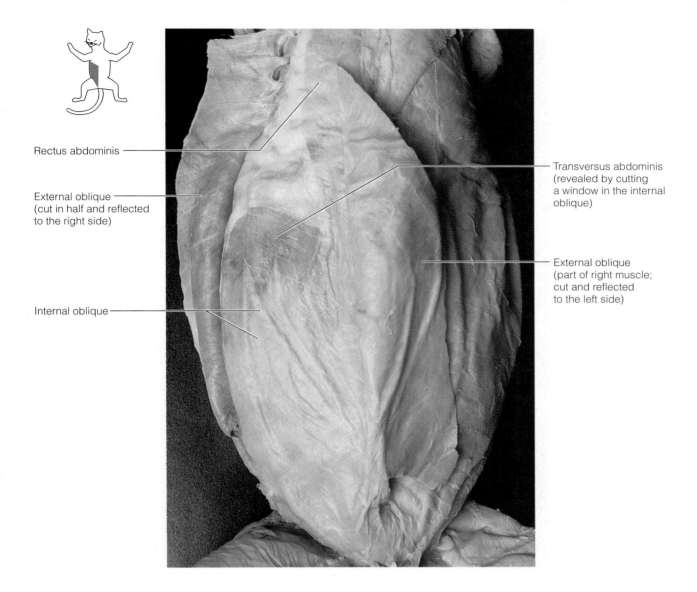

Rectus abdominis

External oblique
(cut in half and reflected
to the right side)

Internal oblique

Transversus abdominis
(revealed by cutting
a window in the internal
oblique)

External oblique
(part of right muscle;
cut and reflected
to the left side)

Figure D1.4   Muscles of the abdominal wall of the cat.

2.   The **external oblique** is a sheet of muscle immediately beside (and running beneath) the rectus abdominis (see Figure D1.4). Carefully free and then transect the external oblique to reveal the anterior attachment of the rectus abdominis and the deeper **internal oblique.** Reflect the external oblique; observe the deeper muscle. Notice which way the fibers run.

How does the fiber direction of the internal oblique compare to that of the external oblique?

_____

_____

3.   Free and then transect the internal oblique muscle to reveal the fibers of the **transversus abdominis,** whose fibers run transversely across the abdomen.

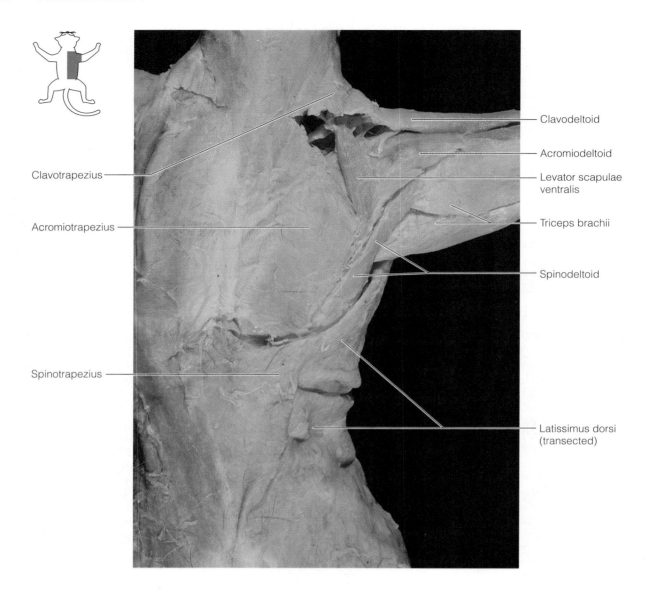

Clavotrapezius

Acromiotrapezius

Spinotrapezius

Clavodeltoid

Acromiodeltoid

Levator scapulae ventralis

Triceps brachii

Spinodeltoid

Latissimus dorsi (transected)

Figure D1.5 Superficial muscles of the anterodorsal aspect of the right shoulder, trunk, and neck of the cat. (Compare to human thorax muscles, posterior view, Figure 15.8a.)

Superficial Muscles of the Shoulder and Dorsal Trunk and Neck

Refer to Figure D1.5 as you dissect the superficial muscles of the dorsal surface of the trunk.

1. Turn your cat on its ventral surface and start your observations with the **trapezius group.** Humans have a single large trapezius muscle, but the cat has three separate muscles—the clavotrapezius, acromiotrapezius, and spinotrapezius—that together perform a similar function. The prefix (clavo-, acromio-, and spino-) in each case reveals the muscle's site of insertion. The **clavotrapezius,** the most superior muscle of the group, is homologous to that part of the human trapezius that inserts into the clavicle. Slip a probe under this muscle and follow it to its apparent origin.

Where does the clavotrapezius appear to originate?

_____

Is this similar to its origin in humans?_____

The fibers of the clavotrapezius are continuous inferiorly with those of the clavicular part of the cat's deltoid muscle (clavodeltoid), and the two muscles work together to extend the humerus. Release the clavotrapezius muscle from adjoining muscles. The **acromiotrapezius** is a large, nearly square muscle easily identified by its aponeurosis, which passes over the vertebral border of the scapula. It originates from the cervical and T_1 vertebrae and inserts into the scapular spine. The triangular **spinotrapezius** runs from the thoracic vertebrae to the scapular spine. This is the most posterior of the trapezius muscles in the cat. Now that you know where they are located, pull on the three trapezius muscles to mimic their action.

Do the trapezius muscles appear to have the same functions in cats as in humans?

_____

2. The **levator scapulae ventralis,** a flat, straplike muscle, can be located in the triangle created by the division of the fibers of the clavotrapezius and acromiotrapezius. Its anterior fibers run underneath the clavotrapezius from its origin at the base of the skull (occipital bone), and it inserts on the vertebral border of the scapula. In the cat it helps to hold the upper edges of the scapulae together and draws them toward the head.

What is the function of the levator scapulae in humans?

_____

_____

_____

3. The **deltoid group:** like the trapezius, the human deltoid muscle is represented by three separate muscles in the cat—the clavodeltoid, acromiodeltoid, and spinodeltoid. The **clavodeltoid,** the most superficial muscle of the shoulder, is a continuation of the clavotrapezius below the clavicle, which is this muscle's point of origin (see Figure D1.5). Follow its course down the forelimb to the point where it merges along a white line with the pectoantebrachialis. Separate it from the pectoantebrachialis, and then transect it and pull it back.

Where does the clavodeltoid insert?_____

What do you think the function of this muscle is?

_____

_____

The **acromiodeltoid** lies posterior to the clavodeltoid and runs over the top of the shoulder. This small triangular muscle originates on the acromion of the scapula. It inserts into the spinodeltoid (a muscle of similar size) posterior to it. The **spinodeltoid** is covered with fascia near the anterior end of the scapula. Its tendon extends under the acromiodeltoid muscle and inserts on the humerus. Notice that its fibers run obliquely to those of the acromiodeltoid. Like the human deltoid muscle, the acromiodeltoid and clavodeltoid muscles in the cat raise and rotate the humerus.

4. The **latissimus dorsi** is a large flat muscle covering most of the lateral surface of the posterior trunk. Its upper edge is covered by the spinotrapezius. As in humans, it inserts into the humerus. But before inserting, its fibers merge with the fibers of many other muscles, among them the xiphihumeralis of the pectoralis group.

Deep Muscles of the Dorsal Trunk and Neck

1. In preparation, transect the latissimus dorsi, the muscles of the pectoralis group, and the spinotrapezius and reflect them back. Be careful not to damage the large brachial nerve plexus, which lies in the axillary space beneath the pectoralis group.

Scalenes

Subscapularis

Serratus ventralis

Rectus abdominis

External oblique

Figure D1.6   Deep muscles of the right inferolateral thorax of the cat.

2.   The **serratus ventralis,** homologous to the serratus anterior of humans, arises deep to the pectoral muscles and covers the lateral surface of the rib cage. It is easily identified by its fingerlike muscular origins, which arise on the first 9 or 10 ribs. It inserts into the scapula. The anterior portion of the serratus ventralis, which arises from the cervical vertebrae, is homologous to the levator scapulae in humans; both pull the scapula toward the sternum. Trace this muscle to its insertion. In general, in the cat, this muscle acts to pull the scapula posteriorly and downward. Refer to Figure D1.6.

3.   Reflect the upper limb to reveal the **subscapularis,** which occupies most of the ventral surface of the scapula. Humans have a homologous muscle.

4.   Locate the anterior, posterior, and middle **scalene** muscles on the lateral surface of the cat's neck and trunk. The most prominent and longest of these muscles is the middle scalene, which lies between the anterior and posterior members. The scalenes originate on the ribs and run cephalad over the serratus ventralis to insert in common on the cervical vertebrae. These muscles draw the ribs anteriorly and bend the neck downward; thus they are homologous to the human scalene muscles, which elevate the ribs and flex the neck. (Notice that the difference is only one of position. Humans walk erect, but cats are quadrupeds.)

5.   Reflect the flaps of the transected latissimus dorsi, spinodeltoid, acromiodeltoid, and levator scapulae ventralis. The **splenius** is a large flat muscle occupying most of the side of the neck close to the vertebrae. As in humans, it originates on the ligamentum nuchae at the back of the neck and inserts into the occipital bone. It functions to raise the head. Refer to Figure D1.7.

6.   To view the rhomboid muscles, lay the cat on its side and hold its forelegs together to spread the scapulae apart. The rhomboid muscles lie between the scapulae and beneath the acromiotrapezius. All the rhomboid muscles originate on the vertebrae and insert on the scapula.

Trapezius
- Clavotrapezius (cut)
- Acromiotrapezius (cut)
- Spinotrapezius (cut)

Supraspinatus

Infraspinatus

Rhomboid capitus

Splenius

Rhomboideus

Figure D1.7 Deep muscles of the superior aspect of the dorsal thorax of the cat.

There are three rhomboids in the cat. The **rhomboid capitis,** the most anterolateral muscle of the group, has no counterpart in the human body. The **rhomboid minor,** located posterior to the rhomboid capitis, is much larger. The fibers of the rhomboid minor run transversely to those of the rhomboid capitis. The most posterior muscle of the group, the **rhomboid major,** is so closely fused to the rhomboid minor that many consider them to be one muscle—the **rhomboideus,** which is homologous to human rhomboid muscles.

7. The **supraspinatus** and **infraspinatus** muscles are similar to the same muscles in humans. The supraspinatus can be found under the acromiotrapezius, and the infraspinatus is deep to the spinotrapezius. Both originate on the lateral scapular surface and insert on the humerus. ■

Activity:
Dissecting Forelimb Muscles

Cat forelimb muscles fall into the same three categories as human upper limb muscles, but in this section the muscles of the entire forelimb are considered together. Refer to Figure D1.8 as you study these muscles.

Muscles of the Lateral Surface

1. The triceps muscle **(triceps brachii)** of the cat is easily identified if the cat is placed on its side. It is a large, fleshy muscle covering the posterior aspect and much of the side of

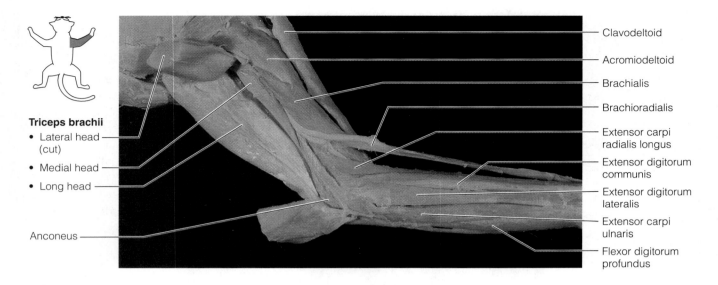

Triceps brachii
- Lateral head (cut)
- Medial head
- Long head

Anconeus

Clavodeltoid

Acromiodeltoid

Brachialis

Brachioradialis

Extensor carpi radialis longus

Extensor digitorum communis

Extensor digitorum lateralis

Extensor carpi ulnaris

Flexor digitorum profundus

Figure D1.8 Lateral surface of the right forelimb of the cat.
The lateral head of the triceps brachii has been transected and reflected.

the humerus. As in humans, this muscle arises from three heads, which originate from the humerus and scapula and insert jointly into the olecranon process of the ulna. Remove the fascia from the superior region of the lateral arm surface to identify the lateral and long heads of the triceps. The long head is approximately twice as long as the lateral head and lies medial to it on the posterior arm surface. The medial head can be exposed by transecting the lateral head and pulling it aside. Now pull on the triceps muscle.

How does the function of the triceps muscle compare in cats and in humans?

_____

Anterior and distal to the medial head of the triceps is the tiny anconeus muscle, sometimes called the fourth head of the triceps muscle. Notice its darker color and the way it wraps the tip of the elbow.

2. The **brachialis** can be located anterior to the lateral head of the triceps muscle. Identify its origin on the humerus, and trace its course as it crosses the elbow and inserts on the ulna. It flexes the cat's foreleg.

Identification of the forearm muscles is difficult because of the tough fascia sheath that encases them, but give it a try.

3. Remove as much of the connective tissue as possible and cut through the ligaments that secure the tendons at the wrist (transverse carpal ligaments) so that you will be able to follow the muscles to their insertions. Begin your identification

of the forearm muscles at the lateral surface of the forearm. The muscles of this region are very much alike in appearance and are difficult to identify accurately unless a definite order is followed. Thus you will begin with the most anterior muscles and proceed to the posterior aspect. Remember to check carefully the tendons of insertion to verify your muscle identifications.

4. The ribbonlike muscle on the lateral surface of the humerus is the **brachioradialis.** Observe how it passes down the forearm to insert on the styloid process of the radius. (If your removal of the fascia was not very careful, this muscle may have been removed.)

5. The **extensor carpi radialis longus** has a broad origin and is larger than the brachioradialis. It extends down the anterior surface of the radius (see Figure D1.8). Transect this muscle to view the **extensor carpi radialis brevis,** which is partially covered by and sometimes fused with the extensor carpi radialis longus. Both muscles have origins, insertions, and actions similar to their human counterparts.

6. You can see the entire **extensor digitorum communis** along the lateral surface of the forearm. Trace it to its four tendons, which insert on the second to fifth digits. This muscle extends these digits. The **extensor digitorum lateralis** (absent in humans) also extends the digits. This muscle lies immediately posterior to the extensor digitorum communis.

7. Follow the **extensor carpi ulnaris** from the lateral epicondyle of the humerus to the ulnar side of the fifth metacarpal. Often this muscle has a shiny tendon, which helps in its identification.

Biceps brachii

Triceps,
medial head

Coracobrachialis

Epitrochlearis

Pronator teres

Brachioradialis

Extensor carpi radialis

Flexor carpi radialis

Flexor digitorum
profundus

Palmaris longus

Flexor carpi ulnaris
(two heads)

Figure D1.9   Medial surface of the right forelimb of the cat.

Muscles of the Medial Surface

1. The **biceps brachii** (refer to Figure D1.9) is a large spin-dle-shaped muscle medial to the brachialis on the anterior surface of the humerus. Pull back the cut ends of the pectoral muscles to get a good view of the biceps. This muscle is much more prominent in humans, but its origin, insertion, and action are very similar in cats and in humans. Follow the muscle to its origin.

Does the biceps have two heads in the cat? _____

2. The broad, flat, exceedingly thin muscle on the postero-medial surface of the arm is the **epitrochlearis.** Its tendon originates from the fascia of the latissimus dorsi, and the muscle inserts into the olecranon process of the ulna. This muscle extends the forearm of the cat; it is not found in humans.

3. The **coracobrachialis** of the cat is insignificant (approx-imately ½ in. long) and can be seen as a very small muscle crossing the ventral aspect of the shoulder joint. It runs be-neath the biceps brachii to insert on the humerus and has the same function as the human coracobrachialis.

4. Referring again to Figure D1.9, turn the cat so that the ventral forearm muscles (mostly flexors and pronators) can be observed. As in humans, most of these muscles arise from the medial epicondyle of the humerus. The **pronator teres** runs from the medial epicondyle of the humerus and declines in size as it approaches its insertion on the radius. Do not bother to trace it to its insertion.

5. Like its human counterpart, the **flexor carpi radialis** runs from the medial epicondyle of the humerus to insert into the second and third metacarpals.

6. The large flat muscle in the center of the medial surface is the **palmaris longus.** Its origin on the medial epicondyle of the humerus abuts that of the pronator teres and is shared

with the flexor carpi radialis. The palmaris longus extends down the forearm to terminate in four tendons on the digits. Comparatively speaking, this muscle is much larger in cats than in humans.

The **flexor carpi ulnaris** arises from a two-headed ori-gin (medial epicondyle of the humerus and olecranon of the ulna). Its two bellies (fleshy parts) pass downward to the wrist, where they are united by a single tendon that inserts into the carpals of the wrist. As in humans, this muscle flexes the wrist. ■

Activity:
Dissecting Hindlimb Muscles

Remove the fat and fascia from all thigh surfaces, but do not cut through or remove the **fascia lata,** which is a tough white aponeurosis covering the anterolateral surface of the thigh from the hip to the leg. If the cat is a male, the cordlike sperm duct will be embedded in the fat near the pubic symphysis. Carefully clear around, but not in, this region.

Posterolateral Hindlimb Muscles

1. Turn the cat on its ventral surface and identify the fol-lowing superficial muscles of the hip and thigh, referring to Figure D1.10. (The deeper hip muscles of the cat will not be identified.) Viewing the lateral aspect of the hindlimb, you will identify these muscles in sequence from the anterior to the posterior aspects of the hip and thigh. Most anterior is the **sartorius.** Approximately 1½ in. wide, it extends around the lateral aspect of the thigh to the anterior surface, where the major portion of it lies. Free it from the adjacent muscles and pass a blunt probe under it to trace its origin and insertion. Homologous to the sartorius muscle in humans, it adducts and rotates the thigh, but in addition, the cat sartorius acts as a knee extensor. Transect this muscle.

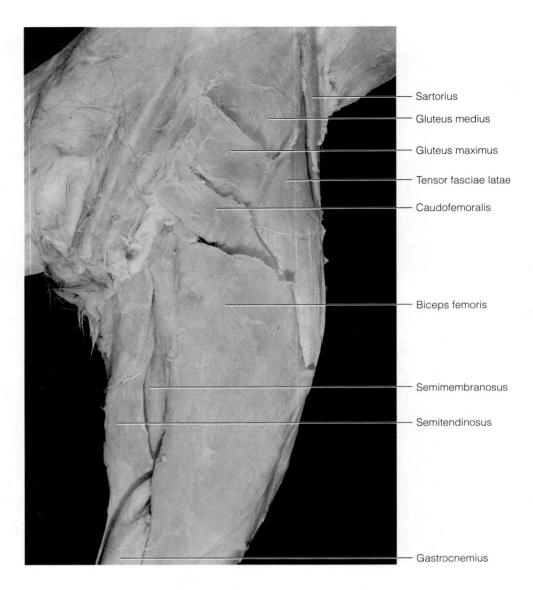

- Sartorius
- Gluteus medius
- Gluteus maximus
- Tensor fasciae latae
- Caudofemoralis
- Biceps femoris
- Semimembranosus
- Semitendinosus
- Gastrocnemius

Figure D1.10   Muscles of the right posterolateral thigh in the cat; superficial view.

2.   The **tensor fasciae latae** is posterior to the sartorius. It is wide at its superior end, where it originates on the iliac crest, and narrows as it approaches its insertion into the fascia lata (or iliotibial band), which runs to the proximal tibial region. Transect its superior end and pull it back to expose the **gluteus medius** lying beneath it. This is the largest of the gluteus muscles in the cat. It originates on the ilium and inserts on the greater trochanter of the femur. The gluteus medius overlays and obscures the gluteus minimus, pyriformis, and gemellus muscles (which will not be identified here).

3.   The **gluteus maximus** is a small triangular hip muscle posterior to the superior end of the tensor fasciae latae and paralleling it. In humans the gluteus maximus is a large fleshy muscle forming most of the buttock mass. In the cat it is only about ½ in. wide and 2 in. long, and is smaller than the gluteus medius. The gluteus maximus covers part of the gluteus medius as it extends from the sacral region and the end of the femur. It abducts the thigh.

4.   Posterior to the gluteus maximus, identify the triangular **caudofemoralis,** which originates on the caudal vertebrae and inserts into the patella via an aponeurosis. There is no homologue to this muscle in humans; in cats it abducts the thigh and flexes the vertebral column.

5.   The **hamstring muscles** of the hindlimb include the biceps femoris, the semitendinosus, and the semimembranosus muscles. The **biceps femoris** is a large, powerful muscle that covers most of the posterolateral surface of the thigh. It is 1½ to 2 in. wide throughout its length. Trace it from its origin on the ischial tuberosity to its insertion on the tibia. Part of the **semitendinosus** can be seen beneath the posterior border of the biceps femoris. Transect and reflect the biceps muscle to reveal the whole length of the semitendinosus. Contrary to what its name implies ("half-tendon"), this muscle is muscular and fleshy except at its insertion. It is uniformly about ¾ in. wide as it runs down the thigh from the ischial tuberosity to the medial side of the ulna. It acts to bend the knee. The

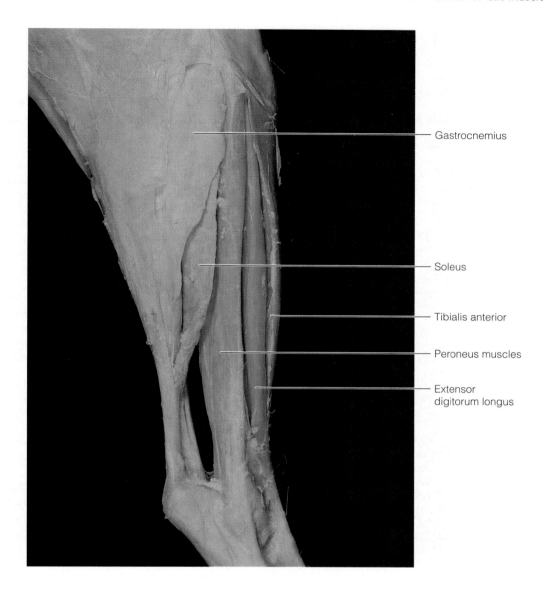

Gastrocnemius

Soleus

Tibialis anterior

Peroneus muscles

Extensor
digitorum longus

Figure D1.11 Superficial muscles of the posterolateral aspect of the right shank (leg).

semimembranosus, a large muscle lying medial to the semi-tendinosus and largely obscured by it, is best seen in an anterior view of the thigh (see Figure D1.12b). If desired, however, the semitendinosus can be transected to view it from the posterior aspect. The semimembranosus is larger and broader than the semitendinosus. Like the other hamstrings, it originates on the ischial tuberosity and inserts on the medial epicondyle of the femur and the medial tibial surface.

How does the semimembranosus compare with its human homologue?

_____

_____

6.   Remove the heavy fascia covering the lateral surface of the shank. Moving from the posterior to the anterior aspect,

identify the following muscles on the posterolateral shank (leg). (Refer to Figure D1.11.) First reflect the lower portion of the biceps femoris to see the origin of the **triceps surae,** the large composite muscle of the calf. Humans also have a triceps surae. The **gastrocnemius,** part of the triceps surae, is the largest muscle on the shank. As in humans, it has two heads and inserts via the calcaneal (Achilles) tendon into the calcaneus. Run a probe beneath this muscle and then transect it to reveal the **soleus,** which is deep to the gastrocnemius.

7.   Another important group of muscles in the leg is the **peroneus muscles** (see Figure D1.11), which collectively appear as a slender, evenly shaped superficial muscle lying anterior to the triceps surae. Originating on the fibula and inserting on the digits and metatarsals, the peroneus muscles flex the foot.

8.   The **extensor digitorum longus** lies anterior to the peroneus muscles. Its origin, insertion, and action in cats are similar to the homologous human muscle. The **tibialis anterior** is anterior to the extensor digitorum longus. The tibialis

anterior is roughly triangular in cross section and heavier at its proximal end. Locate its origin on the proximal fibula and tibia and its insertion on the first metatarsal. You can see the sharp edge of the tibia at the anterior border of this muscle. As in humans, it is a foot flexor.

Anteromedial Hindlimb Muscles

1. Turn the cat onto its dorsal surface to identify the muscles of the anteromedial hindlimb (refer to Figure D1.12). Note once again the straplike sartorius at the surface of the thigh, which you have already identified and transected. It originates on the ilium and inserts on the medial region of the tibia.

2. Reflect the cut ends of the sartorius to identify the **quadriceps** muscles. The **vastus medialis** lies just beneath the sartorius. Resting close to the femur, it arises from the ilium and inserts into the patellar ligament. The small cylindrical muscle anterior and lateral to the vastus medialis is the **rectus femoris.** In cats this muscle originates entirely from the femur.

What is the origin of the rectus femoris in humans?

_____

_____

Free the rectus femoris from the most lateral muscle of this group, the large, fleshy **vastus lateralis,** which lies deep to the tensor fasciae latae. The vastus lateralis arises from the lateral femoral surface and inserts, along with the other vasti muscles, into the patellar ligament. Transect this muscle to identify the deep **vastus intermedius,** the smallest of the vasti muscles. It lies medial to the vastus lateralis and merges superiorly with the vastus medialis. (The vastus intermedius is not shown in the figure.)

3. The **gracilis** is a broad muscle that covers the posterior portion of the medial aspect of the thigh (see Figure D1.12a). It originates on the pubic symphysis and inserts on the medial proximal tibial surface. In cats the gracilis adducts the leg and draws it posteriorly.

How does this compare with the human gracilis?

_____

_____

4. Free and transect the gracilis to view the adductor muscles deep to it. The **adductor femoris** is a large muscle that lies beneath the gracilis and abuts the semimembranosus medially. Its origin is the pubic ramus and the ischium, and its fibers pass downward to insert on most of the length of the femoral shaft. The adductor femoris is homologous to the human adductor magnus, brevis, and longus. Its function is to extend the thigh after it has been drawn forward, and to adduct the thigh. A small muscle about an inch long—the **adductor longus**—touches the superior margin of the adductor femoris. It originates on the pubic bone and inserts on the proximal surface of the femur.

5. Before continuing your dissection, locate the **femoral triangle** (Scarpa's triangle), an important area bordered by the proximal edge of the sartorius and the adductor muscles. It is usually possible to identify the femoral artery (injected with red latex) and the femoral vein (injected with blue latex), which span the triangle (see Figure D1.12). (You will identify these vessels again in your study of the circulatory system.) If your instructor wishes you to identify the pectineus and iliopsoas, remove these vessels and go on to steps 6 and 7.

6. Examine the superolateral margin of the adductor longus to locate the small **pectineus.** It is normally covered by the gracilis (which you have cut and reflected). The pectineus, which originates on the pubis and inserts on the proximal end of the femur, is similar in all ways to its human homologue.

7. Just lateral to the pectineus you can see a small portion of the **iliopsoas,** a long and cylindrical muscle. Its origin is on the transverse processes of T_1 through T_{12} and the lumbar vertebrae, and it passes posteriorly toward the body wall to insert on the medial aspect of the proximal femur. The iliopsoas flexes and laterally rotates the thigh. It corresponds to the human iliopsoas and psoas minor.

Text continues on page 748

(a)

Femoral artery
and vein

Adductors

Spermatic cords

Sartorius

Gracilis

Iliopsoas

Sartorius
(cut and reflected)

Vastus lateralis

Femoral artery

Rectus femoris

Pectineus

Adductor longus

Vastus medialis

Adductor femoris

Semimembranosus

Semitendinosus

Gracilis
(cut and reflected)

(b)

Figure D1.12 **Superficial muscles of the anteromedial thigh.**
(a) Gracilis and sartorius are intact in this superficial view of the right thigh.
(b) The gracilis and sartorius are transected and reflected to show deeper muscles.

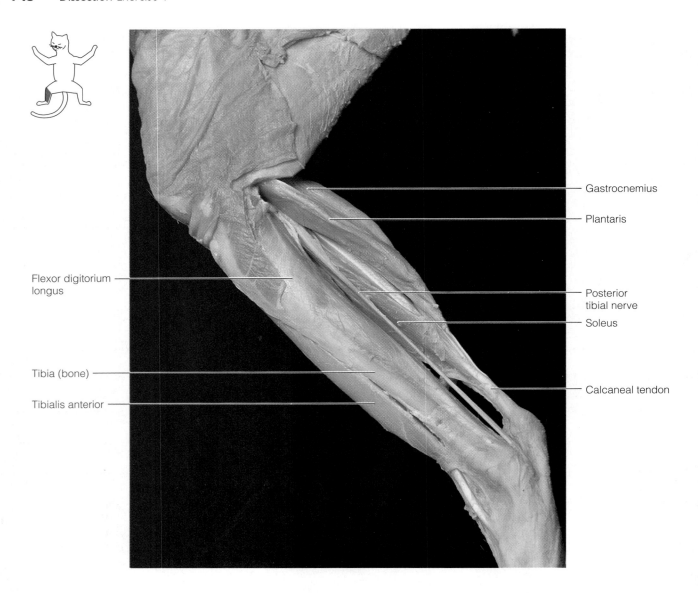

Gastrocnemius

Plantaris

Flexor digitorium longus

Posterior tibial nerve

Soleus

Tibia (bone)

Tibialis anterior

Calcaneal tendon

Figure D1.13 Superficial muscles of the right anteromedial shank (leg) of the cat.

8. Reidentify the gastrocnemius of the shank and then the **plantaris,** which is fused with the lateral head of the gastrocnemius. (See Figure D1.13.) It originates from the lateral aspect of the femur and patella, and its tendon passes around the calcaneus to insert on the second phalanx. Working with the triceps surae, it flexes the digits and extends the foot.

9. Anterior to the plantaris is the **flexor digitorum longus,** a long, tapering muscle with two heads. It originates on the lateral surfaces of the proximal fibula and tibia and inserts via four tendons into the terminal phalanges. As in humans, it flexes the toes.

10. The **tibialis posterior** is a long, flat muscle lateral and deep to the flexor digitorum longus. It originates on the medial surface of the head of the fibula and the ventral tibia. It merges with a flat, shiny tendon to insert into the tarsals.

11. The **flexor hallucis longus** is a long muscle that lies lateral to the tibialis posterior. It originates from the posterior tibia and passes downward to the ankle. It is a uniformly broad muscle in the cat. As in humans, it is a flexor of the great toe. ■

Dissection Review

Many human muscles are modified from those of the cat (or any quadruped) as a result of the requirements of an upright posture. The following questions refer to these differences.

1. How does the human trapezius muscle differ from the cat's?

_____

_____

_____

2. How does the deltoid differ?

_____

_____

_____

3. How do the extent and orientation of the human sartorius muscle differ from its relative position in the cat?

_____

_____

_____

4. Explain these differences in terms of differences in function.

_____

_____

_____

5. The human rectus abdominis is definitely divided by four transverse tendons (tendinous intersections). These tendons are absent or difficult to identify in the cat. How do these tendons affect the human upright posture?

_____

_____

_____

dissection exercise

2

Dissection of Cat Spinal Nerves

Objective

1. To identify on a dissected animal the musculocutaneous, radial, median, and ulnar nerves of the upper limb and the femoral, saphenous, sciatic, common peroneal, and tibial nerves of the lower limb.

Materials

- ❑ Dissecting tray and instruments
- ❑ Disposable gloves or protective skin cream
- ❑ Animal specimen from previous dissections
- ❑ Embalming fluid

The cat has 38 or 39 pairs of spinal nerves (as compared to 31 in humans). Of these, 8 are cervical, 13 thoracic, 7 lumbar, 3 sacral, and 7 or 8 caudal.

A complete dissection of the cat's spinal nerves would be extraordinarily time-consuming and exacting, and is not warranted in a basic anatomy and physiology course. However, it is desirable for you to have some dissection work to complement your study of the anatomical charts. Thus at this point you will carry out a partial dissection of the brachial plexus and lumbosacral plexus and identify some of the major nerves.

See Exercise 21 of this manual for a discussion of human spinal nerves.

Activity:
Dissecting Nerves of the Brachial Plexus

1. Place your cat specimen on the dissecting tray, dorsal side down. Reflect the cut ends of the left pectoralis muscles to expose the large brachial plexus in the axillary region (Figure D2.1). Carefully clean the exposed nerves as far back toward their points of origin as possible.

2. The **musculocutaneous nerve** is the most superior nerve of this group. It splits into two subdivisions that run under the margins of the coracobrachialis and biceps brachii muscles. Trace its fibers into the ventral muscles of the arm it serves.

3. Locate the large **radial nerve** inferior to the musculocutaneous nerve. The radial nerve serves the dorsal muscles of the arm and forearm. Follow it into the three heads of the triceps brachii muscle.

4. In the cat, the **median nerve** is closely associated with the brachial artery and vein. It courses through the arm to supply the ventral muscles of the forearm (with the exception of the flexor carpi ulnaris and the ulnar head of the flexor digitorum profundus). It also innervates some of the intrinsic hand muscles, as in humans.

5. The **ulnar nerve** is the most posterior of the large brachial plexus nerves. Follow it as it travels down the forelimb, passing over the medial epicondyle of the humerus, to supply the flexor carpi ulnaris and the ulnar head of the flexor digitorum profundus (and the hand muscles). ■

(a)

(b)

Figure D2.1   Brachial plexus and major blood vessels of the left forelimb of the cat, ventral aspect. (a) Diagrammatic view. **(b)** Photograph.

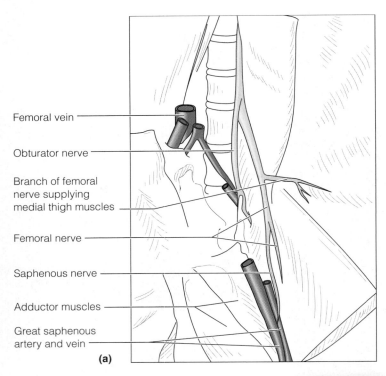

Femoral vein

Obturator nerve

Branch of femoral
nerve supplying
medial thigh muscles

Femoral nerve

Saphenous nerve

Adductor muscles

Great saphenous
artery and vein

(a)

Femoral vein

Obturator nerve

Branch of femoral
nerve supplying
medial thigh muscles

Femoral nerve

Saphenous nerve

Great saphenous
artery and vein

Adductor muscles

**Figure D2.2 Lumbar plexus of
the cat.** Ventral views: **(a)** art,
(b) photograph.

(b)

Activity:
Dissecting Nerves of the
Lumbosacral Plexus

1.   To locate the **femoral nerve** arising from the lumbar
plexus, first identify the right *femoral triangle,* which is bordered by the sartorius and adductor muscles of the anterior

thigh (Figure D2.2). The large femoral nerve travels through
this region after emerging from the psoas major muscle in
close association with the femoral artery and vein. Follow the
nerve into the muscles and skin of the anterior thigh, which it
supplies. Notice also its cutaneous branch in the cat, the
saphenous nerve, which continues down the anterior medial
surface of the thigh (with the great saphenous artery and
vein) to supply the skin of the anterior shank and foot.

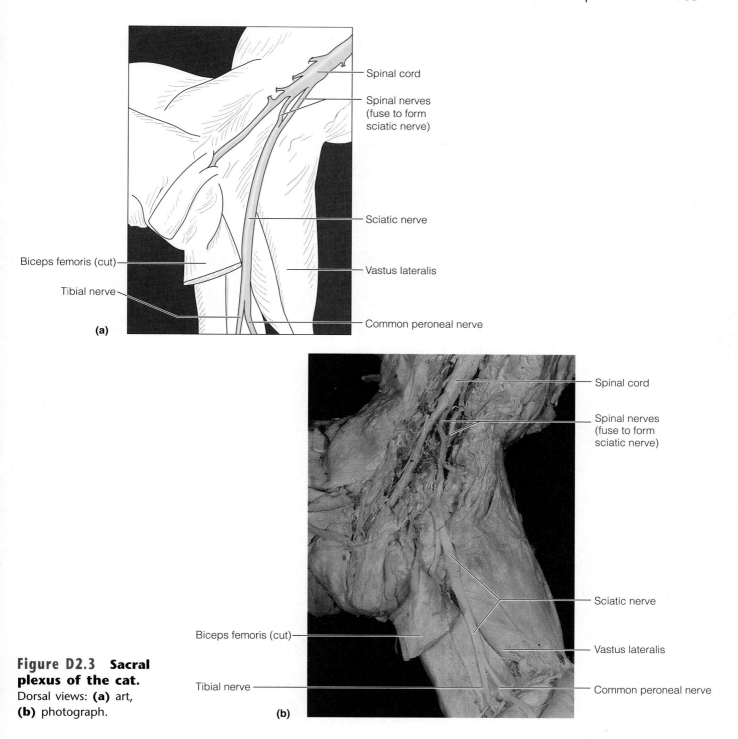

Figure D2.3 Sacral plexus of the cat. Dorsal views: **(a)** art, **(b)** photograph.

2.   Turn the cat ventral side down so you can view the posterior aspect of the lower limb (Figure D2.3). Reflect the ends of the transected biceps femoris muscle to view the large cordlike sciatic nerve. The **sciatic nerve** arises from the sacral plexus and serves the dorsal thigh muscles and all the muscles of the leg and foot. Follow the nerve as it travels down the posterior thigh lateral to the semimembranosus muscle. Note that just superior to the gastrocnemius muscle

of the calf, it divides into its two major branches, which serve the leg.

3.   Identify the **tibial nerve** medially and the **common peroneal nerve,** which curves over the lateral surface of the gastrocnemius.

4.   When you have finished making your observations, wrap the cat for storage and clean all dissecting tools and equipment before leaving the laboratory. ■

Dissection Review

1. From anterior to posterior, put the nerves issuing from the brachial plexus in their proper order (i.e., the median, musculo-cutaneous, radial, and ulnar nerves).

   _____

   _____

2. Which of the nerves named above serves the cat's forearm extensor muscles? _____ Which serves the

   forearm flexors? _____

3. Just superior to the gastrocnemius muscle the sciatic nerve divides into its two main branches, the _____

   and _____ nerves.

4. What name is given to the cutaneous nerve of the cat's thigh? _____

Identification of Selected Endocrine Organs of the Cat

Activity:
Opening the Ventral Body Cavity

1. Don gloves and then obtain your dissection animal. Place the animal on the dissecting tray, ventral side up. Using scissors, make a longitudinal median incision through the ventral body wall. Begin your cut just superior to the midline of the pubic bone and continue it anteriorly to the rib cage. Check the incision guide provided in Figure D3.1 as you work along.

2. Angle the scissors slightly (½ in.) to the right or left of the sternum, and continue the cut through the rib cartilages, just lateral to the body midline, to the base of the throat.

3. Make two lateral cuts on both sides of the ventral body surface, anterior and posterior to the diaphragm, which separates the thoracic and abdominal parts of the ventral body cavity. Leave the diaphragm intact. Spread the thoracic walls laterally to expose the thoracic organs.

4. Make an angled lateral cut on each side of the median incision line just superior to the pubic bone, and spread the flaps to expose the abdominal cavity organs. ∎

Activity:
Identifying Organs

A helpful adjunct to identifying selected endocrine organs of the cat is a general overview of ventral body cavity organs as shown in Figure D3.2. Since you will study the organ systems housed in the ventral body cavity in later units, the objective here is simply to identify the most important organs and those that will help you to locate the desired endocrine organs (marked *). A schematic showing the relative positioning of several of the animal's endocrine organs is provided in Figure D3.3.

See Exercises 27 and 28 of this manual for a discussion of the human endocrine system.

Objectives

1. To prepare the cat for observation by opening the ventral body cavity.

2. To identify and name the major endocrine organs on a dissected cat.

Materials

❑ Dissection animal, tray, and instruments
❑ Embalming fluid
❑ Plastic disposable gloves or protective skin cream

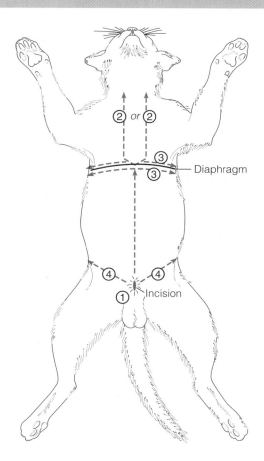

Figure D3.1 Incisions to be made in opening the ventral body cavity of a cat. Numbers indicate sequence.

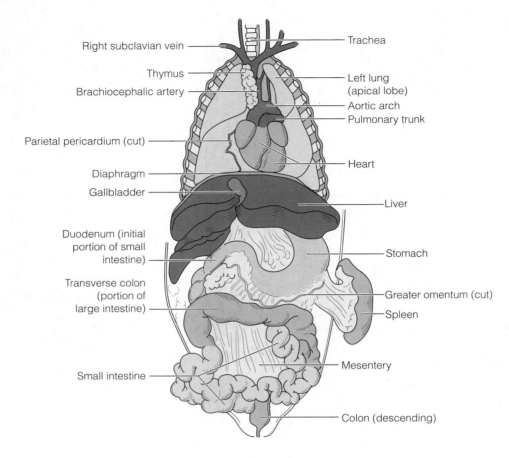

Figure D3.2 Ventral body cavity organs of the cat. Superficial view with greater omentum removed.

Thoracic Cavity Organs

Heart: In the mediastinum enclosed by the pericardium.

Trachea: The windpipe; runs down the midline of the throat and then divides just anterior to the lungs to form the bronchi, which plunge into the lungs on either side.

*Thyroid: Its dark lobes straddle the trachea. This endocrine organ's hormones are the main hormones regulating the body's metabolic rate.

Lungs: Paired organs flanking the heart.

*Thymus: Glandular structure superior to and partly covering the heart. The thymus is intimately involved (via its hormones) in programming the immune system. If you have a young cat, the thymus will be quite large. In old cats, most of this organ has been replaced by fat.

Abdominal Cavity Organs

Liver: Large multilobed organ lying under the umbrella of the diaphragm.

Lift the drapelike, fat-infiltrated greater omentum covering the abdominal organs to expose the following organs:

Stomach: Dorsally located sac to the left side of the liver.

Spleen: Flattened brown organ curving around the lateral aspect of the stomach.

Small intestine: Tubelike organ continuing posteriorly from the stomach.

*Pancreas: Diffuse gland lying deep to and between the small intestine and stomach. Lift the first section of the small intestine with your forceps; you should see the pancreas situated in the delicate mesentery behind the stomach. This gland is extremely important in regulating blood sugar levels.

Large intestine: Taking a U-shaped course around the small intestine to terminate in the rectum.

Push the intestines to one side with a probe to reveal the deeper organs in the abdominal cavity.

Kidneys: Bean-shaped organs located toward the dorsal body wall surface and behind the peritoneum.

*Adrenal glands: Seen above and medial to each kidney, these small glands produce corticosteroids important in preventing stress and abnormalities of water and electrolyte balance in the body.

*Gonads (ovaries or testes): Sex organs producing sex hormones. The location of the gonads is illustrated in Figure D3.3, but their identification is deferred until the reproductive system organs are considered (Exercise 43 and Dissection Exercise 9). ■

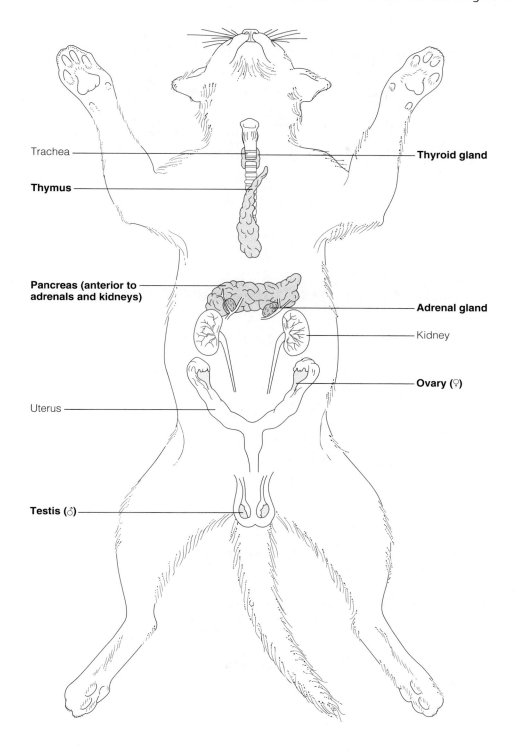

Trachea

Thyroid gland

Thymus

Pancreas (anterior to adrenals and kidneys)

Adrenal gland

Kidney

Ovary (♀)

Uterus

Testis (♂)

Figure D3.3 **Endocrine organs in the cat.**

A c t i v i t y :
Preparing the Dissection Animal for Storage

Before leaving the lab, prepare your animal for storage.

1. Return the animal's skin flaps to their normal position over the ventral body cavity organs.

2. To prevent the internal organs from drying out, soak several paper towels with embalming fluid, and wrap them snugly around the animal's torso. Make sure that the dissected areas are completely enveloped.

3. Place the animal in a plastic storage bag, press out excess air, and securely close the bag with a rubber band or twine.

4. Attach your name tag to the bag for identification.

5. Place the cat in the storage container, and properly dispose of your gloves before leaving the laboratory. ■

Dissection Review

1. How do the locations of the endocrine organs in the cat compare with those in the human?

    _____

    _____

2. Name two endocrine organs located in the throat region: _____ and _____

3. Name three endocrine organs located in the abdominal cavity.

    _____

    _____

4. Given the assumption (not necessarily true) that human beings have more stress than cats, which endocrine organs would you expect to be relatively larger in humans?

    _____

5. Cats are smaller animals than humans. Which would you expect to have a (relatively speaking) more active thyroid gland—

    cats or humans? _____ Why? (We know we are asking a lot with this one, but give it a whirl.)

    _____

    _____

Dissection of the Blood Vessels of the Cat

I f Dissection Exercise 3 was conducted, you have already opened your animal's ventral body cavity and identified many of its organs. In such a case, begin this exercise with the activity "Preparing to Identify the Blood Vessels" on page 761.

Objectives

1. To identify several of the most important blood vessels of the cat.

2. To point out anatomical differences between the vascular system of the human and the laboratory dissection specimen.

Materials

❑ Dissection animal, tray, and instruments

❑ Protective skin cream or disposable gloves

❑ Embalming fluid

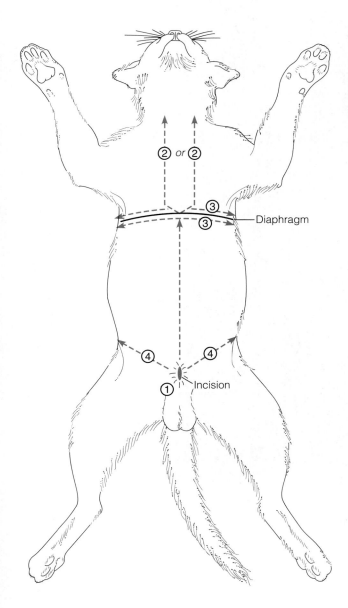

Figure D4.1 Incisions to be made in opening the ventral body cavity of a cat. Numbers indicate sequence.

Activity:
Opening the Ventral Body Cavity

1. Using scissors and Figure D4.1 as a guide, make a longitudinal incision through the ventral body wall. Begin just superior to the midline of the pubic bone, and continue the cut anteriorly to the rib cage.

2. Angle the scissors slightly (½ inch) to the right or left of the sternum, and continue the cut through the rib cartilages just lateral to the body midline, to the base of the throat.

3. Make two lateral cuts on both sides of the ventral body surface, anterior and posterior to the diaphragm. Leave the diaphragm intact. Spread the thoracic walls laterally to expose the thoracic organs.

4. Make an angled lateral cut on each side of the median incision line just superior to the pubic bone and spread the flaps to expose the abdominal cavity organs.

See Exercise 32 of this manual for a discussion of the anatomy of the human blood vessels.

Larynx

Trachea

External jugular vein

Brachiocephalic vein

Superior vena cava (precava)

Vagus nerve

Lobes of lung

Aorta

Heart

Diaphragm (cut)

Liver lobes

Stomach

Spleen

Kidney

Small intestine

Large intestine

Urinary bladder

Figure D4.2 Ventral body cavity organs of the cat. (Greater omentum has been removed.)

Activity:
Preliminary Organ Identification

A helpful prelude to identifying and tracing the blood supply of the various organs of the cat is a preliminary identification of ventral body cavity organs shown in Figure D4.2. Since you will study the organ systems contained in the ventral cavity in later units, the objective here is simply to identify the most important organs. Using Figure D4.2 as a guide, identify the following body cavity organs:

Thoracic Cavity Organs

Heart: In the mediastinum enclosed by the pericardium

Lungs: Flanking the heart

Thymus: Superior to and partially covering the heart

Abdominal Cavity Organs

Liver: Posterior to the diaphragm

Lift the large, drapelike, fat-infiltrated greater omentum covering the abdominal organs to expose the following:

Stomach: Dorsally located and to the left side of the liver

Spleen: A flattened, brown organ curving around the lateral aspect of the stomach

Small intestine: Continuing posteriorly from the stomach

Large intestine: Taking a U-shaped course around the small intestine and terminating in the rectum ∎

Activity:
Preparing to Identify the Blood Vessels

1.   Carefully clear away any thymus tissue or fat obscuring the heart and the large vessels associated with the heart. Before identifying the blood vessels, try to locate the *phrenic nerve* (from the cervical plexus), which innervates the diaphragm. The phrenic nerves lie ventral to the root of the lung on each side, as they pass to the diaphragm. Also attempt to locate the *vagus nerve* (cranial nerve X) passing laterally along the trachea and dorsal to the root of the lung.

2.   Slit the parietal pericardium and reflect it superiorly. Then, cut it away from its heart attachments. Review the structures of the heart. Notice its pointed inferior end (apex) and its broader superior portion. Identify the two *atria,* which appear darker than the inferior *ventricles.*

3.   Identify the **aorta,** the largest artery in the body, issuing from the left ventricle. Also identify the *coronary arteries* in the sulcus on the ventral surface of the heart; these should be injected with red latex. (As an aid to blood vessel identification, the arteries of laboratory dissection specimens are injected with red latex; the veins are injected with blue latex. Exceptions to this will be noted as they are encountered.)

4.   Identify the two large venae cavae—the superior and inferior venae cavae—entering the right atrium. The superior vena cava is the largest dark-colored vessel entering the base of the heart. These vessels are called the **precava** and **postcava,** respectively, in the cat. The caval veins drain the same relative body areas as in humans. Also identify the **pulmonary trunk** (usually injected with blue latex) extending anteriorly from the right ventricle and the right and left pulmonary arteries. Trace the **pulmonary arteries** until they enter the lungs. Locate the **pulmonary veins** entering the left atrium and the ascending aorta arising from the left ventricle and running dorsally to the precava and to the left of the body midline. ∎

Identifying the Arteries of the Cat

Refer to Figure D4.3 and to the summary photo in Figure D4.5 on p. 766 as you study the arterial system of the cat.

1. Reidentify the aorta as it emerges from the left ventricle. As you observed in the dissection of the sheep heart, the first branches of the aorta are the **coronary arteries,** which supply the myocardium. The coronary arteries emerge from the base of the aorta and can be seen on the surface of the heart. Follow the aorta as it arches (aortic arch), and identify its major branches. In the cat, the aortic arch gives off two large vessels, the **right brachiocephalic artery** and the **left subclavian artery.** The right brachiocephalic artery has three major branches, the right subclavian artery and the right and left common carotid arteries. (Note that in humans, the left common carotid artery is a direct branch off the aortic arch.)

2. Follow the **right common carotid artery** along the right side of the trachea as it moves anteriorly, giving off branches to the neck muscles, thyroid gland, and trachea. At the level of the larynx, it branches to form the **external** and **internal carotid arteries.** The internal carotid is quite small in the cat and it may be difficult to locate. It may even be absent. The distribution of the carotid arteries parallels that in humans.

3. Follow the right **subclavian artery** laterally. It gives off four branches, the first being the tiny **vertebral artery,** which along with the internal carotid artery provides the arterial circulation of the brain. Other branches of the subclavian artery include the **costocervical trunk** (to the costal and cervical regions), the **thyrocervical trunk** (to the shoulder), and the **internal mammary artery** (serving the ventral thoracic wall). As the subclavian passes in front of the first rib it becomes the **axillary artery.** Its branches, which supply the trunk and shoulder muscles, are the **ventral thoracic artery** (the pectoral muscles), the **long thoracic artery** (pectoral muscles and latissimus dorsi), and the **subscapular artery** (the trunk muscles). As the axillary artery enters the arm, it is called the **brachial artery,** and it travels with the median nerve down the length of the humerus. At the elbow, the brachial artery branches to produce the two major arteries serving the forearm and hand, the **radial** and **ulnar arteries.**

4. Return to the thorax, lift the left lung, and follow the course of the **descending aorta** through the thoracic cavity. The esophagus overlies it along its course. Notice the paired intercostal arteries that branch laterally from the aorta in the thoracic region.

5. Follow the aorta through the diaphragm into the abdominal cavity. Carefully pull the peritoneum away from its ventral surface and identify the following vessels:

Celiac trunk: The first branch diverging from the aorta immediately as it enters the abdominal cavity; supplies the stomach, liver, gallbladder, pancreas, and spleen. (Trace as many of its branches to these organs as possible.)

Superior mesenteric artery: Immediately posterior to the celiac trunk; supplies the small intestine and most of the large intestine. (Spread the mesentery of the small intestine to observe the branches of this artery as they run to supply the small intestine.)

Adrenolumbar arteries: Paired arteries diverging from the aorta slightly posterior to the superior mesenteric artery; supply the muscles of the body wall and adrenal glands

Renal arteries: Paired arteries running to the kidneys, which they supply

Genital arteries (testicular or ovarian): Paired arteries supplying the gonads

Inferior mesenteric artery: An unpaired thin vessel arising from the ventral surface of the aorta posterior to the genital arteries; supplies the second half of the large intestine

Iliolumbar arteries: Paired, rather large arteries that supply the body musculature in the iliolumbar region

6. The descending aorta ends posteriorly by dividing into three arteries: the two lateral **external iliac arteries,** which continue through the body wall and pass under the inguinal ligament to the hindlimb, and the single median **internal iliac artery.** (There is no common iliac artery in the cat.) The internal iliac persists only briefly before dividing into the right and left internal iliac arteries and the caudal artery, which supply the pelvic viscera.

7. Trace the external iliac artery into the thigh, where it becomes the **femoral artery.** The femoral artery is most easily identified in the **femoral triangle** at the medial surface of the upper thigh. Follow the femoral artery as it courses through the thigh (along with the femoral vein and nerve) and gives off branches to the thigh muscles. (These various branches are indicated on Figure 4.3.) As you approach the knee, the **saphenous artery** branches off the femoral artery to supply the medial portion of the leg. The femoral artery then descends deep to the knee to become the **popliteal artery** in the popliteal region. The popliteal artery in turn gives off two main branches, the **sural artery** and the **posterior tibial artery,** and continues as the **anterior tibial artery.** These branches supply the leg and foot.

Lingual artery

Larynx

Superior thyroid artery

Vertebral artery

Transverse scapular artery

Collateral radial artery

Subscapular artery

Deep brachial artery

Axillary artery

Costocervical trunk

Internal mammary artery

Right subclavian artery

Brachiocephalic artery

Aortic arch

Hepatic artery

Inferior phrenic artery

Edge of diaphragm

Superior mesenteric artery

Lumbar artery

Inferior mesenteric artery

Iliolumbar artery

Right external iliac artery

Circumflex femoral artery

Femoral artery

Proximal caudofemoral artery

Saphenous artery

Sural artery

Popliteal artery

Posterior tibial artery

Anterior tibial artery

External maxillary artery

External carotid artery

Internal carotid artery

Common carotid arteries

Caudal (inferior) thyroid artery

Radial artery

Vertebral artery

Ulnar artery

Thyrocervical trunk

Brachial artery

Ventral thoracic artery

Long thoracic artery

Thoracodorsal artery

Left subclavian artery

Descending thoracic aorta

Left gastric artery (branch of celiac)

Splenic artery (branch of celiac)

Celiac trunk

Adrenal gland

Adrenolumbar artery

Descending abdominal aorta

Renal artery

Genital artery (testicular-ovarian)

Internal iliac artery

Left internal iliac artery

External pudendal artery

Deep femoral artery

Caudal artery

Figure D4.3 **Arterial system of the cat.** (See also Figure 4.5 on p. 766.)

Activity:
Identifying the Veins of the Cat

Refer to Figure D4.4 and to summary Figure D4.5 on p. 766 as you study the venous system of the cat. Keep in mind that the vessels are named for the region drained, not for the point of union with other veins. (As you continue with the dissection, notice that not all vessels shown on Figure D4.4 are discussed.)

1. Reidentify the precava as it enters the right atrium. Trace it anteriorly to identify veins that enter it:

Azygos vein: Passing directly into its dorsal surface; drains the thoracic intercostal muscles

Internal thoracic (mammary) veins: Drain the chest and abdominal walls

Right vertebral vein: Drains the spinal cord and brain; usually enters right side of precava approximately at the level of the internal mammary veins but may enter the brachiocephalic vein in your specimen

Right and left brachiocephalic veins: Form the precava by their union

2. Reflect the pectoral muscles, and trace the brachiocephalic vein laterally. Identify the two large veins that unite to form it—the external jugular vein and the subclavian vein.

3. Follow the **external jugular vein** as it courses anteriorly along the side of the neck to the point where it is joined on its medial surface by the **internal jugular vein.** The internal jugular veins are small and may be difficult to identify in the cat. Notice the difference in cat and human jugular veins. The internal jugular is considerably larger in humans and drains into the subclavian vein. In the cat, the external jugular is larger, and the interior jugular vein drains into it. Identify the *common carotid artery,* since it accompanies the internal jugular vein in this region. Also attempt to find the *sympathetic trunk,* which is located in the same area running lateral to the trachea. Several other vessels drain into the external jugular vein (transverse scapular vein, facial veins, and others). These are not discussed here but are shown on the figure and may be traced if time allows.

4. Return to the shoulder region and follow the course of the **subclavian vein** as it moves laterally toward the arm. It becomes the **axillary vein** as it passes in front of the first rib and runs through the brachial plexus, giving off several branches, the first of which is the **subscapular vein.** The subscapular vein drains the proximal part of the arm and shoulder. The four other branches that receive drainage from the shoulder, pectoral, and latissimus dorsi muscles are shown in the figure but need not be identified in this dissection.

5. Follow the axillary vein into the arm, where it becomes the **brachial vein.** You can locate this vein on the medial side of the arm accompanying the brachial artery and nerve. Trace it to the point where it receives the **radial** and **ulnar veins** (which drain the forelimb) at the inner bend of the elbow. Also locate the superficial **cephalic vein** on the dorsal side of the arm. It communicates with the brachial vein via the median cubital vein in the elbow region and then enters the transverse scapular vein in the shoulder.

6. Reidentify the postcaval vein, and trace it to its passage through the diaphragm. Notice again as you follow its course that the **intercostal veins** drain into a much smaller vein lying dorsal to the postcava, the **azygos vein.**

7. Attempt to identify the **hepatic veins** entering the postcava from the liver. These may be seen if some of the anterior liver tissue is scraped away where the postcava enters the liver.

8. Displace the intestines to the left side of the body cavity, and proceed posteriorly to identify the following veins in order. All of these veins empty into the postcava and drain the organs served by the same-named arteries. In the cat, variations in the connections of the veins to be located are common, and in some cases the postcaval vein may be double below the level of the renal veins. If you observe deviations, call them to the attention of your instructor.

Adrenolumbar veins: From the adrenal glands and body wall.

Renal veins: From the kidneys (it is common to find two renal veins on the right side).

Genital veins (testicular or ovarian veins): The left vein of this venous pair enters the left renal vein anteriorly.

Iliolumbar veins: Drain muscles of the back.

Common iliac veins: Unite to form the postcava.

The common iliac veins are formed in turn by the union of the **internal iliac** and **external iliac veins.** The more medial internal iliac veins receive branches from the pelvic organs and gluteal region whereas the external iliac vein receives venous drainage from the lower extremity. As the external iliac vein enters the thigh by running beneath the inguinal ligament, it receives the **deep femoral vein,** which drains the thigh and the external genital region. Just inferior to that point, the external iliac vein becomes the **femoral vein,** which receives blood from the thigh, leg, and foot. Follow the femoral vein down the thigh to identify the **great saphenous vein,** a superficial vein that courses up the inner aspect of the calf and across the inferior portion of the gracilis muscle (accompanied by the great saphenous artery and nerve) to enter the femoral vein. The femoral vein is formed by the union of this vein and the popliteal vein. The **popliteal vein** is located deep in the thigh beneath the semimembranosus and semitendinosus muscles in the popliteal space accompanying the popliteal artery. Trace the popliteal vein to its point of division into the **posterior** and **anterior tibial veins,** which drain the leg.

9. In your specimen, trace the hepatic portal drainage depicted in Figure D4.6. Locate the **hepatic portal vein** by removing the peritoneum between the first portion of the small intestine and the liver. It appears brown due to coagulated blood, and it is unlikely that it or any of the vessels of this circulation contain latex. In the cat, the hepatic portal vein is

Transverse facial vein

Transverse scapular vein

Right vertebral vein

Axillary vein
Cephalic vein
Subscapular vein

Caudal circumflex
humeral vein

Median cubital vein

Brachial vein

Long thoracic vein

Internal thoracic vein
(mammary vein)

Azygos vein

Phrenic veins

Hepatic veins

Kidney

Renal veins

Genital vein (ovarian-testicular)

Iliolumbar vein

Middle sacral vein

Lateral circumflex
femoral vein

Proximal caudal
femoral vein

Popliteal vein

Great saphenous vein

Sural (caudal
femoral) vein

Anterior tibial vein

Posterior tibial vein

Anterior facial vein

Posterior facial vein

External jugular vein

Internal jugular vein

Inferior thyroid vein

Radial vein

Ulnar vein

Cephalic vein

Left vertebral vein

Subclavian vein

Brachiocephalic veins

Thoracodorsal vein

Costocervical-vertebral trunk

Superior vena cava
(precava)

Inferior vena cava
(postcava)

Diaphragm

Adrenal gland

Adrenolumbar vein

Genital vein
(ovarian-testicular)

Lumbar veins

Common iliac vein

External iliac vein

Internal iliac vein

Deep femoral vein

Femoral vein

Figure D4.4 Venous system of the cat. (See also Figure 4.5 on p. 766.)

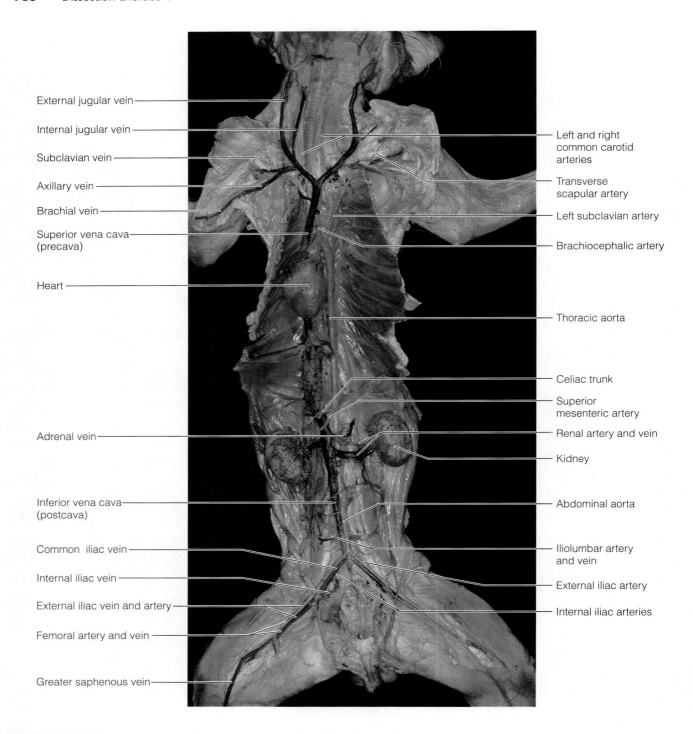

Figure D4.5 Cat dissected to reveal major blood vessels.

formed by the union of the **gastrosplenic** and **superior mesenteric veins.** (In the human, the hepatic portal vein is formed by the union of the splenic and superior mesenteric veins.) If possible, locate the following vessels, which empty into the hepatic portal vein:

Gastrosplenic vein: Carries blood from the spleen and stomach; located dorsal to the stomach.

Superior mesenteric vein: A large vein draining the small and large intestines and the pancreas.

Inferior mesenteric vein: Parallels the course of the inferior mesenteric artery.

Coronary vein: Drains the lesser curvature of the stomach.

Pancreaticoduodenal veins (anterior and posterior): The anterior branch empties into the hepatic portal vein; the posterior branch empties into the superior mesenteric vein. (In humans, both of these are branches of the superior mesenteric vein.)

If the structures of the lymphatic system of the cat are to be studied during this laboratory session, turn to Dissection Exercise 5 for instructions to conduct the study. Otherwise, properly clean your dissecting instruments and dissecting pan, and wrap and tag your cat for storage. ■

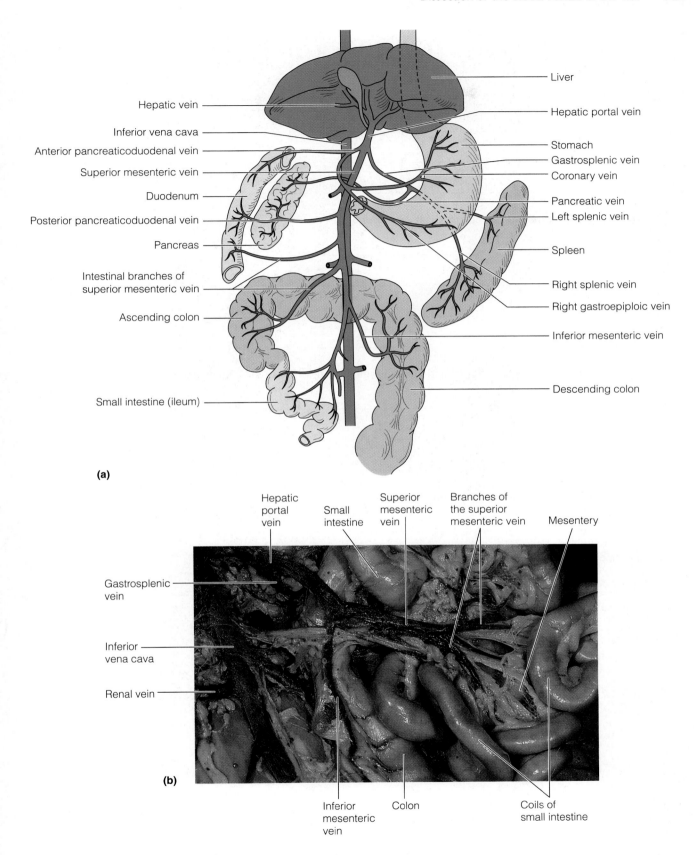

Figure D4.6   Hepatic portal circulation of the cat. (a) Diagrammatic view.
(b) Photograph of hepatic portal system of the cat, midline to left lateral view, just posterior
to the liver and pancreas. Intestines have been pulled to the left side of the cat. The mesentery
of the small intestine has been partially dissected to show the veins of the portal system.

Dissection Review

1. What differences did you observe between the origin of the left common carotid arteries in the cat and in the human?

   _____

   _____

   _____

   Between the origin of the internal and external iliac arteries?

   _____

   _____

   _____

   _____

2. How do the relative sizes of the external and internal jugular veins differ in the human and the cat?

   _____

   _____

   _____

3. In the cat the inferior vena cava is called the

   _____

   and the superior vena cava is referred to as the

   _____

The Main Lymphatic Ducts of the Cat

Activity:
Identifying the Main Lymphatic Ducts of the Cat

1.   Obtain your cat and a dissecting tray and instruments. Because lymphatic vessels are extremely thin-walled, it is difficult to locate them in a dissection. However, the large thoracic duct can be localized and identified.

2.   Move the thoracic organs to the side to locate the **thoracic duct.** Typically it lies just to the left of the mid-dorsal line, abutting the dorsal aspect of the descending aorta. It is usually about the size of pencil lead and red-brown with a segmented or beaded appearance caused by the valves within it. Trace it anteriorly to the site where it passes behind the left brachiocephalic vein and then bends and enters the venous system at the junction of the left subclavian and external jugular veins. If the veins are well injected, some of the blue latex may have slipped past the valves and entered the first portion of the thoracic duct.

3.   While in this region, also attempt to identify the short **right lymphatic duct** draining into the right subclavian vein, and notice the collection of lymph nodes in the axillary region.

Objective

1. To compare and contrast lymphatic structures of the cat to those of a human.

Materials

- ❑ Dissection animal, tray, and instruments
- ❑ Embalming fluid
- ❑ Disposable gloves or protective skin cream

4.   If the cat is triply injected and the lymphatic vessels have been injected with yellow or green latex, trace the thoracic duct posteriorly to identify the cisterna chyli, the saclike enlargement of its distal end. This structure, which receives fat-rich lymph from the intestine, begins at the level of the diaphragm and can be localized posterior to the left kidney.

5.   When you finish identifying these lymphatic structures, clean the dissecting instruments and tray, and properly wrap and return the cat to storage. ■

See Exercise 36 of this manual for a discussion of the human lymphatic system.

Dissection Review

Is the cat's lymphatic drainage pattern basically similar or dissimilar to that of humans? _____

_____

Dissection of the Respiratory System of the Cat

Objective

1. To identify the major respiratory system organs in a dissected animal.

Materials

❑ Dissection animal, tray, and instruments
❑ Disposable gloves or protective skin cream
❑ Embalming fluid
❑ Dissecting microscope

In this dissection exercise, you will be examining both the gross and fine structure of respiratory system organs. Obtain your dissection animal, and dissecting tray and instruments. If desired, apply protective skin cream or don disposable gloves before beginning dissection activities.

Activity:
Identifying Organs of the Respiratory System

1. Examine the external nares, oral cavity, and oral pharynx. Use a probe to demonstrate the continuity between the oral pharynx and the nasal pharynx above.

2. After securing the animal to the dissecting tray, dorsal surface down, expose the more distal respiratory structures by retracting the cut muscle and rib cage. Do not sever nerves and blood vessels located on either side of the trachea if these have not been studied. If you have not previously opened the thoracic cavity, make a medial longitudinal incision through the neck muscles and thoracic musculature to expose and view the thoracic organs (see Figure D4.1, p. 759).

3. Using the orientation Figure D6.1 and the photo in Figure D6.2 as guides, identify the structures named in items 3 through 5. Examine the **trachea,** and determine by finger ex-

See Exercises 37 and 38 of this manual for a discussion of the human respiratory system.

amination whether the cartilage rings are complete or incomplete posteriorly. Locate the *thyroid gland* inferior to the larynx on the trachea. Free the **larynx** from the attached muscle tissue for ease of examination. Identify the **thyroid** and **cricoid cartilages** and the flaplike **epiglottis.** Find the hyoid bone, located anterior to the larynx. Make a longitudinal incision through the ventral wall of the larynx and locate the *true* and *false vocal cords* on the inner wall.

4. Locate the large right and left common carotid arteries and the internal jugular veins on either side of the trachea (see Figure D4.5, p. 766). Also locate a conspicuous white band, the vagus nerve, which lies alongside the trachea, adjacent to the common carotid artery.

5. Examine the contents of the thoracic cavity. Follow the trachea as it bifurcates into two **primary bronchi,** which plunge into the lungs. Note that there are two *pleural cavities* containing the lungs and that each lung is composed of many lobes. In humans there are three lobes in the right lung and two in the left. How does this compare to what is seen in the cat? Identify the pericardial sac containing the heart located in the mediastinum (if it is still present). Examine the pleura, and note its exceptionally smooth texture. Locate the **diaphragm** and the **phrenic nerve.** The phrenic nerve, clearly visible as a white "thread" running along the pericardium to the diaphragm, controls the activity of the diaphragm in breathing. Lift one lung and find the esophagus beneath the parietal pleura. Follow it through the diaphragm to the stomach. ■

Activity:
Observing Lung Tissue Microscopically

Make a longitudinal incision in the outer tissue of one lung lobe beginning at a primary bronchus. Attempt to follow part of the respiratory tree from this point down into the smaller subdivisions. Carefully observe the cut lung tissue (under a dissection scope, if one is available), noting the richness of the vascular supply and the irregular or spongy texture of the lung. ■

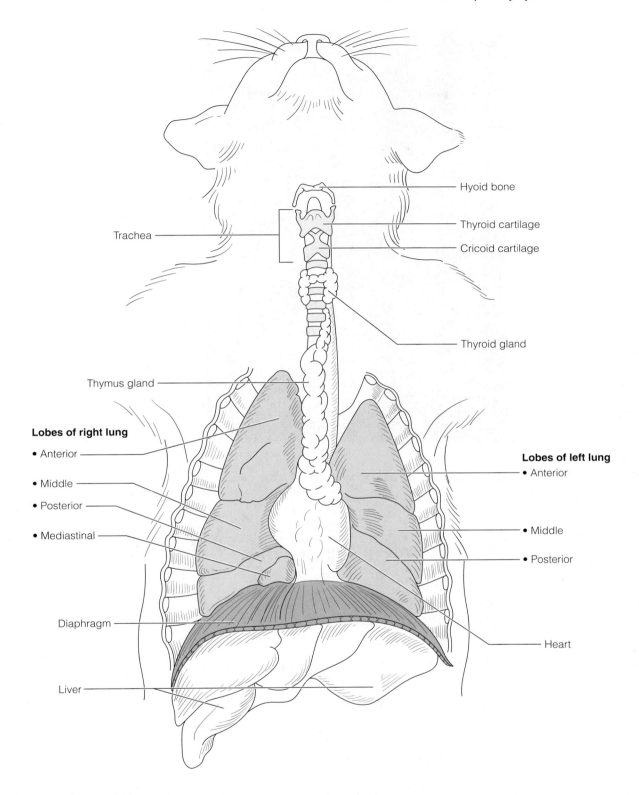

Figure D6.1 **Respiratory system of the cat.** (See also the photo in Figure D6.2.)

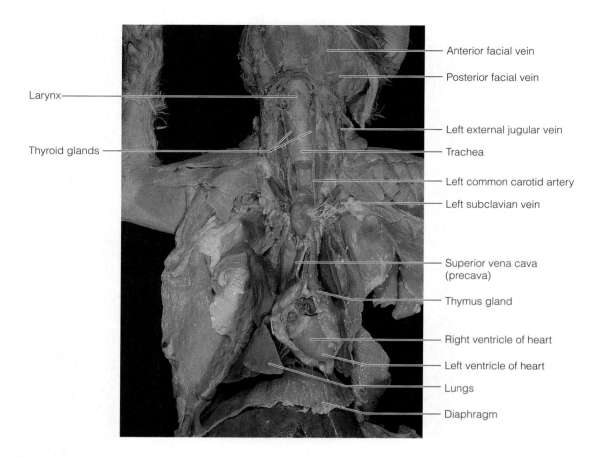

Anterior facial vein

Posterior facial vein

Larynx

Left external jugular vein

Thyroid glands

Trachea

Left common carotid artery

Left subclavian vein

Superior vena cava (precava)

Thymus gland

Right ventricle of heart

Left ventricle of heart

Lungs

Diaphragm

Figure D6.2 Photograph of the respiratory system of the cat.

Dissection Review

1. Are the cartilaginous rings in the cat trachea complete or incomplete?

2. How does the number of lung lobes in the cat compare with the number in humans?

3. Describe the appearance of lung tissue under the dissection microscope.

Dissection of the Digestive System of the Cat

Obtain your cat and secure it to the dissecting tray, dorsal surface down. Obtain all necessary dissecting instruments. If you have completed the dissection of the circulatory and respiratory systems, the abdominal cavity is already exposed and many of the digestive system structures have been previously identified. However, duplication of effort generally provides a good learning experience, so all of the digestive system structures will be traced and identified in this exercise.

Objectives

1. To identify on a dissected animal the organs composing the alimentary canal, and to name their subdivisions if any.
2. To name and identify the accessory organs of digestion in the dissection animal and to indicate their function.

Materials

- ❑ Dissection animal, tray, and instruments
- ❑ Bone cutters
- ❑ Protective hand cream or disposable gloves
- ❑ Embalming fluid
- ❑ Hand lens

If the abdominal cavity has not been previously opened, make a midline incision from the rib cage to the pubic symphysis as shown in Figure D7.1. Then make four lateral cuts—two parallel to the rib cage and two at the inferior margin of the abdominal cavity so that the abdominal wall can be reflected back while you examine the abdominal contents. Observe the shiny membrane lining the inner surface of the abdominal wall, which is the **parietal peritoneum.**

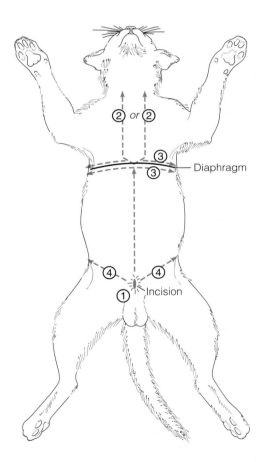

Figure D7.1 Incisions to be made in opening the ventral body cavity of the cat. Numbers indicate sequence.

See Exercise 39 of this manual for a discussion of the human digestive system.

A c t i v i t y :
Identifying Alimentary Canal Organs

1. Using Figure D7.2, locate the abdominal alimentary canal structures.

2. Identify the large reddish brown **liver** just beneath the diaphragm and the greater omentum covering the abdominal contents. The greater omentum assists in regulating body temperature and its phagocytic cells help to protect the body. Notice that the greater omentum is riddled with fat deposits. Lift the greater omentum, noting its two-layered structure and attachments, and lay it to the side or remove it to make subsequent organ identifications easier. Does the liver of the cat have the same number of lobes as the human liver?

_____

3. Lift the liver and examine its inferior surface to locate the **gallbladder,** a dark greenish sac embedded in the liver's ventral surface. Identify the **falciform ligament,** a delicate layer of mesentery separating the main lobes of the liver (right and left median lobes) and attaching the liver superiorly to the abdominal wall. Also identify the thickened area along the posterior edge of the falciform ligament, the _round ligament,_ or _ligamentum teres,_ a remnant of the umbilical vein of the embryo.

4. Displace the left lobes of the liver to expose the **stomach.** Identify the cardiac, fundic, body, and pyloric regions of the stomach. What is the general shape of the stomach?

_____

Locate the **lesser omentum,** the serous membrane attaching the lesser curvature of the stomach to the liver. Make an incision through the stomach wall to expose its inner surface. Can you see the **rugae**? (When the stomach is empty, its mucosa is thrown into large folds called rugae. As the stomach fills, the rugae gradually disappear and are no longer visible.)

5. Lift the stomach and locate the **pancreas,** which appears as a grayish or brownish diffuse glandular mass in the mesentery. It extends from the vicinity of the spleen and greater curvature of the stomach and wraps around the duodenum. Attempt to find the **pancreatic duct** as it empties into the duodenum at a swollen area referred to as the **hepatopancreatic ampulla.** Close to the pancreatic duct, locate the **bile duct** and trace its course superiorly to the point where it diverges into the **cystic duct** (gallbladder duct) and the **common hepatic duct** (duct from the liver). Notice that the duodenum assumes a looped position.

6. Lift the **small intestine** to investigate the manner in which it is attached to the posterior body wall by the **mesentery.** Observe the mesentery closely. What types of structures do you see in this double peritoneal fold?

_____

Other than providing support for the intestine, what other functions does the mesentery have?

_____

_____

Trace the course of the small intestine from its proximal, (duodenal) end to its distal (ileal) end. Can you see any obvious differences in the external anatomy of the small intestine from one end to the other?

_____

With a scalpel, slice open the distal portion of the ileum and flush out the inner surface with water. Feel the inner surface with your fingertip. How does it feel?

_____

Use a hand lens to see if you can see any **villi** and to locate the areas of lymphatic tissue called **Peyer's patches,** which appear as scattered white patches on the inner intestinal surface.

Return to the duodenal end of the small intestine. Make an incision into the duodenum. As before, flush the surface with water, and feel the inner surface. Does it feel any different than the ileal mucosa?

_____

If so, describe the difference. _____

_____

_____

Use the hand lens to observe the villi. What differences do you see in the villi in the two areas of the small intestine?

_____

_____

7. Make an incision into the junction between the ileum and cecum to locate the ileocecal valve. Observe the **cecum,** the initial expanded part of the large intestine. (Lymph nodes may have to be removed from this area to observe it clearly.) Does the cat have an appendix?

_____

8. Identify the short ascending, transverse, and descending portions of the **colon** and the **mesocolon,** a membrane that attaches the colon to the posterior body wall. Trace the descending colon to the **rectum,** which penetrates the body wall, and identify the **anus.**

Identify the two portions of the peritoneum, the parietal peritoneum lining the abdominal wall (identified previously) and the visceral peritoneum, which is the outermost layer of the wall of the abdominal organs (serosa). ∎

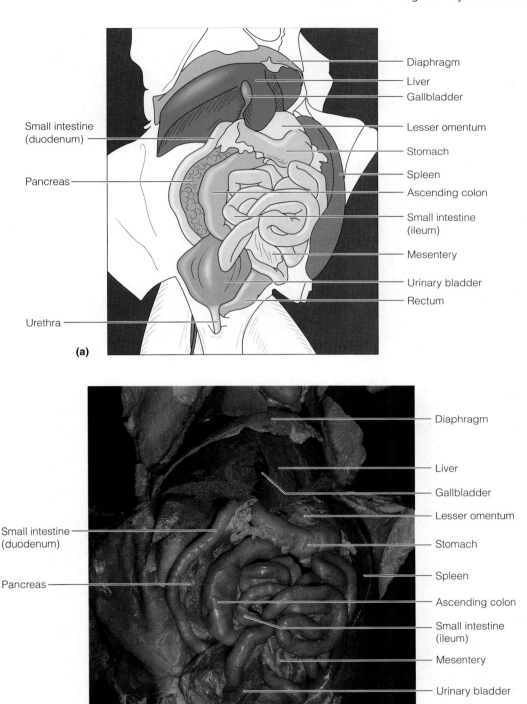

Figure D7.2   Digestive organs of the cat. (a) Diagrammatic view.
(b) Photograph. The greater omentum has been cut from its attachment to the stomach.

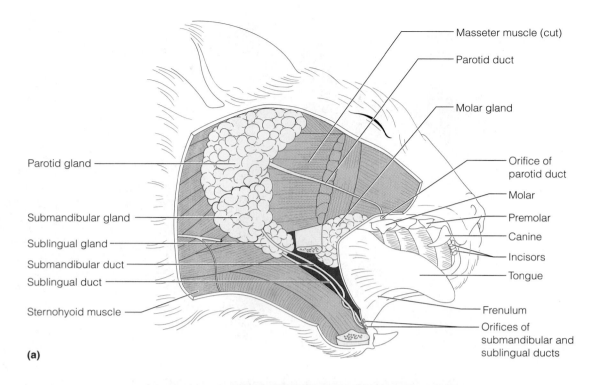

Masseter muscle (cut)

Parotid duct

Molar gland

Parotid gland

Orifice of parotid duct

Molar

Premolar

Submandibular gland

Canine

Sublingual gland

Incisors

Submandibular duct

Sublingual duct

Tongue

Sternohyoid muscle

Frenulum

Orifices of submandibular and sublingual ducts

(a)

Ear (cut)

Lymph node

Parotid gland

Submandibular gland

Masseter muscle

Lymph node

(b)

Figure D7.3 **Salivary glands of the cat.** **(a)** Diagrammatic view. **(b)** Photograph.

Activity:
Exposing and Viewing the Salivary Glands and Oral Cavity Structures

1.   To expose and identify the **salivary glands,** which secrete saliva into the mouth, remove the skin from one side of the head and clear the connective tissue away from the angle of the jaw, below the ear, and superior to the masseter muscle. Many lymph nodes are in this area and you should remove them if they obscure the salivary glands. The cat possesses five pairs of salivary glands, but only those glands described in humans are easily localized and identified (Figure D7.3). Locate the parotid gland on the cheek just inferior to the ear. Follow its duct over the surface of the masseter muscle to the angle of the mouth. The submandibular gland is posterior to the parotid, near the angle of the jaw, and the sublingual gland is just anterior to the submandibular gland within the lower jaw. The ducts of the submandibular and sublingual glands run deep and parallel to each other and empty on the side of the frenulum of the tongue. These need not be identified on the cat.

2.   To expose and identify the structures of the oral cavity, cut through the mandible with bone cutters just anterior to the angle to free the lower jaw from the maxilla.

Identify the **hard** and **soft palates,** and use a probe to trace the hard palate to its posterior limits. Note the transverse ridges, or *rugae,* on the hard palate, which play a role in holding food in place while chewing.

Do these appear in humans? _____

Does the cat have a uvula? _____

Identify the **oropharynx** at the rear of the oral cavity and the palatine tonsils on the posterior walls at the junction between the oral cavity and oropharynx. Identify the **tongue** and rub your finger across its surface to feel the papillae. Some of the papillae, especially at the anterior end of the tongue, should feel sharp and bristly. These are the filiform papillae, which are much more numerous in the cat than in humans. What do you think their function is?

_____

Locate the **lingual frenulum** attaching the tongue to the floor of the mouth. Trace the tongue posteriorly until you locate the **epiglottis,** the flap of tissue that covers the entrance to the respiratory passageway when swallowing occurs. Identify the **esophageal opening** posterior to the epiglottis.

Observe the **teeth** of the cat. The dental formula for the adult cat is as follows:

$$\frac{3,1,3,1}{3,1,2,1} \times 2 = 30$$

What differences in diet and mode of food-getting explain the difference between cat and human dentition?

_____

_____

3.   Prepare your cat for storage by wrapping it in paper towels wet with embalming fluid, return it to the plastic bag, and attach your name label. Wash the dissecting tray and instruments before continuing or leaving the laboratory. ■

Dissection Review

Several differences between cat and human digestive anatomy should have become apparent during the dissection. Indicate the pertinent differences between the human and the cat relative to the following structures:

Structure	Cat	Human
tongue papillae		
number of liver lobes		
appendix		

Dissection of the
Urinary System
of the Cat

Objective

1. To identify on a dissection specimen the urinary system organs and to describe the general function of each.

Materials

❑ Dissection animal, tray, and instruments
❑ Disposable gloves or protective skin cream
❑ Embalming fluid

The structures of the reproductive and urinary systems are often considered together as the *urogenital system,* since they have common embryologic origins. However, the emphasis in this dissection is on identifying the structures of the urinary tract (Figures D8.1 and D8.2) with only a few references to contiguous reproductive structures. The anatomy of the reproductive system is studied in Dissection Exercise 9.

Activity:
Identifying Organs of
the Urinary System

1. Don gloves or apply skin cream. Obtain your dissection specimen, and pin or tie its limbs to the dissection tray. Reflect the abdominal viscera (most importantly the small intestine) to locate the kidneys high on the dorsal body wall. Note that the **kidneys** in the cat, as well as in the human, are retroperitoneal (behind the peritoneum).

2. Carefully remove the peritoneum, and clear away the bed of fat that invests the kidneys. Then locate the adrenal (suprarenal) glands that lie superiorly and medial to the kidneys.

3. Identify the **renal artery** (red latex injected), the **renal vein** (blue latex injected), and the ureter at the hilus region of the kidney. (You may find two renal veins leaving one kidney in the cat but not in humans.)

4. Trace the **ureters** to the **urinary bladder,** a smooth muscular sac located superiorly to the small intestine. If your cat is a female, be careful not to confuse the ureters with the uterine tubes, which lie superior to the bladder in the same general region. See Figure D8.1. Observe the sites where the **ureters** enter the bladder. How would you describe the entrance point anatomically?

_____

_____

5. Cut through the bladder wall, and examine the region of the urethral exit to see if you can discern any evidence of the *internal sphincter.*

6. If your cat is a male, identify the prostate gland (part of the male reproductive system), which encircles the urethra distal to the neck of the bladder (see Figure D8.2). Notice that the urinary bladder is somewhat fixed in position by ligaments.

7. Using a probe, trace the urethra as it exits from the bladder to its terminus in the **urogenital sinus***, a common chamber into which both the vagina and the urethra empty in the female cat, or into the penis of the male. Dissection to expose the urethra along its entire length should not be done at this time because of possible damage to the reproductive structures, which you will study in Dissection Exercise 9.

8. Before cleaning up the dissection materials, observe a cat of the opposite sex. ∎

* In the human female, both the vagina and the urethra have separate external openings.

See Exercise 41 of this manual for a discussion of the human urinary system.

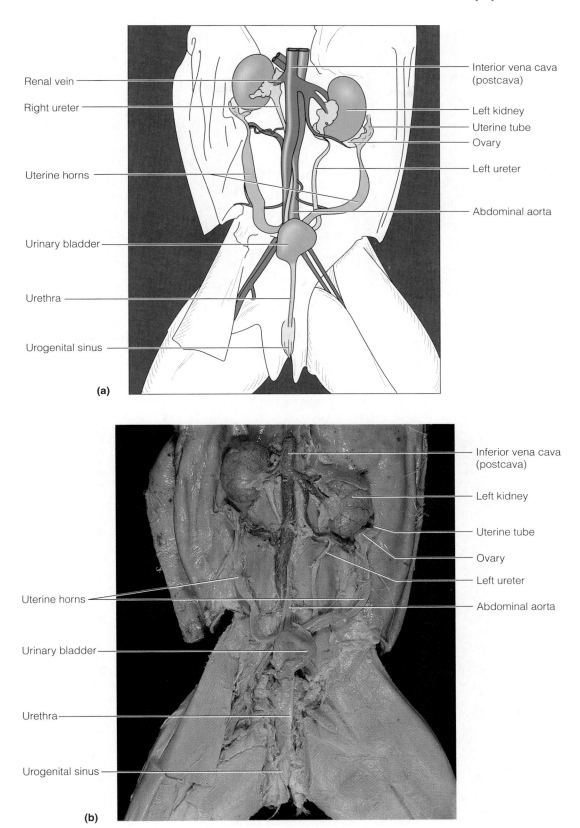

Figure D8.1 Urinary system of the female cat. (Reproductive structures are also indicated.) **(a)** Diagrammatic view. **(b)** Photograph of female urogenital system.

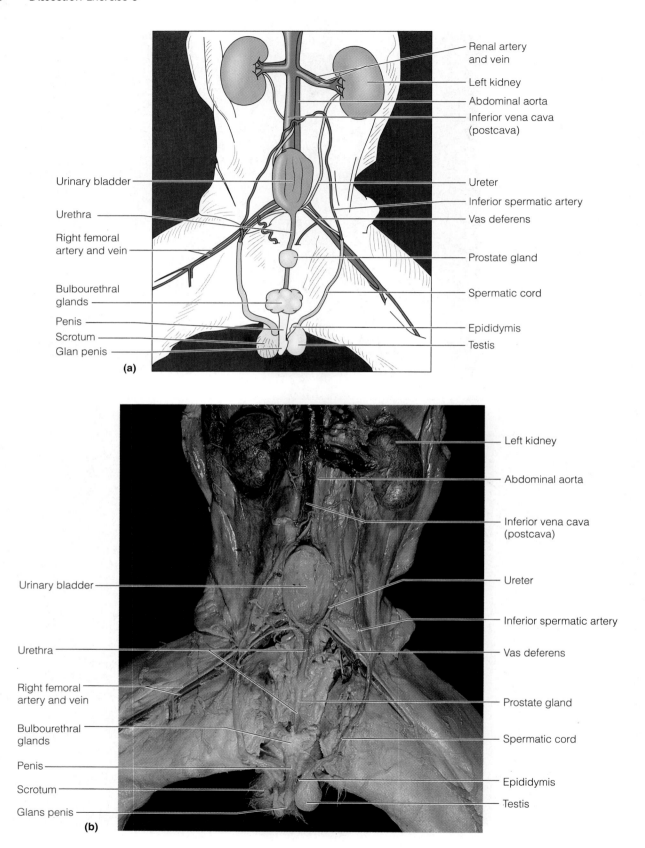

Figure D8.2 Urinary system of the male cat. (Reproductive structures are also indicated.) **(a)** Diagrammatic view. **(b)** Photograph of male urogenital system.

Dissection Review

1. How does the position of the kidneys in the cat differ from their position in humans?

   _____

   _____

   _____

   _____

2. How does the site of urethral emptying in the female cat differ from its termination point in the human female?

   _____

   _____

   _____

3. What gland encircles the neck of the bladder in the male? _____ Is this part of the urinary system?

   _____ What is its function? _____

   _____

   _____

Dissection of the Reproductive System of the Cat

Objective

1. To identify the major reproductive structures of the male and female dissection animal, and to recognize and discuss pertinent differences between the reproductive structures of humans and the dissection animal.

Materials

❑ Dissection animal, tray, and instruments
❑ Disposable gloves or protective skin cream
❑ Embalming fluid
❑ Bone cutters (for female cats)
❑ Small metric rulers (for female cats)

O btain your cat, a dissection tray, the necessary dissecting instruments, and, if desired, protective skin cream or plastic gloves. After you have completed the study of the reproductive structures of your specimen, observe a cat of the opposite sex. (The following instructions assume that the abdominal cavity has been opened in previous dissection exercises.)

Activity:
Identifying Organs of the Male Reproductive System

Refer to Figure D9.1 as you identify the male structures.

1. Identify the **penis** and notice the prepuce covering the glans. Carefully cut through the skin overlying the penis to expose the cavernous tissue beneath, then cross section the penis to see the relative positioning of the three cavernous bodies.

2. Identify the **scrotal sac,** and then carefully make a shallow incision through the scrotum to expose the **testes.** Notice that the scrotum is divided internally.

See Exercise 43 of this manual for a discussion of the human reproductive system.

3. Lateral to the medial aspect of the scrotal sac, locate the **spermatic cord,** which contains the spermatic artery, vein, and nerve, as well as the ductus deferens, and follow it up through the inguinal canal into the abdominal cavity. (It is not necessary to cut through the pelvic bone; a slight tug on the spermatic cord in the scrotal sac region will reveal its position in the abdominal cavity.) Carefully loosen the spermatic cord from the connective tissue investing it, and follow its course as it travels superiorly in the pelvic cavity, loops over the ureter,* and then courses posterior to the bladder and enters the prostate gland.

4. Notice that the **prostate gland** is comparatively smaller in the cat than in the human, and it is more distal to the bladder. (In the human, the prostate is immediately adjacent to the base of the bladder.) Carefully slit open the prostate gland to follow the **ductus deferens** to the urethra, which exits from the bladder midline. The male cat urethra, like that of the human, serves as both a urinary and sperm duct. In the human, the ductus deferens is joined by the duct of the seminal vesicle to form the ejaculatory duct, which enters the prostate. Seminal vesicles are not present in the cat.

5. Trace the **urethra** to the proximal end of the cavernous tissue of the penis. Carefully split the proximal portion of the penis along a sagittal plane to reveal the **bulbourethral glands** lying beneath it.

6. Once again, turn your attention to the testis. Cut it from its attachment to the spermatic cord and carefully slit open the **tunica vaginalis** capsule enclosing it. Identify the **epididymis** running along one side of the testis. Make a longitudinal cut through the testis and epididymis. Can you see the tubular nature of the epididymis and the rete testis portion of the testis with the naked eye? ■

Text continues on page 785

* This position of the spermatic cord and ductus deferens is due to the fact that during fetal development, the testis was in the same relative position as the ovary is in the female. In its descent, it passes laterally and ventrally to the ureter.

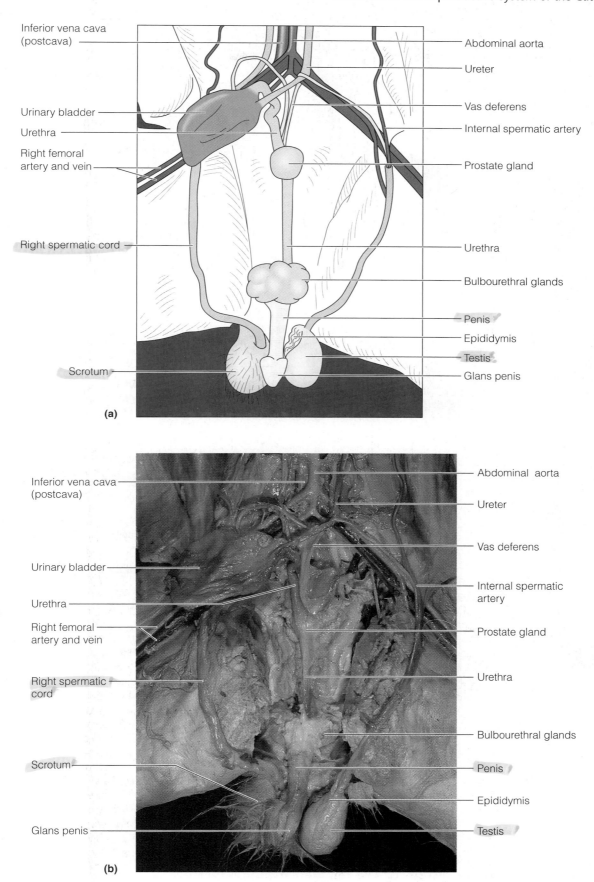

Inferior vena cava
(postcava)

Urinary bladder

Urethra

Right femoral
artery and vein

Right spermatic cord

Scrotum

(a)

Abdominal aorta

Ureter

Vas deferens

Internal spermatic artery

Prostate gland

Urethra

Bulbourethral glands

Penis

Epididymis

Testis

Glans penis

Inferior vena cava
(postcava)

Urinary bladder

Urethra

Right femoral
artery and vein

Right spermatic
cord

Scrotum

Glans penis

(b)

Abdominal aorta

Ureter

Vas deferens

Internal spermatic
artery

Prostate gland

Urethra

Bulbourethral glands

Penis

Epididymis

Testis

Figure D9.1   Reproductive system of the male cat. (a) Diagrammatic
view. **(b)** Photograph.

(a)

(b)

Figure D9.2 **Reproductive system of the female cat. (a)** Diagrammatic view. **(b)** Photograph.

Activity:
Identifying Organs of the Female Reproductive System

Refer to Figure D9.2 showing a dissection of the urogenital system of the female cat as you identify the structures described below.

1. Unlike the pear-shaped simplex, or one-part, uterus of the human, the uterus of the cat is Y-shaped (bipartite or bicornuate) and consists of a **uterine body** from which two **uterine horns** (cornua) diverge. Such an enlarged uterus enables the animal to produce litters. Examine the abdominal cavity and identify the bladder and the body of the uterus lying just dorsal to it.

2. Follow one of the uterine horns as it travels superiorly in the body cavity. Identify the thin mesentery (the *broad ligament*), which helps anchor it and the other reproductive structures to the body wall. Approximately halfway up the length of the uterine horn, it should be possible to identify the more important *round ligament,* a cord of connective tissue extending laterally and posteriorly from the uterine horn to the region of the body wall that would correspond to the inguinal region of the male.

3. Examine the **uterine tube** and **ovary** at the distal end of the uterine horn just caudal to the kidney. Observe how the funnel-shaped end of the uterine tube curves around the ovary. As in the human, the distal end of the tube is fimbriated, or fringed, and the tube is lined with ciliated epithelium. The uterine tubes of the cat are tiny and much shorter than in the human. Identify the *ovarian ligament,* a short thick cord that extends from the uterus to the ovary and anchors the ovary to the body wall. Also observe the *ovarian artery* and *vein* passing through the mesentery to the ovary and uterine structures.

4. Return to the body of the uterus and follow it caudad to the bony pelvis. Use bone cutters to cut through the median line of the pelvis (the pubic symphysis), cutting carefully so you do not damage the urethra deep to it. Expose the pelvic region by pressing the thighs dorsally. Follow the uterine body caudally to the vagina, and note the point where the urethra draining the bladder and the **vagina** enter a common chamber, the **urogenital sinus.** How does this anatomical arrangement compare to that seen in the human female?

_____

_____

5. Observe the **vulva** of the cat, which is similar to the human vulva. Identify the raised **labia majora** surrounding the urogenital opening.

6. To determine the length of the vagina, which is difficult to ascertain by external inspection, slit through the vaginal wall just superior to the urogenital sinus and cut toward the body of the uterus with scissors. Reflect the cut edges, and identify the muscular cervix of the uterus. Approximately how long is the vagina of the cat? (Measure the distance between the urogenital sinus and the cervix.)

_____

7. When you have completed your observations of both male and female cats, clean your dissecting instruments and tray and properly wrap the cat for storage. ∎

Dissection Review

1. The female cat has a _____ uterus; that of the human female is _____.

   Explain the difference in structure of these two uterine types. _____

   _____

   _____

   _____

2. What reproductive advantage is conferred by the feline uterine type?

   _____

   _____

3. Cite differences noted between the cat and the human relative to the following structures:

uterine tubes or oviducts _____

_____

site of entry of ductus deferens into the urethra _____

_____

_____

location of the prostate gland _____

_____

_____

seminal vesicles _____

_____

urethral and vaginal openings in the female

_____

_____

_____

_____

_____

The Metric System

Measurement	Unit and abbreviation	Metric equivalent	Metric to English conversion factor	English to metric conversion factor
Length	1 kilometer (km)	$= 1000\ (10^3)$ meters	1 km = 0.62 mile	1 mile = 1.61 km
	1 meter (m)	$= 100\ (10^2)$ centimeters $= 1000$ millimeters	1 m = 1.09 yards 1 m = 3.28 feet 1 m = 39.37 inches	1 yard = 0.914 m 1 foot = 0.305 m
	1 centimeter (cm)	$= 0.01\ (10^{-2})$ meter	1 cm = 0.394 inch	1 foot = 30.5 cm 1 inch = 2.54 cm
	1 millimeter (mm)	$= 0.001\ (10^{-3})$ meter	1 mm = 0.039 inch	
	1 micrometer (μm) [formerly micron (μ)]	$= 0.000001\ (10^{-6})$ meter		
	1 nanometer (nm) [formerly millimicron (mμ)]	$= 0.000000001\ (10^{-9})$ meter		
	1 angstrom (Å)	$= 0.0000000001\ (10^{-10})$ meter		
Area	1 square meter (m²)	$= 10,000$ square centimeters	1 m² = 1.1960 square yards 1 m² = 10.764 square feet	1 square yard = 0.8361 m² 1 square foot = 0.0929 m²
	1 square centimeter (cm²)	$= 100$ square millimeters	1 cm² = 0.155 square inch	1 square inch = 6.4516 cm²
Mass	1 metric ton (t)	$= 1000$ kilograms	1 t = 1.103 ton	1 ton = 0.907 t
	1 kilogram (kg)	$= 1000$ grams	1 kg = 2.205 pounds	1 pound = 0.4536 kg
	1 gram (g)	$= 1000$ milligrams	1 g = 0.0353 ounce 1 g = 15.432 grains	1 ounce = 28.35 g
	1 milligram (mg)	$= 0.001$ gram	1 mg = approx. 0.015 grain	
	1 microgram (μg)	$= 0.000001$ gram		
Volume (solids)	1 cubic meter (m³)	$= 1,000,000$ cubic centimeters	1 m³ = 1.3080 cubic yards 1 m³ = 35.315 cubic feet	1 cubic yard = 0.7646 m³ 1 cubic foot = 0.0283 m³
	1 cubic centimeter (cm³ or cc)	$= 0.000001$ cubic meter $= 1$ milliliter	1 cm³ = 0.0610 cubic inch	1 cubic inch = 16.387 cm³
	1 cubic millimeter (mm³)	$= 0.000000001$ cubic meter		
Volume (liquids and gases)	1 kiloliter (kl or kL)	$= 1000$ liters	1 kL = 264.17 gallons	1 gallon = 3.785 L 1 quart = 0.946 L
	1 liter (l or L)	$= 1000$ milliliters	1 L = 0.264 gallons 1 L = 1.057 quarts	
	1 milliliter (ml or mL)	$= 0.001$ liter $= 1$ cubic centimeter	1 ml = 0.034 fluid ounce 1 ml = approx. $\frac{1}{4}$ teaspoon 1 ml = approx. 15–16 drops (gtt.)	1 quart = 946 ml 1 pint = 473 ml 1 fluid ounce = 29.57 ml 1 teaspoon = approx. 5 ml
	1 microliter (μl or μL)	$= 0.000001$ liter		
Time	1 second (s)	$= \frac{1}{60}$ minute		
	1 millisecond (ms)	$= 0.001$ second		
Temperature	Degrees Celsius (°C)		$°F = \frac{9}{5}°C + 32$	$°C = \frac{5}{9}(°F - 32)$

Predicted Vital Capacities for Males

Height in centimeters

Age	146	148	150	152	154	156	158	160	162	164	166	168	170	172	174	176	178	180	182	184	186	188	190	192	194
16	3765	3820	3870	3920	3975	4025	4075	4130	4180	4230	4285	4335	4385	4440	4490	4540	4590	4645	4695	4745	4800	4850	4900	4955	5005
18	3740	3790	3840	3890	3940	3995	4045	4095	4145	4200	4250	4300	4350	4405	4455	4505	4555	4610	4660	4710	4760	4815	4865	4915	4965
20	3710	3760	3810	3860	3910	3960	4015	4065	4115	4165	4215	4265	4320	4370	4420	4470	4520	4570	4625	4675	4725	4775	4825	4875	4930
22	3680	3730	3780	3830	3880	3930	3980	4030	4080	4135	4185	4235	4285	4335	4385	4435	4485	4535	4585	4635	4685	4735	4790	4840	4890
24	3635	3685	3735	3785	3835	3885	3935	3985	4035	4085	4135	4185	4235	4285	4330	4380	4430	4480	4530	4580	4630	4680	4730	4780	4830
26	3605	3655	3705	3755	3805	3855	3905	3955	4000	4050	4100	4150	4200	4250	4300	4350	4395	4445	4495	4545	4595	4645	4695	4740	4790
28	3575	3625	3675	3725	3775	3820	3870	3920	3970	4020	4070	4115	4165	4215	4265	4310	4360	4410	4460	4510	4555	4605	4655	4705	4755
30	3550	3595	3645	3695	3740	3790	3840	3890	3935	3985	4035	4080	4130	4180	4230	4275	4325	4375	4425	4470	4520	4570	4615	4665	4715
32	3520	3565	3615	3665	3710	3760	3810	3855	3905	3950	4000	4050	4095	4145	4195	4240	4290	4340	4385	4435	4485	4530	4580	4625	4675
34	3475	3525	3570	3620	3665	3715	3760	3810	3855	3905	3950	4000	4045	4095	4140	4190	4225	4285	4330	4380	4425	4475	4520	4570	4615
36	3445	3495	3540	3585	3635	3680	3730	3775	3825	3870	3920	3965	4010	4060	4105	4155	4200	4250	4295	4340	4390	4435	4485	4530	4580
38	3415	3465	3510	3555	3605	3650	3695	3745	3790	3840	3885	3930	3980	4025	4070	4120	4165	4210	4260	4305	4350	4400	4445	4495	4540
40	3385	3435	3480	3525	3575	3620	3665	3710	3760	3805	3850	3900	3945	3990	4035	4085	4130	4175	4220	4270	4315	4360	4410	4455	4500
42	3360	3405	3450	3495	3540	3590	3635	3680	3725	3770	3820	3865	3910	3955	4000	4050	4095	4140	4185	4230	4280	4325	4370	4415	4460
44	3315	3360	3405	3450	3495	3540	3585	3630	3675	3725	3770	3815	3860	3905	3950	3995	4040	4085	4130	4175	4220	4270	4315	4360	4405
46	3285	3330	3375	3420	3465	3510	3555	3600	3645	3690	3735	3780	3825	3870	3915	3960	4005	4050	4095	4140	4185	4230	4275	4320	4365
48	3255	3300	3345	3390	3435	3480	3525	3570	3615	3655	3700	3745	3790	3835	3880	3925	3970	4015	4060	4105	4150	4190	4235	4280	4325
50	3210	3255	3300	3345	3390	3430	3475	3520	3565	3610	3650	3695	3740	3785	3830	3870	3915	3960	4005	4050	4090	4135	4180	4225	4270
52	3185	3225	3270	3315	3355	3400	3445	3490	3530	3575	3620	3660	3705	3750	3795	3835	3880	3925	3970	4010	4055	4100	4140	4185	4230
54	3155	3195	3240	3285	3325	3370	3415	3455	3500	3540	3585	3630	3670	3715	3760	3800	3845	3890	3930	3975	4020	4060	4105	4145	4190
56	3125	3165	3210	3255	3295	3340	3380	3425	3465	3510	3550	3595	3640	3680	3725	3765	3810	3850	3895	3940	3980	4025	4065	4110	4150
58	3080	3125	3165	3210	3250	3290	3335	3375	3420	3460	3500	3545	3585	3630	3670	3715	3755	3800	3840	3880	3925	3965	4010	4050	4095
60	3050	3095	3135	3175	3220	3260	3300	3345	3385	3430	3470	3500	3555	3595	3635	3680	3720	3760	3805	3845	3885	3930	3970	4015	4055
62	3020	3060	3110	3150	3190	3230	3270	3310	3350	3390	3440	3480	3520	3560	3600	3640	3680	3730	3770	3810	3850	3890	3930	3970	4020
64	2990	3030	3080	3120	3160	3200	3240	3280	3320	3360	3400	3440	3490	3530	3570	3610	3650	3690	3730	3770	3810	3850	3900	3940	3980
66	2950	2990	3030	3070	3110	3150	3190	3230	3270	3310	3350	3390	3430	3470	3510	3550	3600	3640	3680	3720	3760	3800	3840	3880	3920
68	2920	2960	3000	3040	3080	3120	3160	3200	3240	3280	3320	3360	3400	3440	3480	3520	3560	3600	3640	3680	3720	3760	3800	3840	3880
70	2890	2930	2970	3010	3050	3090	3130	3170	3210	3250	3290	3330	3370	3410	3450	3480	3520	3560	3600	3640	3680	3720	3760	3800	3840
72	2860	2900	2940	2980	3020	3060	3100	3140	3180	3210	3250	3290	3330	3370	3410	3450	3490	3530	3570	3610	3650	3680	3720	3760	3800
74	2820	2860	2900	2930	2970	3010	3050	3090	3130	3170	3200	3240	3280	3320	3360	3400	3440	3470	3510	3550	3590	3630	3670	3710	3740

Courtesy of Warren E. Collins, Inc., Braintree, Mass.

Appendix B — Predicted Vital Capacities for Females

Height in centimeters

Age	146	148	150	152	154	156	158	160	162	164	166	168	170	172	174	176	178	180	182	184	186	188	190	192	194
16	2950	2990	3030	3070	3110	3150	3190	3230	3270	3310	3350	3390	3430	3470	3510	3550	3590	3630	3670	3715	3755	3800	3840	3880	3920
17	2935	2975	3015	3055	3095	3135	3175	3215	3255	3295	3335	3375	3415	3455	3495	3535	3575	3615	3655	3695	3740	3780	3820	3860	3900
18	2920	2960	3000	3040	3080	3120	3160	3200	3240	3280	3320	3360	3400	3440	3480	3520	3560	3600	3640	3680	3720	3760	3800	3840	3880
20	2890	2930	2970	3010	3050	3090	3130	3170	3210	3250	3290	3330	3370	3410	3450	3490	3525	3565	3605	3645	3695	3720	3760	3800	3840
22	2860	2900	2940	2980	3020	3060	3095	3135	3175	3215	3255	3290	3330	3370	3410	3450	3490	3530	3570	3610	3650	3685	3725	3765	3800
24	2830	2870	2910	2950	2985	3025	3065	3100	3140	3180	3220	3260	3300	3335	3375	3415	3455	3490	3530	3570	3610	3650	3685	3725	3765
26	2800	2840	2880	2920	2960	3000	3035	3070	3110	3150	3190	3230	3265	3300	3340	3380	3420	3455	3495	3530	3570	3610	3650	3685	3725
28	2775	2810	2850	2890	2930	2965	3000	3040	3070	3115	3155	3190	3230	3270	3305	3345	3380	3420	3460	3495	3535	3570	3610	3650	3685
30	2745	2780	2820	2860	2895	2935	2970	3010	3045	3085	3120	3160	3195	3235	3270	3310	3345	3385	3420	3460	3495	3535	3570	3610	3645
32	2715	2750	2790	2825	2865	2900	2940	2975	3015	3050	3090	3125	3160	3200	3235	3275	3310	3350	3385	3425	3460	3495	3535	3570	3610
34	2685	2725	2760	2795	2835	2870	2910	2945	2980	3020	3055	3090	3130	3165	3200	3240	3275	3310	3350	3385	3425	3460	3495	3535	3570
36	2655	2695	2730	2765	2805	2840	2875	2910	2950	2985	3020	3060	3095	3130	3165	3205	3240	3275	3310	3350	3385	3420	3460	3495	3530
38	2630	2665	2700	2735	2770	2810	2845	2880	2915	2950	2990	3025	3060	3095	3130	3170	3205	3240	3275	3310	3350	3385	3420	3455	3490
40	2600	2635	2670	2705	2740	2775	2810	2850	2885	2920	2955	2990	3025	3060	3095	3135	3170	3205	3240	3275	3310	3345	3380	3420	3455
42	2570	2605	2640	2675	2710	2745	2780	2815	2850	2885	2920	2955	2990	3025	3060	3100	3135	3170	3205	3240	3275	3310	3345	3380	3415
44	2540	2575	2610	2645	2680	2715	2750	2785	2820	2855	2890	2925	2960	2995	3030	3060	3095	3130	3165	3200	3235	3270	3305	3340	3375
46	2510	2545	2580	2615	2650	2685	2715	2750	2785	2820	2855	2890	2925	2960	2995	3030	3060	3095	3130	3165	3200	3235	3270	3305	3340
48	2480	2515	2550	2585	2620	2650	2685	2715	2750	2785	2820	2855	2890	2925	2960	2995	3030	3060	3095	3130	3160	3195	3230	3265	3300
50	2455	2485	2520	2555	2590	2625	2655	2690	2720	2755	2785	2820	2855	2890	2925	2955	2990	3025	3060	3090	3125	3155	3190	3225	3260
52	2425	2455	2490	2525	2555	2590	2625	2655	2690	2720	2755	2790	2820	2855	2890	2925	2955	2990	3020	3055	3090	3125	3155	3190	3220
54	2395	2425	2460	2495	2530	2560	2590	2625	2655	2690	2720	2755	2790	2820	2855	2885	2920	2950	2985	3020	3050	3085	3115	3150	3180
56	2365	2400	2430	2465	2495	2525	2560	2590	2625	2655	2690	2720	2755	2790	2820	2855	2885	2920	2950	2980	3015	3045	3080	3110	3145
58	2335	2370	2400	2430	2460	2495	2525	2560	2590	2625	2655	2690	2720	2750	2785	2815	2850	2880	2920	2945	2975	3010	3040	3075	3105
60	2305	2340	2370	2400	2430	2460	2495	2525	2560	2590	2625	2655	2685	2720	2750	2780	2810	2845	2875	2915	2940	2970	3000	3035	3065
62	2280	2310	2340	2370	2405	2435	2465	2495	2525	2560	2590	2620	2655	2685	2715	2745	2775	2810	2840	2870	2900	2935	2965	2995	3025
64	2250	2280	2310	2340	2370	2400	2430	2465	2495	2525	2555	2585	2620	2650	2680	2710	2740	2770	2805	2835	2865	2895	2925	2955	2990
66	2220	2250	2280	2310	2340	2370	2400	2430	2460	2495	2525	2555	2585	2615	2645	2675	2705	2735	2765	2800	2825	2860	2890	2920	2950
68	2190	2220	2250	2280	2310	2340	2370	2400	2430	2460	2490	2520	2550	2580	2610	2640	2670	2700	2730	2760	2795	2820	2850	2880	2910
70	2160	2190	2220	2250	2280	2310	2340	2370	2400	2425	2455	2485	2515	2545	2575	2605	2635	2665	2695	2725	2755	2780	2810	2840	2870
72	2130	2160	2190	2220	2250	2280	2310	2335	2365	2395	2425	2455	2480	2510	2540	2570	2600	2630	2660	2685	2715	2745	2775	2805	2830
74	2100	2130	2160	2190	2220	2245	2275	2305	2335	2360	2390	2420	2450	2475	2505	2535	2565	2590	2620	2650	2680	2710	2740	2765	2795

Courtesy of Warren E. Collins, Inc., Braintree, Mass.

A.D.A.M. Interactive Anatomy Correlations

Appendix C lists correlations of Dissectible Anatomy images in the A.D.A.M. Interactive Anatomy (AIA) program with exercises in this manual. To start the AIA program, insert the AIA CD in your CD-ROM drive. Double-click the Interactive Anatomy 3.0 application icon located in the Interactive Anatomy folder on your hard drive. From the Introduction screen, click Dissectible Anatomy. From the open dialog, click Open. To change anatomical views, click the View button drop-down menu in the tool bar, and select Anterior, Posterior,

Lateral, or Medial. Click on the Structure List to launch the List Manager (Mac only) or choose from the list of available structures (Win only).

Note that AIA also offers Atlas Anatomy and 3-D Anatomy. Take some time to explore these additional features of AIA.

After you click Open from the open dialog, the following image appears:

Exercise 1
The Language of Anatomy

To illustrate a coronal plane
View: Anterior
Structure: Skull-coronal section

To illustrate a midsagittal plane
View: Medial
Structure: Skin

To illustrate a dorsal body cavity
View: Medial
Structure: Lumbosacral trunk

To illustrate a ventral body cavity
View: Anterior
Structure: Diaphragm

To illustrate serous membranes of the ventral body cavity
View: Anterior
Structure: Pericardial sac, Parietal peritoneum, Peritoneum

To illustrate a parietal pleura
View: Medial
Structure: Parietal pleura

Exercise 8
Classification of
Body Membranes

To illustrate parietal pleural membranes
View: Anterior
Structure: Parietal pleura

To illustrate peritoneal membranes
View: Medial
Structure: Parietal peritoneum

Exercise 10
The Axial Skeleton

To illustrate bones of the cranium and face
View: Anterior
Structure: Skull

To illustrate bones of the cranium
View: Posterior
Structure: Bones—coronal section

To illustrate bones of the cranium and face
View: Lateral
Structure: Skull

To illustrate hyoid bone
View: Lateral
Structure: Hyoid bone

To illustrate paranasal sinuses
View: Anterior
Structure: Mucosa of maxillary, Frontal,
Sphenoidal sinuses, Ethmoidal sinus

To illustrate a vertebral column
View: Anterior
Structure: Intervertebral disc

To illustrate a posterior vertebral column
View: Posterior
Structure: Intervertebral disc

To illustrate a vertebral column and inter-
vertebral discs
View: Medial
Structure: Sacrum

To illustrate a sacrum and lumbar
vertebrae
View: Posterior
Structure: Sacrum

To illustrate a bony thorax
View: Anterior
Structure: Ribs

Exercise 11
The Appendicular Skeleton

To illustrate a pectoral girdle
View: Anterior
Structure: Clavicle

To illustrate a pectoral girdle
View: Posterior
Structure: Scapula

To illustrate a humerus
View: Anterior
Structure: Humerus

To illustrate an ulna
View: Anterior
Structure: Ulna

To illustrate bones of the wrist and hand
View: Anterior
Structure: Bones of hand

To illustrate a female pelvis click on the
Gender button and choose Female
View: Anterior
Structure: Bones—coronal section

To illustrate a male pelvis click on the
Gender button and choose Male
View: Anterior
Structure: Bones—coronal section

To illustrate a femur
View: Anterior
Structure: Femur

To illustrate a tibia
View: Anterior
Structure: Tibia

To illustrate bones of the foot
View: Anterior
Structure: Bones of the foot

Exercise 13
Articulations and
Body Movements

To illustrate a synovial joint of the knee
View: Anterior
Structure: Synovial capsule of knee joint

To illustrate a synovial joint capsule of the
hip
View: Anterior
Structure: Synovial joint capsule of the
hip

Exercise 15
Gross Anatomy of the
Muscular System

To illustrate muscles of the face
View: Anterior
Structure: Orbicularis oculi muscle

To illustrate muscles of mastication
View: Anterior
Structure: Masseter muscle

To illustrate superficial muscles of the
neck
View: Lateral
Structure: Platysma muscle,
Sternocleidomastoid muscle

To illustrate deep muscles of the neck
View: Lateral
Structure: Strap muscles

To illustrate thorax and shoulder muscles
View: Lateral
Structure: Pectoralis major muscle,
Serratus anterior muscle

To illustrate thorax muscles
View: Anterior
Structure: External intercostal muscles,
Internal intercostal muscle

To illustrate abdominal wall
View: Anterior
Structure: Rectus abdominis muscle,
External abdominal oblique muscle,
Internal abdominal oblique muscle

To illustrate thorax muscles
View: Lateral
Structure: Latissimus dorsi muscle

To illustrate posterior muscles of the trunk
View: Posterior
Structure: Rhomboideus muscle

To illustrate muscles associated with the
vertebral column
View: Posterior
Structure: Semispinalis muscle, Splenius
muscle

To illustrate muscles of the humerus that
act on the forearm
View: Posterior
Structure: Brachialis muscle

To illustrate anterior muscles of the
forearm that act on the hand and fingers
View: Anterior
Structure: Flexor carpi radialis muscle,
Palmaris longus muscle, Flexor carpi
ulnaris muscle, Flexor digitorum superficialis muscle

To illustrate deep muscles of the forearm
that act on the hand and fingers
View: Posterior
Structure: Abductor pollicis longus muscle

To illustrate muscles acting on the thigh
View: Anterior
Structure: Sartorius muscle

To illustrate quadriceps
View: Anterior
Structure: Rectus femoris muscle, Vastus
lateralis muscle, Vastus medialis muscle,
Vastus intermedius muscle, Tensor fasciae
latae muscle

To illustrate muscles acting on the thigh
and originating on the pelvis
View: Posterior
Structure: Gluteus maximus muscle,
Gluteus medius muscle, Gluteus minimus
muscle

To illustrate hamstrings
View: Posterior
Structure: Long head of the biceps femoris
muscle, Semitendinosus muscle,
Semimembranosus muscle

To illustrate superficial muscles acting on the foot and ankle
View: Posterior
Structure: Gastrocnemius muscle, Soleus muscle, Popliteus muscle, Tibialis posterior muscle

To illustrate muscles acting on the foot and ankle
View: Anterior
Structure: Tibialis anterior muscle, Extensor digitorum longus muscle, Extensor hallucis longus muscle

Exercise 19
Gross Anatomy of the Brain and Cranial Nerves

To illustrate a cerebrum
View: Anterior
Structure: Skull—coronal section

To illustrate a cerebrum
View: Lateral
Structure: Brain

To illustrate cranial nerves
View: Lateral
Structure: Cranial nerves

To illustrate a vagus nerve
View: Lateral
Structure: Vagus nerve {CN X}

To illustrate phrenic and vagus nerves
View: Anterior
Structure: Phrenic and vagus nerves {CN X}

Exercise 21
Spinal Cord, Spinal Nerves, and the Autonomic Nervous System

To illustrate a spinal cord
View: Posterior
Structure: Spinal cord

To illustrate deep nerve plexuses and intercostal nerves
View: Anterior
Structure: Deep nerve plexuses and intercostal nerves

To illustrate a brachial plexus and branches
View: Anterior
Structure: Brachial plexus

To illustrate a central nervous system and sacral plexus
View: Medial
Structure: Central nervous system and sacral plexus

To illustrate a sciatic nerve and branches
View: Medial
Structure: Sciatic nerve and branches

To illustrate a tibial nerve
View: Posterior
Structure: Tibial nerve and branches

To illustrate autonomic nerve plexuses
View: Medial
Structure: Autonomic nerve plexuses

To illustrate a sympathetic trunk
View: Anterior
Structure: Sympathetic trunk

Exercise 24
Special Senses: Vision

To illustrate muscles and external anatomy of the eye
View: Anterior
Structure: Muscles of the eye

To illustrate eye muscles
View: Lateral
Structure: Eye muscles—medial

Exercise 27
Anatomy and Basic Function of the Endocrine Glands

To illustrate a pituitary gland
View: Lateral
Structure: Pituitary gland

To illustrate a thyroid gland
View: Anterior
Structure: Thyroid gland

To illustrate a suprarenal gland
View: Anterior
Structure: Suprarenal {Adrenal} gland

To illustrate a pancreas
View: Anterior
Structure: Pancreas

To illustrate ovaries click on the Gender button and choose Female
View: Medial
Structure: Ovary

To illustrate testes click on the Gender button and choose Male
View: Anterior
Structure: Testis

To illustrate a thymus gland
View: Anterior
Structure: Thymus gland

Exercise 30
Anatomy of the Heart

To illustrate a heart
View: Anterior
Structure: Heart

To illustrate fibrous pericardium
View: Anterior
Structure: Pericardiacophrenic vein

To illustrate visceral pericardium
View: Anterior
Structure: Epicardium

To illustrate heart chambers and heart valves
View: Anterior
Structure: Heart—cut section

To illustrate coronary arteries
View: Anterior
Structure: Coronary arteries

Exercise 32
Anatomy of Blood Vessels

To illustrate an aortic arch and branches
View: Anterior
Structure: Aortic arch and branches

To illustrate common carotid arteries
View: Anterior
Structure: Common carotid arteries

To illustrate major branches of the descending aorta
View: Anterior
Structure: Celiac trunk and branches

To illustrate a descending thoracic aorta
View: Anterior
Structure: Descending thoracic aorta

To illustrate an abdominal aorta and branches
View: Anterior
Structure: Abdominal aorta and branches

To illustrate an aortic arch and branches
View: Medial
Structure: Aortic arch and branches

To illustrate veins draining into the vena cavae
View: Lateral
Structure: Vena Cavae and tributaries

To illustrate pulmonary circulation
View: Anterior
Structure: Pulmonary arteries

To illustrate arterial supply of the brain
View: Lateral
Structure: Aortic arch and branches

To illustrate hepatic portal circulation
View: Anterior
Structure: Portal vein and tributaries

Exercise 34
Human Cardiovascular Physiology—Blood Pressure and Pulse Determination

To illustrate the position of the heart and valves in the thoracic cavity
View: Anterior
Structure: Heart—cut section

To illustrate pulse points
View: Anterior
Structure: Radial artery, Deep femoral artery

Exercise 36
The Lymphatic System and Immune Response

To illustrate lymphatic vessels and lymphoid organs
View: Anterior
Structure: Lymph vessels

To illustrate a thoracic duct
View: Anterior
Structure: Thoracic duct

To illustrate axillary and cervical lymph nodes
View: Anterior
Structure: Axillary and cervical lymph nodes

To illustrate a thymus gland
View: Anterior
Structure: Thymus gland

To illustrate a spleen
View: Anterior
Structure: Spleen

To illustrate palatine glands and tonsils
View: Anterior
Structure: Palatine glands and tonsils

Exercise 37
Anatomy of the Respiratory System

To illustrate an upper respiratory tract
View: Medial
Structure: Nasal conchae

To illustrate an upper respiratory tract
View: Lateral
Structure: Mediastinal parietal pleura of right pleural cavity

To illustrate a larynx
View: Anterior
Structure: Thyroid gland

To illustrate an epiglottis
View: Anterior
Structure: Epiglottis

To illustrate lower respiratory structures
View: Medial
Structure: Right lung

To illustrate lower respiratory structures
View: Anterior
Structure: Trachea

To illustrate lungs
View: Anterior
Structure: Lungs

To illustrate lungs—coronal section
View: Anterior
Structure: Lungs—coronal section

To illustrate a right lung
View: Lateral
Structure: Right lung

To illustrate a parietal pleura
View: Anterior
Structure: Parietal pleura

To illustrate a diaphragm
View: Anterior
Structure: Diaphragm

To illustrate a trachea
View: Lateral
Structure: Trachea

Exercise 39
Anatomy of the Digestive System

To illustrate an oral cavity
View: Lateral
Structure: Esophagus

To illustrate tonsils
View: Anterior
Structure: Palatine tonsil

To illustrate salivary glands
View: Lateral
Structure: Parotid gland, Parotid duct, Deep salivary glands

To illustrate an esophagus
View: Lateral
Structure: Esophagus

To illustrate a stomach
View: Anterior
Structure: Stomach, Stomach—coronal section

To illustrate a stomach
View: Lateral
Structure: Stomach

To illustrate an ileum
View: Anterior
Structure: Ileum

To illustrate a duodenum
View: Anterior
Structure: Duodenum

To illustrate a transverse colon
View: Anterior
Structure: Transverse colon

To illustrate an ascending and descending colon
View: Anterior
Structure: Ascending colon, Descending colon

To illustrate a colon
View: Medial
Structure: Colon

To illustrate teeth
View: Anterior
Structure: Skull

To illustrate a bile duct, liver, and gallbladder
View: Anterior
Structure: Bile duct {Common bile duct}, Liver—coronal section, Gallbladder

To illustrate a pancreas
View: Anterior
Structure: Pancreas

Exercise 41
Anatomy of the Urinary System

To illustrate kidneys
View: Anterior
Structure: Kidney, Kidney—longitudinal section

To illustrate a ureter
View: Anterior
Structure: Ureter

To illustrate a urinary bladder
View: Anterior
Structure: Urinary bladder

To illustrate a female urethra click on the Gender button and choose Female
View: Medial
Structure: Urethra

To illustrate a male urethra click on the Gender button and choose Male
View: Medial
Structure: Urinary bladder and prostate gland, Urethra

Exercise 43
Anatomy of the Reproductive System

To illustrate a male urethra click on the Gender button and choose Male
View: Medial
Structure: Urinary bladder and prostate gland, Urethra

To illustrate testes click on the Gender button and choose Male
View: Anterior
Structure: Testis

To illustrate a penis click on the Gender button and choose Male
View: Anterior
Structure: Penis

To illustrate a uterus click on the Gender button and choose Female
View: Anterior
Structure: Uterus

To illustrate mammary glands click on the Gender button and choose Female
View: Anterior
Structure: Breast

Credits

ILLUSTRATIONS

Exercise 1
1.2, 1.3, 1.5–1.8: Precision Graphics.

Exercise 3
3.2–3.4: Precision Graphics.

Exercise 4
4.1 and 4.2: Carla Simmons.

Exercise 5
5A.1 and 5A.4: Precision Graphics. 5B.1–5B.5: Steve McEntee.

Exercise 6
6.01–6.03, 6.5, and 6.6: Precision Graphics. 6.4: Kristin Otwell.

Exercise 7
7.1: Elizabeth Morales-Denny. 7.2 and 7. 4: Precision Graphics. 7.5: Barbara Cousins.

Exercise 8
8.1: Precision Graphics.

Exercise 9
Table 9.1: Kristin Mount. 9.1: Laurie O'Keefe. 9.2 and 9.3: Barbara Cousins. 9.4: Precision Graphics.

Exercise 10
Table 10.1: Kristin Otwell. 10.1, 10.2, 10.3a, 10.6, and 10.12: Nadine Sokol. 10.3b, 10.8, and 10.15: Precision Graphics. 10.4 and 10.5: Wendy Hiller-Gee and Precision Graphics. 10.7, 10.13, 10.14, and 10.16: Laurie O'Keefe. 10.9: Vincent Perez. 10.10: Kristin Otwell. 10.11: Kristin Mount.

Exercise 11
Table 11.1: Kristin Otwell. 11.1a, 11.2a, and 11.9b: Precision Graphics. 11.1b and c, 11.2b–d, and 11.3–11.8: Laurie O'Keefe. 11.9a: Kristin Otwell.

Exercise 12
12.1: Nadine Sokol.

Exercise 13
13.1–13.6: Precision Graphics. 13.7 and 13.8: Barbara Cousins.

Exercise 14
14.2, 14.4, and 14.5: Precision Graphics. 14.3: Raychel Ciemma.

Exercise 15
15.1–15.11: Raychel Ciemma. 15.12: Wendy Hiller-Gee. 15.13 and 15.14: Raychel Ciemma.

Exercise 16
16A.1–16A.5: Precision Graphics. 16B.1–16B.4: Steve McEntee.

Exercise 17
17.1–17.3, 17.5, 17.06, and 17.7b: Precision Graphics. 17.7a: Charles W. Hoffman.

Exercise 18
18.1–18.4: Precision Graphics.

Exercise 19
19.1, 19.2, 19.4, 19.5, 19.7–19.9, 19.11, and 19.13: Precision Graphics.

Exercise 20
20.1: Precision Graphics.

Exercise 21
21.2: Stephanie McCann. 21.3–21.7, 21.9, and 21.10: Precision Graphics. 21.8: Kristin Mount.

Exercise 22
22.1, 22.2, and 22.3: Precision Graphics.

Exercise 23
23.2: Precision Graphics.

Exercise 24
24.1a and b: Wendy Hiller-Gee at BioMed Arts. 24.1c, 24.2–24.4, 24.10, and 24.11: Precision Graphics. 24.7–24.9: Shirley Bortoli.

Exercise 25
25.1: Todd A. Buck. 25.2, 25.4, 25.5, and 25.8: Precision Graphics.

Exercise 26
26.1 and 26.2: Precision Graphics.

Exercise 27
27.1 and 27.3: Precision Graphics.

Exercise 28
28.1: Precision Graphics.

Exercise 29
29.2, 29.4 ,29.5, and 29.8: Precision Graphics.

Exercise 30
30.1: Wendy Hiller-Gee. 30.2, 30.3, and 30.6: Barbara Cousins. 30.4 and 30.8: Precision Graphics. 30.5: Kristin Mount.

Exercise 31
31.1: Barbara Cousins. 31.2–31.4: Precision Graphics.

Exercise 32
32. 1 and 32.2: Precision Graphics. 32.3–32.14: Barbara Cousins. 32.15: Kristin Mount.

Exercise 33
33.1 and 33.2: Steve McEntee.

Exercise 34
34.1–34.3: Precision Graphics. 34.4: Shirley Bortoli.

Exercise 35
35A.1–35A.5: Precision Graphics. 35B.1–35B.3: Steve McEntee.

Exercise 36
36.1: Kristin Mount. 36.2: Nadine Sokol. 36.2 and 36.3: Precision Graphics.

Exercise 37
37.1 and 37.2: Raychel Ciemma. 37.3: Wendy Hiller-Gee. 37.4–37.7: Precision Graphics.

Exercise 38
38.1 and 38.5–38.7: Precision Graphics. 38.2: Shirley Bortoli.

Exercise 39
39.1 and 39.10: Nadine Sokol. 39.2–39.9 and 39.11–39.14: Precision Graphics.

Exercise 40
40A.1–40A.2: Precision Graphics. 40B.1–40B.3: Steve McEntee.

Exercise 41
41.1a and 41.3b: Linda McVay. 41.1b: Wendy Hiller-Gee at BioMed Arts. 41.2 and 41.04–41.6: Precision Graphics.

Exercise 42
42.1 and 42.2: Precision Graphics.

Exercise 43
43.1, 43.3, 43.5, and 43.6: Precision Graphics. 43.4: Barbara Cousins. 43.7: Martha Blake.

Exercise 44
44.1, 44.3, 44.5, and 44.7: Precision Graphics.

Exercise 45
45.1: Kristin Mount. 45.2: Precision Graphics.

Exercise 47

47.3: Raychel Ciemma. 47.5, 47.7, 47.7b, 47.9, 47.10, 47.13, and 47.19: Precision Graphics.

Cat Dissection Exercises

D1.1, D3.1 D3.3, D4.1, and D7.1: Precision Graphics. D2.1, D2.3, D3.2, D4.3, D4.4, D4.6, D6.1, D7.2, D7.3, D8.1, D8.2, D9.1, and D9.2 : Kristin Mount.

PHOTOGRAPHS

Special thanks to Christine Eckel, Elena Dorfman, Kirk Wuest, Dennis Strete, Ph.D., Merritt College, and University of California, Berkeley.

Exercise 1

1.1: © Benjamin/Cummings Publishing. Photo by BioMed Arts Associates, Inc. 1.4: © Jenny Thomas/Addison Wesley Longman.

Exercise 2

2.1a, 2.1b, 2.2a, 2.2b, 2.3, 2.4a, 2.4b, 2.4c: © Dr. Robert A. Chase. 2.5: © Carolina Biological Supply Company/Phototake.

Exercise 3

3.1: Courtesy of Leica, Inc. 3.5: © Benjamin/Cummings Publishing. Photo by Victor Eroschencko.

Exercise 4

4.1b: © Dr. Don Fawcett, Science Source/Photo Researchers. 4.3a–f: © Ed Reschke.

Exercise 5

5A.2: © Elena Dorfman/Addison Wesley Longman. 5A.3a: © Keith R. Porter, Science Source/Photo Researchers. 5A.3b: Courtesy of Dr. Mohnadas Narla, Lawrence Berkeley Laboratory. 5A.3c: © David M. Phillips, Science Source/Photo Researchers.

Exercise 6

6.3a–h, 6.5c–e, h, i: © Benjamin/Cummings Publishing Company. Photo by Allen Bell, University of New England. 6.5a, b, f, g, j, and k, 6.6b, and 6.7: © Ed Reschke. 6.6a: © Eric Graves, Science Source/Photo Researchers. 6.6c: © Benjamin/Cummings Publishing Company.

Exercise 7

7.3: © Addison Wesley Longman. 7.6: Courtesy of Marian Rice.

Exercise 9

9.3b: © Benjamin/Cummings Publishing Company. Photo by Allen Bell, University of New England.

Exercise 10

10.4 and 10.5 From A Stereoscopic Atlas of Human Anatomy, by David L. Bassett. 10.9c: © Dr. Robert A. Chase.

Exercise 11

Table 11.1: From A Stereoscopic Atlas of Human Anatomy, by David L. Bassett. 11.5b: Department of Anatomy and Histology, University of California, San Francisco. 11.7a: © Benjamin/Cummings Publishing. Photo by BioMed Arts Associates, Inc.

Exercise 13

13.2: © Dr. Robert A. Chase. 13.7c: © Benjamin/Cummings Publishing Company. Photo by Stephen Spector, courtesy of Dr. Charles Thomas, Kansas University Medical Center. 13.8b: *From A Stereoscopic Atlas of Human Anatomy,* by David L. Bassett.

Exercise 14

14.3b: Courtesy of Marian Rice.

Exercise 15

15.3b, 15.7b, 15.12b, and 15.13b: © Dr. Robert A. Chase. 15.4a and 15.8a: From *A Stereoscopic Atlas of Human Anatomy,* by David L. Bassett. 15.10f: © Benjamin/Cummings Publishing Company. Photo by Stephen Spector, courtesy of Dr. Charles Thomas, Kansas University Medical Center.

Exercise 17

17.2c: © Benjamin/Cummings Publishing. Photo by Victor Eroschencko. 17.4: © Dennis Strete, Ph.D.

Exercise 19

19.2b and c, and 19.4a: © Dr. Robert A. Chase. 19.3: © Benjamin/Cummings Publishing Company. Photo by Mark Nielsen, University of Utah. 19.6a–c and 19.10: From *A Stereoscopic Atlas of Human Anatomy,* by David L. Bassett. 19.11b and 19.12: © Benjamin/Cummings Publishing Company. Photo by Dr. Sharon Cummings, University of California, Davis. 19.13b and 19.14: © Elena Dorfman/Addison Wesley Longman.

Exercise 20

20.1a: © Alexander Tsiaras, Science Source/Photo Researchers.

Exercise 21

21.1b–d: From *A Stereoscopic Atlas of Human Anatomy,* by David L. Bassett.

Exercise 22

22.4, 22.5, and 22.6: © Benjamin/Cummings Publishing Company. Photo by Richard Tauber.

Exercise 24

24.5a and b: © Benjamin/Cummings Publishing Company. Photo by Stephen Spector, courtesy of Dr. Charles Thomas, Kansas University Medical Center. 24.6: © Elena Dorfman/Addison Wesley Longman. 24.12a and b: © Benjamin/Cummings Publishing Company. Photo by Richard Tauber. 24.13: © Don Wong, Science Source/Photo Researchers.

Exercise 25

25.6a–c: © Benjamin/Cummings Publishing Company. Photo by Richard Tauber.

Exercise 29

29.3, 29.6c, and 29.7a–d: © Elena Dorfman/Addison Wesley Longman. 29.8b: © Boehringer Ingleheim International GmbH. Photo by Lennart Nilsson. 29.9: © Benjamin/Cummings Publishing Company. Photo by Jack Scanlan, Holyoke Community College.

Exercise 30

30.3b: © Lennart Nilsson. From *The Body Victorious.* 30.7: © Benjamin/Cummings Publishing Company. Photo by Allen Bell, University of New England. 30.8a, b, and 30.9: © Benjamin/Cummings Publishing Company. Photo by Wally Cash, Kansas State University.

Exercise 36

36.4: © SIU /Peter Arnold.

Exercise 37

37.1b, 37.5a: From *A Stereoscopic Atlas of Human Anatomy,* by David L. Bassett.

Exercise 38

38.3, and 38.4a and b: © Elena Dorfman/Addison Wesley Longman.

Exercise 39

39.7a and c: From *A Stereoscopic Atlas of Human Anatomy,* by David L. Bassett.

Exercise 41

41.3a: © Benjamin/Cummings Publishing Company. Photo by Wally Cash, Kansas State University.

Exercise 43

43.2a: © Ed Reschke.

Exercise 44

44.2b: © Dennis Strete, Ph.D. 44.6a–c: © Benjamin/Cummings Publishing. Photo by Victor Eroschencko.

Exercise 46

46.1a: © Scott Camazine, Science Source/Photo Researchers. 46.1b: Courtesy Jane Cangleton and Reproductive Genetics Center. 46.2.1: © Llewellyn/Uniphoto Picture Agency. 46.2.2: © Marc Martinelli/The Image Bank. 46.2.3 and 46.2.4: © Ogust/Image Works. 46.2.5: © Boisvieux/Explorer/Photo Researchers. 46.2.6, 46.2.7, 46.2.8: © Benjamin/Cummings. Photo by Anthony Loveday.

Exercise 47

47.1, 47.2, 47.3a, 47.4, 47.6a, 47.6b, 47.8, 47.11a, 47.11b, 47.12, 47.13a, 47.14, 47.15a, 47.15b, 47.16, 47.17, 47.18, 47.20, 47.21a, 47.21b, 47.21c, 47.21d, and 47.22: © Paul Waring/BioMed Arts Associates, Inc.

Histology Atlas

PLATES 1, 4, 5, 6, 7, 10, 14, 23, 25, 28, 29, 33, 35, 36, 38, 39, 40, 41, 49, 50: © Dennis Strete, Ph.D.

PLATES 2, 3, 8, 9, 11, 12, 13, 15, 16, 17, 18, 19, 21, 22, 24, 27, 32, 34, 37, 42, 43, 44, 45, 46, 47, 48, 51, 52, 53, 54, 55, 56, 57, 58, 59, 60, 61, 62: © Benjamin/Cummings Publishing. Photo by Victor Eroschencko.

PLATE 20 Carolina Biological Supply/Phototake. PLATE 27: Courtesy of LUMEN Histology, Loyola University Medical Education Network, http://www.lumen/MedEd/histo/frames/ histo©frames.html. PLATE 30: © John Cunningham/Visuals Unlimited. PLATE 31: Courtesy of Marian Rice.

Human Anatomy Atlas

PLATES A and B: © Elena Dorfman/ Addison Wesley Longman. PLATES C-I: © Dr. Robert A. Chase.

Cat Dissection Exercises

D1.2, D1.3 D1.4, D1.7, D1.13, D4.2: Elena Dorfman/Addison Wesley Longman. D1.5, D1.6, D1.8, D1.9, D1.10, D1.11, D1.12 a and b, D2.1b, D2.2b, D2.3b, D4.5, D4.6b, D6.2, D8.1b, D8.2b, D9.1b, D9.2b: © Paul Waring/BioMed Arts Associates, Inc. D7.2b, D7.3b: © Benjamin/Cummings Publishing Company. Photo by Jack Scanlan, Holyoke Community College.

TRADEMARK ACKNOWLEDGEMENTS

Adrenalin is a registered trademark of Warner-Lambert Company.

Albustix, Clinitest, Hemastix, Icotest, Ketostix, and Multistix are registered trademarks of Miles Laboratories, Inc.

Betadine is a registered trademark of The Purdue Frederick Company.

Chemstrip is a registered trademark of Boehringer Mannheim Corporation.

Clinistix is a registered trademark of Ames Company, Inc.

Harleco is a registered trademark of Hartman-Leddon Company, Inc.

Hefty is a registered trademark of Mobil Oil Company.

Landau is a registered trademark of Landau Uniform.

Macintosh and Power Macintosh are registered trademarks of Apple Computer, Inc.

Novocain is a registered trademark of Sterling Drug, Inc.

Parafilm is a registered trademark of American National Can Company.

Seal-Ease, Sedi-stain, and Unopette are registered trademarks of Becton, Dickinson, and Company.

Velcro is a registered trademark of Velcro, U.S.A., Inc.

Wampole is a registered trademark of Henry K. Wampole and Company.

Windows is a registered trademark of Microsoft Corporation.

Index

NEW EXERCISES

Exercise 48B: Respiratory System Mechanics

Exercise 49B: Renal Physiology

Using the Histology Module Instructions

Respiratory System Mechanics: Computer Simulation

Objectives

1. To define *ventilation, inspiration, expiration, forced expiration, tidal volume, vital capacity, expiratory reserve volume, inspiratory reserve volume, residual volume, surfactant, minute respiratory volume, forced expiratory volume,* and *pneumothorax.*

2. To describe the role of muscles and volume changes in the mechanics of breathing.

3. To understand that the lungs do not contain muscle and that respirations are therefore caused by external forces.

4. To explore the effect of changing airway resistance on breathing.

5. To study the effect of surfactant on lung function.

6. To examine the factors that cause lung collapse.

7. To understand the effects of hyperventilation, rebreathing, and breath holding on the CO_2 level in the blood.

Materials

Minimum equipment required:

Windows:

☐ 486/66 MHz or better recommended

☐ Windows 95/98 recommended

☐ SVGA display (256 colors at 640 × 480)

☐ Minimum 16 MB available RAM

☐ Double-speed CD-ROM drive (quad-speed or higher recommended)

☐ Sound card

☐ Speakers or headphones

☐ Printer

Macintosh:

☐ 68040 processor or Power Macintosh recommended

☐ System 7.1 or higher

☐ Minimum 16 MB available RAM

☐ 13-inch or larger color monitor (640 × 480 resolution)

☐ Double-speed CD-ROM drive (quad-speed or higher recommended)

☐ Printer

Software:

☐ Benjamin/Cummings PhysioEx CD-ROM—Respiratory System Mechanics module

The two phases of **pulmonary ventilation** or **breathing** are **inspiration,** during which air is taken into the lungs, and **expiration,** during which air is expelled from the lungs. Inspiration occurs as the external intercostal muscles and the diaphragm contract. The diaphragm, normally a dome-shaped muscle, flattens as it moves inferiorly while the external intercostal muscles between the ribs lift the rib cage. These cooperative actions increase the thoracic volume. Because the increase in thoracic volume causes a partial vacuum, air rushes into the lungs. During normal expiration, the inspiratory muscles relax, causing the diaphragm to rise and the chest wall to move inward. The thorax returns to its normal shape due to the elastic properties of the lung and thoracic wall. Like a deflating balloon, the pressure in the lungs rises, which forces air out of the lungs and airways. Although expiration is normally a passive process, abdominal wall muscles and the internal intercostal muscles can contract to force air from the lungs. Blowing up a balloon is an example where such **forced expiration** would occur.

Getting Started

Begin by making sure you have the computer equipment and software listed in the Materials section on the left.

1. Quit all applications currently running on your computer.

2. Insert the PhysioEx CD-ROM into the CD-ROM drive and keep PhysioEx in the drive during the entire time you use the program.

3. Follow the instructions below for your computer type.

Windows 95/Windows 98
The program will launch automatically when you load the CD-ROM into the drive. If autorun is disabled on your machine follow these instructions:

1. Double-click the **My Computer** icon on the Windows desktop.

2.   Double-click the **PhysioEx** CD icon.

3.   Look for the icon that looks like a tiny movie projector and double-click it.

Macintosh
Double-click the **PhysioEx** CD icon that appears in the window on your desktop to launch the program.

Simulating Spirometry: Measuring Respiratory Volumes and Capacities

This computerized simulation allows you to investigate the basic mechanical function of the respiratory system as you determine lung volumes and capacities. The concepts you will learn by studying this simulated mechanical lung can then be applied to help you understand the operation of the human respiratory system.

Normal quiet breathing moves about 500 ml (0.5 liter) of air (the tidal volume) in and out of the lungs with each breath, but this amount can vary due to a person's size, sex, age, physical condition, and immediate respiratory needs. The terms used for the normal respiratory volumes are defined next. The values are for the normal adult male and are approximate.

Normal Respiratory Volumes

Tidal volume (TV): Amount of air inhaled or exhaled with each breath under resting conditions (500 ml)

Expiratory reserve volume (ERV): Amount of air that can be forcefully exhaled after a normal tidal volume exhalation (1200 ml)

Inspiratory reserve volume (IRV): Amount of air that can be forcefully inhaled after a normal tidal volume inhalation (3100 ml)

Residual volume (RV): Amount of air remaining in the lungs after complete exhalation (1200 ml)

Vital capacity (VC): Maximum amount of air that can be exhaled after a normal maximal inspiration (4800 ml)

$$VC = TV + IRV + ERV$$

Total lung capacity (TLC): Sum of vital capacity and residual volume

An idealized tracing of the various respiratory volumes and their relationships to each other are shown in Figure 38.2, p. 383.

Pulmonary Function Tests

Forced vital capacity (FVC): Amount of air that can be expelled when the subject takes the deepest possible breath and exhales as completely and rapidly as possible

Forced expiratory volume (FEV$_1$): Measures the percentage of the vital capacity that is exhaled during 1 second of the FVC test (normally 75% to 85% of the vital capacity)

Choose **Respiratory System Mechanics** from the main menu. The opening screen for the Respiratory Volumes experiment will appear in a few seconds (Figure 48B.1). The main features on the screen when the program starts are a pair of *simulated lungs within a bell jar* at the left side of the screen, an *oscilloscope* at the upper right part of the screen, a *data display* area beneath the oscilloscope, and a *data control unit* at the bottom of the screen.

The black rubber "diaphragm" sealing the bottom of the glass bell jar is attached to a rod in the pump just below the jar. The rod moves the rubber diaphragm up and down to change the pressure within the bell jar (comparable to the intrapleural pressure in the body). As the diaphragm moves inferiorly, the resulting volume increase creates a partial vacuum in the bell jar because of lowered pressure. This partial vacuum causes air to be sucked into the tube at the top of the bell jar and then into the simulated lungs.

Conversely, as the diaphragm moves up, the rising pressure within the bell jar forces air out of the lungs. The partition between the two lungs compartmentalizes the bell jar into right and left sides. The lungs are connected to an airflow tube in which the diameter is adjustable by clicking the (+) and (−) buttons next to the Radius window in the equipment atop the bell jar. The volume of each breath that passes through the single air-flow tube above the bell jar is displayed in the Flow window. Clicking **Start** below the bell jar begins a trial run in which the simulated lungs will "breathe" in normal tidal volumes and the oscilloscope will display the tidal tracing. When **ERV** is clicked, the lungs will exhale maximally at the bottom of a tidal stroke and the expiratory reserve volume will be displayed in the Exp. Res. Vol. window below the oscilloscope. When **FVC** is clicked, the lungs will first inhale maximally and then exhale fully to demonstrate forced vital capacity. After ERV and FVC have been measured, the remaining lung values will be calculated and displayed in the small windows below the oscilloscope.

The data control equipment in the lower part of the screen records and displays data accumulated during the experiments. When you click **Record Data,** your data is recorded in the computer's memory and is displayed in the data grid. Data displayed in the data grid include the Radius, Flow, TV (tidal volume), ERV (expiratory reserve volume), IRV (inspiratory reserve volume), RV (residual volume), VC (vital capacity), FEV$_1$ (forced expiratory volume—1 second), TLC (total lung capacity), and Pump Rate. Clicking **Delete Line** allows you to discard the data for a single run; clicking **Clear Table** erases the entire experiment to allow you to start over.

If you need help identifying any piece of equipment, choose **Balloons On** from the Help menu and move the mouse pointer onto any piece of equipment visible on the computer's screen. As the pointer touches the object, a pop-up window appears to identify the equipment. To close the pop-up window, move the mouse pointer away from the equipment. Choose **Balloons On** again to turn off this help feature.

A c t i v i t y :
Measuring Respiratory Volumes

Your first experiment will establish the baseline respiratory values.

Figure 48B.1 Opening screen of the Respiratory Volumes experiment.

1. If the grid in the data control unit is not empty, click **Clear Table** to discard all previous data.

2. Adjust the radius of the airways to 5.00 mm by clicking the appropriate button next to the Radius window.

3. Click **Start** and allow the tracing to complete. Watch the simulated lungs begin to breathe as a result of the "contraction and relaxation" of the diaphragm. Simultaneously, the oscilloscope will display a tracing of the tidal volume for each breath. The Flow window atop the bell jar indicates the tidal volume for each breath, and the Tidal Vol. window below the oscilloscope shows the average tidal volume. The Pump Rate window displays the number of breaths per minute.

4. Click **Clear Tracings.**

5. Now click **Start** again. After a second or two, click **ERV,** wait 2 seconds and then click **FVC** to complete the measurement of respiratory volumes. The expiratory reserve volume, inspiratory reserve volume, and residual volume will be automatically calculated and displayed from the tests you have performed so far. Also, the equipment calculates and displays the total lung capacity.

6. Compute the **minute respiratory volume (MRV)** using the following formula (you can use the Calculator in the Tools menu):

$$MRV = TV \times BPM \text{ (breaths per minute)}$$

MRV _____ ml/min

7. Does expiratory reserve volume include tidal volume?

_____

Explain your answer.

_____

_____

8. Now click **Record Data** to record the current experimental data in the data grid. Then click **Clear Tracings.** ∎

Activity:

Examining the Effect of Changing Airway Resistance on Respiratory Volumes

Lung diseases are often classified as obstructive or restrictive. With an obstructive problem, expiratory flow is affected, whereas a restrictive problem might indicate reduced inspiratory volume. Although they are not diagnostic, pulmonary function tests such as FEV_1 can help a clinician determine the difference between obstructive and restrictive problems. FEV_1 is the forced volume exhaled in one second. In obstructive disorders like chronic bronchitis and asthma, airway resistance is increased and FEV_1 will be low. Here you will explore the effect of changing the diameter of the airway on pulmonary function.

1. Do *not* clear the data table from the previous experiment.

2. Adjust the radius of the airways to 4.50 mm by clicking the appropriate button next to the Radius window.

3. Click **Start** to begin respirations.

4. Click **FVC.** As you saw in the previous test, the simulated lungs will inhale maximally and then exhale as forcefully as possible. FEV_1 will be displayed in the FEV_1 window below the oscilloscope.

5. When the lungs stop respiring, click **Record Data** to record the current data in the data grid.

6. Decrease the radius of the airways in 0.50-mm decrements and repeat steps 4 and 5 until the minimum radius (3.00 mm) is achieved. Be sure to click **Record Data** after each trial. Click **Clear Tracings** between trials. If you make an error and want to delete a single value, click the data line in the data grid and then click **Delete Line.**

7. A useful way to express FEV_1 is as a percentage of the forced vital capacity. Copy the FEV_1 and vital capacity values from the computer screen to Chart 1, and then calculate the FEV_1 (%) by dividing the FEV_1 volume by the vital capacity volume. Record the FEV_1 (%) in Chart 1. You can use the Calculator under the Tools menu.

What happened to the FEV_1 (%) as the radius of the airways was decreased?

_____

Explain your answer.

_____

_____ ■

Simulating Factors Affecting Respirations

This part of the computer simulation allows you to explore the action of surfactant on pulmonary function and the effect of changing the intrapleural pressure.

Choose **Factors Affecting Respirations** from the **Experiment** menu. The opening screen will appear in a few seconds (Figure 48B.2). The basic features on the screen when the program starts are the same as in the Respiratory Volumes experiment screen. Additional equipment includes a surfactant dispenser atop the bell jar, and valves on each side of the bell jar. Each time **Surfactant** is clicked, a measured amount of surfactant is sprayed into the lungs. Clicking **Flush** washes surfactant from the lungs to prepare for another run.

Clicking **Open Valve** allows the pressure within that side of the bell jar to equalize with the atmospheric pressure. When **Reset** is clicked, the lungs are prepared for another run.

Data accumulated during a run are displayed in the windows below the oscilloscope. When you click **Record Data,** that data is recorded in the computer's memory and is displayed in the data grid. Data displayed in the data grid include the Radius, Pump Rate, the amount of Surfactant, Pressure Left (pressure in the left lung), Pressure Right (pressure in the right lung), Flow Left (air flow in the left lung), Flow Right (air flow in the right lung), and Total Flow. Clicking **Delete Line** allows you to discard data values for a single run, and clicking **Clear Table** erases the entire experiment to allow you to start over.

Chart 1	FEV₁ as % of VC		
Radius	**FEV₁**	**Vital Capacity**	**FEV₁ (%)**
5.00			
4.50			
4.00			
3.50			
3.00			

Figure 48B.2   Opening screen of the Factors Affecting Respirations experiment.

A c t i v i t y :
Examining the Effect of Surfactant

At any gas-liquid boundary, the molecules of the liquid are attracted more strongly to each other than they are to the air molecules. This unequal attraction produces tension at the liquid surface called *surface tension*. Because surface tension resists any force that tends to increase surface area, it acts to decrease the size of hollow spaces, such as the alveoli or microscopic air spaces within the lungs. If the film lining the air spaces in the lung were pure water, it would be very difficult, if not impossible, to inflate the lungs. However, the aqueous film covering the alveolar surfaces contains **surfactant,** a detergent-like lipoprotein that decreases surface tension by reducing the attraction of water molecules for each other. You will explore the action of surfactant in this experiment.

1.   If the data grid is not empty, click **Clear Table** to discard all previous data values.

2.   Adjust the airway radius to 5.00 mm by clicking the appropriate button next to the Radius window.

3.   If necessary, click **Flush** to clear the simulated lungs of existing surfactant.

4.   Click **Start** and allow a baseline run without added surfactant to complete.

5.   When the run completes, click **Record Data.**

6.   Now click **Surfactant** twice.

7.   Click **Start** to begin the surfactant run.

8.   When the lungs stop respiring, click **Record Data** to display the data in the grid.

How has the air flow changed compared to the baseline run?

_____

_____

Premature infants often have difficulty breathing. Explain why this might be so. (Use your text as needed.)

_____

_____ ■

Investigating Intrapleural Pressure

The pressure within the pleural cavity, **intrapleural pressure,** is less than the pressure within the alveoli. This negative pressure condition is caused by two forces, the tendency of the lung to recoil due to its elastic properties and the surface tension of the alveolar fluid. These two forces act to pull the lungs away from the thoracic wall, creating a partial vacuum in the pleural cavity. Because the pressure in the intrapleural space is lower than atmospheric pressure, any opening created in the thoracic wall equalizes the intrapleural pressure with the atmospheric pressure, allowing air to enter the pleural cavity, a condition called **pneumothorax.** Pneumothorax allows lung collapse, a condition called **atelectasis** (at ″ĕ-lik′tah-sis).

In the simulated respiratory system on the computer's screen, the intrapleural space is the space between the wall of the bell jar and the outer wall of the lung it contains. The pressure difference between inspiration and expiration for the left lung and for the right lung is individually displayed in the Pressure Left and Pressure Right windows.

1. Do *not* discard your previous data.

2. Click **Clear Tracings** to clean up the screen and then click **Flush** to clear the lungs of surfactant from the previous run.

3. Adjust the radius of the airways to 5.00 mm by clicking the appropriate button next to the Radius window.

4. Click **Start** and allow one screen of respirations to complete. Notice the negative pressure condition displayed below the oscilloscope when the lungs inhale.

5. When the lungs stop respiring, click **Record Data** to display the data in the grid.

6. Now click **Open Valve** on the left side of the bell jar above the Start button to open the valve.

7. Click **Start** to begin the run.

8. When the run completes, click **Record Data** again.

What happened to the lung in the left side of the bell jar?

_____

_____

How did the pressure in the left lung differ from that in the right lung?

_____

Explain your reasoning.

_____

_____

How did the total air flow in this trial compare with that in the previous trial in which the pleural cavities were intact?

_____

_____

What do you think would happen if the two lungs were in a single large cavity instead of separate cavities?

_____

_____

9. Now close the valve you opened earlier by clicking it and then click **Start** to begin a new trial.

10. When the run completes, click **Record Data** to display the data in the grid.

Did the deflated lung reinflate? _____
Explain your answer.

_____

_____

11. Click the **Reset** button atop the bell jar. This action draws the air out of the intrapleural space and returns it to normal resting condition.

12. Click **Start** and allow the run to complete.

13. When the run completes, click **Record Data** to display the data in the grid.

Why did lung function in the deflated (left) lung return to normal after you clicked Reset?

_____

_____ ■

Simulating Variations in Breathing

This part of the computer simulation allows you to examine the effects of hyperventilation, rebreathing, and breathholding on CO_2 level in the blood.

Choose **Variations in Breathing** from the **Experiment** menu. The opening screen will appear in a few seconds (Figure 48B.3). The basic features on the screen when the pro-

Figure 48B.3 **Opening screen of the Variations in Breathing experiment.**

gram starts are the same as in the lung volumes screen. The buttons beneath the oscilloscope control the various possible breathing patterns. Clicking **Hyperventilation** causes the lungs to breathe faster than normal. A small bag automatically covers the airway tube when **Rebreathing** is clicked. Clicking **Breath Holding** causes the lungs to stop respiring. Click **Normal Breathing** at any time to resume normal tidal cycles. The window next to the Start button displays the breathing pattern being performed by the simulated lungs.

The windows below the oscilloscope display the PCO_2 (partial pressure of CO_2) of the air in the lungs, Maximum PCO_2, Minimum PCO_2, and Pump rate.

Data accumulated during a run are displayed in the windows below the oscilloscope. When you click **Record Data,** that data is recorded in the computer's memory and is displayed in the data grid. Data displayed in the data grid include the Condition, PCO_2, Max. PCO_2, Min. PCO_2, Pump Rate, Radius, and Total Flow. Clicking **Delete Line** allows you to discard data values for a single run, and clicking **Clear Table** erases the entire experiment to allow you to start over.

Activity:
Exploring Various Breathing Patterns

You will establish the baseline respiratory values in this first experiment.

1. If the grid in the data control unit is not empty, click **Clear Table** to discard all previous data.

2. Adjust the radius of the airways to 5.00 mm by clicking the appropriate button next to the Radius window. Now, read through steps 3–5 before attempting to execute them.

3. Click **Start** and notice that it changes to **Stop** to allow you to stop the respiration. Watch the simulated lungs begin to breathe as a result of the external mechanical forces supplied by the pump below the bell jar. Simultaneously, the oscilloscope will display a tracing of the tidal volume for each breath.

4. After 2 seconds, click the **Hyperventilation** button and watch the PCO_2 displays. The breathing pattern will change to short, rapid breaths. The PCO_2 of the air in the lungs will be displayed in the small window to the right of the Hyperventilation button.

5. Watch the oscilloscope display and the PCO₂ window, and click **Stop** before the tracing reaches the end of the screen.

What happens to PCO₂ during hyperventilation? Explain your answer.

_____

_____

6. Click **Record Data.**
7. Now click **Clear Tracings** to prepare for the next run.

Rebreathing

When Rebreathing is clicked, a small bag will appear over the end of the air tube to allow the air within the lungs to be repeatedly inspired and expired.

1. Click **Start,** wait 2 seconds, and then click **Rebreathing.**
2. Watch the breathing pattern on the oscilloscope and notice the PCO₂ during the course of the run. Click **Stop** when the tracing reaches the right edge of the oscilloscope.

What happens to PCO₂ during the entire time of the rebreathing activity?

_____

_____

Did the depth of the breathing pattern change during rebreathing? (Carefully examine the tracing for rate and depth changes; the changes can be subtle.) Explain.

_____

_____

_____

3. Click **Record Data** and then click **Clear Tracings** to prepare for the next run.

Breath Holding

Breath holding can be considered an extreme form of rebreathing in which there is no gas exchange between the outside atmosphere and the air within the lungs.

1. Click **Start,** wait a second or two, and then click **Breath Holding.**
2. Let the breath-holding activity continue for about 5 seconds and then click **Normal Breathing.**
3. Click **Stop** when the tracing reaches the right edge of the oscilloscope.

What happened to the PCO₂ during breath holding?

_____

What happened to the breathing pattern when normal respirations resume?

_____

_____

4. Click **Record Data.** ■

Using Your Data
Save Your Data

When you have finished all parts of this exercise, you can save your data sets for later review and additional analysis.

1. Select **Save Data to File** from the **File** menu.
2. When the dialog box opens, click a drive and/or directory location.
3. Type a file name into the **File name** box (your instructor may provide you with a name).
4. Click **Save** to complete the save. If you want to review previously saved data, select **Open Data File** from the **File** menu, and choose your file.

Print Your Data

If your instructor requires a printed copy of your experiment results, choose **Print** from the **File** menu.

1. Type your name into the space provided.
2. When you are ready to print, click **OK.** ■

Objectives

1. To define *glomerulus, glomerular capsule, renal corpuscle, renal tubule,* **nephron,** *proximal convoluted tubule, loop of Henle,* and *distal convoluted tubule.*

2. To describe the blood supply to each nephron.

3. To identify the regions of the nephron involved in glomerular filtration and tubular reabsorption.

4. To study the factors affecting glomerular filtration.

5. To explore the concept of carrier transport maximum.

6. To understand how the hormones aldosterone and ADH affect the function of the kidney.

7. To describe how the kidneys can produce urine that is four times more concentrated than the blood.

Materials

Minimum equipment required:

Windows:

❑ 486/66 MHz or better recommended

❑ Windows 95/98 recommended

❑ SVGA display (256 colors at 640 × 480)

❑ Minimum 16 MB available RAM

❑ Double-speed CD-ROM drive (quad-speed or higher recommended)

❑ Sound card

❑ Speakers or headphones

❑ Printer

Macintosh:

❑ 68040 processor or Power Macintosh recommended

❑ System 7.1 or higher

❑ Minimum 16 MB available RAM

❑ 13-inch or larger color monitor (640 × 480 resolution)

❑ Double speed CD-ROM drive (quad-speed or higher recommended)

❑ Printer

Software:

❑ Benjamin/Cummings PhysioEx CD-ROM—Renal System Physiology module

Metabolism produces wastes that must be eliminated from the body. This excretory function is the job of the renal system, most importantly the paired kidneys. Each kidney consists of about one million nephrons that carry out two crucial services, blood filtration and fluid processing.

Microscopic Structure and Function of the Kidney

Each of the million or so **nephrons** in each kidney is a microscopic tubule consisting of two major parts: a **glomerulus** and a **renal tubule.** The glomerulus is a tangled capillary knot that filters fluid from the blood into the lumen of the renal tubule. The function of the renal tubule is to process that fluid, also called the **filtrate.** The beginning of the renal tubule is an enlarged end called the **glomerular capsule,** which surrounds the glomerulus and serves to funnel the filtrate into the rest of the renal tubule. Collectively, the glomerulus and the glomerular capsule are called the **renal corpuscle.**

As the rest of the renal tubule extends from the glomerular capsule, it becomes twisted and convoluted, then dips sharply down to form a hairpin loop, and then coils again before entering a collecting duct. Starting at the glomerular capsule, the anatomical parts of the renal tubule are as follows: the **proximal convoluted tubule,** the **loop of Henle** (nephron loop), and the **distal convoluted tubule.**

Two arterioles supply each glomerulus: an afferent arteriole feeds the glomerular capillary bed and an efferent arteriole drains it. These arterioles are responsible for blood flow through the glomerulus. Constricting the afferent arteriole lowers the downstream pressure in the glomerulus, whereas constricting the efferent arteriole will increase the pressure in the glomerulus. In addition, the diameter of the efferent arteriole is smaller than the diameter of the afferent arteriole, restricting blood flow out of the glomerulus. Consequently, the pressure in the glomerulus forces fluid through the endothelium of the glomerulus into the lumen of the surrounding glomerular capsule. In essence, everything in the blood except the cells and proteins are filtered through the glomerular wall. From the capsule, the filtrate moves into the rest of the renal tubule for processing. The job of the tubule is to reabsorb all the beneficial substances from its lumen while allowing the wastes to travel down the tubule for elimination from the body.

The nephron performs three important functions to process the filtrate into urine: glomerular filtration, tubular reabsorption, and tubular secretion. **Glomerular filtration** is a passive process in which fluid passes from the lumen of the glomerular capillary into the glomerular capsule of the renal tubule. **Tubular reabsorption** moves most of the filtrate back into the blood, leaving principally salt water plus the wastes in the lumen of the tubule. Some of the desirable or needed solutes are actively reabsorbed, and others move passively from the lumen of the tubule into the interstitial spaces. **Tubular secretion** is essentially the reverse of tubular reabsorption and is a process by which the kidneys can rid the blood of additional unwanted substances such as creatinine and ammonia.

The reabsorbed solutes and water that move into the interstitial space between the nephrons need to be returned to the blood, or the kidneys will rapidly swell like balloons. The peritubular capillaries surrounding the renal tubule reclaim the reabsorbed substances and return them to general circulation. Peritubular capillaries arise from the efferent arteriole exiting the glomerulus and empty into the veins leaving the kidney.

Getting Started

Begin by making sure you have the computer equipment and software listed in the Materials section on p. 814.

1.   Quit all applications currently running on your computer.

2.   Insert the PhysioEx CD-ROM into the CD-ROM drive and keep PhysioEx in the drive during the entire time you use the program.

3.   Follow the instructions below for your computer type.

Windows 95/Windows 98
The program will launch automatically when you load the CD-ROM into the drive. If autorun is disabled on your machine follow these instructions:

1.   Double-click the **My Computer** icon on the Windows desktop.

2.   Double-click the **PhysioEx CD** icon.

3.   Look for the icon that looks like a tiny movie projector and double-click it.

Macintosh
Double-click the **PhysioEx** icon that appears in the window on your desktop to launch the program.

Simulating Glomerular Filtration

This computerized simulation allows you to explore one function of a single simulated nephron, glomerular filtration. The concepts you will learn by studying a single nephron can then be applied to understand the function of the kidney as a whole.

Choose **Renal System Physiology** from the main menu. The opening screen for the Simulating Glomerular Filtration experiment will appear in a few seconds (Figure 49B.1). The main features on the screen when the program starts are a simulated blood supply at the left side of the screen, a simulated nephron within a supporting tank on the right side, and a data control unit at the bottom of the display.

The left beaker is the "blood" source representing the general circulation supplying the nephron. The "blood pressure" in the beaker is adjustable by clicking the (+) and (−) buttons on top of the beaker. A tube with an adjustable radius called the *afferent flow* tube connects the left beaker to the simulated glomerulus. Another adjustable tube called the *efferent flow* tube drains the glomerulus. The afferent flow tube represents the afferent arteriole feeding the glomerulus of each nephron, and the efferent flow tube represents the efferent arteriole draining the glomerulus. The outflow of the nephron empties into a collecting duct, which in turn drains into another small beaker at the bottom right part of the screen. Clicking the valve at the end of the collecting duct stops the flow of fluid through the nephron and collecting duct.

The Glomerular Pressure window on top of the nephron tank displays the pressure within the glomerulus. The Glomerular Filt. Rate window indicates the flow rate of the fluid moving from the lumen of the glomerulus into the lumen of the renal tubule.

The concentration gradient bathing the nephron is fixed at 1200 mosm. Clicking **Start** begins the experiment. Clicking **Refill** resets the equipment to begin another run.

The equipment in the lower part of the screen is called the *data control unit*. This equipment records and displays data you accumulate during the experiments. The data set for the first experiment (Afferent) is highlighted in the **Data Sets** window. You can add or delete a data set by clicking the appropriate button to the right of the Data Sets window. When you click **Record Data,** your data is recorded in the computer's memory and is displayed in the data grid. Data displayed in the data grid include the Afferent Radius, Efferent Radius, Beaker Pressure, Glomerular Pressure, Glomerular Filtration Rate, and the Urine Volume. Clicking **Delete Line** allows you to discard data values for a single run, and clicking **Clear Data Set** erases the entire experiment to allow you to start over.

If you need help identifying any piece of equipment, choose **Balloons On** from the Help menu and move the mouse pointer onto any piece of equipment visible on the computer's screen. As the pointer touches the object, a pop-up window appears identifying the equipment. To close the pop-up window, move the mouse pointer away from the equipment. Choose **Balloons On** again to turn off this help feature.

Figure 49B.1 **Opening screen of the Simulating Glomerular Filtration experiment.**

Activity:
Investigating the Effect of Flow Tube Radius on Glomerular Filtration

Your first experiment will examine the effects of flow tube radii and pressures on the rate of glomerular filtration. Click **Start** to see the on-screen action. Continue when you understand how the simulation operates. Click **Refill** to reset the experiment.

1. The **Afferent** line in the Data Sets window of the data control unit should be highlighted in bright blue. If it is not, choose it by clicking the **Afferent** line. The data control unit will now record filtration rate variations due to changing afferent flow tube radius.

2. If the data grid is not empty, click **Clear Data Set** to discard all previous data.

3. Adjust the afferent radius to 0.35 mm, and the efferent radius to 0.40 mm by clicking the appropriate buttons.

4. If the left beaker is not full, click **Refill.**

5. Keep the beaker pressure at 90 mm Hg during this part of the experiment.

6. Click **Start** and watch the blood flow. Simultaneously, filtered fluid will be moving through the nephron and into the collecting duct. The Glomerular Filt. Rate window will display the fluid flow rate into the renal tubule when the left beaker has finished draining.

7. Now click **Record Data** to record the current experiment data in the data grid. Click **Refill** to replenish the left beaker and prepare the nephron for the next run.

8. Increase the afferent radius in 0.05-mm increments and repeat steps 6 through 8 until the maximum radius (0.60 mm) is achieved. Be sure to click **Record Data** after each trial. If you make an error and want to delete a single value, click the data line in the data grid and then click **Delete Line.**

What happens to the glomerular filtration rate as the afferent radius is increased?

_____

_____

Predict the effect of increasing or decreasing the efferent radius on glomerular filtration rate. Use the simulation to reach an answer if you are not sure.

_____

_____ ■

Activity:
Studying the Effect of Pressure on Glomerular Filtration

Both the blood pressure supplying the glomerulus and the pressure in the renal tubule have a significant impact on the glomerular filtration rate. In this activity, the data control unit will record filtration rate variations due to changing pressure.

1.   Click the **Pressure** line in the Data Sets window of the data control unit.

2.   If the data grid is not empty, click **Clear Data Set** to discard all previous data.

3.   If the left beaker is not full, click **Refill.**

4.   Adjust the pressure in the left beaker to 70 mm Hg by clicking the appropriate button.

5.   During this part of the experiment, maintain the afferent flow tube radius at 0.55 mm and the efferent flow tube radius at 0.45 mm.

6.   Click **Start** and watch the blood flow. Filtrate will move through the nephron into the collecting duct. At the end of the run, the Glomerular Filt. Rate window will display the filtrate flow rate into the renal tubule.

7.   Now click **Record Data** to record the current experiment data in the data grid. Click **Refill** to replenish the left beaker.

8.   Increase the pressure in the upper beaker in increments of 10 mm Hg and repeat steps 6 through 8 until the maximum pressure (100 mm Hg) is achieved. Be sure to click **Record Data** after each trial. If you make an error and want to delete a single value, click the data line in the data grid and then click **Delete Line.**

What happened to the glomerular filtration rate as the beaker pressure was increased?

_____

_____

Explain your answer.

_____

_____

_____ ■

Activity:
Assessing Combined Effects on Glomerular Filtration

So far, you have examined the effects of flow tube radius and pressure on glomerular filtration rate. In this experiment you will be altering both variables to explore the combined effects on glomerular filtration rate and to see how one can compensate for the other to maintain an adequate glomerular filtration rate.

1.   Click **Combined** in the Data Sets window of the data control unit.

2.   If the data grid is not empty, click **Clear Data Set** to discard all previous data.

3.   If the left beaker is not full, click **Refill.**

4.   Set the starting conditions at

•   100 mm Hg beaker pressure

•   0.55 mm afferent radius

•   0.45 mm efferent radius

5.   Click **Start.**

6.   Now click **Record Data** to record the current baseline data in the data grid.

7.   Click **Refill.**

You will use this baseline data to compare a run in which the valve at the end of the collecting duct is in the open position with a run in which the valve is in the closed position.

8.   Use the simulation and your knowledge of basic renal anatomy to arrive at answers to the following questions. Be sure to click **Record Data** after each trial. If you make an error and want to delete a single value, click the data line in the data grid and then click **Delete Line.**

Click **Close Valve** on the end of the collecting duct to close it. Click **Start** and allow the run to complete. How does this run compare to the runs in which the valve was open?

_____

_____

Expanding on this concept, what might happen to total glomerular filtration and therefore urine production in a human kidney if all of its collecting ducts were totally blocked?

_____

_____

Would kidney function as a whole be affected if a single nephron was blocked? Explain.

_____

_____ ■

Figure 48B.2 **Opening screen of the Simulating Urine Formation experiment.**

Would the kidney be functioning if glomerular filtration was zero? Explain.

_____

_____

Explain how the body could increase glomerular filtration rate in a human kidney.

_____

_____

_____

_____

If you increased the pressure in the beaker, what other condi-

tion(s) could you adjust to keep the glomerular filtration rate constant?

_____

_____ ∎

Simulating Urine Formation

This part of the computer simulation allows you to explore some aspects of urine formation by manipulating the interstitial solute concentration. Other activities include investigating the effects of aldosterone and ADH (antidiuretic hormone), and the role that glucose carrier proteins play in renal function.

Choose **Simulating Urine Formation** from the **Experiment** menu. The opening screen will appear in a few seconds (Figure 49B.2). The basic features on the screen when the program starts are similar to the glomerular filtration screen. Most of the vascular controls have been moved off-screen to the left because they will not be needed in this set of experiments. Additional equipment includes a *supplies shelf* at the right side of the screen, a *glucose carrier control* located at the top of the nephron tank, and a concentration probe at the bottom left part of the screen.

The maximum concentration of the "interstitial gradient" to be dispensed into the tank surrounding the nephron is adjusted by clicking the (+) and (−) buttons next to the Conc. Grad. window. Click **Dispense** to fill the tank through the jets at the bottom of the tank with the chosen solute gradient. Click **Start** to begin a run. After a run completes, the concentration probe can be clicked and dragged over the nephron to display the solute concentration within.

Hormone is dispensed by dragging a hormone bottle cap to the gray cap button in the nephron tank at the top of the collecting duct and then letting go of the mouse button.

The (+) and (−) buttons in the glucose carrier control are used to adjust the number of glucose carriers that will be inserted into the simulated proximal convoluted tubule when the Add Carriers button is clicked.

Data displayed in the data grid will depend on which experiment is being conducted. Clicking **Delete Line** allows you to discard data values for a single run, and clicking **Clear Data Set** erases the entire experiment to allow you to start over.

Activity:
Exploring the Role of the Solute Gradient on Maximum Urine Concentration Achievable

In the process of urine formation, solutes and water move from the lumen of the nephron into the interstitial spaces. The passive movement of solutes and water from the lumen of the renal tubule into the interstitial spaces relies in part on the total solute gradient surrounding the nephron. When the nephron is permeable to solutes or water, an equilibrium will be reached between the interstitial fluid and the contents of the nephron. Antidiuretic hormone (ADH) increases the water permeability of the distal convoluted tubule and the collecting duct, allowing water to flow to areas of higher solute concentration, usually from the lumen of the nephron into the surrounding interstitial area. You will explore the process of passive reabsorption in this experiment. While doing this part of the simulation, assume that when ADH is present the conditions favor the formation of the most concentrated urine possible.

1. **Gradient** in the Data Sets window of the data control unit should be highlighted in bright blue. If it is not then click Gradient.

2. If the data grid is not empty, click **Clear Data Set** to discard all previous data.

3. Click and hold the mouse button on the ADH bottle cap and drag it to the gray cap at the top right side of the nephron tank. Release the mouse button to dispense ADH onto the collecting duct.

4. Adjust the maximum total solute concentration of the gradient (**Conc. Grad.**) to 300 mosm by clicking the appropriate button. Because the blood solute concentration is also 300 mosm, there is no osmotic difference between the lumen

of the nephron and the surrounding interstitial fluid.

5. Click **Dispense.**

6. Click **Start** to begin the experiment. Filtrate will move through the nephron and then drain into the beaker below the collecting duct.

7. While the experiment is running, click and hold the mouse on the Probe and drag it to the urine beaker. Observe the total solute concentration in the Concentration window.

8. Now click **Record Data** to record the current experiment data in the data grid.

9. Increase the maximum concentration of the gradient in 300-mosm increments and repeat steps 3 through 8 until 1200 mosm is achieved. Be sure to click **Record Data** after each trial. If you make an error and want to delete a single value, click the data line in the data grid and then click **Delete Line.**

What happened to the urine concentration as the gradient concentration was increased?

_____

_____

What factor limits the maximum possible urine concentration?

_____

_____

The solute concentration of the blood is about 300 mosm, and the highest interstitial solute concentration in a human kidney is about 1200 mosm. This means that the maximum urine solute concentration is about four times that of the blood. What would be the maximum possible urine concentration if the maximum interstitial solute concentration were 3000 mosm instead of 1200 mosm? Explain. (Use the simulation to arrive at an answer if you are not sure.)

_____

_____

_____

_____ ■

Activity:
Studying the Effect of Glucose Carrier Proteins on Glucose Reabsorption

Because carrier proteins are needed to move glucose from the lumen of the nephron into the interstitial spaces, there is a limit to the amount of glucose that can be reabsorbed. When all glucose carriers are bound with the glucose they are transporting, excess glucose is eliminated in urine. In this experiment, you will examine the effect of varying the number of glucose transport proteins in the proximal convoluted tubule.

1. Click **Glucose** in the Data Sets window of the data control unit.

2. If the data grid is not empty, click **Clear Data Set** to discard all previous data.

3. Set the concentration gradient (**Conc. Grad.**) to 1200 mosm.

4. Click **Dispense.**

5. Adjust the number of glucose carriers to 100 (an arbitrary figure) by clicking the appropriate button.

6. Click **Add Carriers.** This action inserts the specified number of glucose carrier proteins per unit area into the membrane of the proximal convoluted tubule.

7. Click **Start** to begin the run after the carriers have been added.

8. Click **Record Data** to record the current experiment data in the data grid. Glucose presence in the urine will be displayed in the data grid.

9. Now increase the number of glucose carrier proteins in the proximal convoluted tubule in increments of 100 glucose carriers and repeat steps 6 through 8 until the maximum number of glucose carrier proteins (500) is achieved. Be sure to click **Record Data** after each trial. If you make an error and want to delete a single value, click the data line in the data grid and then click **Delete Line.**

What happened to the amount of glucose present in the urine as the number of glucose carriers was increased?

_____

_____

The amount of glucose present in normal urine is minimal because there are normally enough glucose carriers present to handle the "traffic." Predict the consequence in the urine if there was more glucose than could be transported by the available number of glucose carrier proteins.

_____

_____

Explain why we would expect to find glucose in the urine of a diabetic person.

_____

_____ ■

Activity:
Testing the Effect of Hormones on Urine Formation

The concentration of the urine excreted by our kidneys changes depending on our immediate needs. For example, if a person consumes a large quantity of water, the excess water will be eliminated, producing dilute urine. On the other hand, under conditions of dehydration, there is a clear benefit in being able to produce urine as concentrated as possible, thereby retaining precious water. Although the medullary gradient makes it possible to excrete concentrated urine, urine dilution or concentration is ultimately under hormonal control. In this experiment, you will investigate the effects of two different hormones on renal function, aldosterone produced by the adrenal gland and ADH manufactured by the hypothalamus and stored in the posterior pituitary gland. Aldosterone works to reabsorb Na^+ (and thereby water) at the expense of losing K^+. Its site of action is the distal convoluted tubule. ADH makes the distal tubule and collecting duct more permeable to water, thereby allowing the body to reabsorb more water from the filtrate when it is present.

1. Click **Hormone** in the Data Sets window of the data control unit.

2. If the data grid is not empty, click **Clear Data Set** to discard all previous data.

3. During this part of the experiment, keep the concentration gradient at 1200 mosm.

4. Click **Dispense** to add the gradient and then click **Start** to begin the experiment.

5. Now click **Record Data** to record the current experiment data in the data grid.

You will use this baseline data to compare with the conditions of the filtrate under the control of the two hormones.

6. Keeping all experiment conditions the same as before, do the following:

• Drag the aldosterone bottle cap to the gray cap on the top right side of the nephron tank and release the mouse to automatically dispense aldosterone into the tank surrounding the distal convoluted tubule and collecting duct.

• Click **Start** and allow the run to complete.

• Click **Record Data.**

In this run, how does the volume of urine differ from the previously measured baseline volume?

_____

_____

Explain the difference in the total amount of potassium in the urine between this run and the baseline run.

_____

_____

7. Drag the ADH bottle cap to the gray cap on the top right side of the nephron tank and release it to dispense ADH.

• Click **Start** and allow the run to complete

• Click **Record Data.**

In this run, how does the volume of urine differ from the baseline measurement?

_____

_____

Is there a difference in the total amount of potassium in this run and the total amount of potassium in the baseline run? Explain. (Hint: the urine volume with ADH present is about one-tenth the urine volume when it is not present.)

_____

_____

Are the effects of aldosterone and ADH similar or antagonistic? _____

Consider this situation: we want to reabsorb sodium ions but do not want to increase the volume of the blood by reabsorbing water from the filtrate. Assuming that aldosterone and ADH are both present, how would you adjust the hormones to accomplish the task?

_____

_____

_____

_____

If the interstitial gradient ranged from 300 mosm to 3000 mosm and ADH was not present, what would be the maximum possible urine concentration? _____ ■

A c t i v i t y :
Using Your Data

Save Your Data

When you have finished all parts of this exercise, you can save your data sets for later review and additional analysis.

1. Select **Save Data to File** from the **File** menu.

2. When the dialog box opens, click a drive and/or directory location.

3. Type a file name into the **File name** box (your instructor may provide you with a name).

4. Click **Save** to complete the save. If you want to review previously saved data, select **Open Data File** from the **File** menu, and choose your file.

Print Your Data

If your instructor requires a printed copy of your experiment results, choose **Print** from the **File** menu.

1. Type your name into the space provided.

2. When you are ready to print, click **OK.** ■

H

Using the Histology Module

Materials

Minimum equipment required:

Windows:

- ❏ 486/66 MHz or better recommended
- ❏ Windows 95/98 recommended
- ❏ SVGA display (256 colors at 640 × 480)
- ❏ Minimum 16 MB available RAM
- ❏ Double-speed CD-ROM drive (quad-speed or higher recommended)

Macintosh:

- ❏ 68040 processor or Power Macintosh recommended
- ❏ System 7.1 or higher
- ❏ Minimum 16 MB available RAM
- ❏ 13-inch or larger color monitor (640 × 480 resolution)
- ❏ Double-speed CD-ROM drive (quad-speed or higher recommended)

Software:

- ❏ Benjamin/Cummings PhysioEx CD-ROM—Histology module

Examining a specimen using a microscope accomplishes two goals: first, it gives you an understanding of the cellular organization of tissues, and second, perhaps even more importantly, it hones your observational skills. Because developing these skills is crucial in the understanding and eventual mastery of the way of thinking in science, using this histology module is not intended as a substitute for using the microscope. Instead, use the histology module to gain an overall appreciation of the specimens and then make your own observations using your microscope. The histology module is also an excellent review tool.

Figure H   **Opening screen of the Histology module.**

Getting Started

1. Quit all applications currently running on your computer.

2. Insert the PhysioEx CD-ROM into the CD-ROM drive and keep PhysioEx in the drive during the entire time you use the program.

3. Follow the instructions below for your computer type.

Windows 95/Windows 98
The program will launch automatically when you load the CD-ROM into the drive. If autorun is disabled on your machine follow these instructions:

1. Double-click the **My Computer** icon on the Windows desktop.

2. Double-click the **PhysioEx** CD icon.

3. Look for the icon that looks like a tiny movie projector and double-click it.

Macintosh
Double-click the **PhysioEx** icon that appears in the window on your desktop to launch the program.

Exploring Digital Histology

Choose **Histology** from the main menu. The opening screen for the Histology module will appear in a few seconds (Figure H). The main features on the screen when the program starts are a pair of empty text boxes at the left side of the screen, a large image-viewing window on the right side, and a set of magnification buttons above the image window.

Click **Select an Image** to bring up the image list. Choosing an image from the list displays it in the image-viewing window. The text description of the specimen is displayed in the large text box at the lower left and the image title in the small upper text box. Some of the words in the text are highlighted in color; clicking those words displays another image that you can compare with the main image.

Click and hold the mouse button (left button on a PC) on the image and then drag the mouse to move other areas of the specimen into the image viewing window, much like moving a slide on a microscope stage.

Choose the magnification of the image in the viewing window by clicking one of the magnification buttons above the image window.

To see parts of the slide labeled, click **Labels on**. Click the button again to remove the labels.

Respiratory System Mechanics: Computer Simulation

Define the following terms:

1. Ventilation _____

2. Inspiration _____

3. Expiration _____

Measuring Respiratory Volumes

1. Write the respiratory volume term and the normal value that is described by the following statements:

    Volume of air present in the lungs after a forceful expiration _____

    Volume of air that can be expired forcefully after a normal expiration _____

    Volume of air that is breathed in and out during a normal respiration _____

    Volume of air that can be inspired forcefully after a normal inspiration _____

    Volume of air corresponding to TV + IRV + ERV _____

2. Fill in the formula for minute respiratory volume:

    _____

Examining the Effect of Changing Airway Resistance on Respiratory Volumes

1. Even though pulmonary function tests are not diagnostic, they can help determine the difference between

    _____ and _____ disorders.

2. Chronic bronchitis and asthma are examples of _____ disorders.

3. Describe FEV_1: _____

    _____

4. Explain the difference between FVC and FEV_1: _____

    _____

5. What effect would increasing airway resistance have on FEV_1? _____

Examining the Effect of Surfactant

1. Explain the term *surface tension.*_____

_____

2. Surfactant is a detergent-like _____.

3. How does surfactant work?_____

4. What might happen to ventilation if the watery film lining the alveoli did not contain surfactant?

_____

Investigating Intrapleural Pressure

Complete the following statements.

The pressure within the pleural cavity, __1__, is __2__ than the pressure within the alveoli. This __3__ pressure condition is caused by two forces, the tendency of the lung to recoil due to its __4__ properties and the __5__ of the alveolar fluid. These two forces act to pull the lungs away from the thoracic wall, creating a partial __6__ in the pleural cavity. Because the pressure in the __7__ space is lower than __8__, any opening created in the thoracic wall equalizes the intrapleural pressure with the atmospheric pressure, allowing air to enter the pleural cavity, a condition called __9__. Pneumothorax allows __10__, a condition called __11__.

1. _____

2. _____

3. _____

4. _____

5. _____

6. _____

7. _____

8. _____

9. _____

10. _____

11. _____

12. Why is the intrapleural pressure negative rather than positive?_____

_____

_____

13. Would intrapleural pressure be positive or negative when blowing up a balloon? Explain your answer.

_____

_____

Exploring Various Breathing Patterns

1. Match the term listed in column B with the descriptive phrase in column A. (There may be more than one correct answer.)

    Column A

    _____1.   causes a drop in carbon dioxide concentration in the blood

    _____2.   results in lower blood pH

    _____3.   stimulates an increase respiratory rate

    _____4.   results in a lower respiratory rate

    _____5.   can be considered an extreme form of rebreathing

    _____6.   causes a rise in blood carbon dioxide

    Column B

    a.   rebreathing

    b.   hyperventilation

    c.   breath holding

2. Because carbon dioxide is the main stimulus for respirations, what would happen to respiratory drive if you held your

    breath?_____

    _____

NAME _____ LAB TIME/DATE _____

Renal Physiology—The Function of the Nephron Computer Simulation

Define the following terms:

1. Glomerulus _____

2. Renal tubule _____

3. Glomerular capsule _____

4. Renal corpuscle _____

5. Afferent arteriole _____

6. Efferent arteriole _____

Investigating the Effect of Flow Tube Radius on Glomerular Filtration

1. In terms of the blood supply to and from the glomerulus, explain why the glomerular capillary bed is unusual.

   _____

   _____

2. How would pressure in the glomerulus be affected by constricting the afferent arteriole? Explain your answer.

   _____

   _____

3. How would pressure in the glomerulus be affected by constricting the efferent arteriole? Explain your answer.

   _____

   _____

Assessing Combined Effects on Glomerular Filtration

1. If systemic blood pressure started to rise, what could the arterioles of the glomerulus do to keep glomerular filtration rate

   constant?_____

   _____

2. One of the experiments you performed in the simulation was to close the valve at the end of the collecting duct. Is closin
   that valve more like constricting an afferent arteriole or more like a kidney stone? Explain your answer.

   _____

3. Constricting the efferent arteriole would have the same effect on glomerular filtration as (constricting/dilating) the afferent arteriole.

_____

_____

Exploring the Role of the Solute Gradient on Maximum Urine Concentration Achievable

Complete the following statements.

In the process of urine formation, solutes and water move from the __1__ of the nephron into the __2__ spaces. The passive movement of solutes and water from the lumen of the renal tubule into the interstitial spaces relies in part on the __3__ surrounding the nephron. When the nephron is permeable to solutes or water, and __4__ will be reached between the interstitial fluid and the contents of the nephron. __5__ is a hormone that increases the water permeability of the __6__ and the collecting duct, allowing water to flow to areas of higher solute concentration, usually from the lumen of the nephron into the surrounding interstitial area. __7__ is hormone that causes __8__ reabsorption at the expense of __9__ loss into the lumen of the tubule.

1. _____

2. _____

3. _____

4. _____

5. _____

6. _____

7. _____

8. _____

9. _____

10. Would the passive movement of substances occur if the interstitial solute concentration was the same as the filtrate solute concentration? Explain your answer.

_____

_____

Studying the Effect of Glucose Carrier Proteins on Glucose Reabsorption

1. In terms of the function of the nephron, explain why one might find glucose in the urine exiting the collecting duct.

_____

_____

2. Imagine this scenario: a person has the normal number of glucose carriers in the nephrons yet has glucose in the urine. What could be the cause of this condition? (Hint: think about filtration rate.)

_____

_____

Testing the Effect of Hormones on Urine Formation

Complete the following statements.

The concentration of the __1__ excreted by our kidneys changes depending on our immediate needs. For example, if a person consumes a large quantity of water, the excess water will be eliminated, producing __2__ urine. On the other hand, under conditions of dehydration, there is a clear benefit in being able to produce urine as __3__ as possible, thereby retaining precious water. Although the medullary gradient makes it possible to excrete concentrated urine, urine dilution or concentration is ultimately under __4__ control. In this experiment, you will investigate the effects of two different hormones on renal function, aldosterone produced by the __5__ and ADH manufactured by the __6__ and stored in the __7__. Aldosterone works to reabsorb __8__ (and thereby water) at the expense of losing __9__. Its site of action is the __10__. ADH makes the distal tubule and collecting duct more permeable to __11__, thereby allowing the body to reabsorb more water from the filtrate when it is present.

1. _____

2. _____

3. _____

4. _____

5. _____

6. _____

7. _____

8. _____

9. _____

10. _____

11. _____

12. Match the term listed in column B with the descriptive phrase in column A. (There may be more than one correct answer.)

Column A

_____1. causes production of dilute urine

_____2. results in increased sodium loss

_____3. causes the body to retain more potassium

_____4. will cause water retention due to sodium movement

_____5. causes water reabsorption due to increased membrane permeability

_____6. increases sodium reabsorption

Column B

a. increased ADH

b. increased alsostrene

c. decreased ADH

d. decreased aldostrene